Intelligent Forecasting and Optimization in Electrical Power Systems

Intelligent Forecasting and Optimization in Electrical Power Systems

Editors

Paweł Piotrowski
Grzegorz Dudek
Dariusz Baczyński

MDPI

Basel • Beijing • Wuhan • Barcelona • Belgrade • Novi Sad • Cluj • Manchester

Editors
Paweł Piotrowski
Warsaw University of Technology (WUT)
Warszawa, Poland

Grzegorz Dudek
Czestochowa University of Technology
Czestochowa, Poland

Dariusz Baczyński
Warsaw University of Technology (WUT)
Warszawa, Poland

Editorial Office
MDPI
St. Alban-Anlage 66
4052 Basel, Switzerland

This is a reprint of articles from the Special Issue published online in the open access journal *Energies* (ISSN 1996-1073) (available at: https://www.mdpi.com/journal/energies/special_issues/Intelligent_Forecasting_and_Optimization_in_Electrical_Power_Systems).

For citation purposes, cite each article independently as indicated on the article page online and as indicated below:

Lastname, A.A.; Lastname, B.B. Article Title. *Journal Name* **Year**, *Volume Number*, Page Range.

ISBN 978-3-0365-9080-6 (Hbk)
ISBN 978-3-0365-9081-3 (PDF)
doi.org/10.3390/books978-3-0365-9081-3

Contents

About the Editors

Paweł Piotrowski

Paweł Piotrowski received his PhD in electrical engineering from Warsaw University of Technology (WUT), Poland, in 1994, and habilitation in electrical engineering from Warsaw University of Technology, Poland, in 2014. Currently, he is an associate professor at the Faculty of Electrical Engineering, WUT. He is the author of books concerning forecasting in power engineering, applications of artificial intelligence in power engineering and over 100 scientific papers. His research interests include forecasting and optimization in power engineering. He is Head of Division of Electric Networks and Power Systems in Electrical Engineering Power Institute.

Grzegorz Dudek

Grzegorz Dudek received his PhD in electrical engineering from Czestochowa University of Technology (CUT), Poland, in 2003, habilitation in computer science from Lodz University of Technology, Poland, in 2013, and was granted the title of full professor in 2023. Currently, he is a professor at the Department of Electrical Engineering, CUT. He is the author of four books concerning machine learning for forecasting and evolutionary algorithms for unit commitment and over 130 scientific papers. He came third in the Global Energy Forecasting Competition 2014 (price forecasting track). His research interests include machine learning and artificial intelligence, and their application to practical classification, regression, forecasting, and optimization problems.

Dariusz Baczyński

Dariusz Baczyński received his PhD and habilitation in electrical engineering from Warsaw University of Technology (WUT), Poland, in 1999 and in 2014, respectively. Currently, he is an associate professor at the Faculty of Electrical Engineering, WUT. He also cooperated with IT companies as an expert. His interests include optimization, forecasting, and applications of computer systems in broadly understood electrical power engineering. In particular, he is interested in pareto optimization and modern computational intelligence methods. He is the author of several books and over 50 scientific papers.

energies

MDPI

Editorial

Intelligent Forecasting and Optimization in Electrical Power Systems: Advances in Models and Applications

Grzegorz Dudek [1,*,†], Paweł Piotrowski [2,†] and Dariusz Baczyński [2,†]

[1] Faculty of Electrical Engineering, Czestochowa University of Technology, 42-200 Czestochowa, Poland
[2] Electrical Power Engineering Institute, Warsaw University of Technology, 00-662 Warsaw, Poland; pawel.piotrowski@pw.edu.pl (P.P.); dariusz.baczynski@pw.edu.pl (D.B.)
* Correspondence: grzegorz.dudek@pcz.pl
† These authors contributed equally to this work.

1. Introduction

A modern power system is a complex network of interconnected components, such as generators, transmission lines, and distribution subsystems, that are designed to provide electricity to consumers in an efficient and reliable manner. These systems make use of advanced technologies and control systems to monitor and manage the flow of electricity, including integrating renewable energy sources (RESs), implementing smart grid systems, and using advanced forecasting and optimization techniques to ensure the stability and security of the grid. The aim of modern power systems is to provide a sustainable and reliable source of electricity that meets the needs of the growing population, while minimizing the environmental impact and reducing the costs.

A power system requires forecasts that predict the future electricity demand, the power generation from RESs, and meteorological data that are important regarding consumer demand and the level of generation from RESs. Accurate forecasting enables the effective operation of power systems of all sizes, including microgrids. It is necessary for energy mix optimization, energy storage management, hydro-thermal coordination, fuel reserve planning, electricity import and export planning, and security assessments. It is also crucial in competitive energy markets, as electricity prices are highly influenced by the demand for electricity and energy mixes. Thus, accurate forecasting is financially beneficial for all participants of the energy market.

Citation: Dudek, G.; Piotrowski, P.; Baczyński, D. Intelligent Forecasting and Optimization in Electrical Power Systems: Advances in Models and Applications. *Energies* **2023**, *16*, 3024. https://doi.org/10.3390/en16073024

Received: 4 March 2023
Accepted: 23 March 2023
Published: 26 March 2023

The objective of optimization of power systems is to efficiently utilize available resources to meet a target outcome, such as reducing costs, increasing efficiency, or improving reliability. Optimization of power systems involves finding the optimal operating conditions for a system given constraints such as equipment capacity, energy prices, and system reliability requirements. This requires taking into account a wide range of factors, including energy generation and demand forecasts, load profiles, and the availability of energy storage and other resources. Typical optimization problems in power systems are unit commitment and optimal power flow. Unit commitment is the process of scheduling the available generating units to meet the expected load demand in the most economical way. This involves determining which generators to operate, their power outputs, and their start-up and shut-down schedules over a given time period. Optimal power flow is the process of finding the optimal settings for the controllable devices in the power system, such as generators, transformers, reactive power devices, and switches, to minimize the cost of generating and transmitting electricity in the system while satisfying system constraints, such as power balance, network stability limits, transmission limits, voltage limits, and device operational limits.

This Special Issue explores the latest developments and advancements in the application of artificial intelligence (AI) and machine learning (ML) for forecasting and optimization in the field of power engineering. In recent years, AI and ML have been gaining significant traction and are becoming one of the most important fields in computing. These methods have proven to be effective in solving forecasting and optimization problems in power engineering.

For this Special Issue, we invited researchers to submit original papers and review articles that showcase their latest research results in forecasting and optimization of electrical power systems. Topics of interest include, but are not limited to:

- AI/ML/deep learning for forecasting electricity generation from RESs;
- AI/ML/deep learning for forecasting power demand for electrical power systems;
- Optimization of electrical power systems;
- Forecasting of meteorological data (wind speed and solar radiation) that is important for forecasting electricity generation from RESs;
- Statistical analyses of data for forecasting models (including problems related to big, missing, distorted, and uncertain data);
- Reliability of electrical power systems.

Overall, this Special Issue aims to bring together the latest research and advancements in the application of AI and ML to forecasting and optimization in the field of power engineering and provide a platform for the exchange of ideas and the presentation of new findings.

2. Summary of the Contributions

There were 25 papers submitted to this Special Issue, and 18 papers were accepted. Although each paper covers a different topic, we can identify four categories into which the papers can be classified according to their main focus: electricity demand forecasting, wind power forecasting, photovoltaic power forecasting, and optimization.

2.1. *Electricity Demand Forecasting*

2.1.1. Relevance of the Subject

Demand forecasting in power systems is the process of predicting the future electricity demand of a given area or region. It is an important aspect of power system planning, as it allows utility companies to estimate the amount of energy they will need to supply in the future and to make informed decisions about how to meet that demand. Accurate demand forecasting helps power system operators to avoid both shortages and excess generation, which can be costly and impact the stability of the electrical grid. Forecasting electricity demand can be challenging because it depends on a wide range of factors, including weather patterns, economic trends, and consumer behavior.

In addition to its importance in day-to-day operations, demand forecasting is also critical for mid-term and long-term planning in the power sector. Accurate forecasting assists utility companies in making knowledgeable decisions about investments in new infrastructure, such as power plants and transmission lines, and can aid in optimizing the use of existing resources.

2.1.2. Main Forecasting Problems

There are different types of forecasting problems that can arise in electricity demand forecasting, including:

- Very short-term forecasting, which refers to the prediction of electricity demand or production for a time horizon from seconds to a few hours ahead.
- Short-term forecasting, which refers to predicting electricity demand in the immediate future, usually up to several days ahead.
- Medium-term forecasting, which involves predicting electricity demand for a period of a few weeks to a few months ahead.
- Long-term forecasting, which involves predicting electricity demand for a period several years in advance.

- Peak electricity demand forecasting, which is forecasting of the highest level of electricity consumption within a particular period, typically daily or yearly.
- Seasonal forecasting, which involves predicting the electricity demand for different seasons of the year.
- Special event forecasting, which involves predicting the electricity demand for special events, such as holidays, sports events, or festivals.
- Probabilistic forecasting, which is forecasting of not only a single predicted value, but also a probability distribution or range of potential values with their associated probabilities.
- Uncertainty forecasting, which involves predicting electricity demand in the presence of uncertainty, such as changes in weather patterns, economic conditions, or energy policies.

Each type of forecasting problem requires different data, models, and techniques, and may have different levels of accuracy and uncertainty.

Electricity demand forecasting can be a challenging task due to various factors that can affect the consumption patterns of electricity users:

- Seasonality and trends. There can be significant seasonality and trends in electricity demand, such as increased usage during hot summer months or the growth in the electricity market over time.
- Volatility. Electricity demand can be volatile and subject to unexpected changes due to weather events, economic conditions, or other unforeseen factors.
- Data quality. Electricity demand data may be incomplete or contain errors, which can affect the accuracy of forecasting models.
- Non-linear relationships. There may be non-linear relationships between electricity demand and various factors such as temperature, time of day, and day of the week.
- Uncertainty. The accuracy of forecasting models can be affected by uncertainty around future events or conditions, such as changes in regulations or the introduction of new technologies.

To address these challenges, advanced forecasting techniques such as ML, time series analysis, and statistical modeling are often used to analyze historical data and identify patterns and trends that can help predict future electricity demand. There are many methods used for demand forecasting, including statistical models, AI and ML models, and hybrid models that combine the two. These models use historical data, weather forecasts, economic data, and other factors to make predictions about future electricity demand.

2.1.3. Overview of Article Content

The purpose of [1] is to predict the impact of electric vehicle developments on the Polish power system from 2022 to 2027. The study conducted multi-stage and multi-variant prognostic research by forecasting the number of electric vehicles using seven methods, and then forecasting the annual power demand arising from the operation of these vehicles, both with and without the impact of e-mobility growth, using six methods. The daily profiles of typical days were forecasted with and without e-mobility growth using three methods. To forecast the number of electric vehicles in Poland, a unique growth dynamics model was developed. The researchers also applied an artificial neural network (ANN), specifically the multilayer perceptron (MLP), in the extrapolation of non-linear functions for forecasting the number of electric vehicles and annual power demand without the impact of e-mobility growth. In another innovative proposal, they included two ANN models (MLP and long short-term memory (LSTM)) in an ensemble model for simultaneous extrapolation of 24 non-linear functions to forecast the daily profiles of typical days. The study revealed that e-mobility development in Poland for the next six years (2022–2027) may pose a challenge in terms of the additional demand for electricity. Electric vehicles' largest percentage share of the demand for electricity was in the peak evening time, while the smallest percentage share was during the night. Overall, this study provides important insights for policymakers, energy planners, and stakeholders who need to make informed decisions on how to manage the expected increase in demand for electricity due to the growth of e-mobility in Poland.

Paper [2] investigated the sources of uncertainty in short-term hourly electricity load forecasting and proposed a clustering-based bootstrapping method to increase the accuracy of multi-step ahead point forecasts. The proposed method, called SSA.KM.N, combines singular spectrum analysis and K-means clustering-based generation of Gaussian normal distribution to generate electricity load time series with lower variance and values around the original data. The study compares SSA.KM.N and KM.N using two Malaysian, one Polish, and one Indonesian electricity load time series using four benchmark models for electricity load forecasting: SARIMA, NNAR, TBATS, and DSHW. The results showed that the proposed method improves the accuracy of multi-step ahead forecast values, especially for the SARIMA and NNAR models. The study also noted that the number of bootstrapped series does not seem to affect the forecasting accuracy, and the model suitable for the original series is not necessarily appropriate for all bootstrapped series. The authors suggest combining several models and ensemble learning methods in future research. Overall, the study proposed a novel method for improving the short-term hourly electricity load forecasting accuracy by addressing uncertainty through bootstrap aggregation.

Paper [3] conducted a literature review of autoregressive methods applied to short-term forecasting of power demand, aiming to improve the forecasting efficiency while minimizing the financial costs and time taken. The review analyzed 47 articles and 264 forecasting models, focusing on autoregressive methods, but also including methods with explanatory variables. The analysis included 25 power systems on four continents that were published by 44 different research teams. The paper presents a new approach to developing literature reviews, ranking the forecasting models based on the mean average percentage error (MAPE), and also presenting a flowchart illustrating the process. The most effective models using the autoregressive approach include fuzzy logic, ANNs, wavelet ANNs, adaptive neuro-fuzzy inference systems, genetic algorithms, fuzzy regression, and data envelope analyses. The results of the review constitute an excellent starting point for further tests and pave the way for future research in this area. The paper also discusses the state of research in short-term power demand forecasting, including methods of AI data mining, and big data.

ML ensemble models are the state-of-the-art in forecasting. Paper [4] explored the use of random forest, an ensemble model, for short-term load forecasting, and investigated various data representation and training modes. The study demonstrated that the proposed approach using random forest outperforms both standard statistical models and more sophisticated ML approaches in terms of accuracy for short-term load forecasting. The random forest model is easy to learn and optimize, with a small number of tuning hyperparameters. It has the ability to handle multiple exogenous predictors of different types. The study also shows that the performance of random forest depends significantly on data preprocessing and proper organization of the training process. The proposed approach extends pattern definition and introduces a global mode of training with additional predictors representing calendar data. The proposed model is suitable for forecasting problems with multiple seasonality, nonlinear trends, and varying variance in time series. In future work, the author plans to extend random forest with random data projection and use it for probabilistic forecasting.

A solution for predicting the monthly power demand using statistical methods such as ARIMA, ETS, and Prophet is proposed in [5]. These methods utilize pattern representation of seasonal cycles of the time series to unify the data, filter out a trend, and define longer seasonal cycles. The input and output variables in the pattern space are characterized by a less complex relationship, resulting in a simpler forecasting model. ARIMA and ETS construct global models, while comparative minimum distance methods, such as k-NN, construct local models individually for each query pattern. Outliers in the time series affect the selection of ARIMA and ETS parameters, leading to suboptimal models, while they have a lesser impact on k-NN. Additionally, the statistical models generate forecasts one step ahead, while k-NN predicts the vector representing the entire predicted sequence in one step. A simulation study on monthly electricity load time series for 35 European countries confirmed the high accuracy of the proposed models.

Paper [6] proposed a smart home occupancy prediction technique using environmental variables such as CO_2, noise, and relative temperature via an ML forecasting strategy. The LSTM neural network was used to process time series prediction, and two metaheuristic optimization algorithms (GA and PSO) were used to enhance the performance of the LSTM algorithm. The proposed methods were evaluated using real-world datasets. The results show that GA and PSO can adjust the LSTM model to perform significantly better than benchmark models, including other ML approaches such as basic LSTM. The predicted values were used to determine whether residents were present and control real electrical consumption. The authors suggest a potential field for future research in thermal parameter forecasting using recurrent neural networks for various places such as hospitals, hotels, and public establishments.

2.2. *Wind Power Forecasting*

2.2.1. Relevance of the Subject

The forecasting of power generation in wind farms is a much explored research topic. Five papers devoted to forecasting the energy generation of wind farms have been published in this Special Issue.

Forecasting purposes vary by time horizon. The ability to precisely forecast power generation in the short-term for wind farms (especially large wind farms) is very topical, since such generation is highly unstable and creates problems for distribution and transmission system operators in appropriately preparing the power system for operation. Forecasts of the energy generation of wind farms, especially for the next day, play an important role in this process. They are also utilized in energy market transactions. Even a small improvement in the quality of these generation forecasts translates into an improved security of the system and savings for the economy. High quality forecasts of electrical energy generation are also very important for owners of small wind turbines due to the optimization of energy storage and optimization of the use of various energy carriers (especially in microgrid systems). Medium-term forecasts of wind farm power generation have other purposes: grid integration planning, determining the optimal use of backup power sources, and balancing the supply and demand of electricity. The applications of long-term forecasts of wind farm power generation are, e.g., maintenance scheduling, wind farm design, electricity market restructuring, and optimization of operating costs.

2.2.2. Main Forecasting Problems

For short-term forecasts (more than a few hours), it is not possible to accurately forecast electricity generation from wind farms without using wind speed forecasts. The accuracy of power generation forecasts depends strongly on the quality of wind speed forecasts. For extensive wind farms, an additional problem is the variety in atmospheric conditions (essentially wind speed) in different parts of the farm. The terrain in the vicinity of the wind farm (e.g., forests, hills, and lakes) is another factor that affects the amount of electricity generation. The amount of electricity generated is therefore to some extent dependent on the roughness of the terrain. On the macro scale, it is equally important to select a proper forecasting point from which meteorological variables can be derived. The location of numerical weather prediction (NWP) forecasting points has an impact on the quality of generation forecasts; NWP forecasts at locations far away from the wind farm can generate large forecasting errors.

The problem of the availability (amount of information) of data for the forecasting model is also very important; the more information related to power generation available to be used in the model, the more accurate the generation forecasts will be. Another problem is the use in forecasting models of wind speeds that were forecasted at a much different height than the height of the wind turbines on the wind farm. The final important forecasting problem is that the quality of forecasts decreases as the forecast horizon grows (it is difficult to accurately forecast wind speed for a horizon greater than 6 h).

A fundamental problem for generating a forecast for a specific period of time, for example, for a 1-h period or a 15-min period, is that the instantaneous wind speed forecasts are unknown. Therefore, simplification of the model is necessary, which has an obvious impact on the accuracy of wind farm generation forecasts.

2.2.3. Overview of Article Content

Paper [7] concerns ensemble methods using ML and deep learning for one-day-ahead forecasts of electric energy production in two wind farms. It is worth noting that using two wind farms for forecasting considerably increases the credibility of newly created prediction methods and the conclusions made from them. The authors verified the accuracy of forecasts executed by single methods, hybrid methods, and ensemble methods (for a total of thirteen methods). However, the predictions made by the original ensemble forecasting method, called "Ensemble Averaging without Extremes", had the lowest normalized mean absolute error (nMAE) among all tested methods. A new, original proposal, "Additional Expert Correction", reduced the errors of energy generation forecasts for both wind farms. Using the original skill score (SS) metric proposed by the authors to compare the prediction accuracy proved to be very useful. This metric allows incorporation of both the nMAE and the normalized root mean square error (nRMSE) into the final quality assessment. The results of comparative tests (different sets of inputs to the predictive models) demonstrated that it is better to use NWP point forecasts for hourly lags (−3, −2, −1, 0, 1, 2, 3 (original contribution)) as input data than lags of 0 and −1 that are typically used in such situations. The authors demonstrated that it is better to use forecasts from two different NWP models as input data than from one NWP model. The conclusions drawn from this extensive work can be generalized, at least for Central Europe.

Paper [8] concerns offshore wind power short-term forecasting. ML models are accurate methods of wind power prediction; however, their accuracy depends on the selection of appropriate hyperparameters. The authors proposed a novel optimization algorithm to tune the LSTM model for short-term wind power forecasting. The new Optuna optimization framework was employed to optimize the hyperparameters of the LSTM model, including the number of lag observations, the exposure frequency, the number of nodes, the number of samples in an epoch, and the used difference order, to convert a nonstationary dataset into a stationary dataset. This proposed method improved the wind power prediction accuracy. The method's effectiveness was validated using six distinct datasets, with noted accuracy improvements observed in all cases.

Paper [9] concerns NN-based wind power forecasting models for neuromorphic devices. The authors proposed the use of biologically inspired algorithms adapted to the architecture of neuromorphic devices, such as spiking artificial NNs. They proposed a short-term wind power forecasting model based on spiking artificial NNs adapted to the computational abilities of Loihi (a neuromorphic device developed by Intel). One-step-ahead wind power forecasts were executed using wind power generation data from Ireland. The authors demonstrated that neuromorphic computing offers a new paradigm to create energy efficient, low latency algorithms in contrast to the present state-of-the-art ML/DL strategies, thus potentially reducing the computational cost of training and deploying AI-based forecasting models.

Paper [10] presented a selective review on the recent advancements in long-, short-, and ultra-short-term wind power predictions. A detailed review of recent research achievements and performance and the possible future scope of research are presented. Each category of forecasting methods is divided into four subclasses and a comparative analysis is presented. This review paper also provides future recommendations and discussions on recent development trends in forecasting methods. An analysis of papers showed that hybrid methods are probably the best choice for all three prediction horizons.

Paper [11] concerns an evaluation of the metrics for wind power forecasts. The "Introduction" section provides a valuable and extensive description of the major factors that affect the quality of wind power forecasts. In the "Performance of Forecasting Model" section, a comprehensive inventory of error metrics is presented, which includes both popular and occasionally used metrics, totaling 19 error metrics. This paper conducted a comprehensive review (quantitative analysis) based on more than one hundred papers concerning forecasts of energy generation from wind farms (offshore and onshore). Moreover, the paper includes an extensive statistical analysis of errors (qualitative analysis). In the "Comprehensive Error Analysis" section, the quotients of the nRMSE and nMAE were calculated and a new, unique error dispersion factor (EDF) metric was thus introduced (a combination of two frequently used error metrics). This research presents a unique and novel approach to studying errors in power generation forecasts for wind farms. The EDF shows the average variability of the moduli of error, regardless of the magnitude of the error. The decrease in the EDF with a rise in the forecasting horizon indicates that the variability in the errors decreases with an increasing forecasting horizon. An analysis of the errors and the EDF depending on the class of forecasting methods demonstrated that the variability in the moduli of errors of the best methods (smallest forecasting errors) was usually larger than for the "single method" class (much larger forecasting errors). The moduli of errors in the "single method" class are much larger and much closer in value than in the best (ensemble or hybrid) methods.

2.3. *Photovoltaic Power Forecasting*

2.3.1. Relevance of the Subject

Photovoltaic sources (PV) are perceived by the public as an opportunity for emission-free electricity generation on various scales. Of course, sources of this type are intermittent and often difficult to manage. Depending on the size of the power system and the size of the connected PV system, there is a growing need for increasingly more precise forecasts of energy production. Therefore, the topic of forecasting energy production from PV sources is quite popular, leading to the next three papers in this Special Issue. As it was mentioned above, PV system sizes may vary. For household use, they can be few hundred Watts, for small microgrids, they may be dozens of kW, and for PV farms, they can reach dozens of MW. Electricity produced from a PV source can be utilized in many ways in power systems of different sizes and purposes. Depending on these factors, different prediction horizons and prediction quantization may be chosen. The typical applications of PV forecasts are summarized in the following:

- Control of microgrid elements (sources, receivers, and storage), especially important and difficult in islanded operation mode;
- Energy market participation of PV source operators;
- Control and operation planning of conventional electricity sources (transmission system operator level);
- Control and operation planning of power grids (distribution system operator level).

Control applications usually require ultra-short-term forecasts, with horizons from a number of seconds up to few hours ahead and with quantization from seconds to dozens of minutes. For planning applications, short-term forecasts are used. In this context, short-term refers to horizons from a few hours up to one week It is obvious that without forecasts, most of the businesses and technical processes mentioned above are not plausible. Furthermore, using PV energy forecasts results in substantial economical savings.

2.3.2. Main Forecasting Problems

Electricity production of photovoltaic sources depends strictly on meteorological conditions. This makes these sources similar to wind sources. However, in the case of photovoltaic sources, the geographic location of the source and the season of the year are also important, as these factors affect the maximum insolation during the day. In order to obtain forecasts of electricity production from PV sources of the highest possible quality, it is necessary to use weather forecasts and to take into account seasonal dependencies in predictive models. Most solutions use NWP forecasts. They are indispensable in the case of short-term forecasts and some (longer horizons) ultra-short-term forecasts. In the case of the latter (for shorter horizons), different measurements are utilized to create so-called "nowcasting" meteo forecasts. These measurements may include insolation, energy production of neighboring PV sources, and also images of the sky taken with a camera. Seasonal dependencies are in some way taken into account in NWP forecasts. For example, insolation and temperature are given for the exact time and date. However, there can be some factors, which can grow to considerable depending on the class of prediction models. For physical models, the main problem is the exact determination of PV panel orientation and inclination angle. Machine learning methods generally do not require such data. It is more important to collect and prepare proper datasets for model learning and testing. These datasets should reflect phenomena that can be described by the included parameters. As an example, the influence of various types of precipitation on energy generation (e.g., snow and rain) should be modeled. Another problem is that the soiling of PV panels and their periodic cleaning must be taken into account. Both the problems of precipitation and soiling may influence the energy generation from a few tenths of a percent to several tens of percent.

2.3.3. Overview of Article Content

Paper [12] concerns ultra-short-term forecasting of photovoltaic source power generation. In this case, the forecasting horizon was next step forecasting and the forecast quantization was 5 min. The paper starts with a literature overview and a description of photovoltaic system performance. The data gathered for the presented research are derived from a 3.2 kW PV system. A very detailed statistical analysis of power generation data is presented. On the basis of this analysis, sets of explanatory variables are proposed. There are eight different sets with different numbers of inputs (explanatory variables), varying from one up to fifteen. The authors proposed ten different forecasting models: single (naive, LR, KNNR, MLP, SVR, and IT2FLS), ensemble, and hybrid. Almost every model was tested for more than one set of explanatory variables, giving a total of almost 40 configurations. All the results were evaluated using four quality criteria, i.e., RMSE, nMAPE, nAPEmax, and MBE. The best results were obtained by the hybrid and MLP models when using sets of explanatory variables with higher numbers of variables. The authors presented a detailed analysis of the results.

Paper [13] concerns short-term forecasting with a 1 to 144 h horizon and hourly quantization. The authors presented in detail the dataset used, which includes lagged power production, global horizontal irradiance, NWP forecasts, and regional aggregated solar power predictions. Then, a one-step-ahead model configuration was presented. The authors proposed the use of separate models for (a) the 1st hour ahead, (b) the 2nd to 56th hour ahead, and (c) the 57th to 144th ahead. XGBoost (XGB) and CatBoost (CTB) methods were used to build the prediction models. As evaluation criteria, the RMSE and RMSE scores were selected. The RMSE skill score utilizes the complete history persistence ensemble (CH-PeEn) as a benchmark method. Such a criterion can be used as a comparison for the proposed model with simple forecasting based on historical data. The authors also investigated the model's performance with respect to the development of separate models for each month, for 3 months, or for a universal model. The results of the tests were presented and discussed in detail. The best results were obtained for separate models built for each month of the year.

Paper [14] concerns day-ahead forecasting (24–47 h horizon) with hourly quantization of PV and wind sources. The main idea of this article is the use of multi-task learning (MTL) autoencoders. The authors determined whether MTL autoencoders can be utilized to predict day ahead electricity generation for different sources in non single-task learning. This led to other investigations, such as determining the quality of such predictions and determining whether additional encoder fine-tuning will be necessary. To answer to these questions, the authors used the following datasets: PVOPEN, PVSYN, PVREAL, WINDOPEN, WINDSYN, and WINDREAL. These datasets include over 600 renewable power stations with additional NWP data. The authors tested different autoencoder architectures varying the parameter number by three orders of magnitude. During experiments, the RMSE and nRMSE were used as quality criteria. When considering a multi-task approach, the authors reduced the trainable parameters by up to 203 times. The authors also concluded that the amount of layers requiring fine-tuning depends on the architecture and the model.

2.4. Optimization

2.4.1. Relevance of the Subject

The competitiveness of the economy depends on the ability to save energy and the ability to propose innovative solutions in optimization of power systems. Furthermore, in electrical power engineering, optimization often uses the results of forecasting, creating a synergistic effect. Solving an optimization problem requires several steps, usually problem description, criteria definition, mathematical model construction, objective function definition, optimization method application, and testing. For many problems, these are relatively time-consuming tasks because most of them do not have ready-made toolkits. In this Special Issue, there are four papers concerning different aspects of optimization of power systems.

2.4.2. Overview of Article Content

Paper [15] concerns the optimization of the configuration and operation of a hybrid AC/DC low voltage microgrid. For optimization purposes, the CLONALG algorithm was chosen. The CLONALG algorithm belongs to the family of artificial immune system (AIS) computational intelligence methods. In the presented application, it was equipped with a modified hypermutation operator. The author stated three different optimization tasks: minimization of total active power losses, minimization of costs associated with the operation of the hybrid AC/DC microgrid, and maximization of the level of power generated by the RES. For each task, there is an appropriate mathematical definition of the problem. The test hybrid microgrid consists of AC and DC networks coupled with an electronic power converter. The microgrid supplied a single family housing estate and connects PV, wind, and distributed generation sources. It also included energy storage. The optimization results of the proposed version of the CLONAG algorithm are presented in detail and compared to the evolutionary algorithm. The proposed algorithm achieved better results in most cases.

Paper [16] concerns voltage control in MV networks with distributed generation. Widespread incorporation of distributed generation (DG) (especially renewable) to medium voltage (MV) and low voltage (LV) networks causes many operation problems. One of these problems is the rise in voltage during high energy generation in DG and vice versa. This results, for example, in limitations of PV source generation on sunny days. The main idea of this article is to overcome the voltage problems by appropriately setting the transformer's on-load tap changers and using additional measures such as capacitor banks, reactive power generation in the RES, and energy storage. As an optimization method, the authors used the algorithm of the innovative gunner (AIG). This algorithm, as a computational intelligence method, is generally similar to other methods, especially swarm methods. One feature that distinguishes it from other swarm methods is its method of decision vector modification. Usually, algorithms use additive formulae. However, the AIG uses multiplicative modifications, which makes optimization more dynamic (especially at the beginning). The authors presented the test network and AIG optimization results compared to cuckoo search (CS) and moth-flame optimization (MFO) algorithms.

Paper [17] concerns the reliability of MV distribution networks with distributed generation and ICT infrastructure. Distribution power networks (both MV and LV) were originally designed as hierarchical for unidirectional power flow from generating units connected to higher voltage levels to receivers connected to lower voltage levels. Incorporation of distributed generation (DG) and RESs has changed this operation model. To obtain a better performance of the networks, their structures and operation model must change. Obviously, information and communication technology should be a part of this transition. This article presents a reliability analysis to answer the questions of what the future network structure should be and what additional elements need to be incorporated to obtain the optimal reliability. The authors used several indices (SAIFI, CAIFI, ASAI, ASUI, and EENS) to answer to this question, analyzing five network structures.

Paper [18] concerns the optimization of industrial refrigeration system operation. The authors decided to define this problem as a multi-objective problem with two conflicting objectives: maximization of the effectiveness of the cooling towers and minimization of the overall power requirements of the refrigeration system. The objectives are contradictory because the efficiency of the system increases with the required system power. The structure of the test refrigeration system and objective functions were presented. To solve the optimization problem, the authors proposed and described three different evolutionary algorithms: the non-dominated sorting genetic algorithm (NSGAII), the micro-genetic algorithm (Micro-GA), and the strength Pareto evolutionary algorithm (SPEA2). As determining the optimal solution (in the case of multi-objective optimization) is difficult, the authors introduced a third criterion: the energy efficiency ratio. After many analyses of the obtained results by using this third criterion, the authors proved that the best solution was achieved using the SPEA2 algorithm.

Author Contributions: Conceptualization, G.D., P.P. and D.B.; methodology, G.D., P.P. and D.B.; validation, G.D., P.P. and D.B.; resources, G.D., P.P. and D.B.; writing—original draft preparation, G.D., P.P. and D.B.; writing—review and editing, G.D., P.P. and D.B.; supervision, G.D., P.P. and D.B. All authors have read and agreed to the published version of the manuscript.

Funding: This research received no external funding.

Conflicts of Interest: The authors declare no conflict of interest.

References

1. Piotrowski, P.; Baczyński, D.; Kopyt, M. Medium-Term Forecasts of Load Profiles in Polish Power System including E-Mobility Development. *Energies* **2022**, *15*, 5578. [CrossRef]
2. Sulandari, W.; Yudhanto, Y.; Rodrigues, P.C. The Use of Singular Spectrum Analysis and K-Means Clustering-Based Bootstrap to Improve Multistep Ahead Load Forecasting. *Energies* **2022**, *15*, 5838. [CrossRef]
3. Czapaj, R.; Kamiński, J.; Sołtysik, M. A Review of Auto-Regressive Methods Applications to Short-Term Demand Forecasting in Power Systems. *Energies* **2022**, *15*, 6729. [CrossRef]
4. Dudek, G. A Comprehensive Study of Random Forest for Short-Term Load Forecasting. *Energies* **2022**, *15*, 7547. [CrossRef]
5. Pełka, P. Analysis and Forecasting of Monthly Electricity Demand Time Series Using Pattern-Based Statistical Methods. *Energies* **2023**, *16*, 827. [CrossRef]
6. Mahjoub, S.; Labdai, S.; Chrifi-Alaoui, L.; Marhic, B.; Delahoche, L. Short-Term Occupancy Forecasting for a Smart Home Using Optimized Weight Updates Based on GA and PSO Algorithms for an LSTM Network. *Energies* **2023**, *16*, 1641. [CrossRef]
7. Piotrowski, P.; Baczyński, D.; Kopyt, M.; Gulczyński, T. Advanced Ensemble Methods Using Machine Learning and Deep Learning for One-Day-Ahead Forecasts of Electric Energy Production in Wind Farms. *Energies* **2022**, *15*, 1252. [CrossRef]
8. Hanifi, S.; Lotfian, S.; Zare-Behtash, H.; Cammarano, A. Offshore Wind Power Forecasting— A New Hyperparameter Optimisation Algorithm for Deep Learning Models. *Energies* **2022**, *15*, 6919. [CrossRef]
9. González Sopeña, J.M.; Pakrashi, V.; Ghosh, B. A Spiking Neural Network Based Wind Power Forecasting Model for Neuromorphic Devices. *Energies* **2022**, *15*, 7256. [CrossRef]
10. Sawant, M.; Patil, R.; Shikhare, T.; Nagle, S.; Chavan, S.; Negi, S.; Bokde, N.D. A Selective Review on Recent Advancements in Long, Short and Ultra-Short-Term Wind Power Prediction. *Energies* **2022**, *15*, 8107. [CrossRef]
11. Piotrowski, P.; Rutyna, I.; Baczyński, D.; Kopyt, M. Evaluation Metrics for Wind Power Forecasts: A Comprehensive Review and Statistical Analysis of Errors. *Energies* **2022**, *15*, 9657. [CrossRef]
12. Piotrowski, P.; Parol, M.; Kapler, P.; Fetliński, B. Advanced Forecasting Methods of 5-Minute Power Generation in a PV System for Microgrid Operation Control. *Energies* **2022**, *15*, 2645. [CrossRef]
13. Bezerra Menezes Leite, H.; Zareipour, H. Six Days Ahead Forecasting of Energy Production of Small Behind-the-Meter Solar Sites. *Energies* **2023**, *16*, 1533. [CrossRef]
14. Schreiber, J.; Sick, B. Multi-Task Autoencoders and Transfer Learning for Day-Ahead Wind and Photovoltaic Power Forecasts. *Energies* **2022**, *15*, 8062. [CrossRef]

15. Rokicki, Ł. Optimization of the Configuration and Operating States of Hybrid AC/DC Low Voltage Microgrid Using a Clonal Selection Algorithm with a Modified Hypermutation Operator. *Energies* **2021**, *14*, 6351. [CrossRef]
16. Pijarski, P.; Kacejko, P.; Wancerz, M. Voltage Control in MV Network with Distributed Generation— Possibilities of Real Quality Enhancement. *Energies* **2022**, *15*, 2081. [CrossRef]
17. Parol, M.; Wasilewski, J.; Wojtowicz, T.; Arendarski, B.; Komarnicki, P. Reliability Analysis of MV Electric Distribution Networks Including Distributed Generation and ICT Infrastructure. *Energies* **2022**, *15*, 5311. [CrossRef]
18. Nedjah, N.; de Macedo Mourelle, L.; Lizarazu, M.S.D. Evolutionary Multi-Objective Optimization Applied to Industrial Refrigeration Systems for Energy Efficiency. *Energies* **2022**, *15*, 5575. [CrossRef]

energies

MDPI

Article

Optimization of the Configuration and Operating States of Hybrid AC/DC Low Voltage Microgrid Using a Clonal Selection Algorithm with a Modified Hypermutation Operator

Łukasz Rokicki

Faculty of Electrical Engineering, Warsaw University of Technology, Koszykowa 75 Street, 00-662 Warsaw, Poland; lukasz.rokicki@pw.edu.pl; Tel.: +48-22-234-7951

Citation: Rokicki, Ł. Optimization of the Configuration and Operating States of Hybrid AC/DC Low Voltage Microgrid Using a Clonal Selection Algorithm with a Modified Hypermutation Operator. *Energies* **2021**, *14*, 6351. https://doi.org/10.3390/en14196351

Academic Editors: Luis Hernández-Callejo and Ricardo J. Bessa

Received: 11 August 2021
Accepted: 27 September 2021
Published: 5 October 2021

Abstract: The issue of optimization of the configuration and operating states in low voltage microgrids is important both from the point of view of the proper operation of the microgrid and its impact on the medium voltage distribution network to which such microgrid is connected. Suboptimal microgrid configuration may cause problems in networks managed by distribution system operators, as well as for electricity consumers and owners of microsources and energy storage systems connected to the microgrid. Structures particularly sensitive to incorrect determination of the operating states of individual devices are hybrid microgrids that combine an alternating current and direct current networks with the use of a bidirectional power electronic converter. An analysis of available literature shows that evolutionary and swarm optimization algorithms are the most frequently chosen for the optimization of power systems. The research presented in this article concerns the assessment of the possibilities of using artificial immune systems, operating on the basis of the CLONALG algorithm, as tools enabling the effective optimization of low voltage hybrid microgrids. In his research, the author developed a model of a hybrid low voltage microgrid, formulated three optimization tasks, and implemented an algorithm for solving the formulated tasks based on an artificial immune system using the CLONALG algorithm. The conducted research consisted of performing a 24 h simulation of microgrid operation for each of the formulated optimization tasks (divided into 10 min independent optimization periods). A novelty in the conducted research was the modification of the hypermutation operator, which is the key mechanism for the functioning of the CLONALG algorithm. In order to verify the changes introduced in the CLONALG algorithm and to assess the effectiveness of the artificial immune system in solving optimization tasks, optimization was also carried out with the use of an evolutionary algorithm, commonly used in solving such tasks. Based on the analysis of the obtained results of optimization calculations, it can be concluded that the artificial immune system proposed in this article, operating on the basis of the CLONALG algorithm with a modified hypermutation operator, in most of the analyzed cases obtained better results than the evolutionary algorithm. In several cases, both algorithms obtained identical results, which also proves that the CLONALG algorithm can be considered as an effective tool for optimizing modern power structures, such as low voltage microgrids, including hybrid AC/DC microgrids.

Keywords: hybrid AC/DC microgrid; optimization of configuration and operating states; CLONALG; modified hypermutation operator

1. Introduction

Over the last few years, the development of distributed, renewable energy sources (RES) and growing interest in prosumer installations have been observed. The presence of a large number of generation sources and energy storage devices (ESDs) in low voltage distribution networks promotes the creation of microgrids which are capable of synchronous operation with the remaining part of the power system as well as autonomous island operation.

A significant part of the ESDs and microsources used in microgrids generate DC voltage. Connecting them to the AC network requires the use of DC/AC electronic power converters (EPCs). Some AC microsources, due to the high voltage frequency, require a connection to the microgrid via AC/AC converters. The use of EPCs between microsource or energy storage device and the AC network results in additional power losses and reduces the efficiency of generation units. Increasing the efficiency of the devices included in the microgrid is possible by connecting energy sources and storage devices generating DC voltage to the DC network and units generating AC voltage to the AC network. Both types of network can be connected with each other by means of a single, bidirectional AC/DC converter, thus creating a low voltage hybrid microgrid.

The complexity of a low voltage microgrid, resulting from a large number of microsources, ESDs, EPCs, and controlled loads, requires the development of an appropriate management system for their operation in order to achieve maximum efficiency of RES. Proper selection of elements forming the microgrid, as well as subsequent determination of the operating states of individual devices in such a way that the microgrid as a whole is in the optimal configuration for the problem under consideration is not an easy task. The complexity of the calculations is directly proportional to the number of devices installed. The creation of an effective control system is possible by using appropriate optimization algorithms, including those that use artificial intelligence methods. The division of optimization algorithms that can be used in problems related to microgrids is shown in Figure 1.

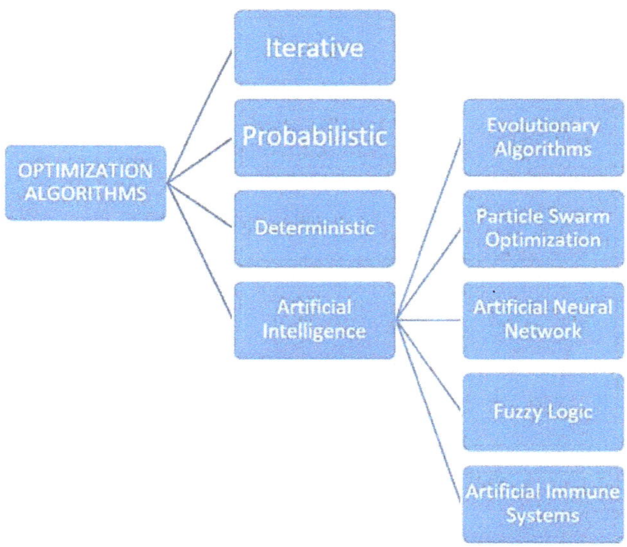

Figure 1. Division of optimization algorithms.

Among the methods of artificial intelligence used in solving optimization tasks, the most popular are evolutionary algorithms (EA) and particle swarm optimization (PSO). In article [1], a memory-based genetic algorithm, which is a type of EA, was used to minimize the power generation costs in the smart grid framework. The proposed method shares optimal power generation in a microgrid through different types of microsources. The authors of [2] use differential evolution algorithm (type of EA) for optimal single-objective economic scheduling and bi-objective environmental-economic scheduling of community microgrids. Another popular artificial intelligence method used to solve optimization tasks is the PSO algorithm. In paper [3], PSO was used to find economically optimal solutions for day-ahead scheduling strategy of a microgrid equipped with CHP microsources. Article [4] concerns the use of the PSO as a management system of microgrids composed of different types of microsources and energy storage devices to minimize total operating costs of the microgrid.

There are many other artificial intelligence methods that can be used for solving different optimizations tasks, such as Artificial Neural Network, Fuzzy Logic or Artificial Immune Systems (AIS) [5–7]. The main objective of the research presented in this paper is to assess whether the AIS, operating on the basis of the CLONALG algorithm, can be used as an effective tool for optimizing the configuration and operating states of low voltage AC/DC hybrid microgrids.

1.1. Review of Knowledge in the Field of Hybrid AC/DC Microgrids

The article [8] presents an overview of microgrids cooperating with AC and DC power grids. Advantages and disadvantages of both technologies were discussed in detail in this publication. The differences in the manner of connecting microsources, ESDs, and receivers to the networks were described, and schematic diagrams of EPCs, protection, and monitoring systems were presented. The article also contains an overview of control and optimization systems aimed at ensuring the quality of electricity and stability of the microgrid. The publication was completed with an economic analysis and examples of operating microgrids all over the world.

Hybrid AC/DC systems constitute a separate group of microgrids. An outline of the structure of hybrid microgrid was presented in [9] and the first research works began in 2010 [10]. The simplest hybrid microgrid is composed of AC and DC networks connected to each other by means of a bidirectional AC/DC EPC [11–14]. In paper [15], the planning process of a hybrid AC/DC microgrid with optimal placement of DC feeders was described. The concept, control paradigm, and implementation of a bus-sectionalized hybrid microgrid was presented in article [16].

Both AC and DC networks in a hybrid microgrid have the same types of microsources and ESDs. The coexistence of AC and DC networks allows for greater efficiency of installed devices than in case of solutions using only one type of voltage The DC network of the hybrid microgrid is a natural place for connecting photovoltaic panels [17,18], fuel cells [19], wind turbine generation sets equipped with DC generators [19], battery energy storage systems [20], and supercapacitors. The ability to integrate AC and DC networks within a hybrid microgrid can also contribute to the development of V2G technology [21–23].

Creating a hybrid microgrid concept allows combining the advantages of DC and AC networks, as well as eliminating some of the disadvantages of these network. The main advantages of using hybrid microgrids are described in [24] and include:

- reduction of energy conversion levels in EPCs and associated power losses,
- increasing the efficiency of RES,
- increasing the level of reliability of energy supply,
- increasing flexibility in regulatory services,
- limiting the impact of higher current and voltage harmonics,
- the ability to simplify the design of some electricity consumers by not having to use a built-in EPCs.

In terms of the method of connecting the hybrid microgrid to the external distribution network, coupled and separated topologies can be distinguished. In coupled topologies, the distribution network and the AC network of the hybrid microgrid are connected directly to each other using an MV/LV transformer. The DC network is connected using a two-way AC/DC converter. This converter can be connected to both low voltage and medium voltage sides of the transformer. In the case of separated topologies, the microgrid DC network is connected to the distribution grid through an AC/DC converter. Depending on the design, this converter can be connected directly to the medium voltage network or using a step-down transformer. In both of these cases, the AC low voltage network does not have a direct connection to the distribution grid. A detailed description of coupled and separated topologies is provided in [25].

As with other types of microgrids, popularization of hybrid microgrids requires the development of appropriate solutions for protection systems, taking into account the specificity of AC and DC networks. Detailed methods for solving problems related to the protection of hybrid microgrids are presented in publications [26–28].

In the field of architecture of hybrid microgrid control systems, the same centralized and distributed control solutions are used as in the case of AC and DC microgrids. It should be noted that the hybrid microgrid control system must provide the ability to control the operating states of the AC/DC converter connecting the networks of both types of voltage in order to properly manage energy exchange between them. An example of a hybrid microgrid centralized control system has been described in [29]. Article [30] presents a coordination control strategy for a hybrid microgrid in standalone mode.

1.2. Objective and Contribution

The main objective of this paper is the assessment of the effectiveness of the AIS in solving tasks related to the optimization of the configuration and operating states of a hybrid AC/DC low voltage microgrid. Below are the contributions of this paper:

- modification of hypermutation operator used in the CLONALG algorithm,
- development of an optimization algorithm of operating states of hybrid AC/DC low voltage microgrid,
- comparison of the obtained results using classic and modified CLONALG algorithms as well as evolutionary algorithms.

The remainder of this paper is organized as follows: Section 2.1 presents the formulated optimization tasks. The mathematical models of this tasks are presented in Section 2.2. Section 3 describes the proposed microgrid optimization algorithm. The case study, including description of the test hybrid AC/DC low voltage microgrid, results of optimization calculations, as well as comparison of the calculation results obtained using the CLONALG algorithm and the evolutionary algorithm and discussion about these results are presented in Section 4. The summary and main conclusions are included in Section 5. The paper ends with a list of references.

2. Optimization Problem Formulation

In order to ensure proper operation of the hybrid AC/DC microgrid as a coherent system, it is necessary to implement appropriate strategies for controlling the operation of individual components of this system. The strategy of centralized two-stage control is considered in this article. Each of the microsources, ESDs, EPCs, and controlled loads should be equipped with a local controller. The task of local controllers is to collect information about the status of individual devices and send them to a central controller, which carries out the process of optimal control of the microgrid. Local controllers also receive signals from the central controller and force the appropriate behavior of the devices they control. The adopted control strategy allows the hybrid microgrid to operate in a synchronous mode with an external distribution grid, or autonomously in island mode. In both of these cases, determining the operating states is necessary to meet the given optimization criteria. The central controller must also distinguish between a number of factors affecting the possible operating states of microgrid components, such as power demand values and generation capacities of RES during the optimization period, acceptable regulatory ranges of individual devices in microgrid, instantaneous energy storage state of charge (SOC) levels, technical data of microgrid components, and mode of microgrid cooperation with an external distribution grid.

2.1. Optimization Tasks

In this paper, three single-criteria optimization tasks are formulated:

- task 1—minimization of total active power losses,
- task 2—minimization of costs associated with the operation of the hybrid AC/DC microgrid,

- task 3—maximization of level of power generated by RES.

In the first task, the control strategy assumes that the individual devices included in a hybrid microsystem will be controlled to obtain the lowest possible values of total active power losses in the optimization period under consideration. The amount of this losses can be described by the following formula:

$$\Delta P_{TOT_T} = \sum_{i=1}^{N_l} \Delta P_{l_i} + \sum_{j=1}^{N_{TR}} \Delta P_{TR_j} + \sum_{k=1}^{N_{EPC}} \Delta P_{EPC_k} \tag{1}$$

The reduction of active power losses is achieved by changing the power flow in the microgrid, resulting from the levels of microsources generation, load level and mode of operation of ESDs, as well as the demand for controlled loads. To minimize active power losses, microgrid central controller must have information about the current power flow in the microgrid and determine the expected power flow for subsequent settings of individual microgrid elements.

In the second task, control strategy assumes determination of such operating states of individual components of hybrid microgrid so that the total costs related to the functioning of this microgrid in the considered optimization period will be as low as possible. In order to implement this strategy, a hybrid microgrid operator (*HMO*) was defined as an intermediary in financial settlements between customers and the distribution system operator (*DSO*). The costs to be minimized can be written using the following formulas:

$$C_{TOT_T} = C_{FIX_T} + C_{VAR_T} \tag{2}$$

$$C_{FIX_T} = C_{FIX_{DSO}} + C_{FIX_{MSHMO}} + C_{FIX_{ESHMO}} - C_{FIX_{MEL}} \tag{3}$$

$$C_{VAR_T} = C_{VAR_{DSO}} + C_{VAR_{MSHMO}} + C_{VAR_{MSL}} + C_{VAR_{ESHMO}} + C_{VAR_{ESL}} \tag{4}$$

In the last task, the control strategy consists of determining the operating states of individual components of the hybrid microgrid so that the sum of power generated in microsources using renewable primary energy resources is as high as possible in the considered optimization period. The level of power generated by *RES* is determined by the following formula:

$$P_{RES_T} = \sum_{i=1}^{N_{RES}} P_{G_i} \tag{5}$$

2.2. Mathematical Models of Formulated Optimization Tasks

In order to solve formulated optimization tasks, appropriate mathematical models are defined for each of them, containing a problem representation, the form of objective function, and a set of constraints.

For each of the formulated tasks, a δ vector is defined, which represents a set of solutions to a given optimization problem. This vector contains a binary sequence coding the operating states of individual components of the hybrid microgrid. On the basis of the data contained in the δ vector, load flow calculations on the hybrid microgrid are performed, and then, depending on the considered optimization criterion, the following are determined: active power losses, costs related to the operation of the hybrid microgrid and RES generation levels. The mathematical notation of the objective functions defined for individual tasks is as follows:

$$F_{O_1} = \min_{\delta}\{\Delta P_{TOT_T}(\delta)\} \tag{6}$$

$$F_{O_2} = \min_{\delta}\{C_{TOT_T}(\delta)\} \tag{7}$$

$$F_{O_3} = \max_{\delta}\{P_{RES_T}(\delta)\} \tag{8}$$

Determining the optimal operating states of a hybrid microgrid requires that the following constraints be met:

- none of the microsources/ESDs connected the hybrid microgrid may operate with output power greater than nominal power of this microsource/ESD:

$$S_i \le S_{n_i} \forall i \in MS_{AC} \quad (9)$$

$$P_i \le P_{n_i} \forall i \in MS_{DC} \quad (10)$$

$$S_{ESD_{AC_i}} \le S_{ESD_{ACn_i}} \forall i \in SD_{AC} \quad (11)$$

$$P_{ESD_{DC_i}} \le P_{ESD_{DCn_i}} \forall i \in SD_{DC} \quad (12)$$

- none of the microsources/ESDs/EPCs connected to the AC part of hybrid microgrid may operate with a power factor $cos(\varphi)$ lower than the nominal power factor of this microsource/ESD/EPC:

$$cos\varphi_{MS_i} \ge cos\varphi_{MS_{n_i}} \forall i \in MS_{AC} \quad (13)$$

$$cos\varphi_{ESD_i} \ge cos\varphi_{ESD_{n_i}} \forall i \in SD_{AC} \quad (14)$$

$$cos\varphi_{EPC_i} \ge cos\varphi_{EPC_{n_i}} \forall i \in PC \quad (15)$$

- current flow in any of the power lines should not be greater than the long-term current carrying capacity of this power line:

$$I_i \le I_{cc_i} \forall i \in L \quad (16)$$

- voltage level at each of the hybrid microgrid nodes may not exceed the maximum or minimum allowable values:

$$U_{min_i} \le U_i \le U_{max_i} \forall i \in N \quad (17)$$

- power flow in the EPC/transformer cannot be greater than the nominal power of this EPC/transformer:

$$S_{EPC_{AC_i}} \le S_{EPC_{ni}} \forall i \in PC \quad (18)$$

$$P_{EPC_{DC_i}} \le P_{EPC_{ni}} \forall i \in PC \quad (19)$$

$$S_{TR_i} \le S_{TR_{ni}} \forall i \in TR \quad (20)$$

- The SOC level of each ESD should be within the limits allowed for that ESD:

$$SOC_{min_i} \le SOC_i \le SOC_{max_i} \forall i \in SD \quad (21)$$

- the synchronous generator acting as a balancing source in the AC part of the hybrid microgrid cannot go into motor operation:

$$P_{SG} \ge 0 \quad (22)$$

3. Description of the Proposed Microgrid Optimization Algorithm

An AIS based on a CLONALG will be used to solve the defined optimization tasks. In order to "build" a properly functioning immune system, it is necessary to determine:

- the method of representation of the optimization task solutions,
- how to create the initial set of antibodies (candidate solutions),

- the method of assessing the solution (formulation of the evaluation function),
- how to adapt the AIS during its operation.

A binary representation of the problem is assumed for each of the defined optimization tasks. The antibody should then be understood as the δ vector encoding the operating states of all devices controlled by the immune system within a hybrid microgrid. The δ vector consists of a binary sequence divided into groups of different lengths, coding individual operating states. The number of groups is equal to the number of operating states determined during the operation of AIS. Determining the number of bits belonging to a single group (length of a group) requires knowledge of the allowable adjustment range of individual devices and the expected accuracy (number of decimal digits).

The initial set of antibodies is created as a $N \times l$ matrix, where N is the number of antibodies in the set and l is the number of bits encoding the given antibody—size of the antibody. Knowing the number of antibodies and their size, the optimization algorithm randomly assigns the values "0" or "1" to individual bits, creating the initial set of antibodies.

After creating the initial set of antibodies, an optimization algorithm determines the operating states of individual devices installed in the microgrid. The next steps performed by the algorithm are the calculation of the power flow, determining the value of the evaluation function appropriate for the optimization task being solved.

The transformation of the objective functions defined in Section 3 into evaluation functions is necessary due to the development of a universal algorithm for solving minimizing and maximizing optimization tasks. In the case of minimizing tasks, the objective and evaluation functions are identical. For the maximization task 3 evaluation function take the following form:

$$eval_3 = C_3 - F_{o_3} \tag{23}$$

Formulation of the evaluation functions in accordance with the above-mentioned description aims at transforming all defined optimization tasks into minimizing tasks.

The load flow calculation performed by the optimization algorithm also allows to check whether the found solution does not violate the constraints. The algorithm enforces compliance with constraints by introducing appropriate penalty functions whose task is to increase the value of evaluation function in case of violation of constraints. The general mathematical notation of penalty functions is as follows:

$$eval_{p_i} = eval_i \cdot \prod_{j=1}^{n} \Psi_j \tag{24}$$

$$\Psi_j = \begin{cases} 1 & in\ the\ absence\ of\ violations \\ a_j + \psi_j{}^{b_j} & if\ violations\ occur \end{cases} \tag{25}$$

The operation of the AIS is based on determining the affinity of the antibody to the presented antigen, whose role in optimization tasks is performed by the antibody encoding the best solution found so far for the given task. The determination of affinity is calculated as follows:

$$AFF_i = \frac{best\_eval}{eval_{p_i}} \tag{26}$$

After determining the affinity, antibodies in the set are sorted in descending order. The next step performed by the algorithm is to select $N1$ antibodies with the highest affinity and create their clones. The number of clones is directly proportional to the affinity of the antibody and calculated as follows:

$$N_{CL_i} = N_{CL_{max}} - \frac{(AFF_{max} - AFF_i) \cdot (N_{CL_{max}} - N_{CL_{min}})}{AFF_{max} - AFF_{min} + \varepsilon}; \quad N_{CL_i} \in \mathbb{N} \tag{27}$$

Created clones are then subjected to a hypermutation process, inversely proportional to the affinity of the antibody. The probability of mutation of an antibody is determined as follows:

$$P_{MUT_i} = P_{MUT_{min}} + \frac{(AFF_{max} - AFF_i)\cdot(P_{MUT_{max}} - P_{MUT_{min}})}{AFF_{max} - AFF_{min} + \varepsilon} \quad (28)$$

In the classical variant of the CLONALG, the mutation operator generates $r = N \cdot l$ pseudo-random numbers between 0 and 1. For binary problem representation, mutation of a single bit in an antibody occurs when the generated pseudo-random number is less or equal to the probability of mutation. The presented scheme of the hypermutation operator operation shows that with a sufficiently high probability value, all bits in the analyzed antibody can undergo mutation. This article presents a modification of the hypermutation operator to enable changing the value of only a single bit in a given antibody. The modification consists of the fact that the number of generated pseudo-random numbers on the basis of which the algorithm decides whether to make a mutation has been limited to $r = N$. In case the generated pseudo-random number is less than the probability of mutation, a second pseudo-random generator is launched, which randomizes an integer ranging from 1 to l. The generated second pseudo-random number is the position of the bit being mutated. The operation diagram of the modified hypermutation operator is shown in Figure 2.

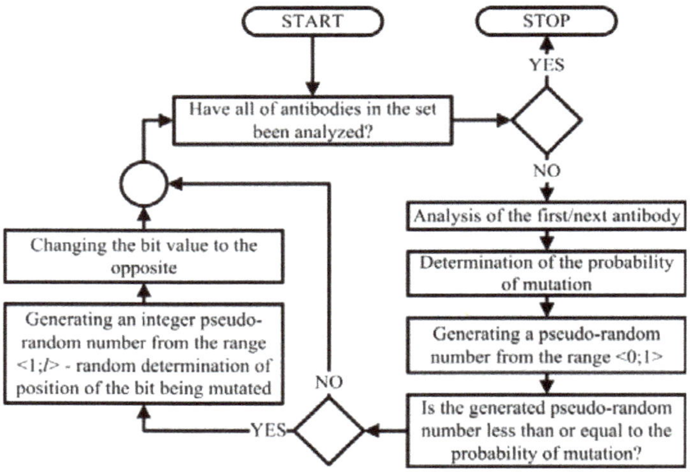

Figure 2. Operation diagram of the modified hypermutation operator.

After the hypermutation operation, the modified clones are added to the antibody set. To prevent excessive growth of the set of antibodies, the algorithm removes N2 antibodies with the lowest affinity, and then complements the free spots in the set with new, randomly generated antibodies. Then the algorithm goes to the next iteration by re-determining the operating states of individual devices installed in the microgrid. The algorithm's operation cycle is repeated until the stop condition is reached. The last step performed by the algorithm is to save the results of the optimization calculations.

The optimization algorithm was implemented using the DPL script language included in the PowerFactory v.15.2 software [31]. Ranges of settings of individual devices in the microgrid and ESDs SOC are loaded once after the script has been started. Generation profiles of RES and power demand profiles of consumers are cyclically loaded for each of the optimization periods considered. All mentioned input data are saved in appropriate text files. Changing the settings of parameters controlling the operation of the algorithm and economic quantities, such as electricity purchase prices or per unit fixed costs, is made directly in the source code of the script implementing the optimization algorithm. The general block diagram of the script implementing the optimization algorithm, taking into account the above description, is presented in Figure 3.

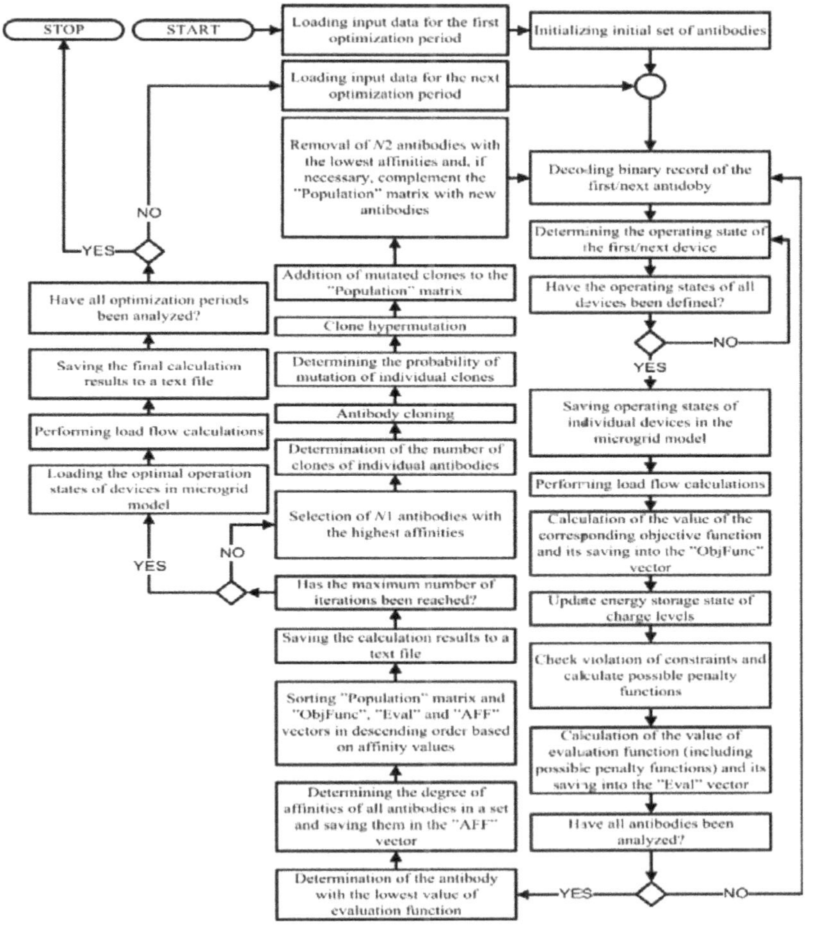

Figure 3. General block diagram of the script implementing the optimization algorithm.

4. Case Study

In order to evaluate the possibility of using the CLONALG with a modified hypermutation operator in the process of optimizing the configuration and operating states of the hybrid microgrid, exemplary calculations were carried out in the test microgrid working synchronously with the distribution power grid and in island mode. The optimization calculations were repeated using the CLONALG with the classic variation of the hypermutation operator and the evolutionary algorithm to compare the obtained results and verify the correct operation of modified CLONALG algorithm. Sample results of the calculations carried out are presented later in this paper.

4.1. Description of Test Hybrid Microgrid

Optimization calculations were carried out for a hybrid test microgrid supplying a single-family house estate. It is a microgrid consisting of AC and DC networks connected to each other with an EPC. Individual nodes of both types of network were connected by overhead lines. DC power lines were built as double-track AsXS 2 × 70 type lines and AC power lines were built as single AsXS 4 × 70 type lines. The AC network is connected to the external distribution grid via MV/LV transformer and AFL6 35 type medium voltage line. The technical data of individual elements of the hybrid test microgrid are given in Table 1. The schematic diagram of the test microgrid is presented in Figure 4.

Table 1. Technical data of individual elements of the hybrid test microgrid.

EPC	S_n [kVA]	U_{AC} [kV]	U_{DC} [kV]	$\cos \varphi_n$ [−]	ΔP_{Load} [kW]	ΔP_{Idle} [kW]
	125	0.4	0.4	0.8	5	0.5
MV/LV Transformer	S_n [kVA]	U_{MV} [kV]	U_{LV} [kV]	ΔU_k [%]	ΔP_{Cu} [kW]	ΔP_{Fe} [kW]
	63	15.75	0.4	4.5	1.2	0.18
AC line AsXS 4 × 70	$R'\left[\frac{\Omega}{km}\right]$	$X'\left[\frac{\Omega}{km}\right]$	I_{dd} [A]			
	0.443	0.083	213			
MV line AFL6 35	$R'\left[\frac{\Omega}{km}\right]$	$X'\left[\frac{\Omega}{km}\right]$	I_{dd} [A]			
	0.852	0.4	145			
DC line AsXS 2 × 70	$R'\left[\frac{\Omega}{km}\right]$	I_{dd} [A]				
	0.443	213				

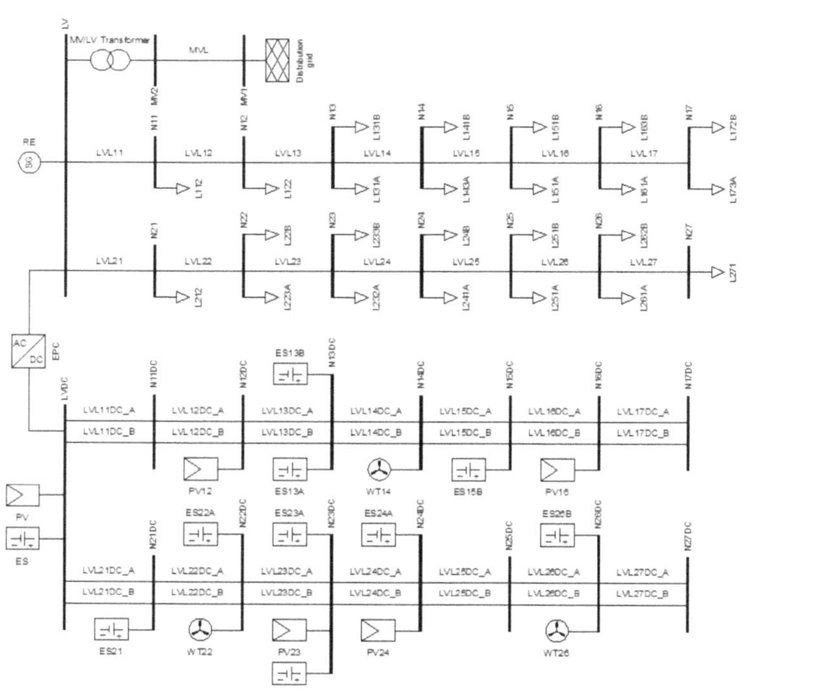

Figure 4. Schematic diagram of the test microgrid.

In a hybrid test microgrid, 24 non-controlled loads connected only to the AC network were modeled. Each of the load was characterized by assigning to it one of three different daily active and reactive power demand characteristics. The total daily power demand characteristics of the test microgrid are presented in Figure 5.

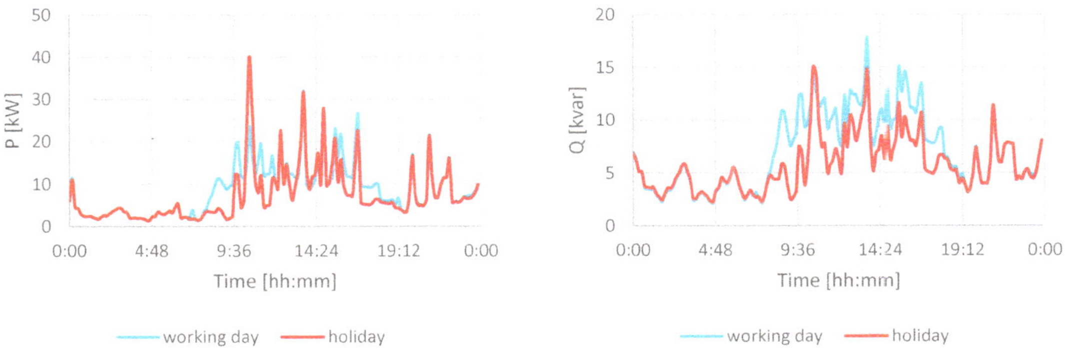

(a) (b)

Figure 5. (**a**) Total daily active power demand characteristics of the test microgrid; (**b**) Total daily reactive power demand characteristics of the test microgrid.

The hybrid test microgrid was equipped with 9 microsources, divided into three categories (1 reciprocating engine (RE) with synchronous generator connected directly to the AC network, 3 wind microturbine generation sets, and 5 photovoltaic sources connected to the DC network via power inverters). Technical data of installed microsources are presented in Table 2. Daily characteristics of photovoltaic sources and wind microturbine generation sets' generation capacity for two selected days of the year are presented in Figure 6.

Table 2. Technical data of the microsources installed in the hybrid test microgrid.

Microsource	Owner	Nominal Power [kW]/[kVA] *	P_{min} [kW]	P_{max} [kW]	Q_{min} [kvar]	Q_{max} [kvar]
AP	HMO	61	0	49	−36.7	36.7
PV	HMO	40	0	It depends	0	0
PV12	Consumer	4	0	on the	0	0
PV16	Consumer	25	0	atmospheric	0	0
PV23	Consumer	6	0	conditions	0	0
PV24	Consumer	10	0	prevailing in	0	0
TW14	Consumer	2	0	a given	0	0
TW22	Consumer	5	0	optimization	0	0
TW26	Consumer	3	0	period	0	0

* apparent power for the microsource connected to the AC network, active power for the microsource connected to the DC network.

For microsources owned by the HMO, fixed costs per unit of 0.0014 $USD/kW/T$ were adopted, where T is an optimization period of 10 min. HMO is not charged with fixed costs resulting from the maintenance of microsources owned by individual consumers. Variable costs per unit of the reciprocating engine were adopted at the level of 0.0279 $USD/kW/T$; it was also assumed that this value constitutes the purchase price of energy generated in microsources owned by individual consumers. Variable costs of the HMOs photovoltaic source are zero. The regulation of the generated power level of photovoltaic sources and wind microturbine generation sets is carried out by detuning the converter connecting the source with the hybrid microgrid from the maximum power operation point (MPP) on the production characteristics of the given source [32].

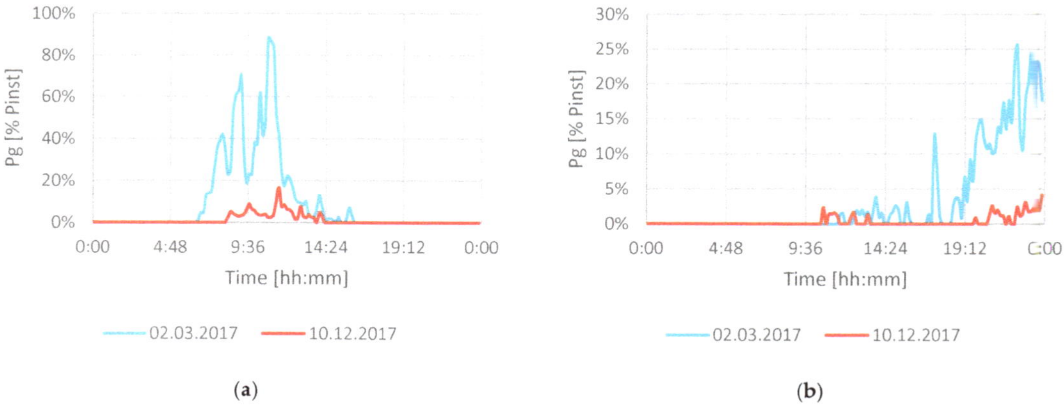

Figure 6. (**a**) Daily generation capacity charecteristics of photovoltaic sources; (**b**) Daily generation capacity characteristics of wind microturbine generation sets.

The test microgrid was equipped with 9 ESDs with a rated power of 50 kW and a capacity of 37 kWh, owned by individual consumers. The HMO also has one energy storage device with a rated power of 40 kW and a capacity of 160 kWh. All ESDs have been connected only to the DC network. In the process of controlling ESDs, it was assumed that they could operate in the full range of power regulation. However, the charge or discharge power may be reduced if the energy level in the storage device is not within acceptable limits. These limits are:

- from 5% to 95% of the storage capacity between 12:00 am to 6:50 pm,
- from 40% to 60% of the storage capacity between 7:00 pm to 11:50 pm.

The greater reduction on the SOC of ESDs in the evening is designed to prepare them for operation during the next day, so that they are able to balance the shortage or surplus of generated power in microsources in relation to the power demand of customers. As in the case of microsources, the HMO is not charged with fixed costs resulting from the maintenance of consumers ESDs. Consumers have full freedom in the choice of energy storage technology and capacity, which forces HMO to maintain the appropriate infrastructure enabling the connection of the storage device to the network. It has been assumed that the costs per unit of maintaining a single connection is 0.004 $USD/kW/T$. The same numerical value is a fixed costs per unit for the energy storage device owned by HMO. HMOs variable costs include per unit costs of discharging energy storage device, amounting to 0.0095 $USD/kW/T$. The price from the sale of energy taken to charge storage devices belonging to the consumers amounts to 0.0322 $USD/kW/T$ and it is HMO revenue.

It should also be noted that the energy storage device owned by HMO is not subject to optimization. The device works as a source balancing the DC network. The use of energy storage as a balancing element of the DC network allows optimization of the operating states of the EPC connecting both DC and AC networks, which translates into control of power flow between both networks. In the case of island operation of the hybrid microgrid, the reciprocating engine was excluded from the optimization process, also to assign the role of balancing source to this device.

4.2. Results of Optimization Calculations Carried out Using the Modified CLONALG Algorithm

In order to obtain the results of optimization calculations, a 24 h microgrid operation simulations were performed for a test microgrid. Simulations were made for both synchronous and island operation. Two different load demand profiles for working day and for holiday were taken into account, as well as two generation profiles for RES for 2 March 2017 and 10 December 2017. In total, eight simulations were carried out for a single optimization task. Each of the simulations was started for the same initial ESDs SOC levels and the following settings of the algorithm control parameters:

- number of antibodies: $N = 400$,
- number of antibodies selected for cloning: $N_1 = 40$,
- number of antibodies replaced by randomly generated new antibodies: $N_2 = 16$,
- maximum probability of mutation: $P_{mut\_max} = 0.53$,
- minimum probability of mutation: $P_{mut\_min} = 0.19$,
- maximum number of clones created for single antibody: $N_{CL_{max}} = 4$,
- minimum number of clones created for single antibody: $N_{CL_{min}} = 2$.

The number of iterations of the optimization algorithm depended on the chosen optimization tasks and microgrid operation mode; detailed values are given in Table 3.

Table 3. Number of iterations of the optimization algorithm.

Optimization Task	Synchronous Operation	Island Operation
Task 1	200	200
Task 2	200	450
Task 3	150	250

Selected results of optimization calculations (for a single optimization period) are presented in Tables 4–6. Exemplary daily changes of optimized values for the adopted generation profile of 2 March 2017 and power demand profile for the working day are presented in Figures 7–9.

Table 4. Selected results of optimization calculations in task 1.

Test Microgrid Operation Mode	Time	RES Generation Profiles Date	Day Type	Optimized Quality Indicator	Other Quality Indicators	
				Active Power Losses [kW]	Costs [USD]	Level of Power Generated in RES [kW]
Synchronous	10:50 am	2 March 2017	working day	1.086	3.565	35.550
			Holiday	0.839	3.766	15.468
		10 December 2017	working day	1.081	2.977	0.754
			holiday	1.421	2.958	1.147
Island	5:30 am	2 March 2017	working day	1.346	3.338	0.000
			holiday	1.161	3.578	0.000
		10 December 2017	working day	1.293	3.204	0.000
			holiday	1.670	3.454	0.000

Table 5. Selected results of optimization calculations in task 2.

Test Microgrid Operation Mode	Time	RES Generation Profiles Date	Day Type	Optimized Quality Indicator	Other Quality Indicators	
				Costs [USD]	Active Power Losses [kW]	Level of Power Generated in RES [kW]
Synchronous	10:50 am	2 March 2017	working day	3.304	1.231	0.000
			holiday	2.707	0.927	4.678
		10 December 2017	working day	3.053	2.515	0.000
			holiday	3.267	2.130	0.000
Island	5:30 am	2 March 2017	working day	2.601	0.886	0.000
			holiday	2.657	0.956	0.000
		10 December 2017	working day	3.094	1.235	0.000
			holiday	2.657	0.956	0.000

Table 6. Selected results of optimization calculations in task 3.

Test Microgrid Operation Mode	Time	RES Generation Profiles Date	Day Type	Optimized Quality Indicator	Other Quality Indicators	
				Level of Power Generated in RES [kW]	Costs [USD]	Active Power Losses [kW]
Synchronous	10:50 am	2 March 2017	working day	29.576	4.681	3.774
			holiday	62.094	5.230	3.148
		10 December 2017	working day	1.702	3.586	1.495
			holiday	1.696	4.525	3.039
Island	5:30 am	2 March 2017	working day	0.000	3.602	1.499
			holiday	0.000	3.602	1.499
		10 December 2017	working day	0.000	3.602	1.499
			holiday	0.000	3.602	1.499

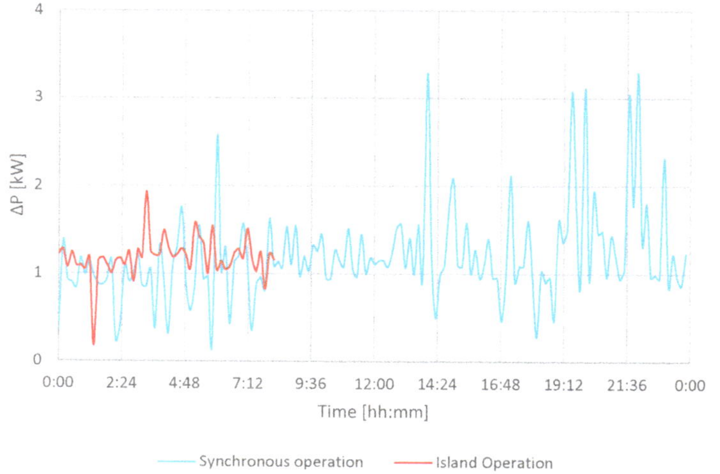

Figure 7. Daily changes of active power losses.

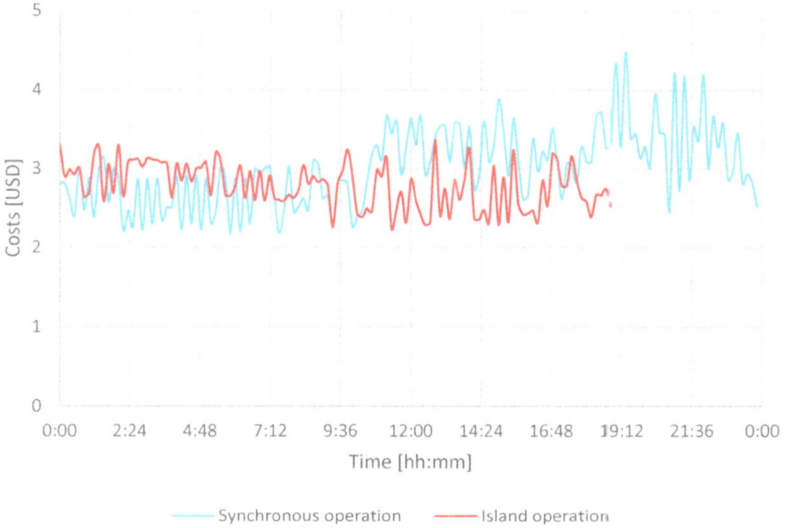

Figure 8. Daily changes of costs of operation of the hybrid test microgrid.

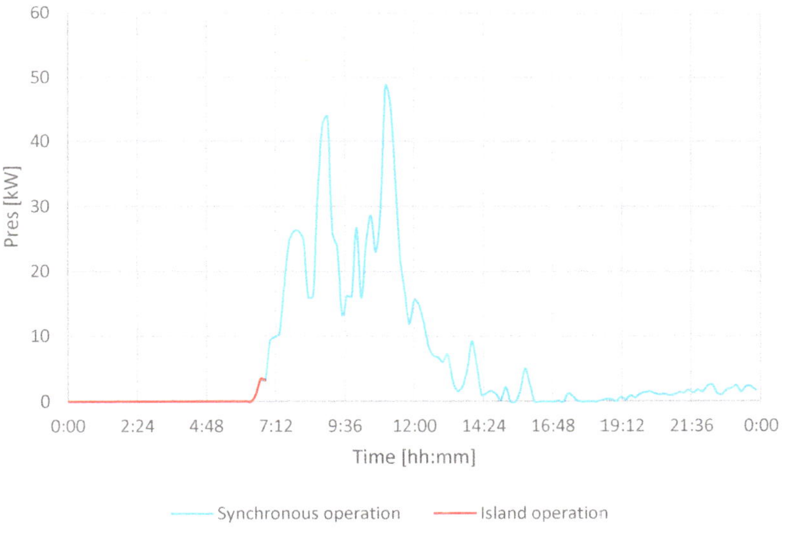

Figure 9. Daily changes of active power generated in RES.

Figure 10 presents the changes in the value of the evaluation function in task 1 for a microgrid operating synchronously with the distribution network depending on the selected RES generation capacity profile, and the power demand profile (the figure shows calculation made at 10:50 am).

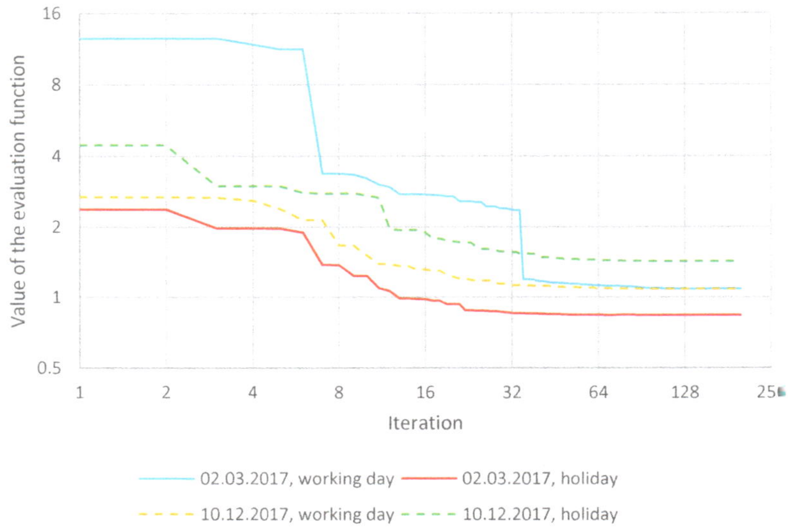

Figure 10. Progress of the optimization process.

Exemplary daily changes of the operating states of a selected microsource for a microgrid operating synchronously with the distribution network depending on the RES generation capacity profile of 2 March 2017 and the power demand profile for a working day are shown in Figure 11.

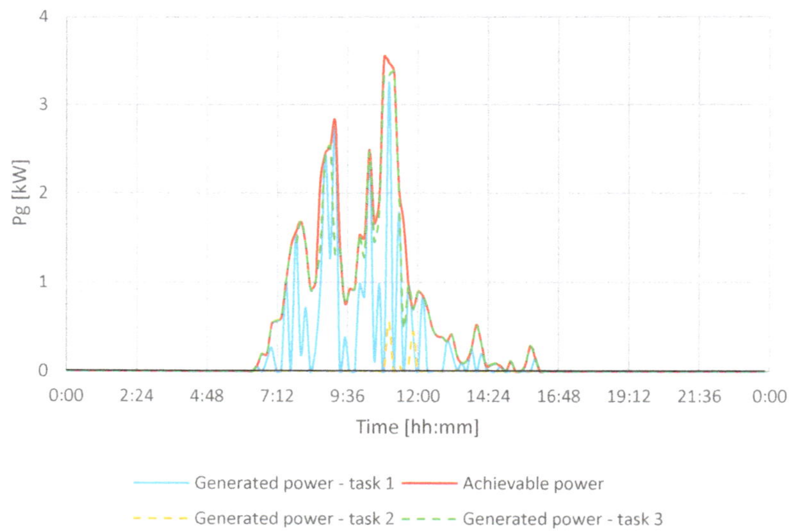

Figure 11. Daily changes of the operating states of the selected microsource (PV12).

4.3. Comparison of Calculation Results Obtained Using the CLONALG Algorithm and the Evolutionary Algorithm

Due to the modification of the hypermutation operator used in the CLONALG and the wish to verify the obtained optimization results, a comparative analysis of these results was carried out with the results obtained using the evolutionary algorithm, which is commonly used to solve optimization tasks in the field of power engineering [33–37]. In the evolutionary algorithm used for comparison, a stochastic sampling with replacement was used as a selection method. A binary representation of the problem, identical to the CLONALG, was also assumed. Within the control parameters of the evolutionary algorithm, a crossover probability of 0.22 and a mutation probability of 0.07 were assumed. The number of chromosomes was equal to the number of antibodies defined in the CLONALG and both algorithms performed the same number of iterations within the considered optimization period. The results of the comparison are shown in Tables 7 and 8.

Table 7. Comparison of results of optimization calculations for the microgrid synchronous operation.

Optimization Task	RES Generation Capacity Profile	Power Demand Profile	Percentage of Optimization Periods at Which: CLONALG Algorithm Achieved a Better Solution	Evolutionary Algorithm Achieved a Better Solution	Both Algorithms Reached an Identical Solution	Number of Analyzed Optimization Periods
Task 1	2 March 2017	working day	83.33%	16.67%	0.00%	144
		holiday	79.86%	20.14%	0.00%	144
	10 December 2017	working day	86.81%	13.19%	0.00%	144
		holiday	79.86%	20.14%	0.00%	144
Task 2	2 March 2017	working day	91.67%	8.33%	0.00%	144
		holiday	91.67%	8.33%	0.00%	144
	10 December 2017	working day	92.36%	7.64%	0.00%	144
		holiday	86.11%	13.89%	0.00%	144
Task 3	2 March 2017	working day	61.81%	2.78%	35.42%	144
		holiday	61.81%	2.78%	35.42%	144
	10 December 2017	working day	36.81%	0.00%	63.19%	144
		holiday	36.81%	0.00%	63.19%	144

Table 8. Comparison of results of optimization calculations for the microgrid island operation.

Optimization Task	RES Generation Capacity Profile	Power Demand Profile	Percentage of Optimization Periods at Which: CLONALG Algorithm Achieved a Better Solution	Evolutionary Algorithm Achieved a Better Solution	Both Algorithms Reached an Identical Solution	Number of Analyzed Optimization Periods
Task 1	2 March 2017	working day	90.00%	10.00%	0.00%	50
		holiday	92.16%	7.84%	0.00%	51
	10 December 2017	working day	93.24%	6.76%	0.00%	74
		holiday	88.41%	11.59%	0.00%	69
Task 2	2 March 2017	working day	100.00%	0.00%	0.00%	114
		holiday	98.77%	1.23%	0.00%	81
	10 December 2017	working day	100.00%	0.00%	0.00%	92
		holiday	98.91%	1.09%	0.00%	92
Task 3	2 March 2017	working day	14.29%	0.00%	85.71%	42
		holiday	11.36%	0.00%	88.64%	44
	10 December 2017	working day	30.77%	0.00%	69.23%	52
		holiday	14.00%	0.00%	86.00%	50

For the selected cases, a comparison was also made with the classic version of the CLONALG algorithm. Calculations were made for:

- all defined optimization tasks,
- synchronous and island operation of the test microgrid,
- total achievable RES active power generation capacity equal to 75.225 kW,
- total load demand for active power equal to 10.704 kW.

The results of the comparison are shown in Table 9. Figure 12 shows the convergence of the optimization process for tasks 1 and 2.

Table 9. Comparison of results of the optimization calculations.

Test Microgrid Operation Mode	Optimized Quality Indicator	Evolutionary Algorithm	CLONALG Algorithm	
			With Modified Hypermutation Operator	With Classic Hypermutation Operator
Synchronous	Active power losses [kW]	**1.220**	1.291	1.254
	Costs [USD]	3.923	**3.091**	3.679
	Level of power generated in RES [kW]	52.082	**61.478**	58.369
Island	Active power losses [kW]	1.180	**1.031**	1.081
	Costs [USD]	3.960	**2.871**	3.556
	Level of power generated in RES [kW]	55.774	**61.957**	57.875

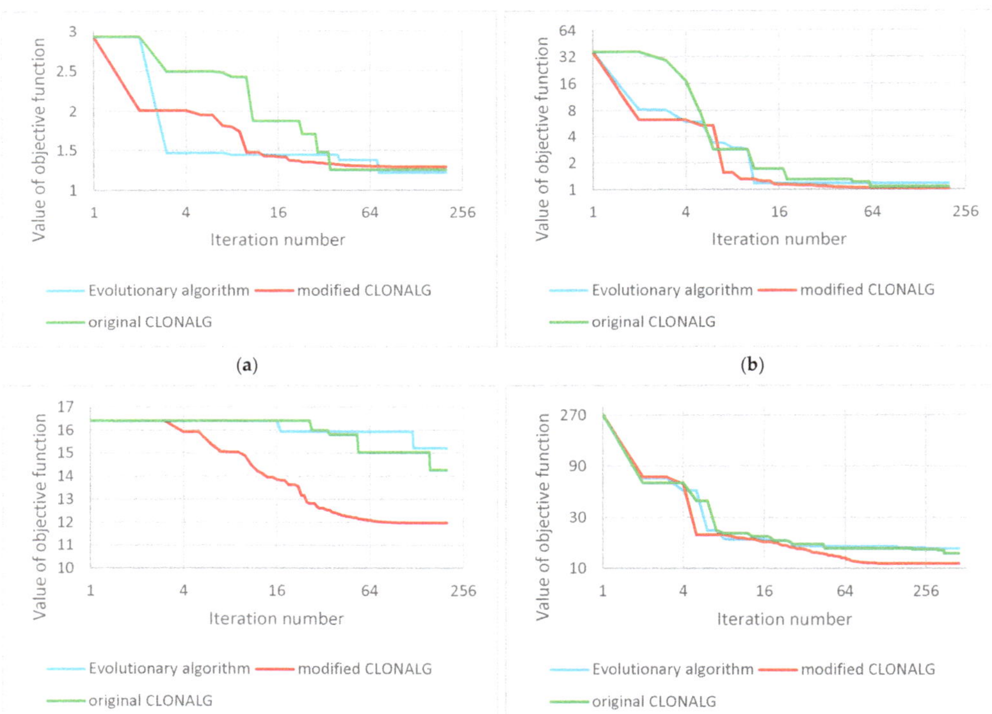

Figure 12. Convergence of optimization process for (**a**) task 1, synchronous operation; (**b**) task 1, island operation; (**c**) task 2, synchronous operation; (**d**) task 2, island operation.

4.4. Discussion

The concept of a hybrid low voltage AC/DC microgrid controlled by AIS could be an interesting way to integrate renewable energy sources, energy storage units, as well as electric vehicles into an efficient and easy to manage power microsystem.

Analyzing the results of optimization calculations constituting a 24 h simulation of the operation of the hybrid microgrid, it should be stated that the AIS, functioning on the basis of a CLONALG, is able to carry out the process of optimizing the configuration and operating states of the hybrid microgrid, working synchronously with the external distribution network. In case of the island operation, for all formulated optimization tasks, the algorithm was not able to ensure the correct operation of the microsystem for 24 h. Premature termination of optimization calculations is not due to the malfunction of the AIS, but to the structures of test microgrid that were not designed for long-term island operation.

The obtained results also depend on the adopted assumptions regarding RES generation profiles and consumer power demand profiles. Analysis of the results of the optimization calculations shows that there is a relationship between the results obtained and the choice of the power demand profile. Similar conclusions can be drawn based on the analysis of various RES generation capacity profiles. In the example of the task of maximizing the level of power generated by RES, there are clear differences between generation capacities in spring and winter.

Analyzing the progress of the optimization process, it should be stated that it proceeded correctly for all considered cases. Subsequent iterations of the algorithm for solving formulated optimization tasks results in a decrease in the value of the evaluation function. The sharp decline in the value of the evaluation function in the initial iterations of the algorithm testifies to the proper functioning of the AIS and the effective elimination of suboptimal solutions. The advantage of the CLONALG over the evolutionary algorithm in the first stage of the optimization process can be due to two mechanisms:

- cloning probability directly proportional to the affinity of the antibody to antigen, combined with inverse proportional hypermutation,
- removal from the population antibodies with the lowest affinity to antigen.

The first mechanism operates by selecting a certain number of antibodies with the highest affinity (the best solution in a given iteration) and subjecting them to the cloning and hypermutation processes. The second mechanism is used to protect the algorithm against an excessive increase in population size, and thus a decrease in its efficiency, by removing the worst solutions of the optimization task and, if necessary, supplementing the population with new randomly generated antibodies. Both mechanisms mentioned above cause that in the initial phase of operation, the AIS rejects the worst solutions faster than the evolutionary algorithm.

While searching for optimal solutions for formulated tasks, the AIS changed the operating states of individual devices in the test hybrid microgrid. The way the selected microgrid element works depends to a large extent on the chosen optimization task and on the input data. For example, in task 2, the analyzed microsource (photovoltaic panel) is switched off practically throughout the simulation, while in task 3, it works with the maximum achievable power.

A comparative analysis of optimization calculations carried out using an AIS based on a CLONALG and an evolutionary algorithm showed that for tasks 1 and 2 in most of analyzed cases more favorable results of calculations were obtained using AIS and thus the proposed optimization method is an effective optimization tool. For task number 3, the advantage of AIS is smaller than in previous cases and for some optimization periods, both algorithms obtained identical results. This is especially visible in the case of optimization of microgrid operating in the island mode; however, it should be noted that the number of analyzed optimization periods is relatively low, compared to the optimization of the synchronously operating microgrid.

Due to the wish to verify the modification of the hypermutation operator introduced in the CLONALG algorithm, a comparison (only for selected optimization periods) was also made with the classic version of this operator. The obtained calculation results indicate that the change in the mode of operation of the hypermutation operator resulted in an improvement of the results obtained in most of the considered cases.

When assessing the convergence of the examined optimization algorithms, it was noticed that the CLONALG algorithm with the modified hypermutation operator in most of analyzed cases gains an advantage over the other algorithms in the first few iterations of the optimization process. Further observations of the behavior of analyzed algorithms allow us to state that in the final stage of the optimization process, the differences in the obtained results are not relatively high and modified CLONALG rarely obtained worst solution. The complexity of the CLONALG algorithm, compared to the evolutionary algorithm will require a longer computation time. This fact theoretically acts as a disadvantage of AIS as an optimization tool, but the observed tendency to remove suboptimal solution quickly in an initial stage of optimization process may be an advantage of the method proposed in this paper.

5. Conclusions

From the obtained results of the optimization calculations, the following conclusions can be made:

- It is possible to use an AIS based on the CLONALG algorithm as an effective optimization tool for hybrid AC/DC microgrids operating on both synchronous and island mode,
- in most of the considered cases, optimization with the use of an AIS resulted in finding a better solution compared to optimization with an evolutionary algorithm,
- the introduced modification of the hypermutation operator contributed to the improvement of the obtained results in relation to the classic version of CLONALG algorithm as well as evolutionary algorithm,
- the modified CLONALG algorithm showed a stronger tendency to reject suboptimal solutions in the early phase of the optimization process than its classic version,
- the artificial immune system, operating on the basis of the CLONALG algorithm, requires greater computational effort that the commonly used evolutionary algorithm; however, the aforementioned tendency to quickly reject suboptimal solutions at an early stage of the optimization process may be an advantage of the AIS,
- both algorithms, due to the way they work, do not guarantee that the solution they find is a global optimum. The analysis of the convergence of these algorithms suggest that the obtained solutions are a quite good approximation of the global optimum.
- the optimization tasks formulated and solved in this paper can be implemented in real installations. However, this process requires the construction of an appropriate telecommunications infrastructure that allows monitoring the state of individual microgrid elements and sending control signals to them.

The author of the article considers it advisable to conduct further research on the possibility of using AISs in solving optimization problems in the field of power engineering, especially in the field of optimization of configuration and operating states of hybrid microgrids. Another important direction of future research is also the optimization of the structure of newly design hybrid microgrids in terms of the selection of the composition of generating units, ESDs, and EPCs coupling AC and DC networks. In addition, from the IT side, research is possible to increase the efficiency of the computational algorithms used.

Funding: This research received no external funding.

Institutional Review Board Statement: Not applicable.

Informed Consent Statement: Not applicable.

Data Availability Statement: Not applicable.

Conflicts of Interest: The authors declare no conflict of interest.

Abbreviations

The following abbreviations are used in this manuscript.

EA	Evolutionary algorithm
PSO	Particle swarm optimization
AIS	Artificial immune systems
AC	Alternating current
DC	Direct current
CLONALG	Clonal selection algorithm
RES	Renewable energy sources
ESD	Energy storage device
EPC	Electronic power converter
V2G	Vehicle to grid
MV	Medium voltage
LV	Low voltage
SOC	State of charge
HMO	Hybrid microgrid operator
DSO	Distribution system operator
RE	Reciprocating engine
MPP	Maximum power operation point

Nomenclature:

ΔP_{TOT_T}	total active power losses in the hybrid microgrid during the optimization period
$N_l/N_{TR}/N_{EPC}$	number of power lines/transformers/EPCs in the hybrid microgrid
$\Delta P_{l_i}/\Delta P_{TR_j}/\Delta P_{EPC_k}$	active power losses in the ith power line/jth transformer/kth EPC belonging to the hybrid microgrid
C_{TOT_T}	total costs related to the operation of hybrid microgrid during the optimization period
C_{FIX_T}/C_{VAR_T}	total fixed/variable costs related to the operation of hybrid microgrid during the optimization period
$C_{FIX_{DSO}}$	fixed costs related to maintaining the connection of hybrid microgrid with external distribution system
$C_{FIX_{MSHMO}}/C_{FIX_{ESHMO}}$	fixed costs related to the maintenance of microsources/ESDs owned by HMO
$C_{FIX_{MEL}}$	fixed costs related to maintaining the infrastructure enabling consumers to connect microsources and ESDs to the microgrid
$C_{VAR_{DSO}}$	variable costs associated with the purchase of energy from DSO
$C_{VAR_{MSHMO}}/C_{VAR_{ESHMO}}$	variable costs related to the operation of microsources/ESDs owned by HMO
$C_{VAR_{MSL}}/C_{VAR_{ESL}}$	variable costs related to the purchase of energy generated in microsources/taken from energy storage units owned by individual customers supplied from hybrid microgrid
P_{RES_T}	total active power generated in RES belonging to hybrid microgrid during the optimization period
N_{RES}	number of RES installed in hybrid microgrid
P_{G_i}	active power generated by ith RES
δ	vector which coding a candidate solution
F_{O_i}	objective function for ith optimization task

S_i/P_i	apparent/active power generated by ith microsource connected to the AC/DC part of hybrid microgrid
S_{n_i}/P_{n_i}	nominal apparent/active power of the ith microsource connected to the AC/DC part of hybrid microgrid
MS_{AC}/MS_{DC}	set of microsources connected to the AC/DC part of hybrid microgrid
$S_{ESD_{AC_i}}/P_{ESD_{DC_i}}$	apparent/active power measured at the connection point of ith ESD connected to AC/DC part of hybrid microgrid
$S_{ESD_{ACn_i}}/P_{ESD_{DCn_i}}$	nominal apparent/active power of the ith ESD connected to AC/DC part of hybrid microgrid
SD_{AC}/SD_{DC}	set of ESDs connected to AC/DC part of hybrid microgrid
$\cos\varphi_{MS_i}/\cos\varphi_{ESD_i}/\cos\varphi_{EPC_i}$	power factor of the ith microsource/ESD/EPC connected to the AC part of hybrid microgrid
$\cos\varphi_{MS_{n_i}}/\cos\varphi_{ESD_{n_i}}/\cos\varphi_{EPC_{n_i}}$	nominal power factor of the ith microsource/ESD/EPC connected to the AC part of hybrid microgrid
$MS_{AC}/SD_{AC}/PC$	set of microsources/ESDs/EPCs connected to the AC part of hybrid microgrid
I_i	current flow in the ith power line
I_{cc_i}	long-term current capacity of the ith power line
L	set of power lines belonging to the hybrid microgrid
U_{min_i}/U_{max_i}	minimum/maximum allowable voltage level for ith node
U_i	voltage measured in ith node
N	set of nodes belonging to the hybrid microgrid
$S_{EPC_{AC_i}}/P_{EPC_{DC_i}}$	apparent/active power flow through the alternating/direct current circuits of the ith EPC
$S_{EPC_{ni}}/P_{EPC_{ni}}$	nominal apparent/active power of the ith EPC
S_{TR_i}	apparent power flow through the ith transformer
$S_{TR_{ni}}$	nominal apparent power of ith transformer
PC/TR	set of EPCs/transformers belonging to the hybrid microgrid
SOC_{min_i}/SOC_{max_i}	minimum/maximum allowable SOC level of the ith ESD
SOC_i	current SOC level of the ith ESD
SD	set of ESDs belonging to the hybrid microgrid
P_{SG}	active power generated by a synchronous generator acting as a balancing source
$eval_i$	evaluation function for ith optimization task
C_3	non-negative constant for task 3
$eval_{p_i}$	evaluation function after taking into account the penalty functions
$eval_i$	evaluation function for ith optimization task
n	number of defined constraints
Ψ_j	general form of the penalty function for jth constraint
ψ_j	detailed form of the penalty function for jth constraint
a_j/b_j	non-negative coefficient of sensitivity of the penalty function significant for minor/major exceedances of jth constraint
AFF_i	affinity of ith antibody
best_eval	value of the evaluation function of the antibody encoding the best solution found so far
N_{CL_i}	number of clones of ith antibody from N1 selected antibodies
$N_{CL_{max}}/N_{CL_{min}}$	maximum/minimum number of clones created for a single antibody
AFF_{max}/AFF_{min}	maximum / minimum affinity of N1 selected antibodies
$\varepsilon \to 0$	constant value
P_{MUT_i}	probability of mutation of ith antibody among the clones created
$P_{MUT_{min}}/P_{MUT_{max}}$	minimum/maximum probability of mutation

References

1. Askarzadeh, A. A memory-based genetic algorithm for optimization of power generation in a microgrid. *IEEE Trans. Sustain. Energy* **2018**, *9*, 1081–1089. [CrossRef]
2. Rana, M.J.; Zaman, F.; Ray, T.; Sarker, R. Economic-environmental scheduling of community microgrid using evolutionary algorithm. In Proceedings of the 2020 IEEE Symposium series on computational intelligence (SSCI), Canberra, Australia, 1–4 December 2020. [CrossRef]
3. Xu, Y.Q.; Sun, K.Y. Economic optimization of a community-scale integrated energy microgrid based on PSO algorithm. In Proceedings of the 2020 12th IEEE PES Asia-Pacific and Energy Engineering Conference (APPEEC), Nanjing, China, 20–23 September 2020. [CrossRef]
4. Radosavljević, J.; Jevtić, M.; Klimenta, D. Energy and operation management of a microgrid using particle swarm optimization. *Eng. Optim.* **2016**, *48*, 811–830. [CrossRef]
5. De Castro, L.N.; Timmis, J.I. Artificial immune systems as a novel soft computing paradigm. *Soft Comput.* **2003**, *7*, 526–544. [CrossRef]
6. Lee, T.F.; Hsiao, Y.C.; Wu, H.Y.; Huang, T.L.; Fang, F.M.; Cho, M.Y. Optimization of reactive power compensation and voltage regulation using artificial immune algorithm for radial transmission networks. In Proceedings of the 2007 International Conference on Intelligent Systems Application to Power Systems, ISAP, Toki Messe, Niigata, Japan, 5–8 November 2007. [CrossRef]
7. Pattanaik, J.K.; Basu, M.; Dash, D.P. Optimal Power flow with FACTS devices Rusing artificial immune systems. In Proceedings of the 2017 International Conference on Technological Advancements in Power and Energy, TAP Energy, Kollam, India, 21–23 December 2017. [CrossRef]
8. Gárate, J.I.; Ibarra, E.; Martinez de Alegria, I.; Planas, E.; Andreu, J. AC and DC technology in microgrids: A review. *Renew. Sustain. Energy Rev.* **2014**, *43*, 726–749.
9. Hybrid AC/DC Microgrids: A Bridge to Future Energy Distribution System. Available online: http://www.acs.eonerc.rwth-aachen.de/cms/E-ON-ERC-ACS/Forschung/Abgeschlossene-Projekte/~{}euwe/HYBRID-AC-DC-MICROGRIDS-A-BRIDGE-TO-FUT/?lidx=1 (accessed on 15 May 2019).
10. Liu, X.; Wang, P.; Loh, P.C. A hybrid AC/DC micro-grid. In Proceedings of the 2010 9th International Power and Energy Conference, IPEC 2010, Singapore, 27–29 October 2010. [CrossRef]
11. Xiao, J.; Nguyen, X.B.; Wang, P.; Huang, J.; Zhou, Q. Implementation of DC/DC converter with high frequency transformer (DHFT) in hybrid AC/DC microgrid. In Proceedings of the 2017 Asian Conference on Energy, Power and Transportation Electrification, ACEPT, Singapore, 24–26 October 2017. [CrossRef]
12. Bandla, K.C.; Prasad Padhy, N. A Multilevel Dual Converter Fed Open End Transformer Configuration for Hybrid AC-DC Microgrid. In Proceedings of the IEEE Power and Energy Society General Meeting, Portland, OR, USA, 5–10 August 2018. [CrossRef]
13. Gu, L.; Jin, K.; Zhu, H.; Wang, C. A single-stage isolated three-phase bidirectional AC/DC converter. In Proceedings of the 2014 IEEE 5th International Symposium on Power Electronics for Distributed Generation Systems, PEDG, Galway, Ireland, 24–27 June 2014. [CrossRef]
14. Lachichi, A.; Junyent-Ferre, A.; Green, T. Power converters design for hybrid LV ac/dc microgrids. In Proceedings of the 2017 6th International Conference on Renewable Energy Research and Applications, ICRERA, San Diego, CA, USA, 5–8 November 2017. [CrossRef]
15. Wu, X.; Wang, Z.; Ding, T.; Li, Z. Hybrid AC/DC Microgrid Planning with Optimal Placement of DC Feeders. *Energies* **2019**, *12*, 1751. [CrossRef]
16. Li, J.; Cai, H.; Yang, P.; Wei, W. A Bus-Sectionalized Hybrid AC/DC Microgrid: Concept, Control Paradigm, and Implementation. *Energies* **2021**, *14*, 3508. [CrossRef]
17. Nasir, M.; Khan, H.A. Solar photovoltaic integrated building scale hybrid AC/DC microgrid. In Proceedings of the 5th IET International Conference on Renewable Power Generation (RPG), London, UK, 21–23 September 2016. [CrossRef]
18. Ma, T.; Cintuglu, M.H.; Mohammed, O. Control of hybrid AC/DC microgrid involving energy storage, renewable energy and pulsem loads. In Proceedings of the 2015 IEEE Industry Applications Society Annual Meeting IAS 2015, Addison, OR, USA, 18–22 October 2015. [CrossRef]
19. Vigneysh, T.; Kumarappan, N.; Arulraj, R. Operation and control of wind/fuel cell based hybrid microgrid in grid connected mode. In Proceedings of the 2013 IEEE International Multi-Conference on Automation, Computing, Control, Communication and Compressed Sensing, (iMac4s), Kattayam, India, 22–23 March 2013. [CrossRef]
20. Debela, T.; Bhattacharya, A. Design and analysis of a DC/AC microgrid with centralized battery energy storage system. In Proceedings of the IECON 2017—43rd Annual Conference of the IEEE Industrial Electronics Society, Beijing, China, 29 October–1 November 2017. [CrossRef]
21. Rahman, M.S.; Hossain, M.J.; Lu, J. Utilization of parked EV-ESS for power management in a grid-tied hybrid AC/DC microgrid. In Proceedings of the Ausralasian Universities Power Engineering Conference, AUPEC 2015, Wollongong, Australia, 27–30 September 2015. [CrossRef]
22. Rahman, M.S.; Hossain, M.J.; Rafi FH, M.; Lu, J. EV charging in a commercial hybrid AC/DC microgrid: Configuration, control and impact analysis. In Proceedings of the 2016 Australasian Universities Power Engineering Conference, AUPEC 2016, Brisbane, Australia, 25–28 September 2016. [CrossRef]

23. Ouramdane, O.; Elbouchikhi, E.; Amirat, Y.; Sedgh Gooya, E. Optimal Sizing and Energy Management of Microgrids with Vehicle-to-Grid Technology: A Critical Review and Future Trends. *Energies* **2021**, *14*, 4166. [CrossRef]
24. Kaushik, R.A.; Pindoriya, N.M. A hybrid AC-DC microgrid: Opportunities & key issues in implementation. In Proceedings of the IEEE International Conference on Green Computing, Communication and Electrical Engineering, ICGCCEE 2014, Coimbatore, India, 6–8 March 2014. [CrossRef]
25. Unamuno, E.; Barrena, J.A. Hybrid ac/dc microgrids—Part I: Review and classification of topologies. *Renew. Sustain. Energy Rev.* **2015**, *52*, 1251–1259. [CrossRef]
26. Mirsaeidi, S.; Dong, X.; Said, D.M. Towards hybrid AC/DC microgrids: Critical analysis and classification of protection strategies. *Renew. Sustain. Energy Rev.* **2018**, *90*, 97–103. [CrossRef]
27. Mirsaeidi, S.; Dong, X.; Shi, S.; Tzelepis, D. Challenges, advances and future directions in protection of hybrid AC/DC microgrids. *IET Renew. Power Gener.* **2017**, *11*, 1495–1502. [CrossRef]
28. Rahman Fahim, S.; Sarker, S.K.; Muyeen, S.M.; Sheikh, M.R.I.; Das, S.K. Microgrid Fault Detection and Classification: Machine Learning Based Approach, Comparison, and Reviews. *Energies* **2020**, *13*, 3460. [CrossRef]
29. Ambia, M.N.; Al-Durra, A.; Muyeen, S.M. Centralized power control strategy for AC-DC hybrid micro-grid system using multi-converter scheme. In Proceedings of the IECON 2011—37th Annual Conference of the IEEE Industrial Electronics Society, Melbourne, Australia, 7–10 November 2011. [CrossRef]
30. Aryani, D.R.; Song, H. Coordination Control Strategy for AC/DC Hybrid Microgrids in Stand-Alone Mode. *Energies* **2016**, *9*, 469. [CrossRef]
31. PowerFactory v.15.2 Manual (DIgSILENT Technical Documentation). Available online: http://www.digsilent.de (accessed on 13 September 2019).
32. Hoke, A.; Maksimović, D. Active power control of photovoltaic power systems. In Proceedings of the 2013 1st IEEE Conference on Technologies for Sustainability, SusTech, Portland, OR, USA, 1–2 August 2013. [CrossRef]
33. Khosa, F.K.; Zia, M.F.; Bhatti, A.A. Genetic algorithm based optimization of economic load dispatch constrained by stochastic wind power. In Proceedings of the ICOSST 2015 International Conference on Open Source Systems and Technologies, Lahore, Pakistan, 17–19 December 2015. [CrossRef]
34. Naimi, D.; Ahmed, S.; Bouktir, T. An efficient optimization method based on genetic algorithm applied to reduce greenhouse gases in power system. In Proceedings of the 2013 International Conference on Control, Decision and Information Technologies, CoDIT, Hammamet, Tunisia, 6–8 May 2013. [CrossRef]
35. Babaei, A.; Razavi, S.; Kamali, S.; Gholami, A. Performance improvement of power system stabilizer be genetic algorithm for one machine infinite bus system. In Proceedings of the 2007 42nd International Universities Power Engineering Conference, Brighton, UK, 4–6 September 2007. [CrossRef]
36. Liang, H.; Gu, X.; Zhao, D. Optimization of system partitioning schemes for power system black-start restoration based on genetic algorithms. In Proceedings of the 2010 Asia-Pacific Power and Energy Engineering Conference, APPEEC, Chengdu, China, 28–31 March 2010. [CrossRef]
37. Atia, R.; Yamada, N. Optimization of a PV-wind-diesel system using a hybrid genetic algorithm. In Proceedings of the 2012 IEEE Electrical Power and Energy Conference, London, ON, Canada, 10–12 October 2012. [CrossRef]

energies

Article

Advanced Ensemble Methods Using Machine Learning and Deep Learning for One-Day-Ahead Forecasts of Electric Energy Production in Wind Farms

Paweł Piotrowski [1,*], Dariusz Baczyński [1], Marcin Kopyt [1] and Tomasz Gulczyński [2]

[1] Electrical Power Engineering Institute, Warsaw University of Technology, Koszykowa 75 Street, 00-662 Warsaw, Poland; dariusz.baczynski@ee.pw.edu.pl (D.B.); marcin.kopyt@ee.pw.edu.pl (M.K.)
[2] Globema Sp. z o. o., Wita Stwosza 22 Street, 02-661 Warsaw, Poland; tomasz.gulczynski@globema.pl
* Correspondence: pawel.piotrowski@ee.pw.edu.pl; Tel.: +48-22-234-7314

Abstract: The ability to precisely forecast power generation for large wind farms is very important, since such generation is highly unstable and creates problems for Distribution and Transmission System Operators to properly prepare the power system for operation. Forecasts for the next 24 h play an important role in this process. They are also used in energy market transactions. Even a small improvement in the quality of these forecasts translates into more security of the system and savings for the economy. Using two wind farms for statistical analyses and forecasting considerably increases credibility of newly created effective prediction methods and formulated conclusions. In the first part of our study, we have analysed the available data to identify potentially useful explanatory variables for forecasting models with additional development of new input data based on the basic data set. We demonstrate that it is better to use Numerical Weather Prediction (NWP) point forecasts for hourly lags: −3, 2, −1, 0, 1, 2, 3 (original contribution) as input data than lags 0, 1 that are typically used. Also, we prove that it is better to use forecasts from two NWP models as input data. Ensemble, hybrid and single methods are used for predictions, including machine learning (ML) solutions like Gradient-Boosted Trees (GBT), Random Forest (RF), Multi-Layer Perceptron (MLP), Long Short-Term Memory (LSTM), K-Nearest Neighbours Regression (KNNR) and Support Vector Regression (SVR). Original ensemble methods, developed for researching specific implementations, have reduced errors of forecast energy generation for both wind farms as compared to single methods. Predictions by the original ensemble forecasting method, called "Ensemble Averaging Without Extremes" have the lowest normalized mean absolute error (nMAE) among all tested methods. A new, original "Additional Expert Correction" additionally reduces errors of energy generation forecasts for both wind farms. The proposed ensemble methods are also applicable to short-time generation forecasting for other renewable energy sources (RES), e.g., hydropower or photovoltaic (PV) systems.

Keywords: wind energy; wind farm; ensemble methods; short-term forecasting; electric energy production; machine learning; deep neural network; swarm intelligence

Citation: Piotrowski, P.; Baczyński, D.; Kopyt, M.; Gulczyński, T. Advanced Ensemble Methods Using Machine Learning and Deep Learning for One-Day-Ahead Forecasts of Electric Energy Production in Wind Farms. *Energies* **2022**, *15*, 1252. https://doi.org/10.3390/en15041252

Academic Editor: Surender Reddy Salkuti

Received: 10 January 2022
Accepted: 6 February 2022
Published: 9 February 2022

1. Introduction

The impact of humanity on climate change is a fact accepted by most scientists and policymakers. Renewable energy sources have become a "natural" alternative to energy sources based on fossil fuels. Obviously, the largest increases in energy production come from wind sources. However, they are known for their basic disadvantage, which is intermittent power generation. A way to overcome this drawback is to develop best possible energy production forecasts and properly prepare the power system for operation by Distribution and Transmission System Operators. Forecasts for the next day play an important role in this process. They are also used in energy market transactions. Even a small improvement in the quality of these forecasts translates into improved security of the system and savings for the economy. Therefore, efforts are made to improve quality by:

- analysing the usefulness of various explanatory data;
- utilizing machine learning;
- preparing forecasts with single, team and hybrid methods;
- analysing the influence of point distribution of Numerical Weather Prediction (NWP) models at large wind farms;
- conducting comparative analysis of forecast quality for various wind farms.

The research presented in this paper concerns two medium-sized wind farms. No real-world wind speed data had been collected, which has made data analysis difficult.

1.1. Related Works

In recent years, ensemble models have become popular to tackle the deficiencies of single prediction models. The concept of ensemble is to achieve data variability to compensate for disadvantages of component models, such as bias, and obtain a solution that is more robust and less susceptible to the errors of NWP models. In their work, Liu, Chen, Lv, Wu and Liu [1] presented different ways of creating an ensemble. One solution (*sol1*) was based on achieving varying training data sets. Bagging and boosting mechanisms were indicated by the authors as a way to create such data, and decision tree-based methods as models using this type of data. Another solution (*sol2*) involved using different prediction models as components of the ensemble. In this case, the same class of prediction tools (different ANN) or their different classes (statistical and machine learning models) were both suggested as viable options. The third way of achieving variability (*sol3*) was to use the same prediction models with different components. MLP networks with different numbers of hidden layers and neurons in them or wavelet networks using different wavelets could be given as an example here. To systematize the papers presented below, they are assigned to the aforementioned groups.

Studies on sol1 have been presented in many works [2–8]. Research of Yildiz, Acikgoz, Korkmaz and Budak [2], Duan, Wang, Ma, Tian, Fang, Cheng, Chang, Y and Liu [3], and Abedinia et al. [4] addressed achieving sol1 by decomposition of input data into IMFs. On the other hand, Memarzadeh and Keynia [5] and Liu, Zhao, Yu, Zhang, and Wang [6] used wavelet decomposition, while Wang, Zhang and Ma [7] used single spectrum analysis instead. Like in the work of Sun, Zhao and Zhang [8], clusterization sometimes followed decomposition.

Literature concerning sol2 offers a plethora of model mixes. Piotrowski et al. [9] analysed different combinations of physical model, kNN regression, MLP and LSTM networks with PSO or BFGS optimization. Other researchers used 2 neural networks of the same type with different Lagrange polynomials in hidden layers [7], different predictive distributions [10], BPNN, ENN, ELM, LSTM [11], ANN-SVR-Gaussian process [12], etc

After data decomposition, Sun, Zhao, and Zhang [8] performed further clustering and created a separate LSTM model for each cluster. Thus, their work could be assigned not only to sol1 but also to the sol3 category. The same applies to the work by Chen and Liu [11], as the authors created the same models for data with different time resolutions. Others authors proposed, among others, using parallel stacked autoencoders [13] and LSTM networks [14–19] with different wavelet activation kernels [14] or with ensemble pruning and combination [15].

Some authors performed comparative analyses. Sun, Zhao, and Zhang [8] compared BP, Elman, and LSTM networks accuracy, Saini, Kumar, Mathur, and Saxena [16] compared RNN, NARX, and LSTM networks and Ahmadi et al. compared different tree models [17]; Kisvari, Lin, and Liu confronted LSTM with GRU [18], while Yildiz et al. [2] compared CNN with other deep learning methods. Although these studies lacked ensemble models as a cherry on top, the performed analyses could be of use when composing ensembles of these models. Semi-ensemble, switchable models would also be a viable alternative: Ouyang, Huang, He, and Tang [19] created models switched by the Markov chain regime, while Sun, Feng, and Zhang [10] created an ensemble with component models accuracy at previous time steps used as a switching condition.

Machine learning models have become frequently used prediction tools, not only as ensemble components, but also as standalone solutions. Decision trees with variants [17], SVR [10,20], and neural networks [8–10] are examples of quite popular predictors. With increasing average PC computational power, deep learning models gained their share of popularity, too. Among them, not only methods such as LSTM [3,6,8,11,14,16,18,21,22], GRU [18,23], or deep ESN [24] have been used in research, but also methods previously associated with image analysis like CNN [1,2,25–27] have been incorporated into studies. In their research, Wang, Li, and Yang [19] proposed an LSTM-based encoder to achieve input attention that understands the importance of variables, Sun, Zhao, and Zhang [8] created different LSTM hybrids for wind power series of multiple time scales, while Niu et al. [23] presented Sequence-to-Sequence GRU Networks as a recurrent method of multi-step ahead prediction.

In the papers reviewed by us, convolution networks were used to extract spatial information from data. In some cases [2,25,26], they were used to add spatial aspect to temporal information. For that purpose, Yin, Ou, Huang, and Meng [28] suggested extracting both temporal and spatial information by cascade of CNN followed by LSTM; in another case [27], extracted spatial information was a replacement for lacking time information.

Data extraction by CNN can be treated as semi-automatic input inference without user involvement. Some authors, however, preferred a different approach, i.e., feature engineering and input selection based on statistical analysis. Lin and Liu [29] presented wind data correction methods according to IEC standards, Medina and Ajenjo [30] presented analysis of optimal time lags for input variables with different time horizons, while other authors presented data cleaning and imputation by Lomnaofski norm [31], extensive sensitivity analysis of input data [9], and analyses of optimal sparsity of NWP model grids [32].

Last but not least, note that all of the mentioned deep learning and ensemble solutions could either use NWP data or be created as a stack of weather forecasting models followed by energy prediction models. Since generated energy prediction accuracy is usually affected by the accuracy of input data, enhanced weather forecasts could lead to improved energy prediction. Better weather forecast could be achieved in multiple ways, e.g., de Mattos Neto et al. [33] proposed in their paper an LSTM-SVR hybrid as a means of obtaining better wind speed forecasts.

1.2. Objective and Contribution

The main objectives of this paper can be summarized as follows:

- Perform extensive statistical analysis of time series of energy generated in two wind farms and perform statistical analysis of potential exogenous explanatory variables;
- Perform very extensive analysis of sensitivity of explanatory variables;
- Verify the accuracy of forecasts conducted by single methods, hybrid methods, and ensemble methods (13 methods in total);
- Develop and verify an original ensemble method, called "Ensemble Averaging Without Extremes" and conduct an original selection of combinations of predictors for ensemble methods;
- Identify the most efficient forecasting methods from among tested methods for data from both wind farms.

Below are listed selected contributions of this paper:

1. This research addresses forecasting for large wind farms. Although this topic frequently appears in literature, this research has its unique values. First of all, an extensive data analysis was performed, including the time series itself, additional data, and two different NWP model parameters (81 inputs in total). Secondly, 13 forecasting methods for two wind farms were tested and compared.
2. Development of an original method, called "Ensemble Averaging Without Extremes". Predictions by this method have yielded the lowest SS metric (Skill Score) and nMAE error among the tested methods; the original "Additional Expert Correction" method yielded additional improvement of forecasts.

3. Construction of a number of different models, data scenarios and parameters resulted in testing more than 400 forecasting models. This makes this research one of the most extensive studies on the topic. The conclusions drawn from this research can be generalized, at least for Central Europe.

The remainder of this paper is organized as follows: Section 2.1 presents statistical analysis of times series and NWP data for two wind farms. The importance of the available basic input data and additional input data is discussed in Sections 2.2 and 2.3. Section 3 describes prediction methods employed and Section 4 gives evaluation criteria for assessment of forecasting quality. Extensive analysis of the results and their discussion is in Section 5. Section 6 summarizes the whole research providing the main conclusions. References are listed at the end of this paper.

2. Data

2.1. Statistical Analysis

For statistical analyses, data acquired for two medium-sized European wind farms (A and B) were used. The range of the acquired data was identical for both farms and spanned from 4 April 2017 to 10 October 2019, with about 29 months in total. Rated powers for Farms A and B were 50 MW and 48.3 MW, respectively.

The following data were available for analysis:

- Wind farm generation time series (forecasted variable);
- Weather forecasts for wind farms location.

Records of actual meteorological parameters were not available; hence, GFS and ECMWF NWP models were used for our research instead. For ECMWF, the archived high-resolution atmospheric model was chosen (HRES) [34]. The GFS model was supplied by the Interdisciplinary Modelling Center, Warsaw University (ICM UW) [35,36]. Both models make it possible to use 4 forecast runs per day (at 0/6/12/18 UTC) with 1 h resolution and maximal horizon of 240 h. Time resolution of HRES changes, however, to a 3 h interval after 90 h horizon and to 6 h after 144 h horizon. For GFS, only the first interval change appears after reaching the 120 h horizon. For each wind farm, only weather forecasts corresponding to the respective spatial point were used. Weather source points for the ECMWF model were chosen as the points nearest to the ones appearing in a dense $1/8 \times 1/8$-degree grid. The same method was applied to GFS with its native spatial resolution of 0.25×0.25 degrees.

Data from both time series of electric energy production (Farm A and Farm B) were normalized separately for anonymization to relative units (1 relative unit is equal to the rated power of the wind farm). However, each time series of NWP forecasts data was normalized using min–max scaling.

Table 1 shows descriptive statistics for time series of hourly electric energy generated by the Wind Farm A and Wind Farm B considered here. Percentage distribution of electric energy generation for both wind farms is shown in Figure 1. The analysis of electric energy generation percentiles shows that values very close to 0 made up more than 25% of both time series samples. Usually, energy generation was within the range of (0–0.1) [p.u.] for both time series samples.

Calculated autocorrelation coefficient (ACF) of hourly generation in both time series shows a little daily periodicity. Autocorrelation coefficients quickly decrease for the following hours of the first day. For both time series, all autocorrelation coefficients are statistically significant (5% significance level) up to 3 days back (72 prior observations). Autocorrelation function (ACF) of the Wind Farm A energy generation time series is presented in Figure 2. However, autocorrelation function (ACF) of the Wind Farm B energy generation time series is presented in Figure 3.

Table 1. Descriptive statistics for hourly electric energy generation.

Descriptive Statistics	Wind Farm A	Wind Farm B
Mean	0.278 [p.u.]	0.288 [p.u.]
Standard deviation	0.284 [p.u.]	0.315 [p.u.]
Minimum	0.000 [p.u.]	0.000 [p.u.]
Maximum	0.990 [p.u.]	0.980 [p.u.]
Coefficient of variation	102.277%	109.164%
The 10th percentile	0.000 [p.u.]	0.000 [p.u.]
The 25th percentile (lower quartile)	0.025 [p.u.]	0.018 [p.u.]
The 50th percentile (median)	0.181 [p.u.]	0.163 [p.u.]
The 75th (upper quartile)	0.459 [p.u.]	0.483 [p.u.]
The 90 percentile	0.742 [p.u.]	0.861 [p.u.]
Variance	0.080	0.099
Skewness	0.906	0.941
Kurtosis	−0.320	−0.468

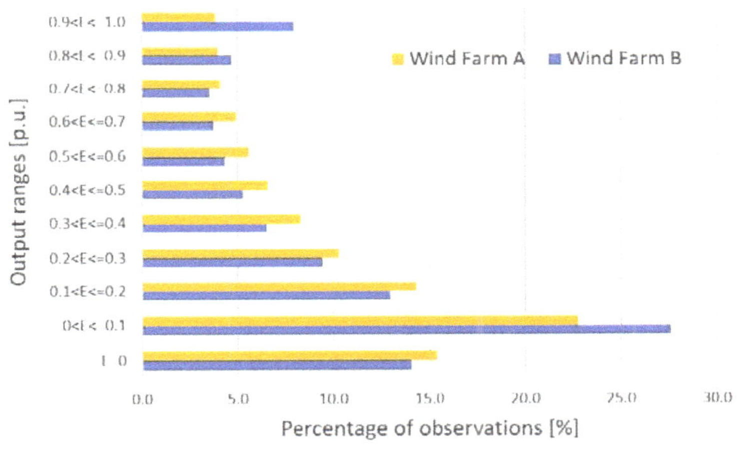

Figure 1. Percentage of time-series observations in particular generation ranges.

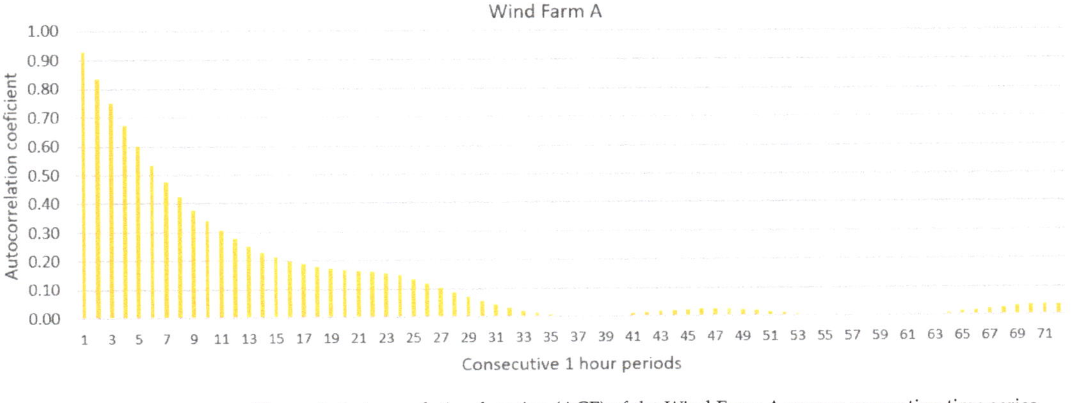

Figure 2. Autocorrelation function (ACF) of the Wind Farm A energy generation time series.

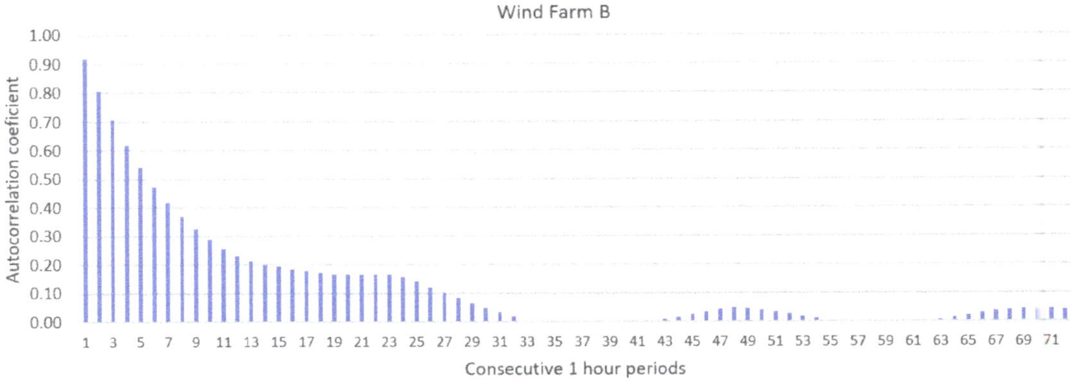

Figure 3. Autocorrelation function (ACF) of the Wind Farm B energy generation time series.

Figure 4 shows daily variability of hourly energy production [p.u.] of Wind Farms A and B. Arithmetic means of hourly energy generations for each hour of the day were calculated based on data span from 4 April 2017 to 28 September 2018 (18 months in total), with omitting test period datetimes—1 October 2018 to 1 October 2019. For the same periods, mean arithmetic hourly generations were calculated for each month, with the averaging of values for the months occurring two times. Pearson linear correlation coefficient between the data is equal to 0.950. Daily variability of electric energy generation of both wind farms is similar.

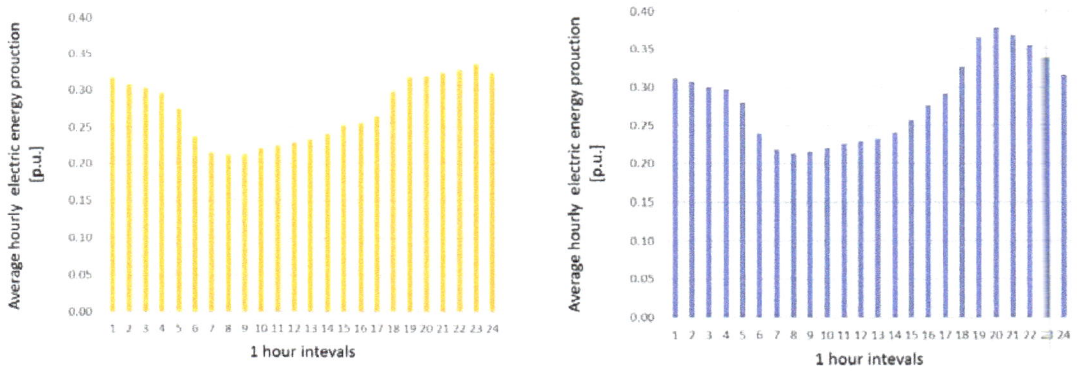

Figure 4. (**a**) Daily variability of hourly energy production of Wind Farm A; (**b**) Daily variability of hourly energy production of Wind Farm B.

Figure 5 shows seasonal variability of electrical energy generation of Wind Farms A and B. Pearson linear correlation coefficient between the data is equal to 0.922. Seasonal variability of electric energy generation of both wind farms is similar.

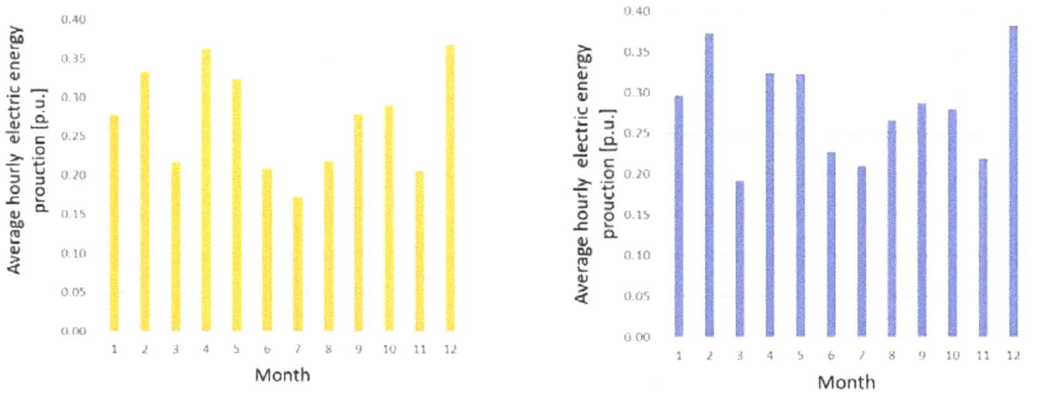

(a) (b)

Figure 5. (**a**) Seasonal variability of electrical energy production of Wind Farm A; (**b**) Seasonal variability of electrical energy production of Wind Farm B.

Figure 6 presents dispersion diagrams—relationships between wind speed forecasts [p.u.] for the beginning of a 1 h period of energy generation and actual production of electrical energy [p.u.] from Wind Farm A for 2 different NWP models (GFS and ECMWF). Figure 7 contains similar diagrams for Wind Farm B. For both figures, points are slightly more concentrated for the ECMWF model (cases **b**). All dispersion diagrams indicate a non-linear relationship between wind speed and the yield of electricity. The exact shape corresponding approximately to the shape of the wind turbine power curve typical for a single turbine cannot be well seen on the diagram due to low concentration of data points. Both of the observed disadvantageous phenomena result probably from the following reasons:

- Data include wind speed forecasts instead of actual, recorded values;
- Wind speed forecasts are momentary values for a given hour and actual wind speed usually changes during a 1 h period;
- Data come from very large wind farms with turbines scattered across vast territory with varying orography. For single wind turbines, data would probably be much more concentrated.

For both wind farms, extreme outliers were treated as unreliable samples, and further removed from data. This can be due to incorrect readings, missing data or scheduled/unscheduled shutdowns of at least a part of the wind farm. Only extreme, rarely occurring outliers were removed from the data, since big errors of wind speed prediction certainly must occur in a 24 h forecast horizon. A scenario with null wind speed forecast and non-zero electricity generation could be given as an example of NWP inaccuracy.

2.2. Analysis of Importance of Available Basic Input Data for Forecasting Methods

A detailed description of the available basic set of potential input variables for forecasting models is presented in Table 2. Figure 8 presents time points (momentary values time lags) of point weather forecasts from GFS and ECMWF models (input data) in relation to periods of electricity generation.

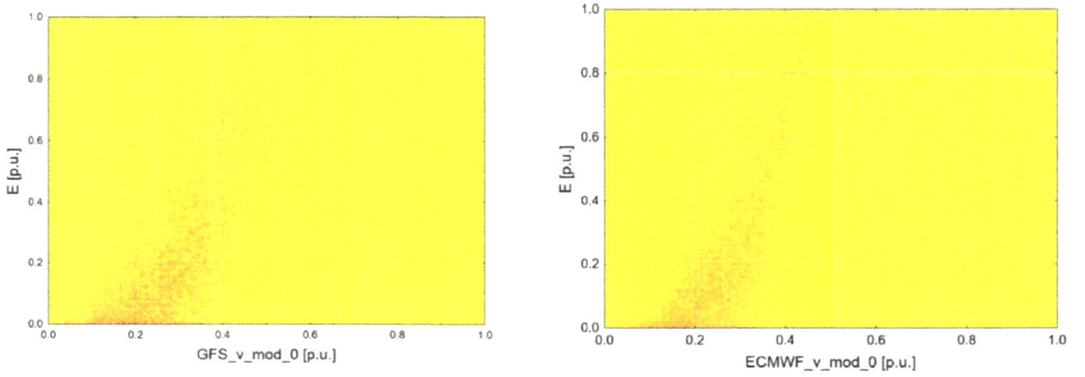

(**a**) (**b**)

Figure 6. (**a**) Relationship between wind speed forecast from GFS NWP model and electricity generation from Wind Farm A; (**b**) Relationship between wind speed forecast from ECMWF NWP model and electricity generation from Wind Farm A.

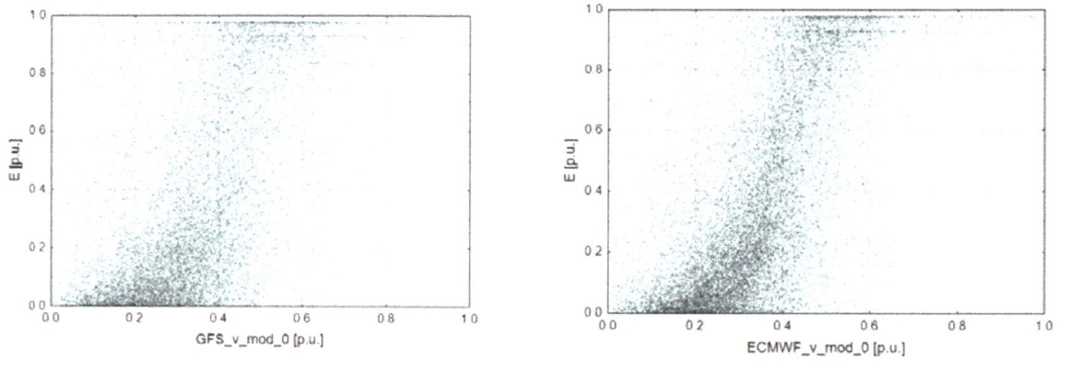

(**a**) (**b**)

Figure 7. (**a**) Relationship between wind speed forecast from GFS NWP model and electricity generation from Wind Farm B; (**b**) Relationship between wind speed forecast from ECMWF NWP model and electricity generation from Wind Farm B.

To identify the most important inputs for prediction models, extensive sensitivity analysis was performed for both wind farms. All of the 68 potential input variables that had been acquired were included. Comparison of the importance of input variables for both farms made it possible to draw general conclusions about the validity of use of given variables in predictions of electricity generation from large wind farms. Figure 9 presents consecutive steps of this analysis. Global Sensitivity Analysis (SA statistics) in the MLP network was performed for 4 models. Each trained model had 68 input variables and 1 output variable (electricity generation), 40, 50, 60, 70, and 80 (5 models) hidden neurons, used the BFGS learning algorithm, hyperbolic tangent hidden layer activation function, and linear output layer activation function. After training each MLP model, GSA was performed and the importance of each input variable was computed. Next, for each input, the overall rating was calculated as the arithmetic mean of 4 results of global sensitivity analysis obtained from each MLP model.

Table 2. Description of available basic input variables for forecasting models.

Input Data Numbers	Input Data Code/Codes	Description of Input Data (Three Categories)
	Category I. Markers of variability of wind farm's daily energy production	
1	hour	Numbers from 0 to 23 refer to the time of the forecast, where 0 refers to power generation from 23:00 to 00:00
2	ave_hour	Arithmetic mean of power generation for the given hour of the day (24 values)
	Category II. Lagged variables of hourly energy generation forecasted time series	
3–5	E-23 h, E-24 h, E-25 h	Energy generation lagged by 23/24/25 h from currently considered timestamp
	Category III. NWP forecasts	
6–12	GFS_t_−3, GFS_t_−2, ..., GFS_t_3	Seven values of air temperature point forecasts from GFS NWP model for respective hourly lags −3, −2, −1, 0, +1 [1], +2, +3 h
13–19	ECMWF_t_−3, ECMWF_t_−2, ..., ECMWF_t_3	Seven values of air temperature point forecasts from ECMWF NWP model for respective hourly lags −3, −2, −1, 0, +1 [1], +2, +3 h
20–26	GFS_p_−3, GFS _p_−2, ..., GFS _p_3	Seven values of atmospheric pressure point forecasts from GFS NWP model for respective hourly lags −3, −2, −1, 0, +1 [1], +2, +3 h
27–33	ECMWF_p_−3, ECMWF_p_−2, ..., ECMWF_p_3	Seven values of atmospheric pressure point forecasts from ECMWF NWP model for respective hourly lags −3, −2, −1, 0, +1 [1], +2, +3 h
34–40	GFS_h_−3, GFS_h_−2, ..., GFS_h_3	Seven values of air humidity point forecasts from GFS NWP model for respective hourly lags −3, −2, −1, 0, +1 [1], +2, +3 h
41–47	GFS_alpha_−3, GFS_alpha_−2, ..., GFS_alpha_3	Seven values of wind direction (0 degree at N, 90 at E) point forecasts from GFS NWP model for respective hourly lags −3, −2, −1, 0, +1 [1], +2, +3 h
48–55	ECMWF_alpha_−3, ECMWF_alpha_−2, ..., ECMWF_alpha_3	Seven values of wind direction (0 degree at N, 90 at E) point forecasts from ECMWF NWP model for respective hourly lags −3, −2, −1, 0, +1 [1], +2, +3 h
55–61	GFS_v_mod_−3, GFS_v_mod_−2, ..., GFS_v_mod_3	Seven values of wind speed (modulus calculated from NS and WE components) point forecasts from GFS NWP model for respective hourly lags −3, −2, −1, 0, +1 [1], +2, +3 h
62–68	ECMWF_v_mod_−3, ECMWF_v_mod_−2, ..., ECMWF_v_mod_3	Seven values of wind speed (modulus calculated from NS and WE components) point forecasts from ECMWF NWP model for respective hourly lags −3, −2, −1, 0, +1 [1], +2, +3 h

[1] 1 h lag refers to the time point for which forecast is to be generated, and energy generated is assigned to the hourly period between the considered time and one hour earlier.

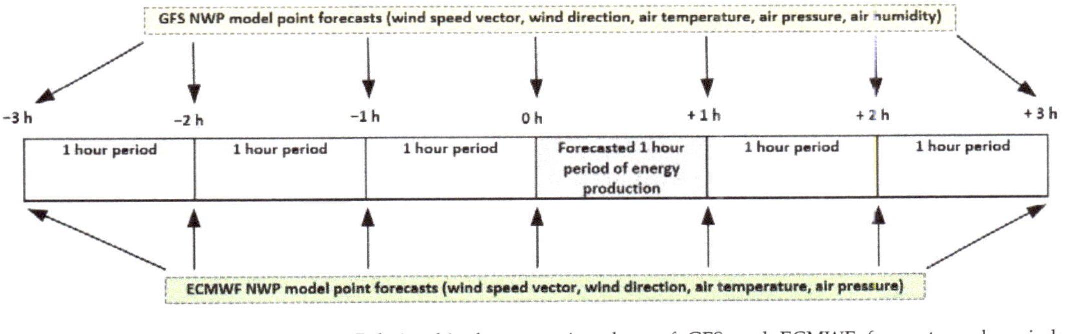

Figure 8. Relationship between time lags of GFS and ECMWF forecasts and periods of electricity generation.

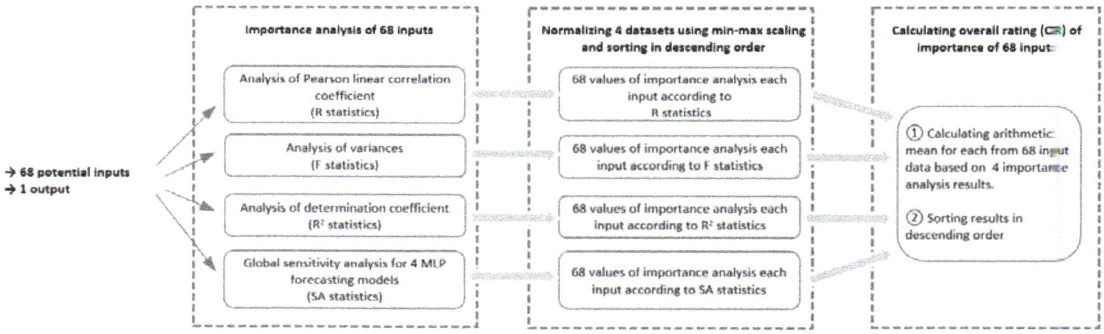

Figure 9. Consecutive steps of potential input data sensitivity analysis of made with 4 different methods and Overall Rating (OR).

Results of sensitivity analysis for potential input variables of prediction models are shown in Figure 10. The most important input variables are definitely wind speed forecasts, notably, the ones closest to the 1 h energy generation period. ECMWF NWP forecasts turned out higher in ranking than GFS forecasts, while the least important input variables were predictions of atmospheric pressure.

The importance of INPUT variables varied between 4 analytic methods used here for both value of metrics and position in importance ranking. The most differing results came from the SA method due to the non-linear modelling (MLP network) used in that method. The remaining analytic methods used linear modelling; hence, their results were similar. Figure 11 contains the interrelationship matrix (Pearson linear correlation coefficients) between 4 analytic methods used to determine the importance of input variables.

2.3. Analysis of Importance of Additional Input Data Created

A detailed description of an additional set of potential input variables for forecasting models, derived from mathematical transformation of basic data, is presented in Table 3. Additional input data are created to verify their potential usefulness and importance in the forecasting process. Wind speed forecasts using either one or both NWP models are averaged to reduce the random component. Percentage differences between averaged wind speed/atmospheric pressure point forecasts for respective pairs of hourly lags are computed to include additional information about the dynamics for wind speed/atmospheric pressure in the model. Physical model (turbine power curve) prognosis is another additional information. A third-order polynomial is used to approximate the power curve, while averaged wind speeds with lag 0 and 1 from time bounds of the predicted periods are inputs to this model.

Figure 12 shows the results of importance analysis of the additional input data created. The analysis was performed according to steps in Figure 3, the same as for the basic input data case, and used 68 basic input variables (described in Table 2) and 13 additional inputs (described in Table 3). Figure 12 presents partial results-40 best input variables out of the total of 81. Studies have shown that additional input data, in particular forecasts from physical models and average values of predictions, are highly valuable as prediction model explanatory data, as additional input data usually rank high in terms of importance (OR metrics). Moreover, similar results for both wind farms show the universality of our procedure of the construction of additional input data. As the next step, it was verified whether additional data can be advantageous for different methods of electricity generation forecasting.

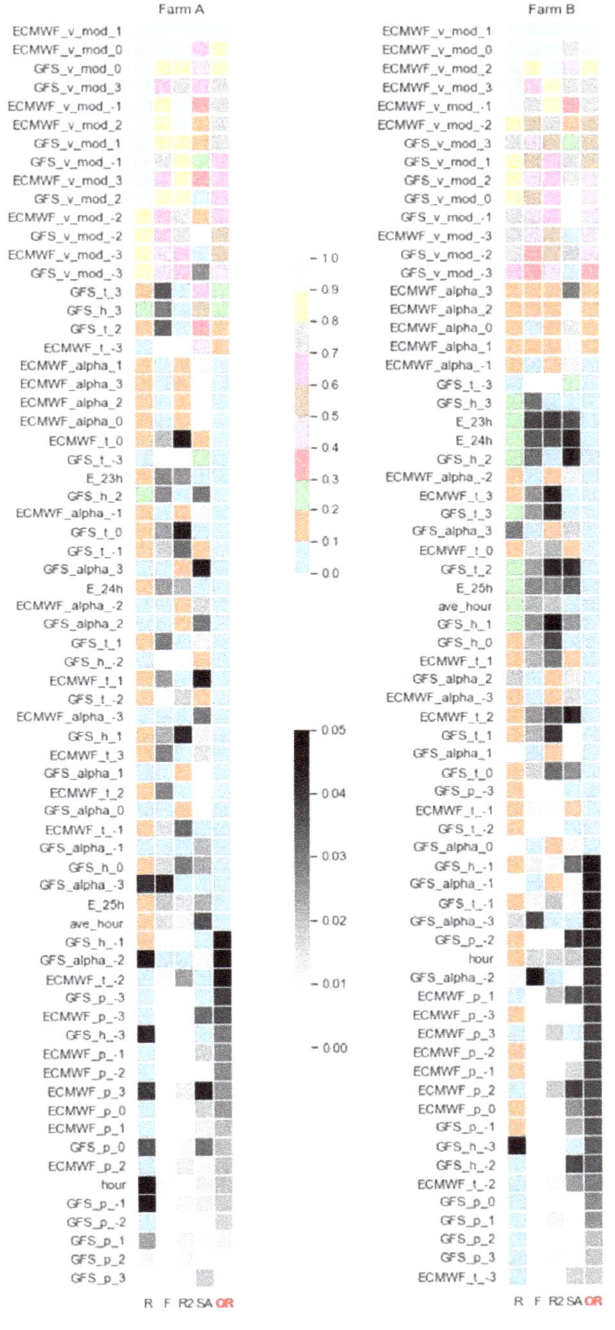

Figure 10. Results of sensitivity analysis of potential input variables for prediction models for Wind Farm A and Wind Farm B: 4 analysis methods and final Overall Rating (OR).

Wind Farm A

	R	F	R2	SA
R	1.000	-	-	-
F	0.973	1.000	-	-
R2	0.981	0.992	1.000	-
SA	0.539	0.564	0.530	1.000

Wind Farm B

	R	F	R2	SA
R	1.000	-	-	-
F	0.939	1.000	-	-
R2	0.948	0.990	1.000	-
SA	0.616	0.753	0.679	1.000

Figure 11. Interrelationship matrix (Pearson linear correlation coefficients) between 4 methods used to determine the importance of input variables.

Table 3. Description of additional input variables created for forecasting models.

Input Data Numbers	Input Data Code(s)	Description of Additional Input Data
1A	ave(GFS_ECMWF)_v_mod_0-1	Arithmetic mean of averaged wind speed point forecasts for hourly lags 0 h and +1 [1] h from ECMWF NWP and GFS NWP models
2A	ave_ECMWF_v_mod_0-1	Arithmetic mean of wind speed point forecasts for hourly lags 0 h and +1 [1] h from ECMWF NWP model
3A	ave_GFS_v_mod_0-1	Arithmetic mean of forecasts of wind speed point for hourly lags 0 h and +1 [1] h from GFS NWP model
4A–7A	differ(GFS_ECMWF)_v_mod_−2_−1 differ(GFS_ECMWF)_v_mod_−1_0 differ(GFS_ECMWF)_v_mod_0_1 differ(GFS_ECMWF)_v_mod_1_2	Percentage difference between averaged wind speed point forecasts for respective pair of hourly lags −2, −1, 0, −1 [1], +2 h from ECMWF NWP and GFS NWP models
8A–11A	differ(GFS_ECMWF)_p_−2_−1 differ(GFS_ECMWF)_p_−1_0 differ(GFS_ECMWF)_p_0_1 differ(GFS_ECMWF)_p_1_2	Percentage difference between averaged atmospheric pressure point forecasts for respective pair of hourly lags −2, −1, 0, +1 [1], +2 h from ECMWF NWP and GFS NWP models
12A	E_from_producer_turbine_power_curve	Forecast of electric energy production calculated based on producer turbine power curve estimated as a polynomial of degree 3, with ave(GFS_ECMWF)_v_mod_0-1 as input data. For input data below and above the cut-in for turbine, the forecast value is equal to zero
13A	E_from_power_curve(scatter_plot)	Forecast electric energy production calculated based on estimated polynomial of degree 3 as turbine power curve with ave(GFS_ECMWF)_v_mod_0-1 as input data. The estimation of power curve was executed based on the scatter plot between electric energy production and wind speed forecasts. For input data below and above the cut-in for turbine, the forecast value is equal to zero

[1] 1 h lag refers to the point of time for which forecast is to be generated, and energy generated is allocated between the considered time and one hour earlier.

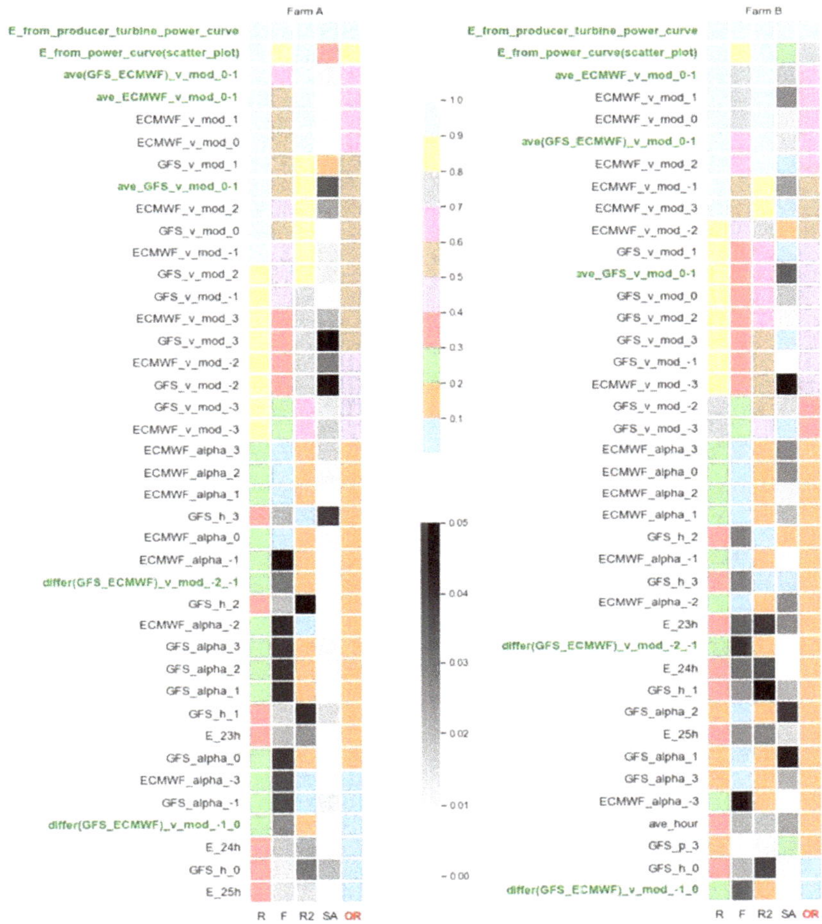

Figure 12. Results of sensitivity analysis of potential input variables including additional input data created for prediction models for Wind Farm A and Wind Farm B: 4 analysis methods and final Overall Rating (OR). Figure contains 40 best inputs out of 81. Names of additional inputs are marked green.

3. Forecasting Methods

This section includes the description of proposed forecasting methods. The research used both single methods as well as advanced ensemble and hybrid methods. Described the Persistence Model is a benchmark for the quality of other, more advanced forecasting methods.

Single methods, using only one individual predictor, are addressed next. The general scheme is presented in Figure 13.

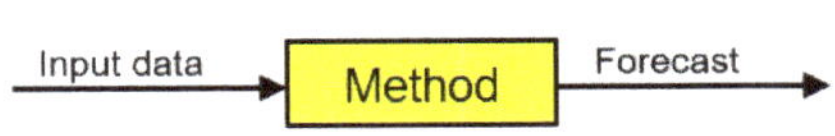

Figure 13. General structure of single method.

Persistence model. The naïve model is the simplest model in forecasting. In the Persistence Model, the forecast generation value is the same as the actual energy generation value from the same hour the day before. Forecasts are calculated by Formula (1):

$$\hat{y}_t = y_{t-24} \quad (1)$$

where $\hat{y}_t$—forecast electric energy generated by wind farm for hour t and $y_{t-24 \cdot n}$—energy generation for period lagged by $t - 24$ from forecast period t.

Physical Model. This forecasting model of generated hourly power is a function of wind speed. The function is in the form of the 3rd-order polynomial. Two different methods were utilized to form 3rd-degree polynomial separately for Wind Farm A and Wind Farm B.

- *Physical model version 1*. The polynomial of degree 3 is estimated based on turbine power curve data from the manufacturer's catalogue (turbine power for wind speeds with 1 m/s steps). For input data below and above cut-in for each turbine, the forecast is equal to zero. The input data depend on variant (ave(GFS_ECMWF)_v_mod_0-1, ave_ECMWF_v_mod_0-1 or ave_GFS_v_mod_0-1).
- *Physical model version 2*. The polynomial of degree 3 is estimated based on the scatter plot between electric energy production and wind speed forecasts (ave(GFS_ECMWF) _v_mod_0-1). For input data below and above cut-in for each turbine, the forecast value is equal to zero. The input data are ave(GFS_ECMWF)_v_mod_0-1.

K-Nearest Neighbours Regression. KNNR is a non-parametric method used for regression problems [37]. The input of the model contains the k-closest training examples in the feature space. The output of KNNR model is the property value for the object. Property value is the average of the values of k-nearest neighbours. Hyperparameter—the value of k (the number of nearest neighbours) needs searching for the appropriate value. The other hyperparameter for tuning is the choice of the distance metric.

Neural Network, Type MLP—Multi-layer Perceptron is a classical type of ANN. Widely used over decades, it proved its applicability as an effective non-linear or linear global approximator [13,38]. It is a feedforward ANN usually with an input layer, one or two hidden layers, and an output layer. Originally, it used the backpropagation algorithm for supervised learning. During years of development, other optimisation algorithms were applied for MLP learning, among them, the BFGS method that was chosen as the learning algorithm in our research. The number of neurons in hidden layer(s) was decided to be the main hyperparameter for tuning.

Support Vector Regression. Support Vector Machine for regression (SVR) transforms the classification task into regression by defining hyperparameter width ε tolerance region around the destination [39]. Hyperparameters of SVR for tuning are the following: regularization constant C, tolerance ε, and parameter s of the Gaussian kernel.

Deep Neural Network Type LSTM. The main difference between LSTM and traditional RNNs is LSTM's internal built format. Its hidden layers contain 3 gates, namely, input, forget, and output gate. This solution allows to control the flow of information and allows to deal with problems such as gradient explosion and vanishing, and taking long-term dependencies into account [10]. A typical LSTM network contains an input layer followed by up to two hidden layers finished by an output layer with dropout layers possible between layers. The dropout mechanism's goal is to prevent overfitting by keeping node in network with Bernoulli distribution probability [40]. The LSTM model contains (among others) the following hyperparameters: the number of hidden layers and neurons in them, activation function in each layer, number of training epochs, batch size, dropout degree, type of model optimizer, and learning rate.

Ensemble methods, using more than one individual predictor and supported by a simple or more complex integration system of individual forecasts, are addressed next. The simplest integration system is weighted averaging of individual predictors. The general scheme of establishing an ensemble of predictors is presented in Figure 14. The ensemble method can use the same type of methods as predictors (e.g., Random Forest, Gradient-

Boosted Trees) or different types of predictors (e.g., single Machine Learning methods as predictors.

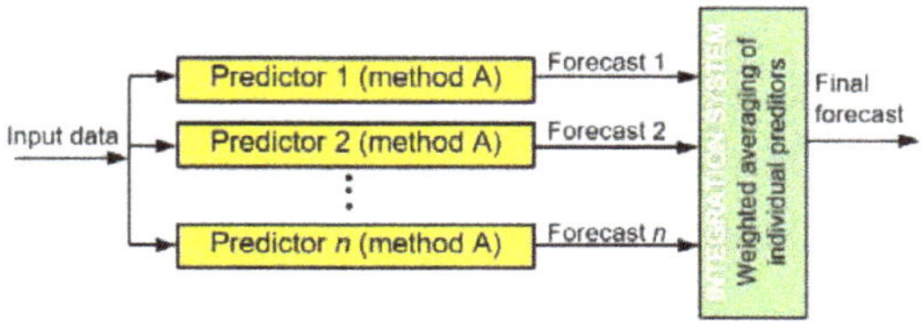

(a)

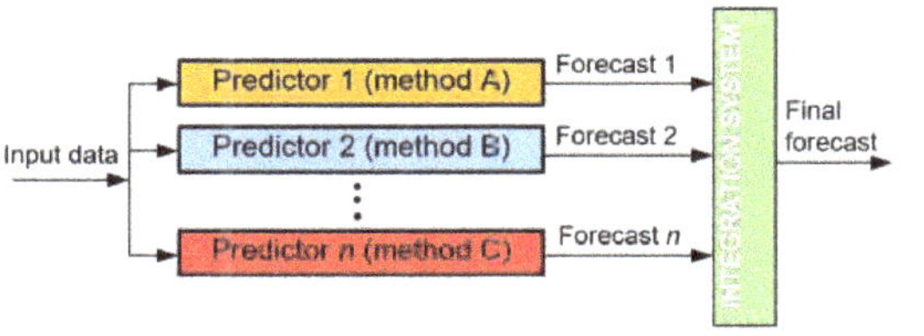

(b)

Figure 14. (**a**) General structure of ensemble method with the same type of methods as individual predictors; (**b**) General structure of ensemble method with different types of methods as individual predictors.

Random Forest Regression. RF is an ensemble method based on many single decision trees (the same type of models). In the regression task, the prediction in a single decision tree is the average target value of all instances associated with the single leaf node [41]. The final prediction is the average value of all n single decision trees. The regularization hyperparameters depend on the algorithm used, but generally restricted are among others: the maximum depth of a single decision tree, maximum number of levels in each decision tree, minimum number of data points placed in a node before the node is split, minimum number of data points allowed in a leaf node and maximum number of nodes. The number of predictors for each of the n single decision trees is made by the random choice of k predictors from all available n predictors [41,42].

Gradient-Boosted Trees for Regression. Gradient boosting refers to an ensemble method that can combine several weak learners into a strong learner [41]. GBT works by sequentially adding predictors (the same type of models) to the ensemble, each one correcting its predecessor. The method tries to fit the new predictor into the residual errors made by the previous predictor. The final prediction is the average value from all n single decision trees. In comparison with random forest, this method has one additional hyperparameter—learning rate, which scales the contribution of each tree [42,43].

Ensemble Averaging Without Extremes. The method developed by the authors of this study involves the deletion of the minimum and maximum forecast from the set of n single predictors (different types of methods) before each calculation of single final forecasts, being an average of forecasts from n-2 single predictors. The deletion is executed 24 times for each forecast separately. The choice predictors in the ensemble is based on the similar levels of forecasting error and mutually independent operation [9]. The final forecast result is calculated by Formula (2).

$$\hat{y}_i = \frac{1}{n-2} \cdot \left(\sum_{k=1}^{s} \hat{y}_i^k - \min\{\hat{y}_i^k\} - \max\{\hat{y}_i^k\}\right) \quad (2)$$

where i is the forecast point, $\hat{y}_i$ is the final forecast value, $\hat{y}_i^k$ is the forecasted value by predictor number k, and n is the number of predictors in the original ensemble before the removal of the outputs of predictors yielding extreme forecasts from the set of results.

Weighted Averaging as an Integrator of Ensemble based on nMAE and R. It integrates the results of selected predictors (different types of methods) into the final verdict of the ensemble. The final forecast is defined as the average of the results generated by all n predictors in the ensemble and is calculated by Formula (3) [9,39]. This method reduces the variance of forecast errors. Predictors are included in the ensemble based on two important elements:

- time series of the residues from forecasts should be most distant from each other (small R values);
- the smallest nMAE errors on the validation subset.

$$\hat{y}_i = \frac{1}{n}\sum_{j=1}^{n}\hat{y}_i^j \quad (3)$$

where i is the prediction point, $\hat{y}_i$ is the final predicted value, $\hat{y}_i^j$ is the predicted value by predictor number j, and n is the number of hybrid predictors in the ensemble.

Hybrid methods, using two or more different methods connected in series, are addressed next.

Machine learning method with additional input data from two Physical models. This hybrid method is a cascade of two different Physical models (version 1 and version 2) with one of the five ML methods (GBT, SVR, KNNR, MLP, or LSTM). ML component uses both forecasts of electric energy production as an additional input. The general scheme of this hybrid method is presented in Figure 15.

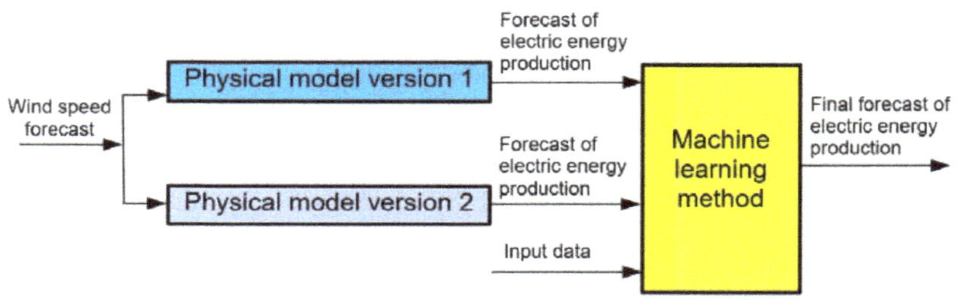

Figure 15. General structure of hybrid method—machine learning method with additional input data from two Physical models.

Physical model version 1 with input data as wind speed forecast from Gradient-Boosted Trees method. This hybrid method consists of the Gradient-Boosted Trees method connected in series with Physical Model Version 1. The GBT method predicts wind speed, while Physical Model Version 1 forecasts electric energy production. Physical Model Version 1 yielded smaller errors than the MLP and GBT methods considered here. The training and testing subsets differ from each other. The training subset uses wind speed based on the manufacturer's reversed turbine power curve (third-order polynomial) as additional input. This allows the method to learn effective wind speed corresponding to actual values of electric energy production. In turn, the testing subset uses ave(GFS_ECMWF)_v_mod_0-1 as its additional input, since electric energy production, and thus effective wind speed, would be unobtainable during the operational work of the models. The concept of this hybrid methods is based on the assumption that GBT will learn better on a training subset containing a precise estimate of wind speed than on one containing wind speed forecasts with a large random component. The general scheme of this hybrid method is presented in Figure 16.

A summary description of thirteen tested forecasting methods is shown in Table 4. The listed methods include four types of ensemble methods, two types of hybrid ones, and seven single methods. Six methods (single/ensemble) are machine learning (ML) methods, including one deep learning method.

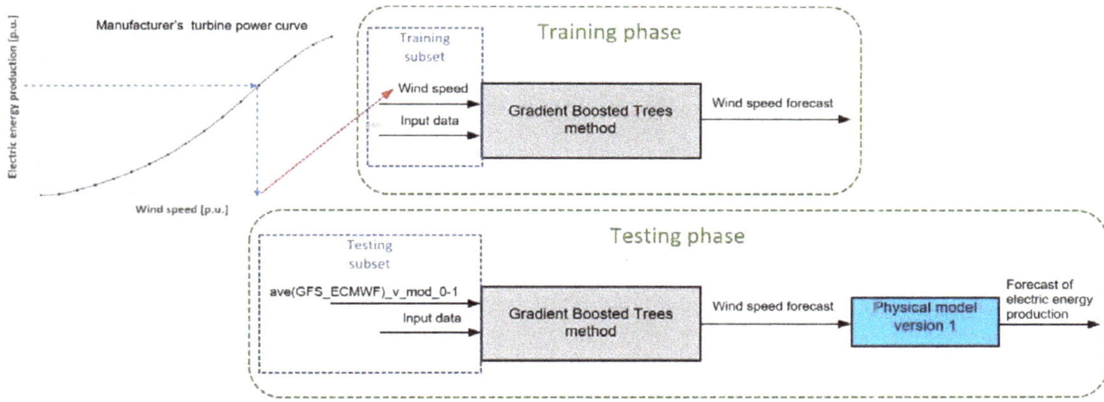

Figure 16. General structure of hybrid method—Physical model version 1 with input data as wind speed forecast from Gradient-Boosted Trees method.

Table 4. Summary description of thirteen tested forecasting methods.

Name of Method	Method Code	Category	Complexity
Persistence	PERSISTENCE	Linear/ non-parametric	Single
Physical model version 1	PHYS_v1	Non-linear/ parametric	Single
Physical model version 2	PHYS_v2	Non-linear/ parametric	Single
K-Nearest Neighbours Regression	KNNR	Non-linear/ non-parametric	Single
Type MLP artificial neural network	MLP	Non-linear/ parametric	Single
Support Vector Regression	SVR	Non-linear/ non-parametric	Single
Deep neural network type LSTM	LSTM	Non-linear/ parametric	Single
Random Forest Regression	RF	Non-linear/ non-parametric	Ensemble (one type of model)
Gradient-Boosted Trees for regression	GBT	Non-linear/ non-parametric	Ensemble (one type of model)
Ensemble Averaging Without Extremes	INT_OUT_EXT [p_1 *, ... , p_n]	Non-linear/ non-parametric	Ensemble (different types of models including hybrid models)
Weighted Averaging as an Integrator of Ensemble based on nMAE and R	INT_AVE [p_1 *, ... , p_n]	Non-linear/ non-parametric	Ensemble (different types of models including hybrid models)
Machine learning method with additional input data from two physical models	PHYS(v1&v2)→ML	Non-linear/ parametric	Hybrid
Physical model version 1 with input data as wind speed forecast from Gradient-Boosted Trees method	GBT→PHYS_v1	Non-linear/ parametric	Hybrid

Remark: * denotes first predictor in ensemble of n predictors.

Additional expert correction of forecasts. Since wind turbines produce no power below the lower and above the upper limits of wind speed, a unique expert correction method is proposed. Obviously, without verification, the use of the correction would be unjustified, as it applies to wind speed forecasts with a large random component instead of real-world wind speeds. Due to that, its effectiveness and validity are verified for the selected group of methods providing best forecasts. A robust wind estimator ave(GFS_ECMWF)_v_mod_0-1 is used as a conditional variable for the method to adjust for bias of singular NWP models. For wind speed forecasts—ave(GFS_ECMWF)_v_mod_0-1 below cut-in and above cut-out wind speeds for wind turbine, forecast electric energy production is corrected to zero. The final prediction with expert correction is calculated by Formula (4).

$$\hat{E}_i = \begin{cases} \hat{E}_i \; for \; v_{min}\langle \hat{v}_i \rangle v_{max} \\ 0 \; for \; \hat{v}_i \le v_{min} \\ 0 \; for \; \hat{v}_i \ge v_{max} \end{cases} \tag{4}$$

where $\hat{E}_i$ is the predicted value (electric energy production), $\hat{v}_i$ is the predicted wind speed (ave(GFS_ECMWF)_v_mod_0-1) and v_{min} and v_{max} are cut-in and cut-out wind speeds of turbine.

4. Evaluation Criteria

Three evaluation criteria are used to test the performance of the methods, including normalized Root Mean Square Error (nRMSE), normalized Mean Absolute Error (nMAE) and normalized Mean Bias Error (nMBE).

Normalized Root Mean Square Error which is sensitive to large error values is calculated by Formula (5):

$$nRMSE = \frac{1}{c_{norm}} \sqrt{\frac{1}{n} \sum_{i=1}^{n} (\hat{y}_i - y_i)^2} \tag{5}$$

where $\hat{y}_i$ is the predicted value (electric energy production), y_i is the actual value, c_{norm} is the normalizing factor (rated power of wind farm), and n is the number of prediction points.

Normalized Mean Absolute Error is calculated by Formula (6). nMAE is a risk metric according to the expected value of the absolute error.

$$nMAE = \frac{1}{n} \sum_{i=1}^{n} \frac{1}{c_{norm}} |\hat{y}_i - y_i| \cdot 100\% \tag{6}$$

Normalized Mean Bias Error (nMBE) captures average bias in prediction and is calculated by Formula (7). The forecasting method overestimates if nMBE > 0 or underestimates if nMBE < 0.

$$nMBE = \frac{1}{n} \sum_{i=1}^{n} \frac{1}{c_{norm}} (\hat{y}_i - y_i) \tag{7}$$

Errors nRMSE and nMAE are basic measures to evaluate the accuracy of proposed models, while nMBE is only auxiliary. In the process of forecasting electric energy production in a wind farm, the changes of nRMSE and nMAE have the same trend, and the smaller the two error values, the more accurate the prediction results. Both show random and systematic errors. A large gap between nMAE and nRMSE for the results of a method indicates that predicted values are extremely distant from the measured data [44,45].

The effectiveness of the forecasting approaches is found by considering the uncertainty and variability of forecasts [46]. For a comparative assessment of the performance test of the analysed methods, the Skill Score (SS) metric was used. The proposed Skill Score metric uses two error metrics—nRMSE and nMAE—and is calculated by Formula (8). Higher SS values are an indication of superior prediction quality.

$$SS = \frac{1}{2}\left[\left(1 - \frac{nMAE_{forecast}}{nMAE_{reference}}\right) + \left(1 - \frac{nRMSE_{forecast}}{nRMSE_{reference}}\right)\right] \tag{8}$$

where $nMAE_{forecast}$ and $nRMSE_{forecast}$ are errors of the analysed method, $nMAE_{reference}$ and $nRMSE_{reference}$ are errors of reference method (persistence method—naive model).

5. Results and Discussion

The range of the acquired data was identical for both wind farms and spanned from 4 April 2017 to 10 October 2019, with about 29 months in total. Data were divided into three subsets—training subset, validation subset, and test subset. The training and validation subsets for the period from 4 April 2017 to 30 September 2018 (17 months) were chosen at random (85% and 15%, respectively). The training subset is used for the estimation of model parameters. The validation subset is used for tuning hyperparameters of parts of methods. The last part of the data (from 1 October 2018 to 1 October 2019—12 months) constituted the test subset used for one-time final evaluation of the quality of specific prediction methods on data for all seasons.

Predictions were conducted sequentially, from single methods with a limited number of input variables to hybrid methods, to ensemble methods. Such procedure allows us to observe differences in the quality of results depending on the complexity of particular methods and the range of input variables used. Research was done in steps in order to verify different hypotheses, and find an optimal input dataset and the best group of prediction methods.

Step 1. Hypotheses verification:

- Are the accuracies of two designed physical models different from each other?
- Is it better to use one or two NWP models with wind speed forecasts for physical models?

Tables 5 and 6 contains results of forecasts for A and B wind farms, respectively. Physical and Persistence (reference) Models were used for predictions.

Table 5. Measures of performance of the proposed Physical Models (test subset) for Wind Farm A.

Method Code	Input Data Codes	nMAE [%]	nRMSE	nMBE
PHYS_v1	ave(GFS_ECMWF)_v_mod_0-1	12.3288	0.1813	0.0349
PHYS_v2	ave(GFS_ECMWF)_v_mod_0-1	12.3700	0.1709	0.0025
PHYS_v1	ave_ECMWF_v_mod_0-1	13.2318	0.1920	0.0663
PHYS_v1	ave_GFS_v_mod_0-1	14.5975	0.2152	−0.0063
PERSISTENCE	E-24 h	28.7790	0.3833	0.0127

Remarks: The best fitting results for each fitting measure are printed in bold in blue. The worst fitting result is printed in red.

Table 6. Measures of performance of the proposed Physical Models (test subset) for Wind Farm B.

Method Code	Input Data Codes	nMAE [%]	nRMSE	nMBE
PHYS_v1	ave_ECMWF_v_mod_0-1	14.6101	0.2246	−0.0173
PHYS_v1	ave(GFS_ECMWF)_v_mod_0-1	15.2718	0.2286	0.0403
PHYS_v2	ave(GFS_ECMWF)_v_mod_0-1	15.9984	0.2196	−0.0025
PHYS_v1	ave_GFS_v_mod_0-1	19.5102	0.2889	−0.0468
PERSISTENCE	E-24 h	29.9931	0.3886	−0.0291

Remarks: The best fitting results for each fitting measure are printed in bold in blue. The worst fitting result is printed in red.

Results of the two Physical Models indicate that PHYS_v1 was better fitted for both wind farms, while results for the NWP models were ambiguous. Although using only the GFS model was clearly the least favourable option, for Wind Farm A it was better to use both NWP models, while for Wind Farm B it was better to use the ECMWF model only. In comparison, the nMAE Persistence Model was twice as good as both Physical Models.

Step 2. Hypotheses verification:

- Is it better to use NWP point forecasts for hourly lags: −3, 2, −1, 0, 1, 2, 3 (original contribution) as input data instead of the typically used lags 0, 1?
- Is it better to use one or two NWP models as input data source?

To verify the above, a strong GBT method was used, recommended by multiple papers. Tables 7 and 8 present the resulting forecasts for Wind Farms A and B using the proposed GBT method with different versions of NWP input data.

Table 7. Measures of performance of the proposed GBT method, with different versions of NWP input data (test subset) for Wind Farm A.

Method Code	Input Data Numbers/Description	nMAE [%]	nRMSE	nMBE
GBT	1−68 (68 inputs)/including NWP forecasts from ECMWF and GFS models (point forecasts for hourly lags: −3, 2, −1, 0, 1, 2, 3)	11.8518	0.1636	0.0006
GBT	1–5, 13–19, 27–33, 48–55, 62–68 (34 inputs)/including NWP forecasts from ECMWF model (point forecasts for hourly lags: −3, 2, −1, 0, 1, 2, 3)	12.5388	0.1701	−0.0019
GBT	1–5, 16–17, 30–31, 51–52, 65–66 (13 inputs)/including NWP forecasts from ECMWF model (point forecasts for hourly lags: 0, 1)	12.9633	0.1760	−0.0006
GBT	1–5, 6–12, 20–26, 34–47, 55–61 (40 inputs)/including NWP forecasts from GFS model (point forecasts for hourly lags: −3, 2, −1, 0, 1, 2, 3)	13.8295	0.1855	0.0037
GBT	1–5, 16–17, 30–31, 51–52, 65–66 (13 inputs)/including NWP forecasts from GFS model (point forecasts for hourly lags: 0, 1)	14.1665	0.1888	0.0026

Remarks: The best fitting results for each fitting measure are printed in bold in blue. The worst fitting result is printed in red.

Table 8. Measures of performance of the proposed GBT method, with different versions of NWP input data (test subset) for Wind Farm B.

Method Code	Input Data Numbers/Description	nMAE [%]	nRMSE	nMBE
GBT	1–68 (68 inputs)/NWP forecasts from ECMWF and GFS models (point forecasts for hourly lags: −3, 2, −1, 0, 1, 2, 3)	14.4555	0.2090	0.0032
GBT	1–5, 13–19, 27–33, 48–55, 62–68 (34 inputs)/including NWP forecasts from ECMWF model (point forecasts for hourly lags: −3, 2, −1, 0, 1, 2, 3)	14.7389	0.2141	0.0066
GBT	1–5, 16–17, 30–31, 51–52, 65–66 (13 inputs)/including NWP forecasts from ECMWF model (point forecasts for hourly lags: 0, 1)	14.9831	0.2161	0.0048
GBT	1–5, 6–12, 20–26, 34–47, 55–61 (40 inputs)/including NWP forecasts from GFS model (point forecasts for hourly lags: −3, 2, −1, 0, 1, 2, 3)	17.8355	0.2397	−0.0013
GBT	1–5, 16–17, 30–31, 51–52, 65–66 (13 inputs)/including NWP forecasts from GFS model (point forecasts for hourly lags: 0, 1)	18.0803	0.2446	−0.0019

Remarks: The best fitting results for each fitting measure are printed in bold in blue. The worst fitting result is printed in red.

Research in step 2 demonstrated that the order of the results obtained with the same combination of input data was the same for both wind farms. Best accuracies were achieved by using both NWP models. The application of the novel and original idea of using point forecasts for hourly lags: −3, 2, −1, 0, 1, 2, 3 yields clearly better results than using typical

0, 1 lags. Like in Physical Models, in this case, forecasts for Wind Farm B were less accurate than for Wind Farm A. Preliminary studies analysing the importance of input data also indicated slightly lesser correlation between NWP forecasts for Wind Farm B than for Wind Farm A. The above findings were used in further research steps; hence, the subsequent versions of forecasts use both NWP models predictions and point forecasts for hourly lags: $-3, 2, -1, 0, 1, 2, 3$.

Step 3. This step is the main, most extensive and labour intensive part of research. Forecasts of energy production were obtained from different single, hybrid and ensemble models, including by original methods. To find proper hyperparameters for them, more than 300 hyperparameter combinations were tested using the Grid Search method. The lowest nMAE score on the validation range was used as the parameter selection criterion. Hyperparameter search ranges and their determined values for chosen methods are summarized in Table A1 in Appendix A. The described determinations were carried out to verify the following:

- Which method group yields the lowest prediction errors (recommended methods) and does the Ensemble Averaging Without Extremes original method developed by us belong to the recommended methods?
- Does the original proposition developed by us—additional input variables (see Table 3)—reduce the prediction error?
- Does the original proposition developed by us—additional expert correction—reduce the prediction error?

Tables 9 and 10 present forecasts for Wind Farms A and B resulting from the proposed single, ensemble and hybrid methods with different sets of input data. For the two best methods, results are shown with and without additional expert correction (see Formula (4)).

Tabular results were ordered by descending SS metric, which was taken as the main determinant of prediction quality, as it takes into account both nMAE and nRMSE errors.

Based on the results from Tables 9 and 10, the following conclusions can be drawn regarding the proposed single, hybrid, and ensemble methods with different sets of input data:

- Original method called "Ensemble Averaging Without Extremes" is the best for both wind farms (SS metric and nMAE error);
- For "Ensemble Averaging Without Extremes", the best fitted solution was to use an ensemble of 5 methods, while a 3-method ensemble was the best for "Weighted Averaging As an Integrator of Ensemble";
- Our original method "Additional Expert Correction" resulted in lower nMAE than for predictions without correction. For "Ensemble Averaging Without Extremes", nMAE decreased by 0.42% for Wind Farm A and 0.92% for Wind Farm B;
- Hybrid methods have worse accuracy measures of nMAE and nRMSE than ensemble methods for both wind farms;
- Deep neural network LSTM is the best single method, MLP is the second best;
- Original hybrid "Physical Model Version 1 With Input Data As Wind Speed Forecast from Gradient-Boosted Trees Method" turned out to be of less advantage than ensemble methods;
- For most methods, using additional input data (numbers: 1A, 2A, 3A) reduced nMAE in comparison with using basic input data only (numbers: 1–68). KNNR was an exception as it yielded the lowest nMAE with a highly reduced number of input data variables;
- Values of nMBE were very low for the analysed methods, which means there was no systematic error in predictions. The lowest nMBE for both wind farms was achieved with the RF Method;
- Prediction errors for Wind Farm B were bigger than for Wind Farm A, which was indicated by results of sensitivity analysis of potential input variables (see Figure 10).

Figures 17–20 provides two forecasts of electric energy generation for Wind Farm A made by the best method with additional expert correction for the two following days of each season (from autumn to summer).

Table 9. Measures of performance of the proposed single, ensemble and hybrid methods with different sets of input data (test subset) for Wind Farm A.

Method Code	Input Data Numbers	SS	nMAE [%]	nRMSE	nMBE
INT_OUT_EXT [GBT, RF, PHYS(v1&v2)→KNNR, MLP, LSTM] with additional expert correction	Different, it depends on predictor in ensemble	**0.5925**	**11.3055**	0.1618	0.0146
INT_AVE [GBT, RF, LSTM] with additional expert correction	Different, it depends on predictor in ensemble	0.5923	11.3387	0.1615	0.0117
INT_OUT_EXT [GBT, RF, PHYS(v1&v2)→KNNR, MLP, LSTM]	Different, it depends on predictor in ensemble	0.5921	11.3527	0.1615	0.0123
INT_AVE [GBT, RF, LSTM]	Different, it depends on predictor in ensemble	0.5910	11.4403	**0.1612**	0.0085
INT_AVE [GBT, RF, PHYS(v1&v2)→KNNR, MLP, LSTM] with additional expert correction	Different, it depends on predictor in ensemble	0.5904	11.3558	0.1627	0.0149
INT_AVE [GBT, RF, PHYS(v1&v2)→KNNR, MLP, LSTM]	Different, it depends on predictor in ensemble	0.5898	11.4174	0.1624	0.0124
LSTM	1–68, 1A, 2A, 3A	0.5842	11.4012	0.1669	0.0252
GBT	1–68	0.5807	11.8518	0.1636	0.0006
RF	1–68, 1A, 2A, 3A	0.5803	11.8847	0.1635	**−0.0004**
PHYS(v1&v2)→GBT	1–68, 1A–13A	0.5791	11.9190	0.1639	0.0022
MLP	1–68, 1A, 2A, 3A	0.5781	11.9211	0.1646	0.0041
GBT→PHYS_v1	1–68, 1A, 2A, 3A	0.5760	11.6604	0.1698	0.0463
PHYS(v1&v2)→MLP	1–68, 1A–13A	0.5694	12.1960	0.1676	−0.0026
PHYS_v1	1A	0.5622	12.3700	0.1709	0.0025
PHYS(v1&v2)→LSTM	1–68, 1A–13A	0.5534	12.2249	0.1796	0.0134
PHYS(v1&v2)→KNNR	1A, 2A, 3A, 12A, 13A	0.5507	12.2899	0.1807	0.0326
PHYS_v1	1A	0.5493	12.3288	0.1813	0.0349
PHYS(v1&v2)→SVR	2A, 12A, 13A	0.5432	13.1031	0.1757	−0.0202
PERSISTENCE	4	0.0000	28.7790	0.3833	0.0127

Remarks: The best fitting results for each fitting measure are printed in bold in blue. The worst fitting result is printed in red.

Figures 21–24 provides two forecasts of electric energy generation for Wind Farm B made by the best method with additional expert correction for two following days of each season (from autumn to summer).

Figures 17–24 show that energy generation of both wind farms in presented days (16 in total) is highly random. For some hours of certain days, generation is periodically close to its rated value, but for other hours generation is very low. There are also few hour periods of null generation. The lowest generations and predictions among the presented 16 days occurred for 4 days of summer months (Figures 20 and 24). It should be noted that generation predictions have periods of both over- and under-forecasting. Most commonly, it can be observed on a few consequent samples of time series. Moreover, time series of generation predictions have slightly smoothened course due to using the ensemble method, as ensemble methods reduce the variance of forecasts. For "Ensemble Averaging Without Extremes", additional removal of extreme forecasts occurs before average forecast calculation, which, in turn, further enhances the smoothening effect for generation prediction time series.

Table 10. Measures of performance of the proposed single, ensemble and hybrid methods with different sets of input data (test subset) for Wind Farm B.

Method Code	Input Data Numbers	SS	nMAE [%]	nRMSE	nMBE
INT_OUT_EXT [GBT, RF, PHYS(v1&v2)→KNNR, MLP, LSTM] with additional expert correction	Different, it depends on predictor in ensemble	**0.5096**	**13.7552**	0.2029	0.0108
INT_OUT_EXT [GBT, RF, PHYS(v1&v2)→KNNR, MLP, LSTM]	Different, it depends on predictor in ensemble	0.5091	13.8199	**0.2025**	0.0075
INT_AVE [GBT, RF, LSTM] with additional expert correction	Different, it depends on predictor in ensemble	0.5087	13.7794	0.2033	0.0061
INT_AVE [GBT, RF, PHYS(v1&v2)→KNNR, MLP, LSTM] with additional expert correction	Different, it depends on predictor in ensemble	0.5078	13.8182	0.2035	0.0128
INT_AVE [GBT, RF, LSTM]	Different, it depends on predictor in ensemble	0.5073	13.8994	0.2028	0.0019
INT_AVE [GBT, RF, PHYS(v1&v2)→KNNR, MLP, LSTM]	Different, it depends on predictor in ensemble	0.5071	13.8928	0.2031	0.0096
RF	1–68, 1A, 2A, 3A	0.4977	14.4916	0.2026	**0.0000**
GBT→PHYS_v1	1–68, 1A, 2A, 3A	0.4944	14.6231	0.2035	0.0053
LSTM	1–68, 1A, 2A, 3A	0.4932	13.9797	0.2127	0.0109
GBT	1–68, 1A, 2A, 3A	0.4909	14.4610	0.2083	−0.0003
MLP	1–68	0.4894	14.6654	0.2068	0.0057
PHYS(v1&v2)→GBT	1–68, 1A–13A	0.4838	14.8258	0.2091	−0.0015
PHYS(v1&v2)→MLP	1–68, 1A–13A	0.4781	15.0611	0.2105	−0.0008
PHYS_v1	2A	0.4675	14.6101	0.2246	−0.0173
SVR	1–68	0.4617	15.9566	0.2116	0.0143
PHYS(v1&v2)→LSTM	1–68, 1A, 2A, 3A, 12A, 13A	0.4574	15.2485	0.2241	0.0111
PHYS(v1&v2)→KNNR	1A, 2A, 3A, 12A, 13A	0.4548	15.1681	0.2272	0.0366
PHYS_v2	1A	0.4508	15.9984	0.2196	−0.0025
PERSISTENCE	4	0.0000	29.9931	0.3886	−0.0291

Remarks: The best fitting results for each fitting measure are printed in bold in blue. The worst fitting result is printed in red.

For both wind farms, additional analysis of nMAE error distribution was made. It concerned hourly periods of prediction using the best forecasting method of "Ensemble Averaging Without Extremes". The goal of analysis was to determine whether error magnitude depends on forecast horizon (from 1 to 24 h) and time of the day. Figure 25 shows the graph of the forecast error (nMAE) depending on the forecast horizon for the test subset for Wind Farm A and Wind Farm B.

nMAE values presented in Figure 25 are visibly greater for Wind Farm B, which complies with the results from Tables 9 and 10. nMAE error equals 11.3055% and 13.7552% for Wind Farm A and B, respectively. The distribution of error values shown in Figure 25 and the distribution of average production of energy in individual hours values shown in Figure 4a,b are very similar for both wind farms. For Wind Farm A, the correlation coefficient is equal to 0.9331, and for Wind Farm B, the correlation coefficient is equal 0.9291. Both autocorrelation coefficients are statistically significant (5% significance level). This phenomenon is related to a strong non-linear relationship between the energy forecast error and the wind speed forecast error. The aforementioned non-linear relationship results from the fact that the generation of energy in the wind source is a third-degree polynomial of the wind speed.

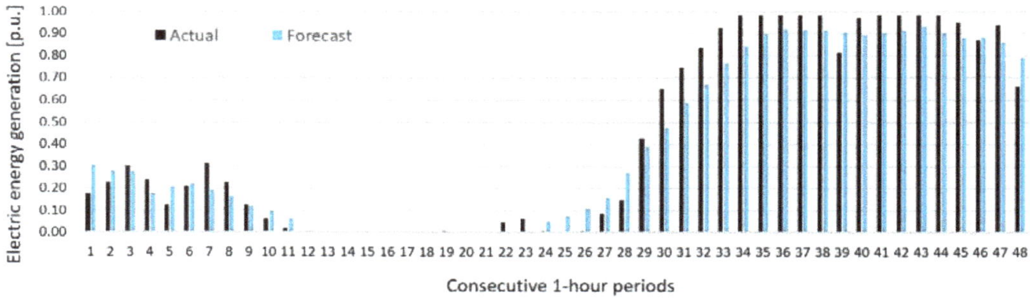

Figure 17. Two forecasts of electric energy generation for Wind Farm A made by INT_OUT_EXT [GBT, RF, PHYS(v1&v2)→KNNR, MLP, LSTM] method with additional expert correction for two consecutive days of an autumn month (November).

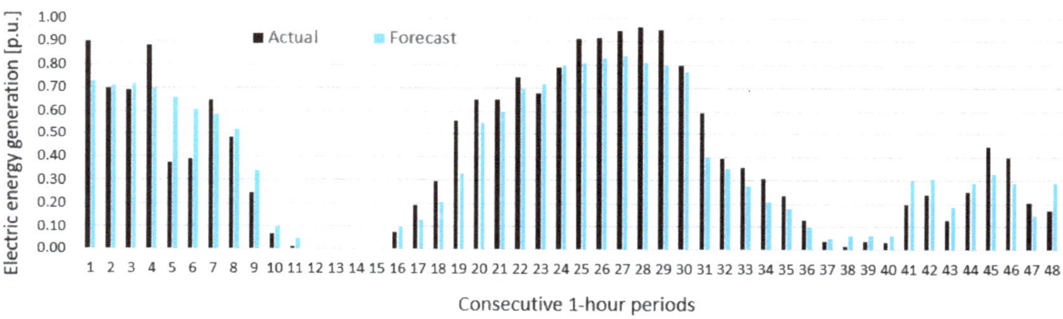

Figure 18. Two forecasts of electric energy generation for Wind Farm A made by INT_OUT_EXT [GBT, RF, PHYS(v1&v2)→KNNR, MLP, LSTM] method with additional expert correction for two consecutive days of the winter month (January).

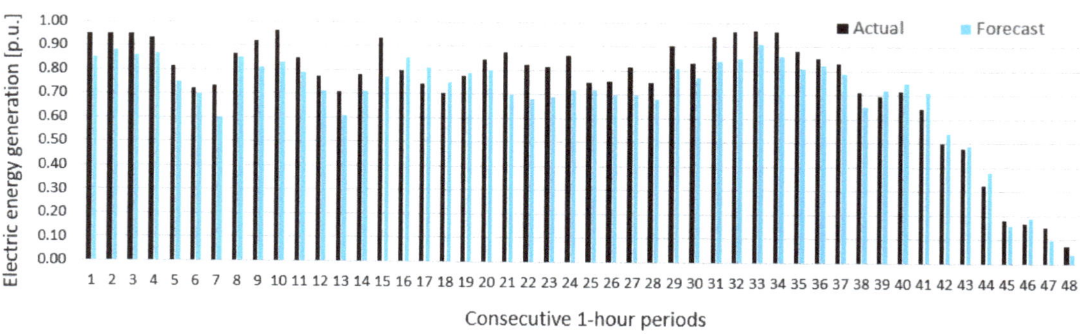

Figure 19. Two forecasts of electric energy generation for Wind Farm A made by INT_OUT_EXT [GBT, RF, PHYS(v1&v2)→KNNR, MLP, LSTM] method with additional expert correction for two consecutive days of the spring month (April).

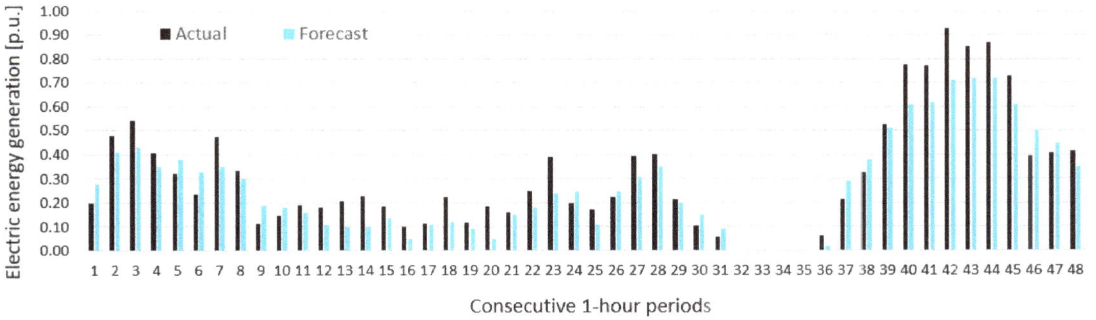

Figure 20. Two forecasts of electric energy generation for Wind Farm A made by INT_OUT_EXT [GBT, RF, PHYS(v1&v2)→KNNR, MLP, LSTM] method with additional expert correction for two consecutive days of the summer month (August).

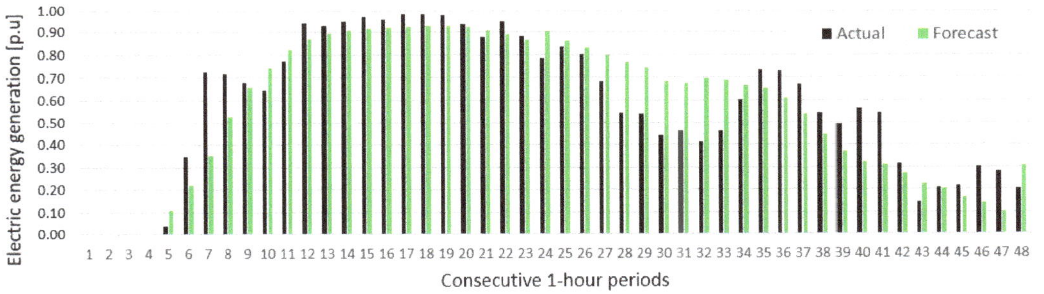

Figure 21. Two forecasts of electric energy generation for Wind Farm B made by INT_OUT_EXT [GBT, RF, PHYS(v1&v2)→KNNR, MLP, LSTM] method with additional expert correction for two consecutive days of an autumn month (November).

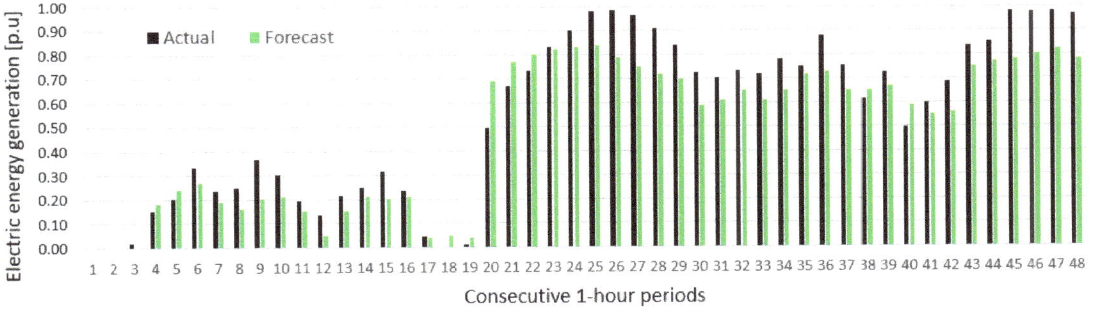

Figure 22. Two forecasts of electric energy generation for Wind Farm B made by INT_OUT_EXT [GBT, RF, PHYS(v1&v2)→KNNR, MLP, LSTM] method with additional expert correction for two consecutive days of the winter month (January).

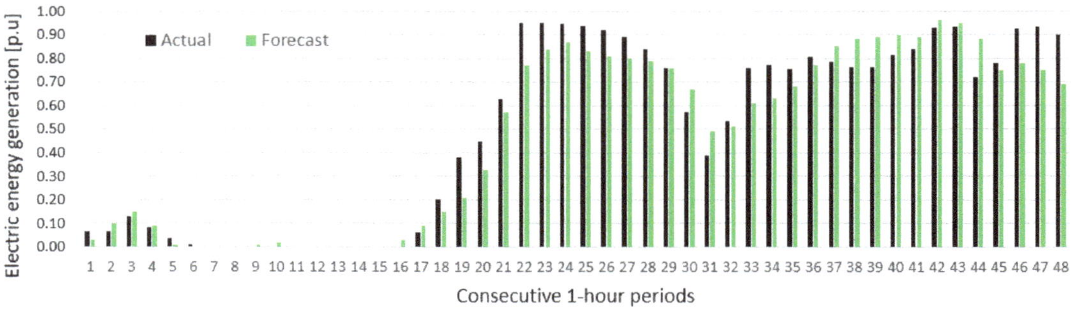

Figure 23. Two forecasts of electric energy generation for Wind Farm B made by INT_OUT_EXT [GBT, RF, PHYS(v1&v2)→KNNR, MLP, LSTM] method with additional expert correction for two consecutive days of the spring month (April).

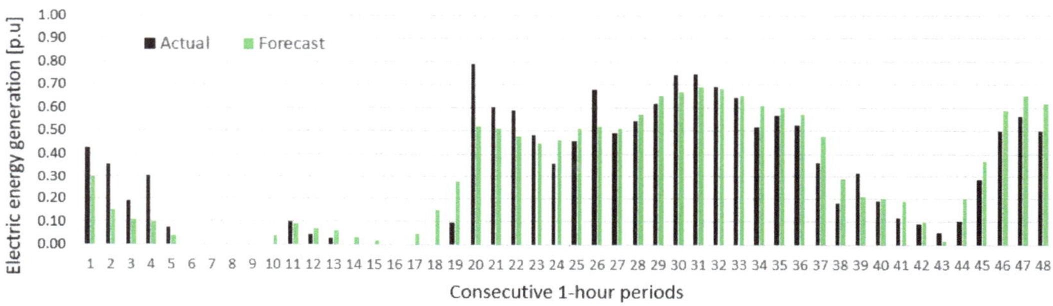

Figure 24. Two forecasts of electric energy generation for Wind Farm B made by INT_OUT_EXT [GBT, RF, PHYS(v1&v2)→KNNR, MLP, LSTM] method with additional expert correction for two consecutive days of the summer month (August).

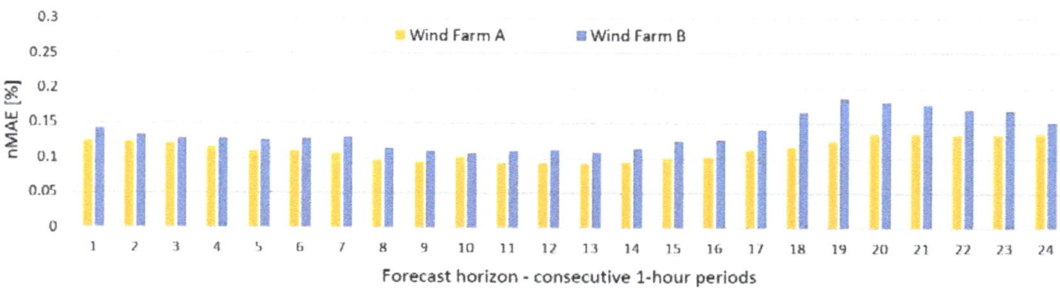

Figure 25. Forecast error depending on the forecast horizon for the test subset for both wind farms.

6. Conclusions

Using two wind farms for statistical analyses and forecasting considerably improves credibility of newly created effective prediction methods and conclusions. The results of the study are summarized below.

Original ensemble methods, developed for researching specific implementations, reduced errors of energy generation forecasts for both wind farms as compared to single methods. The best integration system for ensemble methods for accuracy measure nMAE is a new, original integrator developed for predictions, called "Ensemble Averaging Without Extremes" (method code INT_OUT_EXT), with five methods in the ensemble. The best integration system for ensemble methods for accuracy measure nRMSE is an original integrator developed for predictions called "Weighted Averaging As an Integrator of Ensemble" (method code INT_AVE) with three methods in the ensemble.

A new, original "Additional Expert Correction" reduced errors of energy generation forecasts for both wind farms. Deep neural network LSTM is the best single method, MLP is the second best, while using SVR, KNNR, and Physical model is less favourable for both wind farms. Hybrid methods have worse accuracy measures using nMAE and nRMSE than ensemble methods for both wind farms.

Using meteo forecasts from two NWP models (ECMWF and GFS) as input data yield better results than using a single NWP model. Using NWP point forecasts for hourly lags: −3, −2, −1, 0, 1, 2, 3 (original contribution) as input data is better than using typical lags 0, 1. Using additional input data created, especially input data numbers: 1A, 2A, 3A, reduces prediction errors of most methods in comparison with base input variables (input data numbers: 1–68).

For both wind farms, strong positive correlation was determined between distribution of energy production averages, in particular, hourly periods and distribution of prediction errors (nMAE). Identifying this relationship is valuable, practical information concerning the expected value of prediction error depending on the time of the day. The greater the average generation for a given hour, the greater the prediction error (nMAE) expected. For both analyzed wind farms, the greatest prediction errors are expected during evening hours, while the lowest errors are expected between 08:00 a.m. and 2:00 p.m.

Using original SS metric to compare prediction accuracy is useful, as it allows to incorporate both nMAE and nRMSE into final quality assessment. Both measures are important for the end user of the prediction, as the former is sensitive to reducing the average error, while the latter is sensitive to overforecasting and underforecasting.

More research is needed to verify, among other things, the following:

- Does prediction accuracy depend on using forecasts from more than one spot of a medium-sized wind farm and how the accuracy of forecasts will be affected by other factors, i.e., data with higher time resolution, e.g., 15 min, using real measurements of weather as input data?
- Can measurements and weather predictions be used to create a weather-error sensitive switching regime for models of ensembles?
- Is it possible to reduce the level of high random component in predictions?
- Will the proposed, original method of "Ensemble Averaging Without Extremes" be equally good for different types of RES predictions (e.g., photovoltaic systems and hydropower system)?

Author Contributions: Conceptualization, P.P., D.B.; methodology, P.P., D.B.; investigation, P.P., D.B. and M.K.; supervision, P.P.; validation, M.K., T.G.; writing, P.P., D.B., M.K. and T.G.; visualization M.K., P.P.; project administration, P.P. All authors have read and agreed to the published version of the manuscript.

Funding: This research was funded by The National Centre for Research and Development (Poland), Grant No. ID POIR.01.01.01-00-130/16 (to P.P., D.B., M.K.).

Institutional Review Board Statement: Not applicable.

Informed Consent Statement: Not applicable.

Data Availability Statement: Not applicable.

Conflicts of Interest: The authors declare no conflict of interest.

Abbreviations

The following abbreviations are used in this manuscript:

ACF	Autocorrelation function
ANN	Artificial Neural Network
BFGS	Broyden–Fletcher–Goldfarb–Shanno algorithm
BPNN	Back Propagation Neural Network
CNN	Convolutional Neural Network
DNN	Deep Neural Network
ECMWF	European Center for the Medium-Range Weather Forecast
ELM	Extreme Learning Machine
ENN	Elman Neural Network
ESN	Echo State Network
F	Fisher test
GBT	Gradient-Boosted Trees
GFS	Global Forest System
GRU	Gated Recurrent Unit
HRES	High-resolution atmospheric model
IEC	International Electrotechnical Commission
KNNR	K-Nearest Neighbours Regression
LSTM	Long short-term memory
ML	Machine Learning
MLP	Multi-Layer Perceptron
NARX	Nonlinear Autoregressive Exogenous Model
nMAE	normalized Mean Absolute Error
nMBE	normalized Mean Bias Error
NWP	Numerical Weather Prediction
PSO	Particle Swarm Optimization
p.u.	Per unit
R	Pearson linear correlation coefficient
R^2	Determination coefficient
RES	Renewable Energy Sources
RF	Random Forest
RNN	Recurrent Neural Network
nRMSE	normalized Root Mean Square Error
SA	Sensitivity analysis
SS	Skill Score
SVR	Support Vector Regression
SVM	Support Vector Machine

Appendix A

Table A1. The results of hyperparameters tuning for chosen single, hybrid and ensemble methods for Wind Farm B.

Method Code	Description of Method, the Name and the Range of Values of Hyperparameters Tuning and Selected Values
LSTM	The number of hidden layers: 1–2, selected: 2, the number of neurons in hidden layer: 4–50, selected: 35–20, the activation function in hidden layer: ReLU/sigmoid/tanh, selected hyperbolic tangent, the activation function in output layer: linear, learning algorithms ADAM, RMSprop, selected optimizer: ADAM, lr = 0.001, decay = 1×10^{-5}, epochs: 1500, patience: 100, batch size: 128; shuffle: True. Dropout after each hidden layer: 0/0.2, selected dropout: 0.2
SVR	Regression SVM: Type-1, Type 2, selected: Type-1, kernel type: Gaussian (RBF), the width parameter σ: 0.147, the regularization constant C, range: 1–20 (step 1), selected: 2, the tolerance ε, range: 0.01–1 (step 0.01), selected: 0.05.
PHYS(v1&v2)→KNNR	Distance metrics: Euclidean, Manhattan, Minkowski, selected: Euclidean, the number of nearest neighbours k, range: 1–5[illegible], selected: 4.
MLP	The number of neurons in hidden layer: 10–80, selected 30, learning algorithm: BFGS, the activation function in hidden layer: linear, hyperbolic tangent, selected: hyperbolic tangent, the activation function in output layer: linear.
GBT	Considered max depth: 2/4/6/10, selected depth: 6; trees number: 100/200/400, selected number: 100; learning rate: 0.1/0.01/0.001, selected: 0.1
RF	The number of predictors chosen at random: 30, 35, 40, 45, 50, 55, 60, selected 35 number of decision trees: 1–500, selected: 385. Stop parameters: maximum number of levels in each decision tree: 5, 10, 20, selected 10, minimum number of data points placed in a node before the node is split: 100, 200, 300, selected 200, min number of data points allowed in a leaf node: 10, maximum number of nodes: 100.

References

1. Liu, H.; Chen, C.; Lv, X.; Wu, X.; Liu, M. Deterministic wind energy forecasting: A review of intelligent predictors and auxiliary methods. *Energy Convers. Manag.* **2019**, *195*, 328–345. [CrossRef]
2. Yildiz, C.; Acikgoz, H.; Korkmaz, D.; Budak, U. An improved residual-based convolutional neural network for very short-term wind power forecasting. *Energy Convers. Manag.* **2021**, *228*, 113731. [CrossRef]
3. Duan, J.; Wang, P.; Ma, W.; Tian, X.; Fang, S.; Cheng, Y.; Chang, Y.; Liu, H. Short-term wind power forecasting using the hybrid model of improved variational mode decomposition and Correntropy Long Short-term memory neural network. *Energy* **2021**, *214*, 118980. [CrossRef]
4. Abedinia, O.; Lotfi, M.; Bagheri, M.; Sobhani, B.; Shafiekhah, M.; Catalao, J.P.S. Improved EMD-Based Complex Prediction Model for Wind Power Forecasting. *IEEE Trans. Sustain. Energy* **2020**, *11*, 2790–2802. [CrossRef]
5. Memarzadeh, G.; Keynia, F. A new short-term wind speed forecasting method based on fine-tuned LSTM neural network and optimal input sets. *Energy Convers. Manag.* **2020**, *213*, 112824. [CrossRef]
6. Liu, B.; Zhao, S.; Yu, X.; Zhang, L.; Wang, Q. A Novel Deep Learning Approach for Wind Power Forecasting Based on WD-LSTM Model. *Energies* **2020**, *13*, 4964. [CrossRef]
7. Wang, C.; Zhang, H.; Ma, P. Wind power forecasting based on singular spectrum analysis and a new hybrid Laguerre neural network. *Appl. Energy* **2020**, *259*, 114139. [CrossRef]
8. Sun, Z.; Zhao, S.; Zhang, J. Short-Term Wind Power Forecasting on Multiple Scales Using VMD Decomposition, K-Means Clustering and LSTM Principal Computing. *IEEE Access* **2019**, *7*, 166917–166929. [CrossRef]
9. Piotrowski, P.; Kopyt, M.; Baczyński, D.; Robak, S.; Gulczyński, T. Hybrid and Ensemble Methods of Two Days Ahead Forecasts of Electric Energy Production in a Small Wind Turbine. *Energies* **2021**, *14*, 1225. [CrossRef]
10. Sun, M.; Feng, C.; Zhang, J. Multi-distribution ensemble probabilistic wind power forecasting. *Renew. Energy* **2020**, *148*, 135–149. [CrossRef]
11. Chen, C.; Liu, H. Medium-term wind power forecasting based on multi-resolution multi-learner ensemble and adaptive model selection. *Energy Convers. Manag.* **2020**, *206*, 112492. [CrossRef]
12. Du, P. Ensemble Machine Learning-Based Wind Forecasting to Combine NWP Output With Data From Weather Station. *IEEE Sustain. Energy* **2019**, *10*, 2133–2141. [CrossRef]
13. Chen, J.; Zhu, Q.; Li, H.; Lin, Z.; Shi, D.; Li, Y.; Duan, X.; Liu, Y. Learning Heterogeneous Features Jointly: A Deep End-to-End Framework for Multi-Step Short-Term Wind Power Prediction. *IEEE Trans. Sustain. Energy* **2020**, *11*, 1761–1772. [CrossRef]
14. Shahid, F.; Zameer, A.; Mehmood, A.; Raja, M.A.Z. A novel wavenets long short term memory paradigm for wind power prediction. *Appl. Energy* **2020**, *269*, 115098. [CrossRef]
15. Su, H.-Y.; Huang, C.-R. Enhanced Wind Generation Forecast Using Robust Ensemble Learning. *IEEE Trans. Smart Grid 12.1* **2021**, *2*, 912–915. [CrossRef]
16. Saini, V.K.; Kumar, R.; Mathur, A.; Saxena, A. Short term forecasting based on hourly wind speed data using deep learning algorithms. In Proceedings of the 2020 3rd International Conference on Emerging Technologies in Computer Engineering: Machine Learning and Internet of Things (ICETCE), Jaipur, India, 7–8 February 2020; pp. 1–6. [CrossRef]
17. Ahmadi, A.; Nabipour, M.; Mohammadi-Ivatloo, B.; Amani, A.M.; Rho, S.; Piran, M.J. Long-Term Wind Power Forecasting Using Tree-Based Learning Algorithms. *IEEE Access* **2020**, *8*, 151511–151522. [CrossRef]
18. Kisvari, A.; Lin, Z.; Liu, X. Wind power forecasting A data-driven method along with gated recurrent neural network. *Renew. Energy* **2021**, *163*, 1895–1909. [CrossRef]
19. Ouyang, T.; Huang, H.; He, Y.; Tang, Z. Chaotic wind power time series prediction via switching data-driven modes. *Renew. Energy* **2020**, *145*, 270–281. [CrossRef]
20. Li, L.L.; Zhao, X.; Tseng, M.L.; Tan, R.R. Short-term wind power forecasting based on support vector machine with improved dragonfly algorithm. *J. Clean. Prod.* **2020**, *242*, 118447. [CrossRef]
21. Wang, X.; Li, P.; Yang, J. Short-term Wind Power Forecasting Based on Two-stage Attention Mechanism. *IET Renew. Power Gener.* **2019**, *14*, 297–304. [CrossRef]
22. Yuan, X.; Chen, C.; Jiang, M.; Yuan, Y. Prediction interval of wind power using parameter optimized Beta distribution based LSTM model. *Appl. Soft Comput.* **2019**, *82*, 105550. [CrossRef]
23. Niu, Z.; Yu, Z.; Tang, W.; Wu, Q.; Reformat, M. Wind power forecasting using attention-based gated recurrent unit network. *Energy* **2020**, *196*, 117081. [CrossRef]
24. Hu, H.; Wang, L.; Lv, S.X. Forecasting energy consumption and wind power generation using deep echo state network. *Renew. Energy* **2020**, *154*, 598–613. [CrossRef]
25. Yu, R.; Liu, Z.; Li, X.; Lu, W.; Ma, D.; Yu, M.; Wang, J.; Li, B. Scene learning: Deep convolutional networks for wind power prediction by embedding turbines into grid space. *Appl. Energy* **2019**, *238*, 249–257. [CrossRef]
26. Hu, T.; Wu, W.; Guo, Q.; Sun, H.; Shi, L.; Shen, X. Very short-term spatial and temporal wind power forecasting A deep learning approach. *CSEE J. Power Energy Syst.* **2020**, *6*, 434–443. [CrossRef]
27. Wang, Z.; Zhang, J.; Zhang, Y.; Huang, C.; Wang, L. Short-Term Wind Speed Forecasting Based on Information of Neighboring Wind Farms. *IEEE Access* **2020**, *8*, 16760–16770. [CrossRef]
28. Yin, H.; Ou, Z.; Huang, S.; Meng, A. A cascaded deep learning wind power prediction approach based on a two-layer of mode decomposition. *Energy* **2019**, *189*, 116316. [CrossRef]

29. Lin, Z.; Liu, X. Wind power forecasting of an offshore wind turbine based on high-frequency SCADA data and deep learning neural network. *Energy* **2020**, *201*, 117693. [CrossRef]
30. Medina, S.V.; Ajenjo, U.P. Performance Improvement of Artificial Neural Network Model in Short-term Forecasting of Wind Farm Power Output. *J. Mod. Power Syst. Clean Energy* **2020**, *8*, 484–490. [CrossRef]
31. Tang, B.; Chen, Y.; Chen, Q.; Su, M. Research on Short-Term Wind Power Forecasting by Data Mining on Historical Wind Resource. *Appl. Sci.* **2020**, *10*, 1295. [CrossRef]
32. Piotrowski, P.; Baczyński, D.; Kopyt, M.; Szafranek, K.; Helt, P.; Gulczyński, T. Analysis of forecasted meteorological data (NWP) for efficient spatial forecasting of wind power generation. *Electric Power Systems Research* **2019**, *175*, 105891. [CrossRef]
33. De Mattos Neto, P.S.G.; de Oliveira, J.F.L.; de O. Santos Júnior, D.S.; Siqueira, H.V.; Marinho, M.H.N.; Madeiro, F. An adaptive hybrid system using deep learning for wind speed forecasting. *Inf. Sci.* **2021**, *581*, 495–514. [CrossRef]
34. Atmospheric Model high resolution 10-day forecast (Set I—HRES). Available online: https://www.ecmwf.int/en/forecasts/datasets/set-i (accessed on 15 January 2021).
35. NCEP Products Inventory. Available online: https://www.nco.ncep.noaa.gov/pmb/products/gfs/ (accessed on 15 January 2021).
36. Available online: https://mapy.meteo.pl (accessed on 15 January 2021).
37. Dudek, G.; Pełka, P. Forecasting monthly electricity demand using k nearest neighbor method. *Przegląd Elektrotechniczny* **2017**, *93*, 62–65.
38. Dudek, G. Multilayer Perceptron for Short-Term Load Forecasting: From Global to Local Approach. *Neural Comput. App.* **2019**, *32*, 3695–3707. [CrossRef]
39. Osowski, S.; Siwek, K. Local dynamic integration of ensemble in prediction of time series. *Bull. Pol. Ac. Tech.* **2019**, *67*, 517–525.
40. Piotrowski, A.P.; Napiorkowski, J.J.; Piotrowska, A.E. Impact of deep learning-based dropout on shallow neural networks applied to stream temperature modelling. *Earth-Sci. Rev.* **2020**, *201*, 103076. [CrossRef]
41. Parol, M.; Piotrowski, P.; Kapler, M.; Piotrowski, M. Forecasting of 10-Second Power Demand of Highly Variable Loads for Microgrid Operation Control. *Energies* **2021**, *14*, 1290. [CrossRef]
42. Géron, A. *Hands-on Machine Learning with Scikit-Learn, Keras, and TensorFlow*, 2nd ed.; O'Reilly Media, Inc.: Sevastopol, CA, USA, 2019.
43. Hastie, T.; Tibshirani, R.; Friedman, J.H. *The Elements of Statistical Learning. Data Mining, Inference, and Prediction*, 2nd ed.; Springer: Berlin/Heidelberg, Germany, 2009.
44. Zhao, Y.; Ye, L.; Li, Z.; Song, X.; Lang, Y.; Su, J. A novel bidirectional mechanism based on time series model for wind power forecasting. *Appl. Energy* **2016**, *177*, 793–803. [CrossRef]
45. Shahram, H.; Xiaolei, L.; Zi, L.; Saeid, L. A Critical Review of Wind Power Forecasting Methods—Past, Present and Future. *Energies* **2020**, *13*, 3764. [CrossRef]
46. Barcons, J.; Avila, M.; Folch, A. Diurnal cycle RANS simulations applied to wind resource assessment. *Wind Energy* **2019**, *2*[illegible], 269–282. [CrossRef]

energies

MDPI

Article

Voltage Control in MV Network with Distributed Generation—Possibilities of Real Quality Enhancement

Paweł Pijarski *, Piotr Kacejko and Marek Wancerz

Department of Power Engineering, Faculty of Electrical Engineering and Computer Science, Lublin University of Technology, Nadbystrzycka 38D, 20-618 Lublin, Poland; p.kacejko@pollub.pl (P.K.); m.wancerz@pollub.pl (M.W.)
* Correspondence: p.pijarski@pollub.pl

Abstract: Connecting an increasing number of distributed sources in MV (medium voltage) and LV (low voltage) distribution networks causes voltage problems resulting mainly from periodic power flows towards the HV/MV (HV—high voltage) transformer station. This temporarily changes the nature of distribution networks from receiving to supply networks and causes an increase in the voltage values deep within the network, often above the permissible level. Therefore, it is necessary to search for new voltage control methods that take into account the active participation of distributed sources. The article proposes a concept of such a system in which the control signals are transformer taps in the HV/LV station and the values of reactive powers generated or consumed by RES (renewable energy sources). These values can be determined either by solving the optimisation problem (according to a given quality indicator criterion) or on the basis of appropriately selected settings of the $Q(U)$ characteristics of the inverters and the HV/LV transformer ratio. The article describes both approaches, pointing to the advantages and disadvantages of each of them.

Keywords: voltage control; voltage quality; renewable energy; metaheuristic optimisation; medium voltage; $Q(U)$ characteristics

Citation: Pijarski, P.; Kacejko, P.; Wancerz, M. Voltage Control in MV Network with Distributed Generation—Possibilities of Real Quality Enhancement. *Energies* **2022**, *15*, 2081. https://doi.org/10.3390/en15062081

Academic Editors: Paweł Piotrowski, Grzegorz Dudek, Dariusz Baczyński and Alvaro Luna

Received: 23 January 2022
Accepted: 10 March 2022
Published: 12 March 2022

1. Introduction

The article continues and extends the analysis of problems related to voltage control in MV networks, in which a large number of distributed sources have been installed. The variable power generation of these units due to weather conditions causes frequent changes of voltage values. The most severe are voltage increases above 1.1 U_n, which when transformed into the LV level may damage the receivers, or create conditions for switching off the sources (both on the MV and LV side) by overvoltage protections. The volatility of weather phenomena and the randomness operation of protective devices lead to voltage chaos in the network.

In the previously presented work [1], the authors showed that voltage control is possible, in which not only the HV/MV transformer with OLTC (on load tap changer) is actively involved, but also sources connected to the MV grid. These sources, depending on the voltage conditions (related to the variability of the power generation and voltage changes in the HV grid), can control the values of the generated (or consumed) reactive power on the basis of signals sent from the voltage controller.

The concept of voltage control in the MV network proposed in [1] comes down to on-line solving (for every quarter of an hour) of the OPF (optimal power flow) task after prior estimation of the network state and transmission of control variables determined in the computational process to the actuators (tap changer position and source reactive powers). In the considered OPF problem, the objective function is a voltage quality indicator covering all network nodes (the number of nodes is N). The objective function is described by the Equation (1).

The proven effectiveness of the optimisation task solution (the AIG heuristic algorithm [2] was used) is conditioned, however, by high requirements in terms of accessibility to the network model and its ICT equipment. The network model is the result of the process of estimating its state. The use of estimation algorithms at the MV level is not an easy task, although certainly not as complex as in the case of meshed transmission networks. Similarly, solving the OPF task in real time (the research assumed a discrete control for each time window of one quarter of an hour) requires considerable computational expenditure. Thus, the method of optimal voltage control in MV networks presented in [1], hereinafter referred to as the OPFh-MVt method (optimal power flow heuristic—medium voltage for each time period), can be considered attractive and future-proof, but today it is difficult to convince network operators to wider attempts to implement it.

In the present article, the authors set themselves the goal of searching for an alternative method of voltage control in MV networks with distributed generation, the implementation of which would not be as complicated as in the case of the OPFh-MVt method, while the results would be only slightly worse. The novelty of the proposed approach consists in presenting the optimal method of voltage control in the MV network, the results of which are treated as reference. For practical use, a simplified method is recommended, the results of which have also been positively verified. The novelty of the article also lies in the fact that a very large set of data from real objects was used to verify the presented methods.

The article consists of seven sections. The first section contains an introduction to the subject and the purpose of extending research and analyses related to the considered problem. The second section presents a literature review on voltage control in the MV network. The third section contains the formulation of the optimisation task and the description of the algorithm for its solution. The fourth section includes a description of a simplified method of voltage control using the HV/MV transformer tap changer and control of the reactive power of RES sources with given Q(U) characteristics. Section five presents the IEEE 37 test network. The calculation results showing the effectiveness of the proposed control system are included in section six. Section seven provides a discussion of the results and conclusions.

2. Literature Review

The subject of voltage control in MV networks with distributed generation has been the subject of research in many articles. The authors approach this problem in various ways, trying to demonstrate the effectiveness of the proposed methods of solving it. Generally speaking, there are four main groups of methods presented in the works so far:

- Voltage control using only the on-load tap-changer (and/or possibly a capacitor bank);
- Voltage control with the use of on-load tap changer and reactive power generation in RES;
- Voltage control with the use of on-load tap changer, reactive power generation in renewable energy sources and the use of energy storages connected in selected network nodes;
- Voltage control with the use of on-load tap changer, reactive power generation in RES and the use of electrolyser installations connected in generation nodes.

It is also possible to imagine a comprehensive approach to the problem and apply voltage control using an on-load tap changer, reactive power generation in RES, the use of energy stores connected in selected network nodes and electrolyser installations connected in generation nodes.

A number of works have been prepared in which problems resulting from a radical change in the characteristics of distribution networks, previously considered typical (radial system of operation, unidirectional power flow), are considered. Some of them used both classical and heuristic optimisation methods. The selected voltage control evaluation criteria identified by the authors in other articles are presented below:

- Minimisation of the cost of power and energy losses [3],
- Minimisation of power losses [4,5],
- Minimisation of energy losses [3],
- Minimisation of costs related to active power losses and reactive power flow [6],

- Minimisation of the voltage quality index (while in the literature there are indicators expressed with the use of various dependencies) [7–11],
- Minimisation of the number of tap changer position changes [9].

The above-mentioned selected criteria, ways of solving the outlined problems and many other similar issues can be found, for example, in works [12–24].

The easiest way to adjust is to use only the on-load tap-changer. For example, in [25] the authors used the multi-agent system to find the optimal values of the transformer tap changer in order to minimise the objective function, which is the positive three-phase voltage deviation. This function represents the sum of voltage deviations in the observed nodes.

In article [26], the authors present the results of analyses for the IEEE 13 test network at various load levels. Changing the transformer taps is controlled by the line drop compensator, depending on the required voltage level in the selected network node.

An interesting approach can be found in article [9], where the objective function is the difference between the transformer's taps at two consecutive time points (assuming that there is one transformer in the network). The optimisation task is to minimise the number of tap changer position changes during the day while meeting the constraints.

Study [10] uses a method consisting in adaptive adaptation of the transformer's tap changer to the assumed voltage value in a fictitious node. The electrical distance of this node from the MV busbars in the 110/MV station is also appropriately determined so that the expected voltage value in it influences the quality indicator of the voltage quality in the entire network.

In article [8], seven different objective functions related to the optimal voltage control in the MV and LV distribution networks are considered. HV/MV transformer ratios and MV/LV transformers ratios are addressed as decision variables.

In addition to the transformer tap changer, the ability to generate reactive power in RES is also used for voltage control. A number of works on this subject have been written. An example may be article [6] where the decision variables are the reactive powers of the micro-sources in the LV network at the given transformer ratio. The objective function is the sum of costs related to active power losses and costs related to the reactive power flow. The internal point method is used to solve the optimisation problem.

Decision variables in the form of reactive power generated or consumed by RES are also used in work [27]. The authors consider a three-criteria objective function under the necessary constraints. Weights for individual criteria are determined dynamically.

In [28], a two-criteria objective function is considered, consisting of the sum of the costs of power losses and the costs of switching operations as well as voltage deviations. The objective function contains two criteria, therefore weighting factors were used. The weighting factors are selected by the analytic hierarchy process (AHP), described in article [29].

In [30], a single-criterion objective function is used in the form of a voltage quality indicator. The optimisation task was to minimise the objective function by changing the HV/LV transformer ratio and the reactive power of the sources, but only in a few operating states. The applied method of linear optimisation was locally convergent.

The use of reactive power generation or consumption in RES was also analysed in works [31–36].

The next group of papers are articles devoted to the use of electricity storage to optimise the operation of the distribution network [37–45]. The authors of these studies apply various criteria and methods for solving voltage problems.

In the work [37], the authors consider the medium voltage network and the water supply network, which is a controlled energy storage. Water consumption control (grid load control) is used to control the voltage by changing the electricity consumption. The article [38] presents the Predictive Control (MPC) Model, which consists in the optimal coordination of generation in renewable energy sources, energy storage and the operation of the on-load tap changer. One of the most interesting functions of the objective is included in [39]. The objective function has two normalised criteria (with values ranging from 0 to 1). Each of the two considered criteria is taken into account with an appropriate weighting factor. The first criterion is the voltage deviation, while the second criterion is the total capacity of the energy storage.

In the work [40], in order to solve the voltage problems caused by a large number of photovoltaic installations, a coordinated method of controlling distributed energy storage systems in combination with traditional control (OLTC) has been proposed. A novel charging and discharging system for battery energy storage systems (BESS), which uses real network data, is described in [41]. The article [42] proposes to create an optimal battery charging/discharging schedule in the context of power loss minimisation. Determining the capacity of battery energy storages installed in a grid saturated with photovoltaic installations, in order to control their operation, was proposed in [43]. A review of energy storage technologies and systems and the methods of their application, for example in power grids, have been presented in the works [44,45].

Some authors use the available measurements and also look for the relationship between the voltage values and the power generated in the sources to implement the voltage control process in the distribution network. Some of these methods do not require knowledge of the network model, due to the application of neural solution (deep learning and artificial intelligence. Such attempts can be found, for example, in works [46–51].

The use of voltage value measurements to control the operation of the distribution network without the knowledge of the network topology is presented, for example, in [46]. In the work [47], the authors replace with a linear model the non-linear dependencies between the voltage values in the distribution network nodes and the generated power. Optimal voltage control in a distribution network containing renewable energy sources, which does not require knowledge of its model, was considered in the works [48,49]. In the article [50] a data-driven-based optimisation method for var-voltage sequential control was proposed. An interesting algorithm of voltage control in the distribution network is presented in [51]. The authors also emphasise that the proposed method requires the exchange of information only between neighbouring photovoltaic installations, which significantly reduces the communication complexity.

The applied deep learning algorithms can be combined with optimisation tasks. Examples of such research and analyses are presented, for example, in the works [52–55].

The P2G (power to gas) technology has also been developed for some time, and alkaline water electrolysers (AEL)—[56,57], used for the production of "green hydrogen", are considered the cheapest and the most accessible. From the point of view of voltage control they are controlled active power loads connected at the generation nodes. Within a few years there has been a significant increase in interest in this method of storing surplus electricity from renewable sources [56–64]. Some works concern the optimal size and layout of electrolyser installations. Part of the articles concern the elimination of negative voltage effects in networks saturated with RES installations. The analyses are conducted for both the medium voltage and low voltage distribution networks.

As shown in the literature review, there are many ways to assess the quality of voltage and the effectiveness of its control in power grids. Some of them use a complicated mathematical framework, in others the objective function of the control process is difficult to understand intuitively by a combination of technical and economic indicators. In some solutions it is not necessary to know the network model, but it is necessary to transmit signals from all its nodes and sources. According to the authors of the presented article, only simple voltage quality assessment criteria have a chance for practical use by network operators and sensitive consumers. Therefore, the search for complex alternative criteria was abandoned, assuming that simple criteria such as (1) having a simple physical interpretation (analysis of the deviation from the criterion value) can be treated as the appropriate objective function of a more or less complex optimisation processes.

Comparing the works of other authors with the analyses performed in this article, its originality should be emphasised, consisting in the application of an innovative approach to the problem of voltage control in the considered MV network. It consists in:

- Using the actual annual power generation characteristics by the solar and wind sources under consideration ($P_{gi}(t)$);
- Using the real characteristics of the power received at nodes ($P_{Li}(t)$) of the considered network;
- Using the real waveform of voltage changes in the 110 kV network ($U_{HV}(t)$);
- Determination of the optimal values of the control variables with the use of the proprietary OPFh-MVt algorithm for all the considered grid states and determination of global control assessment indicators;
- Formulation of a simplified method of voltage control using the interaction of the OLTC system and active operation of the source control and its evaluation in relation to the reference results obtained with the OPFh-MVt method.

3. The Method of Voltage Control in the MV Network Using the Results of Cyclic Solving of the OPF Task

The method proposed in this paper is implemented on the basis of the solution of the optimisation task.

The considered objective function was described by the following equation:

$$F(\mathbf{x},\mathbf{y},\mathbf{z}) = \sqrt{\sum_{i=1}^{N}\left(\frac{U_i - U_o}{U_n}\right)^2} = \text{Ind}\,U, \tag{1}$$

while the individual variables of the control process are defined as follows:

$\mathbf{x} = [\vartheta,\ Q_{G1} \ldots Q_{Gk} \ldots Q_{Gp}]$—vector of control variables formed by transformer ratio (ϑ—discrete variable) and reactive power of p sources connected to the MV network;

$\mathbf{y} = [U_{HV},\ P_{L1} \ldots P_{Lm}, Q_{L1} \ldots Q_{Lm}, P_{G1} \ldots P_{Gp}]$—vector of independent variables, formed by: HV network supply voltage, active and reactive power received in m nodes and power generated in p sources, not subject to change during optimisation calculations,

$\mathbf{z} = [U_1 \ldots\ U_j, \delta_1 \ldots \delta_j]$—the vector of state variables containing nodal voltages and their arguments (total number of network nodes $j = p + m$).

The results obtained by that method can be treated as reference for the other simplified solutions. A diagram of the control process is shown in Figure 1.

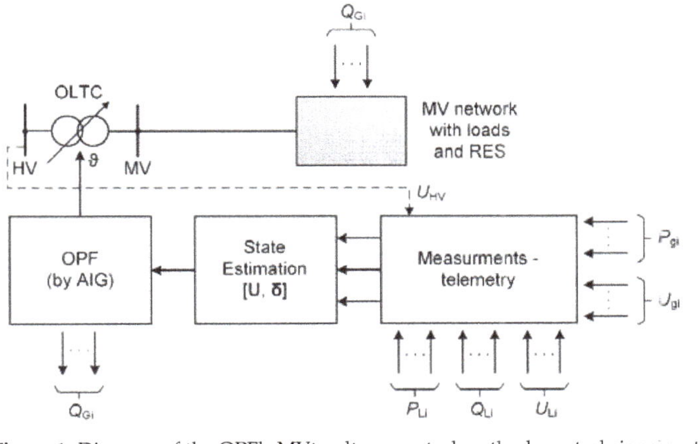

Figure 1. Diagram of the OPFh-MVt voltage control method, control signals—transformer ratio ϑ and reactive power of RES systems.

For the solution of the optimisation task, the original proprietary heuristic algorithm called AIG (Algorithm of Innovative Gunner) was used, described in detail in [2] and successfully tested to solve many technical and mathematical problems—[2,65]. The AIG algorithm is characterised by the fact that the components of the decision vector are subject to "multiplicative" modifications in subsequent iterations, described by the relationship

$$x_l^{(k+1)} = x_l^{(k)} \cdot g_l(\xi) \tag{2}$$

in contrast to "additive" modifications, used in other metaheuristic methods [65–73], described by the relationship

$$x_l^{(k+1)} = x_l^{(k)} + \Delta x_l^{(k)} \tag{3}$$

where k is the next iteration, functions $g_l(\xi)$ and $\Delta x_l^{(k)}$ are a symbolic notation and a characteristic of the heuristic method used.

The innovativeness of the AIG algorithm results from a new method of determining the value of decision variables in subsequent iterations. This means that in each step of the iteration process, the previously obtained solution is corrected by appropriately selected multipliers. This is a fundamental difference compared to other metaheuristic algorithms, in which the process of creating a new solution is based on adding an appropriate component (appropriate for a given method) to the previous solution or searching in its environment. The authors of the article, as the authors of the AIG algorithm, find more and more applications in which its speed and accuracy of calculations are used. It is also used in other applications, even very distant from the power industry [74–77].

In the case of the AIG algorithm, the $g_l(\xi)$ functions have the form of the $\cos\alpha$ function and its inverse $(\cos\alpha)^{-1}$, while α and β are correction angles drawn from the variable interval $(-\alpha_{\max}, \alpha_{\max})$ and $(-\beta_{\max}, \beta_{\max})$ by means of the uniform distribution. A block diagram showing the operation of the AIG algorithm is shown in Figure 2 [2].

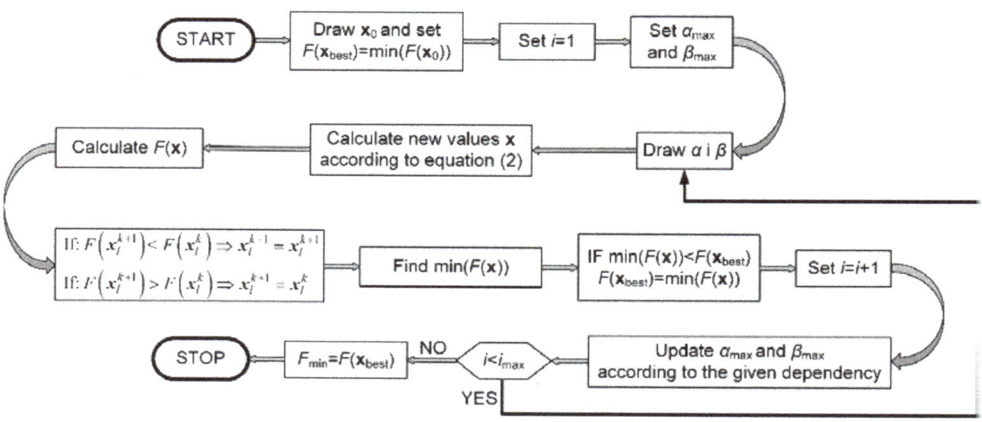

Figure 2. Block diagram of the AIG algorithm (k is the iteration number).

The objective function $F(\mathbf{x})$, which is minimised, is described by Equation (1). The following limitations are checked during the optimisation process:

- Minimum and maximum transformer ratio values (ϑ). The calculations were based on a 10 MVA transformer with 19 operating positions of the tap changer, within the range of ± 9 (plus the tap in the zero position);
- Minimum and maximum reactive power values for each renewable energy source ($Q_{\text{Gmax}}, Q_{\text{Gmin}}$). It was assumed that each RES has the ability to generate/consume a maximum reactive power equal to … P_{nG}; since the maximum power of each power plant is 1 MW, the possible reactive power control is within ± 0.4 Mvar;
- Minimum and maximum voltage values for all network nodes (U_i); the voltage was kept in the range from 0.9 U_{nMV} to 1.1 U_{nMV};

- Permissible values of current carrying capacity of sections of power lines ($I_{l\max}$). The following value was assumed in the calculations:
    - $I_{l\max} = 355$ A for conductors with a cross-section of 120 mm^2,
    - $I_{l\max} = 290$ A for conductors with a cross-section of 70 mm^2,
    - $I_{l\max} = 170$ A for conductors with a cross-section of 50 mm^2,
    - $I_{l\max} = 145$ A for conductors with a cross-section of 35 mm^2,

and the permissible power value o of the transformer (S_{nT}). The calculations assume the rated power of the transformer $S_{nT} = 10$ MVA.

The calculations were performed in Matlab and PowerWorld Simulator, version 22. The main script was written in Matlab, while the power flow calculations were performed in PowerWorld. The connection between the two programs is possible owing to the SimAuto plug-in (included with PowerWorld), which also acts as an interchangeable computing engine that enables data exchange between different applications. The computation process starts with running the script in the Matlab environment. Then, during each iteration, remote connection with the PowerWorld floodlight program is performed, the parameters of the power system elements are changed, and the calculation results are downloaded [1,65,78]. The flow chart of the optimisation process is presented in the general diagram (Figure 3).

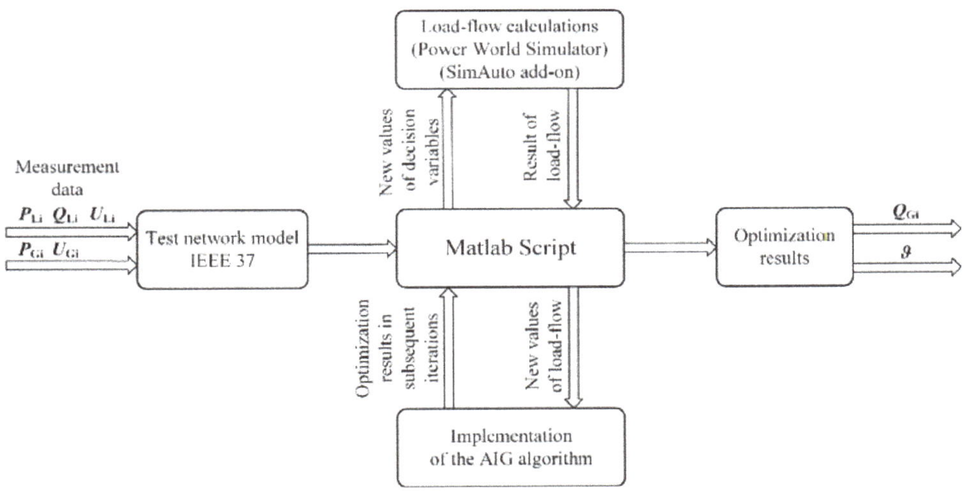

Figure 3. General scheme of the organisation of the computational process used in solving the optimisation task.

Changing the input parameters or downloading the calculation results is done with the use of appropriate commands, appropriate for a given programming environment. After the AIG algorithm is run, optimisation calculations follow, and the results are saved in a file.

4. A Simplified Method of Voltage Control in the MV Network with the Use of the Tap Changer of the HV/MV Transformer and the Active Influence of Distributed Sources

The basic voltage control system in the MV network is shown in Figure 4a. Very often, the role of this system is limited to keeping a constant, set voltage value on the lower side of the HV/MV transformer. The OLTC switches the transformer taps on the HV side, in the considered case their number (up and down) was ± 9, and the voltage change per tap $\Delta U_T = 1.11\%$. These are typical values. At the same time, in many cases, the neutrality of RES in terms of generation (or consumption) of reactive power is sought by setting their power factors to the value $\cos\varphi = 1$. Admittedly, this method of voltage control ensures its set value near the transformer busbars (most often it is 1.05 U_n), but it does not allow for controlling the increase in voltage deep inside the network, which was shown for the test cases. Such a method of control should be assessed negatively.

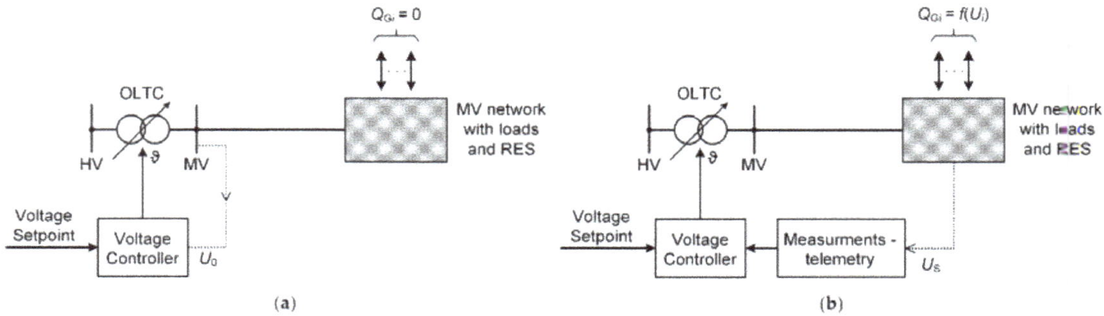

Figure 4. Simplified methods of voltage control in the MV grid with distributed generation: (**a**) traditional regulation—keeping a constant voltage value on the MV busbars, (**b**) keeping a constant voltage value in an optimally selected node deep inside the grid and activating the characteristics of *Q(U)* inverters.

In order to take advantage of the regulation possibilities of the sources, it is possible to consider the way of operating with a defined level of reactive power generation. As the problem is too high voltage values caused by the power flow towards the MV busbars, the method of operation involving reactive power consumption depending on the value of the generated active power, i.e., $Q_G = -0.4\ P_G$, was also considered. This method of voltage control should also be assessed negatively, because in some cases the voltage value is underestimated, and unnecessary reactive power flows increase losses.

The improvement of voltage conditions in the MV network can be achieved also by keeping a constant voltage value not on the transformer busbars, but inside the network—Figure 4b (node *s*). Depending on the possibility of signal transmission from the network to the controller and the method of selecting the set point, the effects of such control may be varied, but they have a significant impact on reducing the negative influence of RES on voltage conditions and improve the efficiency of the OLTC system.

The activation of the characteristics of the *Q(U)* inverters results in a further improvement of voltage conditions in the vicinity of installation of RES units. The required shape of the *Q(U)* characteristic is given in standards [79–81]. Its individual characteristic points can be individually set for each source. Analyses taking into account the characteristics of reactive power as a function of voltage in a network node can be found, inter alia, in the works [7,32,65,82–85]. Figure 5 shows the characteristic that seems to be the most appropriate for a network with a large number of RES—when the voltage reaches the value of 1.1 U_n, the source absorbs the maximum possible value of reactive power.

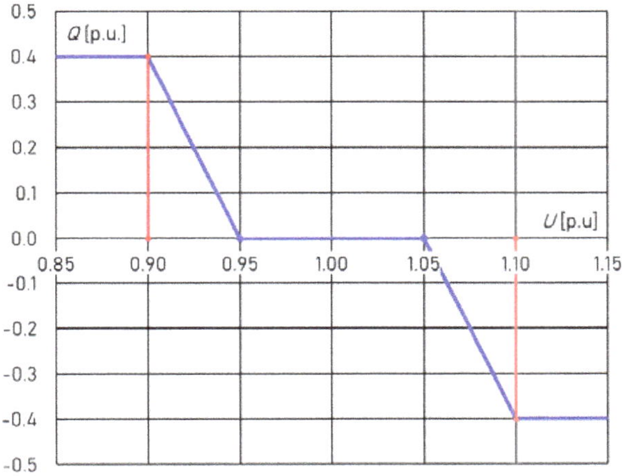

Figure 5. The $Q(U)$ characteristic of the inverter of the RES installation (photovoltaic and wind farm) selected for the analysis of the effectiveness of the voltage regulation in the MV grid.

5. Test Network

The subject of the research was the IEEE 37 network [86], which was assigned a voltage of 15 kV (MV). The supply station has a 10 MVA transformer with a ratio of ϑ = 115/16.5 kV/kV ± 9%. The operation of five sources was considered in this network—three photovoltaic farms and two wind turbines with the same rated power of 1 MW. The diagram of the IEEE 37 network and the location of the sources are shown in Figure 6. A detailed description of the network structure as well as the resistance and reactance of individual branches modelling the lines are presented in Table 1. Table 1 also contains cross-sections and lengths of individual line sections, which show that the network in question is typical for rural areas with an average level of electrification. The network load includes MV/LV transformer substations connected in all nodes (the total number of nodes is m = 37, MV/LV substations are not marked in the Figure 6). Table 1 presents the data of the individual sections of the MV line. Table 1 presents the data of the individual sections of the MV line.

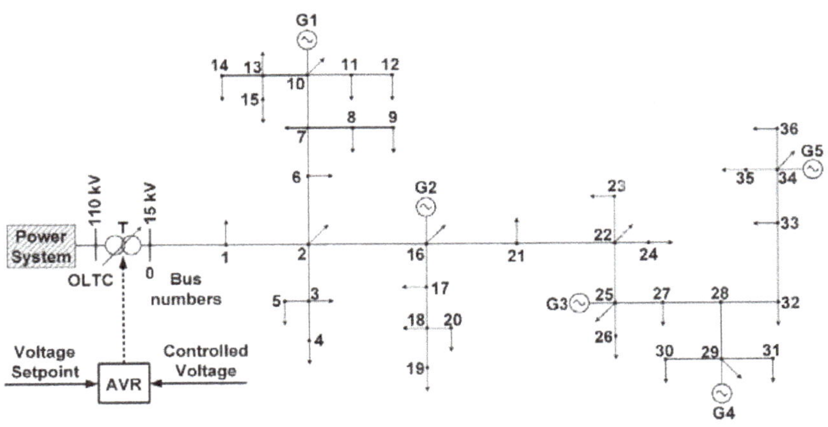

Figure 6. IEEE 37 test network diagram [86].

Table 1. Parameters of the individual sections of the MV line.

Power Line from-to	Line Length l, km	Conductor Cross-Section S, mm^2	Resistance R, Ω	Reactance X, Ω
0–1	1.80	120	0.43	0.18
1–2	1.80	120	0.43	0.18
2–3	0.50	70	0.20	0.20
2–6	0.50	70	0.20	0.20
2–16	2.50	120	0.60	0.25
3–4	0.70	70	0.29	0.28
3–5	0.60	70	0.25	0.24
6–7	1.20	70	0.50	0.48
7–8	1.20	35	0.99	0.48
7–10	2.00	70	0.82	0.80
8–9	0.70	35	0.57	0.28
10–11	0.50	35	0.41	0.20
10–13	1.20	35	0.98	0.48
11–12	1.20	35	0.98	0.48
13–14	0.50	35	0.41	0.20
13–15	0.90	35	0.74	0.36
16–17	1.50	35	1.22	0.60
16–21	2.90	120	0.69	0.29
17–18	0.90	35	0.74	0.36
18–19	0.50	35	0.41	0.20
18–20	0.70	35	0.57	0.28
21–22	2.50	120	0.60	0.25
22–23	1.00	35	0.82	0.40
22–24	1.00	35	0.82	0.40
22–25	1.80	35	1.47	0.72
25–26	0.50	35	0.41	0.20
25–27	2.00	35	1.63	0.80
27–28	1.50	35	1.22	0.60
28–29	0.90	35	0.74	0.36
28–32	1.80	35	1.47	0.72
29–30	1.00	35	0.82	0.40
29–31	0.60	35	0.49	0.24
32–33	1.50	35	1.22	0.60
33–34	1.90	35	1.55	0.76
34–35	0.80	35	0.65	0.32
34–36	0.80	35	0.65	0.32

The authors had hourly measurements of the load and generated power in the MV network and the voltage on the 110 kV (HV) side registered for the entire year, which gives 8760 h. The record of changes in these values is shown in Figure 7. Power generation in wind turbines (Figure 7c) and in photovoltaic farms (Figure 7d) corresponds to real changes resulting from weather conditions (wind speed, solar radiation intensity, cloud cover).

Figure 8 shows the results of the voltage analysis carried out for the tested MV network in the conditions of complete no RES generation. The voltage values determined for 8760 cases form a characteristic multicoloured "band" which, with increasing distance from the MV busbars of the HV/MV transformer, slightly widens and falls downwards. In all cases and for each node, the voltage must be between 1.01 and 1.05 of the rated voltage. Thus, the voltage quality in the state of no generation, even without introducing numerical indicators, can be assessed as good.

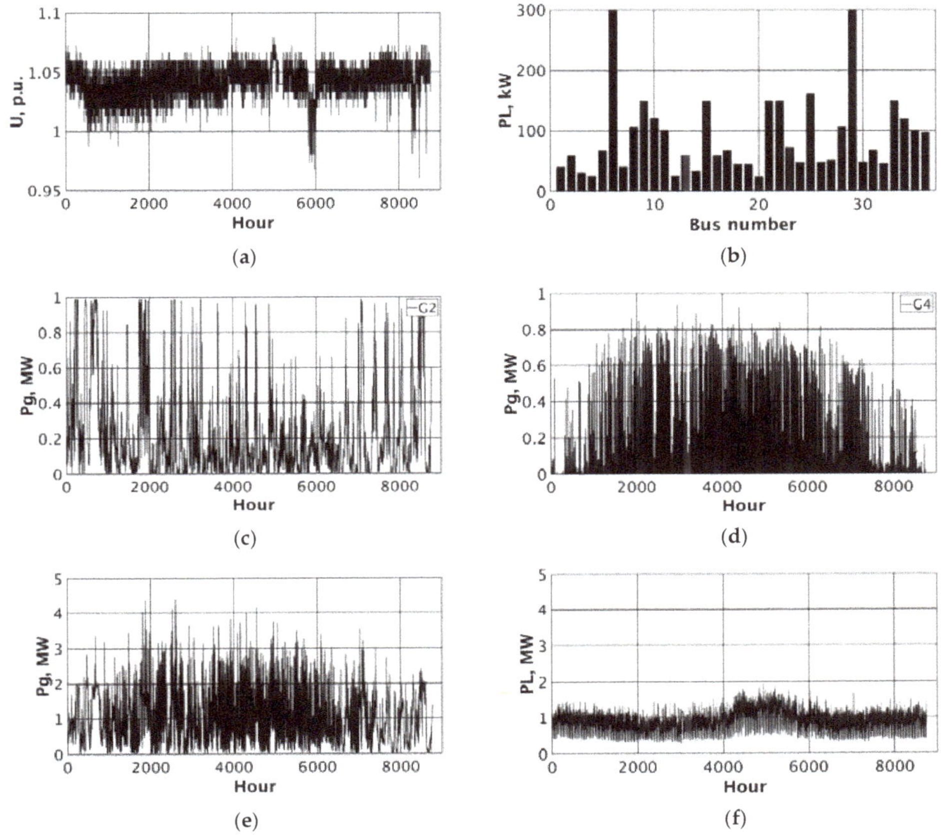

Figure 7. Drawn variable values for subsequent calculation cases (**a**) HV values, (**b**) maximum loads of individual nodes, (**c**) power generated in wind turbine (G2), (**d**) power generated in photovoltaic farm (G4), (**e**) total power generated in renewable energy sources, (**f**) total load power in MV nodes.

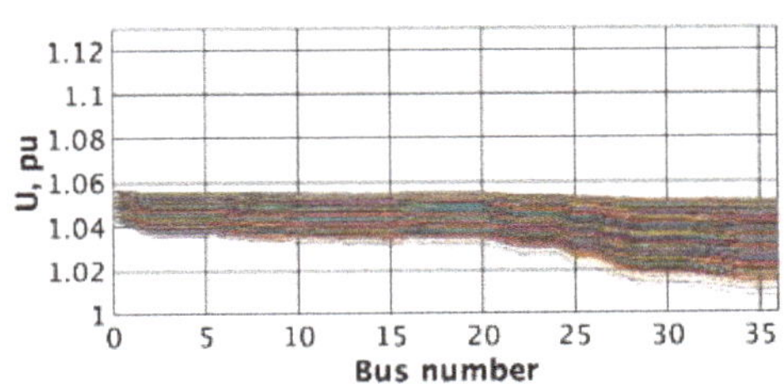

Figure 8. The results of the voltage analysis in the IEEE 37 network without the participation of sources.

6. Calculation Results

Below, the results of the analysis of voltage values in the MV nodes, carried out over a period of one year with the use of the three control methods discussed above, are presented and compared.

6.1. Assessment of Voltage Quality Using a Traditional Circuit

Figure 9 shows the results of the analysis carried out with the assumption that the control system keeps the value of 1.05 U_n on the transformer LV bus (node 0 of IEEE 37 network) by influencing the OLTC. Generators operate with coefficient cos $\varphi = 1$ or absorb reactive power according to defined relation $Q_G = -0.4\ P_G$. As can be seen in Figure 9a,b, the band of voltages clearly widens, exceeding in many cases the critical value of 1.05 U_n. As the voltage drops below 1.02 U_n, under no-generation conditions, it is impossible to ensure stable voltage conditions on the lower side of the MV/LV transformers of the consumers connected at nodes 25 to 37. The values of the voltage quality index, defined by Equation (1), many times exceed the value of 0.1 defined as acceptable (Figure 9c,d). High generation with reactive power absorption slightly reduces the maximum voltage values, but for small generation cases the voltage value drops below the value equal to 1.

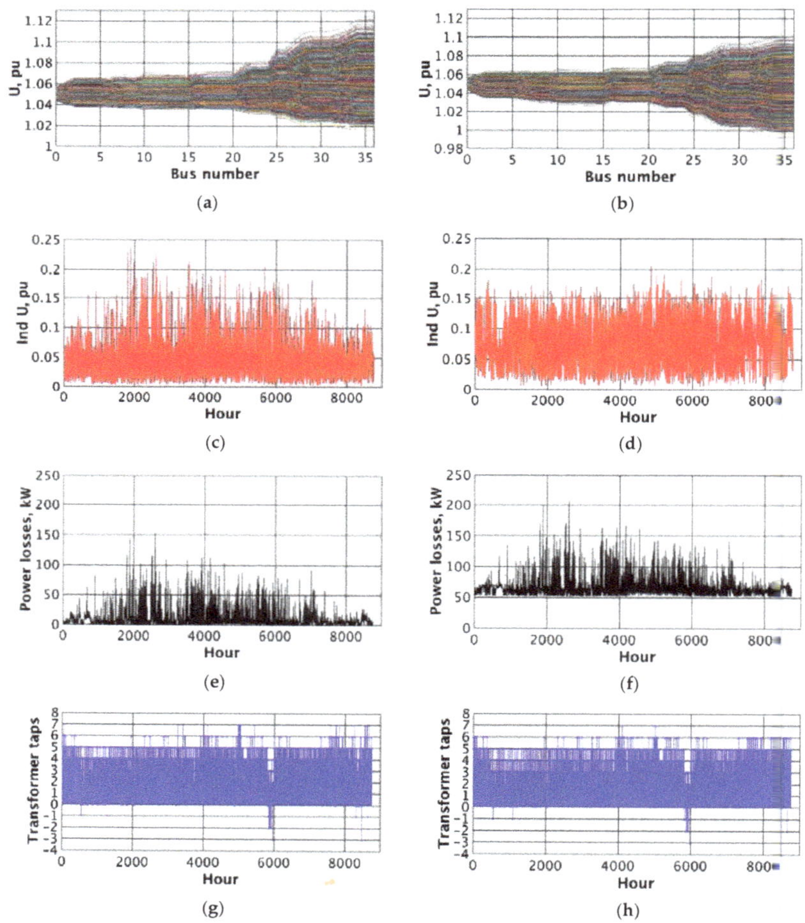

Figure 9. Annual effects of voltage control using the traditional method for two cases of reactive power generation in sources: cos $\varphi = 1$, $Q_G = -0.4\ P_G$ (**a**) voltages in network nodes, cos $\varphi = 1$; (**b**) voltages in network nodes, $Q_G = -0.4\ P_G$; (**c**) voltage quality indicator, cos $\varphi = 1$; (**d**) voltage quality indicator, $Q_G = -0.4\ P_G$; (**e**) power losses, cos $\varphi = 1$; (**f**) power losses, $Q_G = -0.4\ P_G$; (**g**) OLTC position, cos $\varphi = 1$; (**h**) OLTC position, $Q_G = -0.4\ P_G$.

6.2. Voltage Quality Assessment Using the OPFh-MVt Method

Figure 10 shows the results of the analysis carried out with the assumption that the control system operates in accordance with the principles of the OPFh-MVt method. As a result of the optimisation process, repeated in each time window on the basis of data from telemetry and grid state estimation, the HV/MV transformer ratio values and the reactive powers of the sources connected to the grid are determined. As can be seen in Figure 10a, the band of voltages becomes significantly narrower and even at the end of the network it ranges from 1.04 U_n to 1.08 U_n. Moreover, the voltage quality indicator (optimisation task of objective function) decreases in value and in the worst case it practically does not exceed the level of 0.1.

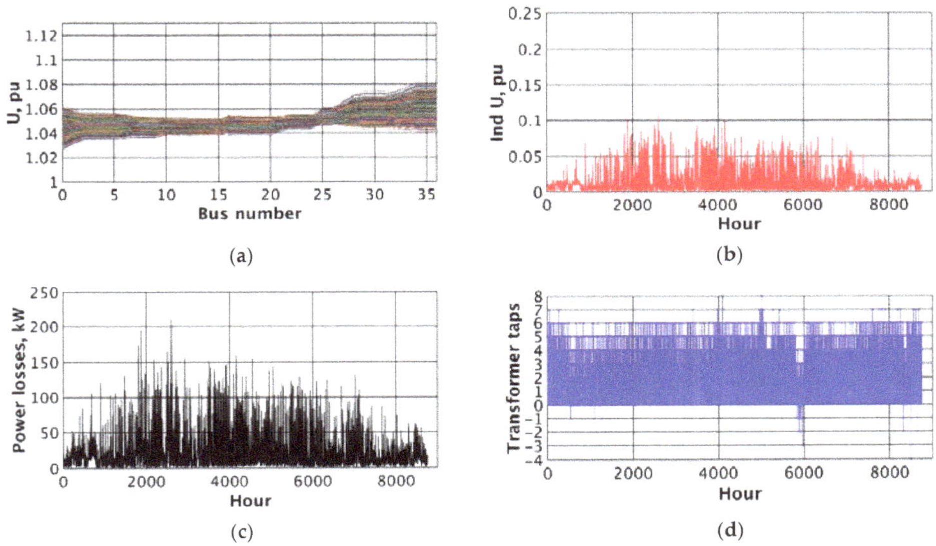

Figure 10. Results of the analysis of the effects of voltage regulation using the OPFh-MVt method: (**a**) voltage values in the network nodes for all hours of the year, (**b**) annual changes in the voltage quality index, (**c**) annual changes in power losses in the network, (**d**) annual changes in the position OLTC.

Generators produce or absorb reactive power so as to minimise the value of the indicator. Changes in the value and direction of reactive power flows take place very rapidly, as they are forced by the course of the optimisation process (Figure 11). The result of high reactive power flows is a significant increase in power losses, which is visible in Figure 10c (compared to Figure 9c). The transformer ratio values changed with OLTC, on the one hand, limit the voltage at the end of the network, but on the other hand, they allow to keep the appropriate voltage value near the station busbar.

Referring to the course of the optimisation process, it should be stated that the AIG algorithm ensures its high convergence and accuracy. Figure 12 shows changes in the best values of the objective function for the selected case.

Figure 12 shows how quickly the AIG algorithm finds the optimal solution. Additionally, for comparison and verification, Figure 12 shows the course of the optimisation process according to the known heuristic algorithms—cuckoo search (CS) and moth-flame optimisation (MFO) compared to the proprietary AIG algorithm. The chart shows that practically 100 iterations are enough to find the optimal solution, so the optimisation process runs efficiently. For AIG, it is even more convergent than for the other tested algorithms.

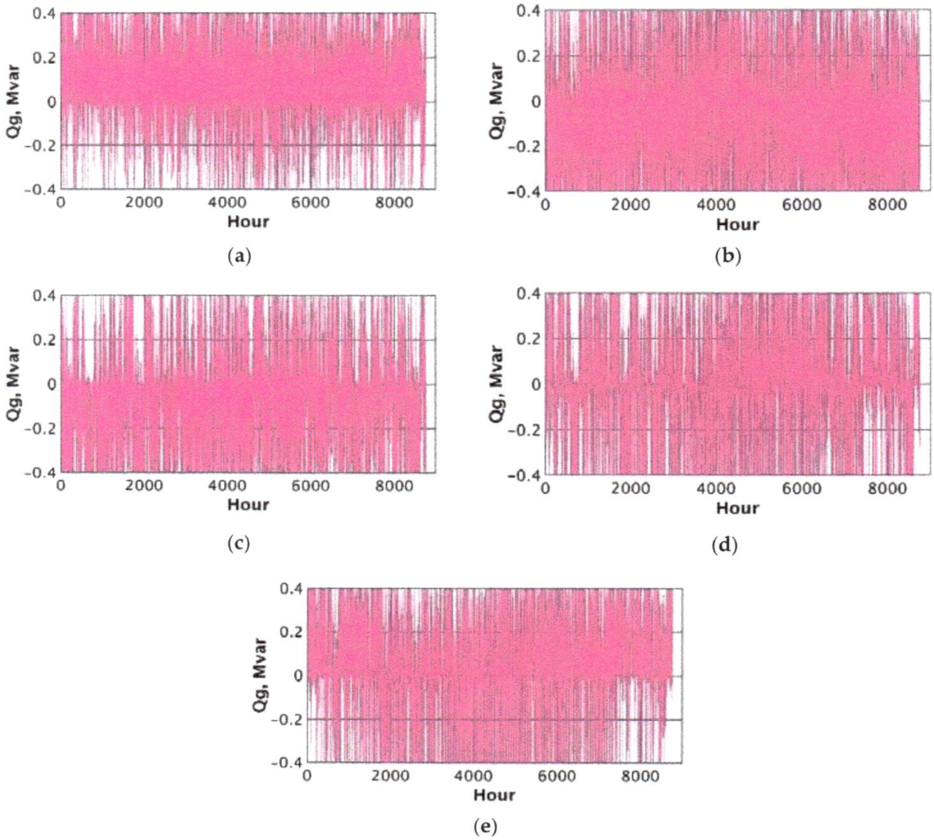

Figure 11. Changes in reactive power generation and consumption of individual RES in the voltage regulation process according to the OPFh-MVt method (**a**) G1, (**b**) G2, (**c**) G3, (**d**) G4, (**e**) G5.

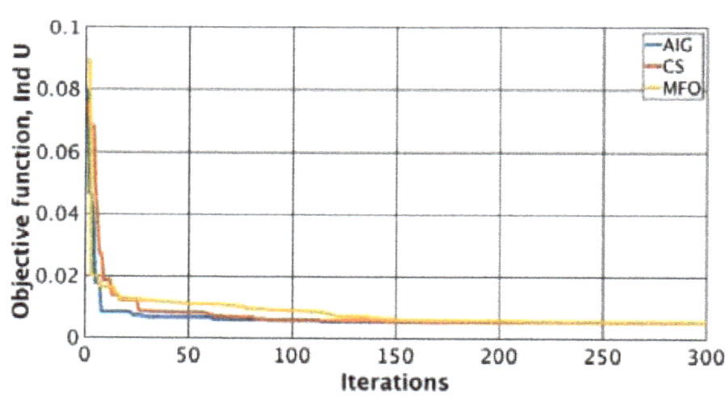

Figure 12. Changes of the best values of the objective function in subsequent iterations for AIG, CS and MFO algorithms.

6.3. Description of the Method Using OLTC Control Related to the Voltage inside the Network with the Simultaneous Use of the Q(U) Characteristics of Individual Sources

Figure 13 presents the results of the analysis carried out with reference to the alternative, simplified method of voltage control, described in point 4. The value of the transformer ratio is determined by the controller to which the voltage is applied from the deep inside of the network. The selection of the node for which the regulator tries to keep the value of 1.05 U_n (internal reference node) is the result of the offline optimisation process, described in the next section. Additionally, for each source, $Q(U)$ characteristics are activated, which ensure local voltage limitation under high-generation conditions. The band of voltages visible in Figure 13a is slightly less coherent than for full optimisation (Figure 10a), but much more favourable than for conditions with traditional control method (Figure 9a,b). The power losses in Figure 13c are clearly smaller than in the case of control OPFh-MVt (Figure 10c). It is a natural consequence of limiting the generation of reactive power in sources only to ensure the appropriate local voltage value, without striving to minimise the global value of the quality indicator Ind U. This is shown in Figure 14—the values of the reactive power absorbed are significantly lower than in the case of the OPFh-MVt control. It can be seen that they do not reach their maximum values and the generation of reactive power does not occur at all.

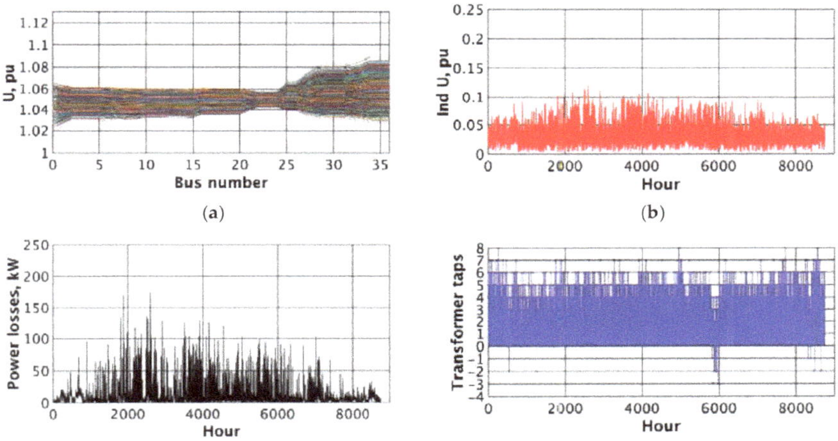

(a) (b) (c) (d)

Figure 13. Graphs for data after optimisation (**a**) voltage values at all MV network nodes, (**b**) voltage indicator, (**c**) power losses, (**d**) transformer tap.

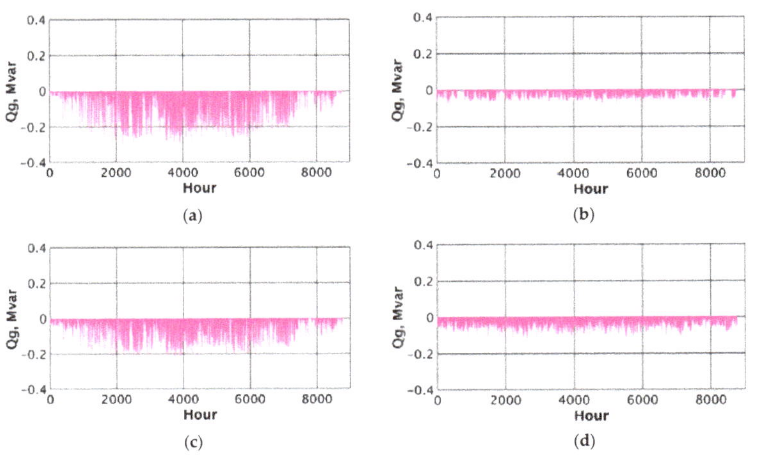

(a) (b) (c) (d)

Figure 14. *Cont.*

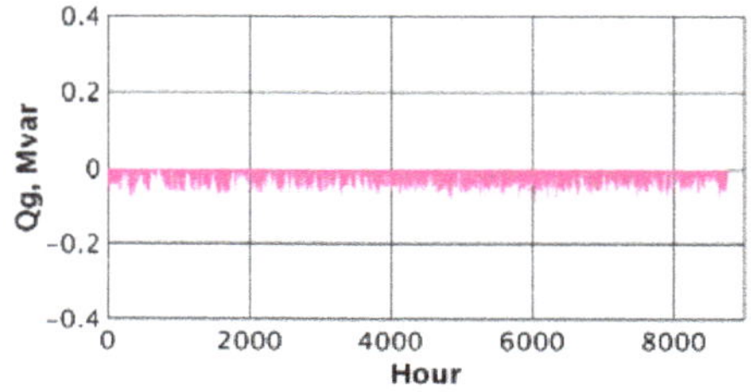

(e)

Figure 14. Changes in reactive power generation and absorption of individual RES in the voltage control process according to the simplified method—OLTC + $Q(U)$ (**a**) G1, (**b**) G2, (**c**) G3, (**d**) G4, (**e**) G5.

6.4. Selection of the Internal Reference Node

The concept of selecting the control reference node not on the HV/MV transformer busbars but inside the network has been known for years. Such a node was called the "centre of gravity of the network load" and it was modelled (without real voltage transmission) by means of a elements (R, X) inside the controller. This solution was called current compensation. With the development of distributed generation, this concept should be modified. As shown in Figure 4b, the reference node for control with OLTC should be appropriately selected, located inside the network and transmission of the voltage value to the controller should be provided. The question is how to select a reference node? The general rule for such a choice can be described as "deep but not too deep". For each of the 8760 h of the year, the effectiveness of the method described in Section 6.3 was simulated, with each of the 37 test network nodes selected as the reference node (in total, calculations were made for 8760 × 37 = 324,120 cases). The following figures show the results of these simulations.

Figure 15 shows the values of the Ind U indicator for the entire IEEE 37 network determined for the simulations described above. A characteristic band of numerical values is visible and despite such a large number of results, it can be clearly seen that the lowest values of the voltage quality index were achieved when node 22 was selected as the reference node. Interestingly, this choice is appropriate for different load conditions, different values of generated power, and different voltage values in the 110 kV network. Placing the reference node too close to the generation sources (deeper into the network, e.g., nodes 28,33,36) results in a significant reduction in the voltage value on the HV/MV transformer busbars and deterioration of the voltage quality for the nodes closer to the transformer and consequently for the entire network. Hence, the rationale for the principle is as defined above (deep but not too deep).

Figure 16 shows the results of three of the 8760 simulations selected for the high generation distributed source state, the medium generation level and for zero generation and high load. The results of the calculations confirm the correctness of choosing node 22 as the reference node. Similar values result from the analysis of the value of the Ind U indicator averaged for the whole year, presented in Figure 17. The choice of node 22 as the reference node minimises the value of this indicator, which confirms the correctness of the selection.

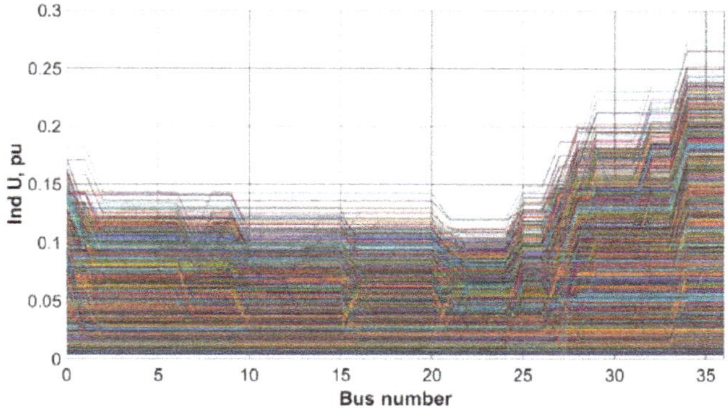

Figure 15. The results of the simulation assessment of the voltage quality indicator Ind U for the IEEE 37 network depending on the selection of the reference node, the voltage of which is maintained at a given level by the OLTC controller.

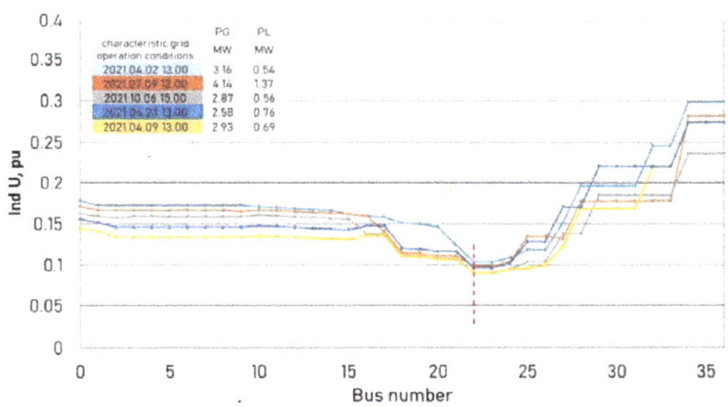

Figure 16. The results of simulations determining the voltage quality index for five characteristic grid operation conditions depending on the selection of the reference node.

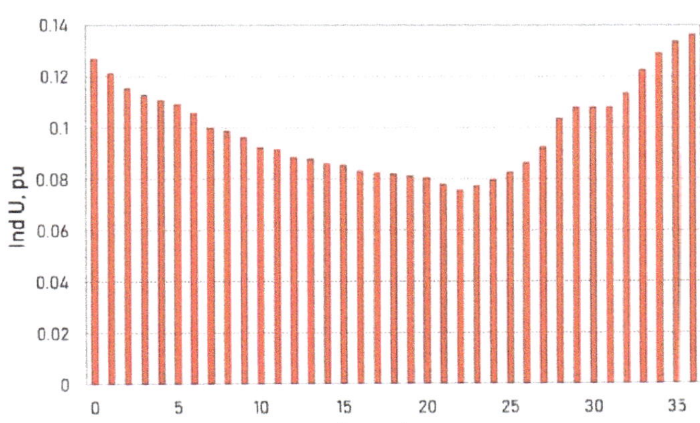

Figure 17. Simulation results determining the voltage quality index averaged for the entire year depending on the selection of the internal reference node.

6.5. Discussion and Comparison of Results for the Analysed Voltage Control Methods

Table 2 summarises and compares the statistical assessment of annual changes in the voltage quality index and relative power losses for the considered cases of voltage control in the considered IEEE 37 network. The rows of the table marked with a superscript [1] refer to network operation without RES generation. The introduction of RES generation with no changes in the voltage control method (transformer with OLTC, zero reactive power—table rows marked with index [2] increases the average value of the voltage quality index from 0.034 to 0.048, while its maximum value increases more than three times (from 0.076 to 0.24). This is a significant deterioration of the voltage quality, with a noticeable increase in relative power losses (on average from 0.740% to 1.208%, maximum from 1.73% to 26.7%, with high generation and very low load).

Table 2. Annual changes in the voltage quality and power loss index.

	$\overline{W}$	σ	Max	Min	Med
Ind U [1]	0.034	0.011	0.076	0.011	0.033
Ind U [2]	0.048	0.035	0.240	0.003	0.038
Ind U [3]	0.011	0.014	0.108	0.001	0.006
Ind U [4]	0.033	0.018	0.121	0.004	0.029
$\Delta P/P_0$ [1] [%]	0.740	0.260	1.730	0.238	0.724
$\Delta P/P_0$ [2] [%]	1.208	2.040	26.70	0.050	0.441
$\Delta P/P_0$ [3] [%]	2.760	3.590	38.30	0.040	1.327
$\Delta P/P_0$ [4] [%]	1.355	2.350	30.57	0.051	0.466

[1] Network with traditional control, the transformer with OLTC keeps the voltage value equal to 1.05 U_n on the MV (bus number 0) buses; no active power generation. [2] Network with traditional control, the transformer with OLTC as describe above, RES variable generation of active power, zero value of reactive power of RES. [3] Voltage control in the MV network using the results of cyclic solving the OPF task. [4] A simplified method of voltage control in the MV network with the use of the tap changer of the HV/MV transformer, keeping the voltage value equal to 1.05 U_n in the depths of the network (bus number 22) and the local influence of reactive power of distributed sources.

The use of voltage control as a solution to the OPF task and the impact on both the transformer ratio (OLTC) and the reactive power of generating sources (RES) significantly improve its quality—Table 2, values with the upper index [3]. The average value of the indicator Ind U decreases four times to the value of 0.011, which is significantly better than in the absence of any generation. Unfortunately, the intensive use of reactive power generation (or absorption) by RES systems leads to a noticeable increase in power losses. Their average relative value increases more than twice (to 2.76%). Thus, to the technical and computational problems related to voltage control based on OPF, there is a doubt related to the clear relationship between the improvement of voltage quality and an increase in power losses.

As stated earlier, a method ensuring relatively easy implementation and a positive impact on voltage quality with a simultaneous limited increase in losses is the use of appropriately selected *Q(U)* characteristics, while striving to keep a constant voltage level in the depths of the network (in Table 2, the rows with the superscript [4]. The voltage quality index is practically the same as for the state with zero generation (0.033), the power losses increase, but to the value of 1.355%, i.e., they are twice lower than in the case of the OPF solution. Thus, the presented results confirm the thesis about the possibility of selecting a relatively easy method of improving voltage conditions in a network with a large number of RES systems.

7. Conclusions

Numerous connections of distributed generation sources to the MV grid cause unfavourable voltage effects, characterised in high-generation conditions by an increase in the voltage values inside the grid, above the permissible level. As the analyses presented in the article showed, the traditional method of regulation with OLTC and keeping a constant voltage value on the MV busbars of the HV/MV transformer does not prevent this phenomenon and it is necessary to look for new solutions.

Undoubtedly, the development of telemetry and software for estimating the state of the MV network allows for the optimisation of its operating conditions, including the optimisation of the voltage control system. The control variables are defined as the result of the optimisation problem—the use of the original AIG heuristic algorithm is shown. Simultaneous control of the HV/MV transformer ratio and influencing the generation or absorption of reactive power by RES units dramatically improves the voltage conditions in the MV network, even with a very high share of distributed generation. Unfortunately, this solution is associated with a significant increase in power losses

Technical difficulties related to the implementation of such an advanced method may be replaced by a compromise by the operation of OLTC on the basis of a measurement signal from the inside of the network and the effect of the activation of the *Q(U)* characteristics of distributed sources. The results of the analyses obtained on the basis of the actual annual HV voltage waveforms and power generated by wind turbines and PV systems indicate that such control can now be treated as a standard for MV grid operation. However, it is justified to continue working on the implementation of more advanced voltage control methods, such as OPFh-MVt described in the article.

Author Contributions: Conceptualisation, P.P., P.K. and M.W.; methodology, P.P., P.K. and M.W.; software, P.P., P.K. and M.W.; validation, P.P., P.K. and M.W.; formal analysis, P.P., P.K. and M.W.; investigation, P.P., P.K. and M.W.; writing—original draft preparation, P.P., P.K. and M.W.; writing—review and editing, P.P., P.K. and M.W.; visualisation, P.P. and M.W.; supervision, P.K. All authors have read and agreed to the published version of the manuscript.

Funding: This research received no external funding.

Institutional Review Board Statement: Not applicable.

Informed Consent Statement: Not applicable.

Data Availability Statement: Not applicable.

Conflicts of Interest: The authors declare no conflict of interest.

References

1. Kacejko, P.; Pijarski, P. Optimal voltage control in MV network with distributed generation. *Energies* **2021**, *14*, 469. [CrossRef]
2. Pijarski, P.; Kacejko, P. A new metaheuristic optimization method: The algorithm of the innovative gunner (AIG). *Eng. Optim.* **2019**, *51*, 2049–2068. [CrossRef]
3. Adamek, S. Optimal Voltage Control in Distributed Power Grid with Dispersed Generation. Ph.D. Thesis, Lublin University of Technology, Lublin, Poland, 2010.
4. Kryonidis, G.C.; Demoulias, C.S.; Papagiannis, G.K. A new voltage control scheme for active medium-voltage (MV) networks. *Electr. Power Syst. Res.* **2019**, *169*, 53–64. [CrossRef]
5. Hashim, T.J.T.; Mohamed, A. Optimal coordinated voltage control in active distribution networks using backtracking search algorithm. *PLoS ONE* **2017**, *12*, e0177507. [CrossRef]
6. Wlodarczyk, P.; Sumper, A.; Cruz, M. Voltage control of distribution grids with multi-microgrids using reactive power management. *Adv. Electr. Comput. Eng.* **2015**, *15*, 83–88. [CrossRef]
7. Farkas, C.; Tóth, A.; Orlay, I. Voltage control methods in the MV grid with a large share of PV. *Int. J. Emerg. Electr. Power Syst.* **2019**, *20*, 1–12. [CrossRef]
8. Szpyra, W.; Kot, A. Optimal voltage control in medium voltage power distribution networks. *Acta Energ.* **2009**, *2*, 89–105.
9. Agalgaonkar, Y.; Pal, B.; Jabr, R. Distribution voltage control considering the impact of PV generation on tap changers and autonomous regulators. *IEEE Trans. Power Syst.* **2014**, *29*, 182–192. [CrossRef]
10. Neagu, B.C.; Grigoras, G. Optimal voltage control in power distribution networks using an adaptive on-load tap changer transformers techniques. In Proceedings of the 2019 IEEE International Conference on Electromechanical and Energy Systems (SIELMEN), Craiova, Romania, 9–11 October 2019; pp. 1–6. [CrossRef]
11. Dib, M.; Ramzi, M.; Nejmi, A. Voltage regulation in the medium voltage distribution grid in the presence of renewable energy sources. *Mater. Today Proc.* **2019**, *13*, 739–745. [CrossRef]
12. Kazmi, S.A.A.; Shahzad, M.K.; Shin, D.R. Multi-objective planning techniques in distribution networks: A composite review. *Energies* **2017**, *10*, 208. [CrossRef]

13. Mahmud, N.; Zahedi, A. Review of control strategies for voltage regulation of the smart distribution network with high penetration of renewable distributed generation. *Renew. Sustain. Energy Rev.* **2016**, *64*, 582–595. [CrossRef]
14. Ziari, I. Planning of Distribution Networks for Medium Voltage and Low Voltage. Ph.D. Thesis, Queensland University of Technology, Brisbane, QLD, Australia, 2011.
15. Jiayi, H.; Chuanwen, J.; Rong, X. A review on distributed energy resources and MicroGrid. *Renew. Sustain. Energy Rev.* **2008**, *12*, 2472–2483. [CrossRef]
16. Kow, K.W.; Wong, Y.W.; Rajkumar, R.K.; Rajkumar, R.K. A review on performance of artificial intelligence and conventional method in mitigating PV grid-tied related power quality events. *Renew. Sustain. Energy Rev.* **2016**, *56*, 334–346. [CrossRef]
17. Kulmala, A. Active Voltage Control in Distribution Networks Including Distributed Energy Resources. Ph.D. Thesis, Tampere University of Technology, Tampere, Finland, 2014.
18. Pathak, A.K.; Sharma, M.P.; Bundele, M. A critical review of voltage and reactive power management of wind farms. *Renew. Sustain. Energy Rev.* **2015**, *51*, 460–471. [CrossRef]
19. Raghav, P.; Sarathy, P. Analysis and Optimization of Medium Voltage Line Voltage Regulators. Master's Thesis, Norwegian University of Science and Technology, Trondheim, Norway, 2018.
20. Zhou, H.; Yan, X.; Liu, G. A review on voltage control using on-load voltage transformer for the power grid. *IOP Conf. Ser. Earth Environ. Sci.* **2019**, *252*, 32144. [CrossRef]
21. Musa, H. An overview on voltage stability indices as indicators of voltage stability for networks with distributed generations penetration. *Int. J. Sci. Technol. Soc.* **2015**, *3*, 244. [CrossRef]
22. Antoniadou-Plytaria, K.E.; Kouveliotis-Lysikatos, I.N.; Georgilakis, P.S.; Hatziargyriou, N.D. Distributed and decentralized voltage control of smart distribution networks: Models, methods, and future research. *IEEE Trans. Smart Grid* **2017**, *8*, 2999–3008. [CrossRef]
23. Provoost, F. *Intelligent Distribution Network Design*; Technische Universiteit Eindhoven: Eindhoven, The Netherlands, 2009; ISBN 978-90-386-1974-3.
24. Małkowski, R.; Izdebski, M.; Miller, P. Adaptive algorithm of a tap-changer controller of the power transformer supplying the radial network reducing the risk of voltage collapse. *Energies* **2020**, *13*, 5403. [CrossRef]
25. Bedawy, A.; Yorino, N.; Mahmoud, K.; Zoka, Y.; Sasaki, Y. Optimal voltage control strategy for voltage regulators in active unbalanced distribution systems using multi-agents. *IEEE Trans. Power Syst.* **2020**, *35*, 1023–1035. [CrossRef]
26. Kersting, W.H. Distribution feeder voltage regulation control. In Proceedings of the 2009 IEEE Rural Electric Power Conference (REPC), Fort Collins, CO, USA, 26–29 April 2009; pp. 1–7.
27. Castro, J.R.; Saad, M.; Lefebvre, S.; Asber, D.; Lenoir, L. Optimal voltage control in distribution network in the presence of DGs. *Int. J. Electr. Power Energy Syst.* **2016**, *78*, 239–247. [CrossRef]
28. Li, P.; Ji, H.; Wang, C.; Zhao, J.; Song, G.; Ding, F.; Wu, J. Coordinated control method of voltage and reactive power for active distribution networks based on soft open point. *IEEE Trans. Sustain. Energy* **2017**, *8*, 1430–1442. [CrossRef]
29. Nick, M.; Cherkaoui, R.; Paolone, M. Optimal allocation of dispersed energy storage systems in active distribution networks for energy balance and grid support. *IEEE Trans. Power Syst.* **2014**, *29*, 2300–2310. [CrossRef]
30. Kacejko, P.; Adamek, S.; Wydra, M. Optimal voltage control in distribution networks with dispersed generation. In Proceedings of the 2010 IEEE PES Innovative Smart Grid Technologies Conference Europe (ISGT Europe), Gothenburg, Sweden, 11–13 October 2010; pp. 1–4.
31. Arefi, A.; Shahnia, F.; Ledwich, G. *Electric Distribution Network Management and Control*; Springer: Singapore, 2018; ISBN 978-981-10-7000-6.
32. Constantin, A.; Lazar, R.; Kjaer, S.B. Voltage regulation (MV) by means of Q(U)—Simulation and measurement on a 12 MVA PV power plant. In Proceedings of the CIRED 2012 Workshop: Integration of Renewables into the Distribution Grid, Lisbon, Portugal, 29–30 May 2012; p. 272. [CrossRef]
33. Dutta, A.; Ganguly, S.; Kumar, C. Model predictive control-based optimal voltage regulation of active distribution networks with OLTC and reactive power capability of PV inverters. *IET Gener. Transm. Distrib.* **2020**, *14*, 5183–5192. [CrossRef]
34. Fandi, G.; Igbinovia, F.O.; Tlusty, J.; Mahmoud, R. Voltage regulation and power losses reduction in a wind farm integrated MV distribution network. *J. Electr. Eng.* **2018**, *69*, 85–92. [CrossRef]
35. Ouali, S.; Cherkaoui, A. Elimination of the impact produced by DG units on the voltage profile of distribution networks. *J. Appl. Math.* **2020**, *2020*, 1395943. [CrossRef]
36. Prochazka, K.; Vaculik, P.; Mezera, D.; Kysnar, F.; Novotny, J. Voltage quality and reactive power flow solution in distribution networks with a high share of renewable energy sources. In Proceedings of the 22nd International Conference and Exhibition on Electricity Distribution (CIRED 2013), Stockholm, Sweden, 10–13 June 2013; p. 1142. [CrossRef]
37. Berkel, F.; Bleich, J.; Bell, M.; Liu, S. A distributed voltage controller for medium voltage grids with storage-containing loads. In Proceedings of the IECON 2018 44th Annual Conference of the IEEE Industrial Electronics Society, Washington, DC, USA, 21–23 October 2018; pp. 3523–3528.
38. Guo, Y.; Wu, Q.; Gao, H.; Chen, X.; Ostergaard, J.; Xin, H. MPC-based coordinated voltage regulation for distribution networks with distributed generation and energy storage system. *IEEE Trans. Sustain. Energy* **2019**, *10*, 1731–1739. [CrossRef]
39. Li, Q.; Zhou, F.; Guo, F.; Fan, F.; Huang, Z. Optimized energy storage system configuration for voltage regulation of distribution network with PV access. *Front. Energy Res.* **2021**, *9*, 641518. [CrossRef]

40. Liu, X.; Aichhorn, A.; Liu, L.; Li, H. Coordinated control of distributed energy storage system with tap changer transformers for voltage rise mitigation under high photovoltaic penetration. *IEEE Trans. Smart Grid* **2012**, *3*, 897–906. [CrossRef]
41. Reihani, E.; Sepasi, S.; Roose, L.R.; Matsuura, M. Energy management at the distribution grid using a Battery Energy Storage System (BESS). *Int. J. Electr. Power Energy Syst.* **2016**, *77*, 337–344. [CrossRef]
42. Teng, J.-H.; Luan, S.-W.; Lee, D.-J.; Huang, Y.-Q. Optimal charging/discharging scheduling of battery storage systems for distribution systems interconnected with sizeable PV generation systems. *IEEE Trans. Power Syst.* **2013**, *28*, 1425–1433. [CrossRef]
43. Kabir, M.N.; Mishra, Y.; Ledwich, G.; Xu, Z.; Bansal, R.C. Improving voltage profile of residential distribution systems using rooftop PVs and Battery Energy Storage systems. *Appl. Energy* **2014**, *134*, 290–300. [CrossRef]
44. Akinyele, D.O.; Rayudu, R.K. Review of energy storage technologies for sustainable power networks. *Sustain. Energy Technol. Assess.* **2014**, *8*, 74–91. [CrossRef]
45. Beaudin, M.; Zareipour, H.; Schellenberglabe, A.; Rosehart, W. Energy storage for mitigating the variability of renewable electricity sources: An updated review. *Energy Sustain. Dev.* **2010**, *14*, 302–314. [CrossRef]
46. Sankur, M.; Dobbe, R.; Von Meier, A.; Arnold, D.B. Model-free optimal voltage phasor regulation in unbalanced distribution systems. *IEEE Trans. Smart Grid* **2020**, *11*, 884–894. [CrossRef]
47. Xu, H.; Dominguez-Garcia, A.D.; Veeravalli, V.V.; Sauer, P.W. Data-driven voltage regulation in radial power distribution systems. *IEEE Trans. Power Syst.* **2020**, *35*, 2133–2143. [CrossRef]
48. Arnold, D.B.; Sankur, M.; Negrete-Pincetic, M.; Callaway, D.S. Model-free optimal coordination of distributed energy resources for provisioning transmission-level services. *IEEE Trans. Power Syst.* **2018**, *33*, 817–828. [CrossRef]
49. Nazaripouya, H.; Pota, H.R.; Chu, C.-C.; Gadh, R. Real-time model-free coordination of active and reactive powers of distributed energy resources to improve voltage regulation in distribution systems. *IEEE Trans. Sustain. Energy* **2020**, *11*, 1483–1494. [CrossRef]
50. Zhang, J.; Chen, Z.; He, C.; Jiang, Z.; Guan, L. Data-driven-based optimization for power system var-voltage sequential control. *IEEE Trans. Ind. Inform.* **2019**, *15*, 2136–2145. [CrossRef]
51. Li, J.; Xu, Z.; Zhao, J.; Zhang, C. Distributed online voltage control in active distribution networks considering PV curtailment. *IEEE Trans. Ind. Inform.* **2019**, *15*, 5519–5530. [CrossRef]
52. Zhang, Y.; Wang, X.; Wang, J.; Zhang, Y. Deep reinforcement learning based volt-VAR optimization in smart distribution systems. *IEEE Trans. Smart Grid* **2021**, *12*, 361–371. [CrossRef]
53. Xu, H.; Dominguez-Garcia, A.D.; Sauer, P.W. Optimal tap setting of voltage regulation transformers using batch reinforcement learning. *IEEE Trans. Power Syst.* **2020**, *35*, 1990–2001. [CrossRef]
54. Wang, W.; Yu, N.; Shi, J.; Gao, Y. Volt-VAR control in power distribution systems with deep reinforcement learning. In Proceedings of the 2019 IEEE International Conference on Communications, Control, and Computing Technologies for Smart Grids (SmartGridComm), Beijing, China, 21–23 October 2019; pp. 1–7. [CrossRef]
55. Duan, J.; Shi, D.; Diao, R.; Li, H.; Wang, Z.; Zhang, B.; Bian, D.; Yi, Z. Deep-reinforcement-learning-based autonomous voltage control for power grid operations. *IEEE Trans. Power Syst.* **2020**, *35*, 814–817. [CrossRef]
56. Brauns, J.; Turek, T. Alkaline water electrolysis powered by renewable energy: A review. *Processes* **2020**, *8*, 248. [CrossRef]
57. Gutiérrez-Martín, F.; Ochoa-Mendoza, A.; Rodriguez-Anton, L.M. Pre-investigation of water electrolysis for flexible energy storage at large scales: The case of the Spanish power system. *Int. J. Hydrogen Energy* **2015**, *40*, 5544–5551. [CrossRef]
58. Hashimoto, K.; Yamasaki, M.; Fujimura, K.; Matsui, T.; Izumiya, K.; Komori, M.; El-Moneim, A.; Akiyama, E.; Habazaki, H.; Kumagai, N.; et al. Global CO2 recycling—Novel materials and prospect for prevention of global warming and abundant energy supply. *Mater. Sci. Eng. A* **1999**, *267*, 200–206. [CrossRef]
59. Hashimoto, K.; Kumagai, N.; Izumiya, K.; Takano, H.; Kato, Z. The production of renewable energy in the form of methane using electrolytic hydrogen generation. *Energy Sustain. Soc.* **2014**, *4*, 17. [CrossRef]
60. El-Taweel, N.A.; Khani, H.; Farag, H.E. Voltage regulation in active power distribution systems integrated with natural gas grids using distributed electric and gas energy resources. *Int. J. Electr. Power Energy Syst.* **2019**, *106*, 561–571. [CrossRef]
61. Estermann, T.; Newborough, M.; Sterner, M. Power-to-gas systems for absorbing excess solar power in electricity distribution networks. *Int. J. Hydrogen Energy* **2016**, *41*, 13950–13959. [CrossRef]
62. Götz, M.; Lefebvre, J.; Mörs, F.; McDaniel Koch, A.; Graf, F.; Bajohr, S.; Reimert, R.; Kolb, T. Renewable power-to-gas: A technological and economic review. *Renew. Energy* **2016**, *85*, 1371–1390. [CrossRef]
63. Mazza, A.; Salomone, F.; Arrigo, F.; Bensaid, S.; Bompard, E.; Chicco, G. Impact of power-to-gas on distribution systems with large renewable energy penetration. *Energy Convers. Manag. X* **2020**, *7*, 100053. [CrossRef]
64. Simonis, B.; Newborough, M. Sizing and operating power-to-gas systems to absorb excess renewable electricity. *Int. J. Hydrogen Energy* **2017**, *42*, 21635–21647. [CrossRef]
65. Pijarski, P.D. *Optymalizacja Heurystyczna w Ocenie Warunków Pracy i Planowaniu Rozwoju Systemu Elektroenergetycznego*; Lublin University of Technology Publishers: Lublin, Poland, 2019; ISBN 978-83-7947-349-6.
66. Kennedy, J.; Eberhart, R. Particle swarm optimization. In Proceedings of the ICNN'95 IEEE International Conference on Neural Networks, Perth, WA, Australia, 27 November–1 December 1995; pp. 1942–1948.
67. Kumar, K.; Zindani, D.; Davim, J.P. *Optimizing Engineering Problems through Heuristic Techniques*; CRC Press: Boca Raton, FL, USA, 2020; ISBN 9781138485365.
68. Mirjalili, S. Moth-flame optimization algorithm: A novel nature-inspired heuristic paradigm. *Knowl. Based Syst.* **2015**, *89*, 228–249. [CrossRef]

69. Mirjalili, S.; Mirjalili, S.M.; Lewis, A. Grey wolf optimizer. *Adv. Eng. Softw.* **2014**, *69*, 46–61. [CrossRef]
70. Hajiabbas, M.P.; Mohammadi-Ivatloo, B. *Optimization of Power System Problems: Methods, Algorithms and MATLAB Codes*; Springer: Cham, Switzerland, 2020; ISBN 978-3-030-34049-0.
71. Pham, D.T.; Ghanbarzadeh, A.; Koç, E.; Otri, S.; Rahim, S.; Zaidi, M. *The Bees Algorithm—A Novel Tool for Complex Optimisation Problems. Intelligent Production Machines and Systems*; Elsevier: Amsterdam, The Netherlands, 2006; pp. 454–459, ISBN 9780080451572.
72. Radosavljević, J. *Metaheuristic Optimization in Power Engineering*; The Institution of Engineering and Technology: London, UK, 2018; ISBN 9781785615467.
73. Yang, X.-S.; Deb, S. Cuckoo search via Lévy flights. In Proceedings of the 2009 IEEE World Congress on Nature & Biologically Inspired Computing (NaBIC), Coimbatore, India, 9–11 December 2009; pp. 210–214.
74. Sarma, S.K. Optimally configured deep convolutional neural network for attack detection in internet of things: Impact of algorithm of the innovative gunner. *Wirel. Pers. Commun.* **2021**, *118*, 239–260. [CrossRef]
75. Dehghani, R.; Poudeh, H.T.; Izadi, Z. The effect of climate change on groundwater level and its prediction using modern meta-heuristic model. *Groundw. Sustain. Dev.* **2022**, *16*, 100702. [CrossRef]
76. Konjaang, J.K.; Xu, L. Meta-heuristic approaches for effective scheduling in infrastructure as a service cloud: A systematic review. *J. Netw. Syst. Manag.* **2021**, *29*, 1–57. [CrossRef]
77. Shafiei, M.; Ghasemi-Marzbali, A. Fast-charging station for electric vehicles, challenges and issues: A comprehensive review. *J. Energy Storage* **2022**, *49*, 104136. [CrossRef]
78. Pijarski, P.; Kacejko, P. Voltage optimization in MV network with distributed generation using power consumption control in electrolysis installations. *Energies* **2021**, *14*, 993. [CrossRef]
79. *CSN EN 50549-1*; Requirements for Generating Plants to Be Connected in Parallel with Distribution Networks—Part 1: Connection to a LV Distribution Network—Generating Plants Up to and including Type B. European Standards S.R.O.: Pilsen, Czech Republic, 2019. Available online: https://www.en-standard.eu/csn-en-50549-1-requirements-for-generating-plants-to-be-connected-in-parallel-with-distribu-tion-networks-part-1-connection-to-a-lv-distribution-network-generating-plants-up-to-and-including-type-b/(accessed on 4 March 2022).
80. *CSN EN 50549-2*; Requirements for Generating Plants to Be Connected in Parallel with Distribution Networks—Part 2: Connection to a MV Distribution Network—Generating Plants Up to and including Type B. European Standards S.R.O.: Pilsen, Czech Republic, 2019. Available online: https://www.en-standard.eu/csn-en-50549-2-requirements-for-generating-plants-to-be-connected-in-parallel-with-distribu-tion-networks-part-2-connection-to-a-mv-distribution-network-generating-plants-up-to-and-including-type-b/(accessed on 4 March 2022).
81. *AS/NZS 4777.2:2015*; Grid Connection of Energy Systems via Inverters, Part 2: Inverter Requirements; EL-042 (Renewable Energy Power Supply Systems & Equipment). Standards Australia: Sydney, Australia, 2015. Available online: https://www.standards.org.au/standards-catalogue/sa-snz/electrotechnology/el-042/as-slash-nzs--4777-dot-2-colon-2015(accessed on 4 March 2022).
82. Pijarski, P.; Kacejko, P. Optimization of the selection of P(U) and Q(U) characteristic parameters of solar microinverters. *Acta Energ.* **2019**, *39*, 6–13. [CrossRef]
83. Juamperez, M.; Yang, G.; Kjær, S.B. Voltage regulation in LV grids by coordinated volt-var control strategies. *J. Mod. Power Syst. Clean Energy* **2014**, *2*, 319–328. [CrossRef]
84. Witzmann, R.; Esslinger, P. Experimental Study on Voltage Dependent Reactive Power Control Q(V) by Solar Inverters in Low-Voltage Networks. In Proceedings of the 22nd International Conference and Exhibition on Electricity Distribution (CIRED 2013), Stockholm, Sweden, 10–13 June 2013; p. 644.
85. Biel, D.; Scherpen, J.M. Active and reactive power regulation in single-phase PV inverters. In Proceedings of the 2018 IEEE 17th European Control Conference (ECC), Limassol, Cyprus, 12–15 June 2018; pp. 782–787. [CrossRef]
86. Valsera-Naranjo, E.; Sumper, A.; Villafáfila-Robles, R.; Martinez-Vicente, D. Probabilistic method to assess the impact of charging of electric vehicles on distribution grids. *Energies* **2012**, *5*, 1503–1531. [CrossRef]

energies

MDPI

Article

Advanced Forecasting Methods of 5-Minute Power Generation in a PV System for Microgrid Operation Control

Paweł Piotrowski [1,*], Mirosław Parol [1], Piotr Kapler [1] and Bartosz Fetliński [2]

1 Institute of Electrical Power Engineering, Faculty of Electrical Engineering, Warsaw University of Technology, Koszykowa 75 Street, 00-662 Warsaw, Poland; miroslaw.parol@pw.edu.pl (M.P.); piotr.kapler@pw.edu.pl (P.K.)
2 Institute of Microelectronics and Optoelectronics, Faculty of Electrical Engineering, Warsaw University of Technology, Koszykowa 75 Street, 00-662 Warsaw, Poland; bartosz.fetlinski@pw.edu.pl
* Correspondence: pawel.piotrowski@pw.edu.pl; Tel.: +48-22-234-7314

Abstract: This paper concerns very-short-term (5-Minute) forecasting of photovoltaic power generation. Developing the methods useful for this type of forecast is the main aim of this study. We prepared a comprehensive study based on fragmentary time series, including 4 full days, of 5 min power generation. This problem is particularly important to microgrids' operation control, i.e., for the proper operation of small energy micro-systems. The forecasting of power generation by renewable energy sources on a very-short-term horizon, including PV systems, is very important, especially in the island mode of microgrids' operation. Inaccurate forecasts can lead to the improper operation of microgrids or increasing costs/decreasing profits for microgrid operators. This paper presents a short description of the performance of photovoltaic systems, particularly the main environmental parameters, and a very detailed statistical analysis of data collected from four sample time series of power generation in an existing PV system, which was located on the roof of a building. Different forecasting methods, which can be employed for this type of forecast, and the choice of proper input data in these methods were the subject of special attention in this paper. Ten various prognostic methods (including hybrid and team methods) were tested. A new, proprietary forecasting method—a hybrid method using three independent MLP-type neural networks—was a unique technique devised by the authors of this paper. The forecasts achieved with the use of various methods are presented and discussed in detail. Additionally, a qualitative analysis of the forecasts, achieved using different measures of quality, was performed. Some of the presented prognostic models are, in our opinion, promising tools for practical use, e.g., for operation control in low-voltage microgrids. The most favorable forecasting methods for various sets of input variables were indicated, and practical conclusions regarding the problem under study were formulated. Thanks to the analysis of the utility of different forecasting methods for four analyzed, separate time series, the reliability of conclusions related to the recommended methods was significantly increased.

Keywords: microgrids; operation control; power generation; PV system; very-short-term forecasting; machine learning; interval type-2 fuzzy logic system

Citation: Piotrowski, P.; Parol, M.; Kapler, P.; Fetliński, B. Advanced Forecasting Methods of 5-Minute Power Generation in a PV System for Microgrid Operation Control. *Energies* **2022**, *15*, 2645. https://doi.org/10.3390/en15072645

Academic Editor: Gabriele Grandi

Received: 12 March 2022
Accepted: 1 April 2022
Published: 4 April 2022

1. Introduction

Microgrids are autonomous energy micro-systems that can operate in both the synchronous (parallel) mode with distribution system operators' grids and the island mode. Control of the microgrid operation in both modes, in particular in the island mode, is a very important issue. Forecasts of power generated from renewable energy sources and forecasts of power demand, in a very-short-term horizon, affect the proper microgrid operation, especially in the island mode. Because of this, forecasts are more and more important. Very-short-term power-generation forecasts, if imprecise, can cause increased costs/decreased profits for microgrids operators or improper operation of energy micro-systems.

It is expected that microgrids will undergo management of electrical power and energy in a very-short-term horizon. All active components of the microgrid, e.g., controllable microsources, energy storage units, and controllable loads, take part in the management process. For the electrical power and energy-management process to proceed correctly, a lot of detailed data is needed. These data include, among others: data on current and forecast loads, data on current and forecast values of power and energy generated by nondispatchable sources (among them, renewable energy sources), and data on the current and forecasted prices of the electrical energy market. These data enable the correct control process of the above-mentioned active components of the microgrid. Obtaining accurate forecasts of power generated in PV systems in a very-short-term horizon is therefore very important from the point of view of power and energy management in the microgrid.

1.1. Related Works

The first part of the literature review refers to the very-short-term forecasting. Within this field, we distinguish between load-demand forecasts and power-generation (wind power and photovoltaic power) forecasts.

The problem of forecasting power demand in a very-short-term horizon is presented in several publications, e.g., in [1–4]. The authors of [4] describe the 10 s forecasting of power demand in the case of highly variable loads. In turn, paper [5] includes a very comprehensive overview of load forecasting methods in short-term and very-short-term horizons. Topics such as the different areas and locations to which this type of forecast can be applied (smart buildings, microgrids, small cities), along with forecasted time horizons, are described in this overview.

Various methods (models) can be applied to prepare forecasts of wind power generation in very short time horizons. In [6], a model of wavelet decomposition and weighted random forests for very-short-term wind power forecasts is presented. The authors of [7] describe hybrid empirical mode decomposition and team empirical mode decomposition models for the needs of wind power forecasts. The authors of [8] present various approaches: neuro fuzzy systems, a support vector regression, and a regression tree in the case of forecasting 1 h wind power. The authors of [9] address, in turn, different approaches for forecasts of wind power in minute horizons. The fuzzy model of Takagi–Sugeno applied to very-short-term forecasts of wind power is presented in [10]. In [11], models based on a discrete-time Markov chain for very-short-term wind power forecasting are described.

Another aspect to be considered is photovoltaic power forecasting in very-short-term or short-term horizons. Two methods, including smart persistence and random forests for the needs of forecasts of PV energy production, are presented in [12]. The authors of [13] address a team model for short-term PV power forecasts. In [14], a complex model for solar power forecasts is described. The model combines wavelet transform, ANFIS, and hybrid firefly and PSO algorithms. The authors of [15] discuss a physical hybrid ANN for 24 h-ahead PV power forecast in microgrids. A very comprehensive review and evaluation of different methods (models) for PV power forecasting are included in [16]. A review of various methods concerning power-generation forecasting in PV systems is also presented in [17]. Paper [18] includes an extensive comparison of different physical models, which can be used for the needs of forecasts concerning PV power generation. In turn, the impact of the availability of design data on the exactness of power-generation forecasts in PV systems, based on physical models, is described in [19].

The second part of the literature review specifically refers to microgrids.

The topic of microgrids was discussed intensively in the literature. In [20,21], a formal definition of microgrids is presented. The idea of microgrids was described in many other publications, e.g., [22,23]. A lot of books and papers address the topic of microgrids' operation control [22–29]. In [24,28,29], a very comprehensive overview of works relating to optimum control (centralized control and decentralized (distributed) control) in microgrids is presented. The authors of [25,28] describe the centralized control logic. In turn, the distributed control logic in microgrids is discussed in [26,28,29]. The authors of [25] present the model of predictive control in microgrids. The operational control in the microgrid island mode is addressed in [25,27]. In [30], a fault detection, localization, and categorization method in the case of a PV-fed DC microgrid is described.

In the analyzed works, different issues concerning photovoltaic power forecasting in a very-short-time horizon and microgrids were considered. The main aim of this paper is to provide a very comprehensive review of the various possible methods of very-short-term photovoltaic power-generation forecasting for the needs of low-voltage microgrids operation, as well as select the best methods among those considered.

1.2. Objective and Contribution

The following are the main objectives of this paper:

- Carry out an analysis of the statistical properties of a time series of the measured values of 5 min power generation in a PV system;
- Verify the usefulness of the available input variables—perform a validity analysis (the time series of solar irradiance, air temperature, PV module temperature, wind direction, and wind speed) using four different methods and select eight sets of input variables to make forecasts using various methods;
- Check the efficiency of 5 min horizon power-generation forecasts by means of ten forecasting methods, including machine learning, hybrid, and ensemble methods (several hundred various models with different set values of parameters/hyperparameters have been verified for this purpose);
- Point out forecasting methods that are the most effective for this 5 min power-generation time series depending on the number of input variables used.
- The selected contributions of this paper are as follows:
- The research concerns unique data—a time series of 5 min power-generation values in a small, consumer PV system. In the case of such small PV systems and such a short forecast horizon (5 min), meteorological forecasts are usually not used in the forecasting process due to difficulties in obtaining them, which makes it problematic to obtain forecasts with very high accuracy;
- We provide a detailed description of the performance of photovoltaic systems regarding the main environmental parameters;
- We performed extensive statistical analyses of the available time series (including an analysis of the importance of the input variables);
- We used tests of ten different prognostic methods (including hybrid and team methods);
- We developed a new, proprietary forecasting method—a hybrid method using three independent, MLP-type neural networks;
- We indicate the most favorable prognostic methods for various sets of input variables (from 3 input variables to 15 input variables) and formulate practical conclusions regarding the problem under study, e.g., from the point of view of microgrids' operation.
- We provide a broad comparative analysis of forecasting methods of a very-short-term horizon for power generation in PV systems that can be connected to low-voltage microgrids.

After completing our studies, we can state that there are efficient, very-short-term forecasting methods for PV power generation, which are suitable for practical use in microgrids' operation.

The organization of this paper is as follows: Section 2 describes the influence of the main environmental parameters on the performance of photovoltaic systems. Section 3.1 includes an analysis of the statistical properties of the time series of PV power generation data investigated in this paper. The analysis leading to the choice of proper input data (explanatory variables) for various prognostic methods is shown in Section 3.2. Section 4 addresses the forecasting methods applied in this paper. In turn, Section 5 discusses criteria employed to evaluate the quality of the forecasting models considered. A broad comparative analysis of forecasting methods of a very short time horizon for power generation in PV systems is presented in Section 6. Section 7 includes the main conclusions resulting from our studies. A list of references ends the paper.

2. Performance of Photovoltaic Systems

The two main environmental parameters affecting the performance of photovoltaic (PV) systems are solar irradiance and cells' temperature [31,32]. Changes in solar irradiance result in a generally proportional shift of the I–V (current–voltage) curve along the current axis, along with a relatively much smaller voltage change. Under low-irradiance conditions, such as those during overcast weather, the maximum power of the PV module tends to be further decreased due to the higher significance of the parallel resistance, which results in a slight decrease in current with an increasing voltage. This effect is highly dependent on PV cells' technology. The current changes resulting from the changing irradiance are instantaneous from the point of view of PV system energy yields. The PV system power output is primarily dependent on the available irradiance.

The PV cell temperature is the second most important factor influencing the energy output of a PV system, as demonstrated by analyses utilizing the performance ratio (PR) parameter to model PV systems' operations [33,34]. The increase in the PV cell temperature results in a decrease in the PV device's open-circuit voltage, along with a minor increase in the short-circuit current. The PV output power temperature coefficients of silicon-based solar cells are of the order of $-0.45\%/K$ [35]. Due to heat capacity of PV modules being heavily dependent not only on the materials and structure of the module itself but also on its mounting structure, tilt angle, and surrounding ground, the rate of response of the module's temperature to the environmental conditions (the irradiance, velocity, and direction of wind and the ambient temperature) varies significantly and must be assumed to be an individual property of the particular system under analysis. The literature provides numerous similar approximations of the influence of temperature on PV systems' efficiency and output power, often using empirically established coefficients [36]. Direct measurement of the temperature of laminated solar cells is difficult, and temperature sensors are usually attached to the rear backsheet of the module. The significant temperature gradient between different points of a single module exposed to sunlight (due to proximity of the frame or mounting structure attachment) makes reliable module temperature challenging, with guidelines suggesting the use of up to four temperature sensors on a single module to model the temperature correctly [37,38]—an effort rarely undertaken, even in research-oriented test systems, and even more so in commercial systems.

Spectral effects, related to the mismatch between the spectral response of a PV module (which primarily depends on PV cell technology) and the spectrum of the incident irradiance (which consists of the direct and diffused component of the solar spectrum and light reflected from the surrounding objects—particularly important for bifacial and multijunction modules), primarily contribute towards varying irradiance effects. However, the spectral mismatch also impacts thermalization and sub-bandgap losses, which result in PV module heating. These factors are difficult to quantify in the analysis of PV systems' performance, as their inclusion would require long-term monitoring of the solar spectra in the location of the system under analysis. Their impact is also highly specific to PV cell technology [39]. The size and layout of a PV system may also impact both the degree and pace of the change of its power output due to external factors, which is particularly important in the case of large-area systems [40].

3. Data

3.1. Statistical Analysis of the Time Series of Power-Generation Data

The installed power of the analyzed PV system is 3.2 kW. The power output of the analyzed system was monitored using the built-in capability of the system's inverter, type SMA Sunnybox SB3000. The built-in measurement system records parameters, such as AC and DC side power, voltages, and currents. The data points are recorded in 5 min intervals. These electrical data are then merged with the data gathered by the meteorological station. Statistical analysis is based on fragmentary time series, including 4 full days. Each day is from a different season. The daily time series includes 288 periods of 5 min. The total number of 5 min periods of power generation in watts is 1152. Before statistical analysis was performed, the data "cleaning" process was performed. Wrong data were identified and replaced with data most relevant to their location (e.g., in the case of non-zero power-generation values in the period between sunset and sunrise or zero-generation values in the period when solar irradiance was non-zero).

Table 1 shows selected statistical measures of the time series of power generation in the PV system. As much as 50% of power generation from the time series was due to small values, below 46.914 (W) (which is more than 68 times less than the installed power of the PV system).

Table 1. Descriptive statistics of time series of power generation.

Statistical Measures	PV System Data
Mean	635.61 (W)
Percentage ratio of mean power to installed power	19.86%
Standard deviation	930.63 (W)
Minimum	0.00 (W)
Maximum	3114.81 (W)
Range	3114.81 (W)
Coefficient of variation	146.41%
The 10th percentile	0.00 (W)
The 25th percentile (lower quartile)	0.00 (W)
The 50th percentile (median)	46.91 (W)
The 75th (upper quartile)	1127.19 (W)
The 90th percentile	2368.10 (W)
Variance	866,074.80 (-)
Skewness	1.23 (-)
Kurtosis	−0.04 (-)

Figure 1 shows the daily time series of power generation for every season of the year (actual measurement data). The whole spring day was cloudless (generation was close to the rated power, very smoothed time series). The opposite of a spring day was a winter day with a much shorter power-generation period and a significantly smaller generation compared to a spring day and a summer day. Dynamic changes in the quantity of generation during the summer and autumn days are evidence of the high variability of cloud cover on these days.

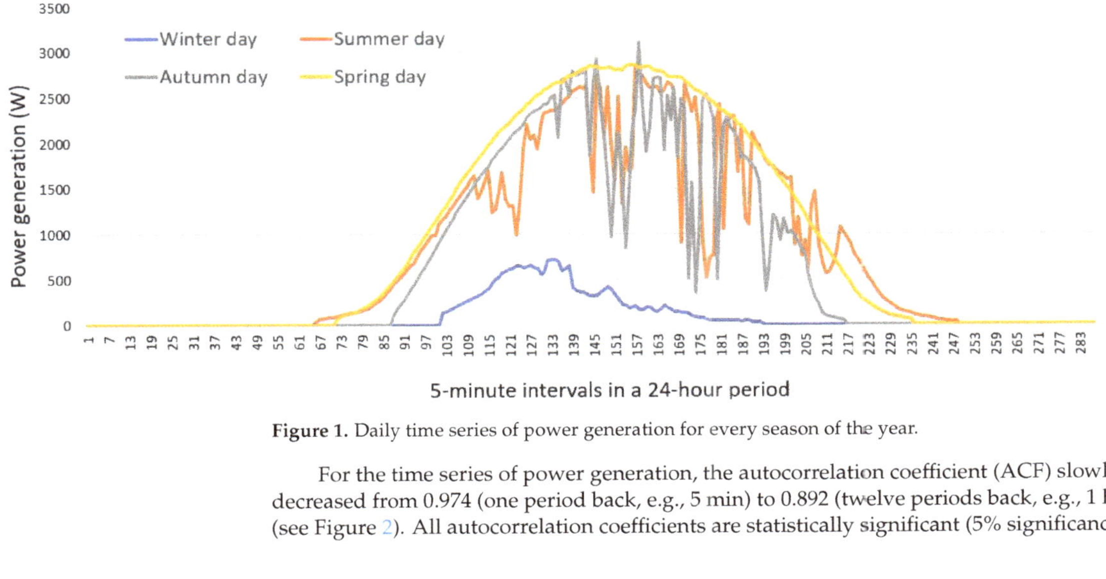

Figure 1. Daily time series of power generation for every season of the year.

For the time series of power generation, the autocorrelation coefficient (ACF) slowly decreased from 0.974 (one period back, e.g., 5 min) to 0.892 (twelve periods back, e.g., 1 h) (see Figure 2). All autocorrelation coefficients are statistically significant (5% significance level). The use of several past values of the forecasted time series of power generation as input data for forecasting models seems justified.

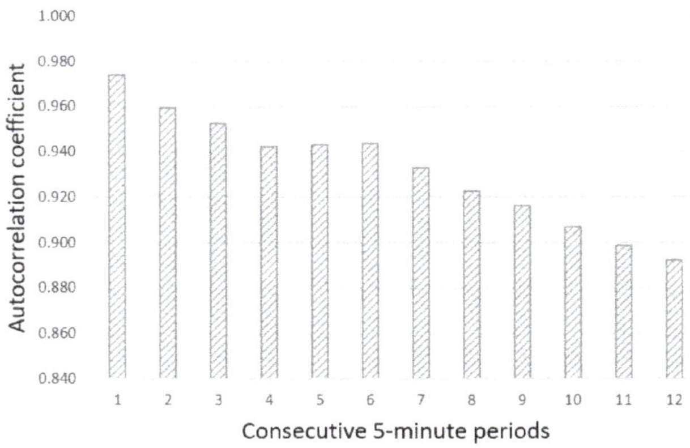

Figure 2. Autocorrelation function (ACF) of the analyzed time series of power generation.

3.2. Analysis of Potential Input Data for Forecasting Methods

The forecasted output is the power generation in the PV system (generation in the DC part of the system). Five additional time series (measured, real values) are available for analysis as potential input data. There are no forecasts of these time series. The following time series are available:

- Solar irradiance (W/m^2);
- Air temperature (°C);
- PV module temperature (°C);
- Wind direction (degrees);
- Wind speed (m/s).

Only the past values of the five exogenous explanatory variables and the past values of the dependent variable (endogenous variable) can be selected as input data for the forecasting methods. Furthermore, a weighted averaging of the time series of power-generation values can be performed. This activity should reduce the random component of this time series. The selected past values of such transformed time series may be a valuable set of input data. They can even potentially replace the past values of the forecasted time series as input data in the forecasting model. The values of the smoothed time series of power generation were calculated from Equation (1).

$$P_t^{smoothed} = P_{t-1} \cdot w_{t-1} + P_{t-2} \cdot w_{t-2} + P_{t-3} \cdot w_{t-3}, \sum_{k=1}^{k=3} w_{t-k} = 1 \quad (1)$$

where $P_t^{smoothed}$ is the smoothed value of power generation for period t and P_{t-k} is the value of power generation for period $t-k$, $w_{t-1} = 0.6$, $w_{t-2} = 0.3$ and $w_{t-3} = 0.1$.

Table 2 shows Pearson linear correlation coefficients (R) between the 5 min power generation and the potential explanatory variables considered. All correlation coefficients are statistically significant (5% level of significance). The number of expertly proposed past values (from one to three withdrawals) for each explanatory variable results from the value of the Pearson correlation coefficient (the higher the value of the Pearson coefficient, the greater the significance of the variable) and the independence of information contained in a given explanatory variable (the small value of the Pearson coefficient comparing the analyzed explanatory variable and other explanatory variables).

Table 2. Values of Pearson linear correlation coefficients between 5 min power generation and the explanatory variables considered.

Code of Variable	Potential Explanatory Variables Considered	R
SPG(T-1)	Smoothed power generation in period $t-1$	0.9756
PG(T-1)	Power generation in period $t-1$	0.9744
PG(T-2)	Power generation in period $t-2$	0.9601
PG(T-3)	Power generation in period $t-3$	0.9536
SI(T-1)	Solar irradiance in period $t-1$	0.9661
SI(T-2)	Solar irradiance in period $t-2$	0.9587
SI(T-3)	Solar irradiance in period $t-3$	0.9385
AT(T-1)	Air temperature in period $t-1$	0.4261
PV_MT(T-1)	PV module temperature in period $t-1$	0.7134
WD(T-1)	Wind direction in period $t-1$	−0.2475
WS(T-1)	Wind speed in period $t-1$	0.1825

The three past values of power generation and the three past values of solar irradiance have very large and similar values of the Pearson's coefficient related to the dependent variable (output data)—power generation. PV module temperature in period $t-1$ has a significantly greater R value than the air temperature in period $t-1$. The smallest R values have wind direction in period $t-1$ and wind speed in period $t-1$. All R values except wind direction in period $t-1$ are positive.

Figure 3 presents dispersion diagrams–relationships between power generation in period t and smoothed power generation in period $t-1$. The relationship is close to linear. The strongest linear relationship is visible for values close to the extremes (power generation close to zero and power generation close to the rated power). The few points significantly deviating from the linear relationship can be interpreted as a change in cloud cover over a period of 5 min. The Pearson linear correlation coefficient between the output data (the power generation in period t) and proposed new input data (the smoothed power generation in period $t-1$) is equal to 0.9756. This R value for the smoothed power generation in period $t-1$ is the biggest of all potential input data.

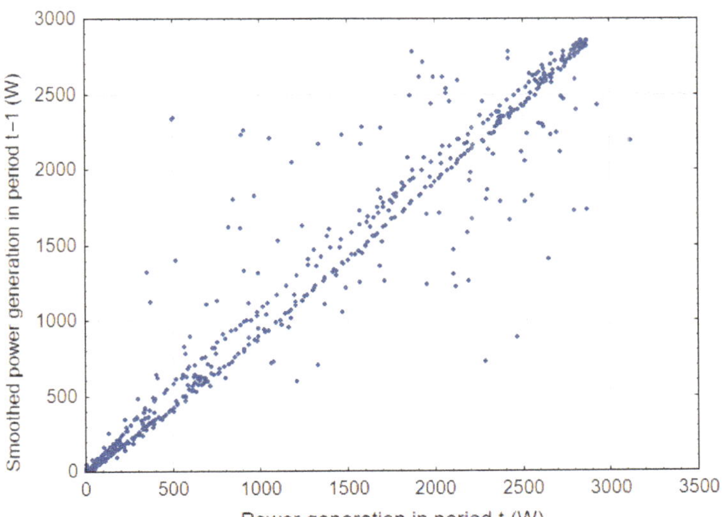

Figure 3. Relationship between power generation in period t and smoothed power generation in period $t-1$.

In order to determine the importance of the potential input data, the following methods of selecting variables were additionally used, using all possible 11 inputs and 1 output:

- C&RT decision trees algorithm for the selection of variables in regression problems—for each potential predictor (input data), the coefficient of determination R^2 is calculated;
- Analysis of variances (F statistics)—this method calculates the quotient of the intergroup variance to the intragroup variance (the dependent variable) in predictor intervals (the number of quantitative predictor classes is determined before the analysis);
- Global Sensitivity Analysis (GSA statistics) for multilayer perceptron (MLP) neural network. A neural network with one hidden layer and four neurons in this layer was used for the analysis. The training algorithm is BFGS, the activation function in the hidden layer is the hyperbolic tangent, and the activation function in the output layer is linear. The value of the importance factor for input data number k is the quotient of the RMSE error of the forecasts of the trained MLP network using the remaining 10 input data and the input data number k is replaced by its mean value from the total data to the RMSE error of the forecasts using all 11 sets of input data. The greater the value of the importance factor for the given input data, the greater their significance. Results below 1 for a given input data mean that these input data can probably be eliminated because the MLP network without these input data has a lower RMSE error in the forecasts;
- The importance of input data using the random forest (RF) algorithm is the many decision trees (DCs). The importance of the given input data is measured by checking to what extent nodes (in all decision trees) using the input data reduce the impurity Gini indicator, with the weight of each node being equal to the number of associated training samples [38]. It was assumed for the analysis that each decision tree would have 6 randomly selected sets of input data from the total of 11.

The results of the input data selection with the C&RT decision tree algorithm are shown in Figure 4. The values of the coefficient of determination are sorted in descending order. The most important explanatory variable according to this method is smoothed power generation in period $t-1$.

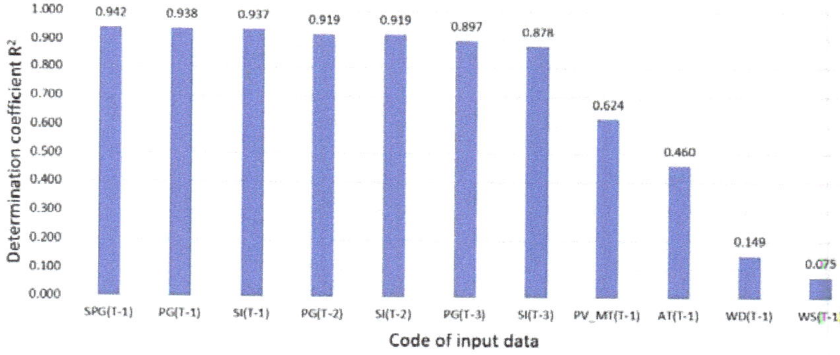

Figure 4. The results of the input data selection with the C&RT decision tree algorithm.

In Figure 5, the results of the input data selection with the use of the analysis of variance (F statistics) are presented. The F values are sorted in descending order. Power generation in period $t-1$ is the most important explanatory variable according to this method.

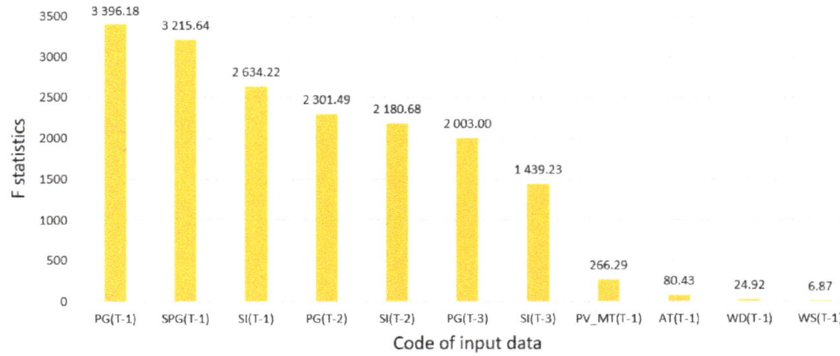

Figure 5. The results of the input data selection using analysis of variance (F statistics).

The results of input data selection using the Global Sensitivity Analysis for the multilayer perceptron (MLP) neural network are shown in Figure 6. The importance factor values are sorted in descending order. The most important explanatory variable according to this method is solar irradiance in period $t-1$.

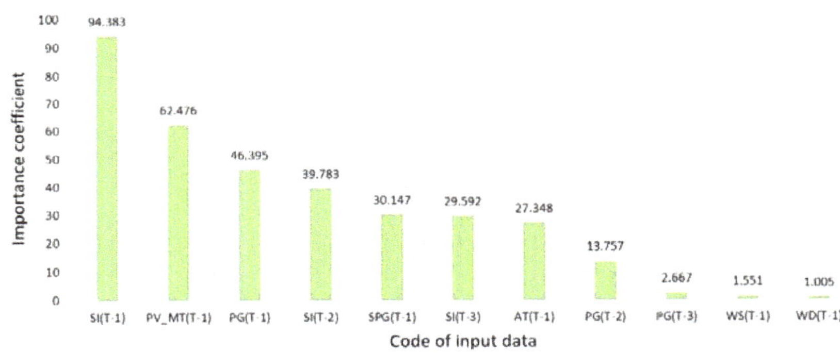

Figure 6. Input data selection results using Global Sensitivity Analysis for MLP-type neural network.

In Figure 7, the results of the input data selection with the use of the random forest algorithm are presented. The importance values are sorted in descending order. Smoothed power generation in period $t-1$ is the most important explanatory variable according to this method.

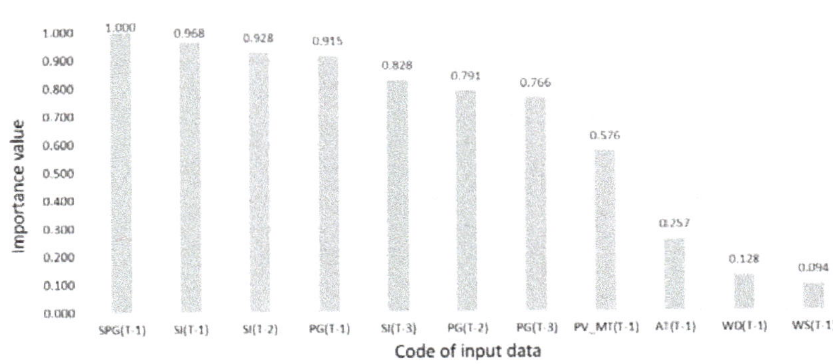

Figure 7. The results of the input data selection using the random forest algorithm.

Based on the analysis of the selection of variables using these four methods, the following conclusions can be drawn:

- For all analyzed methods of selecting variables, the significantly least important input data are wind direction in period $t-1$ and wind speed in period $t-1$. In the vast majority of cases, the last, least-important input data are (somewhat surprisingly) wind speed in period $t-1$;
- The best input data include smoothed power generation in period $t-1$, power generation in period $t-1$, and solar irradiance in period $t-1$.
- The results of the individual selection methods were quite similar, except for the input-data-selection method using global sensitivity analysis for the MLP-type neural network. In this case, the most important input data—solar irradiance in period $t-1$—are significantly more important than the second (surprisingly) in order input data, the PV module temperature in period $t-1$. This method also obtained validity results with the greatest diversification of numerical values;
- For all analyzed methods of selection of input data, the PV module temperatures in period $t-1$ are more important input data than the air temperature in period $t-1$;
- The results of the input-data-selection method with the C&RT decision tree algorithm (values of the coefficient of determination) are very similar to the values of Pearson's linear correlation (Table 2), both in relation to the order of input data in the ranking as well as the values of the coefficients.

Table 3 shows the input datasets that will be applied to forecasts using various methods, including hybrid methods and team methods. The input datasets proposed for the forecast quality tests assume the use of all data nominated on the basis of the selection made using four methods, as well as the use of a limited number of inputs for a given method (e.g., maximum of four sets of input data—this is the limitation of the Interval Type-2 Fuzzy Logic System method due to the computational time consumption). Thanks to the construction of many sets with a different number of input data, it will be possible to verify whether it is reasonable to limit data to those that selection methods indicate as the most important input data or whether it is better to use all available input data that are statistically significant. The persistence model only uses the last known value of the forecast time series for the prediction (set 0 (1 input)). This model is a reference point for other more advanced methods, the forecasts of which should have lower error measures.

Table 3. Sets of input data selected for forecasting methods.

Name of Set	Codes of Input Data and Additional Comments
Set 0 (1 input)	PG(T-1)
SET I (3 inputs)	PG(T-1), PG(T-2), PG(T-3)
SET II A (4 inputs)	PG(T-1), SI(T-1), PV_MT(T-1), AT(T-1)
SET II B (4 inputs)	SPG(T-1), SI(T-1), PV_MT(T-1), AT(T-1)
SET II C (3, 3, 4 inputs)	PG(T-1), PG(T-2), PG(T-3)—inputs for predicting PG $_{forecast}$(T) SI(T-1), SI(T-2), SI(T-3)—inputs for predicting SI $_{forecast}$(T) PG $_{forecast}$(T), SI $_{forecast}$(T), PV_MT(T-1), AT(T-1)—inputs for predicting PG(T)
SET III (11 inputs)	SPG(T-1), PG(T-1), PG(T-2), PG(T-3), SI(T-1), SI(T-2), SI(T-3), AT(T-1), PV_MT(T-1) WD(T-1), WS(T-1)
SET IV (3, 3, 13 inputs)	PG(T-1), PG(T-2), PG(T-3)—inputs for predicting PG $_{forecast}$(T) SI(T-1), SI(T-2), SI(T-3)—inputs for predicting SI $_{forecast}$(T) SPG(T-1), PG(T-1), PG(T-2), PG(T-3), SI(T-1), SI(T-2), SI(T-3), AT(T-1), PV_MT(T-1), WD(T-1), WS(T-1), PG $_{forecast}$(T), SI $_{forecast}$(T)—inputs for predicting PG(T)
SET V (15 inputs)	SPG(T-1), PG(T-1), PG(T-2), PG(T-3), SI(T-1), SI(T-2), SI(T-3), AT(T-1), AT(T-2), AT(T-3), PV_MT(T-1), PV_MT(T-2), PV_MT(T-3), WD(T-1), WS(T-1)

One of the sets (set I (three inputs)) assumes the use of only three retracted values of the forecast time series. This is to compare the quality of forecasts based only on the time series without the use of exogenous input variables with the forecasts using additional exogenous input variables.

Set II C (three, three, and four inputs) and set IV (three, three, and thirteen inputs) are sets for the hybrid method. The first model forecasts power generation in period *t* using the last three values of the time series. The second model forecasts solar irradiance in period *t* using the last three values of the time series. The third model that generates the correct final forecast of power generation in period *t* uses the forecasts from the first model and the second model as input data.

Set V uses all available statistically significant data, including the last three sets' previous values of the following variables: power generation, solar irradiance, PV module temperature, and air temperature.

4. Forecasting Methods

This section describes the methods employed in this paper. Forecasts are made using single methods, ensemble methods, and hybrid methods. In total, ten prognostic methods were used. Figure 8 presents a general diagram of subsequent activities related to the forecasting process.

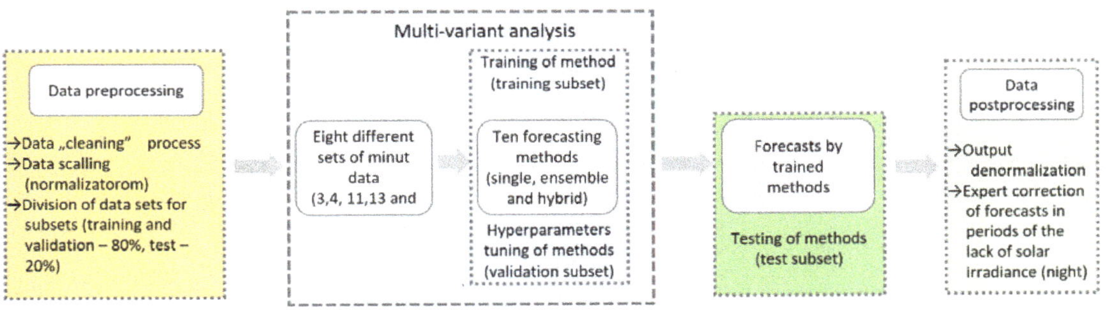

Figure 8. A general diagram of the consecutive steps in the forecasting process.

In the first step, data were preprocessed. In the beginning, before the process of data scaling (normalization) and data processing into the appropriate sets (input data and output data, the process of data "cleaning" was performed. Next, the data from the time series of the PV system's power generation were normalized to relative units (one relative unit is equal to the installed power). The other time series of data (exogenous input variables) were normalized using min–max scaling. The data, including 1152 periods of 5 min, were divided into three subsets: training, validation, and test subsets, respectively. Training and validation subsets consisted of 80% of the time series chosen randomly (division into training and validation parts, which are different depending on the forecasting method used). The test subset comprised the remaining 20% of the time series via random selection. Estimation of model parameters was performed with the training subset. The validation subset was used for tuning the hyperparameters of the selected methods. The last one—test subset—was applied to find the final results of errors in the forecast methods used. The choice of the training and validation subsets from 80% of the data of the time series was made with the usage of gradient-boosted trees (GBT) along with the bootstrap technique. The multiple linear regression model (LR) only used the training subset (80% of the data of the time series) without a validation subset—this model had no hyperparameters, only parameters determined during a one-time parameter-optimization process.

Next, multivariate analysis was performed—using the predictive methods on eight different input datasets in the training subset and the selection of appropriate hyperparameters of the methods in the validation subset. An example of the selected hyperparameters and the scope of their searches for the selected methods is included in Appendix A—Table A1.

Then, the final predictions for the subset test were made for all methods with the selected hyperparameters.

Postprocessing was performed in the last step. The values of the generated forecasts were scaled (de-normalized) to natural values (watts). An expert forecast correction was performed—non-zero power-generation values from the periods between sunset and sunrise were reset (power generation is impossible) in these time periods.

Following is a brief description of the proposed predictive methods. The persistence model was a benchmark for the quality of other, more advanced forecasting methods.

Persistence model. The naive model was the simplest to implement. It assumes that the forecast generation value is equal to the actual power-generation value obtained from the period 5 min before. Forecasts were calculated by Equation (2):

$$\hat{y}_t = y_{t-1} \tag{2}$$

where $\hat{y}_t$—forecast power generated by the PV system in a 5 min period t and y_{t-1}—power generation in a period lagged by $t-1$ from forecast period t.

Multiple linear regression model. This is a linear model that adopts a linear association among the input variables and the single output variable [41,42]. The input data are particular lags of the forecasted output variable. The other input explanatory variables (including their particular lags) are correlated to the output variable. The least-squares approach was used to fit the model.

K-Nearest Neighbors Regression. This technique is a non-parametric method used for regression and classification tasks [42,43]. The input consists of the k nearby training examples from the feature space. When using the KNN regression, the output is the property value for the object. This value represents the average of the values of the k nearest neighbors. The number of nearest neighbors is treated as the main hyperparameter for the tuning process. Models with a very low k value of 1 or 2 are most likely to suffer from overfitting. Along with increasing the value of k, this model should work more efficiently, but it may also lead to an increase in the load on the model and the occurrence of underfitting. The distance metric is the second hyperparameter.

MLP-type artificial neural network. This is a group of feedforward artificial neural networks (ANNs). MLP is an effective and popular linear or non-linear (depending on the kind of activation function in hidden layer/layers and output layer) universal approximator [44,45]. It consists of one input layer, which typically has one or two hidden layers, and one output layer. It often uses the backpropagation algorithm for the supervised learning process. The number of neurons in the hidden layer(s) is usually the main hyperparameter in the tuning task. Another selectable hyperparameter is the activation function in the hidden layer(s) and in the output layer. The Broyden–Fletcher–Goldfarb–Shanno (BFGS) method used for solving unconstrained non-linear optimization problems was chosen as a learning algorithm for the neural network.

Support Vector Regression. SVM for regression of the Gaussian kernel converts the classification process into regression by specifying the width ε tolerance region around the destination [46]. The learning process for SVR is diminished to the quadratic optimization problem and depends on several hyperparameters, such as tolerance ε, regularization constant C, and width parameter s of the Gaussian kernel.

Interval Type-2 Fuzzy Logic System. Type-2 fuzzy sets (T2 FSs) are used in type-2 fuzzy logic systems (T2 FLSs). Type-2 fuzzy sets are an expansion of type-1 fuzzy sets (T1 FSs). Investigations on T2 FSs were performed by Zadeh, Karnik, Mendel, and Liang [47–49]. Membership functions with three dimensions (MFs), including a footprint of uncertainty (FOU), are features of T2 FSs [50]. The structure of T2 FLSs was presented, e.g., in [4]. Typical blocks include the fuzzification block, the fuzzy inference block, the base of fuzzy rules, the type reduction block, and the defuzzification block as components of T2 FLSs. In the type reduction block, the transformation of T2 FS to T1 FS occurs. Usually, for type reduction, the Karnik–Mendel (KM) algorithm is employed [48].

Interval type-2 fuzzy logic systems (IT2 FLSs) (see, e.g., [50]) are often used in practice because of the computational complexity of T2 FLSs [51]. Among the different IT2 FLSs, the IT2 TSK FLS (the IT2 FLS with the inference model of Takagi–Sugeno–Kang [52]), or the IT2 S FLS (the IT2 FLS with the Sugeno inference model), can be distinguished. IT2 TSK FLS and IT2 S FLS require a lower number of model parameters than the standard IT2 FLS. Genetic algorithms (GAs) or PSO algorithms are often used in the training process of the IT2 FLSs (in the determination of their parameters' values).

Random Forest Regression. RF is a collaborative method based on numerous single decision trees (the same type of models). In the regression process, the prediction in a single decision tree consists of the average target value of all instances related to the single leaf node [4]. The final prediction is the average value of all n single decision trees. Random forests are created on the basis of quite deep trees—forecasts using this method are characterized by a low load along with quite a large variance. The regularization hyperparameters depend on the algorithm used but generally restricted, are among others, are factors such as the minimum number of data points placed in a node before the node is split, the maximum number of levels in each decision tree, the maximum depth of a single decision tree, the minimum number of data points allowed in a leaf node, and the maximum number of nodes. The number of predictors for each of the n single decision trees is made by the random choice of k predictors from all available n predictors [4,41]. The overfitting problem, in this case, is usually related to redundant decision trees in the random forest.

Gradient-Boosted Trees for Regression. Gradient boosting refers to an ensemble method that can combine several weak learners into a strong learner [4]. GBT ensures the minimization of variance and bias in relation to single prognostic models. On the other hand, the algorithm is more susceptible to outliers than, for example, simple decision tree models. The GBT algorithm sequentially adds predictors (the same type of models) to the ensemble, each one correcting its own predecessor. This technique tries to fit the new predictor into the residual errors made by the previous predictor. The final prediction consists of the average value from all n single decision trees. In comparison with random forest, GBT has one additional hyperparameter—the learning rate, which is used for scaling the contribution of each tree [41,52]. The problem of overfitting is most often associated with too many trees in the ensemble.

Weighted Averaging Ensemble. This is an integration of the results of selected predictors into the final verdict of the ensemble. The final forecast is defined as the average of the results produced by all n predictors organized in an ensemble [42,46]. The final prediction result is calculated by Equation (3).

$$\hat{y}_i = \frac{1}{n}\sum_{j=1}^{n}\hat{y}_i^j \quad (3)$$

where i is the prediction point, $\hat{y}_i$ is the final predicted value, $\hat{y}_i^j$ is the value predicted by predictor number j, and n is the number of predictors in the ensemble. Note: all weights are equal to $1/n$ in this case.

This formula makes use of the stochastic distribution of the predictive errors. The process of averaging reduces the final error of forecasting. The averaging of the forecast results is an established method of reducing the variance of forecast errors. An important condition for including the predictor in the ensemble is independent operation from the others and a similar level of prediction error [42,46]. The choice of predictors (forecasting methods) is based on the smallest RMSE error on the validation subset, and only predictors of different types are selected for the ensemble.

Hybrid method—connection of three MLP models. As an element of the prognostic problem decomposition, separate forecasts of selected exogenous variables for the forecast of the power-generation period can be made. This procedure creates new input explanatory exogenous variables (forecasts) that may be valuable for power generation in PV system forecasting methods. In the first step, MLP no. 1 forecasts power generation in period t. On the other hand, MLP no. 2 forecasts solar irradiance in period t. In the second step, neural network MLP no. 3 forecasts the final value of power generation in period t based on forecasts from the neural network of MLP no. 1 and no. 2 and other endogenous and exogenous variables (4 or 13 depending on the variant). For each of the three MLP neural networks, their appropriate hyperparameters are selected (the number of neurons in the hidden layer and activation functions in the hidden layer as well as in the output layer). Figure 9 shows a general diagram of the developed, proprietary hybrid method.

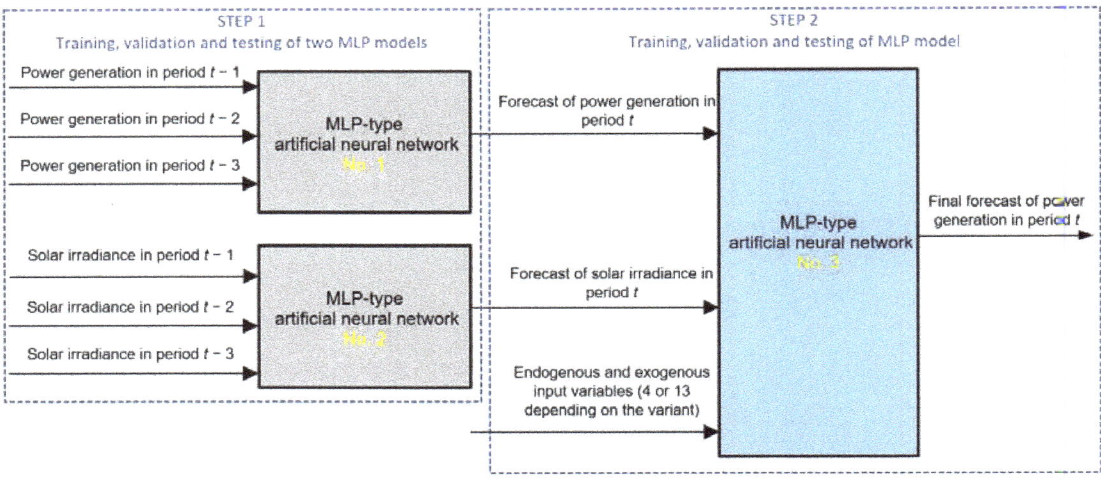

Figure 9. General scheme of the developed hybrid method with the use of three MLP neural network models.

Table 4 shows tested input datasets for each method and the codes of the methods. One reason for organizing data into such sets was to verify the influence of the type and number of variables on the forecast accuracy.

Table 4. Tested input datasets for each method and the codes of the methods.

The Name of Method	The Method Code	Complexity of Method/Type	Tested Sets of Input Data
Persistence model	NAIVE	Single/linear	Set 0
Multiple linear regression model	LR	Single/linear	SET I, SET II A, SET II B, SET III, SET V
K-Nearest Neighbors Regression	KNNR	Single/non-linear	SET I, SET II A, SET II B, SET III, SET V
MLP-type artificial neural network	MLP	Single/non-linear	SET I, SET II A, SET II B, SET III, SET V
Support Vector Regression	SVR	Single/non-linear	SET I, SET II A, SET II B, SET III, SET V
Interval Type-2 Fuzzy Logic System	IT2FLS	Single/non-linear	SET I, SET II A, SET II B
Random forest regression	RF	Ensemble/non-linear	SET I, SET II A, SET II B, SET III, SET V
Gradient-Boosted Trees for regression	GBT	Ensemble/non-linear	SET I, SET II A, SET II B, SET III, SET V
Weighted Averaging Ensemble	WAE (p1 *, . . . , pm)	Ensemble/non-linear	SET I, SET IIB, SET III
Hybrid method—connection of three MLP models	MLP&MLP→MLP	Hybrid/non-linear	SET II C, SET IV

Remark: * denotes first predictor in ensemble of m predictors.

5. Evaluation Criteria

In order to have a broader view of the quality of individual forecasting models, four evaluation criteria were used, including RMSE, nMAPE, nAPEmax, and MBE. The RMSE error was adopted as the most important measure due to the greater sensitivity to large partial errors. In all three tables (presented later) with performance measures of proposed methods, the results are sorted by this error measure. On the other hand, the second measure in the order of importance is the nMAPE error. The nAPEmax and MBE measures, in turn, are only auxiliary.

The Root Mean Square Error is calculated by Equation (4). The RMSE measure is typically used for power-generation forecasts from RES, including PV systems.

$$\mathrm{RMSE} = \sqrt{\frac{1}{n}\sum\nolimits_{i=1}^{n}(y_i - \hat{y}_i)^2} \quad (4)$$

where $\hat{y}_i$ is the predicted value, y_i is the actual value, and n is the number of prediction points.

The Normalized Mean Absolute Percentage Error is determined by Equation (5). Due to the zero values occurring in the power-generation time series, it is impossible to use the popular and recommended measure of the MAPE error. Therefore, the nMAPE measure was used, in which the real power-generation value presented in the denominator of the MAPE formula was replaced with the value representing the normalizing factor (the installed power of PV system).

$$\mathrm{nMAPE} = \frac{1}{n}\sum_{i=1}^{n}\frac{1}{c_{norm}}|y_i - \hat{y}_i|\cdot 100\% \quad (5)$$

where c_{norm} is the normalizing factor (installed power).

The Normalized Maximum Absolute Percentage Error is calculated by Equation (6). The nAPEmax error is the largest partial error of all individual n nAPE errors.

$$\mathrm{nAPEmax} = \max_{i=1,\ldots,n}\frac{1}{c_{norm}}|y_i - \hat{y}_i|\cdot 100\% \quad (6)$$

The Mean Bias Error (MBE) captures the average bias in the prediction and is defined by Equation (7). The forecasting method underestimates values if the nMBE < 0 or overestimates values if the nMBE > 0. The MBE error of a properly functioning prognostic method should be equal to or very close to zero.

$$\mathrm{MBE} = \frac{1}{n}\sum\nolimits_{i=1}^{n}(y_i - \hat{y}_i) \quad (7)$$

6. Results and Discussion

This section presents a wide comparative analysis of very-short-term forecasting methods for power generation in PV systems.

Table 5 shows performance measures of the proposed methods (on test subset) using three sets of input data—SET I (three inputs). This is the most basic set of input data using only the last three retracted values of the forecast time series of power generation. The study was completed to verify the quality of the forecasts; in this case, it was worse compared to forecasting methods that also use exogenous input variables with a similar amount of input data. Furthermore, the Table 5 shows forecast errors for the simplest reference method—persistence methods (NAIVE), using only one set of input data. Tabular results are ordered by ascending RMSE error values. Table A1 in Appendix A shows the results of hyperparameter tuning for the proposed methods using only three sets of input data.

Table 5. Performance measures of proposed methods (on test subset) using three input data.

Method Code	Input Data Set	RMSE (W)	nMAPE (%)	nAPEmax (%)	MBE (W)
MLP	SET I (3 inputs)	122.558	1.474	32.781	−4.539
WAE [MLP,RF]	SET I (3 inputs)	129.491	1.527	30.847	−2.623
RF	SET I (3 inputs)	133.931	1.674	28.439	−6.832
IT2FLS	SET I (3 inputs)	135.965	1.773	29.808	−6.536
KNNR	SET I (3 inputs)	137.828	1.533	33.802	−5.291
LR	SET I (3 inputs)	140.989	1.617	34.214	9.71[illegible]
SVR	SET I (3 inputs)	142.441	1.481	34.426	−3.364
GBT	SET I (3 inputs)	154.257	1.948	29.019	6.427
NAIVE *	SET 0 (1 input)	165.783	1.975	40.199	−9.030

Remarks: The best fitting result for each fitting measure is printed in bold in blue. The worst fitting result is printed in red. * Reference model.

Based on the results from Table 5, the following preliminary conclusions can be drawn regarding the proposed methods using only three sets of input data:

- The smallest RMSE and nMAPE errors were obtained by the MLP method among the seven tested methods (including two group methods), and this method can be considered the preferred one. The RF and IT2FLS methods obtained a slightly higher RMSE error;
- The difference in the quality of forecasts between the best MLP method and the worst NAIVE is quite large;
- The SVR method obtained an RMSE error significantly higher than the best MLP method (according to the RMSE measure), while the nMAPE error was almost identical to the MLP method.

Table 6 shows performance measures of the proposed methods (on test subsets) using four sets of input data (SET II A (four inputs), SET II B (four inputs), and SET II C (three, three, and four inputs)). In this case, the amount of input data is limited to only the most relevant input data, both endogenous and exogenous. The study was completed to verify the quality of the forecasts; in this case, it was worse compared to the prognostic methods using all available statistically significant endogenous and exogenous input variables. Another goal of this research was to verify which of the two sets of input data (SET IIA, SET IIB) obtains smaller forecasting errors using different forecasting methods. In addition, the quality of the proprietary hybrid model was verified in relation to other forecasting methods. Furthermore, the table shows forecast errors for the simplest reference method—the persistence method (NAIVE), using only one set of input data. Tabular results are ordered by ascending RMSE error values.

Based on the results from Table 6, the following preliminary conclusions can be drawn regarding the proposed methods using only four different sets of input data (including exogenous variables):

- The use of exogenous variables for forecasts made it possible to reduce the RMSE error of all the methods used;
- The use of smoothed power generation in period $t-1$ (in SET 2B) as an input variable instead of power generation in period $t-1$ (in SET 2A) turned out to be beneficial—all tested methods obtained a lower RMSE error;
- The smallest RMSE error and nMAPE error were obtained by the original, proprietary hybrid method (MLP&MLP->MLP). On the other hand, the MLP method obtained RMSE and nMAPE errors that were slightly higher;
- The largest RSME error, significantly greater than other methods, was obtained by the reference method—the NAIVE method, while the GBT method was the method with the second-greatest RSME error;

- The SVR method using four sets of input data (including exogenous variables) significantly reduced the RMSE error compared to the forecasts using three sets of input data (only the last two withdrawn values of the forecast process)—see Table 5.

Table 6. Performance measures of proposed methods (on test subset) using four input data.

Method Code	Input Data Set	RMSE (W)	nMAPE (%)	nAPEmax (%)	MBE (W)
MLP&MLP→MLP	SET II C (3, 3, 4 inputs)	**98.113**	**1.336**	19.110	−0.192
MLP	SET 2B (4 inputs)	99.761	1.393	18.100	−6.335
WAE[SVR,MLP]	SET 2B (4 inputs)	101.825	1.423	17.743	**−0.098**
SVR	SET 2B (4 inputs)	104.942	1.506	**16.329**	−0.638
WAE[IT2FLS,MLP]	SET 2B (4 inputs)	105.274	1.535	18.784	−0.116
IT2FLS	SET 2B (4 inputs)	110.461	1.540	19.223	4.956
KNNR	SET 2B (4 inputs)	116.924	1.837	19.328	6.615
MLP	SET 2A (4 inputs)	116.949	1.715	26.040	−2.637
RF	SET 2B (4 inputs)	117.383	1.632	23.832	−0.142
KNNR	SET 2A (4 inputs)	118.985	1.506	27.016	4.145
IT2FLS	SET 2A (4 inputs)	121.590	1.661	27.173	7.511
LR	SET 2B (4 inputs)	123.803	1.961	19.513	−8.993
SVR	SET 2A (4 inputs)	124.312	1.695	30.977	6.775
GBT	SET 2B (4 inputs)	124.747	1.914	17.422	−9.287
RF	SET 2A (4 inputs)	129.469	1.676	28.638	11.728
LR	SET 2A (4 inputs)	132.943	2.076	27.148	−2.051
GBT	SET 2A (4 inputs)	134.457	1.914	28.318	−4.637
NAIVE *	SET 0 (1 input)	165.783	1.975	40.199	−9.030

Remarks: The best fitting result for each fitting measure is printed in bold in blue. The worst fitting result is printed in red. * Reference model.

Table 7 shows, in turn, performance measures of the proposed methods (on test subsets) using 11, 13, and 15 sets of input data. This study aimed to verify whether the use of as many available and statistically significant endogenous and exogenous input variables would improve the quality of forecasts compared to a limited number of input data (three or four sets). In addition, the quality of the proposed proprietary hybrid model and the original "Weighted Averaging Ensemble" models compared to other forecast methods was verified. Furthermore, the Table 7 shows forecast errors for the simplest reference method—the persistence method (NAIVE), using only one set of input data. Tabular results are ordered by ascending RMSE error values.

Table 7. Performance measures of proposed methods (on test subset) using 1, 11, 13, and 15 sets of input data.

Method Code	Input Data Set	RMSE (W)	nMAPE (%)	nAPEmax (%)	MBE (W)
MLP&MLP→MLP	SET IV (3,3,13 inputs)	**61.633**	**0.805**	13.918	1.196
MLP	SET V (15 inputs)	63.092	0.848	**12.375**	−3.498
MLP	SET III (11 inputs)	64.794	0.809	16.173	3.664
WAE (SVR,MLP)	SET V (15 inputs)	65.391	0.832	12.397	−0.053
SVR	SET V (15 inputs)	71.618	0.824	12.629	−6.088
WAE (LR,MLP)	SET III (11 inputs)	71.884	0.826	15.038	**−0.048**
SVR	SET III (11 inputs)	90.379	1.416	16.065	3.570
LR	SET V (15 inputs)	90.491	1.015	14.982	−0.049
LR	SET III (11 inputs)	91.674	1.030	21.215	−3.760
KNNR	SET V (15 inputs)	104.505	1.360	18.228	−4.819
RF	SET V (15 inputs)	111.269	1.577	22.362	−2.483
RF	SET III (11 inputs)	116.497	1.619	23.848	3.406
KNNR	SET III (11 inputs)	118.490	1.565	23.348	7.507
GBT	SET V (15 inputs)	118.547	1.523	20.386	−0.179
GBT	SET III (11 inputs)	122.569	1.587	25.624	−2.750
NAIVE *	SET 0 (1 input)	165.783	1.975	40.199	−9.030

Remarks: The best fitting result for each fitting measure is printed in bold in blue. The worst fitting result is printed in red. * Reference model.

Based on the results from Table 7, the following preliminary conclusions can be drawn regarding the proposed methods using different numbers of sets of input data ranging from 11 to 15 (including exogenous variables):

- The use of a larger number of input data, from 11 to 15 for forecasts (including exogenous variables), allowed for a significant reduction in the RMSE error of all methods used compared to the use of only four sets of input data;
- The smallest RMSE error and nMAPE error were obtained by the original, proprietary hybrid method (MLP&MLP->MLP), and it is the recommended method. On the other hand, the MLP method obtained RMSE and nMAPE errors that were slightly higher;
- The largest RSME error, significantly greater than other methods, was obtained by the reference method—the NAIVE method, while the GBT method was the method with the second-greatest RSME error;
- The SVR method using 15 sets of input data (including exogenous variables) was one of the best methods, but the use of 11 sets of input data proved to be less favorable;
- Team methods with different types of predictors in the team (WAE (SVR,MLP) and WAE (LR,MLP)) were also among the best methods—the RMSE error of these methods was slightly greater than the second MLP method in the list;
- For all tested methods, it was more advantageous to use 15 sets of input data than 11 sets of input data.

Figure 10 shows the RMSE error, for each of the eight tested datasets, obtained by the best prognostic method for the test range. The MLP neural network method (yellow) is definitely the most common method for various input datasets. On the other hand, the smallest RMSE error (green) was obtained by the proprietary developed hybrid model (MLP&MLP->MLP). The highest RMSE error (gray) was achieved by the persistence (naïve) model as the simplest one, using only one set of input data. It should be noted that the quality of forecasts increases significantly with the increasing number of input data used. Thus, it can be concluded that by providing the predictive model at the input with more information related to the predicted process, in particular, with more than just one retracted value of a given explanatory variable (both exogenous and endogenous), smaller forecast errors can be expected.

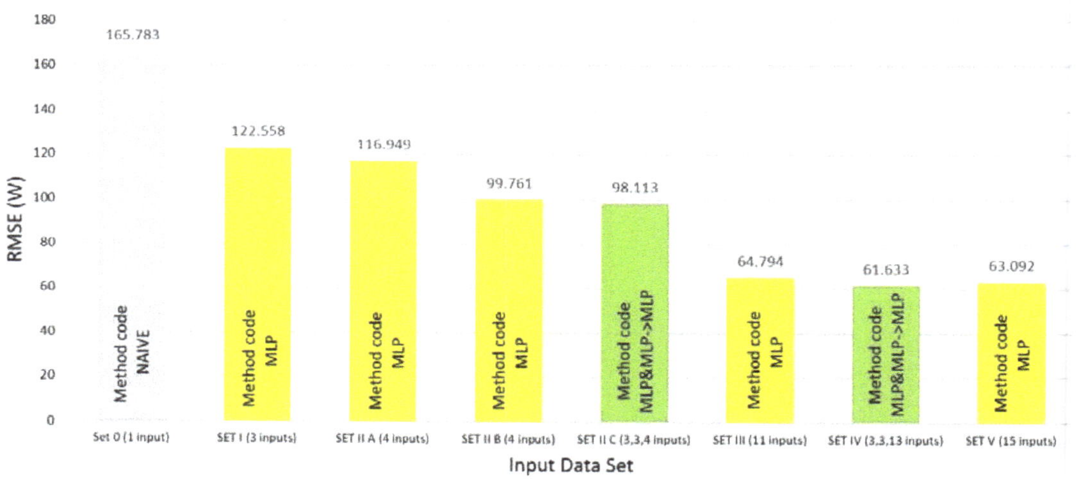

Figure 10. Summary of RMSE error values for the best predictive method depending on the dataset.

Figure 11 shows a scatter plot between the actual power-generation values and the values obtained from the forecast using the best method—a proprietary hybrid model (MLP&MLP->MLP) for the test range. From the graph, it can be observed that the accuracy of forecasts was the highest for small power-generation values below 750 W (where the installed power of a PV system is equal to 3200 W).

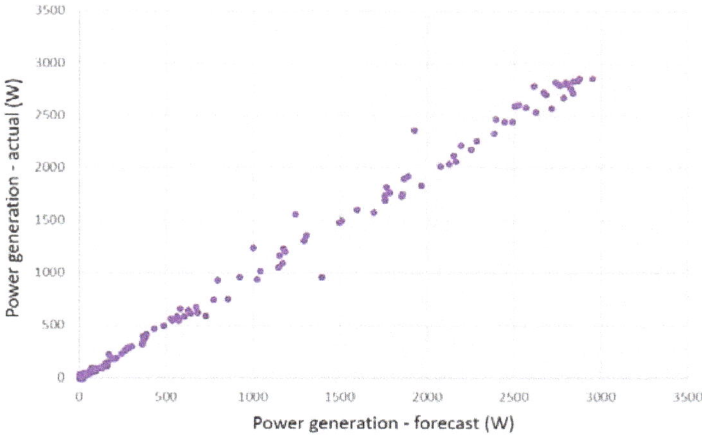

Figure 11. The scatter plot of the real power-generation values and the values obtained from the forecast with the best hybrid model.

7. Conclusions

The analysis of the available input variables with the use of four different methods of selecting input variables for forecasting models allowed us to identify the most important input variables. The most important input data include smoothed power generation in period $t-1$, power generation in period $t-1$, and solar irradiance in period $t-1$. The significantly least-important input data are wind direction in period $t-1$ and wind speed in period $t-1$.

The influence of the type and number of input variables on the quality of forecasts was investigated. The use of only three withdrawn values of power generation showed that this is the least-effective solution. Additionally, the use of other available exogenous variables (the selected historical values of solar irradiance, PV module temperature, wind direction, and wind speed) allowed us to reduce the RSME error of forecasts. An additionally valuable input variable is the smoothed value of power generation (see Equation (1)), a value calculated on the basis of the reverted values of the forecast process. The smallest forecast errors (RMSE) were obtained using a set of SET IV and SET V input variables, i.e., sets with the largest number of input variables.

The effectiveness of many prognostic methods, both single as well as team and hybrid, was verified. The smallest RMSE and nMAPE errors were obtained for the original, developed hybrid method using three MLP neural networks (method code MLP & MLP-MLP) using a set of SET IV input variables. Compared to the reference method (method code NAIVE), the hybrid method obtained an RMSE error 62.8% lower. However, compared to the best single method (the MLP method code) using the SET V input variable set, the RMSE error of the hybrid method was 2.3% lower. In the case of the number of input variables limited to four, the proprietary hybrid method also obtained the smallest RMSE error. Compared to the method code MLP, the RMSE error for the hybrid method was 1.7% lower. Among the single prognostic methods, the MLP neural network was the best method. Other machine learning techniques (RF, SVR, KNNR, and GBT) obtained slightly larger RMSE errors. The most advantageous of these four machine learning techniques was the SVR method with the SET V set of input variables. It is also advantageous to use the collective method (method code WAE (SVR, MLP)), which obtained an RMSE error slightly greater than the best method single (MLP).

In the authors' opinion, some of the forecasting methods investigated are effective and promising tools for practical applications, e.g., for very-short-term PV generation power forecasting. In turn, forecasts of this type are very useful for the needs of low-voltage microgrid operation control.

Research may be continued and expanded in the future. The proposed research directions include:

- Increasing the forecast horizon to 1 h (4 forecasts for consecutive 15 min periods);
- Using various techniques for decomposing the prognostic problem and examining their impact on the quality of forecasts (in the case of obtaining data from a period of several years);
- Examining the distribution of forecast errors during the day—verifying whether there is a relationship between the RMSE error rate and the time of day;
- Quality-testing forecasting models using additional solar irradiance, wind speed, and wind direction forecasts (in the case of obtaining such meteorological forecasts).

Author Contributions: Conceptualization, P.P. and M.P.; formal analysis, P.P.; methodology, P.P. and P.K.; investigation, P.P., P.K., M.P., and B.F.; supervision, M.P.; validation, P.P.; writing, P.P., P.K., M.P., and B.F.; visualization P.P.; project administration, P.P. All authors have read and agreed to the published version of the manuscript.

Funding: This research received no external funding.

Institutional Review Board Statement: Not applicable.

Informed Consent Statement: Not applicable.

Data Availability Statement: Not applicable.

Acknowledgments: The dataset deployed for calculations in this paper was gathered by the meteorology station of the Photovoltaic Laboratory at the Institute of Microelectronics and Optoelectronics of Warsaw University of Technology. The authors thank this institute heartily for sharing these data.

Conflicts of Interest: The authors declare no conflict of interest.

Abbreviations

The following abbreviations are used in this manuscript:

ACF	Autocorrelation Function
ANFIS	Adaptive Neuro-Fuzzy Inference System
ANN	Artificial Neural Network
BFGS	Broyden–Fletcher–Goldfarb–Shanno algorithm
C&RT	Classification and Regression Trees algorithm
DCs	Decision Trees
EIASC	Enhanced iterative algorithm with stop condition
GA	Genetic algorithm
GBT	Gradient-Boosted Trees
GSA	Global sensitivity analysis
IASC	Iterative algorithm with stop condition
IT2FLS	Interval Type-2 Fuzzy Logic System
KM	Karnik–Mendel
KNNR	K-Nearest Neighbors Regression
LR	Linear Regression
MBE	Mean Bias Error
MLP	Multi-Layer Perceptron
$nAPE_{max}$	Normalized Maximum Absolute Percentage Error
nMAPE	Normalized Mean Absolute Percentage Error
PR	Performance ratio
PSO	Particle Swarm Optimization

PV	Photovoltaic
R	Pearson linear correlation coefficient
RES	Renewable Energy Sources
R^2	Determination coefficient
RF	Random forest
RMSE	Root Mean Square Error
SVM	Support Vector Machine
SVR	Support Vector Regression

Appendix A

Table A1 shows the results of hyperparameter tuning for the proposed methods using three sets of input data.

Table A1. Results of hyperparameter tuning for proposed methods using three sets of input data.

Method Code	Description of Method, Name, and Range of Values of Hyperparameters' Tuning and Selected Values
SVR	Regression SVM: Type-1, Type 2, selected: Type-1; kernel type: Gaussian (RBF); width parameter σ: 0.333; regularization constant *C*, range: 1–50 (step 1), selected: 2; tolerance ε, range: 0.01–0.2 (step 0.01), selected: 0.02.
KNNR	Number of nearest neighbours *k*, Distance metrics: Euclidean, Manhattan, Minkowski, selected: Euclidean; range: 1–50, selected: 13.
MLP	Learning algorithm: BFGS; the number of neurons in hidden layer: 2–10, selected: 3; activation function in hidden layer: linear, hyperbolic tangent, selected: hyperbolic tangent; activation function in output layer: linear.
IT2FLS	Interval Type-2: Sugeno FLS, Mamdani FLS, selected: Sugeno FLS; learning and tuning algorithm: GA, PSO, selected: PSO; initial swarm span: 1500–2500, selected: 2000; minimum neighborhood size: 0.20–0.30, selected: 0.25; inertia range: from [0.10–1.10] to [0.20–2.20], selected: [0.50–0.50]; number of iterations in the learning and tuning process: 5–20, selected: 20; type of the membership functions: triangular, Gauss, selected: Gauss; the number of output membership functions: 3–81, selected: 81; defuzzification method: Centroid, Weighted average of all rule outputs, selected: Weighted average of all rule outputs; AND operator type: min, prod, selected: min; OR operator type: max, probor, selected: probor; implication type: prod, min, selected: min; aggregation type: sum, max, selected: sum; the k-Fold Cross-Validation value: 1–4, selected: 4; window size for computing average validation cost: 5–10, selected: 7; maximum allowable increase in validation cost: 0.0–1.0, selected: 0.1; the type-reduction methods: KM, IASC, EIASC, selected: KM.
RF	The number of decision trees: 2–50, selected: 5; the number of predictors chosen at random: 1, 2, selected 2. Stop parameters: maximum number of levels in each decision tree: 5, 10, 20, selected 10; minimum number of data points placed in a node before the node is split: 10, 20, 30, 40, 50, selected 20; minimum number of data points allowed in a leaf node: 10; maximum number of nodes: 100.
GBT	Considered max depth: 2/4, selected depth: 2; trees number: 50/100/150/200/250, selected number: 100; learning rate: 0.1/0.01/0.001, selected: 0.1.

References

1. Shamsollahi, P.; Cheung, K.W.; Chen, Q.; Germain, E.H. A neural network based very short term load forecaster for the interim ISO New England electricity market system. In Proceedings of the 2001 Power Industry Computer Applications Conference, Sydney, Australia, 20–24 May 2001; pp. 217–222.
2. Parol, M.; Piotrowski, P. Very short-term load forecasting for optimum control in microgrids. In Proceedings of the 2nd International Youth Conference on Energetics (IYCE 2009), Budapest, Hungary, 4–6 June 2009; pp. 1–6
3. Parol, M.; Piotrowski, P. Electrical energy demand forecasting for 15 minutes forward for needs of control in low voltage electrical networks with installed sources of distributed generation. *Przegląd Elektrotechniczny Electr. Rev.* **2010**, *86*, 303–309. (In Polish)
4. Parol, M.; Piotrowski, P.; Kapler, P.; Piotrowski, M. Forecasting of 10-Second Power Demand of Highly Variable Loads for Microgrid Operation Control. *Energies* **2021**, *14*, 1290. [CrossRef]

5. Hernandez, L.; Baladron, C.; Aguiar, J.M.; Carro, B.; Sanchez-Esguevillas, A.J.; Lloret, J.; Massana, J. A Survey on Electric Power Demand Forecasting: Future Trends in Smart Grids, Microgrids and Smart Buildings. *IEEE Commun. Surv. Tutor.* **2014**, *16*, 1460–1495. [CrossRef]
6. Niu, D.; Pu, D.; Dai, S. Ultra-Short-Term Wind-Power Forecasting Based on the Weighted Random Forest Optimized by the Niche Immune Lion Algorithm. *Energies* **2018**, *11*, 1098. [CrossRef]
7. Bokde, N.; Feijóo, A.; Villanueva, D.; Kulat, K. A Review on Hybrid Empirical Mode Decomposition Models for Wind Speed and Wind Power Prediction. *Energies* **2019**, *12*, 254. [CrossRef]
8. Adnan, R.M.; Liang, Z.; Yuan, X.; Kisi, O.; Akhlaq, M.; Li, B. Comparison of LSSVR, M5RT, NF-GP, and NF-SC Models for Predictions of Hourly Wind Speed and Wind Power Based on Cross-Validation. *Energies* **2019**, *12*, 329. [CrossRef]
9. Würth, I.; Valldecabres, L.; Simon, E.; Möhrlen, C.; Uzunoğlu, B.; Gilbert, C.; Giebel, G.; Schlipf, D.; Kaifel, A. Minute-Scale Forecasting of Wind Power—Results from the Collaborative Workshop of IEA Wind Task 32 and 36. *Energies* **2019**, *12*, 712. [CrossRef]
10. Liu, F.; Li, R.; Dreglea, A. Wind Speed and Power Ultra Short-Term Robust Forecasting Based on Takagi–Sugeno Fuzzy Model. *Energies* **2019**, *12*, 3551. [CrossRef]
11. Carpinone, A.; Giorgio, M.; Langella, R.; Testa, A. Markov chain modeling for very-short-term wind power forecasting. *Electr. Power Syst. Res.* **2015**, *122*, 152–158. [CrossRef]
12. Tato, J.H.; Brito, M.C. Using Smart Persistence and Random Forests to Predict Photovoltaic Energy Production. *Energies* **2018**, *12*, 100. [CrossRef]
13. Zhu, R.; Guo, W.; Gong, X. Short-Term Photovoltaic Power Output Prediction Based on k-Fold Cross-Validation and an Ensemble Model. *Energies* **2019**, *12*, 1220. [CrossRef]
14. Abdullah, N.A.; Rahim, N.A.; Gan, C.K.; Adzman, N.N. Forecasting Solar Power Using Hybrid Firefly and Particle Swarm Optimization (HFPSO) for Optimizing the Parameters in a Wavelet Transform-Adaptive Neuro Fuzzy Inference System (WT-ANFIS). *Appl. Sci.* **2019**, *9*, 3214. [CrossRef]
15. Nespoli, A.; Mussetta, M.; Ogliari, E.; Leva, S.; Fernández-Ramírez, L.; García-Triviño, P. Robust 24 Hours ahead Forecast in a Microgrid: A Real Case Study. *Electronics* **2019**, *8*, 1434. [CrossRef]
16. Ahmed, R.; Sreeram, V.; Mishra, Y.; Arif, M.D. A review and evaluation of the state-of-art in PV solar power forecasting: Techniques and optimization. *Renew. Sust. Energ. Rev.* **2020**, *124*, 109792. [CrossRef]
17. Antonanzas, J.; Osorio, N.; Escobar, R.; Urraca, R.; Martinez-de-Pison, F.J.; Antonanzas-Torres, F. Review of photovoltaic power forecasting. *Sol. Energy* **2016**, *136*, 78–111. [CrossRef]
18. Mayer, M.J.; Gróf, G. Extensive comparison of physical models for photovoltaic power forecasting. *Appl. Energy* **2021**, *283*, 116239. [CrossRef]
19. Mayer, M.J. Influence of design data availability on the accuracy of physical photovoltaic power forecasts. *Sol. Energy* **2021**, *227*, 532–540. [CrossRef]
20. Study Committee: C6; CIGRÉ Working Group C6.22. *Microgrids 1: Engineering, Economics, & Experience-Capabilities, Benefits, Business Opportunities, and Examples-Microgrids Evolution Roadmap*; Technical Brochure 635; CIGRE: Paris, France, 2015.
21. Marnay, C.; Chatzivasileiadis, S.; Abbey, C.; Iravani, R.; Joos, G.; Lombardi, P.; Mancarella, P.; von Appen, J. Microgrid evolution roadmap. Engineering, economics, and experience. In Proceedings of the 2015 International Symposium on Smart Electric Distribution Systems and Technologies (EDST15), CIGRE SC C6 Colloquium, Vienna, Austria, 8–11 September 2015.
22. *Microgrids: Architectures and Control*; Hatziargyriou, N.D. (Ed.) Wiley-IEEE Press: Hoboken, NJ, USA, 2014.
23. Baczynski, D.; Ksiezyk, K.; Parol, M.; Piotrowski, P.; Wasilewski, J.; Wojtowicz, T. *Low Voltage Microgrids. Joint Publication Edited by Mirosław Parol*; Publishing House of the Warsaw University of Technology: Warsaw, Poland, 2013. (In Polish)
24. Li, Y.; Nejabatkhah, F. Overview of control, integration and energy management of microgrids. *J. Mod. Power Syst. Clean Energy* **2014**, *2*, 212–222. [CrossRef]
25. Olivares, D.E.; Canizares, C.A.; Kazerani, M. A Centralized Energy Management System for Isolated Microgrids. *IEEE Trans. Smart Grid* **2014**, *5*, 1864–1875. [CrossRef]
26. Morstyn, T.; Hredzak, B.; Agelidis, V.G. Control strategies for microgrids with distributed energy storage systems: An overview. *IEEE Trans. Smart Grid* **2018**, *9*, 3652–3666. [CrossRef]
27. Lopes, J.A.P.; Moreira, C.L.; Madureira, A.G. Defining Control Strategies for MicroGrids Islanded Operation. *IEEE Trans. Power Syst.* **2006**, *21*, 916–924. [CrossRef]
28. Parol, M.; Rokicki, Ł.; Parol, R. Towards optimal operation control in rural low voltage microgrids. *Bull. Pol. Ac. Tech.* **2019**, *67*, 799–812.
29. Parol, M.; Kapler, P.; Marzecki, J.; Parol, R.; Połecki, M.; Rokicki, Ł. Effective approach to distributed optimal operation control in rural low voltage microgrids. *Bull. Pol. Ac. Tech.* **2020**, *68*, 661–678.
30. Zakir, M.; Sher, H.A.; Arshad, A.; Lehtonen, M. A fault detection, localization, and categorization method for PV fed DC-microgrid with power-sharing management among the nano-grids. *Int. J. Electr. Power Energy Syst.* **2022**, *137*, 107858. [CrossRef]
31. Sharma, V.; Chandel, S.S. Performance and degradation analysis for long term reliability of solar photovoltaic systems: A review. *Renew. Sust. Energ. Rev.* **2013**, *27*, 753–767. [CrossRef]
32. Makrides, G.; Norton, M.; Georghiou, G.E. Performance of Photovoltaics Under Actual Operating Conditions. *Third Gener. Photovolt.* **2012**, 201–232. [CrossRef]

33. Dierauf, T.; Growitz, A.; Kurtz, S.; Cruz, J.L.B.; Riley, E.; Hansen, C. *Weather-Corrected Performance Ratio*; Technical Report; NREL/TP-5200-57991; National Renewable Energy Lab.(NREL): Golden, CO, USA, April 2013.
34. Nordmann, T.; Clavadetscher, L. Understanding temperature effects on PV system performance. In Proceedings of the 3rd IEEE World Conference on Photovoltaic Energy Conversion, Osaka, Japan, 11–18 May 2003; pp. 2243–2246.
35. Virtuani, A.; Pavanello, D.; Friesen, G. Overview of Temperature Coefficients of Different Thin Film Photovoltaic Technologies. In Proceedings of the 25th EUPVSEC Conference, Valencia, Spain, 6–10 September 2010. [CrossRef]
36. Skoplaki, E.; Palyvos, J.A. On the temperature dependence of photovoltaic module electrical performance: A review of efficiency/power correlations. *Sol. Energy* **2009**, *83*, 614–624. [CrossRef]
37. Makrides, G.; Theristis, M.; Bratcher, J.; Pratt, J.; Georghiou, G.E. Five-year performance and reliability analysis of monocrystalline photovoltaic modules with different backsheet materials. *Sol. Energy* **2018**, *171*, 491–499. [CrossRef]
38. Taylor, N. *Traceable Performance Measurements of PV Devices*; European Union: Luxembourg, 2010. [CrossRef]
39. Polo, J.; Alonso-Abella, M.; Ruiz-Arias, J.A.; Balenzategui, J.L. Worldwide analysis of spectral factors for seven photovoltaic technologies. *Sol. Energy* **2017**, *142*, 194–203. [CrossRef]
40. Rahmann, C.; Vittal, V.; Ascui, J.; Haas, J. Mitigation Control Against Partial Shading Effects in Large-Scale PV Power Plants. *IEEE Trans. Sustain. Energy* **2016**, *7*, 173–180. [CrossRef]
41. Géron, A. *Hands-on Machine Learning with Scikit-Learn, Keras, and TensorFlow*, 2nd ed.; O'Reilly Media, Inc.: Sevastopol, CA, USA, 2019.
42. Piotrowski, P.; Baczyński, D.; Kopyt, M.; Gulczyński, T. Advanced Ensemble Methods Using Machine Learning and Deep Learning for One-Day-Ahead Forecasts of Electric Energy Production in Wind Farms. *Energies* **2022**, *15*, 1252. [CrossRef]
43. Dudek, G.; Pełka, P. Forecasting monthly electricity demand using k nearest neighbor method. *Przegląd Elektrotechniczny Electr. Rev.* **2017**, *93*, 62–65.
44. Piotrowski, P.; Baczyński, D.; Kopyt, M.; Szafranek, K.; Helt, P.; Gulczyński, T. Analysis of forecasted meteorological data (NWP) for efficient spatial forecasting of wind power generation. *Electr. Power Syst. Res.* **2019**, *175*, 105891. [CrossRef]
45. Dudek, G. Multilayer perceptron for short-term load forecasting: From global to local approach. *Neural Comput. Appl.* **2019**, *32*, 3695–3707. [CrossRef]
46. Osowski, S.; Siwek, K. Local dynamic integration of ensemble in prediction of time series. *Bull. Pol. Ac. Tech* **2019**, *67*, 517–525.
47. Karnik, N.; Mendel, J.; Liang, Q. Type-2 fuzzy logic systems. *IEEE Trans. Fuzzy Syst.* **1999**, *7*, 643–658. [CrossRef]
48. Karnik, N.N.; Mendel, J.M. Centroid of a type-2 fuzzy set. *Inf. Sci.* **2001**, *132*, 195–220. [CrossRef]
49. Mendel, J.; John, R. Type-2 fuzzy sets made simple. *IEEE Trans. Fuzzy Syst.* **2002**, *10*, 117–127. [CrossRef]
50. Khosravi, A.; Nahavandi, S.; Creighton, D.; Srinivasan, D. Interval Type-2 Fuzzy Logic Systems for Load Forecasting: A Comparative Study. *IEEE Trans. Power Syst.* **2012**, *27*, 1274–1282. [CrossRef]
51. Liang, Q.; Mendel, J.M. Interval type-2 fuzzy logic systems: Theory and design. *IEEE Trans. Fuzzy Syst.* **2000**, *8*, 535–550. [CrossRef]
52. Hastie, T.; Tibshirani, R.; Friedman, J.H. *The elements of Statistical Learning. Data Mining, Inference, and Prediction*, 2nd ed.; Springer: Berlin/Heidelberg, Germany, 2009.

energies MDPI

Article

Reliability Analysis of MV Electric Distribution Networks Including Distributed Generation and ICT Infrastructure

Miroslaw Parol [1], Jacek Wasilewski [2], Tomasz Wojtowicz [1], Bartlomiej Arendarski [3,*] and Przemyslaw Komarnicki [4]

1 Institute of Electrical Power Engineering, Faculty of Electrical Engineering, Warsaw University of Technology, Koszykowa 75 Street, 00-662 Warsaw, Poland; miroslaw.parol@pw.edu.pl (M.P.); tomasz.wojtowicz@pw.edu.pl (T.W.)
2 PSE S.A., Warszawska 165 Street, 05-520 Konstancin-Jeziorna, Poland; jacek.wasilewski@pse.pl
3 Fraunhofer Institute for Factory Operation and Automation IFF, D-39106 Magdeburg, Germany
4 Department of Engineering and Industrial Design, University of Applied Sciences Magdeburg-Stendal, D-39114 Magdeburg, Germany; przemyslaw.komarnicki@h2.de
* Correspondence: bartlomiej.arendarski@iff.fraunhofer.de; Tel.: +49-391-4090-145

Abstract: In recent years, the increased distributed generation (DG) capacity in electric distribution systems has been observed. Therefore, it is necessary to research existing structures of distribution networks as well as to develop new (future) system structures. There are many works on the reliability of distribution systems with installed DG sources. This paper deals with a reliability analysis for both present and future medium voltage (MV) electric distribution system structures. The impact of DG technology used and energy source location on the power supply reliability has been analyzed. The reliability models of electrical power devices, conventional and renewable energy sources as well as information and communications technology (ICT) components have been proposed. Main contribution of this paper are the results of performed calculations, which have been analyzed for specific system structures (two typical present network structures and two future network structures), using detailed information on DG types, their locations and power capacities, as well as distribution system automation applied (automatic stand-by switching on—ASS and automatic power restoration—APR). The reliability of the smart grid consisting of the distribution network and the coupled communications network was simulated and assessed. The observations and conclusions based on calculation results have been made. More detailed modeling and consideration of system automation of distribution grids with DG units coupled with the communication systems allows the design and application of more reliable MV network structures.

Keywords: distribution of electric power; distributed storage and generation; smart grids; power distribution reliability; information and communication technology

Citation: Parol, M.; Wasilewski, J.; Wojtowicz, T.; Arendarski, B.; Komarnicki, P. Reliability Analysis of MV Electric Distribution Networks Including Distributed Generation and ICT Infrastructure. *Energies* **2022**, *15*, 5311. https://doi.org/10.3390/en15145311

Academic Editor: Abu-Siada Ahmed

Received: 6 June 2022
Accepted: 14 July 2022
Published: 21 July 2022

1. Introduction

The increasing penetration of distributed energy generation (DG) from renewable energy sources (RES) contributes to a decrease in greenhouse gases emission and reduces the dependency on fossil energy sources. At the same time, however, this trend means the electric power networks cannot continue to operate as before. The power grids were originally designed for the classical, hierarchical system with a unidirectional power flow from the central generation, through transmission and distribution level up to the loads. DGs nowadays largely feed directly to the distribution networks, which were not designed for this purpose. Therefore, the planning, operation and maintenance of distribution networks need to be changed.

In power system planning and operation, effective reliability analysis and assessment are key aspects. The reliability of the electric power system is usually expressed as a measure of the ability of the system to provide the customers with a sufficient supply. Continuous energy supply is one of the most important success criteria of a power system. However, the occurrence of major outages can have a significant economic impact on electricity suppliers and the end users who lose electrical service. Competition on the power market forces utilities to reduce costs through, for example, postponing preventive maintenance or replacing equipment only when it has already broken down [1].

There are many studies on reliability of distribution systems with DG units (including RESs) which are connected to them. Large part of those studies concerns the evaluation of the reliability of distribution networks with distributed generation (DG) units installed. A reliability model for distributed generations and an analytical probabilistic approach to investigate impacts of DG units on reliability of electric power distribution grid is proposed in [2]. An analytical technique using explicit expressions for this purpose is studied in [3]. In turn, paper [4] describes the impact of DG units on the radial distribution grid reliability using the analytical as well as Monte Carlo Simulation methods. A probabilistic technique for the evaluation of the distribution network reliability by means of some specific methods used for the estimation of wind speed profile is presented in [5]. Paper [6] describes a Monte Carlo method for the needs of a reliability assessment of distribution systems with distributed generation sources installed with the use of parallel computing. Different scenarios concerning the impact of photovoltaic systems on performance of a test system are analyzed. In paper [7], optimal coordination of distributed generation sources, energy storage and demand management techniques, in the context of a reliability assessment of distribution grids, is presented. The main goal of this action is to maximize the network reliability. Paper [8] addresses the reliability assessment of distribution systems with renewable energy sources (wind and PV units) installed in order to minimize power losses in the systems. An integrated approach for the needs of assessing the influence distributed energy sources, including PV installations, on the reliability performance of power grids is presented in [9]. The modified Monte Carlo method is used for this purpose. Paper [10] presents the problem of the optimization of a hybrid photovoltaic—battery system sizing. A genetic algorithm is used for addressing the reliability in considered grids.

Reliability analyses concerning distribution systems also appear in other various issues. For example, this analysis can be a part of the electric power distribution grids planning process, as it was presented in [11]. In turn, paper [12] describes an approach allowing for the evaluation of reliability indices of a distribution grid for some specific operation practices, i.e., use of telecontrolled switches and islanded operation mode. Paper [13] presents the problem of distribution system reconfiguration optimization in a multi-criteria category utilizing a set of well-known reliability indices for this purpose. Another issue is an extension of the distribution grid reliability evaluation by including electric vehicles in different modes of grid operation [14]. The reliability issue of the information and power terminal to be used in disaster scenarios as a small-scale microgrid, which includes PV generation, battery storage, loads, electric vehicle and ICT components is considered in [15]. In paper [16], a comprehensive review on the smart grid research is presented. The recent achievements in the field of network reliability are described. Paper [17] presents a deep neural network ensemble model for the needs of estimation of outages in an overhead power distribution grid. The neural networks creating the ensemble are trained by a novel algorithm.

Many works are devoted to reliability evaluations in microgrids. An analytical method for the evaluation of the customer's supply reliability in a microgrid, which includes DG units, is presented in [18]. The optimal operation control based on centralized control logic in microgrids functioning in synchronous and islanded mode are introduced in [19], which can have an impact on improvement of supply reliability for consumers connected to these microgrids. Paper [20] describes the impact of operating conditions and protection systems on the microgrid reliability indices. In paper [21], an efficient control to manage power in microgrids with energy storage is proposed. The control system, developed in Real Time Digital Simulator, improves the reliability and resiliency of the microgrid consisting of photovoltaic installations, battery storage, diesel generator and controllable loads.

There are several papers concerning possible cooperation of distribution systems and microgrids in the context of reliability. In paper [22], the influence of microgrids on the distribution grid reliability has been discussed. An analytical method for the evaluation of reliability of the distribution grid in a network environment of multi-microgrids is discussed in [23]. Paper [24] describes a novel method for determining the optimal location and size of micro-grid systems to improve the continuity of supply in radial distribution networks in rural areas. The microgrids are used for reducing the non-served energy, taking into account the reliability and investment costs. In turn, paper [25] presents a method allowing for evaluating the reliability of active distribution grids with multiple microgrids using a Monte Carlo approach. A review and classification of the state-of-the-art of reliability assessment in the case of microgrids connected to distribution grids is presented in [26].

A very important issue for network reliability studies is having accurate models of DG units, particularly models of renewable energy sources. A model used for the purpose of wind farms probabilistic representation for reliability investigations is described in [27]. In paper [28], a review of thirteen wind turbine reliability studies is presented. Paper [29] presents a model allowing for evaluation of generation availability in the case of small hydro power plants.

In the grids with a large share of distributed generation, mainly renewable sources, the additional information and communications technology (ICT) to monitor, control and protect these power system components is applied. This additional ICT smooths the transition from conventional power systems to smart grids. However, it increases the complexity of such integrated systems, thus necessitating new methods for the planning and the optimal integration of advanced communication systems in electric power grids.

A comprehensive overview on smart grids and their technical, management, security, and optimization aspects is given in [30]. In addition to the definition of electrical components, much emphasis is placed on communication, protocols, architecture and security as well as optimization using cloud computing infrastructure, web application scheme as well as information flows and agent clusters. The impact of automation and communication technology on the reliability of the electric distribution systems is given in [31,32]. This facilitates analysis and modeling of coexisting ICT infrastructures on power grid reliability [33,34] and on smart grids altogether. Cooperation between the communication layer and the electrical network and the resulting coupled subsystem, along with the proposal of a multi-agent system for cooperative control of microgrids are mathematically modeled in [35].

A reliability perspective of the smart grid and critical overview of the reliability impacts of major smart grid resources, such as renewables, demand response and storage are given in [36]. This article provides a grid-wide IT architectural framework to meet the reliability challenges that are further enhanced by the ideal mix of these resources leading to a flatter net demand. An optimal control of smart grid including distributed generation and telecommunications and, in particular, smart power substations for improving the network parameters and reliability is given in [37].

The issue of cyber security in networks using the SCADA system are considered in [38], where four attack scenarios for cyber components, which may trip breakers of physical components, are analyzed. In [39], models of cascading failures and uncertainty on the supply side are proposed, followed by an assessment of the reliability of cyber-physical power systems. Cyberpower grids based on IEEE 14-bus and 39-bus system with control centers and corresponding communication networks are tested for false data injection attacks and defense mechanisms in [40].

Communication requirements, specifications, functions and applications in advanced electric power grids are summarized in [41]. An overview of communication standards and protocols, available technologies, data transfer methods, and future development trends is given in [42,43]. The ICT is used for bi-directional data transmission from the monitoring and control of devices to the control center where an operator with an appropriate computer application and algorithms can analyze these data and perform effective monitoring, control and protection of the system [44]. The performance of IEC 61,850 messages in LTE communication for reactive power management in a microgrid is analyzed in [45]. The ICT also provides communication between markets, forecast applications and web services for the customers, which supports the management of the demand and supply process [46].

The issue of reliability of electric distribution systems with DG sources installed in them was broadly discussed in many publications. Quite often there is a lack of any detailed information on distribution network structures (parameters of distribution transformers, data on overhead lines and underground cables) and on DG source types, as well as their power capacities and locations, considered in the existing papers is observed. Moreover, the details on the reliability parameters of distribution system components being considered in these publications are often missing. We intend to present such details in this paper. We are convinced there is still a research space to present different, more detailed studies on reliability of electric distribution systems with integrated DG sources (for various network structures and data describing them) as well as ICT components.

This paper concentrates on the reliability analysis for both present and future electric distribution system structures. Two present electric distribution system structures are considered: a typical urban distribution network (UDN) and typical rural distribution network (RDN). Moreover, two future electric distribution system structures are analyzed: the urban distribution network with connected microgrids (DNMG) and active managed distribution network (AMDN). The impact of DG technology used, energy source locations, and their power capacity on the power supply reliability have been analyzed. The reliability models of electrical power devices, conventional and renewable energy sources, as well as information and communication technology (ICT) components have also been proposed.

The main contribution of this paper is investigating the analyzed subject in a more thorough way, that is: giving detailed data on considered distribution networks structures; on reliability parameters of distribution network components; on DG source types, as well as their power capacities and locations, for which the reliability calculations have been made, taking into account distribution system automation (automatic stand-by switching on (ASS) and automatic power restoration (APR)); presenting the results (seven commonly known reliability indices) achieved from the carried out computations and discussing the results (indices). The impact of DG type on these reliability indices has been investigated. It is worth noting, the reliability assessment of smart grid, i.e., electric power network coupled with the communication network, has also been done. We would like to highlight that this paper relates to MV distribution grids, for which reliability indices are the worst among all electric power distribution networks, as it is reported e.g., in [47–49]. Therefore, the importance of this paper on practical applications can be seen.

This paper evaluates reliability of four electric distribution system structures (two present ones and two future ones) and presents the reliability indices obtained for these structures. In our opinion, more detailed modeling and consideration of system automation of the distribution grids with DG units coupled with the communication systems allows for the design and application of such MV network structures for which the best reliability indices can be obtained.

2. Problem Statement

The main goal of this research was to analyze the impact of a type and location of DG units in present and future distribution network structures on the power supply reliability. The distribution system automation (ASS and APR) has also been considered in the studies. The analysis performed by the authors of this paper has been done for the benchmark structures of the MV distribution networks with the connected DG sources. The benchmark structures (shown in Figures A1, A3 and A4) have also been developed by the authors of this paper. The following reliability indices have been calculated with the use of DIgSILENT PowerFactory software [50,51]:

- SAIFI is the System Average Interruption Frequency Index, which provides the average number of interruptions, above 3 min, in the system that a customer experiences during the observation period, mostly in one year. The index is a dimensionless number and can be calculated as follows [52,53]:

$$\text{SAIFI} = \frac{\text{Total number of customer interruptions}}{\text{Total number of customers served}} = \frac{\sum_i \text{N}_i}{\text{N}_\text{T}} [1/\text{yr}] \quad (1)$$

where N_i is the number of customers interrupted by *i*-th outage in the observation period and N_T is the total number of customers in considered system.

- CAIFI (Customer Average Interruption Frequency Index)—total number of all interruptions, above 3 min, divided by the total number of consumers affected by an interruption in the analyzed system. CAIFI can be calculated as follows [53,54]:

$$\text{CAIFI} = \frac{\text{Total number of customer interruptions}}{\text{Total number of customers affected}} = \frac{\sum_i \text{N}_i}{\text{CN}} [1/\text{yr}] \quad (2)$$

where CN is the total number of consumers, which experienced one or more outages.

- SAIDI is the System Average Interruption Duration Index, and it measures the total duration of an interruption, above 3 min, for the average customer during a given time period. It is normally calculated for the period of one year and presents customer minutes or hours of interruption. Mathematical representation of SAIDI is given in Equation (3) [52,53]:

$$\text{SAIDI} = \frac{\sum \text{Customer interruption durations}}{\text{Total number of customers served}} = \frac{\sum_i \text{r}_i \text{N}_i}{\text{N}_\text{T}} [\text{hr}/\text{yr}] \quad (3)$$

where r_i is restoration time and failure duration in the case of consumers interrupted by *i*-th outage.

- CAIDI, the Customer Average Interruption Duration Index represents the average time required to restore service after an outage occurs, which indicates how long an average interruption, above 3 min, lasts. It measures the duration of time that the customer is de-energized per interruption. To calculate the index Equation (4) can be used [52,53]:

$$\text{CAIDI} = \frac{\sum \text{Customer interruption durations}}{\text{Total number of customers interruptions}} = \frac{\sum_i \text{r}_i \text{N}_i}{\sum_i \text{N}_i} [\text{hr}] \quad (4)$$

- ASAI (Average Service Availability Index)—the probability of having all loads supplied. The index is often expressed in a percentage, and it can be calculated from Equation (5) [52,53]:

$$\text{ASAI} = \frac{\text{Customer hours service availability}}{\text{Customer hours service demand}} = \frac{\text{N}_\text{T} \cdot (T) - \sum_i \text{r}_i \text{N}_i}{\text{N}_\text{T} \cdot (T)} [\text{pu}] \quad (5)$$

where T is the observation time period, usually one year, and in a non-leap year is equal to 8760 h.

- ASUI (Average Service Unavailability Index)—the probability of having one or more loads interrupted, which can be calculated as follows [54]:

$$\text{ASUI} = \frac{\text{Customer hours service unavailability}}{\text{Customer hours service demand}} = 1 - \text{ASAI } [\text{pu}] \quad (6)$$

- EENS (Expected Energy Not Supplied)—the total amount of energy which is expected not to be delivered to loads. The index can be calculated from the Equation (7) [53,55]:

$$\begin{aligned} \text{EENS} &= \sum (\text{Customer annual outage time} \cdot \text{connected power}) \\ &= \sum_i \text{r}_i \text{P}_{\text{ave},i} \ [\text{MWh}/\text{yr}] \end{aligned} \quad (7)$$

where $\text{P}_{\text{ave},i}$—the average active power of customers which is interrupted by *i*-th outage.

A further aim of this research was to analyze the impact of ICT components integrated with the power system on the overall reliability of the smart grid supply. Therefore, a basic distribution power supply system was proposed, for which simulations using the sequential Monte Carlo method were carried out. The following reliability indices have been calculated along with the distribution of the results with the use of Matlab software: SAIDI, SAIFI, CAIDI, EENS, ASAI and ASUI.

3. Reliability Models of Electric Distribution System Components

The operation of an electric power system component can be described as a stochastic process $\{X_t\colon t \in T \wedge X_t\colon \Omega \to S\}$, where T is the life cycle time (continuous value), Ω is the space of coexisted events with the operating process of the system element and $S = \{s_1, s_2, \ldots, s_m\}$ is the finite set of discrete operational states of the system component [56]. According to the element operation types, the states can be functional (full or partial one), stand-by and nonfunctional (failure or planned repair mode, etc.). The transitions between the component states may be caused by random events (failures and repairs), deterministic events (preventive repairs in a scheduled time) and random-deterministic events (conditional realization of preventive repairs).

There are many types of recommended mathematical techniques used in the reliability analysis [50]. Among them one can find as follows:

- Minimal cut-set,
- Zone branch,
- Fault tree,
- Discrete event simulation (Monte Carlo),
- Boolean algebra,
- Failure mode effects and criticality analysis (FMECA),
- Markov and semi-Markov models,
- Stochastic Petri nets.

The information about the accuracy and applicability of the aforementioned techniques can be found in many meaningful publications, e.g., [57].

3.1. Electrical Power Devices

The elements of an electric distribution system, such as lines, transformers, power switches, busbar of switchgears, protection and control elements are modeled as objects, which can be functional or in failure state. The time between these states is represented as a random variable described by an adequate type of probability distribution. In reliability analyses, the following probability distribution types are most often used [57,58]:

- Exponential EXP (λ), $\lambda > 0$, λ—the rate parameter;
- Weibull WEI (λ, β), $\lambda > 0$, $\beta > 0$, λ—the scale parameter, β—the shape parameter;
- Gamma GAM (b, p), $b > 0$, $p > 0$, b—the rate (scale) parameter, p—the shape parameter;
- Normal NOR (μ, σ), $\mu \geq 0$, $\sigma > 0$, μ—the expected value, σ—the standard deviation;
- Pareto PAR (b, δ), $b > 0$, $\delta > 0$, b—the scale parameter, δ—the shape parameter;
- Gumbel GUM (b_0, t_0), $b_0 > 0$, $t_0 \geq 0$, b_0—the scale parameter, t_0—the location parameter;
- Log-normal LNOR (μ_0, σ_0), $\mu_0 \geq 0$, $\sigma_0 > 0$, μ_0—the expected value of natural logarithm, σ_0—the standard deviation of natural logarithm.

Additionally, one can take into consideration the third state that is a preventive repair state with the average annual maintenance duration. A two-state reliability model can be assumed for the MV networks belonging to Polish distribution system operators. Table 1 presents the reliability models of different types of electrical power devices. All the probability distribution parameters have been estimated based on the observations of Polish power distribution systems [59,60].

It is also necessary to determine the adequate reliability model of an equivalent point supplying an analyzed distribution system. The reliability characteristics of that point can be found by an assessment on the power transmission system level or statistical research.

Some reliability analyses take into account the separate characteristics of protection devices (fuses, relays, releases, etc.) and automation equipment (automatic reclosing, stand-by switching on and others) [61].

Table 1. Reliability parameters of selected types of MV distribution system elements; elaborated on the basis of [59,60].

Distribution System Component	Probability Distribution of Time to Failure			Probability Distribution of Time to Repair		
	Type	Parameters		Type	Parameters	
Overhead lines (100 km)	EXP	$\lambda = 25.0$ 1/a		EXP	$\lambda = 0.071$ 1/h	
Underground cable with paper-insulation used in urban networks (100 km)	WEI	$\lambda = 12.0$ 1/a	$\beta = 2.24$	EXP	$\lambda = 0.69$ 1/h	
Underground cables with polythene-insulation used in urban networks (100 km)	WEI	$\lambda = 12.0$ 1/a	$\beta = 1.85$	EXP	$\lambda = 0.72$ 1/h	
Underground cables with paper-insulation used in rural networks (100 km)	WEI	$\lambda = 13.0$ 1/a	$\beta = 2.33$	EXP	$\lambda = 0.30$ 1/h	
Underground cables with polythene-insulation used in rural networks (100 km)	WEI	$\lambda = 7.2$ 1/a	$\beta = 1.20$	EXP	$\lambda = 0.30$ 1/h	
110 kV/MV transformers	EXP	$\lambda = 0.06$ 1/a		EXP	$\lambda = 0.08$ 1/h	
Circuit-breakers	EXP	$\lambda = 0.132$ 1/a		NOR	$\mu = 5.5$ h	$\sigma = 1.0$ h
Disconnectors	EXP	$\lambda = 0.0055$ 1/a		EXP	$\lambda = 0.115$ 1/h	
Busbars in MV switchgear (bay)	EXP	$\Lambda = 0.003$ 1/a		EXP	$\lambda = 0.10$ 1/h	
Current transformers	EXP	$\Lambda = 0.009$ 1/a		EXP	$\lambda = 0.047$ 1/h	

3.2. Distributed Generation Sources

From the point of view of a modeling and reliability assessment, DG sources can be divided into two classes:

- Sources based on conventional energy carriers, such as diesel oil, gas, biogas, etc.,
- Sources based on renewable energy carriers, such as wind, solar radiation, water, etc.

In the first case, the availability of the energy source for generation is highly probable. On the other hand, the availability of the renewable energy resources (second group) requires considering more appropriate probabilistic models [62].

3.2.1. Conventional Energy Sources

The conventional electric energy sources are:

- Engine-driven generators,
- Turbine-driven generators,
- Microturbines,
- Fuel cells.

Depending on the type of service, the aforementioned energy sources can be modeled as the following Markov chain:

- Two-state model—in case of continuous service; determined by the failure rate λ and the repair rate μ,
- Four-state model—in case of peak service; shown in Figure 1

Both the standby anticipation rate ρ and operation rate ν should be determined individually depending on the analyzed electric distribution system. The reliability parameters of engine-driven generators (EDG) and turbine-driven generators (TDG) can be found in Table 2. The values of parameters ν and ρ have been arbitrarily selected.

One can observe a much lower failure rate λ for generation units (both EDG and TDG) operating in peak service. This fact obviously results from less wear of individual components of a generation unit. Therefore, the possibility of failure is decreased. The TDG exhibit the lowest failure rates of units in peak service. Simultaneously, the lower repair rate is observed in comparison to other cases. It is attributable to a relatively small number of long-duration events.

In recent years, a new type of gas turbine, microturbine (MT), has become a fully developed technology. As MT's have only relatively recently been used as the commercial generation sources, there is not wide access to reliability data obtained from a long-time operation of this DG type. The same problem concerns the fuel cells (FC) as a relatively new technology in an industrial and commercial usage.

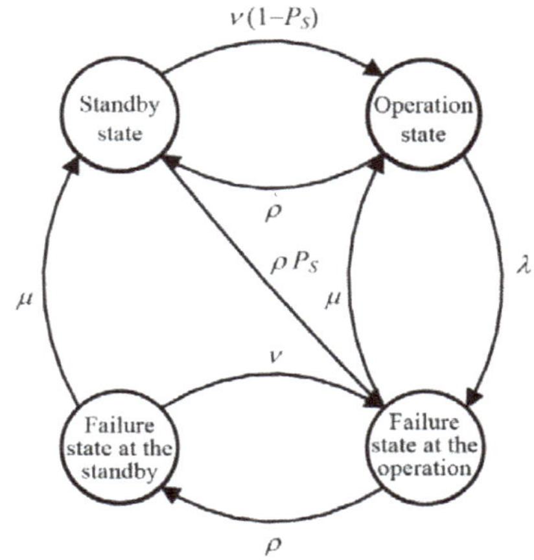

Figure 1. Four-state reliability model of a conventional generation unit operating in peak service. P_S is a probability value of unsuccessful unit starting, ν is an operation rate, ρ is a standby rate, λ is a failure rate and μ is a repair rate [63].

Table 2. Reliability parameters of EDG and TDG [64].

Parameter	Continuous Service		Peak Load Service	
	EDG	**TDG**	**EDG**	**TDG**
λ [1/a]	4.30	4.50	0.90	0.30
μ [1/h]	0.15	0.14	0.26	0.009
$(1 - P_S)$ [-]	-	-	0.912	0.912
ν [1/a]	1.33	1.33	1.33	1.33
ρ [1/a]	4.0	4.0	4.0	4.0

As in case of EDG and TDG, MT and FC can be modeled using a two-state or four-state Markov chain with the failure rate λ and repair rate μ as well as a probability of unsuccessful DG unit starting P_S. For the reliability calculation purpose λ, μ and P_S have been obtained from the manufacturers data available only for peak service. The same values of parameters ν and ρ as for EDG and TDG have been assumed. All of these are presented in Table 3.

Table 3. Reliability parameters of MT and FC.

Parameter	MT	FC
λ [1/a]	0.40	0.80
μ [1/h]	0.013	0.05
$(1 - P_S)$ [-]	0.961	0.976
ν [1/a]	1.33	1.33
ρ [1/a]	4.0	4.0

The reliability parameters presented in Table 3 have been obtained based on the data given by different manufacturers (catalogues and brochures). It is necessary to treat these values a little distrustfully. These reliability parameters come from laboratory research, which cannot reflect the real conditions in an operating process.

3.2.2. Renewable Energy Sources

Among the most popular renewable energy sources in electric distribution systems, there are small hydro power plants (SHPP), small wind-turbine power plants (WTPP) and photovoltaic power plants (PVPP).

The parameters of different energy carriers (i.e., a river flow, wind speed and solar radiation) can be modeled as the homogenous Markov chain with the states representing different intervals of available energy and the transition rates between λ_{ij} (transition rate from state i to state j). The general reliability model of a renewable generation system is shown in Figure 2.

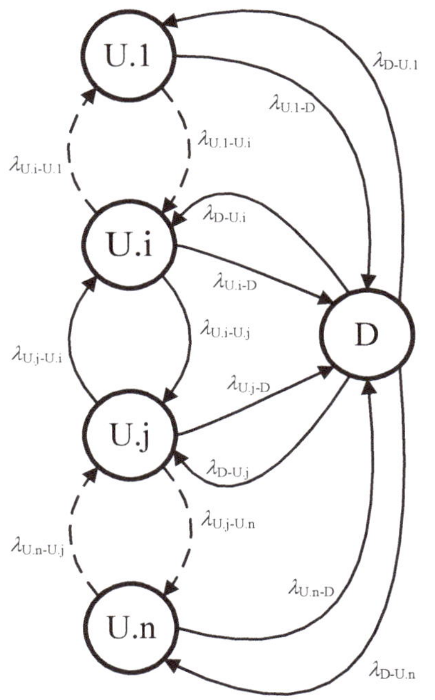

Figure 2. N + 1-state reliability model of a renewable generation system. *U.i* is an up unit state (normal operation) with *i*-th of *n* level of energy carrier $\{1, \ldots, i, j, \ldots, n\}$ and *D* is a down unit state (failure).

In order to represent the reliability of particular types of renewable energy sources, the authors of this paper have found an exemplary number of states, fraction of DG rated apparent power corresponding to the state and values of the transition rates between the states [29,62,65]. All of these are presented in Tables 4–6.

Table 4. Reliability parameters of WTPP [62].

	Fraction of DG Rated Apparent Power				
	0%	**60%**	**100%**	**0%**	**0%**
FROM\TO	U.1	U.2	U.3	U.4	D
U.1	0	200 1/a	0	0	4 1/a
U.2	20 1/a	0	400 1/a	0	4 1/a
U.3	0	300 1/a	0	10 1/a	4 1/a
U.4	0	0	500 1/a	0	24 1/a
D	90 1/a	90 1/a	90 1/a	24 1/a	0

Table 5. Reliability parameters of SHPP [29].

	Fraction of DG Rated Apparent Power					
	38%	**54%**	**66%**	**80%**	**100%**	**0%**
FROM\TO	U.1	U.2	U.3	U.4	U.5	D
U.1	0	11.9 1/a	0	0	0	1.8 1/a
U.2	1.7 1/a	0	9.4 1/a	0	0	1.8 1/a
U.3	0	4.4 1/a	0	5.2 1/a	0	1.8 1/a
U.4	0	0	27.2 1/a	0	3.6 1/a	1.8 1/a
U.5	0	0	0	27.7 1/a	0	1.8 1/a
D	88.2 1/a	88.2 1/a	88.2 1/a	88.2 1/a	88.2 1/a	0

Table 6. Reliability parameters of PVPP [65].

	Fraction of DG Rated Apparent Power			
	100%	**20%**	**0%**	**0%**
FROM\TO	U.1	U.2	U.3	D
U.1	0	1250 1/a	0	0.1 1/a
U.2	2250 1/a	0	2250 1/a	0.1 1/a
U.3	0	1250 1/a	0	0.1 1/a
D	146 1/a	146 1/a	146 1/a	0

It is necessary to mention the values of fraction the DG rated apparent power given in Table 4. The state U.1 concerns a situation when a wind speed is less than the cut-in wind speed of a wind turbine and the power generation equals 0. No power generation is also in the state U.4. In this case, the wind speed is greater than the cut-out wind speed when a wind turbine is switched off.

3.3. Information and Communication Devices

To implement future smart grid functions, the information and communication technology (ICT) is needed. The ICT devices integrated in a power system collect, process and transfer data within the infrastructure. This requires robust communication channels to ensure reliable data flow. For that they use different sorts of communication media, such as Power Line Communication (PLC), Digital Subscriber Lines (xDSLs), fiber optics, IEEE 802.11 (WLAN), IEEE 802.16 (WiMAX), GSM/GPRS, IEEE 802.15.4 (Zigbee), depending on application, technical characteristics and feasibility [41,66]. Several types of devices are installed in integrated communication networks in smart grids, such as phasor measurement unit (PMU), remote terminal unit (RTU), programmable logic controller (PLC), gateway, router, modem, Digital Protective Relay (DPR), Digital Fault Recorder (DFR), PQ meter and smart meter. The types of these devices depend on their application and tasks such as measuring the electrical parameters, controlling the automation systems, transferring collected data and resaving control signals from the control center applications. The comparison of ICT devices, protocols and typical functions in power system is given in Table 7.

Table 7. Comparison of ICT equipment in OSI model [67–69].

ICT Devices	Protocols	Data Type	Function	DOD Model	Layers	OSI Layers
Gateway RTU	HTTP, FTP, Telnet, SMB, SMTP, SNMP, DNS, DHCP, NFS	User Data	Provides services directly to the user applications; identifies communication participant; dedicates quality of service; verify user authentication and privacy and determines adequate resources availability	Process/Application	Application layers	7 Application
Gateway	HTTP, FTP, Telnet, SMTP, AFP, TDI	Encoded User Data	Performs data translations with encoding, formatting, compression and encryption services; provides a common interface for user applications			6 Presentation
Gateway	TCP, UDP, SPX, NetBEUI	Sessions	Establishes, maintains and terminates connections between applications; enable two applications to communicate over a network by opening a session and synchronizing the participating hosts			5 Session
Gateway RTU	IP, IPX, NWLink, NetBEUI	Datagram/Segments	Provides reliable data transfer between hosts including flow control, segmentation, multiplexing, error control, transmission confirmation; maintains quality of service functions	Host to host	Data flow layers	4 Transport
Gateway Router Relay device	IP, IPX, NWLink, NetBEUI	Datagram/Packets	Establishes, controls and terminates network connections; structures and manages multi-node networks including traffic control, addressing, routing, switching, forwarding, error handling and packet sequencing; translates logical network address into physical machine address	Internet		3 Network
Gateway Router RTU	IEEE802.2, LLC, Ethernet, FDDI, Token Ring, IEEE802.11, (WLAN, WiFi), PPP, DLC	Frames	Performs data transfer between nodes connected by a physical layer; divided into two sublayers: Media Access Control (MAC) layer provides data access and transmission, Logical Link Control (LLC) layer controls frames synchronization and error checking	Network Access		2 Data link
Gateway Router Modem RTU, PMU	IEEE802, IEEE802.2, ISO 2110, Ethernet, FDDI Token Ring	Bits	Defines physical specifications of connection, communication medium, transmission mode, connection protocol; controls transmission and reception of data bit streams over the physical medium			1 Physical

Communication network equipment can fail causing interruptions in data transfer, information exchange and other corresponding services. Both hardware and software can be affected for various reasons, impacting the reliability of the communication network. The reliability parameters like mean time between failure (MTBF) and mean time to repair (MTTR) for chosen devices: phasor measurement unit (PMU), remote terminal unit (RTU), programmable logic controller (PLC), gateway and router are given in Table 8. These values are calculated based on literature research on simulation models, laboratory tests and vendors' data presented in [70–72].

Table 8. Reliability of selected communication components; elaborated on the basis of [70–72].

Type	Modell	MTBF (yr)	MTTR (h)	A	U
PMU	Simulation	1.46350	23.1168	0.99820	0.0018
RTU	Calculation	11.0000	4.2000	0.99952	0.00048
PLC	Calculation	17.000	2.8000	0.99968	0.00032
Gateway	G650 Media	16.0009	8.4100	0.99994	0.00006
Router	Cisco 2811	34.2466	1.0000	0.99999	0.00001

where: A—availability, U—unavailability.

3.4. *Interdependencies Modeling of Coupled Electric Power System and ICT Infrastructures*

With the rise of smart grid technologies, the interdependencies of communication technologies and electric power systems become an important aspect in the development of both networks [73]. Modeling such interdependencies will be even more complex in the planning and future operation of multi-energy systems (MES) integrating various energy converters and sources of different physical nature [74].

Infrastructures interdependencies are based on physical and functional relationships among individual components both within and between systems. To characterize the effects of failure propagation from the single component or system to mutually dependent interconnected systems the structure modeling of complex infrastructures can be used [75]. The individual operating conditions of the component in the system can be analyzed and the fault propagation can be reduced by having fast recognition of threats, redundancy design and alternative modes of operation [76]. The concept of complex networks theory [77], which is based on the graph theory, can be used to describe and analyze critical infrastructures on a large scale with multifaceted topologies [78]. The interdependency modeling techniques of coupled infrastructures for integrating ICT within the electric power system (EPS) are offered in [33,34].

A graph can represent a network with its set of components and connections between them. Applying graph representation to the coupled EPS and ICT infrastructures, the vertices indicate system components such as buses, gateways and routers while edges correspond to the power lines, cables and communication links. In order to characterize the interdependencies between the infrastructures they can be classified as follows:

- Connection of an electric node with another electric node, which characterize typical power flow in the electric power grid;
- Connection of a communication node with another communication node that corresponds to a data flow;
- Connection of an electric node to a communication node that represents an electricity supply for the ICT infrastructure;
- Connection of a communication node to an electric node which is responsible for sending and requesting information to and from the power system component.

3.5. *Tool for Reliability Analysis*

The reliability assessment in electric power distribution systems has been carried out with the use of DIgSILENT PowerFactory software (PF). This software enables an assessment of different reliability indices for power systems in a generation area (hierarchical level HL I) as well as in transmission and distribution system (hierarchical levels HL II and HLIII adequately).

The procedure of a reliability assessment in PF is shown in Figure 3, according to [51].

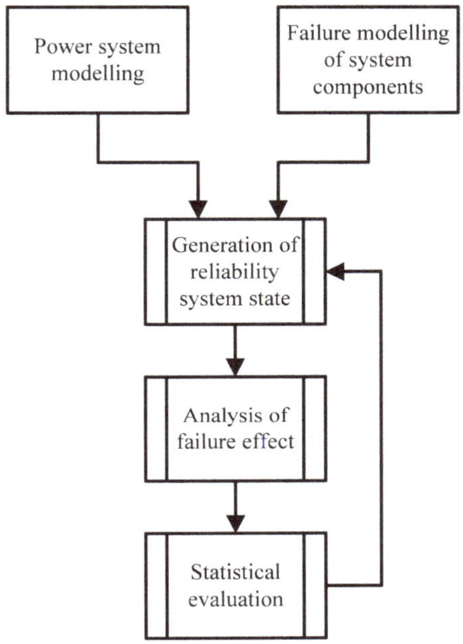

Figure 3. Flow diagram of the reliability assessment in PF software.

The first step is modeling an electric power network structure where technical requirements are met (no overloading, acceptable voltage deviations, etc). For all main network components, the failure models are defined by giving a description of the appropriate probability distributions.

The next stage of the reliability assessment is to generate a list of system states relevant with the failure models and load models. In other words, it is a combination of one or more simultaneous faults and a specific load condition. For each system state, some defined power system reactions are analyzed such as:

- Automatic stand-by switching on (ASS),
- Fault clearance by using protection equipment,
- Automatic power restoration (APR) by opening separating switches (fault separation) and closing normally open switches.

Finally, the system state generation combined with the failure effect analysis updates the calculation of statistic indices. The detailed description of the used algorithm in PF can be found in [79].

4. Reliability Analysis of Electric Power Network

4.1. Assumptions and Limitations

The urban and rural structures of present electric distribution networks (UDN, RDN) with distributed generation have been investigated. Different structures of future distribution networks with embedded generation have been also analyzed, such as: active managed MV distribution network (AMDN) and MV distribution network with connected LV microgrids (DNMG).

The diagrams and parameters of all the basic structures (as a starting point of analysis) are shown in detail in Appendices A–D.

In Poland any part of the electric distribution system (both MV and LV network) controlled by distribution system operators cannot operate in islanded mode. The autonomous operation is admitted only for power networks and installations belonging to a consumer. For that restriction, the islanded operation of UDN and RDN is not admitted.

In the loop structures, i.e., UDN and DNMG, the ASS as well as the APR have been considered in the reliability assessment. All the MV distribution system models do not take into consideration a possibility of power supply reserve (e.g., with the use of ASS) on the level of LV distribution network

For all investigated DG sources, the reliability models described in Section 3.2 have been assumed. The reliability models of the energy storages (chemical battery and flywheel) have been not considered in this analysis.

4.2. Results of Test Calculations

Before the reliability indices have been calculated, the load flow analysis had been carried out for all investigated distribution network structures. Branch overloading and the excess of permissible voltage deviation in nodes has not been observed.

All the calculated system reliability indices for all considered present and future distribution networks are presented in Tables 9–12.

Table 9. Reliability system indices for UDN structure (Figure A1).

Variants: UDN_1, UDN_2, UDN_3, ASS—Busbars 1–2, 14–16, 15–17	
Index	**Case_1, Case_2, Case_3**
SAIFI [1/a]	0.731679
CAIFI [1/a]	0.731679
SAIDI [h]	2.684
CAIDI [h]	3.668
ASAI [-]	0.9996936566
ASUI [-]	0.0003063434
EENS [MWh/a]	42.363
Variants: UDN_1, UDN_2, UDN_3, ASS and APR—all busbars and terminals	
SAIFI [1/a]	0.380772
CAIFI [1/a]	0.380772
SAIDI [h]	1.603
CAIDI [h]	2.191
ASAI [-]	0.9998169754
ASUI [-]	0.0001830246
EENS [MWh/a]	25.449

Table 10. Reliability system indices for RDN structure (Figure A3).

Variants: RDN_1, RDN_2, RDN_3	
Index	**Case_1, Case_2, Case_3**
SAIFI [1/a]	0.806104
CAIFI [1/a]	0.806104
SAIDI [h]	8.306
CAIDI [h]	10.304
ASAI [-]	0.9990518242
ASUI [-]	0.0009481758
EENS [MWh/a]	72.132

Table 11. Reliability system indices for DNMG structure (Figure A4).

	Variant: DNMG_1		
Index	**Case_1**	**Case_2**	**Case_3**
SAIFI [1/a]	0.352252	0.352234	0.352209
CAIFI [1/a]	0.354118	0.354100	0.354075
SAIDI [h]	2.183	2.293	8.046
CAIDI [h]	6.197	6.509	22.844
ASAI [-]	0.9997507987	0.9997382643	0.9990815262
ASUI [-]	0.0002492013	0.0002617357	0.0009184738
EENS [MWh/a]	33.829	35.533	124.821
	Variant: DNMG_2		
Index	Case_1	Case_2	Case_3
SAIFI [1/a]	0.346583	0.346587	0.346654
CAIFI [1/a]	0.348419	0.348423	0.348490
SAIDI [h]	2.393	2.472	8.992
CAIDI [h]	6.904	7.133	25.938
ASAI [-]	0.9997268479	0.9997177990	0.9989735626
ASUI [-]	0.0002731521	0.0002822010	0.0010264374
EENS [MWh/a]	37.077	38.306	139.453
	Variant: DNMG_3		
Index	Case_1	Case_2	Case_3
SAIFI [1/a]	0.358422	0.358436	0.358511
CAIFI [1/a]	0.360315	0.360330	0.360405
SAIDI [h]	2.474	2.558	9.438
CAIDI [h]	6.903	7.136	26.324
ASAI [-]	0.9997175397	0.9997079937	0.9989226588
ASUI [-]	0.0002824603	0.0002920063	0.0010773412
EENS [MWh/a]	38.492	39.795	147.018

Table 12. Reliability system indices for AMDN structure (Figure A5).

			Variant: AMDN_1		
Index	**Case_1**	**Case_2**	**Case_3**	**Case_4**	**Case_5**
SAIFI [1/a]	0.087041	0.081571	0.074728	0.074746	0.076239
CAIFI [1/a]	0.087041	0.081571	0.074728	0.074746	0.076239
SAIDI [h]	0.661	0.661	0.661	0.661	0.661
CAIDI [h]	7.593	8.103	8.845	8.842	8.669
ASAI [-]	0.9999245511	0.9999245511	0.9999245511	0.9999245511	0.9999245511
ASUI [-]	0.0000754489	0.0000754489	0.0000754489	0.0000754489	0.0000754489
EENS [MWh/a]	9.037	9.037	9.037	9.037	9.037
			Variant: AMDN_2		
Index	Case_1	Case_2	Case_3	Case_4	Case_5
SAIFI [1/a]	0.081363	0.080969	0.080475	0.080476	0.079877
CAIFI [1/a]	0.081363	0.080969	0.080475	0.080476	0.079877
SAIDI [h]	0.715	0.715	0.715	0.715	0.715
CAIDI [h]	8.788	8.831	8.885	8.885	8.952
ASAI [-]	0.9999183769	0.9999183769	0.9999183769	0.9999183769	0.9999183769
ASUI [-]	0.0000816231	0.0000816231	0.0000816231	0.0000816231	0.0000816231
EENS [MWh/a]	9.376	9.376	9.376	9.376	9.376
			Variant: AMDN_3		
Index	Case_1	Case_2	Case_3	Case_4	Case_5
SAIFI [1/a]	0.073720	0.076901	0.080882	0.080871	0.080003
CAIFI [1/a]	0.213266	0.222469	0.233984	0.233954	0.231441
SAIDI [h]	0.700	0.700	0.700	0.700	0.700
CAIDI [h]	9.498	9.105	8.657	8.658	8.752
ASAI [-]	0.9999200708	0.9999200708	0.9999200708	0.9999200708	0.9999200708
ASUI [-]	0.0000799292	0.0000799292	0.0000799292	0.0000799292	0.0000799292
EENS [MWh/a]	9.283	9.283	9.283	9.283	9.283

Three variants of DG unit location are considered. Based on UDN_1 variant as a basic UDN structure (see Figure A1), a node including the PVPP and BES connected changes from no. 5 to no. 1 (UDN_2) and no. 8 (UDN_3). In each variant, three cases of different power generation levels of the considered PVPP and BES are analyzed, i.e., Case_1—51 kW, Case_2—100 kW and Case_3—510 kW. The other DG sources do not change the location in all variants and the power generation values in all cases.

There are also three variants of DG unit location to be considered. Based on RDN_1 variant as a basic RDN structure (see Figure A3), a node the WTPP and BES are connected to changes from no. 72 to no. 43 (RDN_2) and no. 71 (RDN_3). In each variant, three cases of different power generation levels of the considered WTPP and BES are analyzed, i.e., Case_1—2.4 MW, Case_2—1.6 MW and Case_3—0.8 MW. The other DG sources do not change the location in all variants and the power generation values in all cases.

Three variants of MG location are considered. Based on DNMG_1 variant as a basic DNMG structure (see Figure A4), a MG is connected to changes from no. 6 to no. 3 (DNMG_2) and no. 1 (DNMG_3). In the second variant (DNMG_2) a load equivalent is shifted from node no. 3 to node no. 6. In the third variant (DNMG_3) an additional load equivalent is connected to node no. 6 (P = 100 kW, Q = 20 kvar). In each variant the change of only one of two microsources in considered MG is analyzed in three cases: Case_1—WTPP (170 kW); MT (30 kW), Case_2—WTPP (170 kW); FC (30 kW), Case_3—WTPP (170 kW); PVPP (30 kW). In all the cases reactive power generated in microsources is equal to 0 kvar.

Three variants of DG unit location are considered. Based on AMDN_1 variant as a basic AMDN structure (see Figure A5), a node the DG unit is connected to changes from no. 4 to no. 2 (AMDN_2) and no. 1 (AMDN_3). In each variant, five cases of different types of the considered DG unit are analyzed, i.e., Case_1—WTPP, Case_2—PVPP, Case_3—EDG, Case_4—TDG, Case_5—SHPP. The change of power generation level of the DG unit is not considered.

For AMDN_1 variant an impact of automatic on-load tap changer at the 110 kV/MV transformer on the maximum active and reactive power generated by a DG unit has been analyzed as well. The branch power capacity and permissible voltage deviation (±10%) was the criterion used to determine the maximum power generation. The first investigated case assumes the peak load and 110 kV/MV transformer operation without on-load tap changer. Maximum values of active and reactive power generated by the DG source are P = 14.3 MW and Q = 4.3 Mvar adequately at +10% voltage deviation. In the second case (peak-off load and transformer operation without on-load tap changer) the DG unit can generate only P = 1.8 MW and Q = 0.54 Mvar. The last case assumes the automatic on-load tap changer at the transformer as well as the peak-off load, the DG source can generate the power up to the cable load capacity (the line between nodes no. 3 and 4).

4.3. Observations

Based on the test calculation results the following observations have been made:

- Looking at the reliability indices calculated for UDN and RDN one can certainly state that location and type of DG source does not directly affect the power supply reliability. The APR application in UDN definitely improves all the investigated reliability indices, i.e., SAIFI, CAIFI, SAIDI, CAIDI, EENS, ASAI, ASUI).
- Three considered variants of DG source location in AMDN have been compared. Looking at ASAI index, the highest value is for "AMDN_1" and the lowest one is for "AMDN_2". There is also an impact of DG types on both the interruption frequency and interruption duration.
- An impact of DG type on reliability indices is also noticeable for DNMG. For all considered DNMG variants, the lowest ASAI value is observed in "Case_3" with WTPP and PVPP connected to a microgrid. Analyzing three variants of microgrid locations, the highest ASAI value is observed for "DNMG_1". In addition, "DNMG_3" characterizes the lowest value of ASAI index.
- Both the interruption frequency (SAIFI, CAIFI) and interruption duration (SAIDI, CAIDI) for UDN are lower than for RDN.

- The interruption frequency (SAIFI, CAIFI) calculated for RDN is about tenfold higher than for AMDN with DG unit connected (except the variant AMDN_3 in which the ratio is about four). An interesting observation about interruption durations is made. The values of CAIDI index are almost comparable to one another, but the SAIDI for AMDN is over twelvefold lower than for RDN structure. This observation shows the strongly irregular distribution of the repair duration in the analyzed AMDN structure.
- DNMG with connected LV microgrids does not significantly improve the power supply reliability in comparison to UDN with APR.

5. Smart Grid Reliability Assessment

5.1. Model Structure

A simple distribution system structure was created to analyze the reliability assessment of an electric power network coupled with a communications network [80,81], see Figure A6 in Appendix E. Integrated communication allows for monitoring all nodes in the network, and thus faster detection of the location of failures in the power network and taking corrective actions. The component aging is disregarded and only the constant failure rate related to their useful life is analyzed. Since failures in the power system usually occur randomly, the sequential Monte Carlo method was employed to simulate and assess a smart grid's reliability over time. The method produces a distribution of possible outcomes rather than a single expected value.

The artificial operating/failure histories of the relevant smart grid elements are generated. The period during which the element is operating is called time to failure (TTF). The period during which the element fails is called time to repair (TTR).

The parameters TTF, TTR constitute random variables and may have different probability distributions. Exponential distribution is used here to assess the reliability of both the electric power distribution system and the communications network. The exponential distribution's probability distribution function is described as follows [80,82]:

$$f(t) = \begin{cases} \lambda e^{-\lambda t}, & 0 < t < \infty \\ 0, & otherwise \end{cases} \tag{8}$$

The method for generating an artificial failure history of a component is presented in Figure 4. Each time interval is computed with different random numbers. This simulates contingencies occurring in a real system realistically.

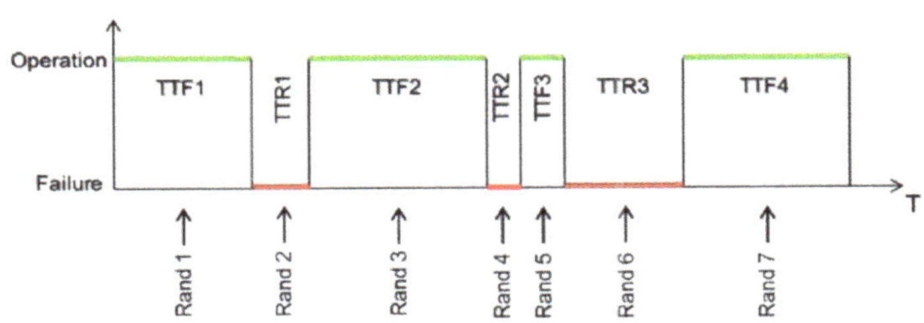

Figure 4. Method of failure history generation [80,81].

TTF and TTR are calculated for a given failure rate and repair rate from Table 13 with the formulas [83]:

$$TTF_i = -\frac{1}{\lambda}\ln(u_i) \tag{9}$$

$$TTR_j = -\frac{1}{\mu}\ln(u_j) \tag{10}$$

where u_i, u_j are random numbers uniformly distributed in the range of 0–1 and λ and μ are the failure rate and the repair rate, respectively.

Table 13. Values of the reliability parameters used for the simulations.

Components	Failure Rate (λ)	Mean Time to Repair (MTTR) [h]	MTTR$_{SG}$ with com Network [h]
Distribution lines	0.8	50	40
Transformers	0.1	100	80

The communications infrastructure in smart grid supplies additional information on power system states, thus improving overall network performance. It enables faster detection of and response to failures or even prevents their occurrence. Accessing information faster facilitates earlier dispatching of service teams or faster responses by different system operators to potential imbalances in power systems, for instance.

By extension, communications shorten interruption times. This method, thus, entails shortening interruption times by shortening the time to repair for combined system. The shorter time to repair is denoted as TTR$_{SG}$ (see Figure 5) and simulated. Since communications are assumed to improve distribution system reliability, their absence or failure do not diminish an electric power system's (EPS) performance, as shown on the right in Figure 5.

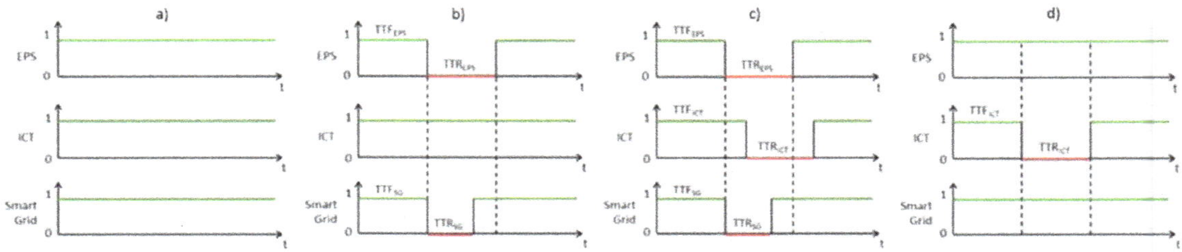

Figure 5. Methodology of smart grid co-simulation: (**a**) EPS, ICT and Smart Grid in operation, (**b**) outage in EPS, ICT in operation, shortening the failure in Smart Grid, (**c**) failure in EPS, failure in ICT, shortening the failure in Smart Grid, (**d**) EPS in operation, failure in ICT, Smart Grid in operation [80,81].

5.2. The Algorithm Used

The reliability simulations are based on the time sequential Monte Carlo technique. This method has been adapted into the proposed approach of Smart Grid reliability assessment. The algorithm used to compute reliability indices of electric power distribution systems (EPS), communications networks (ICT) and integrated Smart Grid (SG) system is presented in a block diagram in Figure 6.

The program starts with definition of system input data such as network topology with location of the components, failure and repair rates and connected loads. The number of sample years (N) and simulation period (T) are also entered in this step. The simulation begins with generating random numbers [0,1] for each element in the system and converting them into time to failure (TTF) using equation 4. In the next step, the element with the shortest TTF is determined, i.e., the component that will fail first. In the conditional block, it is then checked to see if the found minimal TTF value matches within one year. If this is not a case, it means that the TTF is longer than 8760 h and within this year no failure occurs. Further steps will be skipped and random numbers of TTF are computed again for all elements. If the TTF is shorter than 8760 h, the time to repair (TTR) will be computed for that element which indicates its out of operation state duration. Moreover, its location in the interconnection matrix as well as the location of any load nodes which can be influenced by the given component is also determined. After that, a new random number is generated for that component and converted into new TTF. Simulation time has to be updated for each element according to Equation (11).

$$TTF = t + TTR + TTF\_new \tag{11}$$

The updated TTF value indicates subsequent time of failure occurrence and is compared with previously generated TTFs of other components. After that, the TTR for other elements can be computed. These procedures are repeated in a loop for each element until the simulation period (e.g., one year) is completed and all of the simulation sequences comprising the defined number of years are finished. Then, the reliability parameters of each component, such as failure and repair rate as well as unavailability, are computed. Finally, based on these parameters, the reliability indices for the whole system are calculated for the total sample number of years (N).

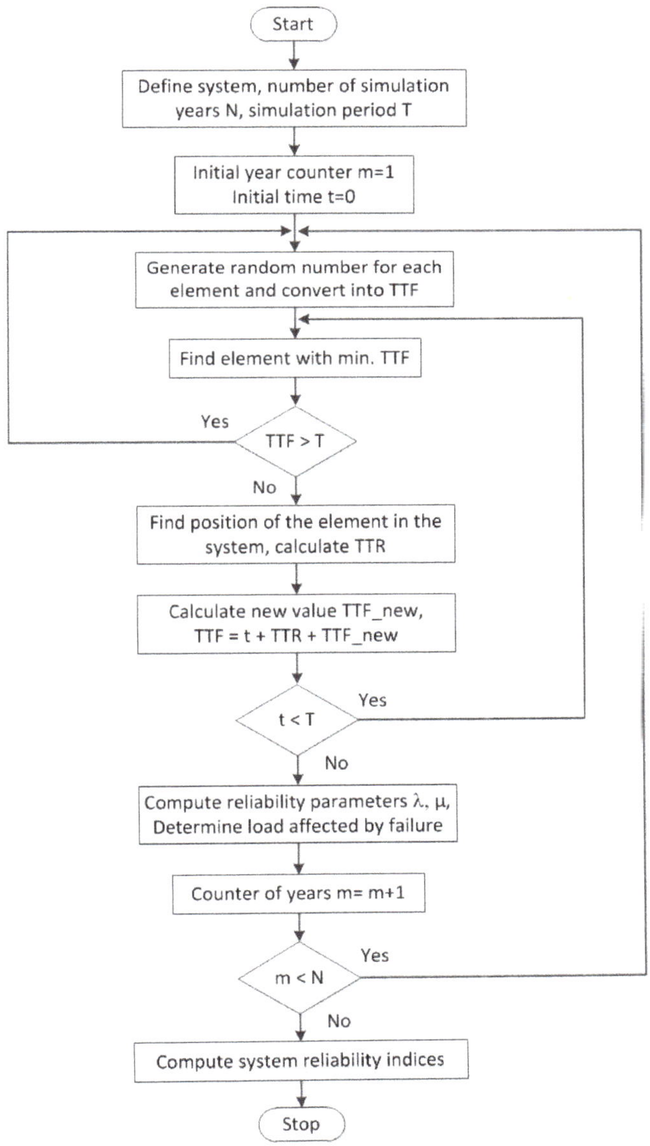

Figure 6. Sequential Monte Carlo algorithm [81,82].

5.3. Simulation Results

The simulations of an integrated smart grid system were run with the structure illustrated in Figure A6. In the simulations, component failures and the existence of communication were considered, and generation parameters were omitted. In order to calculate the relevant reliability indicators, the number of customers has been taken into account.

Simulations of N = 100 years with the step-in sequence of one year (8760 h) with the resolution of one hour were run using the input data presented in Table 13. Faster responses to failures, shorten the time to repair for the smart grid. The results of reliability parameters obtained with time sequential simulation are strongly influenced from the failure, repair rates and system structure. The distribution of reliability indices significantly depends on the number of simulated years.

Simulation results are presented in Table 14 and Figure 7. The average system indices show that the electric power system with a communications infrastructure is more reliable. The presence of ICT shortens interruption times, represented by the SAIDI index.

Table 14. Average system indices for the electric power system analyzed.

Network Structure	SAIDI [h/yr]	SAIFI [fail./yr]	CAIDI [h/fail.]	EENS [MWh/yr]	ASAI [%]
Without communication	9.352	0.105	38.101	0.094	99.893
With communication	7.482	0.104	30.608	0.075	99.915

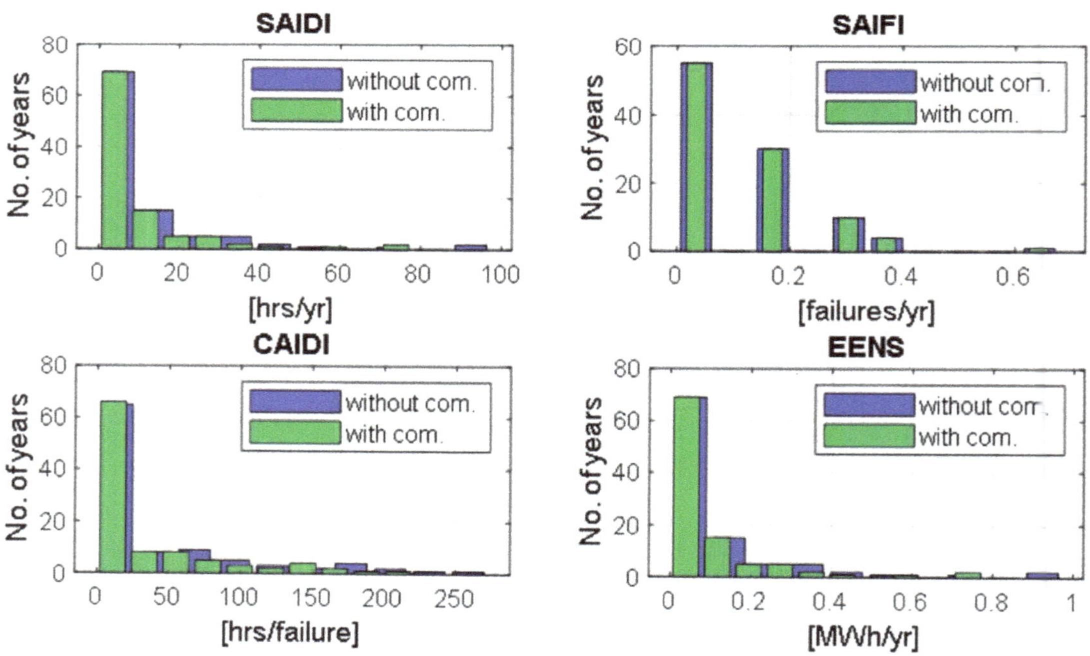

Figure 7. *Cont.*

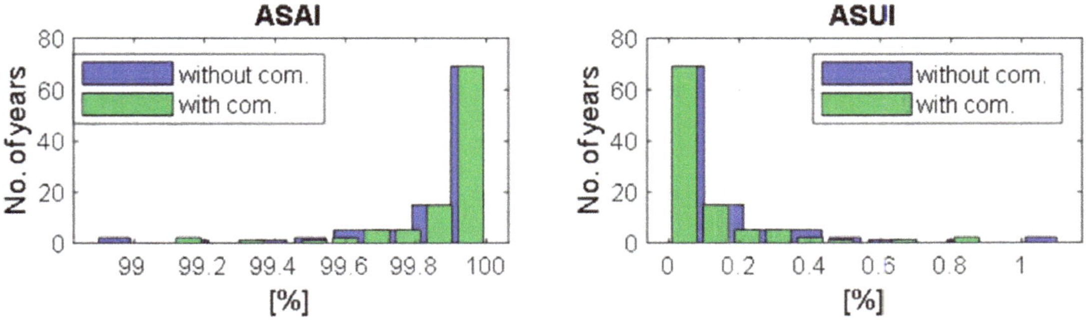

Figure 7. Index distributions obtained from the simulation of an electric power system with and without a communications infrastructure.

Moreover, the range of index distributions shifts toward zero in systems with communications, reflecting improved reliability. Distributions of the SAIFI index representing the number of interruptions in one year are identical for systems with and without communications because only the durations of interruptions change in both scenarios but the number of interruptions in one year remain constant.

The CAIDI index representing the average interruption duration is significantly smaller in electric power systems with ICT than in systems without communications. This is due to the shorter break times.

6. Conclusions

Electrical component failures in distribution systems have been proven the cause of the majority of power interruptions in electric power system.

The number, location, and type of DG sources in existing (conventional) company-owned distribution networks that may not operate in islanded mode have no direct impact on reliability indices. However, a DG unit connection to a distribution network may cause a load alleviation in lines, transformers, etc. and this phenomenon varies the form of the risk function of power system components. The DG sources connected to a distribution network indirectly improves electric service reliability for consumers. DG units may increase voltage in busbars and terminals and short-circuit currents in a distribution system.

Power supply is chiefly improved by providing power redundancy and using remote control switches with distribution system automation, such as ASS, APR and AR (automatic reclosing). The reliability calculations corroborate this. Power interruption frequency and power interruption duration are lower in urban looped distribution networks than in rural distribution networks with feeders supplied from one point. The continued growth of DG capacity in distribution systems requires research and development of new (future) distribution network structures, e.g., actively managed networks (smart grids), microgrids, clustered networks, etc. All of these networks are assumed to be capable of operating autonomously (i.e., islanded and unconnected to the main grid) and to be equipped with distribution system automation (e.g., ASS, APR and AR).

The calculations confirm that the future distribution network structures have higher electrical service reliability than existing distribution networks. Future distribution systems have lower interruption frequencies and durations.

Actively managed distribution networks appear to be a promising idea [84]. Assuming the voltage limitation on network busbars and terminals, the impact of the active on-load tap changer in the 110 kV/MV transformer on the maximum active and reactive power generated by a DG source has been analyzed. This study has demonstrated that the automatic voltage regulator (AVR) at the transformer allows the increase of installed capacity of DG unit connected to the distribution network.

Considering the impact of DG type and locations in future electric distribution networks on the power supply reliability, the results of reliability assessment allow the formulation of the following remarks:

- Power supply continuity is higher when DG sources are based on conventional fuels (gas, diesel oil, etc.). The installation of renewable DG units (energy carriers that are hard to forecast) delivers worse results.
- Types of microsources in distribution networks with connected LV microgrids affect the reliability indices. This is noticeable in installed microsources based on energy carriers dependent on weather conditions.
- The location of DG sources in distribution systems also affects electrical service reliability. The most favorable reliability indices are obtained when the potential maximum number of energy consumers are double-feed (loop: transformer substation—DG unit). In case of aggregation of supplying sources at the one network busbar or terminal, worse reliability indices are observed.
- The appropriate level of DG power generation is a significant issue. The DG power available ensures peak loads are covered. The level of power generation is often limited by branch power capacity and maximum allowable voltage deviation. Active voltage regulation in power transformers can resolve voltage level problems.

The authors intend to focus on optimizing future distribution system structures and devising an optimal development strategy for existing distribution networks in future studies. This will require the determining of accurate reliability models of electric power equipment, protection and automation systems, DG sources and energy storage systems in different types of network structures.

The use of information and communications technology to monitor, control and protect power systems is an important way to meet the challenges of continuously developing electric power grids. The installation of measurement sensors, automated control systems and communication devices will increase the complexity of such integrated systems, thus requiring new methods for designing and optimal integrating of advanced communications systems in electric power grids.

A reliability assessment of smart grids consisting of an electric power distribution system and an integrated communications network based on Monte Carlo simulation was developed and tested in this study. The simulation algorithm delivers the distributions and average values of reliability indices for smart grids, electric power systems and communications networks. This enabled analyzing the influence of the coexistent ICT infrastructure on the power distribution system's reliability and, thus, the entire smart grid. Although some assumptions were made in the methodology to model the systems, the algorithm developed delivers valuable results for the assessment of reliability when designing and optimizing systems. Widespread use of a reliable information and communications infrastructure will improve smart grids' functionality and reliability.

Since this study concentrates on the monitoring of smart grids with advanced ICT, future studies ought to analyze their control and protection. Applying reliable control and protection schemes to the system will help minimize outages and their impact on overall system operation. Future studies ought to examine more complex models of ICT network operation, including several levels of communication performance, e.g., full communication of all components, full communication of all components with limited quality of service (QoS) and limited communication.

Author Contributions: Conceptualization, M.P., J.W. and B.A.; methodology, J.W. and B.A.; investigation, J.W. and B.A.; supervision, M.P. and. P.K.; writing, M.P., J.W., T.W. and B.A.; visualization, J.W. and T.W. All authors have read and agreed to the published version of the manuscript.

Funding: This study was funded in part by the Polish Ministry of Education and Science in the research project "National Electrical Power Security" (PBZ-MEiN-1/2/2006; tasks no. 5.3.5 and 5.3.6).

Informed Consent Statement: Not applicable.

Data Availability Statement: Not applicable.

Acknowledgments: Three of four test MV distribution power network structures used in this paper has been developed under the research project "National Electrical Power Security" supported by the Polish Ministry of Education and Science. Similarly, reliability calculations in the test networks have been carried out under the research project. The smart grid reliability research was carried out within the SECVER project supported by the Federal Ministry of Economic Affairs and Energy (BMWi) in Germany.

Conflicts of Interest: The authors declare no conflict of interest.

Abbreviation

The following abbreviations are used in this manuscript:

AMDN	Active managed distribution network
APR	Automatic power restoration
AR	Automatic reclosing
ASAI	Average service availability index
ASUI	Average service unavailability index
ASS	Automatic stand-by switching on
AVR	Automatic voltage regulator
BES	Battery energy storage
BTPP	Biogas-turbine power plant
CAIDI	Customer average interruption duration index
CAIFI	Customer average interruption frequency index
DG	Distributed generation
DGU	Distributed generation unit
DNMG	Urban distribution network with connected microgrids
EDG	Engine-driven generator
EENS	Expected energy not supplied
EPS	Electric power system
ET	Electric traction substation
EXP	Exponential
FC	Fuel cell
FMECA	Failure mode effects and criticality analysis
GAM	Gamma
GT	Grounding transformer
GTPP	Gas-turbine power plant
GUM	Gumbel
HL	Hierarchical level
ICT	Information and communication technology
IEC	International Electrotechnical Commission
IP	Industrial park
IT	Information technology
LNOR	Log-normal
LTE	Long term evolution
LV	Low voltage
MG	Microgrid
MG(G)	Microgrid generation
MG(L)	Microgrid load
MT	Microturbine
MTBF	Mean time between failure
MTTR	Mean time to repair
MV	Medium voltage
MVUS	MV urban distribution substation
NOR	Normal
PAR	Pareto
PF	PowerFactory software
PFCB	PFC capacitor bank
PLC	Programmable logic controller

PMU Phasor measurement unit
PS Power system
PV Photovoltaic
PVPP Photovoltaic power plant
QoS Quality of service
RDN Rural distribution network
RES Renewable energy source
RTU Remote terminal unit
R&PU Residential and public utility
SCADA Supervisory control and data acquisition
SAIDI System average interruption duration index
SAIFI System average interruption frequency index
SG Smart grid
SHPP Small hydro power plant
TDG Turbine-driven generator
TTF Time to failure
TTR Time to repair
UDN Urban distribution network
WEI Weibull
WTPP Wind-turbine power plant

Appendix A. UDN Structure

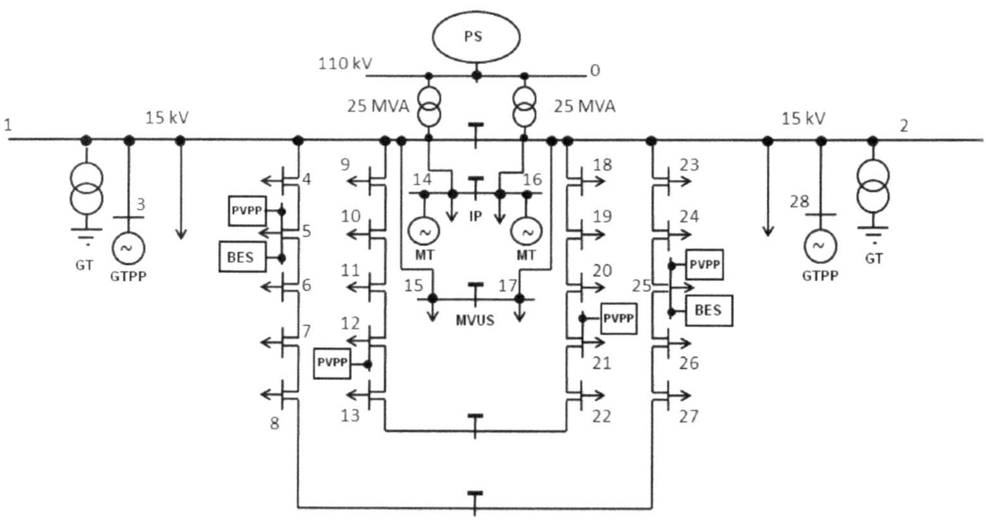

Figure A1. MV urban distribution network with the distributed generation sources (PS—power system, PVPP—photovoltaic power plant, BES—battery energy storage, MT—micro-turbine, GTPP—gas-turbine power plant, GT—grounding transformer, MVUS—MV urban distribution substation, IP—industrial park).

Table A1. Distributed generation sources and energy consumers connected to the urban MV distribution network.

Node No.	P_{LOAD}/P_{GEN} [kW]	Q_{LOAD} [kvar]	Load/DG Type	Node No.	P_{LOAD}/P_{GEN} [kW]	Q_{LOAD} [kvar]	Load/DG Type
1	5000	1500	R&PU	15	4700	1700	R&PU
2	5000	1500	R&PU	16	200	-	MT
3	1200	-	GTPP	16	1080	420	IP
4	510	185	R&PU	17	4650	1835	R&PU
5	530	210	R&PU	18	545	215	R&PU
5	30	-	BES	19	310	120	R&PU
5	51	-	PVPP	20	315	115	R&PU
6	305	120	R&PU	21	335	110	R&PU
7	300	110	R&PU	21	51	-	PVPP
8	530	210	R&PU	22	540	195	ET
9	320	115	R&PU	23	495	180	R&PU
10	540	210	R&PU	24	535	195	R&PU
11	495	180	R&PU	25	325	105	R&PU
12	340	110	R&PU	25	51	-	PVPP
12	51	-	PVPP	25	30	-	BES
13	300	120	R&PU	26	325	105	R&PU
14	1060	415	IP	27	540	195	R&PU
14	200	-	MT	28	1200	-	GTPP

Types of energy consumers: R&PU—residential and public utility, IP—industrial (park), ET—electric traction substation. The nodes the DG sources are connected to are marked as gray background.

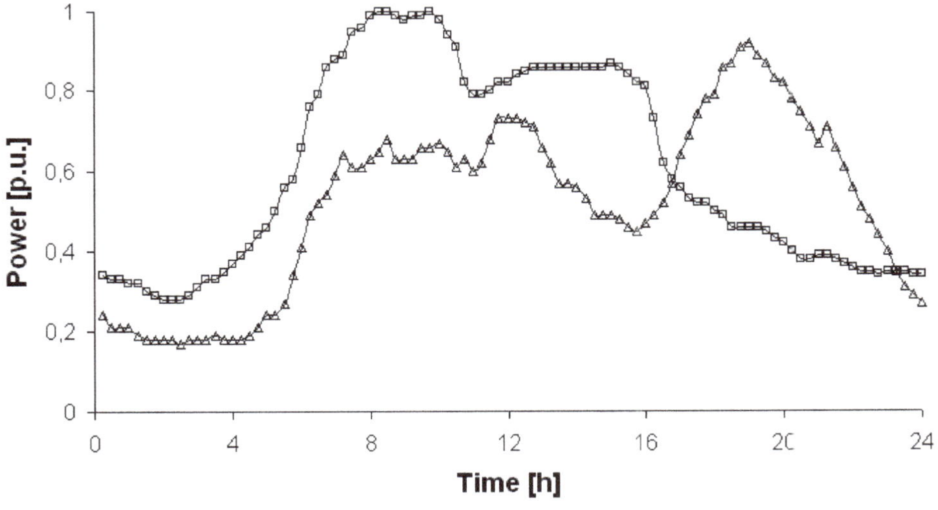

Figure A2. Typical daily load profile (15-min intervals) for residential consumers (a line with the triangles) as well as industrial parks and electric traction substation (a line with the quadrants); based on [85].

Table A2. Line parameters of the urban MV distribution network.

From Node No.	To Node No.	Line Type	Length [m]	From Node No.	To Node No.	Line Type	Length [m]
1	3	3xXUHAKXS1 × 120	2000	1	15	HAKnFty 3 × 240	3000
1	4	3xYHAKXS1 × 120	110	2	16	HAKnFty 3 × 120	1600
4	5	3xYHAKXS1 × 120	250	2	17	HAKnFty 3 × 240	3000
5	6	3xYHAKXS1 × 120	100	2	28	3xYHAKXS 1 × 120	190
6	7	3xYHAKXS1 × 120	130	18	19	3xYHAKXS 1 × 120	100
7	8	3xYHAKXS1 × 120	160	19	20	3xYHAKXS 1 × 120	110
8	27	3xYHAKXS1 × 120	170	20	21	3xYHAKXS 1 × 120	130
1	9	3xYHAKXS1 × 120	200	21	22	3xYHAKXS 1 × 120	80
9	10	3xYHAKXS1 × 120	120	2	23	3xYHAKXS 1 × 120	120
10	11	3xYHAKXS1 × 120	140	23	24	3xYHAKXS 1 × 120	200
11	12	3xYHAKXS1 × 120	170	24	25	3xYHAKXS 1 × 120	80
12	13	3xYHAKXS1 × 120	80	25	26	3xYHAKXS 1 × 120	140
13	22	3xYHAKXS1 × 120	140	26	27	3xYHAKXS 1 × 120	90
1	14	HAKnFty 3 × 120	1600	2	28	3xXUHAKXS1 × 120	2500

XUHAKXS—single-Al core cable, radial field, polythene-coated, polythene sheath, YHAKXS—single-Al core cable, radial field, polythene-coated, polyvinyl chloride sheath, HAKnFty—triple-Al core cable, radial field, paper-coated, steel armor, polyvinyl chloride sheath.

Appendix B. RDN Structure

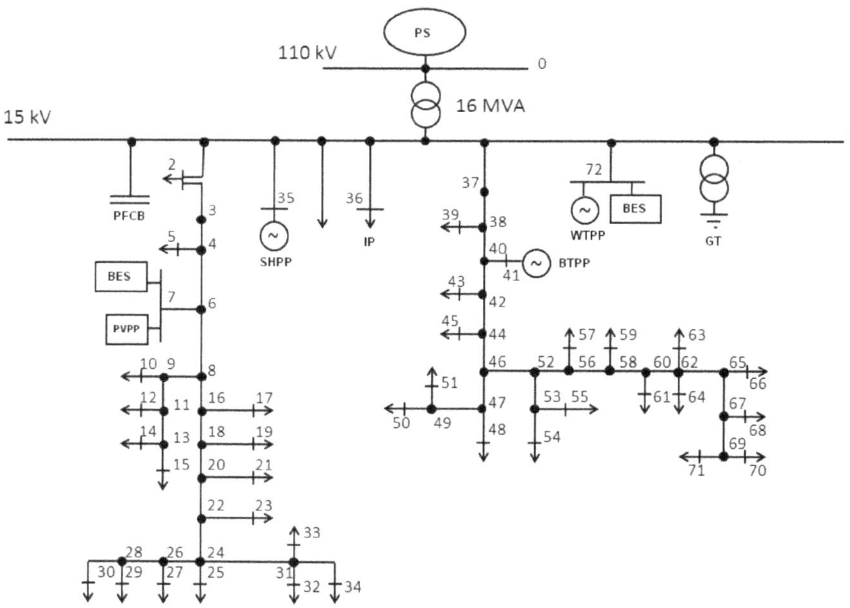

Figure A3. Rural MV distribution network with the distributed generation sources (PS—power system, PVPP—photovoltaic power plant, BES—battery energy storage, SHPP—small hydropower plant, BTPP—biogas-turbine power plant, WTPP—wind-turbine power plant, PFCB—PFC capacitor bank, GT—grounding transformer, IP—industrial park).

Table A3. Distributed generation sources and energy consumers connected to the rural MV distribution network.

Node No.	P_{LOAD}/P_{GEN} [kW]	Q_{LOAD} [kvar]	Load/DG Type	Node No.	P_{LOAD}/P_{GEN} [kW]	Q_{LOAD} [kvar]	Load/DG Type
1	4000	1400	R&PU	36	1595	580	IP
2	565	185	R&PU	39	335	120	R&PU
5	210	75	R&PU	41	1200	-	BTPP
7	85	30	R&PU	43	135	50	R&PU
7	200	-	PVPP	45	140	45	R&PU
7	200	-	BES	48	75	30	R&PU
10	145	45	R&PU	50	135	50	R&PU
12	135	50	R&PU	51	200	75	R&PU
14	125	40	R&PU	54	190	75	R&PU
15	190	70	R&PU	55	210	75	IP
17	205	80	R&PU	57	135	50	R&PU
19	125	50	R&PU	59	75	30	R&PU
21	125	45	IP	61	210	75	R&PU
23	75	30	R&PU	63	215	75	R&PU
25	200	75	R&PU	64	140	45	R&PU
27	225	75	R&PU	66	200	70	R&PU
29	120	45	R&PU	68	125	50	R&PU
30	125	50	R&PU	70	140	45	R&PU
32	120	50	R&PU	71	85	30	R&PU
33	130	45	R&PU	72	2400	-	WTPP
34	90	30	R&PU	72	500	-	BES
35	2000	-	SHPP	-	-	-	-

The nodes the DG sources are connected to are marked as gray background.

Table A4. Line parameters of the rural MV distribution network.

From Node No.	To Node No.	Line Type	Length [m]	From Node No.	To Node No.	Line Type	Length [m]
1	2	3xYHAKXS 1 × 120	500	1	37	3xYHAKXS1 × 120	450
2	3	3xYHAKXS 1 × 120	150	37	38	70 AFL	200
3	4	70 AFL	160	38	39	50 AFL	100
4	5	50 AFL	130	38	40	70 AFL	150
4	6	70 AFL	200	40	41	70 AFL	90
6	7	50 AFL	50	40	42	70 AFL	180
6	8	70 AFL	120	42	43	50 AFL	70
8	9	50 AFL	310	42	44	70 AFL	180
9	10	50 AFL	60	44	45	50 AFL	120
9	11	50 AFL	210	44	46	70 AFL	210
11	12	50 AFL	80	46	47	50 AFL	200
11	13	50 AFL	160	47	48	50 AFL	90
13	14	50 AFL	50	47	49	50 AFL	150
13	15	50 AFL	120	49	50	50 AFL	90
8	16	70 AFL	160	49	51	50 AFL	110
16	17	50 AFL	100	46	52	70 AFL	140
16	18	70 AFL	180	52	53	50 AFL	200
18	19	50 AFL	90	53	54	50 AFL	60
18	20	70 AFL	210	53	55	50 AFL	190
20	21	50 AFL	160	52	56	70 AFL	210
20	22	70 AFL	130	56	57	50 AFL	120
22	23	50 AFL	110	56	58	70 AFL	130
22	24	70 AFL	120	58	59	50 AFL	70

Table A4. *Cont.*

From Node No.	To Node No.	Line Type	Length [m]	From Node No.	To Node No.	Line Type	Length [m]
24	25	50 AFL	140	58	60	70 AFL	200
24	26	50 AFL	180	60	61	50 AFL	130
26	27	50 AFL	160	60	62	70 AFL	160
26	28	50 AFL	150	62	63	50 AFL	200
28	29	50 AFL	70	62	64	50 AFL	130
28	30	50 AFL	290	62	65	50 AFL	140
24	31	50 AFL	200	65	66	50 AFL	210
31	32	50 AFL	120	65	67	50 AFL	140
31	33	50 AFL	60	67	68	50 AFL	30
31	34	50 AFL	300	67	69	50 AFL	120
1	35	3xYHAKXS 1 × 120	4000	69	70	50 AFL	90
1	36	3xYHAKXS 1 × 120	2500	69	71	50 AFL	200
24	31	50 AFL	200	1	72	3xYHAKXS1 × 120	3000

AFL—steel-cored aluminum conductor.

Appendix C. DNMG Structure

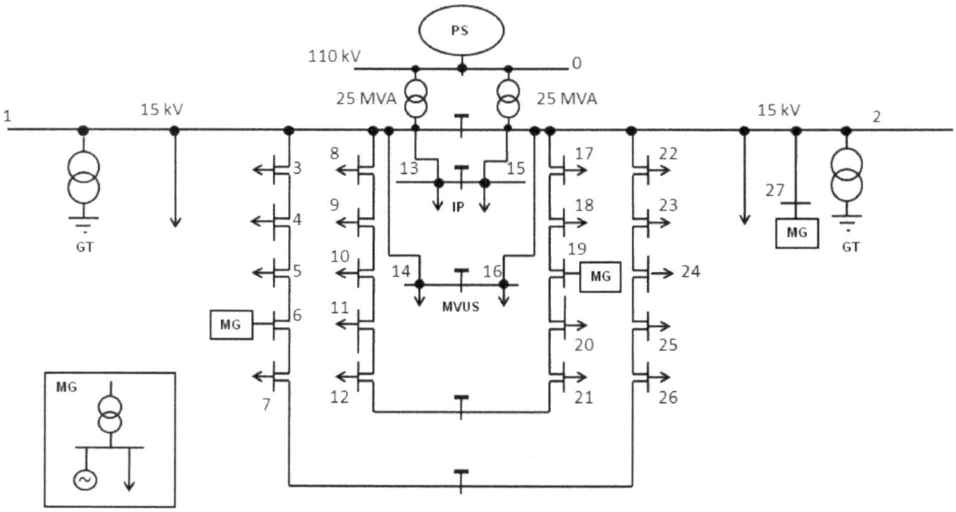

Figure A4. MV distribution network with connected LV microgrids (PS—power system, MG—microgrid, IP—industrial park, MVUS—MV urban distribution substation, GT—grounding transformer).

Table A5. Distributed generation sources and energy consumers connected to the MV distribution network with LV microgrids.

Node No.	P_{LOAD}/P_{GEN} [kW]	Q_{LOAD} [kvar]	Load/DG Type	Node No.	P_{LOAD}/P_{GEN} [kW]	Q_{LOAD} [kvar]	Load/DG Type
1	5000	1500	R&PU	16	4650	1835	R&PU
3	510	185	R&PU	17	545	215	R&PU
4	530	210	R&PU	18	310	120	R&PU
5	305	120	R&PU	19	580	145	MG(L)
6	580	145	MG(L)	19	200	-	MG(G)
6	200	-	MG(G)	20	335	110	R&PU
7	530	210	R&PU	21	540	195	R&PU
8	320	115	R&PU	22	495	180	R&PU
9	540	210	R&PU	23	535	195	R&PU
10	495	180	R&PU	24	325	105	R&PU
11	340	110	R&PU	25	325	105	R&PU
12	300	120	R&PU	26	540	195	R&PU
13	1060	415	IP	2	5000	1500	R&PU
14	4700	1700	R&PU	27	580	145	MG(L)
15	1080	420	IP	27	200	-	MG(G)

Energy consumers and power generation types: MG(L)—microgrid load, MG(G)—microgrid generation. The nodes the MG generations are connected to are marked as gray background.

Table A6. Line parameters of the MV distribution network with LV microgrids.

From Node No.	To Node No.	Length [m]	From Node No.	To Node No.	Length [m]
1	3	110	2	15	1600
3	4	250	2	16	3000
4	5	100	2	17	190
5	6	130	17	18	100
6	7	160	18	19	110
7	26	170	19	20	130
1	8	200	20	21	80
8	9	120	2	22	120
9	10	140	22	23	200
10	11	170	23	24	80
11	12	80	24	25	140
12	21	140	25	26	90
1	13	1600	2	27	2500
1	14	3000			

The MV distribution network with connected LV microgrids consists entirely of underground cables 3x(YHAKXS 1 × 240).

Appendix D. AMDN Structure

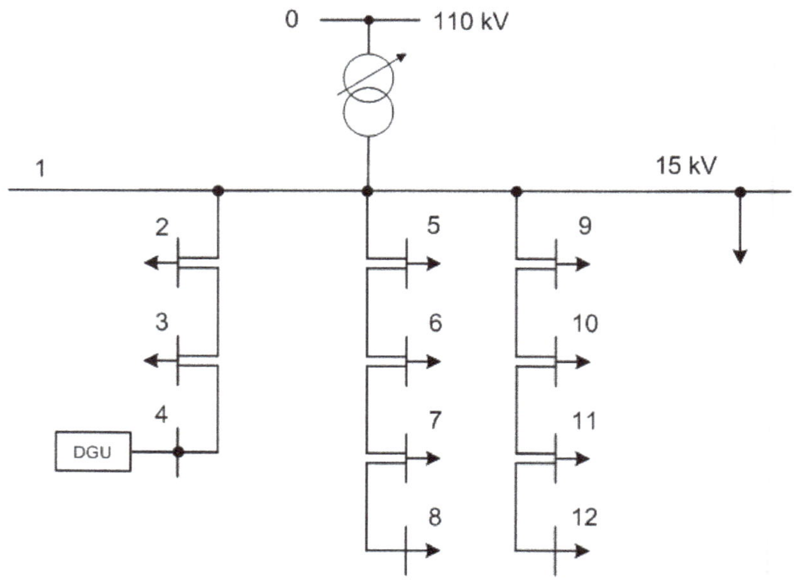

Figure A5. Actively managed MV distribution network with the DG unit (based on [84]).

Table A7. Energy consumers connected to the actively managed MV distribution network.

Node No.	P_{LOAD}/P_{GEN} [kW]	Q_{LOAD} [kvar]	Load/ DG Type	Node No.	P_{LOAD}/P_{GEN} [kW]	Q_{LOAD} [kvar]	Load/ DG Type
1	7000	2100	R&PU	8	400	130	R&PU
2	400	130	R&PU	9	450	150	R&PU
3	500	165	R&PU	10	550	165	R&PU
5	450	135	R&PU	11	1600	640	IP
6	2000	800	IP	12	400	120	R&PU
7	550	165	R&PU				

Table A8. Line parameters of the actively managed MV distribution network.

From Node No.	To Node No.	Length [m]	From Node No.	To Node No.	Length [m]
1	2	500	7	8	140
2	3	150	1	9	300
3	4	250	9	10	140
1	5	400	10	11	180
5	6	150	11	12	120

The actively managed MV distribution network consists entirely of underground cables 3x(YHAKXS 1 × 240).

Appendix E. Smart Grid Structure

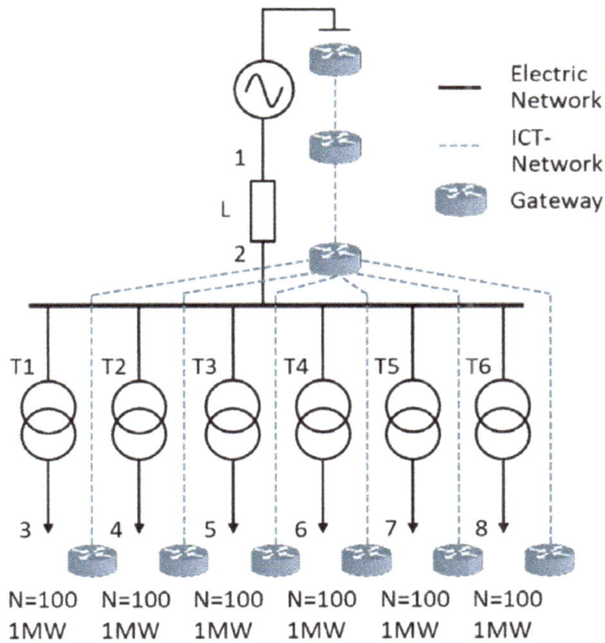

Figure A6. A proposed benchmark system for coupled electric power system and communications network (N—number of customers) (based on [80,81]).

References

1. Chowdhury, A.; Koval, D. Power Distribution System Reliability. In *Practical Methods and Applications*; John Wiley & Sons, Inc. Publication: Hoboken, NJ, USA, 2009.
2. Fotuhi-Firuzabad, M.; Rajabi-Ghahnavie, A. An Analytical Method to Consider DG Impacts on Distribution System Reliability. In Proceedings of the 2005 IEEE/PES Transmission & Distribution Conference & Exhibition: Asia and Pacific, Dalian, China, 18 August 2005.
3. Bae, I.-S.; Kim, J.-O. Reliability Evaluation of Distributed Generation Based on Operation Mode. *IEEE Trans. Power Syst.* **2007**, *22*, 785–790. [CrossRef]
4. Hlatshwayo, M.; Chowdhury, S.; Chowdhury, S.P.; Awodele, K.O. Reliability Enhancement of Radial Distribution Systems with DG Penetration. In Proceedings of the 45th International Universities Power Engineering Conference UPEC2010, Cardiff, UK, 31 August–3 September 2010.
5. Atwa, Y.M.; El-Saadany, E.F. Reliability Evaluation for Distributed System with Renewable Distributed Generation During Islanded Mode of Operation. *IEEE Trans. Power Syst.* **2009**, *24*, 572–581. [CrossRef]
6. Martinez-Velasco, J.A.; Guerra, G. Reliability Analysis of Distribution Systems with Photovoltaic Generation Using a Power Flow Simulator and a Parallel Monte Carlo Approach. *Energies* **2016**, *9*, 537. [CrossRef]
7. Escalera, A.; Castronuovo, E.D.; Prodanović, M.; Roldán-Pérez, J. Reliability Assessment of Distribution Networks with Optimal Coordination of Distributed Generation, Energy Storage and Demand Management. *Energies* **2019**, *12*, 3202. [CrossRef]
8. Kumar, S.; Sarita, K.; Vardhan, A.S.S.; Elavarasan, R.M.; Saket, R.K.; Das, N. Reliability Assessment of Wind-Solar PV Integrated Distribution System Using Electrical Loss Minimization Technique. *Energies* **2020**, *13*, 5631. [CrossRef]
9. Ndawula, M.B.; Djokic, S.Z.; Hernando-Gil, I. Reliability Enhancement in Power Networks under Uncertainty from Distributed Energy Resources. *Energies* **2019**, *12*, 531. [CrossRef]
10. Kebede, F.; Olivier, J.-C.; Bourguet, S.; Machmoum, M. Reliability Evaluation of Renewable Power Systems through Distribution Network Power Outage Modelling. *Energies* **2021**, *14*, 3225. [CrossRef]
11. Verbo, P.; Järventausta, P.; Kiviko, K.; Pylvänäinen, J.; Partanen, J.; Lassila, J.; Honkapuro, S.; Kaipia, T. Applying Reliability Analysis in Evaluation of Life-Cycle Costs of Alternative Network Solutions. In Proceedings of the FPS 2005 International Conference on Future Power Systems, Amsterdam, The Netherlands, 16–18 November 2005; pp. 1–4.

12. Conti, S.; Rizzo, S.A.; El-Saadany, E.F.; Essam, M.; Atwa, Y.M. Reliability Assessment of Distribution Systems Considering Telecontrolled Switches and Microgrids. *IEEE Trans. Power Syst.* **2014**, *29*, 598–607. [CrossRef]
13. Paterakis, N.G.; Mazza, A.; Santos, S.F.; Erdinç, O.; Chicco, G.; Bakirtzis, A.G.; Catalão, J.P. Multi-Objective Reconfiguration of Radial Distribution Systems Using Reliability Indices. *IEEE Trans. Power Syst.* **2016**, *31*, 1048–1062. [CrossRef]
14. Xu, N.Z.; Chung, C.Y. Reliability Evaluation of Distribution Systems Including Vehicle-to-Home and Vehicle-to-Grid. *IEEE Trans. Power Syst.* **2016**, *31*, 759–768. [CrossRef]
15. Lombardi, P.; Hänsch, K.; Arendarski, B.; Komarnicki, P. Information and power terminals: A reliable microgrid infrastructure for use in disaster scenarios. *Int. J. Crit. Infrastruct. Prot.* **2017**, *19*, 49–58. [CrossRef]
16. Alotaibi, I.; Abido, M.A.; Khalid, M.; Savkin, A.V. A Comprehensive Review of Recent Advances in Smart Grids: A Sustainable Future with Renewable Energy Resources. *Energies* **2020**, *13*, 6269. [CrossRef]
17. Das, S.; Kankanala, P.; Pahwa, A. Outage Estimation in Electric Power Distribution Systems Using a Neural Network Ensemble. *Energies* **2021**, *14*, 4797. [CrossRef]
18. Bae, I.-S.; Kim, J.-O. Reliability Evaluation of Customers in a Microgrid. *IEEE Trans. Power Syst.* **2008**, *23*, 1416–1422.
19. Parol, M.; Rokicki, Ł.; Parol, R. Towards optimal operation control in rural low voltage microgrids. *Bull. Pol. Ac Tech.* **2019**, *67*, 799–812.
20. Xu, X.; Mitra, J.; Wang, T.; Mu, L. Evaluation of Operational reliability of a Microgrid Using a Short-term Outage Model. *IEEE Trans. Power Syst.* **2014**, *29*, 2238–2247. [CrossRef]
21. Worku, M.Y.; Hassan, M.A.; Abido, M.A. Real Time Energy Management and Control of Renewable Energy based Microgrid in Grid Connected and Island Modes. *Energies* **2019**, *12*, 276. [CrossRef]
22. Costa, P.M.; Matos, M.A. Assessing the contribution of microgrids to the reliability of distribution networks. *Electr. Power Syst. Res.* **2009**, *79*, 382–389. [CrossRef]
23. Conti, S.; Nicolosi, R.; Rizzo, S.A. An Analytical Formulation to Assess Distribution System Reliability in Presence of Conventional and Renewable Distributed Generators. In Proceedings of the CIGRE Symposium "The Electric Power System of the Future–Integrating Supergrids and Microgrids", Bologna, Italy, 13–15 September 2011.
24. Marcos, F.P.; Domingo, C.M.; San Román, T.G.; Arín, R.C. Location and Sizing of Micro-Grids to Improve Continuity of Supply in Radial Distribution Networks. *Energies* **2020**, *13*, 3495. [CrossRef]
25. Bie, Z.; Zhang, P.; Li, G.; Hua, B.; Meehan, M.; Wang, X. Reliability Evaluation of Active Distribution Systems Including Microgrids. *IEEE Trans. Power Syst.* **2012**, *27*, 2342–2350. [CrossRef]
26. López-Prado, J.L.; Vélez, J.I.; Garcia-Llinás, G.A. Reliability Evaluation in Distribution Networks with Microgrids: Review and Classification of the Literature. *Energies* **2020**, *13*, 6189. [CrossRef]
27. Leite Andréa, P.; Borges Carmen, L.T.; Falcão Djalma, M. Probabilistic Wind Farm Generation Model for Reliability Studies Applied to Brazilian Sites. *IEEE Trans. Power Syst.* **2006**, *21*, 1493–1501. [CrossRef]
28. Artigao, E.; Martín-Martínez, S.; Honrubia-Escribano, A.; Gómez-Lázaro, E. Wind turbine reliability: A comprehensive review towards effective condition monitoring development. *App. Energy* **2018**, *228*, 1569–1583. [CrossRef]
29. Borges Carmen, L.T.; Pinto Roberto, J. Small Hydro Power Plants Energy Availability Modeling for Generation Reliability Evaluation. *IEEE Trans. Power Syst.* **2008**, *23*, 1125–1135. [CrossRef]
30. Ardito, L.; Procaccianti, G.; Menga, G.; Morisio, M. Smart Grid Technologies in Europe: An Overview. *Energies* **2013**, *6*, 251–281. [CrossRef]
31. Celli, G.; Ghiani, E.; Pilo, F.; Soma, G. Impact of ICT on the Reliability of Active Distribution Networks. In Proceedings of the CIRED Workshop, Lisbon, Portugal, 29–30 May 2012.
32. Girón, C.; Rodríguez, F.J.; Giménez de Urtasum, L.; Borroy, S. Assessing the contribution of automation to the electric distribution network reliability. *Int. J. Electr. Power Energy Syst.* **2018**, *97*, 120–126. [CrossRef]
33. Sanchez, J.; Caire, R.; HadjSaid, N. ICT and Electric Power Systems Interdependencies Modeling. In Proceedings of the Internationaler ETG-Kongress in Berlin, Berlin, Germany, 5–6 November 2013.
34. Sanchez, J.; Caire, R.; Hadjsaid, N. ICT and Power Distribution Modeling using Complex Networks. In Proceedings of the IEEE PES Powertech Conference, Grenoble, France, 16–20 June 2013.
35. Smith, E.; Robinson, D.; Agalgaonkar, A. Cooperative Control of Microgrids: A Review of Theoretical Frameworks, Applications and Recent Developments. *Energies* **2021**, *14*, 8026. [CrossRef]
36. Moslehi, K.; Kumar, R. A Reliability Perspective of the Smart Grid. *IEEE Trans. Smart Grid* **2010**, *1*, 57–64. [CrossRef]
37. Pavon, W.; Inga, E.; Simani, S.; Nonato, M. A Review on Optimal Control for the Smart Grid Electrical Substation enhancing Transition Stability. *Energies* **2021**, *14*, 8451. [CrossRef]
38. Zhang, Y.; Wang, L.; Xiang, Y.; Ten, C.-W. Power System Reliability Evaluation with SCADA Cybersecurity Considerations. *IEEE Trans. Smart Grid* **2015**, *6*, 1707–1721. [CrossRef]
39. Chen, L.; Zhao, N.; Cheng, Z.; Gu, W. Reliability Evaluation of Cyber–Physical Power Systems Considering Supply- and Demand-Side Uncertainties. *Energies* **2022**, *15*, 118. [CrossRef]
40. Liu, R.; Mustafa, H.M.; Nie, Z.; Srivastava, A.K. Reachability-Based False Data Injection Attacks and Defence Mechanisms for Cyberpower System. *Energies* **2022**, *15*, 1754. [CrossRef]
41. Jewell, W.; Namboodiri, V.; Aravinthan, V.; Karimi, B.; Kezunovic, M.; Dong, Y. *Communication Requirements and Integration Options for Smart Grid Deployment*; Power Systems Engineering Research Center (PSERC): Tempe, AZ, USA, 2012.

42. Güngör, V.; Sahin, D.; Kocak, T.; Ergüt, S.; Buccella, C.; Cecati, C.; Hancke, G. Smart Grid Technologies: Communication Technologies and Standards. *IEEE Trans. Ind. Inform.* **2011**, *7*, 529–539. [CrossRef]
43. Molokomme, D.N.; Chabalala, C.S.; Bokoro, P.N. A Review of Cognitive Radio Smart Grid Communication Infrastructure Systems. *Energies* **2020**, *13*, 3245. [CrossRef]
44. Powalko, M.; Rudion, K.; Komarnicki, P.; Blumschein, J. Observability of the distribution system. In Proceedings of the 20th International Conference and Exhibition on Electricity Distribution, CIRED 2009, Prague, Czech Republic, 8–11 June 2009.
45. Hussain, S.S.; Aftab, M.A.; Ustun, T.S. Performance Analysis of IEC 61850 Messages in LTE Communication for Reactive Power Management in Microgrids. *Energies* **2020**, *13*, 6011. [CrossRef]
46. VDE; DKE. *The German Roadmap E-Energy/Smart Grids 2.0. Smart Grid Standardization Status, Trends and Prospects*; VDE Association of Electrical Electronic & Information Technologie, DKE German Commission for Electrical, Electronic & Information Technologies of DIN and VDE, Eds.; English Version; H. Heenemann GmbH & Co.: Berlin, Germany, 2013.
47. Bundesnetzagentur. Monitoringbericht 2019. Available online: https://www.bundesnetzagentur.de/SharedDocs/Mediathek/Monitoringberichte/Monitoringbericht_Energie2019.pdf?__blob=publicationFile&v=6 (accessed on 16 March 2022).
48. *CEER Benchmarking Report 6.1 on the Continuity of Electricity and Gas Supply. Ref: C18-EQS-86-03*; Council of European Energy Regulators Asbl: Brussels, Belgium, 2018.
49. Parol, M. Analysis of indexes concerning interruptions in delivery of electricity in distribution networks. *Prz. Elektrotechniczny (Electr. Rev.)* **2014**, *90*, 122–126. (In Polish)
50. *IEEE Std 493-2007*; IEEE Recommended Practice for the Design of Reliable Industrial and Commercial Power Systems. IEEE: New York, NY, USA, 2007.
51. DIgSILENT. *DIgSILENT PowerFactory Manual*; DIgSILENT GmbH: Gomaringen, Germany, 2017.
52. *IEEE Std 1366TM-2012 (Revision of IEEE 1366-2003)*; IEEE Guide for Electric Power Distribution Reliability Indices. IEEE: New York, NY, USA, 2012.
53. Common T&D Reliability Indices. Available online: https://www.ewh.ieee.org/r6/san_francisco/pes/pes_pdf/Reliability_and_Artificial_Intelligence/Common_T&D_Reliability_Indices.pdf (accessed on 9 July 2022).
54. Billinton, R.; Allan, R.N. *Reliability Evolution of Power Systems*, 2nd ed.; Springer: New York, NY, USA, 1996.
55. Choi, S. Assessment of Reliability in the Distribution System of an Industrial Complex. *J. Electr. Eng. Technol.* **2007**, *2*, 201–207. [CrossRef]
56. CIGRE. CIGRE. CIGRE Task Force 38-03-10. Power System Reliability Analysis. In *Composite Power System Reliability Evaluation, Paris*; CIGRE: Paris, France, 1992.
57. Brown, R.E. *Electric Power Distribution Reliability*, 2nd ed.; CRC Press: Boca Raton, FL, USA, 2009.
58. Paska, J. *Reliability of Electric Power Systems*; Publishing House of the Warsaw University of Technology: Warsaw, Poland, 2005. (In Polish)
59. Stepien, J.C. Probabilistic reliability models of 15 kV cable lines. *Prz. Elektrotechniczny (Electr. Rev.)* **2005**, *3*, 245–249. (In Polish)
60. Bargiel, J.; Goc, W.; Sowa, P.; Tejchman, B.; Paska, J. Reliability of Medium Voltage Power Networks. In Proceedings of the VI Scientific-Technical Conference Power Networks in Industry and Power Engineering, Szklarska Poreba, Poland, 10–12 September 2008; pp. 359–364. (In Polish).
61. Bollen, M.H.J.; Sun, Y.; Ault, G.W. Reliability of distribution networks with DER including intentional islanding. In Proceedings of the FPS 2005 International Conference on Future Power Systems, Amsterdam, The Netherlands, 16–18 November 2005.
62. Sayas, F.C.; Allan, R.N. Generation availability assessment of wind farms. *IEEE Proc.-Gener Transm. Distrib* **1996**, *143*, 507–518. [CrossRef]
63. IEEE Task Force on Models for Peaking Service Units. A four State Model for Estimate Outage Risk for Units in Peaking Service. *IEEE Trans. Power Appar. Syst.* **1972**, *91*, 618–627.
64. Smith, A.C.; Donovan, M.D.; Bartos, M.J. Reliability survey of 600 to 1800 kW diesel and gas-turbine generating units. *IEEE Trans. Ind Appl.* **1990**, *26*, 741–755. [CrossRef]
65. Cha, S.T.; Jeon, D.H.; Bae, I.S.; Lee, I.R. Reliability Evaluation of Distribution System Connected Photovoltaic Generation Considering Weather Effects. In Proceedings of the International Conference on Probabilistic Methods Applied to Power Systems, Ames, IA, USA, 12–16 September 2004; pp. 451–456.
66. Usman, A.; Shami, S. Evolution of Communication Technologies for Smart Grid Applications. *Renew. Sust. Energ. Rev* **2013**, *19*, 191–199. [CrossRef]
67. Medhi, D.; Ramasamy, K. *Network Routing: Algorithms, Protocols, and Architectures*, 2nd ed.; Elsevier: Amsterdam, The Netherlands, 2018.
68. Wei, M.; Chen, Z. Communication Systems and Study Method for Active Distribution Power Systems. In Proceedings of the 9th Nordic Electricity Distribution and Asset Management Conference, Aalborg, Denmark, 6–7 September 2010.
69. Ali, M. Reliability of information and communication technology equipment in power system–study review. *Nontechnical Res. Proj.–Masters Otto-Von-Guericke-Univ. Magdebg.* **2014**, 1–67. [CrossRef]
70. Zhang, P.; Chan, K. Reliability Evaluation of Phasor Measurement Unit Using Monte Carlo Dynamic Fault Tree Method. *IEEE Trans. Smart Grid* **2012**, *3*, 1235–1243. [CrossRef]
71. Dolezilekm, D.J. *Choosing Between Communications Processors, RTUs, and PLCs as Substation Automation Controllers*; Schweitzer Engineering Laboratories, Inc.: Pullman, WA, USA, 2000; White Paper.

72. Momken, B. *High Availability Solutions, Avaya Servers and Media Gateways. Avaya Aura™ Communication Manager (Avaya CM) Software*; Avaya Inc.: Durham, NC, USA, 2010; Issue 6.0.
73. Müller, S.; Georg, H.; Rehtanz, C.; Wietfeld, C. Hybrid Simulation of Power Systems and ICT for Real-Time Applications. In Proceedings of the Innovative Smart Grid Technologies Europe (ISGT Europe), 3rd IEEE PES, Berlin, Germany, 14–17 October 2012.
74. Suslov, K.; Piskunova, V.; Gerasimov, D.; Ukolova, E.; Akhmetshin, A.; Lombardi, P.; Komarnicki, P. Development of the methodological basis of the simulation modelling of the multi-energy systems. In Proceedings of the International Scientific and Technical Conference Smart Energy Systems, E3S Web of Conferences, Kazan, Russia, 18–20 September 2019.
75. Zio, E.; Sansavini, G. Modeling Interdependent Network Systems for Identifying Cascade-Safe Operating Margins. *IEEE Trans. Reliab.* **2011**, *60*, 94–101. [CrossRef]
76. Rinaldi, S. Modeling and Simulating Critical Infrastructures and Their Interdependencies. In Proceedings of the 37th Hawaii International Conference on System Sciences, Big Island, HI, USA, 5–8 January 2004.
77. Boccaletti, S.; Latora, V.; Moreno, Y.; Chavez, M.; Hwang, D.-U. Complex networks: Structure and dynamics. *Phys. Rep.* **2006**, *424*, 175–308. [CrossRef]
78. Tranchita, C.; Hadjsaid, N.; Viziteu, M.; Rozel, B.; Caire, R. ICT and Powers Systems: An Integrated Approach. In *Chapter 5: Securing Electricity Supply in the Cyber Age: Exploring the Risks of Information and Communication Technology in Tomorrow's Electricity Infrastructure*; Springer: New York, NY, USA, 2010.
79. Casteren, J.; van Bollen, M.; Schmieg, M. Reliability assessment in electrical power systems: The Weibull-Markov stochastic model. *IEEE Trans. Ind Appl.* **2000**, *36*, 911–915. [CrossRef]
80. Trojan, P. Reliability Assessment of Smart Grid. Master's Thesis, Otto-von-Guericke-Universität Magdeburg, Magdeburg, Germany, August 2013.
81. Trojan, P.; Arendarski, B.; Komarnicki, P. Reliability assessment of smart grid. In Proceedings of the Tagungsband Zum Power and Energy Student Summit 2014 in Stuttgart. Kongress PESS, Stuttgart, Germany, 23–28 January 2014; pp. 131–136.
82. Godha, N.R.; Deshmukh, S.R.; Dagade, R.V. Time sequential Monte Carlo Simulation for Evaluation of Reliability Indices of Power Distribution System. In Proceedings of the 2012 IEEE Symposium on Computers and Informatics-ISCI 2012, Penang, Malaysia, 18–20 March 2012.
83. Billinton, R.; Allan, R.N. *Reliability Evaluation of Engineering Systems–Concepts and Techniques*, 2nd ed.; Plenum Press: New York, NY, USA, 1992.
84. Djapic, P.; Ramsay, C.; Pudjianto, D.; Strbac, G.; Mutale, J.; Jenkins, N.; Allan, R. Taking an Active Approach. *IEEE Power Energy Mag.* **2007**, *5*, 68–77. [CrossRef]
85. Rudion, K.; Styczynski, Z.A.; Hatziargyriou, N.; Papathanasiou, S.; Strunz, K.; Ruhle, O.; Orths, A.; Rozel, B. Development of Benchmarks for Low and Medium Voltage Distribution Networks with High Penetration of Dispersed Generation. In Proceedings of the International Symposium "Modern Electric Power Systems–MEPS'06", Wroclaw, Poland, 6–8 September 2006.

energies

MDPI

Article

Evolutionary Multi-Objective Optimization Applied to Industrial Refrigeration Systems for Energy Efficiency

Nadia Nedjah [1,*,†], Luiza de Macedo Mourelle [2,†] and Marcelo Silveira Dantas Lizarazu [1,†]

1 Department of Electronics Engineering and Telecommunications, State University of Rio de Janeiro, Rio de Janeiro 20550-900, Brazil; lizarazu@eng.uerj.br
2 Department of Systems Engineering and Computation, State University of Rio de Janeiro, Rio de Janeiro 20550-900, Brazil; ldmm@eng.uerj.br
* Correspondence: nadia@eng.uerj.br
† These authors contributed equally to this work.

Abstract: Refrigeration systems based on cooling towers and chillers are widely used equipment in industrial buildings, such as shopping centers, gas and oil refineries and power plants, among many others. Cooling towers are used to recover the heat rejected by the refrigeration system. In this work, the refrigeration is composed of cooling towers dotted with ventilators and compression chillers. The growing environmental concerns and the current scenario of scarce water and energy resources have lead to the adoption of actions to obtain the maximum energy efficiency in such refrigeration equipment. This backs up the application of computational intelligence to optimize the operating conditions of the involved equipment and cooling processes. In this context, we utilize multi-objective optimization algorithms to determine the optimal operational setpoints of the cooling system regarding the cooling towers, its fans and the included chillers. We use evolutionary multi-objective optimization to provide the best trade-offs between two conflicting objectives: maximization of the effectiveness of the cooling towers and minimization of the overall power requirement of the refrigeration system. The optimization process respects the constraints to guarantee the correct and safe operation of the equipment when the evolved solution is implemented. In this work, we apply three evolutionary multi-objective algorithms: Non-dominated Sorting Genetic Algorithm (NSGA-II), Micro-Genetic Algorithm (Micro-GA) and Strength Pareto Evolutionary Algorithm (SPEA2). The results obtained are analyzed under different scenarios and models of the cooling system's equipment, allowing for the selection of the best algorithm and best equipment's model to achieve energy efficiency of the studied refrigeration system.

Keywords: energy efficiency; cooling towers; chillers; evolutionary multi-objective optimization

Citation: Nedjah, N.; de Macedo Mourelle, L.; Lizarazu, M.S.D. Evolutionary Multi-Objective Optimization Applied to Industrial Refrigeration Systems for Energy Efficiency. *Energies* **2022**, *15*, 5575. https://doi.org/10.3390/en15155575

Academic Editors: Paweł Piotrowski, Grzegorz Dudek and Dariusz Baczyński

Received: 27 June 2022
Accepted: 26 July 2022
Published: 1 August 2022

1. Introduction

The technical and scientific community is moving fast towards adopting premises and drastic measures that allow the achievement of a maximal level of energy efficiency of industrial installations. This is due to the ever growing environmental concerns regarding the inefficient electrical power usage and its ever growing demand, as well as to the misuse of water resources. So, in order to achieve energy efficiency in industrial refrigeration systems, we require the utilization of modern mechanisms and methodologies that allow yielding a good or maybe the best possible solution for a process. Many industrial processes generate unwanted heat. So, this heat often must be somehow dissipated. In this case, water is generally used. The returning water in refrigeration systems is often at higher temperatures. It can be discarded or cooled down for further usage. However, the disposal of water is an environmentally unsustainable practice. Furthermore, the disposal of water, which comes at a high temperature would have a very negative impact on the local underwater flora and fauna. Hence, modern sustainable refrigeration system must be designed, configured and operated to reuse water. It is noteworthy to point out that there are more advanced refrigeration systems that are based on the usage of cryogenic fluids [1]. These kind of systems also aim at achieving high degrees of energy efficiency as required in critical systems, such as spaceships and nuclear stations [2]. An interesting survey of refrigeration methods can be found in [3].

Cooling towers are the basic equipment of industrial refrigeration systems. They are intended whenever there are large cooling demands. Moreover, cooling towers offer a clean and economical solution to water reuse in the cooling process. A cooling tower operates together with other equipment such as fans, chillers and pumps to ensure water circulation in the system [4,5]. A coordinated configuration of all the equipment composing the cooling system must be guaranteed. This is because a modification of some parameter in one of these equipment items can impact either positively or negatively the performance of the others parts of the system. When the cascading effects are unsatisfactory to the refrigeration system, a reduction of energy efficiency is often observed.

In this work, we propose to exploit computational intelligence techniques to optimize the energy requirement and effectiveness of an industrial refrigeration system composed of cooling towers, tower ventilators and chillers. For this purpose, quantitative and qualitative data are required to achieve good results. These data are usually collected from field data and data-sheets provided by the equipment manufacturers.

The attainable energy efficiency of a cooling tower is intrinsically dependent on that of the heat exchange process between the returning hot water and the air volume induced in counter-flow to this in the tower via ventilators. It is also influenced by climatic and operational aspects. This optimization is a complex process, and is mainly dependent on the precision of the model used for the equipment of the overall system.

The multi-objective optimization is two-fold. It aims at maximizing the efficiency of the heat exchange performed by the cooling tower while minimizing the global energy requirement of the refrigeration system. The optimization takes into account all the equipment necessary for the correct and safe operation of the refrigeration system. In this work, three evolutionary multi-objective optimization algorithms are applied: NSGA-II, Micro-GA and SPEA2. These algorithms will deliver the optimal settings of the system's parameters to configure the composing cooling towers, tower fans and chillers. Mainly, the variables for which the optimization process will answer for are the cooling tower fan speed setpoint and the water temperature setpoint to be provided by the chillers. It is needless to state that the proposed optimization respects the restrictions imposed for a proper and safe operation of all the involved equipment composing the refrigeration system. The restrictions are set as provided by the equipment suppliers. The cooling system used in this work is based on compression chillers. Herein, such chillers are modeled in two different ways: a simple model wherein only one variable is considered and a more complete one wherein two variables are taken into account. The results yielded from both models are compared in terms of accuracy with respect to the field data. The two models provided for the chillers are used to set up the two objective functions for the optimization process. We also explore two different scenarios regarding the stopping criteria of the optimization algorithms. The performance results using different models and stopping criteria are compared, allowing the selection of the best algorithm for each scenario and the best model for the application.

This paper is structured into six sections. First, in Section 2, we briefly introduce the structure of the studied refrigeration system. Then, in Section 3, we provide a review of related research works. In the sequel, in Section 4, we define the objective functions and operational restrictions. After that, in Section 5, we describe the methodology behind each of the optimization algorithms applied in this work. Then, in Section 6, we analyze the evolved results for different algorithms, stopping scenarios and system models. Subsequently, in Section 7, we compare the effectiveness and efficiency of the used algorithm regarding the achievement of the main objective, which is the energy efficiency of the refrigeration system. Finally, in Section 8, we draw some conclusions and point out some promising directions for future work.

2. System's Structure

The refrigeration system to be optimized is composed of chillers and cooling towers. This configuration is commonly used in commercial buildings and industrial facilities to ensure the thermal comfort of the transiting people and adequate equipment cooling and electrical rooms. The configuration of the cooling system considered in this work is presented in Figure 1. It includes two cooling towers, each composed of three elementary cells. Each cell includes a fan operating with an electric motor. Considering all the components composing the cooling tower, only the fans allow speed variation, through the use of frequency converters, while the others always remain operating at a fixed speed and equal to the nominal one.

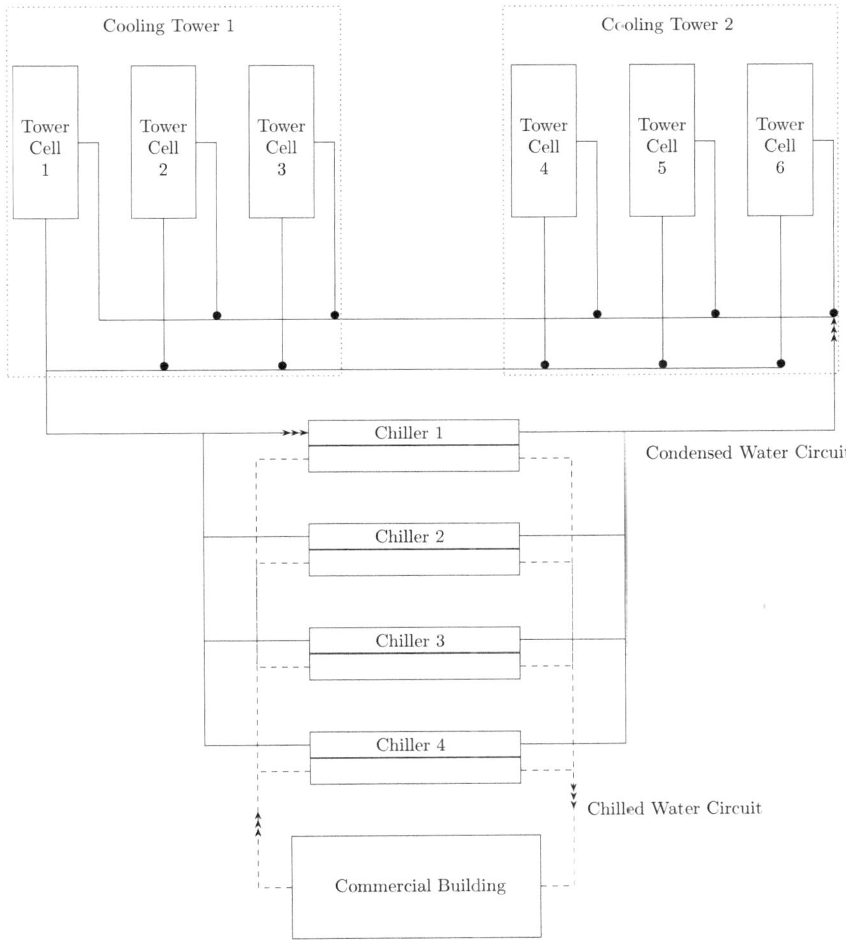

Figure 1. Refrigeration system's configuration.

In the case under study, the number of condensation water lift pumps in operation must be equal to the number of chillers in operation. Hence, the total number of cells in operation in the cooling towers can also be obtained based on the number of chillers in operation.

Among the equipment that composes the refrigeration system considered in this work, only the tower fans allow speed variation, through the use of frequency converters Lift pumps and chillers operate at fixed speed, which is equal to the rated speed. Thus, as the condensed water pumps are not influenced by the speed variation of the tower fans nor by the variation in the temperature of the water passing through the chillers, both in the condenser and in the evaporator, the required energy cannot be taken into account in the optimization process. Therefore, the optimization will be dedicated to the electrical energy demand of the fans and the chillers.

3. Related Works

In [6], the energy efficiency of the refrigeration system is achieved through a control strategy based on extreme search. The proposed control system is based on the global energy requirements, composed of chillers and tower fans. It attempts to reach energy efficiency exploiting variation of the fan speed setpoint. In [7], an extreme search strategy very similar to that presented in [6] is presented. The variable manipulated by the control system is the cooling tower output temperature, in contrast with the work reported in [6]. It exploits the tower ventilators. The achieved improvements vary in function of the chiller's thermal load.

In [8], a control strategy called Optimum Approach Temperature (OAT) is proposed for the energy optimization of the cooling tower. The approach concept represents the difference between the condensing water temperature and the wet bulb temperature The OAT strategy is an optimization that can only be applied to cooling towers.

In [9], an optimal control strategy for a chiller-based refrigeration system is presented. In this work, the equipment model precision is ensured via an online updating process of the underlying parameters. It relies on the recursive least squares method. A genetic algorithm is used as a global optimization tool. The used cost function, which must be minimized, models the global energy as required by the chillers, fans and condensed water pumps.

In [10], an energy optimization system based on simulation for the refrigeration system is proposed. Therein, the chillers are driven by frequency converters, and the tower fans and condensing water pumps operate at predefined velocity. The optimization system uses evolutionary computing. The cost function, which must be minimized, models the energy demand of the refrigeration system regarding the chiller's load, cooling tower ventilators and water pumps. The optimization process considers three kinds of restrictions. The first one guarantees that at any time, the tower thermal capacity must be higher than the chillers' cooling load. The second restriction upholds the minimal and maximal thresholds for the water temperature. The third one allows to maintain water flow within the prescribed minimal and maximal threshold.

In [11], a model that is based on prior experiments is proposed. It allows to simultaneously optimize the available performance parameters while ensuring a minimum energy consumption from an induced draft cooling tower operating under a given set of conditions. It is claimed that the proposed model for the cooling tower performance is suitable for on-line optimization. The objective function is formulated dependent on several performance parameters such as the approach, tower characteristic ratio, effectiveness and evaporation rate, air and water flow rates.

In [12], an overview of the research and development of optimization approaches for water-cooled refrigeration systems is presented. This work survey allows to understand the new significant directions and innovative results in this field. Therein, a taxonomy of the existing optimization approaches is proposed.

4. Problem Formalization

The effectiveness of the cooling tower is defined as its operational efficiency, and is related to the efficiency of the heat exchange between the hot water coming from the process and the air mass induced in the tower in counter-current, through fans. This efficiency is influenced by several factors, which are explained in the modeling of the cooling tower [13]. Among the factors that influence the effectiveness of the tower, we have the relationship between the water and air flows inside the tower and climatic factors, defined by external and wet bulb temperatures. In this work, the water flow that reaches the tower cells only varies as a function of the number of pumps that are in operation, i.e., as a function of the number of operating chillers. On the other hand, the air flow in each cell can vary continuously through the variation of the fan speed. The external temperature influences the thermal load to be served by the chillers, and the wet bulb temperature influences the efficiency of the thermal exchange of the tower, as it represents the lowest possible outlet temperature to be reached. Thus, this work aims to explore multi-objective optimization in order to solve the problem composed of the following conflicting objectives:

- Maximizing the effectiveness of the cooling tower;
- Minimizing the overall energy consumption of the refrigeration system.

To this end, the process variables are collected in the field from the instrumentation already installed in the cooling towers. Local weather conditions are provided by a weather station installed and integrated into the cooling system. So, based on the process data provided by the existing Supervisory Control and Data Acquisition (SCADA) system, the following variables are provided as inputs to the optimization system proposed: the number of chillers that are in operation; the temperature of the hot water reaching the cooling tower; the wet bulb temperature on site; the flow of water that reaches the cooling tower; and the water flow that leaves each chiller.

In this work, the model considers adjustments in the speed of the tower fans as well as adjustments in the chilled water temperature leaving the chillers. This modeling deals with two conflicting variables.

In the studied refrigeration system, the cooling tower operates in conjunction with compression chillers. These occasion the highest energy consumption. The condensed water and chilled water circulation pumps always operate at a fixed speed. So, the inclusion of these into the calculation of the overall energy required by the cooling system does not provide any advantage, as the objective is to evaluate the energy efficiency as achieved after application of the optimization algorithms. Thus, only the consumption of the chillers and tower fans are considered in the implementation of the proposed energy optimization system.

As a premise for the implementation, we consider that the optimal output values of the optimization system must be obtained based on the best compromise between the objectives established above, respecting the operational limits and restrictions defined for the equipment that compose the cooling system. The objective is to obtain, at each predefined interval of one hour, the best setpoint of speed for the tower fans and/or the best setpoint of the temperature of the chilled water leaving the chiller, depending on the modeled scenario. The optimization simulations will be performed using the improved version of three evolutionary algorithms: Strength Pareto Evolutionary Algorithm, Non-Dominated Sorting Genetic Algorithm and Micro-Genetic Algorithm. Note that an explanation of the dynamics of the used optimization algorithms will be provided in Section 5.

4.1. Objective Functions

In this work, we optimize two conflicting objective to solve the energy efficiency problem. The first objective function, $\mathbb{F}_1$, estimates the tower's effectiveness while the second objective function, $\mathbb{F}_2$, approximates the required power of the refrigeration system. Thus, finding the solution that maximizes function $\mathbb{F}_1$ allows the maximization of the heat exchange efficiency of the cooling tower. As we intend to use multi-objective optimization, the found solution will also minimize function $\mathbb{F}_2$, allowing the minimization of the power consumption of the cooling system.

Objective function $\mathbb{F}_1$, which evaluates the efficiency of the heat exchange of the cooling tower is defined in Equation (1):

$$\begin{aligned} \max \mathbb{F}_1 &= \epsilon_a \\ &= c_0 + c_1\left(\frac{\dot{m}_a}{\dot{m}_w}\right) + c_2(T_{w_i} - T_b) + c_3\left(\frac{\dot{m}_a}{\dot{m}_w}\right)^2 + c_4(T_{w_i} - T_b)^2 + c_5\left(\frac{\dot{m}_a}{\dot{m}_w}\right)(T_{w_i} - T_b) \end{aligned} \quad (1)$$

wherein ϵ_a represents the effectiveness of the cooling tower, $\dot{m}_a$ and $\dot{m}_w$ represent the mass flow of air and water and T_{w_i} and T_b represent the temperature of inlet water and that of the bulb. For details about the model's variables, see [13]. The objective function $\mathbb{F}_2$, which evaluates the power required by the system composed of chillers and cooling tower fans is defined in Equation (2):

$$\begin{aligned} \min \mathbb{F}_2 &= n_1 P_v + n_2 P_{ch} \\ &= n_1 \sqrt{3} V_n I_n \left(d_0 \left(\frac{\dot{m}_a}{\dot{m}_{an}}\right)^3 + d_1 \left(\frac{\dot{m}_a}{\dot{m}_{an}}\right)^2 + d_2 \frac{\dot{m}_a}{\dot{m}_{an}} + d_3 \right) \\ &\quad + n_2 Q_{ch_{no}} E_{no} Z_C(T_{ae_{co}}, T_{ae_{ev}}) Z_E(T_{ae_{co}}, T_{ae_{ev}}), \end{aligned} \quad (2)$$

where n_1 and n_2 are discrete variables, representing the number of fans and chillers that must operate in order to meet the requested thermal demand and the commitment to lower energy consumption, respectively. Moreover, P_v and P_{ch} represent the electrical power demanded by fans and chillers, respectively. Recall that the number of fans in operation corresponds to the number of tower cells required in order to guarantee its operational limits. In this problem, we have $n_1 = n_2 + 1$. Moreover, terms Z_C and Z_E of Equation (2) are defined as in Equation (3):

$$\begin{aligned} Z_C = \; & b_0 + b_1 \Delta T_{ag} + b_2 \Delta T_{ag}^2 + b_3 T_{ae_{co}} + b_4 T_{ae_{co}}^2 + b_5 \Delta T_{ag}^2 T_{ae_{co}} + b_6 \Delta T_{ag} T_{ae_{co}}^2; \\ Z_E = \; & a_0 + a_1 T_{ae_{ev}} + a_2 T_{ae_{ev}}^2 + a_3 T_{ae_{co}} + a_4 T_{ae_{co}}^2 + a_5 T_{ae_{ev}} T_{ae_{co}}, \end{aligned} \quad (3)$$

wherein we have $\Delta T_{ag} = T_{ae_{ev}} - T_{as_{ev}}$ [14]. It is noteworthy to emphasize that all the aforementioned variables are fully defined herein or in the model descriptions of the cooling tower and fans [13] and/or of the chillers [14]. The coefficients $a_0 \dots a_5$, $b_0 \dots b_6$, $c_0 \dots c_5$, $d_0 \dots d_3$ are obtained using the Levemberg–Marquardt method as a non-linear regression technique [15]. Their values are given in Table 1. The precision and faithfulness of the resulting models are validated using real field data as proven in [13,14].

Table 1. Model's coefficients to evaluate the system's effectiveness and the power required by the refrigeration system.

Z_E	Value	Z_C	Value	ϵ_a	Value	P_v	Value
a_0	−1.0405	b_0	−0.1177	c_0	+0.0262	d_0	+0.7931
a_1	+0.1379	b_1	+0.3381	c_1	+0.4935	d_1	+0.0330
a_2	−0.0090	b_2	−0.0513	c_2	+0.14350	d_2	+0.0557
a_3	+0.0840	b_3	−0.0276	c_3	−0.0289	d_3	+0.0039
a_4	−0.0022	b_4	+0.0022	c_4	−0.0129		
a_5	+0.0033	b_5	+0.0030	c_5	−0.0533		
		b_6	−0.0006479				

Restrictions

For the optimization problem, four operational constraints related to the considered refrigeration system are required to guarantee correct system operation. The first constraint $\mathbb{G}_1$ concerns the lowest possible value to be reached by the cooling tower outlet temperature. It cannot be lower than the local instantaneous wet bulb temperature due to the saturation of the air leaving the tower after heat transfer and mass with the hot water that reaches the tower. The wet bulb temperature varies throughout the day and can be calculated as a function of ambient temperature and relative humidity. Therefore, the first restriction is defined as in Equation (4):

$$\mathbb{G}_1 : T_{as} \geq T_b, \quad (4)$$

wherein T_{as} represents the cooling tower leaving water temperature and T_{BU} represents the wet bulb temperature.

The second constraint G_2 models the operational conditions of the chiller considered in this work. The manufacturer of the chiller establishes in [16] a restriction regarding the temperature difference between the water inlet and outlet of the condenser. The surge curve of the chiller can be found [17], where it is possible to observe two operating zones for the chiller: with or without surge. The operation in the surge zone of the chiller compressor causes a series of inconveniences, such as vibrations and load oscillations, generating mechanism wear and unexpected performance of the electrical protection in cases of overload. Furthermore, in this operating condition there is a considerable reduction in the coefficient of performance (COP) of the equipment. The COP of a chiller represents the relationship between the cooling capacity ($kW_{thermal}$) and the electrical power required ($kW_{electric}$) for its operation. So, the chiller should preferably operate in the zone below the surge line. It represents the maximum admissible limit for the temperature difference between the inlet and outlet of water in the condenser as a function of the chiller load. Based on this, the second restriction can be defined as in Equation (5):

$$\mathbb{G}_2 : \Delta T_{co} \leq 7,3c_t - 0.3, \quad \text{with} \quad \Delta T_{co} = T_{ae} - T_{as}, \tag{5}$$

wherein c_t is the chiller load factor, with $c_t \in [0, 15, 1]$, T_{ae} is the temperature of the water that leaves the chiller condenser and leaves towards the cooling tower, and T_{as} is the temperature of the water leaving the cooling tower and going towards the condenser inlet. Note that the manufacturer does not recommend operating the chiller with a load below 15% [16]. So, we have to consider a third constraint. It is defined as in Equation (6):

$$\mathbb{G}_3 : 15\% \leq c_{t\%} \leq 100\%. \tag{6}$$

Moreover, the nominal design temperature of the cooling tower is 36.4 °C [17]. Therefore, temperatures above this value should be avoided. So, we must impose a fourth restriction, which concerns the maximum limit of the water inlet temperature in the cooling tower. We define this constraint as in Equation (7):

$$\mathbb{G}_4 : T_{ae} \leq 36.4. \tag{7}$$

5. Evolutionary Algorithms for Multi-Objective Optimization

There are several evolutionary algorithms for multi-objective optimization. The main and more efficient ones are based on the Pareto dominance concept [18,19]. Techniques based on the Pareto concept can be classified into non-elitist techniques and elitist techniques [20]. Multiple Objective Genetic Algorithm (MOGA) [21], Non-Dominated Sorting Genetic Algorithm (NSGA) [22] and Niched Pareto Genetic Algorithm (NPGA and NPGA-II) [23] are examples of non-elitist techniques. Pareto Archived Evolution Strategy (PAES) [24], Memetic Pareto Archived Evolution Strategy (M-PAES), Pareto Envelope-Based Selection Algorithm (PESA and PESA-II), Strength Pareto Evolutionary Algorithm (SPEA and SPEA2) [25], Non-Dominated Sorting Genetic Algorithm II (NSGA-II) [26] and Multiobjective Messy Genetic Algorithm (MOMGA and MOMGA-II) [27] are examples of elitist techniques.

The implementation of elitism in genetic algorithms can significantly accelerate performance [28]. It prevents premature loss of good solutions, according to results presented in [29,30]. The first approach uses elitism is SPEA in [29]. There follows PESA [31], PAES [24], MOMGA [27] and NSGA-II [26]. Since then, elitism is used systematically.

More recently, some elitist algorithms for multi-objective optimization problems are presented with improvements to some of the already established methods, such as SPEA2, NSGA-II and PESA-II. Aiming at these improved algorithms, we have SPEA2+ [32], Chaotic-NSGA-II [33], IPESA-II [34] and NSGA-III [35]. However, there are still no records of a significant number of applications of these algorithms. The purpose of these improved methods is to obtain greater diversity and greater speed of convergence, in order to solve extremely complex problems.

Among the most recently proposed algorithms, NSGA-III stands out, which is an improvement on NSGA-II for applications with many objectives (from four objectives). This algorithm is based on the concept of reference point, emphasizing non-dominated individuals close to a set of reference points provided and updated throughout the iterations. In this way, the maintenance of diversity is achieved through the adaptive update of the reference points distributed in the search space. In NSGA-III, the crowding distance operator, used in NSGA-II, is replaced by the clustering operator, which operates based on distributed reference points. In [35], the NSGA-III is compared to the MOEA/D algorithm, showing satisfactory results.

For the application of energy optimization proposed in this work, only multi-objective algorithms based on the Pareto concept that implement elitism will be used. This follows from the bibliographic study carried out. We found out that these strategies present a better performance in most applications. In addition, due to the fact that the proposed work regards an engineering application that involves a feasibility study for the implementation, the exploitation of multi-objective optimization algorithms already applied to engineering problems must be prioritized. This same consideration is carried out in [36].

The Micro-GA algorithm is a good option for the application at hand, since the operational restrictions of the equipment that compose the cooling system limit the search space to a relatively small region. Therefore, in this work, the multi-objective evolutionary algorithms chosen for the solution of the proposed optimization problem are: SPEA2 [37], NSGA-II [26] and Micro-GA [21]. In the sequel, we give a brief description of the optimization strategies adopted in each of the applied algorithms.

5.1. SPEA2

The main steps of SPEA2 are sketched in Algorithm 1. This algorithm was developed as an improvement of SPEA, and incorporates techniques that should improve the efficiency of the optimization process. It requires variables N, $\overline{N}$ and T, which represent the population size, the external population size (file) and the maximum number of generations, respectively. It returns the set of non-dominated individuals A that establish the best compromise with the defined objectives and constraints.

The methodology implemented in SPEA2 can be explained through the following steps [37]:

Step 1 *Initialization:* Initially, two populations are generated: a random initial population P_0 and an initial external population, termed *file*, such that $\overline{P_0} = \varnothing$. Variable t is defined and set to 0, which must be incremented with each new generation of new non-dominated individuals.

Step 2 *Fitness evaluation:* Each solution in the current populations P_t and $\overline{P_t}$ is evaluated with respect to the objective functions. Then, it is evaluated with respect to dominance relationships. So, each individual is evaluated in relation to the individuals that it dominates and to those that dominate it. When this step is performed for the first time, only individuals from population P_t will be evaluated. Therefore, each individual i of population P_t and in file $\overline{P_t}$ will be assigned a value called *strength*, represented by $S(i)$. The strength of individual i coincides with the number of solutions individual i actually dominates, and it is defined as in Equation (8):

$$S(i) = |\{ j | j \in P_t \cup \overline{P_t} \wedge i \succ j\}|. \tag{8}$$

Moreover, each individual is associated with a value called *raw fitness* that is equivalent to the sum of the strengths of all the individuals that dominate the individual under analysis, both in the population and in the file, as defined in Equation (9):

$$R(i) = \sum_{j \in P_t \cup \overline{P_t}, j \succ i} S(j). \tag{9}$$

Note that the strength of a given individual i will be higher when more individuals are dominated by i, and its raw fitness will be lower when less individuals dominate i. Although the raw fitness provides assignments to individuals based on Pareto dominance, if there are many individuals with identical raw fitness values, this mechanism may fail. Therefore, SPEA2 uses neighborhood density information to effectively guide the search. An adaptation of the kth-nearest neighbor method is used, wherein the density at any point is a function of the distance to the kth-nearest neighbor. In this case, SPEA2 simply takes the inverse of the distance to the kth-nearest neighbor as an estimate of the density. The most accurate way to estimate neighborhood density is to calculate the Euclidean distance in the feasible region from an individual i to each individual j in the file and in the population, and store the obtained values in a list. Another possible way is to consider the term $k = \sqrt{N + \overline{N}}$ as a common point and list the results obtained for all individuals. After sorting the list in ascending order. The kth neighbor will be the one that gives the smallest distance sought, denoted by σ_i^k. Therefore, the density $D(i)$, corresponding to the individual i, is defined as in Equation (10):

$$D(i) = (\sigma_i^k + 2)^{-1}. \tag{10}$$

Note that constant 2 is added to the denominator in order to ensure that its value is greater than zero, and that the density is always less than 1. Finally, the fitness value of the individual is simply defined by $F(i) = R(i) + D(i)$. It is noteworthy to mention that the lower the value of an individual's fitness, the more apt it is, and hence the more chances it will have to propagate over generations and disseminate its characteristics to other individuals.

Algorithm 1 Main steps of SPEA2.

Require: $N, \overline{N}, T$
Ensure: A
generate P_0 randomicallt, with $|P_0| = N$
generate $\overline{P_0} = \varnothing$
$t := 0$;
while true **do**
  compute *Fitness* in P_t and $\overline{P_t}$
  copy non-dominated solutions in P_t and $\overline{P_t}$ to $\overline{P_{t+1}}$
  if $|\overline{P_{t+1}}| > |\overline{N}|$ **then**
    repeat
      reduce $|\overline{P_{t+1}}|$ Using slicing algorithm
    until $|\overline{P_{t+1}}| = |\overline{N}|$
  else if $|\overline{P_{t+1}}| < |\overline{N}|$ **then**
    repeat
      complete $\overline{P_{t+1}}$ with P_t and $\overline{P_t}$
    until $|\overline{P_{t+1}}| = |\overline{N}|$
  end if
  if $t \geq T$ **then**
    save in A the set of non-dominated solution of $\overline{P_{t+1}}$
    halt
  else
    apply selection binary operator with reposition in $\overline{P_{t+1}}$
    apply recombination operator
    apply mutation operator
    save in P_{t+1} the genetic operators' results
    $t := t + 1$
  end if
end while

Step 3 *Contextual selection:* In this step, all non-dominated individuals from population P_t and file $\overline{P_t}$ are copied to next generation file $\overline{P_{t+1}}$. If the size of $\overline{P_{t+1}}$ exceeds $\overline{N}$, it must reduce use of the *slicing algorithm*. If the size of $\overline{P_{t+1}}$ is smaller than $\overline{N}$, $\overline{P_{t+1}}$ must be completed using the best dominated individuals in P_t and $\overline{P_t}$. The slicing algorithm is an iterative process that eliminates, at each iteration, the individual with the smallest Euclidean distance to the nearest neighbor. In the case of a tie, the second smallest Euclidean distance is verified, and so on. The iterative process ends when the population dimension of $\overline{P_{t+1}} = \overline{N}$.

Step 4 *Finalization:* If $t \geq T$, or any other used stopping criterion is satisfied, A is defined as the set of non-dominated individuals that represent the best solution in $\overline{P_{t+1}}$ and for the optimization process. If the stopping conditions are not yet met, proceed with the selection at Step 5.

Step 5 *Selection:* In this step, individuals are selected through the selection operator by a tournament, whose winners are the individuals with the lowest fitness value.

Step 6 *Crossover and mutation:* In this step, the selected individuals are recombined using crossover and mutation operators, thus generating the new individuals of population P_{t+1}. Then, the generation counter is incremented ($t = t + 1$) and the fitness calculation at Step 2 is to be returned to.

5.2. NSGA-II

The main steps of NSGA-II are sketched in Algorithm 2. Initially, NSGA-II generates a random population P_0, with $|P_0| = N$. This initial population is ordered based on solution non-domination. Thus, in this first iteration, a fitness value is calculated for each solution, which makes it possible to determine its respective level of dominance.

Algorithm 2 Main steps of NSGA-II.

Require: T, N
Ensure: Q_{t+1}
$P_0 := Q_0 := 0$; Generate P_0 randomically with $|P_0| = N$; $t := 0$
Apply tournament selection
Apply crossover, recombining solutions; Apply mutation; Generate Q_0
while $t < T$ **do**
  $R_t := P_t \cup Q_t$; Sort R_t using non-dominance; $P_{t+1} := 0$; $i := 1$
  while $|P_{t+1}| \leq N$ **do**
    Compute crowding distance for N_i
    if $|N_i| > |(N - P_{t+1})|$ last spots in P_{t+1} **then**
      Sort N_i regarding crowding operator ($\prec_{obj}$)
      $P_{t+1} := P_{t+1} \cup N_i[1 : (|N| - |P_{t+1}|)]$
    else
      $P_{t+1} := P_{t+1} \cup N_i$
    end if
    $i := i + 1$
  end while
  Apply crossover, recombining solutions; Apply mutation; Generate Q_{t+1}
  $t := t + 1$
end while

In order to choose the best solution, tournament selection is used. Then, recombination and mutation operators are applied to generate solution offspring. The first population of descendants is named Q_0, with $|Q_0| = N$. Then, both the initial populations P_0 and Q_0 are pooled into a single population $R_0 = P_0 \cup Q_0$, with $|R_0| = 2N$. This is the procedure used to generate the initial population R_0 in the first iteration.

In the following t iterations, where $t = 1, 2, 3, \cdots, T$, with T representing the maximum number of iterations, a population R_t ordered by non-dominance is handled. Elitism is guaranteed by combining the previous and current populations in R_t. After sorting, non-dominated solutions are ranked at the level (or boundary) N_1, and these come to play a leading role during the process. The remaining solutions are ranked at one of the levels N_2, N_3 and so on, up to the last level N_d, so that all individuals belong to a certain level of domination. If the size of N_1 is smaller than N, the algorithm considers that all its individuals form the new population P_{t+1}. The remaining space in this new population, that is, $|N| - |N_1|$ spots, must be filled in by the individuals of the subsequent non-dominated levels, using the crowding distance-based comparison operator to select the last remaining spots in P_{t+1}.

In NSGA-II, the fitness of each individual i is called $rank_j$, and depends on the boundary or dominance level to which it belongs and the operator based on the crowding operator, generally represented by $\prec_m$. The latter, in turn, depends on the value of crowding distance $dist_i$ of the evaluated individual i regarding a given objective. In this way, each individual i is compared to an individual j in order to choose which one of them should belong to the new population P_{t+1}.

Crowding operator $\prec_m$ for objective m helps in the algorithm selection process, in order to allow the convergence to the Pareto optimal front. The *crowded comparison* defines that the individuals selected for the new population P_{t+1} will be those with a lower value of *rank*. Therefore, an individual j will be chosen if it has a *rank* less than an individual $p \neq j$, i.e., $rank_j < rank_p$). If the individuals j and p have the same rank, the one associated with the highest value of crowding distance will be selected. That is, if $rank_j = rank_p$, we choose j if $dist_j > dist_p$. Otherwise, the individual p is chosen.

Algorithm 3 shows the procedure to compute the crowding distance, where ℓ is the number of individuals (solutions) contained in the set T, $f_{obj}(i)$ is the value of the *obj*-th objective function for solution i. The terms $f_{obj_{max}}$ and $f_{obj_{min}}$ represent, respectively, the maximum and minimum values obtained for each objective, considering the set ℓ of individuals. The use of the crowding distance allows the most scattered individuals to occupy the last available spots for the formation of the new population P_{t+1}, guaranteeing the diversity of solutions. According to [38], it is important to maintain a good spread in the solutions of the boundaries already found, in order to better explore the search space.

Algorithm 3 Crowding distance procedure.

Require: n_i, f_{obj}
Ensure: $dist_i$
$dist_0 := \infty$;
$dist_\ell := \infty$
for $i := 1 \rightarrow \ell - 2$ **do**
  $dist_i := 0$
end for
for each obj **do**
  Sort f_{obj} regarding objective obj
  $dist_0 := \infty$;
  $dist_\ell := \infty$
  for $i := 1 \rightarrow \ell - 2$ **do**
    $dist_i := dist_i + \frac{f_{obj}(i+1) - f_{obj}(i-1)}{f_{obj_{max}} - f_{obj_{min}}}$
  end for
end for

5.3. MicroGA

The main steps of Micro-GA are sketched in Algorithm 4, where N represents the population size, P the population, P_i the initial Micro-GA population, M the population memory, E the external memory, *iter* the current iteration, $iter_{max}$ the maximum number of iterations and N_{RC} the number of iterations between two replacement cycles.

Algorithm 4 Main steps of Micro-GA.

```
Require: iter_max, N_RC
Ensure: E
generate initial population P randomically, with |P| = N
distribute P between the two portions of M
iter := 0;
while iter < iter_max do
    choose initial P_i for the Micro-GA from M
    repeat
        /* Micro-GA cycle */
        perform binary tournament selection based on dominance relationship
        apply recombination operator
        apply mutation operator
        apply elitism keeping only one non-dominated solution
    until nominal convergence is reached
    copy two non-dominated solutions from P_i to the external memory E
    if E is full when trying to insert non-dominated solution into P_i then
        apply the adaptive grid
    end if
    copy two non-dominated solutions from P_i to M, (replaceable portion)
    if iter mod N_RC then
        apply the replacement cycle
    end if
    iter := iter + 1
end while
```

Micro-GA is a genetic algorithm that uses a very small population during a reset process. In fact, this reset process is the Micro-GA performed in conjunction with the use of an external file to store the non-dominated solutions obtained during the iterations. This algorithm is able to obtain the Pareto front with a reduced number of iterations [21]. The basic idea is suggested from theoretical results, where a population size equal to 3 is proven sufficient for the convergence of the genetic algorithm, regardless of the chromosome length [39]. Micro-GA uses two memories: the population memory, which is used to obtain diversity, and the external memory, used to store the solutions of the Pareto-optimal set. The population memoryis divided into two parts: one called the replaceable portion and the other the non-replaceable portion. The percentages of each of the portions can be determined in advance. Initially, a random population is generated, which is distributed between the replaceable and non-replaceable portions of the population memory. The non-replaceable portion will never be modified during the process, and has the function of providing diversity to the algorithm. The initial population of Micro-GA at the beginning of each of its cycles is taken from both portions of population memory.

During each cycle, Micro-GA implements the conventional genetic operators: tournament selection, two-point recombination, uniform mutation and elitism. Regardless of the number of non-dominated solutions in the population, only one is arbitrarily chosen at each iteration to be used in the next generation. A Micro-GA cycle ends when the nominal convergence is reached. This happens when the difference between the average fitness and the maximum fitness converges to a value less than or equal to 5%. Nominal convergence can also be defined in terms of a certain (usually low) number of generations, ranging from 2 to 5. At the end of a cycle, two non-dominated solutions from the current population obtained (the first and the last) are chosen, which will be compared with the solutions stored in the external memory, initially empty. If one or both of the chosen solutions remain non-dominated after the comparison, they will be included in external memory. Then, the dominated solutions from the external memory are discarded. These two chosen solutions are also compared with two distinct solutions of the replaceable portion of the population memory, so that the non-dominated ones will remain. Thus, during the process, the replaceable portion of the population memory will tend to have more non-dominated solutions, some of which will be used in the initial Micro-GA population of the following iterations, i.e., in the next cycles.

The Micro-GA approach allows for the use of three types of elitism. The first is based on the fact that the non-dominated solutions produced in each cycle of the Micro-GA are stored; therefore, no value information of the evolutionary process is lost. The second type of elitism is based on the fact that the best solutions found after the nominal convergence replace some elements of population memory. This allows gradual convergence to obtain the best solutions, provided that the genetic operators of recombination and mutation yield diversity and spread. The third type of elitism is applied after a pre-established number of iterations, and is called the replacement cycle. The replacement cycle is a process in which some solutions in various regions of the front obtained so far are removed, in order to use them to fill in the replaceable portion of the population memory. Depending on the size defined for this memory, as many solutions as necessary are chosen to guarantee a good distribution.

In order to maintain diversity on the Pareto front, an approach similar to adaptive grid, presented in [24], is applied. Once the file that stores the non-dominated solutions reaches its limit, the search space covered is then divided, indicating a set of coordinates for each solution. From then on, each new non-dominated solution will only be accepted if it belongs to a geometric space that is less populated than the denser regions previously mapped. Thus, preference is given to solutions that appear in less populated regions, thus favoring the scattering of individuals on the Pareto front. So, the adaptive grid aims to divide the search space explored by the solutions stored in the file into h hypercubes, establishing a set of coordinates for each solution. The hypercubes are resized as new solutions and extrapolate the limits of solutions already found in the explored search space. Each hypercube can be interpreted as a small space that contains a certain number of solutions. The number of dimensions of the hypercubes corresponds to the number of search variables in the problem. So, the application of adaptive grid allows to obtain well-distributed Pareto fronts [21]. The adaptive grid requires two parameters: estimated size for the Pareto front and the number of solutions into which the search space will be divided for each objective. The first parameter coincides with the size of the external memory. For the second parameter, the usages of values between 15 and 25 are prescribed [21]. Thus, when the external memory is full, the adaptive grid is used to decide which non-dominated solutions will be eliminated.

6. Performance Results

This section is organized into five sections. First, in Section 6.1, we give all the equipment data and settings of the refrigeration system as used in its model. Then, in Section 6.2, we motivate the two stopping criteria exploited to terminate the optimization processes. Subsequently, in Section 6.3, we present the selection method of the preferred solution among those in the obtained Pareto front. After that, in Sections 6.4–6.6, we introduce the parameter settings and performance results of each of the three applied algorithms: SPEA2, NSGA-II and Micro-GA, respectively.

6.1. System Parameters

The configuration of the cooling system considered in this work, as presented in Figure 1, has the following characteristics. The cooling towers have a capacity of 2500 TR each (the TR unit represents tons of refrigeration and is commonly used in refrigeration systems. One TR corresponds to the power that provides the heat required to melt a ton of ice in 24 h. We have 1 TR = 3.5168 kW). The fan's motor has a nominal power of 30 HP (the HP unit represents horse power. We have 1 HP = 735.5 W). The two cooling towers must guarantee the thermal requirements of four chillers of 1000 TR. The rated power of the chiller compressor's motor is 586 kW while that of condensed water lift pump and chilled water circulation pump motors is 120 HP.

The daily thermal load is guaranteed using two chillers only. The third chiller is available as a sporadic ally in the case of an additional thermal load. The fourth chiller would only operate in a rotational situation, in which there is periodical alternation of operating chillers. Moreover, the alternation allows the avoidance of excessive equipment wear or failure. Therefore, the situation of operation with two chillers is the most common for the cooling system to be optimized, as considered in this work. The compression chillers used are from the manufacturer York®, model YKLKLLH9-CZFS, with rated voltage of 4.16 kV, thermal capacity of 1000 TR, rated electrical power of the compressor motor of 586 kW [40].

In the studied refrigeration system, the efficiency optimization of heat exchange, occurring in each tower cell is provided after determining the best trade-off between the water and air flows. Each lift pump of condensed water operates with a nominal flow of 505 m^3/h. The nominal flow in the Chiller's condenser is 496.8 m^3/h [40]. Hence, the number of pumps must coincide with the number of operating chillers to guarantee nominal flow. Note that The number of operating tower cells is dependent on that of operating lift pumps of the condensed water. This is decided so that the input flows into the tower cells are always in their operating thresholds. These limits are 30% smaller and 20% greater than the nominal input flow. This nominal value is 404 m^3/h [17]. It follows that the input flow into each tower cell must be in the range [282.8, 484.8]. The flow is given in m^3/h.

Table 2 indicates the possible scenarios with up to two operating chillers. The number of water lift pumps and that of cells are the ones that must operate to guarantee the minimal and maximal flow thresholds for the equipment. In Table 2, the indicated flows are in m^3/h. The configurations showing the placeholder were impossible, since according to the respective theoretical values, the real flow would be beyond the cell's required limits. As indicated in Table 2, in the case under study, for n operating chillers, we set the refrigeration system with $n+1$ tower cells. This ensures that the cells will always operate within its inlet flow prescribed interval.

Table 2. Inlet flows for the cooling tower cells in m^3/h.

#Chillers	#Pumps	#Cells (Theoretical)					#Cells (Real)				
		1	2	3	4	5	1	2	3	4	5
1	1	505.00	252.50	168.30	-	-	550.00	280.00	170.00	-	-
2	2	1010.0	505.00	336.70	252.50	-	-	485.00	330.00	-	-

6.2. Stopping Criteria

In this work, we investigate the effectiveness of two stopping criteria for the optimization processes. One criterion is based on a simple overall number of iterations used in the optimization algorithm and the other is based on an overall lapsed optimization time.

Regarding the first stopping criterion, the number of iterations to the finalization of the optimization process is determined experimentally, during the algorithm calibration stage. We verify that 50 iterations is sufficient to obtain a Pareto front with good distribution and a sufficient number of points to choose the best solution to be applied onto the refrigeration system's cooling towers, fans and chillers. So, the first stopping criterion is 50 iterations.

Regarding the second stopping criterion, the lapsed time till the termination of the optimization process is defined based on the transport delay of the refrigeration system being optimized. The transport delay is the time interval required to achieve system stability after defining a new setpoint. For the real system under consideration, we could discover that after setting a new speed setpoint for the tower fans, the system requires 15 min on average to establish a new temperature value for water condensation. Considering the transport delay of the thermal system is quite high, we deemed it important that the optimization process should take the shortest possible period of time to yield the optimal solution to be applied. Note that in the case this time value is close to the system transport delay, the selected solution to be applied may no longer be the best alternative. For instance, assuming that the optimization system obtains the optimal solution in a time interval equal to the transport delay, only after 30 min would we be able to configure a new setpoint for the fan speed. Furthermore, due to a possible variation of the thermal load after this time interval, the speed setpoint obtained could no longer yield the optimal solution at that instant. Hence, we arbitrated that the solution to be applied must be available no later than the equivalent of 10% of the transport delay, i.e., after 90 s. So, the second stopping criterion is 90 s.

6.3. Preferred Solution Selection

Multi-objective algorithms return a set of solutions that guarantee a good trade-off between the optimization conflicting objectives. Therefore, we need a criterion to identify the adequate solutions to be applied in the real application at hand. There are several possible selection criteria [41]. In this work, we select the solution in the Pareto front that provides the lowest mean square of the normalized objectives.

In this application, the overall power consumption of the refrigeration system is in the order of hundreds of thousands of Watts while the effectiveness of the cooling tower varies between 0 and 1. Thus, the objective values must be normalized to avoid giving preference to solutions on the Pareto front that minimize the power consumption over those that maximize effectiveness. For this purpose, we normalize the system's effectiveness metric using Equation (11) and to normalize the power consumption values, we apply Equation (12):

$$\epsilon_n = \frac{\epsilon_e - \epsilon_{min}}{\epsilon_{max} - \epsilon_{min}}, \tag{11}$$

$$P_n = \frac{P_g - P_{min}}{P_{max} - P_{min}}, \tag{12}$$

wherein ϵ_n stands for the normalized effectiveness, ϵ_{min} and ϵ_{max} for the minimum and maximum effectiveness, respectively, considering all the Pareto front solutions. Likewise, P_n stands for the normalized overall power consumption, and P_{min} and P_{max} for the minimum and maximum powers, respectively, considering all the Pareto front solutions.

Recall that the energy efficiency of the application, as modeled in this work, consists of maximizing the system's effectiveness while minimizing its power consumption. So, the criterion defined for choosing the optimal solution is defined formally as in Equation (13):

$$S^* = \min_F \left(\sqrt{\frac{0.5}{\epsilon_n^2} + 0.5P_n^2} \right), \tag{13}$$

wherein S^* represents the solution selected from the Pareto front F. In this work, we consider that the two defined objectives are equally important to achieve the system's energy efficiency. So, both objectives have the same weight.

In Figure 2, we show that the minimum point of the mean square curve of the normalized objectives can be used as a separator between the regions that favor one objective over the other. The solutions towards the left of the minimum point of this curve give preference to maximizing the system's effectiveness, which is achieved by increasing the system's power requirements. On the other side, the solutions towards the right of the minimum point of this curve give preference for minimizing the system's power requirements, which is achieved by decreasing effectiveness, i.e., increasing its inverse $1/\epsilon_a$.

The data used to evaluate the performance of applied optimization algorithms were collected in the field using the existing Supervisory Control and Data Acquisition (SCADA). The dataset includes 21,385 operational points at a rate of 1 point every 5 s. So the overall dataset was collected over 29 h and 42 min of operation of different days and times, so that we could contemplate different conditions of thermal load and different weather conditions. It is noteworthy to mention that the wet bulb temperatures were obtained from the database of the *Instituto Nacional de Pesquisas Espaciais* (INPE), available at [42], as recorded by the meteorological station at Santos Dummont airport in Rio de Janeiro/Brazil.

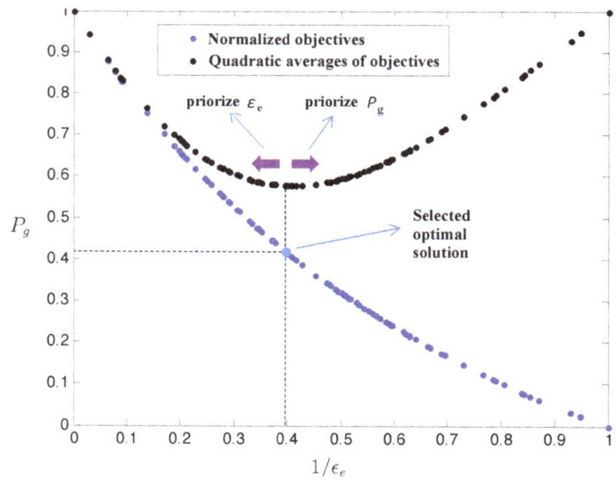

Figure 2. Illustration used to motivate the usage of the selection criterion of the best solution.

For the sake of synthesis and without loss of generality, the analysis presented in the sequel considers the results and Pareto fronts obtained by the applied algorithms, only for 3 points, namely 8, 16 and 26, of the 35 operational points of the whole dataset [43]. These points depict very different load situations. Table 3 presents the data for the illustrative operational points. During the period of time in which the field data are collected, a maximum of two chillers are used. Note that this does not impact the evaluation conducted herein, since the dataset includes 21,385 points, and was also used to validate the tower's and chiller's mathematical models [13,14].

Table 3. Collected data for the operational points used to discuss the performance of the optimization process.

Point	#Chillers	$\dot{m}_{water_{in}}$ (kg/s)	$T_{water_{in}}$ (°C)	T_{BU} (°C)	$\dot{m}_{water_{ev}}$ (kg/s)	$T_{ae_{ev}}$ (°C)	$T_{as_{ev}}$ (°C)
8	2	87.02	29.68	22.94	130.53	10.41	6.01
16	2	70.71	29.40	24.49	106.07	11.16	6.27
26	1	51.65	27.94	23.31	154.96	9.72	6.05

6.4. SPEA2's Performance Results

For the SPEA2 algorithm, the combined MATLAB/C++ implementation available in [44] is used. The parameters' settings used are as follows: population equal to 100, probability of recombination equal to 5% and probability of mutation equal to 15%. In addition, the tournament is used for selecting the best individuals. The choice of these parameters was validated experimentally after repeated tests with several possible sets of parameters. We could verify that populations greater than 100 and recombination and mutation rates above the mentioned values only increased the execution time of the algorithm, not providing significant changes in the results nor in the quality of the obtained Pareto frontiers.

Table 4 presents the selected solutions, as evolved by SPEA2, together with the corresponding values of the objective functions for the 3 operational points (the results for all the 35 operational points used in the optimization are available in Appendix A of [43]). In this table, n_{best} stands for the solution, ϵ_a the effectiveness of the cooling tower, P_g the global power required by fans and chillers and ec the savings in terms of energy consumption.

Table 4. Optimal solutions obtained by SPEA2 for the 4 operational points.

Point	SPEA2—50 Iterations					SPEA2—90 s				
	n_{best} (Hz)	T_{best} (°C)	ϵ_a	P_{global} (kW)	ec (%)	n_{best} (Hz)	T_{best} (°C)	ϵ_a	P_{global} (kW)	ec (%)
8	44.28	6.27	0.5192	857.44	9.55	43.48	6.25	0.5178	858.30	9.45
16	60.00	6.91	0.7452	934.74	11.34	59.97	6.91	0.7451	935.46	11.27
26	58.22	6.80	0.7359	371.46	16.03	59.23	6.83	0.7402	372.43	15.79

Table 5 exhibits the parameters of the optimal solutions obtained that guarantee the established restrictions for the 3 operational points (the results for all the 35 operational points used in the optimization are available in Appendix B of [43]). As before, $T_{ae_{co}}$ stands for the predicted temperature for the water of the condenser circuit that leaves the cooling tower and travels towards the chillers, and $T_{as_{co}}$ the predicted temperature for the water in the condenser circuit, that leaves the chillers and goes towards the cooling tower.

Table 5. Verification of compliance of SPEA2 with the operational restrictions of the equipment for the 3 operational points.

Point	SPEA2—50 Iterations					SPEA2—90 s				
	n_{best} (Hz)	T_{best} (°C)	$T_{ae_{co}}$ (°C)	ΔT (°C)	$T_{as_{co}}$ (°C)	n_{best} (Hz)	T_{best} (°C)	$T_{ae_{co}}$ (°C)	ΔT (°C)	$T_{as_{co}}$ (°C)
8	44.28	6.27	26.18	3.24	30.91	43.48	6.25	26.19	3.25	30.94
16	60.00	6.91	25.74	1.25	31.80	59.97	6.91	25.74	1.25	31.81
26	58.22	6.80	24.53	1.22	28.71	59.23	6.83	24.51	1.20	28.68

Comparing the results obtained with the two stopping criteria, i.e., after 50 iterations and after 90 s, we verify that after a number of iterations greater than 50, operational points 8 and 16 converge to solutions that provide a reduction in both savings and in the effectiveness of the tower, compared to the result obtained for 50 iterations. This is due to the fact that the algorithm's execution with stopping criterion after 90 s is not a continuation of that after 50 iterations, i.e., these are different executions, and due to the stochastic character of the algorithm, it cannot be guaranteed that the solutions obtained in different executions are identical, but rather they represent very close points. Points 8 and 16 show reductions in global energy savings of 0.1% and 0.07%, respectively, and reductions in effectiveness of 0.14% and 0.01%, respectively, after new executions with a number of iterations greater than 50. We observe that this does not rule out the optimal solutions presented for these points, since they ensure a good trade-off between the established objectives. The variations in the achieved results for the different stopping criteria are negligible in practical terms.

Operational point 26, after execution with a number of iterations greater than 50, shows a reduction of 0.24% in global energy savings, in order to obtain an increase of 0.43% in the tower's effectiveness.

Figure 3 shows the Pareto front obtained for the stopping condition of 50 iterations for the 3 operational points indicated in Tables 4 and 5. Figure 4 shows the Pareto front obtained for the stopping conditions of 90 s for the 3 operational points indicated in Tables 4 and 5. In both fronts, the circled points represent the chosen optimal solution. Note that the Pareto fronts obtained when using both stopping conditions are identical, verifying the correct convergence of the algorithm.

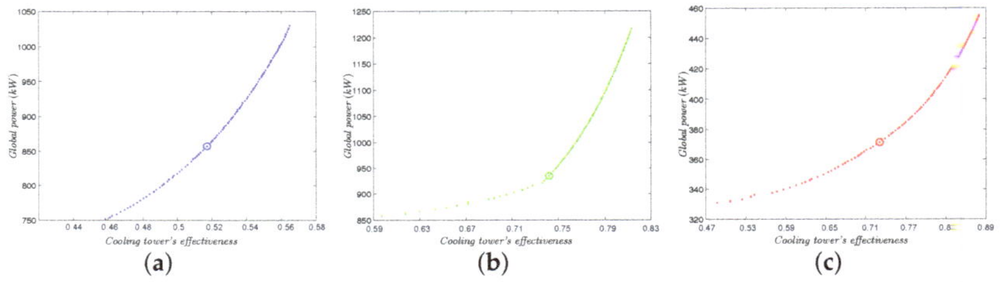

Figure 3. Pareto fronts and selected optimal solutions for 3 of the 35 operating points used in the implementation of the optimization with SPEA2 with stopping criterion after 50 iterations. (**a**) Operational point 8; (**b**) Operational point 16; (**c**) Operational point 26.

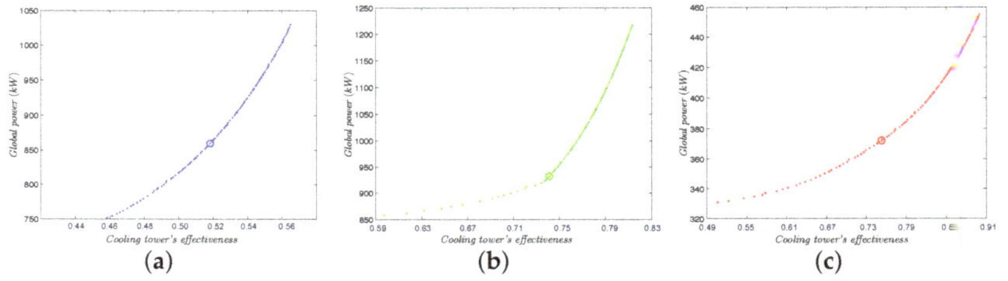

Figure 4. Pareto fronts and selected optimal solutions for 3 of the 35 operating points used in the implementation of the optimization with SPEA2 with stopping criterion after 90 s. (**a**) Operational point 8; (**b**) Operational point 16; (**c**) Operational point 26.

6.5. NSGA-II's Performance Results

For the NSGA-II algorithm, the MATLAB implementation available in [45] is used. The parameters' settings used are as follows: population equal to 100, probability of recombination equal to 0.8 and probability of mutation equal to 0.3. In addition, the binary tournament is used to select the best individuals. Once again, the parameter values are chosen based on tests carried out in order to reduce the execution time and obtain a Pareto front with good distribution and sufficient number of solutions.

Table 6 presents the selected solutions, as evolved by NSGA-II, together with the corresponding values of the objective functions for the 3 operational points (the results for all the 35 operational points used in the optimization are available in Appendix A of [43]). In this table, n_{best} stands for the solution, ϵ_a the effectiveness of the cooling tower, P_g the global power required by fans and chillers and ec the savings in terms of energy consumption.

Table 6. Optimal solutions obtained by NSGA-II for the 4 operational points.

Point	NSGA-II—50 Iterations					NSGA-II—90 s				
	n_{best} (Hz)	T_{best} (°C)	ϵ_a	P_{global} (kW)	ec (%)	n_{best} (Hz)	T_{best} (°C)	ϵ_a	P_{global} (kW)	ec (%)
8	44.46	6.26	0.5196	859.03	9.37	46.04	6.28	0.5223	859.98	9.26
16	59.88	6.95	0.7448	928.58	11.96	59.78	6.93	0.7444	930.36	11.78
26	56.87	6.71	0.7301	372.54	15.76	57.37	6.73	0.7323	373.09	15.62

Table 7 exhibits the parameters of the optimal solutions obtained that guarantee the established restrictions for the 3 operational points (the results for all the 35 operational points used in the optimization are available in Appendix B of [43]). Recall that $T_{ae_{co}}$ stands for the predicted temperature for the water of the condenser circuit that leaves the cooling tower and travels towards the chillers, and $T_{as_{co}}$ the predicted temperature for the water in the condenser circuit that leaves the chillers and goes towards the cooling tower.

Table 7. Verification of compliance of NSGA-II with the operational restrictions of the equipment for the 3 operational points.

Point	NSGA-II—50 Iterations					NSGA-II—90 s				
	n_{best} (Hz)	T_{best} (°C)	$T_{ae_{co}}$ (°C)	ΔT (°C)	$T_{as_{co}}$ (°C)	n_{best} (Hz)	T_{best} (°C)	$T_{ae_{co}}$ (°C)	ΔT (°C)	$T_{as_{co}}$ (°C)
8	44.46	6.26	26.18	3.24	30.92	46.04	6.28	26.16	3.22	30.89
16	59.88	6.95	25.74	1.25	31.76	59.78	6.93	25.74	1.25	31.78
26	56.87	6.71	24.56	1.25	28.79	57.37	6.73	24.55	1.24	28.78

Comparing the results obtained for the stopping criteria after 50 iterations and after 90 s, it appears that, after a number of iterations greater than 50, for the operational points, indicated in Table 6, the optimization converged to solutions that reduce the overall energy savings while achieving a better or similar value for tower effectiveness. In this case, the optimization regarding operating point 8 presents a reduction of 0.11% in energy savings for an increase of 0.27% in the effectiveness of the tower. The optimization regarding point 26 shows a 0.14% reduction in overall energy savings for a 0.22% increase in effectiveness. Unlike the others, for point 16 the optimization exhibits a reduction in both consumption and effectiveness, respectively, of 0.18% and 0.04%, and this is due to the fact that the execution with stopping criterion of 90 s is not a continuation of that of 50 iterations. As noted before, for the stochastic character of the algorithms, it cannot be guaranteed that they will converge to exactly the same solution, but rather to very close points.

Figure 5 shows the Pareto front obtained when using the stopping condition of 50 iterations for the 3 operational points indicated in Tables 6 and 7. Figure 6 presents the Pareto front achieved for the stopping criterion of 90 s for the 3 operational points indicated in Tables 6 and 7. The circled points represent the selected optimal solution. Note that the Pareto fronts obtained for the two stopping criteria are practically identical, verifying the proper convergence of the algorithm.

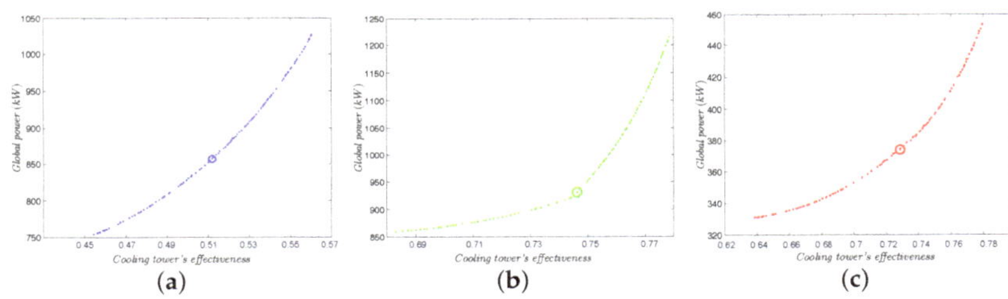

Figure 5. Pareto fronts and selected optimal solutions for 3 of the 35 operating points used in the implementation of the optimization with NSGA-II with stopping criterion after 50 iterations. (**a**) Operational point 8; (**b**) Operational point 16; (**c**) Operational point 26.

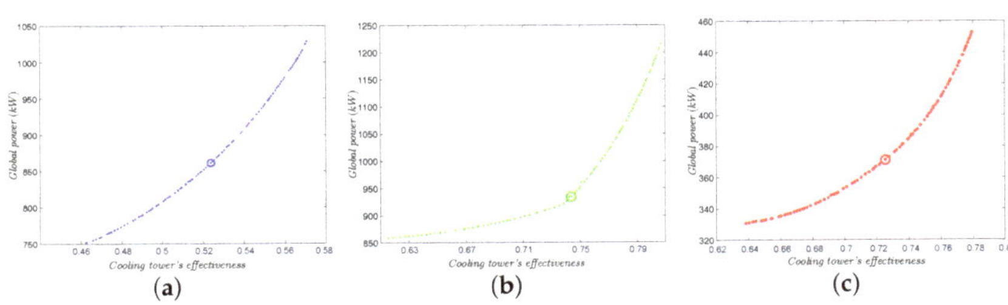

Figure 6. Pareto fronts and selected optimal solutions for 3 of the 35 operating points used in the implementation of the optimization with NSGA-II with stopping criterion after 90 s. (**a**) Operational point 8; (**b**) Operational point 16; (**c**) Operational point 26.

6.6. Micro-GA's Performance Results

For the Micro-GA algorithm, the Toolbox SGALAB from MATLAB [46] is used. The parameters' settings used are as follows: population memory equal to 100, external memory equal to 100, percentage of non-replaceable memory equal to 20%, internal Micro-GA population equal to 6, recombination rate equal to 0.8, mutation rate equal to 0.2, number of Micro-GA iterations until achieving nominal convergence equal to 4 and a replacement cycle of 15 iterations. The binary tournament is used for selecting the best individuals. These values are obtained based on the recommended values in [21] and through experiments in order to obtain a Pareto boundary with good distribution with a fast possible convergence. It is noteworthy to point out that higher values for the mutation rate and for the initial population only increased the algorithm convergence time, leaving the results practically unchanged. Differently from what is indicated in [21], where it is suggested that values for the internal population should be set as 3 to 4, we notice, in this case, that the use of an internal population equal to 6 allowed us to further reduce the algorithm convergence time.

Table 8 presents the selected solutions, as evolved by Micro-GA, together with the corresponding values of the objective functions for the 3 operational point s(The results for all the 35 operational points used in the optimization are available in Appendix A of [43]). In this table, n_{best} stands for the solution, ϵ_a the effectiveness of the cooling tower, P_g the global power required by fans and chillers and ec the savings in terms of energy consumption.

Table 8. Optimal solutions obtained by Micro-GA for the 4 operational points.

Point	Micro-GA—50 Iterations					Micro-GA—90 s				
	n_{best} (Hz)	T_{best} (°C)	ϵ_a	P_{global} (kW)	ec (%)	n_{best} (Hz)	T_{best} (°C)	ϵ_a	P_{global} (kW)	ec (%)
8	43.50	6.26	0.5179	857.17	9.58	43.54	6.20	0.5179	865.90	8.58
16	56.67	6.77	0.7322	946.05	10.21	53.24	6.81	0.7181	929.14	11.90
26	57.29	6.74	0.7320	372.23	15.83	58.30	6.65	0.7363	378.59	14.25

Table 9 exhibits the parameters of the optimal solutions obtained that guarantee the established restrictions for the 3 operational points (the results for all the 35 operational points used in the optimization are available in Appendix B of [43]). As before, $T_{ae_{co}}$ stands for the predicted temperature for the water of the condenser circuit that leaves the cooling tower and travels towards the chillers, and $T_{as_{co}}$ the predicted temperature for the water in the condenser circuit that leaves the chillers and goes towards the cooling tower.

Table 9. Verification of compliance of Micro-GA with the operational restrictions of the equipment for the 3 operational points.

Point	Micro-GA—50 Iterations					Micro-GA—90 s				
	n_{best} (Hz)	T_{best} (°C)	$T_{ae_{co}}$ (°C)	ΔT (°C)	$T_{as_{co}}$ (°C)	n_{best} (Hz)	T_{best} (°C)	$T_{ae_{co}}$ (°C)	ΔT (°C)	$T_{as_{co}}$ (°C)
8	43.50	6.26	26.19	3.25	30.93	43.54	6.20	26.19	3.25	30.98
16	56.67	6.77	25.81	1.32	32.03	53.24	6.81	25.87	1.38	32.05
26	57.29	6.74	24.55	1.24	28.77	58.30	6.65	24.53	1.22	28.80

Comparing the results obtained after executing the Micro-GA with different stopping criteria, i.e., after 50 iterations and after 90 s, we note that the optimization regarding operating point 8 offered the same effectiveness for both criteria, varying only in the achieved global energy savings. After 90 s, a reduction of 1.00% is achieved. Recall that the different stopping criteria occasion different executions, and due to the stochastic character of the optimization algorithms, there could be a deviation between the results. Nonetheless, a convergence confirmation of the algorithm regarding the region containing the optimal solutions is apparent. The optimization regrading operational point 16 shows a reduction of 1.41% in the tower's effectiveness in order to obtain a 1.72% increase in global energy savings. For point 26, the optimization reaches a reduction of 1.58% in the overall energy savings in order to obtain an increase of 0.43% in the tower's effectiveness.

As observed for operational points 26, the result obtained after 90 s is below that obtained after 50 iterations, since the reduction in energy savings is lower than the increase in effectiveness in the tower. This is due to the fact that the criterion for choosing the optimal solution adopted does not verify whether the new optimal solution obtained in a new execution is better or worse than the one obtained in the previous optimization, with 50 iterations. The stopping criteria are applied in two different executions of the algorithm. Thus, after reaching the stopping criterion, the optimal point is simply chosen based on the lowest mean square of the normalized objectives, without evaluating whether the result obtained with the stopping criterion after 90 s is better or worse than after 50 iterations. In this work, the comparison between the optimal solutions obtained for each stopping criterion is performed in a stage after the execution of the algorithm.

Figure 7 presents the Pareto fronts obtained for the stopping criterion of 50 iterations for the 3 operational points indicated in Tables 8 and 9. Figure 8 shows the Pareto fronts obtained for the stopping criterion of 90 s for the 3 operational points indicated in Tables 8 and 9. The circled points represent the preferred optimal solution. Note that the reached Pareto fronts do not have satisfactory solution distribution, as was the case for SPEA2 and NSGA-II. It can also be observed that there is a visible displacement in the optimal solutions obtained after 90 s, which is not satisfactory. Even so, the obtained results confirm the convergence of the algorithm, since the variations verified for the objectives are very small and the solutions for the two stopping criteria are very close.

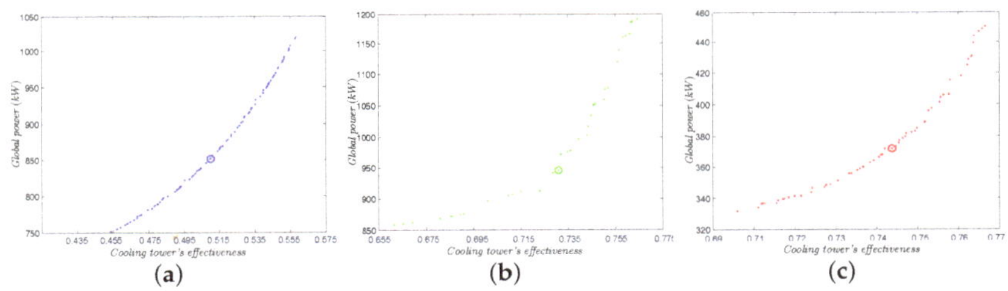

Figure 7. Pareto fronts and selected optimal solutions for 3 of the 35 operating points used in the implementation of the optimization with Micro-GA with stopping criterion after 50 iterations. (**a**) Operational point 8; (**b**) Operational point 16; (**c**) Operational point 26.

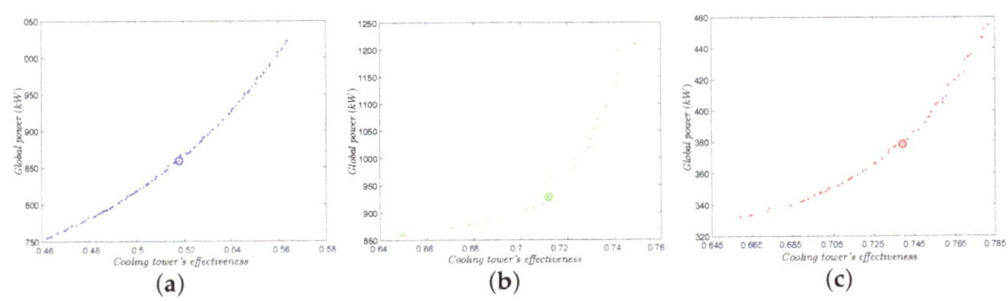

Figure 8. Pareto fronts and selected optimal solutions for 3 of the 35 operating points used in the implementation of the optimization with Micro-GA with stopping criterion after 90 s. (**a**) Operational point 8; (**b**) Operational point 16; (**c**) Operational point 26.

7. Performance Comparison

We now compare the results obtained by the optimization processes when using multi-objective algorithms SPEA2, NSGA-II and Micro-GA. First of all, the obtained results are compared to collected field data to evaluate the gains obtained in terms of energy savings and cooling tower effectiveness. Then, the results achieved by the algorithms are compared with each other in order to choose the most adequate algorithm for the application. Three metrics are used in the comparison and selection process:

- The average, minimum and maximum savings obtained in terms of power consumption by the refrigeration system;
- The average, minimum and maximum effectiveness reached for the cooling tower;
- The ratio between the average savings in terms of overall power consumption and the corresponding reduction in terms of average effectiveness of the cooling tower;

where the average values are computed by applying the optimization results to the 21,385 collected field data regarding the 35 operational points. In this work, the third metric will be termed Energy Efficiency Ratio (EER). It is computed using Equation (14):

$$\mathrm{EER} = \frac{\mathrm{PS}_{avg}}{\Delta\epsilon_{avg}} \tag{14}$$

Table 10 presents the evaluated metrics results of SPEA2, NSGA-II and Micro-GA, regarding both stopping criteria. The values indicated refer to the application of the results obtained for the 35 operational points as presented in Appendix A of [43] to the 21,385 actual field data collected from the real refrigeration system. The execution time indicates the average time spent by the implementation of the considered algorithm with the stopping criterion of 50 iterations. This time duration is given in seconds. For the 90 s case, we report the number of required iterations instead, as the execution time is fixed, i.e., 90 s.

Table 10. Metrics evaluation for the three applied algorithms regarding both stopping criteria.

Metric	After 50 Iterations			After 90 s		
	SPEA2	**NSGA-II**	**Micro-GA**	**SPEA2**	**NSGA-II**	**Micro-GA**
PS_{avg} (%)	8.48	8.28	8.43	8.50	8.32	8.43
PS_{min} (%)	−3.72	−4.09	−4.64	−3.72	−4.26	−3.[illegible]6
PS_{max} (%)	26.07	25.36	25.67	25.27	26.16	25.46
ϵ_{avg}	0.6232	0.6219	0.6183	0.6247	0.6200	0.6159
ϵ_{min}	0.4843	0.4826	0.4818	0.4833	0.4818	0.4816
ϵ_{max}	0.8474	0.8470	0.8350	0.8473	0.8466	0.8235
Time (s)	15.70	69.82	77.92	90	90	90
#Iterations	50	50	50	548	63	78

In Table 10, we can observe that the algorithms implemented in MATLAB (NSGA-II and Micro-GA) require a longer execution time for 50 iterations compared to algorithms implemented in C++ (SPEA2). This result is expected. However, it is noteworthy that the execution time in a dedicated implementation for real usage purposes will depend on the characteristics of the running processor and available memory resources. Moreover, a more efficient codification of the selected algorithm can always be achieved. For both stopping criteria, we can also observe that the algorithm that achieved the best average power savings is SPEA2 followed Micro-GA by NSGA-II. Figure 9 allows a visual comparison of the improvement yielded in terms of average power savings for both stopping criteria (PS_{avg}—50 i; and PS_{avg}—90 s). It is noteworthy to point out that SPEA2 provides a solution that offers a greater average power saving in the case of the 90 s based stopping criterion.

Moreover, note that for both stopping criteria, SPEA2 presents the best average effectiveness, but in this case followed by NSGA-II then Micro-GA. For the first stopping criterion, the optimization time for SPEA2 is the lowest but in the case of the second stopping criterion, the number of iterations required by SPEA2 is the highest. Notably, there are records of negative values of power energy savings, which occur at points wherein the fan speed in the field collected data is 30 Hz. In these cases, the optimization also suggests increasing their speed in order to increase the tower's effectiveness, with a consequent increase in the power energy consumption of the system. This is consistent and matches the expected solution for the proposed optimization system. Figure 10 allows a visual comparison of the improvement yielded in terms of average effectiveness of the tower for both applied algorithms for both stopping criteria (ϵ_{avg}—50 i; and ϵ_{avg}—90 s). Once again, it is noteworthy to point out that SPEA2 provides a solution that offers a greater average cooling tower effectiveness in the case of the 90 s based stopping criterion.

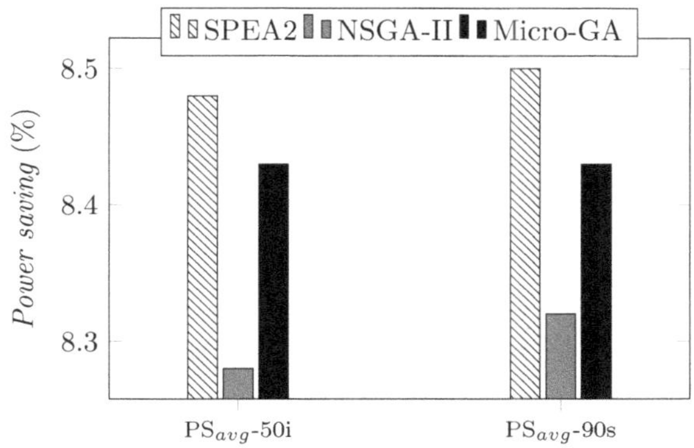

Figure 9. Comparison of average power savings for both stopping criteria as obtained by the applied algorithms.

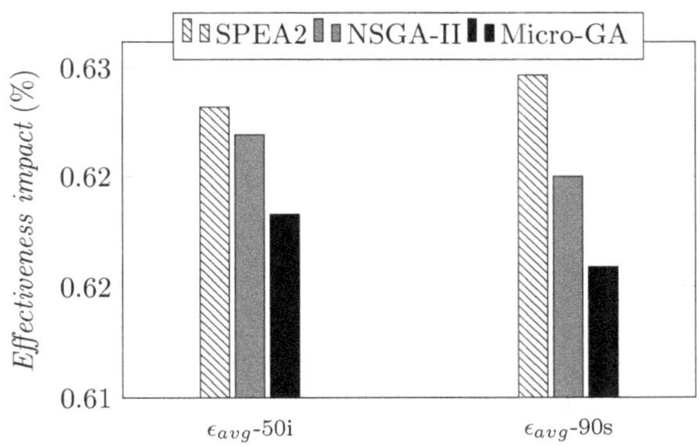

Figure 10. Comparison of average effectiveness improvements for both stopping criteria as obtained by the applied algorithms.

It is known that the system's effectiveness depends not only on the setpoints of the cooling tower operation, but also on external factors, such as ambient temperature and wet bulb temperature. Thus, the reference value for evaluating the algorithms must be at least the average effectiveness obtained by applying the 21,385 operational points collected for the cooling tower modeling, which is 0.6761. Hence, the best algorithm for the application must be the one that achieves the highest average global power savings, with the least possible detriment to the average effectiveness of the cooling tower.

It is noteworthy to point out that, in Table 11, for all three algorithms, the value of EER is greater than 1, which is quite satisfactory. This means that the power savings achieved outweigh the reduction in effectiveness of the cooling tower. So, for both stopping criteria, we note that we have In decreasing order of performance: SPEA2, NSGA-II then Micro-GA. SPEA2 offers the highest value of EFR, which corresponds to 1.60 and 1.65, respectively. So, for the second stopping criterion, SPEA2 achieves a power savings of about to 1.65 times the reduction in the tower effectiveness.

Table 11. Results for the selection of the best algorithm considering both stopping criteria.

Metric	After 50 Iterations			After 90 s		
	SPEA2	NSGA-II	Micro-GA	SPEA2	NSGA-II	Micro-GA
PS_{avg} (%)	8.48	8.28	8.43	8.50	8.32	8.43
ϵ_{avg} (%)	62.32	62.19	61.83	62.47	62.00	61.59
$\Delta\epsilon_{avg}$ (%)	5.29	5.42	5.78	5.14	5.61	6.02
EER	1.60	1.53	1.46	1.65	1.48	1.40

A reduction in the performance of the NSGA-II and Micro-GA algorithms can be seen when comparing the values of FER obtained with the stopping criteria after 50 iterations and after 90 s. For NSGA-II, this factor reduces from 1.53 to 1.48, and for Micro-GA, from 1.46 to 1.40. This is mainly due to the criterion used to choose the optimal solution, which is impacted by the Pareto front distribution. In this case, after 90 s, NSGA-II and Micro-GA added points to the Pareto front that led the adopted decision criterion to choose optimal solutions that favored an increase in terms of average effectiveness of the cooling tower. Figure 11 allows a visual comparison of the improvement yielded in terms of average effectiveness of the tower for both applied algorithms for both stopping criteria (ϵ_{avg}—50 i; and ϵ_{avg}—90 s). Note that, as expected, the solution provided by SPEA2 offers a greater energy efficiency ration in the case of the 90 s based stopping criterion.

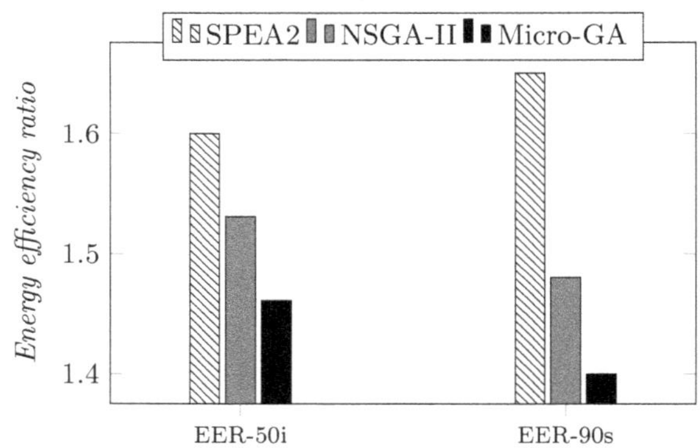

Figure 11. Comparison of energy efficiency ratio achieved by the best solutions for both stopping criteria as obtained by the applied algorithms.

So, it is now safe to conclude that SPEA2 is the best algorithm for the studied application and that the 90 s based stopping criterion is more adequate as it allows for a more interesting trade-off between average power saving and average tower effectiveness, yielding a better ration regarding energy efficiency.

8. Conclusions

The proposed work analyzes the feasibility of applying a multi-objective optimization to the operation of refrigeration systems based on cooling towers and chillers, in order to obtain the operational setpoints that meet the best compromise between two conflicting objectives: reduction of energy consumption and increasing of the tower's effectiveness. This allows obtaining the maximum energy efficiency possible for the whole refrigeration system. For this purpose, it is necessary to formally model the main equipment involved in the considered refrigeration system. Precise and faithful models for the cooling towers and its fans and for the chiller have been developed previously. We also conducted a preliminary survey to select evolutionary multi-objective optimization algorithms to be applied. Algorithms SPEA2, NSGA-II and Micro-GA are chosen so as to investigate their performance regarding the energy efficiency optimization.

We conducted a thorough analysis of the Pareto fronts yielded by the usage of the chosen algorithms. This is performed based on two optimization scenarios with regards to the stopping criterion to be used: either a fixed number of iterations (50 iterations) or a fixed time interval (90 s). We considered these two possibilities so as to obtain the optimal solution to be applied to the real refrigeration system, hence yielding the expected energy efficiency. These iteration and time thresholds are thus set to meet the requirements of the application and to verify the performance impact of the solution reached by the optimization process. After analyzing the obtained global performance results, we conclude that the results obtained with SPEA2 when combined with the stopping criterion of after 90 s should be adopted.

There are several directions to carry on this work aiming at improving the analysis. The used models can be made more sophisticated to offer support for other kind of chillers. In addition, it would be interesting to compare the performance of the chosen algorithms by varying the speed of the condensed and chilled water pumps. The frequency converters could be considered in the optimization process. In this case, the variation of the speed of the cooling tower's fans would have to be taken into account. Furthermore, in the present work, the increase in terms of water consumption of the refrigeration system is not considered in contrast to the reduction of the cooling tower's effectiveness. Thus, developing a model that estimates the system's water consumption in terms of the tower's effectiveness would be interesting. There is also the possibility to explore the usage of other kinds of multi-objective optimization algorithms, such as those based on swarming strategies as apposed to the evolutionary strategy. Among these algorithms, we can mention the work in progress exploring multi-objective particle swarm optimization and multi-objective tribe optimization. Another possible direction could be the study of the effects of cryogenic fluids on the system's energy efficiency.

Author Contributions: Data curation, M.S.D.L.; Formal analysis, L.d.M.M.; Investigation, M.S.D.L.; Methodology, N.N.; Software, M.S.D.L.; Supervision, N.N. and L.d.M.M.; Writing—original draft, N.N.; Writing—review & editing, L.d.M.M. All authors have read and agreed to the published version of the manuscript.

Funding: This work is supported by Conselho Nacional de Desenvolvimento Científico e Tecnológico (CNPq-Brazil) and by Fundação Carlos Chagas Filho de Amparo à Pesquisa do Estado do Rio de Janeiro (FAPERJ-Brazil) with grant numbers 203.111/2018 and 201.013/2022. We are most grateful for their continuous financial support.

Conflicts of Interest: The authors declare no conflict of interest.

References

1. Dondapati, R.S.; Saini, V.; Verma, K.N.; Usurumarti, P.R. Computational prediction of pressure drop and heat transfer with cryogen based nanofluids to be used in micro-heat exchangers. *Int. J. Mech. Sci.* **2017**, *130*, 133–142. [CrossRef]
2. Dondapati, R.S.; Rao, V. Entropy generation minimization (EGM) to optimize mass flow rate in dual channel cable-in-conduit conductors (CICCs) used for fusion grade magnets. *Fusion Eng. Des.* **2014**, *89*, 837–846. [CrossRef]
3. Chakravarthy, V.; Shah, R.; Venkatarathnam, G. A Review of Refrigeration Methods in the Temperature Range 4–300 K. *J. Therm. Sci. Eng. Appl.* **2011**, *3*, 020801. [CrossRef]
4. Yu, K.; Hung, Y.; Hsieh, S.; Hung, S. Chiller Energy Saving Optimization Using Artificial Neural Networks. *J. Appl. Sci.* **2011**, *16*, 3008–3014. [CrossRef]
5. Pegallapati, A.S.; Ramgopal, M. Effect of heat transfer area distribution on frosting performance of refrigerator evaporator. *Int. J. Heat Mass Transf.* **2021**, *175*, 121317. [CrossRef]
6. Li, X.; Li, Y.; Seem, J.E.; Li, P. Extremum seeking control of cooling tower for self-optimizing efficient operation of chilled water systems. In Proceedings of the 2012 American Control Conference, Montreal, QC, Canada, 27–29 June 2012; pp. 3396–3401. [CrossRef]
7. Tyagi, V.; Sane, H.; Darbha, S. An extremum seeking algorithm for determining the set point temperature for condensed water in a cooling tower. In Proceedings of the 2006 American Control Conference, Minneapolis, MN, USA, 14–16 June 2006; pp. 1127–1131. [CrossRef]
8. Liu, C.; Chuah, Y. A study on an optimal approach temperature control strategy of condensing water temperature for energy saving. *Int. J. Refrig.* **2011**, *34*, 816–823. [CrossRef]

9. Ma, Z.; Wang, S. Supervisory and optimal control of central chiller plants using simplified adaptive models and genetic algorithm. *Appl. Energy* **2011**, *88*, 198–211. [CrossRef]
10. Kie, P.L.T.; Theng, L. Intelligent Control of Heating, Ventilating and Air Conditioning Systems. In *Advances in Neuro-Information Processing*; Lecture Notes in Computer Science; Springer: Heidelberg/Berlin, Germany, 2009; Volume 5507, pp. 927–934.
11. Singh, K.; Das, R. Simultaneous optimization of performance parameters and energy consumption in induced draft cooling towers. *Chem. Eng. Res. Des.* **2017**, *123*, 1–13. [CrossRef]
12. Jia, L.; Wei, S.; Liu, J. A review of optimization approaches for controlling water-cooled central cooling systems. *Build. Environ.* **2021**, *203*, 108100. [CrossRef]
13. Nedjah, N.; Mourelle, L.M.; Lizarazu, M.S.D. Mathematical Modeling of Cooling Tower-based Refrigeration Systems for Energy Efficiency Optimization. In Proceedings of the International Conference on Computational Science and Its Applications, Malaga, Spain, 4–7 July 2022.
14. Nedjah, N.; Mourelle, L.M.; Lizarazu, M.S.D. Mathematical Modeling of Chiller-based Refrigeration Systems for Energy Efficiency Optimization. In Proceedings of the International Conference on Computational Science and Its Applications, Malaga, Spain, 4–7 July 2022.
15. More, J.J. The Levenberg–Marquardt algorithm: Implementation and theory. *Numer. Anal.* **1978**, *630*, 105–116.
16. YORK. *Curva de Surge das Unidades Resfriadoras UR-S2O3, UR-S2O4, UR-S2O5 e RU-S2O6*; Technical Report; YORK: Sorocaba, SP, Brazil, 2009.
17. Alpina. *Memória de cálculo das torres de resfriamento TR-2501 e TR-2502*; Technical Report; Alpina S/A Indústria e Comércio: São Bernardo do Campo, SP, Brazil, 2013.
18. Goldberg, D.E. Genetic Algorithms in Search, Optimization and Machine Learning. *Addion Wesley* **1989**, *1989*, 36.
19. Nedjah, N.; Mourelle, L.M. *Real World Multi-Objective System Engineering*; Intelligent System Engineering Series; Nova Science Publishers: New York, NY, USA, 2005.
20. Nedjah, N.; de Macedo Mourelle, L. Evolutionary multi-objective optimisation: A survey. *Int. J. Bio Inspired Comput.* **2015**, *7*, 1–25. [CrossRef]
21. Coello, C.A.C.; Pulido, G.T. A Micro-Genetic Algorithm for Multiobjective Optimization. In *First International Conference on Evolutionary Multi-Criterion Optimization*; Number 1993 in Lecture Notes in Computer Science; Springer: Zurich, Switzerland, 2001; pp. 126–140.
22. Srinivas, N.; Deb, K. Muiltiobjective Optimization Using Nondominated Sorting in Genetic Algorithms. *Evol. Comput.* **1994**, *2*, 221–248. [CrossRef]
23. Horn, J.; Nafpliotis, N.; Goldberg, D. A niched Pareto genetic algorithm for multiobjective optimization. In Proceedings of the First IEEE Conference on Evolutionary Computation, IEEE World Congress on Computational Intelligence, Orlando, FL, USA, 27–29 June 1994; Volume 1, pp. 82–87. [CrossRef]
24. Knowles, J.D.; Corne, D.W. Approximatting the Non-Dominated Front Using the Pareto Archived Evolution Strategy. *Evol. Comput. J.* **2000**, *8*, 149–172. [CrossRef] [PubMed]
25. Knowles, J.; Corne, D. M-PAES: A memetic algorithm for multiobjective optimization. In Proceedings of the 2000 Congress on Evolutionary Computation, CEC00 (Cat. No.00TH8512), La Jolla, CA, USA, 16–19 July 2000; Volume 1, pp. 325–332. [CrossRef]
26. Deb, K.; Agrawal, S.; Pratap, A.; Meyarivan, T. A Fast Elitist Non-dominated sorting genetic algorithm for multi-objective optimization: NSGA-II. In Proceedings of the Parallel Problem Solving from Nature VI Conference, Paris, France, 18–20 September 2000; Volume 1917, pp. 849–858.
27. Veldhuizen, D.A.V.; Lamont, G.B. Multiobjective Optimization with Messy Genetic Algorithms. In Proceedings of the 2000 Symposium on Applied Computing, Como, Italy, 19–21 March 2000; pp. 470–476. [CrossRef]
28. Laumanns, M.; Thiele, L.; Deb, K.; Zitzler, E. Combining Convergence and Diversity in Evolutionary Multi-objective Optimization. *Evol. Comput.* **2002**, *10*, 263–282. [CrossRef] [PubMed]
29. Zitzler, E.; Thiele, L. *An Evolutionary Algorithm for Multiobjective Optimization: The Strength Pareto Approach*; Technical Report 43; Computer Engineering and Communication Networks Lab (TIK), Swiss Federal Intitute of Technology (ETH): Zurich, Switzerland, 1998.
30. Rudolph, G. *Evolutionary Search under Partially Ordered Sets*; Tech. Rep. CI-67/99; Department of Computer Science/LS11, University of Dortmund: Dortmund, Germany, 1999.
31. Corne, D.; Knowles, J.; Oates, M. The Pareto Envelope-Based Selection Algorithm for Multi- Objective optimization. In Proceedings of the Sixth International Conference on Parallel Problem Solving from Nature, PPSN-VI, Paris, France, 18–20 September 2000; pp. 839–848.
32. Kim, M.; Hiroyasu, T.; Miki, M.; Watanabe, S. SPEA2+: Improving the Performance of the Strength Pareto Evolutionary Algorithm 2. In Proceedings of the International Conference on Parallel Problem Solving from Nature 2004, Birmingham, UK, 18–24 September 2004; pp. 742–751.
33. Guo, D.; Wang, J.; Huang, J.; Han, R.; Song, M. Chaotic-NSGA-II: An Effective Algorithm to Solve Multi-objective Problems. In Proceedings of the International Conference on Intelligent Computing and Integrated Systems (ICISS), Guilin, China, 22–24 October 2010; pp. 20–23. [CrossRef]
34. Li, M.; Yang, S.; Liu, X.; Wang, K. IPESA-II: Improved Pareto Envelope-Based Selection Algorithm II. *Evol. Multi-Criterion Optim. Lect. Notes Comput. Sci.* **2013**, *7811*, 143–155. [CrossRef]

35. Deb, K.; Jain, H. An evolutionary many-objective optimization algorithm using reference-point-based nondominated sorting approach, part I: Solving problems with box constraints. *Evol. Comput. IEEE Trans.* **2014**, *18*, 577–601. [CrossRef]
36. Sulieman, D.; Jourdan, L.; Talbi, E.G. Using multiobjective metaheuristics to solve VRP with uncertain demands. In Proceedings of the IEEE Congress on Evolutionary Computation, Barcelona, Spain, 18–23 July 2010; pp. 1–8. [CrossRef]
37. Zitzler, E.; Laumanns, M.; Thiele, L. *SPEA2: Improving the Strength Pareto Evolutionary Algorithm*; TIK-Report 103; Swiss Federal Institute of Technology: Zürich, Switzerland, 2001.
38. Deb, K. *Multi-Objective Optimization Using Evolutionary Algorithms*; Wiley and Sons: New York, NY, USA, 2001.
39. Goldberg, D.E. Sizing Populations for Serial and Parallel Genetic Algorithms. In Proceedings of the Third International Conference on Genetic Algorithms, Fairfax, WV, USA, 4–7 June 1989; Schaffer, J.D., Ed.; Morgan Kaufmann: San Mateo, CA, USA, 1989; pp. 70–79.
40. YORK. *Print-out de seleção das Unidades Resfriadoras UR-S2O3, UR-S2O4, UR-S2O5 e UR-S2O6*; Technical Report; YORK: Sorocaba, SP, Brazil, 2006.
41. Sayyaadi, H.; Nejatolahi, M. Multi-objective optimization of a cooling tower assisted vapor compression refrigeration system. *Int. J. Refrig.* **2011**, *34*, 243–256. [CrossRef]
42. INPE. Dados Observacionais do Centro de Previsão de Tempo e Estudos Climáticos. 2014. Available online: http://bancodedados.cptec.inpe.br (accessed on 26 June 2022).
43. Lizarazu, M.S.D. Otimização Multiobjetivo Aplicada à Efciência Energética de Torres de Resfriamento. Master's Thesis, Universidade do Estado do Rio de Janeiro, Rio de Janeiro, Brazil, 1996. Available online: https://www.pel.uerj.br/bancodissertacoes/Dissertacao_Marcelo_Lizarazu.pdf (accessed on 26 June 2022).
44. Popov, A. *Combined Matlab—C Implementation of SPEA2*; MathWorks: Natick, MA, USA, 2007.
45. Heris, S.M.K. A Version of the NSGA-II in Matlab. 2010. Available online: http://delta.cs.cinvestav.mx/~ccoello/EMOO/EMOOsoftware.html (accessed on 26 June 2022).
46. Chen, L. *SGALAB 1003 Beta 5.0.0.8 (Matrix Variable Inputs): Genetic Algorithm Toolbox for Multi-objective Problems with Fuzzy Logic Controller Applications*; MathWorks: Natick, MA, USA, 2004.

energies

MDPI

Article

Medium-Term Forecasts of Load Profiles in Polish Power System including E-Mobility Development

Paweł Piotrowski, Dariusz Baczyński * and Marcin Kopyt

Electrical Power Engineering Institute, Warsaw University of Technology, Koszykowa 75 Street, 00-662 Warsaw, Poland; pawel.piotrowski@pw.edu.pl (P.P.); marcin.kopyt@pw.edu.pl (M.K.)
* Correspondence: dariusz.baczynski@pw.edu.pl; Tel.: +48-22-234-7255

Abstract: The main objective of this study was to conduct multi-stage and multi-variant prognostic research to assess the impact of e-mobility development on the Polish power system for the period 2022–2027. The research steps were as follows: forecast the number of electric vehicles (using seven methods), forecast annual power demand arising solely out of the operation of the forecast number of electric vehicles, forecast annual power demand with and without the impact of e-mobility growth (using six methods), forecast daily profiles of typical days with and without the impact of e-mobility growth (using three methods). For the purpose of this research, we developed a unique Growth Dynamics Model to forecast the number of electric vehicles in Poland. The application of Multi-Layer Perceptron (MLP) to the extrapolation of non-linear functions (to the forecast number of electric vehicles and forecast annual power demand without the impact of e-mobility growth) is our original, unique proposal to use the Artificial Neural Network (ANN). Another unique, innovative proposal is to include Artificial Neural Networks (Multi-Layer Perceptron and Long short-term memory (LSTM)) in an Ensemble Model for simultaneous extrapolation of 24 non-linear functions to forecast daily profiles of typical days without taking e-mobility into account. This research determined the impact of e-mobility development on the Polish power system, both in terms of annual growth of demand for power and within particular days (hourly distribution) for two typical days (summer and winter). Under the (most likely) balanced growth variant of annual demand for power, due to e-mobility, such demand would grow by more than 4%, and almost 7% under the optimistic variant. Percentage growth of power demand in terms of variation according to time of day was determined. For instance, for the balanced variant, the largest percentage share of e-mobility was in the evening "peak" time (about 6%), and the smallest percentage was in the night "valley" (about 2%).

Keywords: mid-term forecast; e-mobility; electric vehicles (EVs); power system demand; load profile forecast; machine learning (ML)

Citation: Piotrowski, P.; Baczyński, D.; Kopyt, M. Medium-Term Forecasts of Load Profiles in Polish Power System including E-Mobility Development. *Energies* **2022**, *15*, 5578. https://doi.org/10.3390/en15155578

Academic Editor: Hongseok Kim

Received: 12 July 2022
Accepted: 29 July 2022
Published: 1 August 2022

1. Introduction

The last two decades have seen tremendous change in how electric power is used. On the one hand, certain users have reduced their demand for power. As a result of various regulations and technological progress, equipment (e.g., lighting, refrigerators, TV sets) has become increasingly energy efficient. On the other hand, growing population wealth has increased the quantities and diversities of power-consuming equipment (e.g., computers, air conditioners, heat pumps). In the Polish electric power system, a very characteristic symptom of these developments has been significantly increased demand for power in the summer months, especially during heat waves. At the same time, the hitherto typical winter peak of demand for electric power has been decreasing over the years. This requires changes to the planning of how the electric power system operates. Sufficient generation reserves and power transmission capacity should be ensured. To some extent, it leads to problems with upgrades and maintenance works in generation units and power transmission lines.

To operate the power system correctly, appropriate demand forecasts should be prepared, with various time horizons. Annual and monthly quantities of power are both important here, as are the aspects related to peak power and minimum power and the shape, or profile, of the demand curve. Power demand forecasts allow one to properly optimise the composition of the generating units and anticipated potential contingencies.

Electric vehicle owners are a new group of consumers that significantly affect the shape of the power demand curve. The research presented here aimed to answer the question as to the extent of the impact of electric vehicles on the demand profile of the Polish system in the medium term i.e., until 2027.

Over the period 2014–2021, the share of electric vans and cars in new vehicle sales in Poland rose from 0.04% to 2.86%. In that period, the number of EVs increased more than 80-fold [1]. The Polish Alternative Fuels Association forecasts that, by 2024, the market share (sales) of Battery Electric Vehicles (BEVs) will have increased by as much as 14 times, to 10% of the entire market for new vehicles in Poland. In 2025, the cumulative number of registered EVs (BEV and Plug-in Hybrid Electric Vehicles (PHEVs)) in Poland is forecast to be more than 516,000 vehicles, and more than 1.6 million by 2030 [1].

1.1. Related Works

In recent years, dynamic transition to electric vehicles (EVs) has become a major challenge facing the Green Transition. The predicted zero-emission future entails the need to anticipate the effects of progress in vehicle electrification. This involves a number of analyses, regarding both forecasts of the dynamics of development of EVs and their impact on the electric power system and its stability. The following five categories of studies on the topic have been identified in recent literature: forecasts of annual demand on the national level [2–9], forecast number of EVs [7–10], analyses of EV impacts on the power system [11,12], forecast power demand profiles [13–19], and studies combining these particular aspects [20–26].

The identified papers addressing annual demand applied to various parts of the world. Nayyar Hussain Mirjat et al. [2] used the Long-range Energy Alternatives Planning System (LEAP) model to analyse the effect of energy policies on Pakistan's demand until 2050. Similar research using the same system, but for Ethiopia, was conducted by Gebremeskel, Ahlgren and Beyene [3]. The former compared four scenarios, including ones that address maximisation of energy efficiency, Renewable Energy Sources (RESs) penetration, or clean coal technologies. The latter determined scenarios depending on the level of economic development, electrification and urbanisation. Unlike the former, the latter provided for replacement of traditional cars with EVs, which was assumed to be achieved by 2050. Other research in this segment focused on traditional improvement of stability or quality of forecasts. Angelopoulos, Siskos and Psarras [4] proposed, for the Greek system, a disaggregation framework aimed at achieving a robust additive model. He, Wang, Guang and Zhao [5] presented the Simulated Annealing Chicken Swarm Optimization (SACWO) method to optimise the weights adopted in forecast models, and Manowska [6] presented the application of LSTM to power demand by user groups (residential, commercial, transport, industry, and agriculture).

Determination of annual national demand is used for finding a baseline for changes in energy demand as a result of progressing electrification of vehicle fleets. Another contribution of research on EV impacts is to determine the scale of this development by using certain assumptions [3,10] or forecasts [7–9]. This can be done in various ways. Wu & Chen [7] proposed Principal Component Analysis—General regression neural network (PCA-GRNN) as the prediction method. Rietmann, Hügler & Lieven [8] applied a logistic growth model, and Ding and Li [9] applied seven varied models, including ones based on the Grey Model. Some papers [7,8] addressed forecasts of the number of cars sold rather than the number of vehicles actually present in the market. Although such an approach allows for a clearer overview of the situation, based on economic factors, it fails to answer the question of how many cars there are on the market at the same time, which makes it necessary to adopt additional assumptions in longer forecast horizons. Unlike their predecessors, Viri, Makinen and Liimatainen [10] applied a scenario model to analyse the baseline and ±30% larger increase in the number of EVs. The baseline scenario was defined using a Suomen alueellinen autokantamalli, Finnish regional car fleet model (SALAMA). This model allowed for the inclusion of factors such as car age, age of car retirement, user's age group, etc.

After the future number of vehicles is determined, the impact of that number of vehicles on the system can be defined. Such two-step analyses were conducted by Liu and Liu [20], Nogueira, Sousa and Alves [21], Wörner et al. [22], and Brdulak, Chaberek and Jagodziński [23]. The aspects addressed by them included analysis of the impact of vehicles on the peak and off-peak grid balance [21] and analysis of the sufficiency and development needs of charging infrastructure [22]. Other topics included the definition of changes in annual and monthly peak power [20] and the effect of Personal light electric vehicles (PLEVs), such as scooters, on the power grid load. However, the number of vehicles was not always necessary in EV impact studies. The paper by Galvin [11] attempted, instead, to determine how changes in specifications, such as weight and EV motor power output, affect consumption of energy. Feng et al. [12] focused on forecasts of load of vehicle charging stations.

The shape of future power demand profiles can be useful in determining the trends of change in relation to the traditional process of power delivery to users. It makes it possible to determine how the currently-used system balancing solutions might potentially evolve. In the literature considered here, research addressing forecast profiles of power demand has been quite diverse. Kalhori, Emami, Fallahi and Tabarzadi [13] presented a fuzzy logic system for demand with temperature uncertainty; Carmo, Souza and Barbosa [14] proposed a bottom-up approach to creating scenarios for daily curves based on demand, divided into Residential, Tertiary and Industry segments. A different approach was used in papers by Brodowski, Bielecki and Filocha [15] and Hinde, Verdejo and Martínez-Ramón [16], since they focused on creating a hybrid forecasting system. The former approach was based on using Principal Component Analysis (PCA) and clustering of data using Fuzzy C-Means to create a set of hierarchical demand estimators. The latter approach integrated regression and clustering. A secondary objective of the latter approach was to obtain feedback on how automatic division of demand into clusters has been achieved in hourly, daily and monthly intervals. The remaining identified papers also included analyses of the effect of COVID-19 on energy consumption and daily demand curve [17], standardisation of the modelling of load profiles for Europe [18] or forecasts of EV charging profiles [19]. The latter can also be found in Liu and Liu [20], with typical daily curves for subsequent years.

Some studies, classified as combining certain aspects, included the determination of the number and impact of EVs on the electric power system [20–23]. The remaining papers were more extensive, or featured slightly different characteristics. Piotrowski et al. [24] additionally analysed changes in daily profiles over the years. The research by Zou et al. [25] combined EV charging scenarios with effect on the sufficiency of charging infrastructure. Bibak and Tekiner-Mogulkoc [26] focused on EV control in various scenarios using Vehicle-to-grid (V2G) and its impact on daily profiles. The factors affecting the acceptance of that mechanism were presented by Heuveln et al. [27].

1.2. Objective and Contribution

The purpose of this paper was to conduct multi-stage and multi-variant prognostic studies (multiple secondary objectives) with the final objective being to determine the magnitude of impact of e-mobility on the Polish electric power system by 2027 (annual power demand figures). In addition, this paper determined the effect of e-mobility development on hourly profiles of power demand for typical winter and summer days. An important indirect objective was to develop methods to forecast the relative profiles for typical winter and summer days without taking into account the effect of e-mobility.

The research presented in this paper, compared to other studies on the impact of e-mobility on the Polish electric power system, is distinguished by the most comprehensive approach allowing the obtaining of more accurate results than studies using certain simplifications. These simplifications applied either to assumptions in relation to forecasts or the forecasts concerning only one research aspect (e.g., forecasts of annual electricity demand without the impact of e-mobility [6,28], forecasts of annual energy demand resulting from e-mobility [29]). As examples of the comprehensiveness of our presented approach, two elements can be mentioned. The first element is the division of the forecasted number of EVs into different categories of vehicles, thanks to which the estimation of energy demand is more accurate than in studies that do not take into account different EV categories. The second is the inclusion of the increase in the annual demand for electricity resulting from factors other than e-mobility in forecasting the shapes of daily profiles of typical days. It is worth adding that the proposed methodology of forecasting changes in the shape of daily profiles of typical days is a unique, innovative research. The most similar studies, but using simplified (linear) methods of forecasting of the shapes of typical days' daily profiles, are described in [24]. The linear model is unable to represent the non-linear shape of the variability of power demand in the respective hour over consecutive years. Furthermore, to apply linear regression, one needs to build 24 independent models for the given typical day, whereas a single neural network (non-linear model) simultaneously generates 24 values of the typical day profile in a single step.

Below are listed the selected contributions of this paper:

1. Application of the MLP artificial neural network for the extrapolation of non-linear function (forecast number of electric vehicles in Poland from 2022 to 2027), which is our original, unique proposal to apply ANN.
2. Development of an original, unique Growth Dynamics Model, using forecasts of the ratio of annual growth rate to the forecast number of electric vehicles in Poland from 2022 to 2027.
3. Application of the MLP artificial neural network for extrapolation of non-linear function (forecast annual power demand in Poland from 2022 to 2027, excluding the development of e-mobility in Poland), which is our original, unique proposal to use ANN.
4. Application of combined artificial neural networks (MLP and LSTM) in an Ensemble Model for simultaneous extrapolation of 24 non-linear functions, which is our original, unique proposition to use ANN in forecasting the shapes of daily profiles of typical days for 2022–2027 without e-mobility.

The remainder of this paper is organised as follows: Section 2 presents the characteristics of the applied data time series. Section 3 specifies forecasting methods used in this paper and results of stepwise, multi-stage forecasts (Sections 3.2 and 3.3), the final objectives of which are to forecast daily profiles of energy demand in the Polish electric power system, taking into account the development of e-mobility. Evaluation criteria used for the assessment of forecasting quality are presented in Section 3.1. Discussion is in Section 4. Finally, the main conclusions of our studies are summarised in Section 5, and references are listed at the end of this paper.

2. Data

Different time series from Poland were applied to multi-stage research which has as its primary end objective medium-term forecasts of the shape of daily profiles of power demand in the Polish electric power system, taking into account the development of e-mobility. The next research steps used a total of six different types of time series.

The first time series were annual values for electric power demand in Poland from 1990 to 2021 (a total of 32 values). The time series was used for forecasting annual values of power demand within a six years' horizon (2022–2027). Figure 1 presents historical data for annual power demand. The process generally displayed a growing multi-annual trend, with temporary disturbances (drops in energy demand) due to economic situation (e.g., financial crisis 2007–2009, COVID-19 pandemic in 2019–2020). The growing trend was markedly more dynamic since Poland joined the European Union (EU) in 2004.

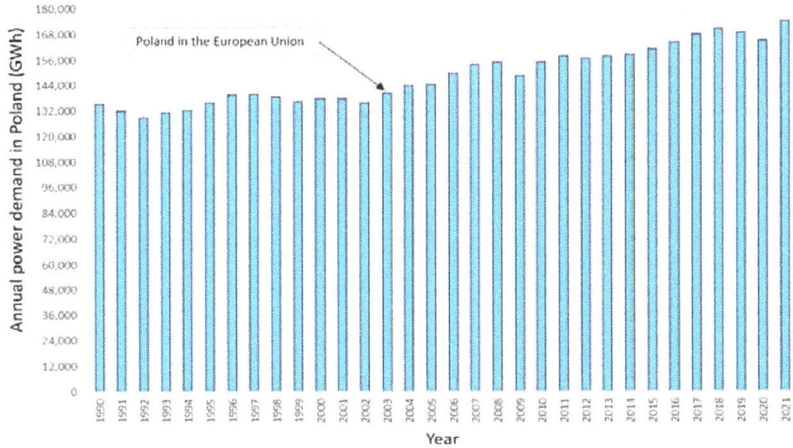

Figure 1. Historical data for annual power demand in Poland.

The second time series were cumulative values of the number of EVs in Poland since 2011 through to 2021 (a total of 11 values). The time series was used for forecasting the cumulative number of EVs in a six years' horizon (2022–2027). Figure 2 presents the total number of EVs in Poland (2011–2021). The process involved strongly non-linear growth, particularly evident in the last three years.

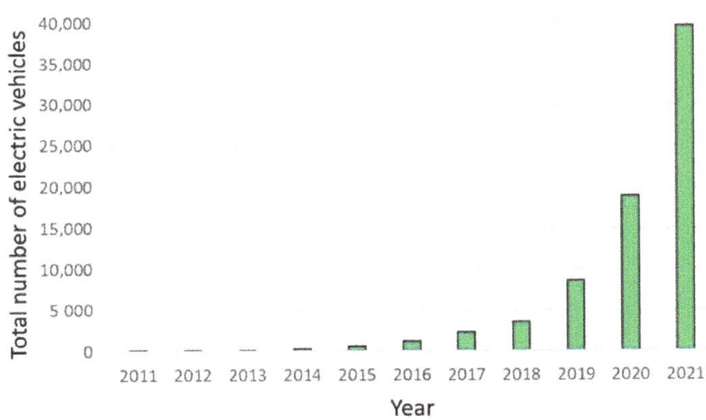

Figure 2. Total number of EVs in Poland.

The third type of time series were six-year forecasts (2022–2027) of power demand due to e-mobility development in Poland, in three variants (optimistic, balanced and pessimistic). The time series were calculated based on the forecast number of EVs in the six years' horizon (2022–2027) and various EV statistical figures (including estimates of annual power demand per EV).

The fourth type of time series were hourly values of power demand in the national electric power system from 2009 to 2021 (a total of 13 years of hourly values). This time series was used to construct profiles of typical days for each of the 13 years. The third Wednesday of January and the third Wednesday of July are "typical days" in the Polish

Power System, representing the winter and the summer business days, respectively [24]. Two daily profiles of typical days were calculated for each year of 2009–2021. The hourly values of the profile were computed as an arithmetic average of hourly values from five business Wednesdays. The five business Wednesdays were the following: the typical day (the third Wednesday of January or the third Wednesday of July), two prior business Wednesdays and two following business Wednesdays. This exercise evened out the profiles and reduced the random component resulting from single days.

Upon building the profiles of typical days for 13 years (2009–2021), two time series of the fifth type were established, containing 24 hourly values for the typical day profiles, for each of the 13 years. Hourly values of both profiles were normalised in each year separately. Normalisation was achieved by dividing the values of profiles from each hour by average hourly power demand during the year [24]. Normalisation enabled us to the track changes in the profiles in 2009–2021, ignoring any profile change resulting from multi-annual growing trends of power demand in the subsequent years. Figure 3 presents relative values of daily profiles of electric energy demand for the typical winter day and summer day in Poland from 2009 to 2021. The charts show that the profile of typical days changed over subsequent years. For the profile of the typical winter day, power demand was decreasing in subsequent years, albeit unevenly for particular hours. For the profile of the typical summer day, power demand has been growing unevenly. Based on these two time series, both profiles with six years' horizons (2022–2027) were forecast, without taking into account an increase in power demand due to e-mobility development.

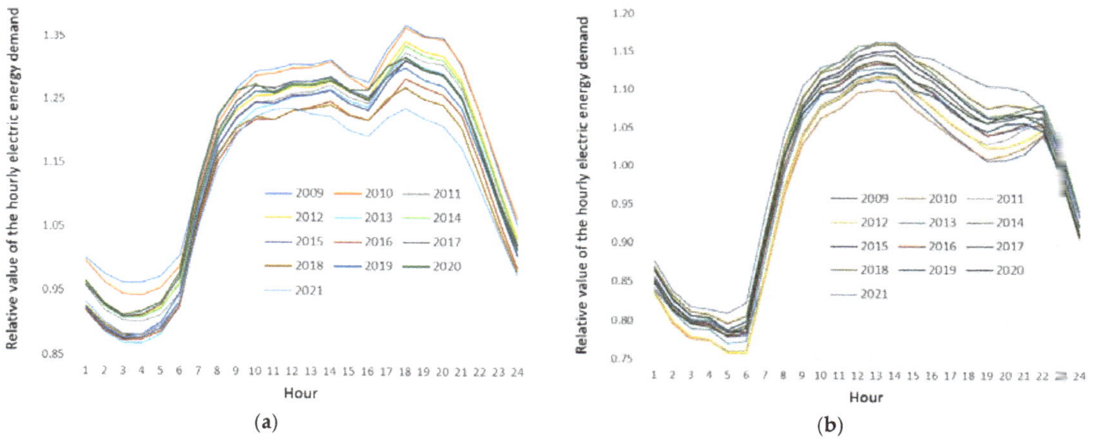

Figure 3. (**a**) Daily profiles of power demand for the winter typical day in Poland; (**b**) Daily profiles of power demand for the summer typical day in Poland.

The sixth type of times series were daily profiles of power demand for various EV types (BEV, PHEV, electric buses, electric heavy trucks and electric delivery vans and two different charging methods (slow charging and rapid charging). These profiles were expertly developed and uniquely based on different statistical data. The profiles were used for the final calculation of forecast profiles of typical days in the power system with the six years' horizon (2022–2027), taking into account e-mobility development. In addition, this process required forecasts of annual EV numbers (2022–2027) and forecasts of annual power demand without taking into account e-mobility development (2022–2027).

3. Methods and Results

This section and its subsections describe forecasting methods and results as particular forecasts of various kinds were performed, until the final objective was achieved that being forecast shapes of profiles until 2027 taking into account e-mobility development. The following models and methods were applied to particular forecasts, including: trend extrapolation models, methods based on time series, methods based on deterministic chaos theory, artificial neural networks (MLP and LSTM), as well as ensemble methods. The general diagramme of the studies described in this paper is shown in Figure 4.

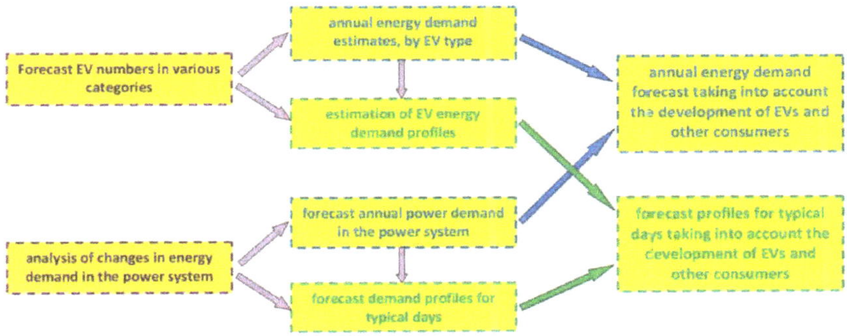

Figure 4. General diagramme of studies described in this paper.

3.1. Evaluation Criteria

To assess the quality of particular forecasting models within their parameter estimation ranges (availability of observed and forecast values), the following five evaluation criteria were used: Root Mean Square Error (RMSE) and Mean Absolute Percentage Error (MAPE) as the main evaluation criteria, Mean Bias Error (MBE), Pearson coefficient of linear correlation (R) and R-squared (R^2) as three auxiliary evaluation criteria. It is worthwhile to note that the least RMSE and MAPE errors in the parameter estimation range do not mean that the model would generate the most accurate "ex ante" (forward-looking) forecasts. On the one hand, relatively small errors within the parameter estimation range are desirable (it would mean that the process has been well framed as a function of time). On the other hand, extremely small errors would mean that the model was unable to generalise (forecasts matching observed values too tightly, which would result in lower prognostic potential of such a model). Expert assessment of the magnitude of error was therefore required to select the preferred prognostic models.

Root Mean Square Error was calculated by Formula (1). RMSE is sensitive to large errors and is more useful when large errors are particularly undesirable.

$$RMSE = \sqrt{\frac{1}{n}\sum_{i=1}^{n}(y_i - \hat{y}_i)^2} \quad (1)$$

where, $\hat{y}_i$ is the predicted value, y_i is the observed value, and n is the number of prediction points.

Mean Absolute Percentage Error is calculated by Formula (2).

$$MAPE = \frac{1}{n}\sum_{i=1}^{n}\left|\frac{y_i - \hat{y}_i}{y_i}\right| \cdot 100\% \quad (2)$$

Mean Bias Error captures the average bias in the prediction, and is calculated by Formula (3)

$$MBE = \frac{1}{n}\sum_{i=1}^{n}(y_i - \hat{y}_i) \quad (3)$$

Pearson coefficient of linear correlation between observed and predicted data was calculated by Formula (4). The forecasting method overestimates values if MBE < 0 or underestimates values if MBE > 0. The MBE error of a properly functioning prognostic method should be equal or very close to zero.

$$R = \frac{C_{y\hat{y}}}{std(y) \cdot std(\hat{y})} \tag{4}$$

where, $C_{y\hat{y}}$ s the covariance between the observed and predicted data, and *std* denotes standard deviation of the variable.

The bigger the error R (range from −1 to 1), the more accurate the prediction results. R-squared was calculated by Formula (5).

$$R^2 = 1 - \left(\left(\sum_{i=1}^{n} (\hat{y}_i - y_i)^2 \right) / \left(\sum_{i=1}^{n} (y_i - \overline{y})^2 \right) \right) \tag{5}$$

where, $\overline{y}$ is the mean of the observed load values.

The R-squared formula describes the difference between the goodness of fit of perfectly fitting model and models the sum of squared errors related to the sum of squared deviations of measured values from the mean value. The bigger R-squared is (range from 0 to 1) the better the model's fit is and the more the process is explained by it. R-squared value gets lower with increasing concentration of the observed data around the mean value.

3.2. Forecast Number of Electric Vehicles in POLAND from 2022 to 2027

A very short time series of the cumulative number of EV registrations in Poland (period 2010–20121) increases the uncertainty of forecasts and justifies the use of an ensemble model based on several models. The process was assumed to be in its inception phase. The process growth dynamics (cumulative number of registered EVs) was strongly non-linear. A similar trend is also evident in other countries.

The methods used for forecasting EV numbers can also be grouped as follows: methods with control of the process growth ceiling (logistic function and a Model According to Prigogine) and methods without control of the process growth ceiling (other methods). It is worthwhile noting that the process reviewed here would not be growing indefinitely. At some point, the process would reach its ceiling. This is due to the fact that the number of vehicles (regardless of their power source) in a country would not grow indefinitely, and would strongly depend on the size of the population. Table 1 shows the grouping of the methods used for forecasting the number of electric vehicles.

Table 1. Groups of methods used for forecasting the number of electric vehicles.

Name of Method	Equation No.	Model Parameter Optimisation Method	Minimised Error	Step Forecasts
Category I. Methods of direct trend extrapolation				
Logistic function *	(6)	Differential Evolution and Particle Swarm Optimisation algorithm (DEPS)	Sum of Squared Errors (SSE)	No
Exponential function	(7)	Method of Least Squares (MLS)	SSE	No
MLP artificial neural network	-	Broyden–Fletcher–Goldfarb–Shanno algorithm (BFGS)	SSE	No
Category II. Methods based on deterministic chaos theory				
Model According to Prigogine *	(8)	DEPS	SSE	Yes
Category III. Methods using time series				
Grey Model GM(1,1)	(9)	DEPS	SSE	Yes

Table 1. *Cont.*

Name of Method	Equation No.	Model Parameter Optimisation Method	Minimised Error	Step Forecasts
Growth Dynamics Model	(10)	MLS	SSE	Yes
Category IV. Method using more than one model				
Ensemble Model	(12)	-	-	No

Remark: * growth ceiling control models.

The extrapolation model of the logistic function was described by Formula (6) [24].

$$y(t) = \frac{a}{1 + b \cdot e^{-c \cdot t}} \tag{6}$$

where, t is the sample index in the time series of the process, $a > 0$ is the saturation level and $b > 0$, $c > 0$ are the parameters. The saturation level is adopted to be exactly 15 million—in Poland, more than 24 million vehicles of all power-source types are registered (about 750 vehicles per 1000 inhabitants).

The extrapolation model of the exponential function was described by Formula (7).

$$y(t) = a \cdot e^{b \cdot t} \tag{7}$$

where, a, b are the parameters.

The application of the MLP artificial neural network to the extrapolation exercise is our original, unique proposition to use ANN. MLP is typically used in regressive (including forecasting [30,31]) and classification problems [32] and requires a large number of learning modules. In this case, MLP was first used for the construction of non-linear function, or the approximation exercise. In this case there was no explicit function formula, rather it was embedded in the architecture of the neural network, in the weights (parameters) and functions of activation of particular layers of the neural network. Next, MLP was applied to forecast out-of-range values (extrapolation). Figure 5 presents a diagram of subsequent actions using an MLP neural network to obtain the forecast number of EVs.

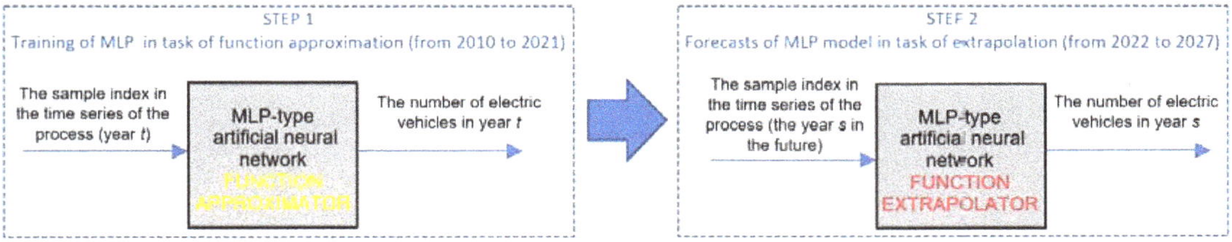

Figure 5. Application of MLP to forecasts of EV numbers.

Tests for various hyperparameters were performed to select appropriate models.

The tested number of hidden neurons ranged from 1 to 4. The number of learning epochs was tested for the following values: 10, 20, 50, 100, 150 and 200. The number of learning epochs had a large influence on the level of "smoothing out" of the function being approximated. With too large a number of learning epochs, MLP learned the values too strictly, thus, losing the capacity to generalise. For the hidden layer, the Linear and Exponential activation functions were tested. In the baseline layer, the exponential activation function was adopted as the one appropriate for the process studied. Such a choice ensured that the MLP neural network was capable of extrapolating (able to predict outside of the learning range). As a result of extensive tests, outcomes of the forecasts of two MLP models were finally selected. Both models had one input and one output. To optimise the weights (model parameters), BFGS optimisation algorithm was applied.

The first of the selected models had 2 hidden neurons and an exponential activation function in the hidden layer and in the output layer (MLP 1-2-1 (exp/exp)).

The other selected model had 2 hidden neurons and a linear activation function in the hidden layer, and exponential activation function in the output layer (MLP 1-2-1 (linear/exp)). Both models were learning for 100 learning epochs, with weights being updated following each learning epoch.

The Model According to Prigogine was described by Formula (8) [28,29].

$$y(t) = y(t-1) \cdot \left[1 + r \cdot \left(1 - \frac{y(t-1)}{K}\right)\right] \tag{8}$$

where $y(t)$ is the population size in period t, $r > 0$ is the population growth rate, $K > 0$ is the development ceiling (forecast population growth in the future). The development ceiling was assumed to be 15 milion.

Grey Model GM (1,1) was described by Formula (9) [24]. In this model, the order of the Grey Differential Equation and the number of variables are equal to 1. This model is recommended by literature [33], especially for very short time series and where the process evolution is in its initial phase.

$$\begin{aligned} \hat{y}(t) &= \hat{y}^{(1)}(t) - \hat{y}^{(1)}(t-1), \\ \hat{y}^{(1)}(t) &= \left[y^{(1)}(1) - \tfrac{u}{a}\right] \cdot e^{(-a(t-1))} + \tfrac{u}{a} \\ \hat{y}^{(1)}(t) &= \sum_{i=1}^{t} y(i),\ t = 1, 2, \ldots, n \end{aligned} \tag{9}$$

where $n \geq 4$ is the length of time series, a is the evolution parameter, u is the grey variable and $\hat{y}(t)$ is the forecast in period t.

The Growth Dynamics Model is our original, unique proposal for a model. In Step One, annual growth rates were calculated for 2011–2021 as the rate of the number of EVs in the year to the number of EVs in the prior year. In Step Two, annual growth rates were approximated to a linear function. Figure 6 presents the variability of annual growth rates.

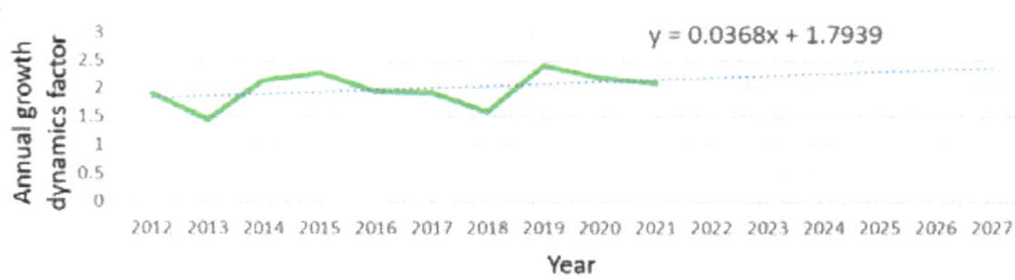

Figure 6. Variability of annual growth rates.

Formula (10) presents a linear function equation with calculated parameter values.

$$c_{growth}(t) \ = \ 0.0368 * t + 1.7939 \tag{10}$$

In Step Three, annual growth rates for 2022–2027 were forecast, using extrapolation of the linear function onto subsequent periods (forward-looking). In Step Four, appropriate stepwise forecasts of the number of EVs were conducted for 2022–2027, using the calculated forecast annual growth rates. Forecast number of EVs was calculated for each year according to Formula (11)

$$y(t) = y(t-1) * \hat{c}_{growth}(t) \tag{11}$$

where, $y(t)$—forecast number of EVs in period t, $\hat{c}_{growth}(t)$—forecast annual growth rate for period t, calculated as extrapolation of the linear function.

The Ensemble Model was described by Formula (12) [24]. The forecast in the Ensemble Model was the weighted arithmetic average of forecasts from several models. To construct such a model, seven single prognostic models were used (logistic function, exponential function, MLP 1-2-1 (exp/exp), MLP 1-2-1 (linear/exp), Model According to Prigogine, Grey Model GM(1,1) and Growth Dynamics Model). Averaged results of forecasts from different models should increase reliability of forecasting.

$$\hat{y}_t = \frac{\sum_{i=1}^{k} \hat{y}_t^i \cdot w_i}{\sum_{i=1}^{k} w_i}, \quad w_i = \frac{1}{k} \tag{12}$$

where k is the number of forecasting models and $\hat{y}_t^i$ is the forecast in period t generated by the model number i.

Table 2 presents summary results (quality assurance metrics) for the model's parameter estimation range (2010–2021). Figure 7 presents the results of forward-looking (2022–2027) forecasts of the total number of EVs in Poland for the eight methods.

Table 2. Performance metrics in the model's parameter estimation range for the forecast number of EVs.

Method Name	RMSE	MAPE [%]	MBE	R	R^2
MLP 1-2-1 (exp/exp)—pessimistic variant	4.13	15.26	54.53	0.99983	0.99962
Logistic function	15.61	16.83	39.34	0.99982	0.99963
Exponential function	16.06	16.67	38.33	0.99982	0.99962
MLP 1-2-1 (linear/exp)	18.97	9.28	20.34	0.99982	0.99964
Ensemble model—balanced variant	34.99	9.83	31.15	0.99975	0.99950
Model According to Prigogine	93.41	12.62	−17.70	0.99898	0.99797
Grey Model GM(1,1)	249.29	11.08	131.74	0.99980	0.99907
Growth Dynamics Model—optimistic variant	346.09	10.78	−47.22	0.99884	0.99729

Remarks: The best fitting results for each fitness metric are printed in bold in blue. The worst fitted result is printed in red.

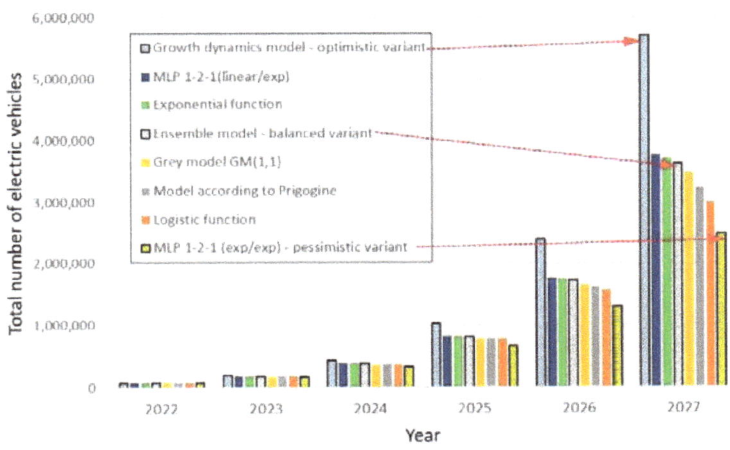

Figure 7. Results of forecasts of the total number of EV in Poland from 2022 to 2027 obtained by eight methods.

The best fitting model for historical data was MLP 1-2-1 (exp/exp) model. It was selected as the pessimistic variant (the lowest "ex ante" forecast values (2022–2027)). The Growth Dynamics Model was the least fitting model in the parameter estimation range. This model was selected as the optimistic variant (largest "ex ante" forecast values (2022–2027)). The Ensemble Model was selected as the most credible model for a balanced model.

The pessimistic and optimistic variants were the models that differed the most from the remaining models, in terms of forecast values. This effect was particularly evident for the 2027 forecasts.

3.3. Forecast Annual Power Demand in Poland from 2022 to 2027 Excluding the Development of E-Mobility in Poland

Forecasts of annual demand for power with the exclusion of e-mobility were conducted by six methods.

The first model, a modified Holt's model, is presented in detail in [34]. Model parameters for the data from the estimation range (1991–2021) were selected using optimisation by the DEPS method. The minimum SSE was sought. Forecasts were conducted by a Stepwise Method (2022–2027).

The second model, the Model According to Prigogine was described by Formula (8). Model parameters for the data from the estimation range (1991–2021) were selected using optimisation by the DEPS method. The minimum SSE was sought. Forecasts were conducted by a Stepwise Method (2022–2027).

The third model, the Method of Constant Annual Growth, Version 1, was described by Formula (13). Annual growth was the average annual growth rate based on historical data of the forecasting exercise. Forecasts were conducted by a Stepwise Method (2022–2027).

$$\hat{y}_t = \hat{y}_{t-1} + \frac{\sum_{j=2}^{k}(y_j - y_{j-1})}{k-1} \quad (13)$$

where k is the number of the data points in the time series and y_{j-1} is the previous value (or forecast) from the time series.

The fourth model, the Method of Constant Annual Growth, Version 2, was described by Formula (14). Annual growth was equal to the slope A from the linear function used for the approximation of the trend line (1991–2021) of annual power demand. The parameter value was A = 1382.30. Forecasts were conducted by a Stepwise Method (2022–2027).

$$\hat{y}_t = \hat{y}_{t-1} + A \quad (14)$$

where, A—parameter from the linear function used for the approximation of the trend line.

The fifth model was an original, unique proposed model, MLP Artificial Neural Network. The model was described (conceptually) in Section 3.2. To select the appropriate model, tests for various hyperparameters were conducted. The tested number of hidden neurons ranged from 1 to 3. The number of learning epochs was tested for the following values: 5, 10, 20 and 50. The number of learning epochs had a large influence on the level of "smoothing out" of the function being approximated. For the hidden layer, the Linear and Hyperbolic Tangent activation functions were tested. In the output layer, a Linear activation function was adopted as the one appropriate for the process studied here, which was due to the variability of the forecast process. Such a choice ensured that the MLP neural network was capable of extrapolating (able to predict outside of the learning range). To optimise the weights (model parameters), the BFGS optimisation algorithm was applied. The selected final mode had two hidden neurons and Hyperbolic Tangent activation function in the hidden layer, and a Linear activation function in the output layer (MLP 1-2-1 (tangh/linear). This model learned for 10 learning epochs, with weights updated following each learning epoch.

The sixth model, the Ensemble Model, was described by Formula (12). The following methods were selected for the ensemble model: Modified Holt's Model, Model According to Prigogine, Constant Annual Growth Method, Version 1, Constant Annual Growth Method, Version 2, and MLP 1-2-1 (tangh/linear).

Table 3 presents summary results (quality assurance metrics) for the model's parameter estimation range (1991–2021). Figure 8 presents the time series of the observed annual power demand figures in Poland and results of forecasts from 2022 to 2027, obtained by the six methods.

Table 3. Performance metrics for the forecast values of annual power demand in Poland within the model's parameter estimation range.

Method Name	RMSE	MAPE [%]	MBE	R	R^2
MLP 1-2-1 (tangh/linear)	509.04	1.574	−112.04	0.9776	0.9556
Ensemble model	1402.67	1.571	152.27	0.9753	0.9510
Constant Annual Growth Method—Version 2	1658.21	1.755	−114.49	0.9693	0.9393
Constant Annual Growth Method—Version 1	1665.56	1.751	-5.6×10^{-12}	0.9693	0.9393
Model according to Prigogine	1665.76	1.751	−2.18	0.9694	0.9394
Modified Holt's Model	1692.48	1.776	990.05	0.9681	0.9253

Remarks: The best fitting results for each fitness metric are printed in bold in blue. The worst fitted result is printed in red.

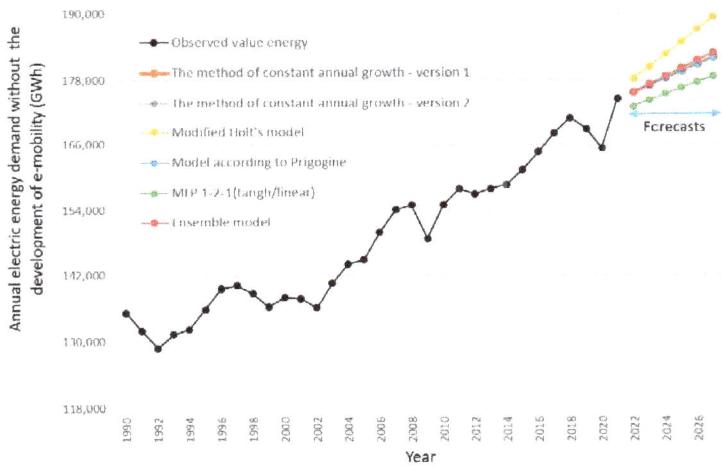

Figure 8. Time series of observed values of annual electric energy demand in Poland and results of forecasts from 2022 to 2027 obtained by six the methods.

MLP 1-2-1 (tangh/linear) was the model that best fit the historical data, and at the same time it generated the lowest values of "ex ante" forecasts (2022–2027). The least fitting model in the parameter estimation range was the Modified Holt's model, and at the same time it generated the largest "ex ante" forecast values (2022–2027). Results of forecasts from the Ensemble Model were selected for further analyses.

3.4. Forecast Annual Power Demand in Poland from 2022 to 2027 Solely due to the Operation of the Forecast Number of EVs

The algorithm had four steps. Figure 9 shows the details. Table 4 contains input data and summary calculation results for 2027 (forecasts with six years' horizon).

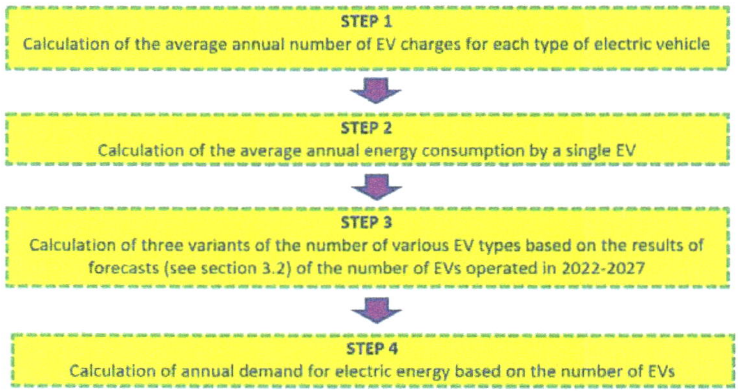

Figure 9. Diagram of calculation of forecasts of annual power demand resulting solely from the operation of the forecast number of electric vehicles.

Table 4. Input data and calculation results for 2027.

Description	BEVs	PHEVs	Electric Buses	Electric Trucks	Electric Delivery Vans
			Input data for calculations		
Average annual mileage [km]	15,000	10,000	80,000	80,000	80,000
Battery capacity in a single EV [kWh]	49.7	12.2	167.5	276.6	48.2
Average EV driving range [km]	343.6	61.2	175.0	252.2	260.0
			Results of calculation of average values		
Average annual number of charges	44	163	457	317	308
Average annual power consumption by a single EV [kWh]	2169.7	1993.5	76,571.4	87,727.2	14,830.8
			Results of calculation of the number of EVs in 2027		
Pessimistic variant [pcs]	1,217,911	1,262,651	1126	709	2849
Balanced variant [pcs]	1,777,612	1,842,912	1501	945	3799
Optimistic variant [pcs]	2,801,269	2,904,173	1876	1181	4749
			Outcome of calculation of power demand for charging the forecast number of EVs in 2027		
Pessimistic variant [GWh]	2642.5	2517.1	86.2	62.2	42.3
Balanced variant [GWh]	3856.8	3673.8	114.9	82.9	56.3
Optimistic variant [GWh]	6077.8	5789.4	143.7	103.6	70.4

Battery capacity and average driving ranges of BEVs and PHEVs were determined as averages calculated for 38 and 65 different vehicle models, respectively, costing less than PLN 0.25 million. For electric vans, these variables were determined as average figures for 17 vehicles of that type. For electric buses, the range and battery capacity were adopted in accordance with [24].

For forecast number of electric trucks of various sizes and the number of electric buses in the six years' horizon, constant growth of the number of vehicles was assumed (average growth from the last several years), such was the observed dynamics of both processes. Forecast numbers of BEVs and PHEVs in subsequent years were calculated as 49% and 51%, respectively, of the forecast values for the given year of the number of EVs of all categories (having deducted forecast number of vehicles from the remaining three categories).

The analysis of results from Table 4 showed that BEVs and PHEVs, or mainly passenger transport, would have by far the biggest impact on annual power demand in Poland in the next six years. For buses, the level of electrification would be about 16% of the fleet (currently, there are about 12,000 buses with various power-source variants). Electric trucks would have no significant impact on power demand, despite large battery capacities, due to the fact that their number in the next six years was not predicted to be very large. Electric delivery vans would probably be more significantly electrified, as the manufacturers' offer gets bigger every year. For that reason, the calculations assumed a 75% share of those vehicles in the category of electric trucks of various sizes.

Figure 10 shows three variants of the forecast annual demand for power in Poland in 2022–2027, resulting solely from the operation of the forecast number of electric vehicles.

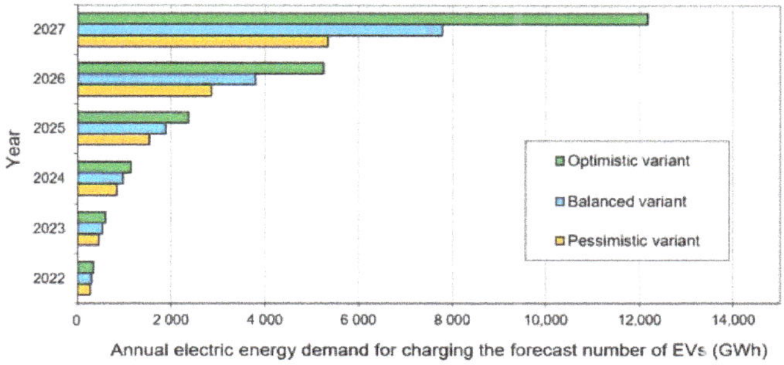

Figure 10. Forecast annual power demand in Poland in 2022–2027 solely due to the operation of the forecast number of electric vehicles.

3.5. Forecast Annual Power Demand in Poland from 2022 to 2027 Taking into Account the Development of E-Mobility in Poland

The forecast value for the year was the sum of the forecast power demands, excluding e-mobility, for the year (for the result obtained from the Ensemble Model, details in Section 3.3) and the three variants of forecast power demand resulting from the operation of electric vehicles in the year (details in Section 3.2). The results of calculations are presented in Table 5. Figure 11 shows, for the three variants the forecast percentage growth of power demand due to e-mobility in Poland from 2022 to 2027.

Table 5. Results of calculations of power demand with e-mobility, in three variants.

	Power Demand Without e-Mobility	Power Demand with E-Mobility		
Year	Ensemble Model [GWh]	Pessimistic Variant [GWh]	Balanced Variant [GWh]	Optimistic Variant [GWh]
2022	175,799	176,063	176,098	176,135
2023	177,272	177,734	177,795	177,865
2024	178,733	179,566	179,702	179,873
2025	180,180	181,718	182,067	182,557
2026	181,615	184,481	185,415	186,882
2027	183,038	188,389	190,824	195,224

The results in Table 5 and Figure 11 show that, for the initial period (forecasts for 2022–2024), the impact of e-mobility on the Polish electric power system was negligible. In subsequent years (forecasts for 2023–2027), the impact of e-mobility on the Polish power system grew dynamically, reaching almost 7% for the optimistic variant. Such an extra annual amount of power (more than 12 [TWh]) is a big challenge for the Polish electric power system, especially in the context of the energy crisis (energy deficit). On the other hand, the mechanisms of the "Fit for 55" package (phasing out manufacturing of petrol or diesel vehicles) means that significantly larger quantities of power need to be generated to meet the e-mobility demand (EV charging). Obviously, these will be covered by RES to some extent.

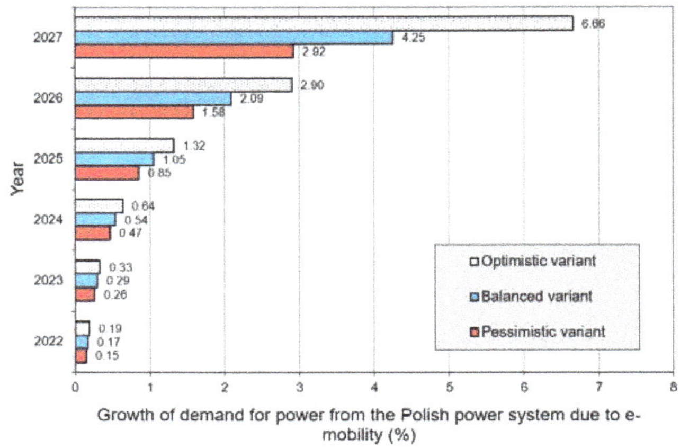

Figure 11. Forecast percentage growth of power demand due to e-mobility in Poland from 2022 to 2027.

3.6. Forecast Daily Profiles of Typical Days in 2022–2027 without E-Mobility

The third Wednesday of July and the third Wednesday of January are "typical days" in the Polish Power System, representing the summer and the winter business days, respectively. Forecasts for both relative daily profiles (normalised hourly values) were conducted for both typical days with a horizon of six years (2022–2027). The normalisation procedure is described in Section 2. Normalisation enabled us to the track changes in the profiles over 2009–2021, ignoring any profile change resulting from a multi-annual growing trend of power demand in the subsequent years.

Forecast relative values of the profiles in the subsequent years, conducted separately for each hour, used the following methods: MLP-type Artificial Neural Network, LSTM-type Deep Neural Network (DNN) and Ensemble Model (final forecast).

Both the MLP and LSTM were used in Step One for simultaneous approximation of 24 different functions (variability of demand for power for the respective hour of the day between 2009 and 2021) in a single model of artificial neural network with 24 outputs (function values) and 1 input (sample index in the time series of the process). In the next step, relative values of both profiles were forecast by extrapolating the function onto six consecutive periods (from 2022 to 2027). The use of an artificial neural network for extrapolation of the function is our original, unique ANN application proposal. Figure 12 conceptually presents our unique method using two different models of neural networks.

The first model used in the Ensemble Model was MLP. The model is described (conceptually) in Section 3.2. Tests for various hyperparameters were performed to select appropriate models. The tested number of hidden neurons ranged from 3 to 12. The number of learning epochs was tested for the following values: 10, 20, 30, 40 and 50. The number of learning epochs had a large influence on the level of "smoothing out" of the function being approximated. For the hidden layer, a hyperbolic tangent was selected as the activation function. Linear activation function in the output layer was adopted as the appropriate one for the process studied here. Such a choice ensured that the MLP neural network was capable of extrapolating (able to predict outside of the learning range). To optimise the weights (model parameters), the BFGS optimisation algorithm was applied. The finally selected models for forecasting the typical winter and summer day profiles had six neurons in the hidden layer (MLP 1-6-24 (tangh/linear)). Both models were learning for 20 learning epochs, with weights being updated following each learning epoch.

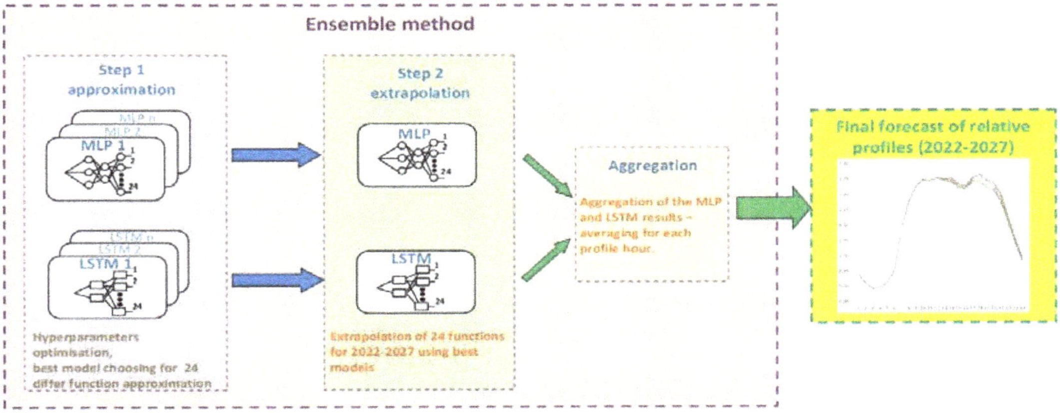

Figure 12. Conceptual graph of the proposed method for relative profile forecasting.

The other model applied in the Ensemble Model was LSTM. The model is described in [35]. Tests for various hyperparameters were performed to select appropriate models. The number of hidden neurons was tested in the range between 3 and 50. The number of learning epochs was tested for the following values: 50, 100, 200, 300 and 500. For the hidden layer, Sigmoid, Relu, and Hyperbolic Tangent activation functions were tested. Linear function was applied in the output layer. Adaptive Moment (Adam) and Root Mean Square Propagation (RMSprop) Optimisation Algorithms were tested to optimise the weights. In addition, the Dropout technique, with the value of 0.1 along the hidden layer, was applied, and the absence of that mechanism was tested. Both final models, selected by expert choice, were taught for 200 epochs, with the Sigmoid function in the hidden layer without the Dropout technique, and with the Adam (winter profile) and RMSPprop (summer profile) optimisation algorithms. The finally selected models for forecasting the profiles of typical days in January (LSTM 1-3-24 (sigm/linear)) and in July (LSTM 1-4-24 (sigm/linear)) had 3 and 4 neurons in the hidden layer, respectively.

The appropriate models (MLP, LSTM) were selected by expert choice based on the size of errors in the estimation range, and the observation of the level of "smoothing out" of forecasts within the model's parameter estimation range (the model should preserve a non-linear shape of forecasts with simultaneous avoidance of overestimating).

The application of two independent non-linear models was much more accurate than a simple linear regression model, for example, in terms of extrapolating forecasts independently for each hour. The linear model was unable to reflect the non-linear shape of the variability of power demand in the respective hour over consecutive years. To apply linear regression, one needs to build 24 independent models for the given typical day, whereas a single neural network generates 24 values of the typical day profile simultaneously in a single step. Such a simplified linear approach to the forecast profiles was applied in [24]. Figure 13 presents differences in the operation of the linear model of regression with trend extrapolation, and the proposed non-linear Ensemble Model. The shapes of forecast curves of the Ensemble Model were clearly non-linear (especially for the estimation of model parameters). The Linear Regression (LR) Model clearly underestimated forecasts in 2022–2027, as compared to the Ensemble Model, as it failed to incorporate a significant change in the downward trend in the last years of the model's parameter estimation range (2014–2021). Process figures from recent years should weigh more since they reflect the most current status of the process trend line. The figures presented in Figure 13 indicated that the process had stabilised (cessation of the downward trend).

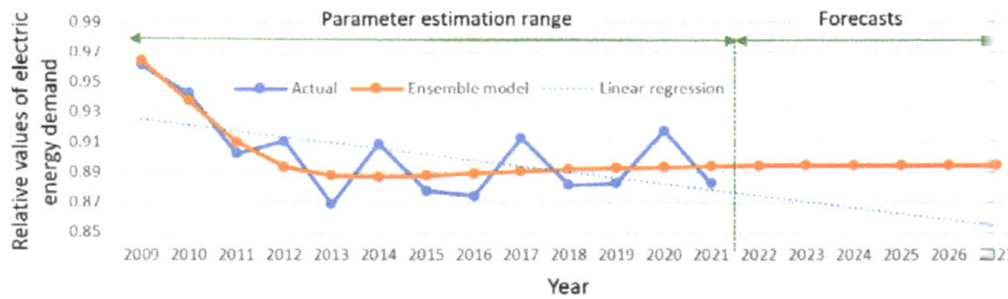

Figure 13. Comparison of forecast outcomes from the linear regression model and non-linear Ensemble Model at 4:00 am on the winter typical day.

Table 6 presents aggregate results of relative forecast profiles of typical days for the three models (error metrics) in the models' parameter estimation range (2009–2021). Error metrics were average values calculated on errors obtained separately for each of the 24 h of the given profile. The error metrics thus obtained indicated that, within the estimation range, the models were better fitted, in terms of relative profile of the typical day, for the summer rather than for the winter. Detailed summary of the results (five error metrics) for each separate hour is provided in Table A1 in Appendix A.

Table 6. Summary of forecast relative profiles of typical days for three models (error metrics) within the models' parameter estimation range (2009–2021).

Method Name	RMSE	MAPE [%]	R	R^2
Relative profile of the winter typical day				
MLP 1-6-24 (tangh/linear)	0.0172	1.300	0.7447	0.5649
LSTM 1-3-24 (sigm/linear)	0.0197	1.423	0.7094	0.4180
Ensemble Model	0.0179	1.311	0.7345	0.5285
Relative profile of the summer typical day				
MLP 1-6-24 (tangh/linear)	0.0110	0.922	0.7726	0.5091
LSTM 1-4-24 (sigm/linear)	0.0128	0.970	0.7191	0.5116
Ensemble model	0.0116	0.931	0.7532	0.5865

Figure 14 presents forecast power demand in particular hours of the typical winter and summer days (2022–2027), and observed values (2009–2021). Figure 15 presents forecast relative profiles of power demand for the typical winter and summer days. The trend of change in relative demand varied by hour. For the profile of the winter typical day, relative power demand slightly fell in particular hours between 01:00 pm and 12:00 pm between the last year of the forecast (2027) and the first year of the forecast (2022). In the night "valley", profile changes were minimal.

For the profile of the summer typical day, there were both slight increases in demand for power in the night "valley" of daily demand, and in the evening "peak" of power demand. In the morning "peak", power demand fell slightly.

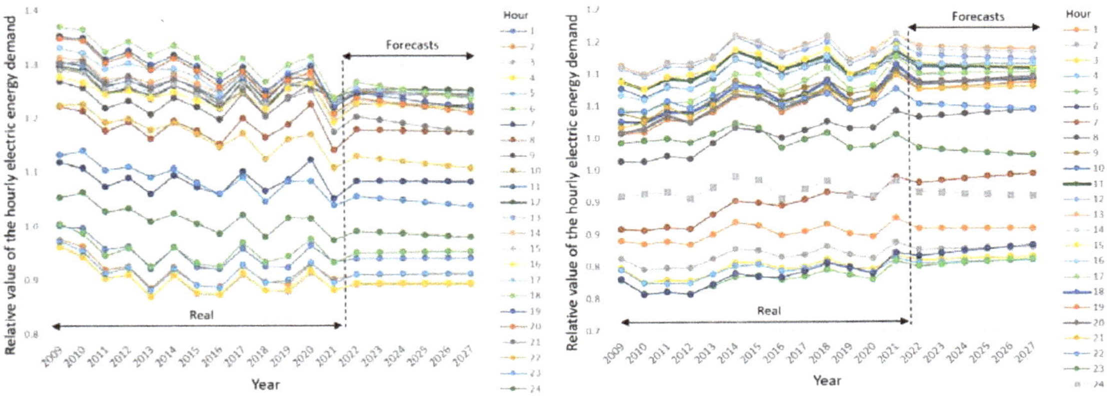

(a) (b)

Figure 14. (**a**) Forecast relative power demand in particular hours of the winter typical day; (**b**) Forecast relative power demand in particular hours of the summer typical day.

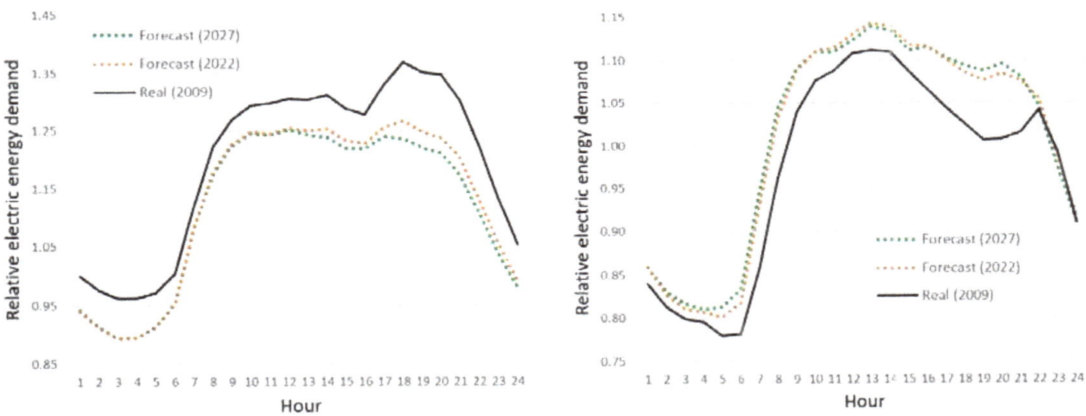

(a) (b)

Figure 15. (**a**) Forecast relative profiles of power demand for the winter typical day; (**b**) Forecast relative profiles of power demand for the summer typical day.

3.7. Forecast Daily Profiles of Typical Days in 2022–2027 with E-Mobility

In Step One, calculations were conducted to transform the forecast values of relative power demand profiles (2022–2027) into absolute values of power demand. Relative forecast values of both profiles (2022–2027) (details in Section 3.6) were recalculated to absolute values [GWh]. However, forecast annual power demand in Poland (2022–2027), as detailed in Section 3.3, was recalculated to obtain average hourly figures of power demand during that year. This method incorporated both the growing trend of the annual power demand and how relative profiles evolved over the subsequent years.

In Step Two, daily profiles of power demand for EV charging were calculated. Calculations of hourly figures from daily profiles of power demand, due solely to e-mobility, used the relative profiles developed for a business day for EVs and daily power demand from four EV categories calculated on annual values. The subsequent calculations assumed that BEVs and PHEVs belonged to a single category, cars with the same relative profiles. The methodology of construction of relative hourly profiles for EV power demand and the relative profiles alone are described in detail in [36]. A total of six relative profiles were used. Each of the three EV categories had two profiles, power drawn from rapid charging stations and from slow charging stations. Slow charging was assumed to be the following: 70% for electric cars, 20% for electric buses and 35% for electric heavy trucks and electric delivery vans, and the remaining power was drawn from rapid charging stations.

Calculations were performed for the three variants of forecast EV numbers in each EV category separately, for 2022–2027. In the next step, combined daily power demand for charging EVs of any type was calculated. Figure 16 shows the outcome of calculations for the balanced variant of the forecast number of EVs in 2027.

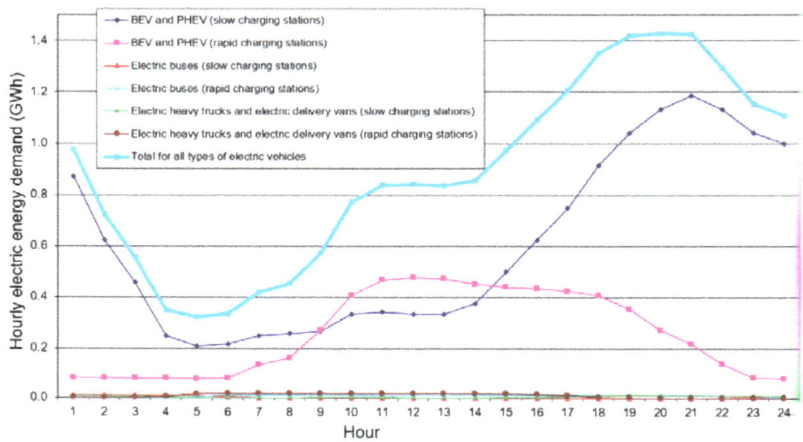

Figure 16. Daily profiles of power demand for charging EVs—balanced variant of the forecast number of EVs in 2027.

In the third and last step, power demand for the typical winter day profile and typical summer day profile were calculated taking into account the development of e-mobility. Figure 17 presents daily power demand profiles for the typical winter and summer days with and without e-mobility for the balanced variant of the number of EVs in 2027.

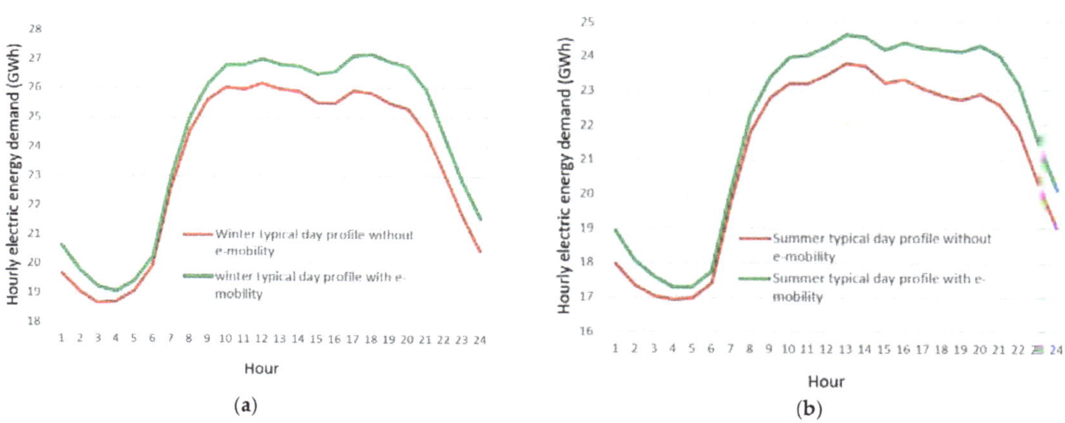

Figure 17. (**a**) Daily power demand profiles for the winter typical day with and without e-mobility for the balanced variant of the number of EVs in 2027; (**b**) Daily power demand profiles for the summer typical day with and without e-mobility for the balanced variant of the number of EVs in 2027.

Percentage growth of power demand due to e-mobility for the balanced variant varied by the time of the day. For the typical winter and summer day profiles, the distribution of percentages within a day were very similar. During the "evening" peak, the percentage share of e-mobility was largest, whereas during the night "valley", the percentage share was the lowest. For the summer typical day profile, percentage shares of e-mobility for all times of the day were slightly higher than for the winter typical day profile. Figure 18 shows the percentage growth of power demand due to e-mobility for the balanced variant of forecast number of EVs in 2027, for the typical winter and summer days. This was due to the fact that, in the winter period, power demand was slightly more than in the summer period, and power demand values due to e-mobility were adopted to be the same for both seasons of the year.

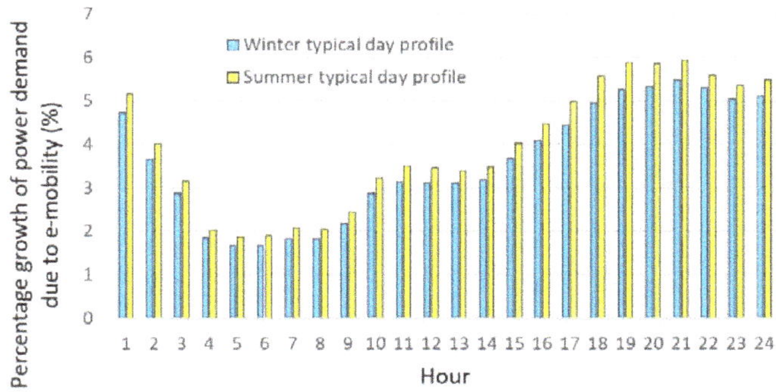

Figure 18. Percentage growth of power demand due to e-mobility for the balanced variant of forecast number of EVs in 2027 for the typical winter and summer days.

4. Discussion

Our observation is that, for the forecast number of EVs and for the forecast demand for power from the Polish electric power system, the models with the best fit, within the models' parameter estimation ranges (RMSE metric), generated at the same time the smallest "ex-ante" (forward-looking) forecast values of all methods. The opposite was noted for models with the worst fit to the observed values within the models' parameter estimation range (RMSE metric). These models, at the same time, generated "ex ante" (forward-looking) forecasts with the highest values of all methods. The conclusion could be, therefore, that well-fitting models would tend to underestimate forecast values, and models with a relatively poor fit would tend to overestimate the forecasts.

Regarding the forecast impact of e-mobility on the Polish electric power system, forecasts (pessimistic variant, balanced variant, and optimistic variant) could be noted to vary more with growing forecast horizon. Particularly wide differences occurred for six years' advance (2027). Therefore, the conclusion could be that uncertainty of forecasts for that horizon was relatively large. The increase in electric power demand for 2027 in the optimistic variant was almost 7%, which is a significant warning signal as to potential problems with meeting power requirements in Poland.

Our study forecasted on average 0.85 million EVs and 1.54–2.38 TWh corresponding load for the year 2025. Comparatively, previous studies determined the EV number to be 3.64 million and energy to be 6.11 TWh [37] or 0.021–0.176 million EVs and 0.19–1.5 TWh, respectively [25]. For the first study, the number of EVs was quadruple and amount of energy was more than twice to almost quadruple greater than in our forecasts. The difference could be attributed to the first study using a simplified procedure of calculation, not decomposing EVs into categories, and the short period used for forecast parameters estimation. Although the second study decomposed EVs into categories, it also used pre-2019 data. This period concerns time when EVs were treated like a novelty in Poland rather than a valid conventional car alternative. It can be noted that from that period of pioneering EVs in Poland the dynamic of process has changed steeply and EVs have started to be bought on a much larger scale, so more recent data has better accuracy.

Our demand forecasts excluding e-mobility determined that demand in 2025 would equal ca.180 TWh (on different model average). Comparatively, another study forecast 149 TWh with quasi-linear dynamics of growth [7]. We deem the results to be roughly comparable, as flattening of the curve can be attributed to the referred authors using ca. 20 years for training phase of the forecasting model instead of using less, but fresher data, thus obtaining a more conservative, averaged forecast.

The pace of growth in the number of electric trucks, especially heavy ones, is a big unknown. Currently, this segment of e-mobility is in its inception phase, and the momentum of this process is unknown. For electric buses, the growth ceiling, and, therefore, the current impact on the Polish electric power system is quite low, due to the relatively small total number of buses used in Poland (slightly more than 12,000 pcs). Even so, financial inequalities between Polish regions make the future transformation process unequal. Due to the cost of acquisition of vehicles and loading infrastructure one can expect that the biggest cities will note the greatest increase in the number of electric buses. Study concerning the dynamics of growth for the Polish capital city showed tripling of the number of electric buses over 2021–2022, for instance [38]. Although the study referred to determined lack of problems for charging infrastructure with the increase of number of electric buses, the result could potentially vary with region and situation. Current socio-political factors, such as petrol prices or increasing inflation, could also affect interest of customers in using public transport, and, in turn, further increase number of electric buses. Another valid method of eco-transport, especially in big cities, are individual and shared e-scooters. This vehicle type could potentially optimize energy spent per capita for routes without good direct connection by electric buses, and reduce traffic. Facilitating movement of this vehicle requires, however, adoption of proper legislation to ensure safety of drivers, especially in the face of the increasing popularity of this solution [39].

Meanwhile, the potential for rising numbers of electric trucks is very large. Currently there are more than 3.6 million such vehicles registered in Poland. If all of those vehicles became electric, their impact on power demand would be huge. Assuming that the share of electric vans is 75% and of electric heavy trucks 25% out of 3.6 million of electric trucks overall, annual power demand resulting only from electric trucks would be, according to our calculations, about 118,000 TWh, i.e., 33% more than the current (2021) total annual power production in Poland.

The analysis of the results of the forecasts of relative power demand profiles shows that the dynamics of change in both relative profiles decrease significantly over 2022–2027 as compared to changes after 2009. For both relative profiles, the largest changes between 2009 and 2027 (forecast) are visible during the evening "peak" of power demand. In the winter season, the evening "peak" of power demand would have decreased over time, while in the summer period, the evening "peak" of power demand would have increased over time. Since 2009 to about 2016, the dynamics of change in relative profiles was high. In the subsequent years since 2017 and to 2021 the changes became less dynamic, the same applies to forecasts from 2022 to 2027.

Analyses of the impact of e-mobility on the Polish power system until 2027 show that, for the profiles of the typical winter and summer days the percentage share of e-mobility is the highest during the evening "peak", which is very unfavourable for the electric power system. This could be partially alleviated by changing the habits of EV users so that they begin slow charging of their EVs just after midnight rather than from afternoon (after returning from work). However, this requires incentives for EV users, such as a significantly reduced electricity price at night. To avoid local grid overload, remote control of the charging process seems to be necessary for users charging their vehicles at night. The controller could collaborate with the controllers of other cars, thus, coordinating the charging process in the respective part of the network, and with the power meter, thus, increasing the automation of the process, or by deploying demand management software. Recent studies show, however, that user acceptance of external management of their cars is rather low in Poland [40]. In light of this, more effort should be put into either policy or incentives creation, as solutions such as V2G could decrease the negative impact of increased load caused by electric vehicles [26]. Albeit not all factors influencing the customers are easily transferable between countries, the quality of quick-charging infrastructure and simplicity of use could be named as universal factors impacting users' decisions to join such programs [27].

The work on hourly electricity profiles, excluding EV, can be compared with the work of Brodowski et al. [15], where the authors predicted mean hourly load profile in the Polish Electric Grid starting from 11 am in the first variant and 8 am in the second one. In order to compare accuracy of the studies, our summer and wind profiles were averaged over hours, and presented in Figure 19. The comparison determined that both studies resulted in a similar magnitude of error, though the referred study showed higher deviation from average from both sides of the average. For early morning (hourly periods 2–8) our model showed more accuracy, while for the rest of the periods both study models were comparable. It can be noted that 8/11 models demonstrated lowest error directly after moments of start, with rapid increase shortly after. Our model, in turn, was more stable in all analysed periods. It must be, however, emphasised that, due to the difference in tested period ranges (year 2004 for 8/11 models, years 2009–2021 in our model) the above comparison could only be roughly done. Other studies, pertaining to profile creation with decomposition into summer and winter day profiles, resulted in 2.8% MAPE over the year 2016 [18]. In view of the nature of our study, concerning forward extrapolation, both of the above studies cannot be directly compared with our study, as it had no direct test data equivalent.

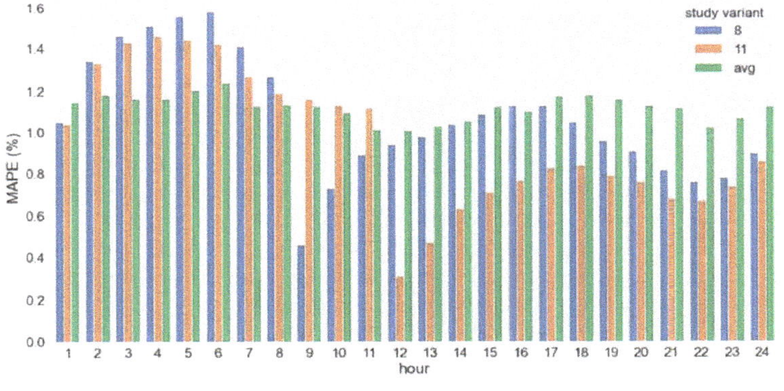

Figure 19. Comparison of hourly energy demand profile excluding EV between our study averaged profile and profiles started to be predicted at 8/11 am.

5. Conclusions

As a result of multi-step and multi-variant forecasts, the impact of e-mobility on the Polish electric power system was determined in terms of annual growth of power demand and on a daily basis (times of the day) for two typical days (summer and winter ones). This impact varied by e-mobility development variant. For the balanced (i.e., the most likely) growth variant, annual power demand would grow by almost 7% due to e-mobility. However, the percentage growth of power demand due to e-mobility for the balanced variant varied by the time of the day. For typical winter and summer day profiles, the distribution of percentages in different times of the day was very similar. During the "evening" peak, the percentage share of e-mobility was the largest, whereas during the night "valley", the percentage share was the lowest.

The outcomes of forecast power demand amounts in particular times of typical winter and summer days (2022–2027) without e-mobility indicated that, depending on the time of the day, the trends of changes in relative demand were different. For the profile of the typical winter day, relative power demand fell slightly in particular hours between 01:00 pm and midnight between the last year of the forecast (2027) and the first year of the forecast (2022). In the night "valley", profile changes were minimal.

This research shows that the development of e-mobility in Poland for the horizon of 6 years (2027) may cause a problem regarding covering the additional demand for electricity. The problem concerns both the value of the total annual energy demand, but also the "evening" peak of the typical summer and winter days, in which the impact of e-mobility on the demand for electricity is greatest.

The proposed unique methods developed by the authors proved to be effective. An MLP artificial neural network was applied for non-linear extrapolation of a single function (forecast number of electric vehicles in Poland from 2022 to 2027 and forecast annual power demand in Poland from 2022 to 2027, without the development of e-mobility in Poland). Ensemble Methods (MLP and LSTM) were applied to conduct simultaneous extrapolation of 24 non-linear functions (forecast daily profiles of typical days in 2022–2027 without e-mobility).

A novel, original Growth Dynamics Model was developed that used forecast annual growth ratios to forecast the number of electric vehicles in Poland from 2022 to 2027.

This research has some limitations which should be pointed out in order to ensure the integrity of scientific research. The main limitation is the use of only the time series of forecasted processes in forecasting models. For this reason, the proposed methods of EV number forecasting can be considered appropriate only for the medium-term horizon (up to several years ahead) due to the relatively short time series of historical data and the dependence of the forecasted process on many factors that may undergo dynamic changes in the future (electricity price, incentives supporting e-mobility, dynamic development of hydrogen-powered electric cars, FCV (Fuel Cell Vehicle)). In the case of forecasts of profiles of typical days, for forecast horizons greater than 6 years, one should expect more and more errors in forecasts as the forecast horizon grows. In the future, there may be various additional factors influencing the shape of the typical day profiles. A factor that may affect the shape of the daily load profiles is, for example, the development of RES. Other factors include climate change and the introduction of dynamic tariffs.

We intend to expand future research on e-mobility development to include forecasts of the development of the number of charging stations, discriminating between rapid and slow charging stations, and forecast the development of e-mobility in as disaggregated a manner as possible (separate forecasts by EV type, including electric bikes and scooters). Regarding the research on the impact of e-mobility development on the electric power system, changes in profiles of typical days due to RES development in Poland (wind farms, photovoltaic systems, and energy storage) can be taken into account or studied in addition. An important element planned in future research will be the use of exogenous explanatory variables (input data) in prognostic models (historical values and forecasts), in addition to the withdrawn values of the explained variable.

Author Contributions: Conceptualization, P.P., D.B.; Methodology, P.P., D.B., M.K.; Investigation, P.P., D.B. and M.K.; Supervision, P.P.; Validation, P.P. and M.K.; Writing, P.P., D.B. and M.K.; Visualization P.P. and D.B.; Project administration, P.P. All authors have read and agreed to the published version of the manuscript.

Funding: This research was funded by the 2021 edition of the competition for grant of the Scientific Council for the Discipline Automatic Control, Electronics and Electrical Engineering of the Warsaw University of Technology (to P.P., D.B., M.K.).

Institutional Review Board Statement: Not applicable.

Informed Consent Statement: Not applicable.

Data Availability Statement: Not applicable.

Conflicts of Interest: The authors declare no conflict of interest.

Abbreviations

The following abbreviations are used throughout this manuscript:

adam	Adaptive Moment (Estimation)
ANN	Artificial Neural Network
BEVs	Battery Electric Vehicles
BFGS	Broyden–Fletcher–Goldfarb–Shanno algorithm
DEPS	Differential Evolution and Particle Swarm Optimisation algorithm
DNN	Deep Neural Network
EVs	Electric vehicles
EU	European Union
FCV	Fuel Cell Vehicle
LEAP	Long-range Energy Alternatives Planning System
LR	Linear Regression
LSTM	Long short-term memory
ML	Machine Learning
MLP	Multi-Layer Perceptron
MLS	Method of Least Squares
MAPE	Mean Absolute Percentage Error
MBE	Mean Bias Error
PCA-GRNN	Principal Component Analysis—General regression neural network
PHEVs	Plug-in Hybrid Electric Vehicles
PLEVs	Personal light electric vehicles
R	Pearson linear correlation coefficient
R^2	Determination coefficient
RES	Renewable Energy Sources
RMSE	Root Mean Square Error
RMSProp	Root Mean Square Propagation
SACWO	Simulated Annealing Chicken Swarm Optimization
SALAMA	(Suomen alueellinen autokantamalli, Finnish regional car fleet model).
SSE	Sum of Squared Errors
V2G	Vehicle-to-grid

Appendix A

Table A1. Forecast accuracy metrics of the MLP + LSTM Ensemble Method for particular times of the day.

	Relative Profile of the Winter Typical Day					Relative Profile of the Summer Typical Day				
Hour *	RMSE	MAPE [%]	MBE	R	R^2	RMSE	MAPE [%]	MBE	R	R^2
1	0.0165	1.383	0.00430	0.8054	0.6212	0.0094	0.907	0.00142	0.6934	0.4680
2	0.0151	1.410	0.00245	0.8350	0.6877	0.0090	0.950	0.00185	0.7131	0.4849
3	0.0149	1.356	0.00349	0.8468	0.7004	0.0098	0.971	0.00249	0.6526	0.3751
4	0.0158	1.394	0.00396	0.8208	0.6501	0.0092	0.928	0.00134	0.6770	0.4251
5	0.0159	1.412	0.00270	0.7932	0.6176	0.0098	0.994	0.00143	0.7337	0.5241
6	0.0167	1.446	0.00296	0.7397	0.5315	0.0116	1.029	0.00251	0.7878	0.5914
7	0.0196	1.450	0.00362	0.5105	0.2327	0.0092	0.805	0.00055	0.9346	0.8718
8	0.0221	1.521	0.00423	0.5153	0.2367	0.0092	0.752	0.00084	0.9314	0.8662
9	0.0200	1.393	0.00306	0.5647	0.3020	0.0116	0.862	0.00134	0.8591	0.7341
10	0.0187	1.308	0.00459	0.6331	0.3616	0.0118	0.886	0.00198	0.8074	0.6417

Table A1. *Cont.*

	Relative Profile of the Winter Typical Day					Relative Profile of the Summer Typical Day				
Hour *	**RMSE**	**MAPE [%]**	**MBE**	**R**	**R^2**	**RMSE**	**MAPE [%]**	**MBE**	**R**	**R^2**
11	0.0171	1.141	0.00619	0.7159	0.4386	0.0123	0.885	−0.00063	0.7491	0.5597
12	0.0165	1.135	0.00353	0.6888	0.4470	0.0120	0.882	0.00030	0.7581	0.5744
13	0.0168	1.131	0.00408	0.6769	0.4210	0.0127	0.931	−0.00081	0.7715	0.5928
14	0.0177	1.131	0.00048	0.6820	0.4646	0.0132	0.982	0.00058	0.7567	0.5714
15	0.0189	1.261	0.00542	0.6262	0.3375	0.0133	0.990	−0.00022	0.7603	0.5727
16	0.0186	1.194	0.00486	0.5862	0.2955	0.0134	1.013	−0.00007	0.8209	0.6734
17	0.0201	1.283	0.00210	0.7187	0.5094	0.0142	1.067	0.00102	0.8241	0.6771
18	0.0200	1.296	0.00234	0.8283	0.6810	0.0153	1.064	0.00298	0.8001	0.6249
19	0.0190	1.253	0.00006	0.8441	0.7125	0.0152	1.067	0.00187	0.8191	0.6614
20	0.0197	1.312	−0.00033	0.8511	0.7242	0.0125	0.947	0.00141	0.8778	0.7651
21	0.0197	1.370	0.00407	0.8343	0.6818	0.0114	0.871	0.00128	0.8521	0.7226
22	0.0176	1.279	0.00388	0.8565	0.7189	0.0108	0.769	−0.00049	0.5679	0.3095
23	0.0164	1.284	0.00578	0.8479	0.6775	0.0105	0.858	0.00135	0.4894	0.4010
24	0.0152	1.322	0.00327	0.8063	0.6322	0.0105	0.930	0.00047	0.4407	0.3820

* Hour corresponds to the point in time of measurement/end of hourly period, e.g., hour = 12 corresponds to 11:00–12:00 period.

References

1. The Polish Alternative Fuels Association. Available online: https://pspa.com.pl (accessed on 15 May 2022).
2. Mirjat, N.H.; Uqaili, M.A.; Harijan, K.; Walasi, G.D.; Mondal, M.A.H.; Sahin, H. Long-term electricity demand forecast and supply side scenarios for Pakistan (2015e2050): A LEAP model application for policy analysis. *Energy* **2018**, *165*, 512–526. [CrossRef]
3. Gebremeskel, D.H.; Ahlgren, E.O.; Beyene, G.B. Long-term evolution of energy and electricity demand forecasting: The case of Ethiopia. *Energy Strategy Rev.* **2021**, *36*, 100671. [CrossRef]
4. Angelopoulos, D.; Siskos, Y.; Psarras, J. Disaggregating time series on multiple criteria for robust forecasting: The case of long-term electricity demand in Greece. *Eur. J. Oper. Res.* **2019**, *275*, 252–265. [CrossRef]
5. He, Y.; Wang, M.; Guang, F.; Zhao, W. Research on the method of electricity demand analysis and forecasting: The case of China. *Electr. Power Syst. Res.* **2020**, *187*, 106408. [CrossRef]
6. Manowska, A. Using the LSTM Network to Forecast the Demand for Electricity in Poland. *Appl. Sci.* **2020**, *10*, 8455. [CrossRef]
7. Wu, M.; Chen, W. Forecast of Electric Vehicle Sales in the World and China Based on PCA-GRNN. *Sustainability* **2022**, *14*, 2206. [CrossRef]
8. Rietmann, N.; Hügler, B.; Lieven, T. Forecasting the trajectory of electric vehicle sales and the consequences for worldwide CO_2 emissions. *J. Clean. Prod.* **2020**, *261*, 121038. [CrossRef]
9. Ding, S.; Li, R. Forecasting the sales and stock of electric vehicles using a novel self-adaptive optimized grey model. *Eng. Appl. Artif. Intell.* **2021**, *100*, 104148. [CrossRef]
10. Viri, R.; Mäkinen, J.; Liimatainen, H. Modelling car fleet renewal in Finland: A model and development speed-based scenarios. *Transp. Policy* **2021**, *112*, 63–79. [CrossRef]
11. Galvin, R. Are electric vehicles getting too big and heavy? Modelling future vehicle journeying demand on a decarbonized US electricity grid. *Energy Policy* **2022**, *161*, 112746. [CrossRef]
12. Feng, J.; Yang, J.; Li, Y.; Wang, H.; Ji, H.; Yang, W.; Wang, K. Load forecasting of electric vehicle charging station based on grey theory and neural network. *Energy Rep.* **2021**, *7*, 487–492. [CrossRef]
13. Kalhori, M.R.N.; Emami, I.T.; Fallahi, F.; Tabarzadi, M. A data-driven knowledge-based system with reasoning under uncertain evidence for regional long-term hourly load forecasting. *Appl. Energy* **2022**, *314*, 118975. [CrossRef]
14. Carmo, D.L.; Souza, R.C.; Barbosa, C.R.H. Partial decomposition approach to generate load curve forecasting scenarios. *Electr. Power Energy Syst.* **2020**, *115*, 105436. [CrossRef]
15. Brodowski, S.; Bielecki, A.; Filocha, M. A hybrid system for forecasting 24-h power load profile for Polish Polish electric grid. *Appl. Soft Comput.* **2017**, *58*, 527–539. [CrossRef]
16. Hinde, Ó.G.; Verdejo, V.G.; Martínez-Ramón, M. Forecast-informed power load profiling: A novel approach. *Eng. Appl. Artif. Intell.* **2020**, *96*, 103948. [CrossRef]
17. Yukseltan, E.; Kok, A.; Yucekaya, A.; Bilge, A.; Agca Aktunc, E.; Hekimoglu, M. The impact of the COVID-19 pandemic and behavioral restrictions on electricity consumption and the daily demand curve in Turkey. *Util. Policy* **2022**, *76*, 101359. [CrossRef]
18. Behm, C.; Nolting, L.; Praktiknjo, A. How to model European electricity load profiles using artificial neural networks. *Appl. Energy* **2020**, *277*, 115564. [CrossRef]
19. Arias, M.B.; Bae, S. Electric vehicle charging demand forecasting model based on big data technologies. *Appl. Energy* **2016**, *183*, 327–339. [CrossRef]

20. Liu, Z.; Liu, Z. Load characteristics forecasting of Hubei power grid up to year 2030 with the development of electric vehicles. *Energy Rep.* **2022**, *8*, 259–268. [CrossRef]
21. Nogueira, T.; Sousa, E.; Alver, G.R. Electric vehicles growth until 2030: Impact on the distribution network power. *Energy Rep.* **2022**, *8*, 145–152. [CrossRef]
22. Wörner, R.; Morozova, I.; Cao, D.; Schneider, D.; Neuburger, M.; Mayer, D.; Körner, C.; Kagerbauer, M.; Kostorz, N.; Blesl, M.; et al. Analysis and Prediction of Electromobility and Energy Supply by the Example of Stuttgart. *World Electr. Veh. J.* **2021**, *12*, 78. [CrossRef]
23. Brdulak, A.; Chaberek, G.; Jagodziński, J. Determination of Electricity Demand by Personal Light Electric Vehicles (PLEVs): An example of e-Motor Scooters in the Context of Large City Management in Poland. *Energies* **2020**, *13*, 194. [CrossRef]
24. Piotrowski, P.; Baczyński, D.; Robak, S.; Kopyt, M.; Piekarz, M.; Polewaczyk, M. Comprehensive forecast of electromobility mid-term development in Poland and its impacts on power system demand. *Bull. Pol. Acad. Sci. Tech. Sci.* **2020**, *68*, 697–709. [CrossRef]
25. Zou, Y.; Zhao, J.; Gao, X.; Chen, Y.; Tohidi, A. Experimental results of electric vehicles effects on low voltage grids. *J. Clean.* **2020**, *255*, 120270. [CrossRef]
26. Bibak, B.; Tekiner-Mogulkoc, H. The parametric analysis of the electric vehicles and vehicle to grid system's role in flattening the power demand. *Sustain. Energy Grids Netw.* **2022**, *30*, 100605. [CrossRef]
27. Heuveln, K.; Ghotge, R.; Annema, J.A.; Bergen, E.; Wee, B.; Pesch, U. Factors influencing consumer acceptance of vehicle-to-grid by electric vehicle drivers in the Netherlands. *Travel Behav. Soc.* **2021**, *24*, 34–45. [CrossRef]
28. Parol, M.; Piotrowski, P. Long-term forecasting method of annual electrical energy demand in electric distribution companies. *Przegląd Elektrotech.* **2010**, *86*, 182–186.
29. Piotrowski, P.; Baczyński, D.; Kapler, P. Multi-variant forecasts of the number of electric vehicles in Poland until 2025 and their impact on annual electric energy demand. *Przegląd Elektrotech.* **2020**, *96*, 138–141. [CrossRef]
30. Piotrowski, P.; Baczyński, D.; Kopyt, M.; Szafranek, K.; Helt, P.; Gulczyński, T. Analysis of forecasted meteorological data (NWP) for efficient spatial forecasting of wind power generation. *Electr. Power Syst. Res.* **2019**, *175*, 105891. [CrossRef]
31. Dudek, G. Multilayer perceptron for short-term load forecasting: From global to local approach. *Neural Comput. Appl.* **2019**, *32*, 3695–3707. [CrossRef]
32. Géron, A. *Hands-on Machine Learning with Scikit-Learn, Keras, and TensorFlow*, 2nd ed.; O'Reilly Media Inc.: Sevastopol, CA, USA, 2019.
33. Liu, S.; Lin, Y. Grey Systems. In *Theory and Applications*; Springer: Berlin/Heidelberg, Germany, 2010.
34. Piotrowski, P. Forecasting in Power Engineering in different time horizons. In *Prace Naukowe Politechniki Warszawskiej "Elektryka"*; Oficyna Wydawnicza Politechniki Warszawskiej: Warsawz, Poland, 2013; Volume 144.
35. Duan, J.; Wang, P.; Ma, W.; Tian, X.; Fang, S.; Cheng, Y.; Chang, Y.; Liu, H. Short-term wind power forecasting using the hybrid model of improved variational mode decomposition and Correntropy Long Short -term memory neural network. *Energy* **2021**, *214*, 118980. [CrossRef]
36. Kłos, M.; Paska, J.; Baczyński, D.; Piotrowski, P.; Robak, S.; Rosłaniec, S.; Helt, P.; Błędzińska, M.; Bielas, R.; Kapler, P.; et al. *Energy Efficiency through the Development of Electromobility in Poland, Research Project Co-Financed by the Norwegian Financial Mechanism 2009–2014 under the Bilateral Cooperation Fund*; Norwegian Financial Mechanism: Warsaw, Poland, 2017.
37. Kłos, M.; Marchel, P.; Paska, J.; Bielas, R.; Błędzińska, M.; Michalski, L.; Wróblewski, K.; Zagrajek, K. Forecast and Impact of Electromobility Development on the Polish Electric Power System. *E3S Web Conf.* **2019**, *84*, 1005. [CrossRef]
38. Zagrajek, K.; Paska, J.; Kłos, M.; Pawlak, K.; Marchel, P.; Bartecka, M.; Michalski, Ł.; Terlikowski, P. Impact of Electric Bus Charging on Distribution Substation and Local Grid in Warsaw. *Energies* **2020**, *13*, 1210. [CrossRef]
39. Turoń, K.; Czech, P. The Concept of Rules and Recommendations for Riding Shared and Private E-Scooters in the Road Network in the Light of Global Problems. In *Modern Traffic Engineering in the System Approach to the Development of Traffic Networks*; TSTP 2019; Advances in Intelligent Systems and Computing; Macioszek, E., Sierpiński, G., Eds.; Springer: Cham, Switzerland, 2020; Volume 1083. [CrossRef]
40. Zagrajek, K. A Survey Data Approach for Determining the Probability Values of Vehicle-to-Grid Service Provision. *Energies* **2021**, *14*, 7270. [CrossRef]

energies

MDPI

Article

The Use of Singular Spectrum Analysis and K-Means Clustering-Based Bootstrap to Improve Multistep Ahead Load Forecasting

Winita Sulandari [1,*], Yudho Yudhanto [2] and Paulo Canas Rodrigues [3]

[1] Study Program of Statistics, Universitas Sebelas Maret, Surakarta 57126, Indonesia
[2] Informatics Engineering, Vocational School, Universitas Sebelas Maret, Surakarta 57126, Indonesia
[3] Department of Statistics, Federal University of Bahia, Salvador 40170-110, Brazil
* Correspondence: winita@mipa.uns.ac.id

Abstract: In general, studies on short-term hourly electricity load modeling and forecasting do not investigate in detail the sources of uncertainty in forecasting. This study aims to evaluate the impact and benefits of applying bootstrap aggregation in overcoming the uncertainty in time series forecasting, thereby increasing the accuracy of multistep ahead point forecasts. We implemented the existing and proposed clustering-based bootstrapping methods to generate new electricity load time series. In the proposed method, we use singular spectrum analysis to decompose the series between signal and noise to reduce the variance of the bootstrapped series. The noise is then bootstrapped by K-means clustering-based generation of Gaussian normal distribution (KM.N) before adding it back to the signal, resulting in the bootstrapped series. We apply the benchmark models for electricity load forecasting, SARIMA, NNAR, TBATS, and DSHW, to model all new bootstrapped series and determine the multistep ahead point forecasts. The forecast values obtained from the original series are compared with the mean and median across all forecasts calculated from the bootstrapped series using the Malaysian, Polish, and Indonesian hourly load series for 12, 24, and 36 steps ahead. We conclude that, in this case, the proposed bootstrapping method improves the accuracy of multistep-ahead forecast values, especially when considering the SARIMA and NNAR models.

Keywords: electricity load forecasting; bootstrap aggregating; singular spectrum analysis; time series forecasting; calendar variation

Citation: Sulandari, W.; Yudhanto, Y.; Rodrigues, P.C. The Use of Singular Spectrum Analysis and K-Means Clustering-Based Bootstrap to Improve Multistep Ahead Load Forecasting. *Energies* **2022**, *15*, 5838. https://doi.org/10.3390/en15165838

Academic Editors: Paweł Piotrowski, Grzegorz Dudek and Dariusz Baczyński

Received: 12 July 2022
Accepted: 8 August 2022
Published: 11 August 2022

1. Introduction

Electricity load forecasting plays a critical role in controlling the balance between power demand and supply. Sometimes, the energy demand exceeds the energy supply and vice versa, which results in financial losses. An important aspect of a smart grid system is determining an accurate load forecasting model. Electricity load forecasting provides information that will simplify the work of planning consumption, generation, distribution, and other essential tasks of the smart grid system [1,2].

Much work has been performed to develop models and strategies to improve the electricity load forecasting accuracy. Generally, an hourly load series shows three relationships, i.e., between the observations for consecutive hours on a particular day, between the observations for the same hour on consecutive days, and between the observations for the same hour on the same day in successive weeks. In certain countries, the hourly load series may become more complex due to calendar variations [3]. The effect of calendar variation is usually considered by including a dummy variable in the model [4–7]. In countries with four seasons, the temperature is often included in the load forecasting model [8,9]. For countries with two seasons, such as Malaysia, it is also possible to include temperature information to improve the forecasts' accuracy [10]. Many models, from simple to complex, have been proposed and developed by researchers and practitioners around the world to improve the accuracy of electricity load forecasting, e.g., regression and seasonal autoregressive integrated moving average (SARIMA) models [4,11,12], exponential smoothing [3,6,13–15], neural network (NN) [16–19], singular spectrum analysis (SSA) [20–22], wavelets [23,24], fuzzy systems [10,25,26], support vector machine [1,21,27,28], among others. However, the most suitable model for electricity load forecasting in a given country may not be the best to model the data in another country because of different consumption and behavioral characteristics.

In this study, we discuss the implementation of the bootstrap aggregating method to improve the accuracy of multistep ahead load forecast. Bootstrap aggregation, which is known by the acronym "bagging", was proposed by [29] to reduce the variance of the predictor. It works by generating replicated bootstrap samples of the training data and using them to obtain the aggregated predictor. Bagging aims at improving the point forecast by considering sources of uncertainty, namely, the parameter estimates, the appropriate model determination, and the noise. In 2016, [30] successfully applied the development of this method in the field of time series forecasting by using the moving block bootstrap (MBB). Further, [31] explored how bagging improves point forecasts and showed that model selection as a solution to model uncertainty was the most influential on the success of bagging in time series. As described in [30], MBB bagging methods first apply the Box–Cox transform to the original series and then decompose it into a trend, seasonal, and noise using STL (Seasonal and Trend decomposition using Loess). STL is a decomposition method developed by [32]. In MBB, the noise is bootstrapped and added back to the trend and seasonal components. The new transformed bootstrapped series are then inverted and modeled. However, MBB is more appropriate for bootstrapping stationary time series. When the original data are not stationary, the bootstrapped series may be very noisy and do not fluctuate as the original series [30].

Recently, [33] proposed three clustering-based bootstrap aggregating methods, i.e., Smoothed MBB (S.MBB), K-means clustering based (KM), and K-means clustering based-generated from Gaussian normal distribution (KM.N), which perform better under noisy and fluctuating data. In adapting the fluctuating data, the S.MBB method smooths the noise using simple exponential smoothing before applying MBB. Meanwhile, KM and KM.N methods adapt to the noisy series by first implementing the K-means cluster. The original series clusters into K groups and then creates new time series based on the clusters. The difference between KM and KM.N is in how they generate the bootstrap series. In KM, a new time series is created directly by sampling values of clusters, while in KM.N, it is created by generating values based on the parameters of the Gaussian normal distribution of clusters. Both the KM and KM.N methods succeeded in making the bootstrapped time series have low variance between each other. Based on the experimental study of the electricity load series with multiple seasonal and calendar effects, KM.N performed better than the KM method [33]. However, this method creates a bootstrapped series based on the original data without sorting out signal and noise. Thus, in more complex series where the calendar effect may not be visible clearly in the original data, it will produce a noisier bootstrapped series at specific points, especially at times affected by calendar variation.

Inspired by [30,33], this study proposes an SSA–clustering-based method named SSA.KM.N as a modification of KM.N. Our proposed method combines singular spectrum analysis (SSA) as an alternative to the STL method in MBB and KM.N to generate new series from the remainder of the SSA decomposition. Literature shows that SSA is powerful in decomposing time series with complex seasonal patterns ([5,20]). SSA plays a role in breaking down time series, which have trends, multiple seasonal components, and are affected by calendar variation, into signal and noise, which generally contain extreme values representing calendar effects in more detail. By taking advantage of the unique strengths of SSA and KMN, our methodology can better adapt to fluctuating time series related to the effects of calendar variation. Bootstrapping the noise using KM.N and adding it to the signal is expected to produce a bootstrapped time series with low variance and values around the original series.

In this work, the proposed method compares with KM.N in its application to bootstrap two Malaysian load time series with different sample sizes and different time periods; one from Poland and another from Indonesia. We evaluate the impact and benefits of applying SSA.KM.N and KM.N in overcoming the source of uncertainty in time series forecasting and their success in increasing the accuracy of multistep ahead point forecast obtained by standard models such as SARIMA, NNAR, TBATS, and DSHW models.

The rest of the paper is organized as follows. Section 2 describes the methods used in this paper, starting from forecasting methods, decomposition, bagging, and ensemble methods. We also present the procedure of our proposed approach in this section. Section 3 reports the application of KM.N and SSA.KM.N to the four electricity load time series and shows the error evaluation for 12, 24, and 36 steps-ahead point forecasts obtained from SARIMA, NNAR, DSHW, and TBATS for further investigation and assessment. Conclusions are found in Section 4.

2. Materials and Methods

This section contains a brief overview of the methods used for time series modeling and forecasting, the decomposition method, the ensemble learning, and the proposed approach.

2.1. Forecasting Methods

SARIMA, and exponential smoothing (i.e., TBATS and DSHW), are popular approaches to forecast trend and seasonal time series. On the other hand, NNAR is a powerful method for capturing nonlinear relationships in time series data. These four methods are frequently used in modeling load series, and their forecast accuracy is used as benchmarks for other proposed methods [4,10,22]. For example, the Spanish Transmission System Operator uses autoregressive (AR) and NN models [7].

The seasonal ARIMA model, notated as SARIMA $(p,d,q)(P,D,Q)_S$, is an extension of ARIMA model that accommodates the seasonal component of the time series [34,35], and can be written as follows:

$$\phi_p(\mathrm{B})\psi_P\left(\mathrm{B}^S\right)\nabla^d\nabla_S^D z_t = \theta_q(\mathrm{B})\vartheta_Q\left(\mathrm{B}^S\right)a_t \tag{1}$$

where z_t is observation at time t, S is seasonal period, p, P, q, and Q are the orders of autoregressive, seasonal autoregressive, moving average, and seasonal moving average, respectively. Superscript d and D notate the regular and seasonal differentiation, while $\nabla_S = 1 - \mathrm{B}^S$ is a backshift operator, and $\phi_p(\mathrm{B})$, and $\theta_q(\mathrm{B})$ are polynomials in B of degree p and q, respectively. Notations $\psi_P\left(\mathrm{B}^S\right)$, and $\vartheta_Q\left(\mathrm{B}^S\right)$ are polynomials in B^S of degrees P and Q, and a_t is white noise. The orders of p, q, P, and Q can be determined from the correlogram and partial correlogram. Oftentimes, the identification of these orders is not an easy task and the user experience is required [36]. The automatic algorithms discussed in [37] with the "auto.arima" function of the R software can be used to help handle this problem [31]. However, other researchers may prefer to estimate the parameters manually instead of using automated packages [38]. In this case, we use "Arima" function included in the package "forecast" in the software R [39].

NNAR is a feedforward neural network that consists of lagged input neurons, one hidden layer with nonlinear function, and one output neuron [40,41]. NNAR $(p,P,k)_S$ model can be represented as in Equation (2),

$$z_t = b_1 + \sum_{j=1}^{k} v_j f_j\left(b_0 + \sum_{i=1}^{p} z_{t-i}w_{ij} + \sum_{m=1}^{P} z_{t-mS}w_{mj}\right) \tag{2}$$

where b_0 and b_1 are biased, k is the number of neurons in the hidden layer, p is the order of the non-seasonal component, while P is the order of the seasonal component. The sigmoid function at the jth neuron in the hidden layer, f_j, is defined in Equation (3),

$$f_j(u) = \frac{1}{[1 + \exp(-u)]}, \quad (3)$$

where $u = b_0 + \sum_{i=1}^{p} z_{t-i}w_{ij} + \sum_{m=1}^{P} z_{t-mS}w_{mj}$. In its implementation we use the "nnetar" function in the R package "forecast" ([39]). Later in the experimental study, P is set to be 1, p is the optimal number of lags for the linear model fitted to the seasonally adjusted data, and k is determined by the rounded value of $(P + p)/2$. The final forecast values are obtained by averaging 20 networks with different random starting weights.

TBATS and DSHW are modifications of the exponential smoothing to handle trends and multiple seasonal patterns in time series forecasting [42]. DSHW, proposed by [13], accommodates two seasonal patterns where one cycle may be nested within another. Meanwhile, TBATS, proposed by [3], can handle a more complex seasonal pattern in time series forecasting. The term "complex" means that the time series has a trend, and multiple seasonal patterns with integer or non-integer periods, and this may include dual calendar seasonal effects. Success studies of the use of DSHW and TBATS in modeling electricity load time series can be found in [3,13,43]. Detail of these models can be found in [3,13,42,44]. In this paper, TBATS and DSHW are fitted by using the "tbats" and "dshw" functions included in the R package "forecast" ([39]).

2.2. SSA Decomposition Method

SSA is a technique in the field of time series analysis that has a vast range of applicability in decomposition [20], missing value imputation [45], and forecasting [46]. In this study, we focus on the use of the SSA algorithm to decompose time series into the following two components: signal and unstructured noise. SSA consists of four steps, namely, embedding, singular value decomposition (SVD), grouping, and diagonal averaging [47,48].

In embedding, we transform a time series $z_t = \{z_1, z_2, \ldots, z_{n_{tr}}\}$ into a trajectory matrix as in (4) as follows:

$$Z = \begin{bmatrix} z_1 & z_2 & z_3 & \cdots & z_{c_2} \\ z_2 & z_3 & z_4 & \cdots & z_{c_2+1} \\ \vdots & \vdots & \vdots & \cdots & \vdots \\ z_{c_1} & z_{c_1+1} & z_{c_1+2} & \cdots & z_{n_{tr}} \end{bmatrix} \quad (4)$$

where c_1 is the window length, n_{tr} the number of training data sets, $c_2 = n_{tr} - c_1 + 1$ and $c_1 \leq c_2$.

The matrix Z is then decomposed by SVD and expressed as follows:

$$Z = Z^{(1)} + Z^{(2)} = \sum_{l=1}^{r_1} \sqrt{\lambda_l} u_l v_l' + \sum_{l=r_1+1}^{r_2} \sqrt{\lambda_l} u_l v_l' \quad (5)$$

where λ_l are the eigenvalues of matrix Z, u_l and v_l are left and right singular vectors of the matrix Z corresponding to the eigenvalues λ_l, respectively. We need to determine r_1, the number of signal components used for reconstruction in the grouping stage. Finally, we can obtain signal and noise by the diagonal averaging procedure, i.e., the anti-diagonals that map the matrices of the signal and noise components back to time series. The original time series can then be expressed as follows:

$$z_t = \tilde{z}_t^{(1)} + \tilde{z}_t^{(2)} \quad (6)$$

where $\tilde{z}_t^{(1)}$ is the signal and $\tilde{z}_t^{(2)}$ is the noise.

2.3. Bagging and Ensemble Methods

The K-means clustering-based bagging method, proposed by [33], clusters a univariate time series into K groups by the K-means method. Each cluster has its average value and standard deviation as parameters of the cluster. A Gaussian normal distribution according to these parameters is then used to generate a random number as a new value of the bootstrap time series. For example, suppose z_t is a value of the original time series at time t that belongs to the ith cluster. In that case, we can obtain the new value of the bootstrap time series at time t by generating a random value based on the Gaussian normal distribution of the particular cluster. The code for bootstrapping time series by K-means clustering-based-generated from Gaussian normal distribution (KM.N) can be found in [49].

2.4. Proposed Approach

The proposed SSA.KM.N bagging method is a modification of KM.N where the first stage decomposes the original time series using SSA (Figure 1; blue cells). As shown in [20,50], SSA can be used to decompose complex time series into several simple pattern components.

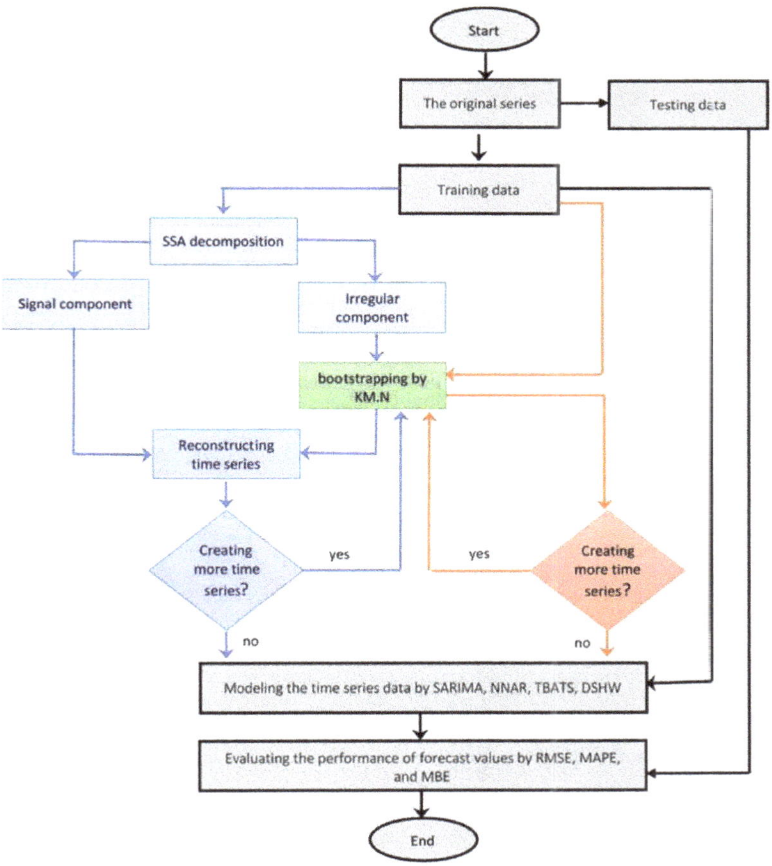

Figure 1. Procedure for generating bootstrapped time series by the SSA-KM.N method.

Step 1. Divide the series into the following two parts: training and testing datasets;
Step 2. Generate $\mathcal{B}$ new series from the original training data by the KM.N method;
Step 3. Generate $\mathcal{B}$ new series from the original training data by the SSA.KM.N method;

a. Apply SSA to the original training data to define the signal and the irregular component;

b. Generate $\mathcal{B}$ new series from the irregular component obtained in Step 3a by the KM.N method;

c. Sum each new irregular component obtained from Step 3b with the signal component so that we obtain $\mathcal{B}$ bootstrapped series of the original training data;

Step 4. Model the original training data, and each bootstrapped series by SARIMA, NNAR, TBATS, and DSHW;

Step 5. Calculate up to M-steps-ahead forecast values by each model obtained in Step 4:

a. Define the M-steps-ahead forecast values from the SARIMA, NNAR, TBATS, and DSHW models obtained from the original training data series;

b. Apply mean and median to calculate the final forecast of the bootstrap series determined in Step 2 and Step 3 for the first $n_{\mathcal{B}}$ bootstrap series;

Step 6. Evaluate the forecast accuracy based on root mean square error (RMSE) and mean absolute percentage error (MAPE).

The two accuracy measures considered in this study can be defined as follows. MAPE, calculated by Equation (7), is frequently used in evaluating load forecasting accuracy since it is a scale-independent error that may compare forecast performance between different data sets [44,51]. Meanwhile, RMSE, calculated using Equation (8), is a scale-dependent error that can be used to compare the accuracy performance of several models on the same data set [51].

$$\text{MAPE} = \frac{100\%}{H}\sum_{h=1}^{H}\left|e_{n_{tr}+h}/y_{n_{tr}+h}\right| \tag{7}$$

$$\text{RMSE} = \left(\frac{\sum_{h=1}^{H} e^2_{n_{tr}+h}}{H}\right)^{1/2} \tag{8}$$

In addition, we also evaluate the model using mean bias error (MBE) as defined in Equation (9). It provides information whether there is a positive or negative bias [52]. We can calculate MBE by the following:

$$\text{MBE} = \frac{\sum_{h=1}^{H} e_{n_{tr}+h}}{H} \tag{9}$$

where

$$e_{n_{tr}+h} = y_{n_{tr}+h} - \hat{y}_{n_{tr}+h}$$

$\hat{y}_{n_{tr}+h}$ and $y_{n_{tr}+h}$ are the predicted value and the actual value at time $(n_{tr}+h)$, respectively. H is the number of observations included in the calculation and n_{tr} is the size of training data.

In this study, each bootstrapped series is modeled separately, and the final forecast for time t is obtained by the following two ensemble methods: the mean and the median across all forecast values at time t, calculated from the bootstrap series. In this study, we obtain the mean and the median of the predicted values generated from the first $n_{\mathcal{B}}$ (between 1 to 100) bootstrapped series to investigate whether the number of generated time series affects the accuracy of the forecast results.

3. Results and Discussion

In this section, we discuss two hourly electricity loads in Johor, Malaysia, and two other electricity load datasets from Poland and Indonesia. We decided to use these four data sets to show the generality of our work for electricity load forecasting.

3.1. Application to the Hourly Electricity Load of Johor, Malaysia

This subsection focuses on short-term forecasting of hourly electricity load with application to Malaysian data. We consider two datasets with different sizes that can be accessed in [53]. The first is the hourly load series from 1 January to 31 December 2009, and the second is the hourly load series from 1 January to 31 July 2010, which are depicted in Figure 2 (see Figure 2a,b, respectively).

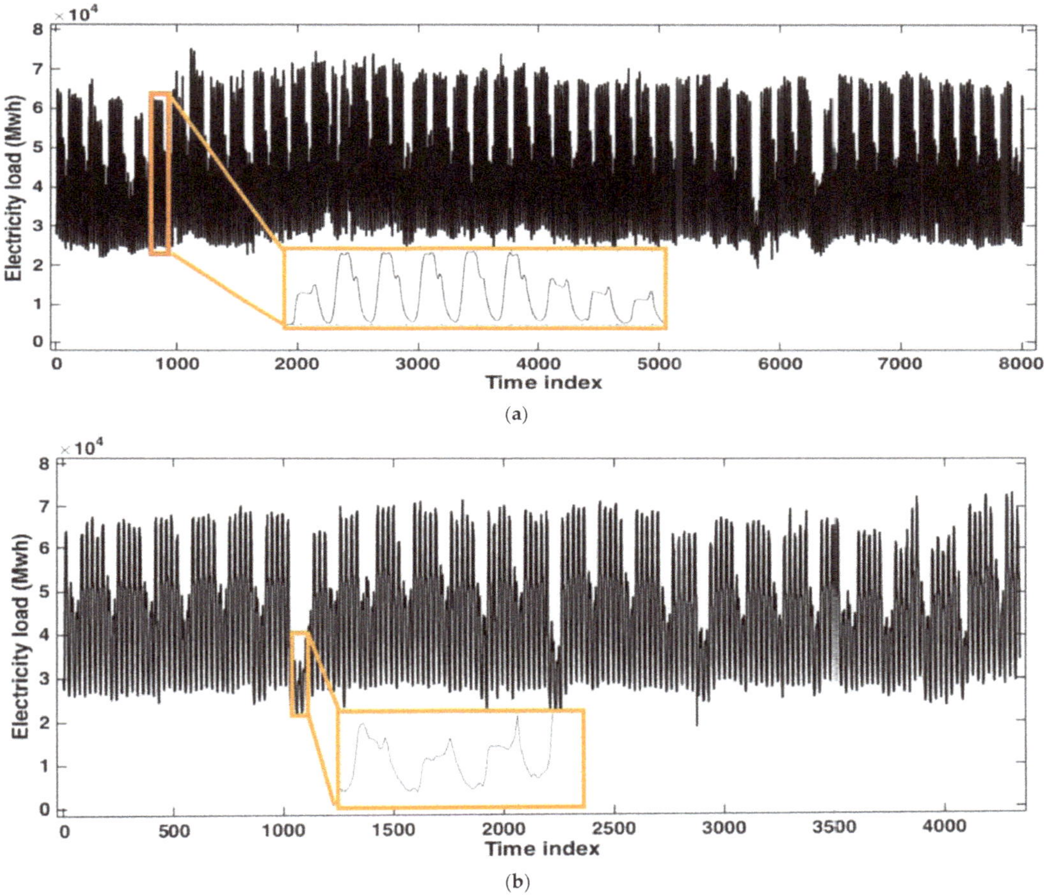

Figure 2. Hourly load series of Johor, Malaysia: (**a**) 1 January time 00:00, to 30 November 2009, time 23:00; (**b**) 1 January, time 00:00, to 30 June 2010, time 23:00.

The period from 1 January, time 00:00, to 30 November 2009, time 23:00, and the period from 1 January, time 00:00, to 30 June 2010, time 23:00, were used for estimation purposes as the training data. The remainder was used to evaluate the forecast performance of the models. These periods are summarized in Table 1.

Our analysis generated 100 bootstrap time series using the KM.N and the SSA.KM.N methods. Note that the original series is included in those 100-bootstrap series. Figure 3 shows the original time series and a realization of the bootstrap series by each of the two methods.

Table 1. Training and testing datasets of hourly load series used in the experimental study.

	1st Data Set	2nd Data Set
Training	01/01/2009 00:00–30/11/2009 23:00	01/01/2010 00:00–30/06/2010 23:00
Testing (for up to h-step ahead):		
$h = 1$	01/12/2009 00:00–31/12/2009 23:00	01/07/2010 00:00–31/07/2010 23:00
$h = 12$	01/12/2009 00:00–01/12/2009 11:00	01/07/2010 00:00–01/07/2020 11:00
$h = 24$	01/12/2009 00:00–01/12/2009 23:00	01/07/2010 00:00–01/07/2020 23:00
$h = 36$	01/12/2009 00:00–01/12/2009 11:00	01/07/2010 00:00–02/07/2020 11:00

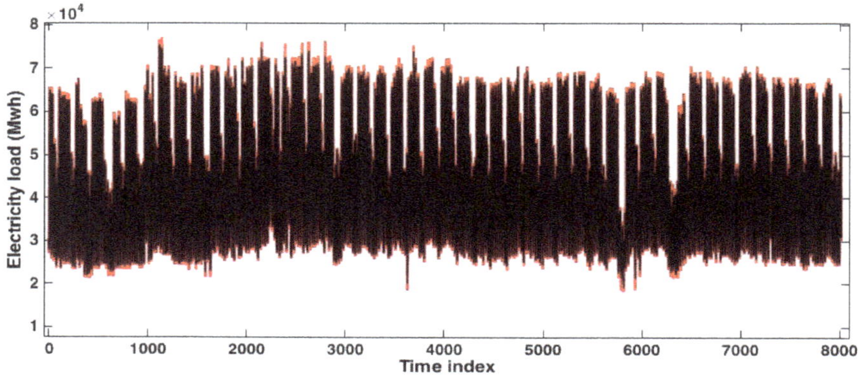

(**a**)

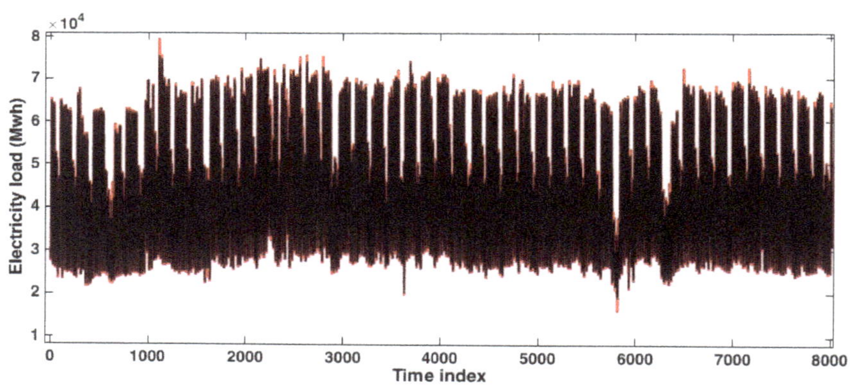

(**b**)

Figure 3. The original series (in black) and the bootstrap time series (in red) obtained by the (**a**) KM.N method; (**b**) SSA.KM.N method.

From Figure 3, we can see that both the KM.N and SSA.KM.N methods produce bootstrap series (in red) with almost the same pattern as the original series (in black). Even in certain parts, where the data have lower or higher values than other times as a result of the calendar variation, the SSA.KM.N can generate series closer to the original time series than the KM.N, visually. As illustrations, we zoom the load on the time period influenced by the Prophet's birthday (Figure 4a) and Eid al-Fitr (Figure 4b). Figure 4 shows that the variance of bootstrapped series obtained by KM.N (left) is larger than those obtained by SSA.KM.N (right).

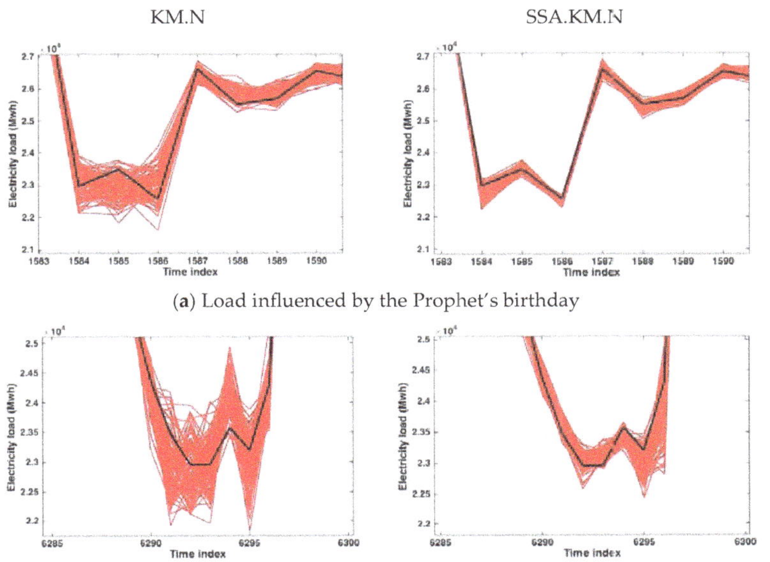

(**a**) Load influenced by the Prophet's birthday

(**b**) Load influenced by Eid al-Fitr

Figure 4. The original series (in black) and the bootstrap time series (in red) obtained by the KM.N method (left) and SSA.KM.N (right) (**a**) for time period influenced by the Prophet's birthday; (**b**) for time period influenced by the Aid al-Fitr.

The performance of the one-step ahead forecast accuracy of SARIMA, NNAR, TBATS, and DSHW is shown in Table 2. All these calculations were performed in the R software. From the analysis of the correlogram and the partial correlogram, the model SARIMA(2,0,3)(2,1,2)$_{24}$ was chosen for the first data set and the SARIMA(2,0,0)(3,1,0)$_{24}$ for the second data set. The most appropriate NNAR, TBATS, and DSHW models were reconstructed and chosen automatically by the "nnetar", "tbats", and "dshw" functions in R. Based on Table 2, we can see that both for the first and second data sets, the NNAR and the DSHW produce smaller RMSE and MAPE than SARIMA and TBATS in the case of one-step ahead forecasting.

Table 2. RMSE and MAPE for one-step-ahead forecasts for the testing data of the two data sets of hourly electricity load in Malaysia, obtained by SARIMA, NNAR, TBATS, and DSHW.

Method	Testing of 1st Data Set (1–31 December 2009)		Testing of 2nd Data Set (1–31 July 2010)	
	RMSE	**MAPE**	**RMSE**	**MAPE**
SARIMA	1168.02	1.90	1719.68	2.54
NNAR	749.92	1.21	1182.68	1.75
TBATS	1584.47	2.91	1466.33	2.37
DSHW	839.18	1.29	1035.14	1.46

Furthermore, we investigate how these four models work for multistep ahead load forecasting with and without bagging implementation. The comparative values of RMSE and MAPE for 12, 24, and 36 steps ahead for SARIMA, NNAR, TBATS, and DSHW are presented in Tables 3–6, respectively. We also present in Tables 3–6 the RMSE and MAPE obtained from the forecast values of each model with four different numbers of bootstrap time series, to infer whether the number of bootstrap samples interferes with the accuracy of the forecasts.

Table 3. RMSE and MAPE of *h*-step ahead forecast obtained by SARIMA model from the original series, KM.N and SSA.KM.N bootstrap series.

	1st Data Set						2nd Data Set					
	RMSE			MAPE			RMSE			MAPE		
h	12	24	36	12	24	36	12	24	36	12	24	36
original	466.51	1408.20	2118.07	0.88	1.94	3.13	1139.08	1268.58	1324.19	2.27	2.48	2.97
Bagging + Mean												
KM.N												
$n_B = 25$	512.57	**1404.68**	**2035.79**	0.97	1.94	**3.00**	1345.29	1405.54	1366.41	**2.25**	2.57	**2.87**
$n_B = 50$	530.99	1419.45	2049.64	0.99	1.96	3.03	1320.97	1393.69	1366.42	2.28	2.59	2.91
$n_B = 75$	544.23	1429.02	2053.97	1.02	1.99	3.05	1323.97	1394.82	1367.79	2.29	2.60	2.92
$n_B = 100$	556.73	1438.17	2062.04	1.09	2.03	3.08	1329.88	1397.27	1371.85	2.32	2.61	2.93
SSA. KM.N												
$n_B = 25$	490.52	1368.94	2029.14	1.04	**1.94**	2.96	**1067.12**	**1234.85**	**1286.55**	**2.11**	**2.40**	**2.87**
$n_B = 50$	494.71	1357.47	2018.93	1.05	**1.94**	**2.95**	1069.56	1235.92	1288.94	2.13	2.41	2.88
$n_B = 75$	498.23	**1356.26**	**2015.31**	1.07	**1.94**	**2.95**	1068.79	1235.36	1288.80	2.13	2.41	2.88
$n_B = 100$	495.76	1360.24	2018.48	1.06	1.95	**2.95**	1072.37	1236.66	1289.22	2.14	2.41	2.88
Bagging + Median												
KM.N												
$n_B = 25$	548.32	1415.38	**2045.61**	1.03	1.98	**3.04**	1330.61	1388.67	1348.25	**2.12**	2.49	**2.80**
$n_B = 50$	543.85	1424.97	2055.74	1.02	1.98	3.05	1306.42	1383.85	1355.37	2.18	2.53	2.86
$n_B = 75$	549.99	1430.50	2056.88	1.05	2.00	3.06	1322.84	1390.97	1360.40	2.22	2.55	2.88
$n_B = 100$	566.06	1444.64	2066.11	1.09	2.04	3.09	1325.92	1392.35	1365.12	2.27	2.58	2.90
SSA.KM.N												
$n_B = 25$	501.40	1372.29	2029.04	1.10	1.98	2.98	1072.91	1239.46	1290.89	**2.13**	2.41	**2.88**
$n_B = 50$	514.83	**1355.26**	2012.78	1.13	1.97	**2.95**	1076.47	1239.43	1292.00	2.15	2.42	2.89
$n_B = 75$	514.08	1355.95	**2011.37**	1.13	1.97	**2.95**	**1069.09**	**1234.93**	**1289.34**	**2.13**	**2.41**	**2.88**
$n_B = 100$	508.08	1365.38	2018.37	1.11	1.97	2.96	1074.03	1238.00	1290.43	2.15	2.42	**2.88**

Green cells represent the RMSE and MAPE values of the SARIMA model obtained from the bootstrap series lower than those obtained from the original series. Bold values represent the lowest value in a column of each bagging method with green cells.

Based on Table 3, we can see that the SSA.KM.N performed better than the KM.N in reducing the RMSE and MAPE of forecasts for 24- and 36 steps-ahead, respectively, obtained by the SARIMA model. The green cells in Table 3 represent the RMSE and MAPE values of the SARIMA model obtained from the bootstrap series, which are lower than those obtained from the original series. Bold values represent the lowest value in a column of each bagging method with green cells.

Moreover, in Table 3, it can be seen that for the first dataset, SARIMA provided high accuracy values for forecasting one day ahead (next 24 h), indicated by the MAPE values of less than two. For the second dataset, the MAPE values were between two and three. For each bagging method, there is no significant difference between the forecast results obtained by the mean and the median ensemble.

Based on the analysis for multistep-ahead forecasting by the SARIMA model with the bagging methods, it cannot be concluded that the more bootstrapped series used in the calculation, the more accurate the forecasting results will be. As we can see, the values in bold (see Table 3) are not in the $n_B = 100$ row, being some in the $n_B = 25$ row.

The comparative forecast results obtained by the NNAR model reconstructed from the original and bootstrap series are presented in Table 4. Based on the analysis of this table, the NNAR tends to produce larger RMSE than SARIMA. This is not in line with the results for predicting one step ahead (see Table 2). However, both the KM.N and SSA.KM.N methods can improve the accuracy performance of the forecasts. The 36-steps-ahead forecast values obtained from the bootstrap series using SSA.KM.N produced a larger RMSE than those obtained from the original time series, but the MAPE value showed the opposite direction.

Similar to the results shown in Table 3, in this case, a greater number of bagging samples does not necessarily result in a better performance in terms of forecasting accuracy.

Table 4. RMSE and MAPE of h-step ahead forecast obtained by NNAR model from the original series, KM.N and SSA.KM.N bootstrap series.

	1st Data Set						2nd Data Set					
	RMSE			MAPE			RMSE			MAPE		
h	12	24	36	12	24	36	12	24	36	12	24	36
Original	1761.72	2365.26	4393.21	2.12	3.18	5.81	2474.07	2797.86	5179.86	3.62	4.24	6.15
Bagging + Mean												
KM.N												
$n_B = 25$	**960.09**	**1723.56**	**2881.12**	**1.10**	**2.35**	**4.01**	2367.77	2062.45	4128.65	3.50	3.32	5.19
$n_B = 50$	964.93	1730.89	2894.54	1.12	2.37	4.03	2433.15	2096.23	4154.82	3.58	3.34	5.22
$n_B = 75$	996.61	1747.33	2949.97	1.16	2.39	4.08	**2358.62**	**2044.60**	**4084.19**	**3.48**	**3.28**	**5.13**
$n_B = 100$	1007.27	1766.48	2972.19	1.16	2.42	4.12	2388.40	2067.72	4120.69	3.51	3.30	5.17
SSA. KM.N												
$n_B = 25$	973.69	1661.79	2756.24	1.11	2.29	3.93	2279.29	**1958.57**	5344.05	**3.41**	**3.15**	**6.04**
$n_B = 50$	**954.97**	**1618.73**	**2701.08**	1.11	**2.23**	**3.84**	2349.54	1997.22	5451.24	3.52	3.18	6.14
$n_B = 75$	971.00	1634.74	2732.76	1.11	2.24	3.86	2325.34	1980.90	5443.80	3.47	3.16	6.13
$n_B = 100$	968.26	1628.32	2716.57	1.11	2.24	3.85	**2311.58**	1974.44	5452.81	3.45	3.16	6.13
Bagging + Median												
KM.N												
$n_B = 25$	913.18	1695.23	**2836.71**	1.07	2.34	**3.96**	**2364.00**	2069.78	**4029.04**	**3.52**	3.35	5.14
$n_B = 50$	**912.72**	**1690.94**	2861.26	**1.05**	**2.32**	3.98	2469.39	2110.04	4136.53	3.63	3.36	5.21
$n_B = 75$	948.08	1697.71	2883.83	1.13	2.35	4.01	2399.13	**2068.57**	4035.98	3.53	3.31	5.11
$n_B = 100$	964.86	1713.66	2892.47	1.12	2.36	4.02	2418.19	2080.95	4088.09	3.55	3.32	5.15
SSA.KM.N												
$n_B = 25$	978.42	1654.41	2785.53	1.13	2.28	3.95	**2248.30**	**1952.46**	5361.38	**3.37**	3.14	**6.05**
$n_B = 50$	979.69	1633.47	2657.16	1.13	2.26	3.81	2270.89	1954.88	5411.12	3.39	3.14	6.09
$n_B = 75$	**975.65**	1642.36	2683.66	**1.12**	2.26	3.83	2315.22	1980.65	5473.96	3.44	3.16	6.15
$n_B = 100$	959.00	**1612.46**	**2631.46**	1.13	**2.23**	**3.78**	2307.14	1977.87	5438.96	3.43	3.15	6.12

Green cells represent the RMSE and MAPE values of the SARIMA model obtained from the bootstrap series lower than those obtained from the original series. Bold values represent the lowest value in a column of each bagging method with green cells.

Table 5 shows that, for the first dataset, bagging did not improve the forecast accuracy obtained by the TBATS model. For the second dataset, the SSA.KM.N enhanced the performance of forecasting accuracy, but this did not apply to the KM.N. Based on the RMSE, the forecast values calculated from the TBATS model were more accurate than those obtained from the NNAR model.

Contrary to the results shown in Table 5, bagging implementation improved the forecasting accuracy of the DSHW model for the first dataset but not for the second dataset (see Table 6). In this case, the KM.N bagging performed better than the SSA.KM.N in reducing the forecasting error. Table 6 shows that the MAPE values obtained by the DSHW model for the first dataset are on average 2–3 times higher than those obtained by the DSHW model for the second dataset. However, in this case, the application of KM.N was able to reduce the MAPE value for 12-step ahead by approximately 36%.

By implementing the SSA.KM.N in the hourly load forecasting of Malaysia up to 36 steps ahead, the RMSE was able to be reduced by 4.97% and 40% when using SARIMA and NNAR, respectively. Meanwhile, KM.N was able to reduce the RMSE value by up to 3.8% for SARIMA and up to 35.43% for NNAR. Furthermore, although in one case, the SSA.KM.N bagging implementation for predicting up to 36 steps ahead using TBATS and DSHW can decrease the RMSE by more than 10%, in another case, it may behave differently. Similar conclusions were obtained when analyzing the MAPE.

For further evaluation, we consider MBE to see the direction of the models and present the results in Tables 7 and 8. The MBE is supposed to provide information on the long-term performance of the model. Based on Tables 7 and 8, the interpretation of the model performance is consistent with that based on RMSE and MAPE. The directions of the bias generated by the models with and without bagging are the same, except for the forecasting of 12 steps ahead by SARIMA (see Table 7). It may be related to the weakness of MBE where the positive and negative errors can cancel each other, and high individual errors can result in low MBE values. However, we can see that bagging methods, both KM.N and SSA.KM.N, reduces MBE values obtained from the NNAR model compared with those obtained from the original time series. In the case of modeling the second data set by TBATS, SSA, KM.N yields lower MBEs than the KM.N bagging method.

Table 5. RMSE and MAPE of h-step ahead forecast obtained by TBATS model from the original series, KM.N and SSA.KM.N bootstrap series.

	1st Data Set						2nd Data Set					
	RMSE			MAPE			RMSE			MAPE		
h	12	24	36	12	24	36	12	24	36	12	24	36
original	1613.36	2057.21	2146.92	4.03	3.72	4.08	1574.78	2771.37	2685.71	3.87	5.54	5.54
Bagging + Mean												
KM.N												
$n_B = 25$	1740.64	2333.25	2360.33	3.99	3.98	3.91	1410.82	3122.42	2864.93	3.50	5.93	5.86
$n_B = 50$	1668.46	2245.73	2282.53	3.89	3.86	3.84	1296.39	3045.17	2805.20	3.18	5.67	5.63
$n_B = 75$	1664.15	2246.18	2290.12	**3.86**	3.85	3.84	1280.05	3047.59	2806.13	3.14	5.67	5.62
$n_B = 100$	1732.41	2325.22	2370.24	3.91	3.93	3.90	**1255.27**	3002.58	2775.60	**3.10**	5.59	5.55
SSA. KM.N												
$n_B = 25$	1966.14	2547.93	2560.87	4.61	4.44	4.26	**1371.73**	1974.32	2286.35	**3.43**	4.03	4.61
$n_B = 50$	2023.97	2609.35	2634.36	4.60	4.48	4.28	1447.05	1939.14	2316.25	3.51	3.99	4.68
$n_B = 75$	2011.41	2580.08	2615.16	4.60	4.45	4.28	1503.40	**1861.19**	**2247.76**	3.56	**3.87**	4.61
$n_B = 100$	1994.50	2572.82	2597.21	4.59	4.45	4.27	1468.42	1879.36	2259.16	3.52	3.90	**4.60**
Bagging + Median												
KM.N												
$n_B = 25$	1937.56	2611.75	2574.41	4.19	4.27	4.16	1336.97	3256.25	2953.00	3.30	6.03	5.91
$n_B = 50$	1719.88	2336.30	2337.81	3.86	3.90	**3.87**	1297.49	3240.04	2933.74	3.19	5.96	5.83
$n_B = 75$	1818.78	2406.98	2410.71	4.01	4.03	4.00	1297.06	3239.03	2934.09	3.19	5.96	5.83
$n_B = 100$	2010.25	2642.92	2639.85	4.19	4.31	4.21	**1287.17**	3203.36	2909.66	**3.16**	5.90	5.78
SSA.KM.N												
$n_B = 25$	1980.32	2604.85	2589.40	4.56	4.46	4.16	1594.16	2542.15	2846.89	3.99	5.13	5.72
$n_B = 50$	2021.70	2674.56	2675.42	4.51	4.50	4.19	1482.30	2157.74	2548.26	3.71	4.46	5.16
$n_B = 75$	2043.60	2661.86	2678.06	4.59	4.53	4.27	1476.01	2070.70	2462.00	3.65	4.28	5.01
$n_B = 100$	2040.19	2629.41	2651.92	4.59	4.50	4.25	1415.00	**2012.09**	**2390.52**	**3.54**	**4.14**	**4.88**

Green cells represent the RMSE and MAPE values of the SARIMA model obtained from the bootstrap series lower than those obtained from the original series. Bold values represent the lowest value in a column of each bagging method with green cells.

3.2. Application to the Hourly Electricity Load of Poland

Figure 5 shows the hourly electricity load of Poland, in Megawatts (Mwh), from 26 October, at 01:00 to 16 December 2020 at 00:00. The data were accessed from https://www.pse.pl/obszary-dzialalnosci/krajowy-system-elektroenergetyczny/zapotrzebowanie-kse (accessed on 21 January 2021). This data set contains the linear trend and multiple seasonal patterns with daily and weekly periods. There was a slight pattern change around time index 400 (11 November 2020) due to the influence of the National Independence Day holiday (shown by the orange rectangle in Figure 5). We fit the model using the first 1212 observations and evaluated the forecasting accuracy performance using the last 36 observations.

In this experimental study, we generate 50 bootstrapped series from the original electricity load of Poland using KM.N and SSA.KM.N. We then model each generated time series by SARIMA, NNAR, TBATS, and NNAR, in the same way as for the Malaysian data. The accuracy evaluation was based on RMSE, MAPE, and MBE for 12, 24, and 36 steps ahead, and is summarized in Tables 9–11.

Based on Tables 9 and 10, we can see that SSA.KM.N can improve the accuracy of 24 and 36 steps ahead of electricity load forecasting for the Polish data using the NNAR model, while KM.N fails to improve forecasting accuracy for this model. On the other hand, the KM.N bagging method works well on the DSHW model, while SSA.KM.N does not perform so well. However, both of them succeeded in increasing the accuracy of forecasting for the ARIMA model.

Table 6. RMSE and MAPE of h-step ahead forecast obtained by DSHW model from the original series, KM.N and SSA.KM.N bootstrap series.

	1st Data Set						2nd Data Set					
	RMSE			MAPE			RMSE			MAPE		
h	12	24	36	12	24	36	12	24	36	12	24	36
original	3448.13	4802.00	4207.31	4.19	6.97	6.39	1458.59	1231.29	1148.92	2.77	2.05	2.00
Bagging + Mean												
KM.N												
$n_{\mathcal{B}} = 25$	**2183.44**	3644.90	3236.11	**2.67**	**5.39**	**5.32**	1809.37	1902.75	1718.99	3.29	3.06	2.75
$n_{\mathcal{B}} = 50$	2230.95	3640.88	3233.82	2.83	5.46	5.40	1890.92	2012.37	1801.60	3.43	3.24	2.89
$n_{\mathcal{B}} = 75$	2194.68	**3590.45**	**3183.48**	2.81	5.40	5.34	1918.15	2065.24	1842.82	3.47	3.33	2.96
$n_{\mathcal{B}} = 100$	2268.22	3660.82	3241.71	2.91	5.50	5.40	1913.90	2052.31	1828.07	3.49	3.32	2.96
SSA. KM.N												
$n_{\mathcal{B}} = 25$	3168.62	4611.09	4018.48	3.76	6.62	5.99	2361.26	2782.86	2349.97	3.68	4.33	3.82
$n_{\mathcal{B}} = 50$	3157.42	**4596.48**	4006.13	3.74	6.60	5.97	2552.00	2945.08	2475.15	4.06	4.56	3.95
$n_{\mathcal{B}} = 75$	3156.84	4598.52	4007.59	3.74	6.60	5.97	2544.55	2946.87	2487.05	4.03	4.59	4.05
$n_{\mathcal{B}} = 100$	**3153.79**	4597.77	**4005.71**	**3.73**	6.60	5.97	2521.97	2911.70	2455.59	4.02	4.55	3.99
Bagging + Median												
KM.N												
$n_{\mathcal{B}} = 25$	2556.48	4295.01	3778.13	3.18	6.27	5.93	1860.06	1967.00	1771.43	3.33	3.12	2.79
$n_{\mathcal{B}} = 50$	**2542.87**	4273.27	3758.95	**3.12**	6.23	5.90	1922.97	2026.09	1817.49	3.47	3.23	2.88
$n_{\mathcal{B}} = 75$	2574.91	**4263.89**	**3750.11**	3.15	**6.21**	**5.88**	1934.91	2060.74	1843.21	4.48	3.30	2.93
$n_{\mathcal{B}} = 100$	2658.87	4320.54	3801.11	3.21	6.28	5.94	1935.22	2046.94	1830.22	3.52	3.29	2.92
SSA.KM.N												
$n_{\mathcal{B}} = 25$	3184.94	4621.30	4025.46	3.77	6.63	5.99	2372.52	2855.73	2427.66	3.58	4.50	4.06
$n_{\mathcal{B}} = 50$	3183.66	4607.60	**4014.75**	3.77	6.62	5.98	2602.98	2938.16	2472.20	4.11	4.46	3.93
$n_{\mathcal{B}} = 75$	**3178.47**	**4606.99**	4014.94	**3.76**	6.62	5.98	2517.25	2841.47	2404.81	4.01	4.38	3.91
$n_{\mathcal{B}} = 100$	3184.52	4610.00	4016.52	3.77	6.62	5.98	2521.24	2812.68	2374.88	4.03	4.33	3.84

Green cells represent the RMSE and MAPE values of the SARIMA model obtained from the bootstrap series lower than those obtained from the original series. Bold values represent the lowest value in a column of each bagging method with green cells.

We can see from Table 10 that implementing the bagging method on Polish data does not reduce the MAPE values in the case of the TBATS model. Still, it makes the MBEs smaller (in absolute values) than those obtained from the original data (Table 11). Furthermore, although the RMSE and MAPE values of the DSHW model decreased with bagging, the results were the opposite when analyzing the MBE values. SSA.KM.N gives better outcomes for the NNAR model, while KM.N is better for the DSHW model.

3.3. *Application to the Hourly Electricity Load of Java-Bali, Indonesia*

To show the generality of the implementation of bagging methods in electricity load forecasting, we also discuss the hourly electricity load of Java-Bali, Indonesia. The data consists of 1464 observations, from 1 October to 30 November 2015. Figure 6 shows that the data has no trend but has double seasonal patterns. It is relatively stable except at time points around 312–336 (14 October 2015) due to the influence of the Hijriyah New Year holiday (shown by the orange rectangle in Figure 6). Moreover, this data set was also discussed in [54].

We generate 50 bootstrapped time series for this case based on 1428 observations (1 October at 01.00 to 29 November at 12.00). The error evaluation in terms of RMSE, MAPE, and MBE obtained from the SARIMA, NNAR, TBATS, and DSHW models are summarized in Tables 12–14, respectively. The overall results shown in Tables 12–14 for this application are similar to those of the previous applications for Malaysian and Polish electricity load data.

Table 7. MBEs of h-step ahead forecast for the first data set obtained by SARIMA, NNAR, TBATS, and DSHW models.

	SARIMA			NNAR			TBATS			DSHW		
h	12	24	36	12	24	36	12	24	36	12	24	36
original	145.99	907.89	1446.04	751.37	1574.98	2308.25	−69.78	598.89	607.95	2233.79	3829.69	3311.54
Bagging + Mean												
KM.N												
$n_B = 25$	167.58	**905.80**	**1390.65**	443.29	**1232.64**	1968.24	290.35	1200.32	1246.45	**1356.39**	**2890.41**	2518.15
$n_B = 50$	214.87	937.02	8881.84	**441.63**	1236.06	**1269.93**	384.09	1287.13	**470.02**	1447.27	2924.45	**1012.76**
$n_B = 75$	230.20	950.19	1425.00	460.41	1246.86	2003.83	371.32	1252.58	1318.35	1436.04	2891.67	2510.35
$n_B = 100$	257.88	969.21	1441.16	473.65	1266.20	2025.98	348.74	1242.51	1298.32	1490.21	2951.16	2554.25
SSA. KM.N												
$n_B = 25$	−137.62	728.93	1264.851	411.07	1170.29	1895.15	269.68	1112.43	1132.33	1921.13	3601.54	3093.45
$n_B = 50$	−150.17	714.81	**869.88**	**386.29**	**1130.32**	**1210.20**	221.89	1042.36	**334.78**	1919.84	3593.18	**1010.49**
$n_B = 75$	−164.26	**705.98**	1243.53	402.64	1142.61	1863.60	238.15	1053.85	1069.63	1919.89	3595.02	3087.34
$n_B = 100$	−154.71	714.48	1250.64	391.24	1133.87	1851.27	323.68	1139.93	1162.73	**1913.61**	**3591.93**	3083.82
Bagging + Median												
KM.N												
$n_B = 25$	198.26	925.21	**1407.78**	**413.57**	1213.51	**1938.24**	447.49	1340.14	1307.19	**1533.13**	**3350.10**	2965.00
$n_B = 50$	230.70	946.82	1426.14	413.63	1209.42	1948.60	297.71	1130.41	1117.38	1553.86	3350.17	**2963.54**
$n_B = 75$	234.08	951.99	1428.00	420.36	**1207.34**	1954.58	375.90	1203.43	1178.55	1593.23	3355.65	2964.73
$n_B = 100$	260.12	972.47	1444.30	440.88	1227.32	1970.68	558.25	1415.63	1394.91	1645.62	3403.65	3003.21
SSA.KM.N												
$n_B = 25$	−152.57	722.76	1258.23	387.98	1150.80	1894.71	343.25	1289.54	1343.09	**1932.04**	3610.62	3097.90
$n_B = 50$	−172.75	699.00	1237.00	378.60	1130.51	1820.84	431.02	1376.17	1441.36	1935.88	**3603.03**	**3092.51**
$n_B = 75$	−177.18	**696.20**	**1233.93**	382.15	1136.42	1835.54	421.09	1348.28	1416.42	1934.80	3603.84	3094.52
$n_B = 100$	−168.17	708.37	1244.94	**367.05**	**1113.81**	**1801.46**	413.72	1325.61	1396.90	1939.26	3606.19	3095.18

Green cells represent the MBE values obtained from the bootstrap series for the first data set lower (in absolute values) than those obtained from the original series. Bold values represent the lowest value in a column of each bagging method with green cells.

Table 8. MBEs of h-step ahead forecast for the second data set obtained by SARIMA, NNAR, TBATS, and DSHW models.

	SARIMA			NNAR			TBATS			DSHW		
h	12	24	36	12	24	36	12	24	36	12	24	36
original	287.06	−228.44	−591.45	2319.17	1146.12	2903.47	−1266.71	−2379.66	−1457.27	−1178.78	−805.60	−534.08
Bagging + Mean												
KM.N												
$n_B = 25$	616.40	**−3.69**	**−389.82**	1733.21	**472.40**	1818.49	−1191.85	−2641.61	−1815.55	−1496.71	−1538.48	−935.39
$n_B = 50$	568.38	−38.61	−582.74	1784.60	530.70	**1375.18**	−1071.33	−2541.56	**−431.21**	−1568.82	−1638.40	**0.61**
$n_B = 75$	566.62	−40.03	−423.90	**1729.60**	476.55	1807.17	−1041.47	−2524.44	−1688.93	−1591.59	−1687.65	−[illegible]044.28
$n_B = 100$	559.38	−43.89	−428.30	1744.63	493.21	1827.51	**−1009.37**	−2480.44	−1640.25	−1592.14	−1679.05	−[illegible]047.71
SSA. KM.N												
$n_B = 25$	291.41	−224.24	**−571.59**	**1672.25**	**481.19**	2374.84	**−97.56**	−869.941	−314.26	−1783.46	−2256.37	−[illegible]713.42
$n_B = 50$	283.98	−225.54	−606.41	1728.23	557.90	**1976.58**	−149.71	−914.66	**−18.19**	−1952.89	−2375.25	**−335.45**
$n_B = 75$	283.69	−226.25	−574.95	1708.84	537.57	2457.88	−125.375	**−832.99**	−348.08	−1943.21	−2389.85	−[illegible]849.06
$n_B = 100$	**285.39**	**−222.95**	−571.83	1695.97	518.81	2448.55	−209.48	−940.32	−418.47	−1931.83	−2363.00	−[illegible]804.90
Bagging + Median												
KM.N												
$n_B = 25$	653.55	**10.43**	**−376.27**	**1730.47**	**422.54**	1737.09	−1138.90	−2723.60	−1888.97	−1529.41	−1578.47	−957.24
$n_B = 50$	600.71	−16.70	−403.8	1811.03	532.83	**1586.62**	−1102.34	−2699.53	−1860.87	−1591.57	−1636.36	−995.47
$n_B = 75$	604.67	−19.03	−405.87	1751.48	487.22	1788.34	−1095.80	−2690.14	−1849.37	−1600.02	−1674.19	−1024.61
$n_B = 100$	578.09	−35.40	−419.54	1763.27	504.38	1817.09	**−1083.54**	−2654.90	−1816.41	−1609.89	−1663.54	−1023.59
SSA.KM.N												
$n_B = 25$	289.37	−221.65	−571.52	**1652.67**	**470.11**	**2375.43**	**−3.50**	**−755.85**	**−200.04**	−1746.97	−2341.34	−[illegible]867.79
$n_B = 50$	285.59	−223.41	−574.00	1668.23	499.54	2419.34	−227.46	−1074.75	−432.69	−1977.57	−2334.53	−[illegible]784.67
$n_B = 75$	**283.23**	−224.74	−575.07	1694.84	517.58	2458.64	−267.19	−1001.17	−421.91	−1921.88	−2271.71	−[illegible]797.42
$n_B = 100$	284.76	**−221.12**	**−571.49**	1690.07	514.03	2440.18	−350.40	−1103.52	−506.35	−1925.08	−2246.15	−[illegible]737.53

Green cells represent the MBE values obtained from the bootstrap series for the first data set lower (in absolute values) than those obtained from the original series. Bold values represent the lowest value in a column of each bagging method with green cells.

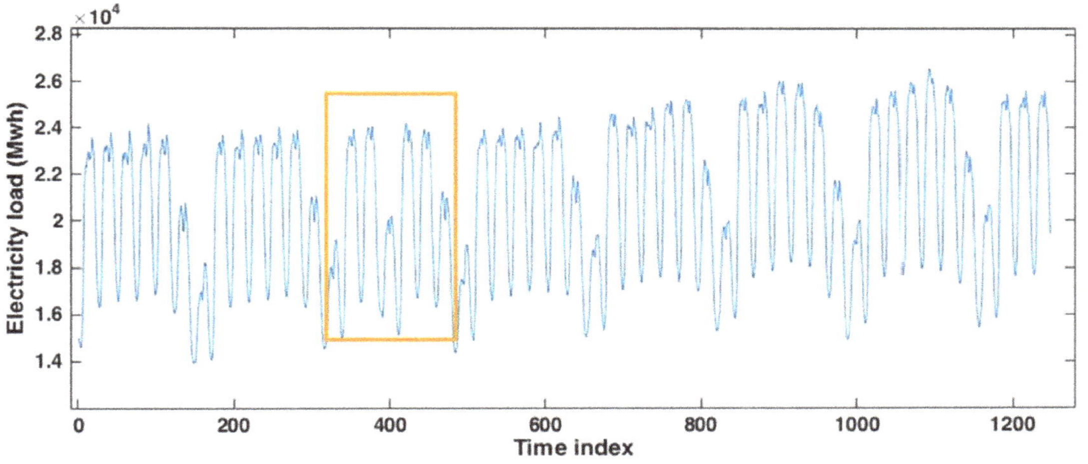

Figure 5. The hourly electricity load of Poland between 26 October and 16 December 2020.

Table 9. RMSEs of h-step ahead forecast for the hourly electricity load of Poland obtained by SARIMA, NNAR, TBATS, and DSHW models.

	SARIMA			NNAR			TBATS			DSHW		
h	12	24	36	12	24	36	12	24	36	12	24	36
Original	2120.28	2436.07	2540.46	107.18	308.45	382.81	259.27	437.82	466.29	351.51	289.03	284.54
Bagging + Mean												
KM.N												
$n_B = 10$	298.01	254.32	229.17	133.53	343.25	416.31	309.15	456.14	468.23	323.29	258.16	254.41
$n_B = 20$	276.62	237.69	220.48	135.79	319.32	390.97	307.57	457.72	467.18	326.77	255.35	255.16
$n_B = 25$	266.20	231,54	218.07	132.76	305.93	371.25	308.52	459.61	469.26	305.33	240.89	240.29
$n_B = 50$	300.58	257.61	230.87	146.32	316.44	383.40	304.79	455.58	467.42	318.51	246.98	243.43
SSA. KM.N												
$n_B = 10$	361.10	304.85	256.78	130.42	254.16	290.95	291.24	433.47	441.28	365.75	298.55	291.71
$n_B = 20$	356.84	302.42	254.80	128.66	228.89	262.25	296.73	439.51	444.58	369.77	298.33	291.90
$n_B = 25$	353.73	299.78	252.83	132.14	234.19	268.70	296.91	442.01	447.02	370.41	300.02	291.14
$n_B = 50$	353.14	299.24	252.50	131.97	231.66	267.86	299.30	435.43	437.14	374.79	300.84	290.49
Bagging + Median												
KM.N												
$n_B = 10$	323.16	270.39	243.86	127.12	330.51	401.92	319.28	463.28	470.74	324.95	261.77	253.61
$n_B = 20$	287.77	243.37	222.76	135.81	312.82	385.15	312.11	461.49	470.19	330.34	257.77	256.02
$n_B = 25$	265.14	227.19	213.62	136.70	298.21	370.86	312.14	461.90	470.74	316.96	249.81	245.81
$n_B = 50$	302.26	257.68	228.44	150.87	313.86	381.42	310.11	460.61	470.76	336.98	261.51	257.76
SSA.KM.N												
$n_B = 10$	364.58	306.80	258.27	137.93	286.20	326.78	296.29	448.13	448.54	366.90	303.00	297.67
$n_B = 20$	362.98	307.99	259.25	132.48	245.93	287.50	297.94	449.62	451.81	376.13	303.83	298.76
$n_B = 25$	364.04	308.75	260.17	137.58	247.58	284.35	300.09	453.73	456.32	373.93	304.04	295.95
$n_B = 50$	306.64	258.38	1.44	133.83	232.00	267.26	299.78	450.28	450.92	378.22	305.12	294.39

Green cells represent the RMSE values obtained from the bootstrap series for the Polish data set lower than those obtained from the original series.

Table 10. MAPEs of h-step ahead forecast for the hourly electricity load of Poland obtained by SARIMA, NNAR, TBATS, and DSHW models.

	SARIMA			NNAR			TBATS			DSHW		
h	12	24	36	12	24	36	12	24	36	12	24	36
original	7.09	9.99	10.20	0.36	0.84	1.19	1.01	1.52	1.59	1.34	1.13	1.11
Bagging + Mean												
KM.N												
$n_B = 10$	1.14	0.99	0.86	0.42	0.98	1.29	1.15	1.89	1.84	1.23	0.98	0.97
$n_B = 20$	1.02	0.90	0.82	0.43	0.91	1.21	1.13	1.89	1.84	1.19	0.91	0.93
$n_B = 25$	0.97	0.88	0.81	0.41	0.86	1.40	1.13	1.90	1.85	1.13	0.87	0.89
$n_B = 50$	1.12	0.99	0.86	0.48	0.92	1.20	1.14	1.88	1.84	1.14	0.86	0.87
SSA. KM.N												
$n_B = 10$	1.44	1.22	0.86	0.42	0.75	0.91	1.10	1.76	1.71	1.37	1.16	1.13
$n_B = 20$	1.42	1.21	0.86	0.42	0.69	0.83	1.10	1.80	1.73	1.38	1.15	1.12
$n_B = 25$	1.41	1.21	0.86	0.43	0.71	0.84	1.10	1.81	1.75	1.38	1.16	1.12
$n_B = 50$	1.41	1.20	0.86	0.43	0.70	0.84	1.10	1.79	1.71	1.39	1.15	1.11
Bagging + Median												
KM.N												
$n_B = 10$	1.22	1.05	0.91	0.39	0.93	1.24	1.18	1.93	1.87	1.22	1.01	0.98
$n_B = 20$	1.06	0.92	0.83	0.411	0.90	1.20	1.15	1.91	1.86	1.18	0.90	0.92
$n_B = 25$	0.96	0.86	0.79	0.41	0.84	1.14	1.15	1.91	1.86	1.14	0.89	0.90
$n_B = 50$	1.12	0.98	0.84	0.49	0.93	1.20	1.16	1.91	1.86	1.20	0.91	0.92
SSA.KM.N												
$n_B = 10$	1.45	1.23	0.96	0.46	0.83	1.01	1.10	1.84	1.75	1.38	1.19	1.17
$n_B = 20$	1.44	1.24	0.96	0.44	0.74	0.90	1.09	1.85	1.77	1.40	1.17	1.15
$n_B = 25$	1.45	1.24	0.97	0.46	0.74	0.89	1.09	1.87	1.79	1.39	1.17	1.14
$n_B = 50$	1.44	1.22	0.86	0.44	0.70	0.84	1.09	1.85	1.76	1.40	1.18	1.13

Green cells represent the MAPE values obtained from the bootstrap series for the Polish data set lower than those obtained from the original series.

Table 11. MBEs of h-step ahead forecast for the hourly electricity load of Poland obtained by SARIMA, NNAR, TBATS, and DSHW models.

	SARIMA			NNAR			TBATS			DSHW		
h	12	24	36	12	24	36	12	24	36	12	24	36
original	1561.97	411.79	962.12	86.73	198.18	286.42	−117.26	79.96	170.71	135.94	0.51	40.35
Bagging + Mean												
KM.N												
$n_B = 10$	273.31	174.83	76.82	104.49	227.21	309.50	−120.87	−39.71	66.39	154.18	44.32	67.89
$n_B = 20$	244.56	147.74	53.40	104.45	211.54	289.94	−129.99	−49.57	53.46	187.92	84.30	107.71
$n_B = 25$	232.10	136.46	42.70	97.04	198.57	270.37	−132.55	−51.01	53.80	168.66	72.60	95.20
$n_B = 50$	268.19	174.37	81.24	118.23	215.67	287.11	−118.26	−35.53	66.27	195.48	92.75	110.85
SSA. KM.N												
$n_B = 10$	344.43	247.28	163.03	99.88	163.18	202.15	−150.01	−41.64	45.34	171.18	15.69	58.17
$n_B = 20$	340.06	234.96	158.87	101.06	147.35	178.70	−150.37	−52.88	33.91	178.98	23.50	65.33
$n_B = 25$	337.43	241.26	155.77	105.18	152.51	183.96	−147.70	−54.21	33.98	179.56	23.77	63.67
$n_B = 50$	336.97	240.48	154.86	104.43	150.84	183.92	−156.03	−64.71	16.50	189.36	25.20	65.02
Bagging + Median												
KM.N												
$n_B = 10$	290.28	194.92	91.57	92.97	214.79	−288.01	−136.15	−61.41	47.86	140.73	33.83	50.60
$n_B = 20$	253.80	158.19	66.38	100.62	207.48	−228.01	−131.90	−56.81	48.96	195.85	91.18	109.55
$n_B = 25$	231.23	138.56	14.34	97.07	193.80	−228.01	−129.80	−53.97	50.75	180.31	76.99	96.29
$n_B = 50$	267.03	173.02	85.13	121.50	216.16	−228.01	−120.45	−44.30	59.69	206.92	101.26	120.60
SSA.KM.N												
$n_B = 10$	347.60	248.26	162.68	111.64	189.09	231.51	−146.64	−61.85	28.45	154.73	−3.77	45.28
$n_B = 20$	346.04	249.75	164.58	107.36	160.93	201.89	−145.21	−62.12	30.76	174.09	15.25	61.85
$n_B = 25$	347.75	250.10	166.04	112.56	162.74	198.52	−147.18	−67.27	27.96	173.74	18.04	59.33
$n_B = 50$	345.41	248.56	162.74	106.97	151.23	182.94	−151.43	−72.11	18.52	187.02	19.27	61.72

Green cells represent the MBE values obtained from the bootstrap series for the Polish data set lower (in absolute values) than those obtained from the original series.

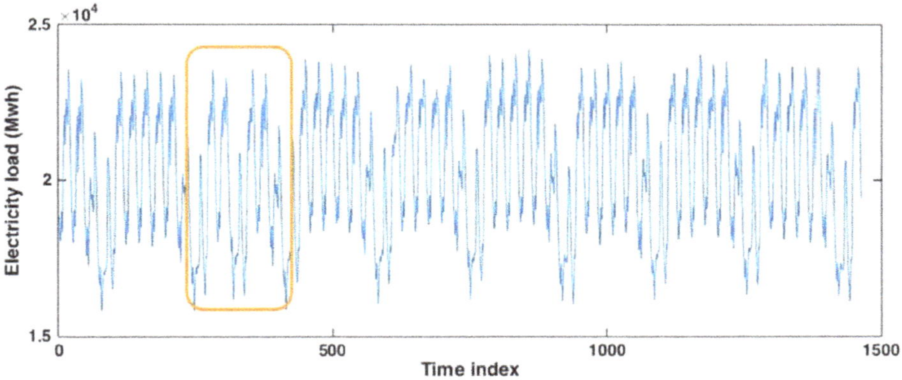

Figure 6. The hourly electricity load of Indonesia between 1 October and 30 November 2015.

Table 12. RMSE of h-step ahead forecast for the hourly electricity load of Indonesia obtained by SARIMA, NNAR, TBATS, and DSHW models.

	SARIMA			NNAR			TBATS			DSHW		
h	12	24	36	12	24	36	12	24	36	12	24	36
original	1236.72	2440.28	3119.10	193.12	216.95	234.57	611.23	702.52	609.34	167.51	265.62	351.18
Bagging + Mean												
KM.N												
$n_B = 10$	140.46	148.61	182.96	192.32	200.15	211.26	581.86	681.02	593.49	138.18	231.80	339.36
$n_B = 20$	137.65	148.79	182.42	201.59	201.79	207.81	581.67	678.01	592.20	137.31	224.45	318.91
$n_B = 25$	136.02	149.93	184.64	197.91	198.36	204.31	581.49	677.87	591.87	139.20	227.65	327.53
$n_B = 50$	136.02	149.93	184.64	194.08	200.23	208.27	580.00	677.27	590.68	147.81	245.81	362.49
SSA. KM.N												
$n_B = 10$	181.30	178.87	194.99	136.43	177.76	208.49	605.75	704.10	608.36	170.08	270.61	364.78
$n_B = 20$	186.56	184.79	199.14	135.42	177.59	208.45	600.50	700.37	605.79	167.54	277.51	385.43
$n_B = 25$	185.58	183.79	198.45	134.86	178.64	209.78	597.76	698.29	604.43	167.74	277.13	383.10
$n_B = 50$	193.68	189.13	201.59	146.47	184.91	213.14	602.30	699.76	605.76	150.49	260.36	370.83
Bagging + Median												
KM.N												
$n_B = 10$	139.51	144.76	183.00	190.81	199.99	210.46	581.67	680.87	593.92	153.44	245.70	352.49
$n_B = 20$	138.10	146.75	183.39	198.30	200.90	205.40	581.72	680.30	593.77	162.20	247.44	344.32
$n_B = 25$	139.17	150.71	187.80	195.26	196.90	202.43	584.21	681.03	594.27	160.89	242.23	344.98
$n_B = 50$	136.72	146.71	184.83	192.91	199.18	206.05	585.00	682.61	595.21	161.88	258.46	372.81
SSA.KM.N												
$n_B = 10$	180.44	178.15	194.23	135.73	174.27	206.11	605.40	700.65	606.85	166.57	268.12	358.32
$n_B = 20$	187.25	184.52	198.52	132.52	176.76	207.01	604.35	700.16	606.58	168.90	267.43	360.99
$n_B = 25$	186.72	183.84	198.25	135.54	178.50	208.79	598.77	696.99	604.16	168.29	268.13	366.47
$n_B = 50$	193.01	188.59	200.85	149.27	186.48	213.25	607.15	701.46	607.49	155.88	258.63	359.15

Green cells represent the RMSE values obtained from the bootstrap series for the Indonesian data set lower than those obtained from the original series.

Table 13 shows that SSA.KM.N produces lower MAPE than KM.N for the NNAR model. Compared with that obtained from the original data, the MAPE of the NNAR model was able to be reduced by up to 31.38%, 24.27%, and 17% for 12, 24, and 36 steps ahead of forecast values, respectively. Meanwhile, KM.N failed to lower the MAPE value for 12 steps-ahead, and it only declined approximately 8.74% and 11% for 24 and 36 steps-ahead forecast values, respectively.

In addition, for the NNAR model, MBE presented in Table 14 shows the application of the SSA.KM.Ns bagging method provides less bias for 12 and 24 steps-ahead forecast values than without bagging. However, this does not apply to KM.N.

Based on the experimental findings of the four data sets, bagging implementation can work well to improve the forecasting accuracy of the SARIMA and NNAR models.

However, the TBATS and DSHW did not yield the same behavior. The success of this implementation is thought to be influenced by the uncertainty of the models. In this experimental study, we found that some bootstrapped series failed to be modeled by TBATS and DSHW, affecting the final forecast results calculated based on the mean and median across all the forecast values.

Table 13. MAPE of h-step ahead forecast for the hourly electricity load of Indonesia obtained by SARIMA, NNAR, TBATS, and DSHW models.

	SARIMA			NNAR			TBATS			DSHW		
h	12	24	36	12	24	36	12	24	36	12	24	36
original	5.83	10.87	13.45	0.95	1.03	1.00	2.78	3.25	2.59	0.77	1.10	1.40
Bagging + Mean												
KM.N												
$n_B = 10$	0.67	0.70	0.76	0.96	0.96	0.92	2.67	3.15	2.53	0.64	0.92	1.29
$n_B = 20$	0.65	0.71	0.76	1.01	0.98	0.92	2.66	3.15	2.53	0.61	0.90	1.22
$n_B = 25$	0.63	0.71	0.77	0.99	0.96	0.91	2.66	3.14	2.53	0.63	0.91	1.25
$n_B = 50$	0.63	0.71	0.77	0.97	0.96	0.91	2.65	3.14	2.53	0.68	0.96	1.37
SSA. KM.N												
$n_B = 10$	0.87	0.86	0.85	0.66	0.78	0.84	2.76	3.24	2.58	0.79	1.13	1.45
$n_B = 20$	0.91	0.90	0.88	0.66	0.78	0.83	2.73	3.23	2.57	0.78	1.12	1.51
$n_B = 25$	0.90	0.89	0.88	0.66	0.78	0.84	2.71	3.22	2.56	0.79	1.12	1.50
$n_B = 50$	0.94	0.92	0.89	0.73	0.83	0.87	2.73	3.22	2.57	0.68	1.03	1.43
Bagging + Median												
KM.N												
$n_B = 10$	0.67	0.69	0.75	0.95	0.96	0.92	2.66	3.15	2.54	0.71	0.98	1.35
$n_B = 20$	0.65	0.70	0.76	0.99	0.97	0.91	2.66	3.26	2.54	0.75	1.00	1.33
$n_B = 25$	0.66	0.72	0.78	0.97	0.94	0.89	2.68	3.16	2.54	0.74	0.97	1.32
$n_B = 50$	0.65	0.71	0.76	0.96	0.95	0.90	2.68	3.17	2.55	0.76	1.02	1.43
SSA.KM.N												
$n_B = 10$	0.87	0.86	0.85	0.67	0.78	0.83	2.73	3.24	2.58	0.77	1.12	1.43
$n_B = 20$	0.91	0.90	0.88	0.65	0.78	0.83	2.75	3.23	2.58	0.76	1.11	1.43
$n_B = 25$	0.90	0.89	0.87	0.67	0.79	0.84	2.72	3.22	2.57	0.77	1.10	1.44
$n_B = 50$	0.95	0.89	0.92	0.75	0.84	0.88	2.76	3.23	2.58	0.70	1.06	1.41

Green cells represent the MAPE values obtained from the bootstrap series for the Indonesian data set lower than those obtained from the original series.

Table 14. MBE of h-step ahead forecast for the hourly electricity load of Indonesia obtained by SARIMA, NNAR, TBATS, and DSHW models.

	SARIMA			NNAR			TBATS			DSHW		
h	12	24	36	12	24	36	12	24	36	12	24	36
original	760.77	1914.50	2658.76	−171.70	−73.39	−7.27	−247.66	−394.84	−210.28	13.77	53.09	184.57
Bagging + Mean												
KM.N												
$n_B = 10$	−110.53	−107.27	−74.87	−176.59	−90.30	−17.23	−196.70	−361.33	−186.34	30.98	63.46	194.51
$n_B = 20$	−107.48	−109.13	−77.58	−185.87	−104.32	−29.48	−192.40	−354.91	−178.92	14.41	46.96	170.86
$n_B = 25$	−104.71	−110.04	−79.76	−181.50	−104.71	−30.37	−192.71	−355.37	−179.71	25.91	58.99	184.20
$n_B = 50$	−104.71	−110.04	−79.76	−176.85	−99.07	−24.80	−199.37	−358.92	−184.30	52.43	89.04	224.39
SSA. KM.N												
$n_B = 10$	−157.22	−147.24	−104.51	−121.14	−33.90	26.25	−242.95	−399.82	−220.51	9.29	53.06	193.08
$n_B = 20$	−163.43	−154.65	−111.92	−120.07	−32.97	27.19	−237.07	−394.13	−214.88	12.78	68.00	216.31
$n_B = 25$	−162.16	−153.24	−110.25	−119.85	−32.31	28.96	−232.08	−390.43	−211.79	14.10	68.99	215.21
$n_B = 50$	−170.78	−158.66	−113.53	−133.25	−41.03	22.80	−241.80	−395.38	−215.15	8.55	60.63	206.89
Bagging + Median												
KM.N												
$n_B = 10$	−109.90	−105.26	−73.05	−174.03	−91.36	−18.07	−197.11	−361.31	−185.10	36.11	72.84	204.91
$n_B = 20$	−109.00	−107.40	−75.75	−182.41	−105.51	−31.60	−196.47	−359.60	−182.80	29.83	62.95	191.74
$n_B = 25$	−108.90	−111.56	−80.54	−177.65	−105.33	−31.90	−199.31	−360.76	−183.54	32.22	71.68	199.89
$n_B = 50$	−105.43	−105.82	−73.99	−174.42	−100.32	−26.06	−203.93	−364.34	−187.13	57.59	99.10	234.13
SSA.KM.N												
$n_B = 10$	−156.02	−146.64	−104.14	−119.95	−35.59	25.79	−239.17	−392.13	−211.83	5.88	46.37	185.00
$n_B = 20$	−164.52	−153.32	−109.83	−118.08	−32.99	27.89	−239.19	−391.83	−211.09	−1.97	46.70	186.84
$n_B = 25$	−162.60	−152.57	−109.06	−121.50	−34.02	28.08	−231.34	−387.20	−207.83	1.60	52.27	193.65
$n_B = 50$	−171.08	−158.53	−113.30	−136.04	−42.54	21.33	−242.47	−393.92	−212.83	4.19	48.66	190.21

Green cells represent the MBE values obtained from the bootstrap series for the Indonesian data set lower (in absolute values) than those obtained from the original series.

The number of bootstrap series does not seem to affect the forecasting accuracy calculated by the mean and median ensemble. In some cases, the SSA.KM.N was able to improve the multistep-ahead forecasting accuracy, but in other cases, the KM.N provided better results.

In this case, the selection of the model is an important step to be considered. The application of bagging with the right forecasting model will increase the accuracy of multi-step ahead forecast values. Further development of the hybrid model, i.e., FFORMA [55] and exponential smoothing-neural network [56] or other combinations depending on the pattern of the data, can be considered to help overcome the uncertainty of the models [30].

4. Conclusions

In this study, we evaluated the impact and benefit of applying the existing KM.N and our proposed clustering-based bootstrap method, SSA.KM.N, in overcoming the uncertainty in time series multistep-ahead point forecasts. We focused on time series with a trend, seasonal, and affected by calendar variation and considered two Malaysian, one Polish, and one Indonesian electricity load time series as illustrative examples.

KM.N is considered an appropriate method for bootstrapping data with complex seasonal patterns, such as electrical load data. In the proposed method, we combined SSA and KM.N with the hope of producing bootstrap values that are more similar to the original data. We considered the SSA method to decompose the load series into signal and noise. By SSA, the observed values influenced by the calendar variation appear more clearly in the noise component than in the original data. Bootstrapping this residual value and adding it to the signal will result in the bootstrap series values around the original data.

Furthermore, we applied the following four models, usually used as benchmark models in forecasting electricity load time series: SARIMA, NNAR, TBATS, and DSHW. These four models are applied to all bootstrapped series to obtain up to 36-steps ahead of forecast values. The final forecast at time t is obtained by the following two ensemble methods: the mean and the median across all forecast values at time t. Based on the experimental results, we note that the number of bootstrapped series does not seem to affect the forecasting accuracy calculated by the mean and the median ensemble. We also found that the model suitable for the original series is not necessarily good for all bootstrapped series. We note that the accuracy of multiple-step-ahead forecasting values can be improved when the model, with different parameters, is appropriate for both the original and bootstrap data. Thus, combining several models and ensemble learning methods can be the direction of future research.

Author Contributions: Conceptualization, methodology, analysis, writing-original draft preparation, W.S.; investigation, software, visualization, and project administration, W.S. and Y.Y.; writing-review and editing, supervision, validation, W.S. and P.C.R. All authors have read and agreed to the published version of the manuscript.

Funding: This research was funded by the Ministry of Education, Culture, Research, and Technology Indonesia with source from the DIPA DIKTI RISTEK (Direktorat Riset, Teknologi, dan Pengabdian Kepada Masyarakat, Direktorat Jenderal Pendidikan Tinggi, Riset, dan Teknologi) 2022, Number SP DIPA-023.17.1.690523/2022 (second revision on 22 April 2022), in the scheme of National Competitive Basic Research (Penelitian Dasar Kompetitif Nasional) with Contract Number 096/E5/PG.02.00.PT/2022.

Institutional Review Board Statement: Not applicable.

Data Availability Statement: Data were obtained from [53] and are available at https://data.mendeley.com/datasets/f4fcrh4tn9/1 (accessed on 5 March 2021).

Acknowledgments: We thank the three anonymous reviewers for their valuable comments and suggestions to improve the quality of this paper. W. Sulandari and Y. Yudhanto acknowledge support from LPPM (Lembaga Penelitian dan Pengabdian kepada masyarakat) Universitas Sebelas Maret and thanks Subanar for his guidance. P.C. Rodrigues acknowledges financial support from the Brazilian national council for scientific and technological development (CNPq) grant "bolsa de produtividade PQ-2" 305852/2019-1.

Conflicts of Interest: The authors declare no conflict of interest. The funders had no role in the design of the study; in the collection, analyses, or interpretation of data; in the writing of the manuscript, or in the decision to publish the results.

Abbreviations

Abbreviation	Definition
DSHW	Double Seasonal Holt-Winters
KM	K-means Clustering-Based
KM.N	K-means Clustering-Based Generated from Gaussian Normal Distribution
MAPE	Mean Absolute Percentage Error
MBE	Mean Bias Error
MBB	Moving Block Bootstrap
NN	Neural Network
NNAR	Neural Network Autoregressive
RMSE	Root Mean Square Error
SARIMA	Seasonal Autoregressive Integrated Moving Average
S.MBB	Smoothed Moving Block Bootstrap
SSA	Singular Spectrum Analysis
SSA.KM.N	Singular Spectrum Analysis, K-means Clustering-Based Generated from Gaussian Normal Distribution
STL	Seasonal and Trend decomposition using Loess
SVD	Singular Value Decomposition
TBATS	Trigonometric, Box-Cox transform, ARMA errors, Trend, and Seasonal Components

References

1. Ahmad, W.; Ayub, N.; Ali, T.; Irfan, M.; Awais, M.; Shiraz, M.; Glowacz, A. Towards Short Term Electricity Load Forecasting Using Improved Support Vector Machine and Extreme Learning Machine. *Energies* **2020**, *13*, 2907. [CrossRef]
2. Ayub, N.; Irfan, M.; Awais, M.; Ali, U.; Ali, T.; Hamdi, M.; Alghamdi, A.; Muhammad, F. Big Data Analytics for Short and Medium-Term Electricity Load Forecasting Using an AI Techniques Ensembler. *Energies* **2020**, *13*, 5193. [CrossRef]
3. De Livera, A.M.; Hyndman, R.J.; Snyder, R.D. Forecasting Time Series with Complex Seasonal Patterns Using Exponential Smoothing. *J. Am. Stat. Assoc.* **2011**, *106*, 1513–1527. [CrossRef]
4. Soares, L.J.; Medeiros, M.C. Modeling and Forecasting Short-Term Electricity Load: A Comparison of Methods with an Application to Brazilian Data. *Int. J. Forecast.* **2008**, *24*, 630–644. [CrossRef]
5. Sulandari, W.; Subanar, S.; Suhartono, S.; Utami, H.; Lee, M.H.; Rodrigues, P.C. SSA-Based Hybrid Forecasting Models and Applications. *Bull. Electr. Eng. Inform.* **2020**, *9*, 2178–2188. [CrossRef]
6. Bernardi, M.; Petrella, L. Multiple Seasonal Cycles Forecasting Model: The Italian Electricity Demand. *Stat. Methods Appl.* **2015**, *24*, 671–695. [CrossRef]
7. López, M.; Valero, S.; Sans, C.; Senabre, C. Use of Available Daylight to Improve Short-Term Load Forecasting Accuracy. *Energies* **2021**, *14*, 95. [CrossRef]
8. Hong, T.; Wang, P.; White, L. Weather Station Selection for Electric Load Forecasting. *Int. J. Forecast.* **2015**, *31*, 286–295. [CrossRef]
9. Luo, J.; Hong, T.; Fang, S.-C. Benchmarking Robustness of Load Forecasting Models under Data Integrity Attacks. *Int. J. Forecast.* **2018**, *34*, 89–104. [CrossRef]
10. Sadaei, H.J.; de Lima e Silva, P.C.; Guimarães, F.G.; Lee, M.H. Short-Term Load Forecasting by Using a Combined Method of Convolutional Neural Networks and Fuzzy Time Series. *Energy* **2019**, *175*, 365–377. [CrossRef]
11. Cabrera, N.G.; Gutiérrez-Alcaraz, G.; Gil, E. Load Forecasting Assessment Using SARIMA Model and Fuzzy Inductive Reasoning. In Proceedings of the 2013 IEEE International Conference on Industrial Engineering and Engineering Management, Bangkok, Thailand, 10–13 December 2013; pp. 561–565. [CrossRef]
12. Chikobvu, D.; Sigauke, C. Regression-SARIMA Modelling of Daily Peak Electricity Demand in South Africa. *J. Energy South. Afr.* **2012**, *23*, 23–30. [CrossRef]
13. Taylor, J.W. Short-Term Electricity Demand Forecasting Using Double Seasonal Exponential Smoothing. *J. Opera. Tional. Res. Soc.* **2003**, *54*, 799–805. [CrossRef]
14. Taylor, J.W. Triple Seasonal Methods for Short-Term Electricity Demand Forecasting. *Eur. J. Oper. Res.* **2010**, *204*, 139–152. [CrossRef]
15. Arora, S.; Taylor, J.W. Short-Term Forecasting of Anomalous Load Using Rule-Based Triple Seasonal Methods. *IEEE Trans. Power Syst.* **2013**, *28*, 3235–3242. [CrossRef]
16. Bakirtzis, A.G.; Petridis, V.; Kiartzis, S.J.; Alexiadis, M.C.; Maissis, A.H. A Neural Network Short Term Load Forecasting Model for the Greek Power System. *IEEE Trans. Power Syst.* **1996**, *11*, 858–863. [CrossRef]

17. Charytoniuk, W.; Chen, M.-S. Very Short-Term Load Forecasting Using Artificial Neural Networks. *IEEE Trans. Power Syst.* **2000**, *15*, 263–268. [CrossRef]
18. Mordjaoui, M.; Haddad, S.; Medoued, A.; Laouafi, A. Electric Load Forecasting by Using Dynamic Neural Network. *Int. J. Hydrogen Energy* **2017**, *42*, 17655–17663. [CrossRef]
19. Leite Coelho da Silva, F.; da Costa, K.; Canas Rodrigues, P.; Salas, R.; López-Gonzales, J.L. Statistical and Artificial Neural Networks Models for Electricity Consumption Forecasting in the Brazilian Industrial Sector. *Energies* **2022**, *15*, 588. [CrossRef]
20. Sulandari, W.; Subanar, S.; Suhartono, S.; Utami, H. Forecasting Time Series with Trend and Seasonal Patterns Based on SSA. In Proceedings of the 2017 3rd International Conference on Science in Information Technology (ICSITech) Theory and Applicattion of IT for Education, Industry and Society in Big Data Era, Bandung, Indonesia, 25–26 October 2017; pp. 694–699. [CrossRef]
21. Zhang, X.; Wang, J.; Zhang, K. Short-Term Electric Load Forecasting Based on Singular Spectrum Analysis and Support Vector Machine Optimized by Cuckoo Search Algorithm. *Electr. Power Syst. Res.* **2017**, *146*, 270–285. [CrossRef]
22. Sulandari, W.; Subanar, S.; Lee, M.H.; Rodrigues, P.C. Time Series Forecasting Using Singular Spectrum Analysis, Fuzzy Systems and Neural Networks. *MethodsX* **2020**, *7*, 101015. [CrossRef]
23. Tao, D.; Xiuli, W.; Xifan, W. A Combined Model of Wavelet and Neural Network for Short Term Load Forecasting. In Proceedings of the International Conference on Power System Technology, Kunming, China, 13–17 October 2002; Volume 4, pp. 2331–2335. [CrossRef]
24. Moazzami, M.; Khodabakhshian, A.; Hooshmand, R. A New Hybrid Day-Ahead Peak Load Forecasting Method for Iran's National Grid. *Appl. Energy* **2013**, *101*, 489–501. [CrossRef]
25. Pandian, S.C.; Duraiswamy, K.; Rajan, C.C.A.; Kanagaraj, N. Fuzzy Approach for Short Term Load Forecasting. *Electr. Power Syst. Res.* **2006**, *76*, 541–548. [CrossRef]
26. Al-Kandari, A.M.; Soliman, S.A.; El-Hawary, M.E. Fuzzy Short-Term Electric Load Forecasting. *Int. J. Electr. Trical. Power Energy Syst.* **2004**, *26*, 111–122. [CrossRef]
27. Chen, Y.; Yang, Y.; Liu, C.; Li, C.; Li, L. A Hybrid Application Algorithm Based on the Support Vector Machine and Artificial Intelligence: An Example of Electric Load Forecasting. *Appl. Math. Model.* **2015**, *39*, 2617–2632. [CrossRef]
28. Li, Y.; Fang, T.; Yu, E. Study of Support Vector Machines for Short-Term Load Forecasting. *Proc. CSEE* **2003**, *23*, 55–59.
29. Breiman, L. Bagging Predictors | SpringerLink. *Mach. Learn.* **1996**, *24*, 123–140. [CrossRef]
30. Bergmeir, C.; Hyndman, R.J.; Benítez, J.M. Bagging Exponential Smoothing Methods Using STL Decomposition and Box–Cox Transformation. *Int. J. Forecast.* **2016**, *32*, 303–312. [CrossRef]
31. Petropoulos, F.; Hyndman, R.J.; Bergmeir, C. Exploring the Sources of Uncertainty: Why Does Bagging for Time Series Fore-casting Work? *Eur. J. Oper. Res.* **2018**, *268*, 545–554. [CrossRef]
32. Laurinec, P.; Lóderer, M.; Lucká, M.; Rozinajová, V. Density-Based Unsupervised Ensemble Learning Methods for Time Series Forecasting of Aggregated or Clustered Electricity Consumption. *J. Intell. Inf. Syst.* **2019**, *53*, 219–239. [CrossRef]
33. Cleveland, R.B.; Cleveland, W.S.; Terpenning, I. STL: A Seasonal-Trend Decomposition Procedure Based on Loess. *J. Off. Stat.* **1990**, *6*, 3.
34. Box, G.E.; Jenkins, G.M.; Reinsel, G.C.; Ljung, G.M. *Time Series Analysis: Forecasting and Control*; John Wiley & Sons: Hoboken, NJ, USA, 2015.
35. Wei, W.W.-S. *Time Series Analysis: Univariate and Multivariate Methods*, 2nd ed.; Pearson Addison-Wesley: Boston, MA, USA, 2006.
36. Huang, S.-J.; Shih, K.-R. Short-Term Load Forecasting via ARMA Model Identification Including Non-Gaussian Process Considerations. *IEEE Trans. Power Syst.* **2003**, *18*, 673–679. [CrossRef]
37. Hyndman, R.J.; Khandakar, Y. Automatic Time Series Forecasting: The Forecast Package for R. *JSS* **2008**, *27*, 1–23. [CrossRef]
38. Mohamed, N.; Ahmad, M.H.; Ismail, Z. Improving Short Term Load Forecasting Using Double Seasonal Arima Model. *World Appl. Sci. J.* **2011**, *15*, 223–231.
39. Hyndman, R.J.; Athanasopoulos, G.; Bergmeir, C.; Caceres, G.; Chhay, L.; O'Hara-Wild, M.; Petropoulos, F.; Razbash, S.; Wang, E.; Yasmeen, F. Forecast: Forecasting Functions for Time Series and Linear Models. In R Package Version 8.15. 2021. Available online: https://pkg.robjhyndman.com/forecast/ (accessed on 18 December 2021).
40. Hyndman, R.J.; Athanasopoulos, G. *Forecasting: Principles and Practice*, 2nd ed.; OTexts: Melbourne, Australia, 2018.
41. Zhang, G.; Patuwo, B.E.; Hu, M.Y. Forecasting with Artificial Neural Networks: The State of the Art. *Int. J. Forecast.* **1998**, *14*, 35–62. [CrossRef]
42. Sulandari, W.; Suhartono; Subanar; Rodrigues, P.C. Exponential Smoothing on Modeling and Forecasting Multiple Seasonal Time Series: An Overview. *Fluct. Noise Lett.* **2021**, *20*, 2130003. [CrossRef]
43. Sulandari, W.; Subanar, S.; Suhartono, S.; Utami, H. Forecasting Electricity Load Demand Using Hybrid Exponential Smoothing-Artificial Neural Network Model. *Int. J. Adv. Intell. Inform.* **2016**, *2*, 131–139. [CrossRef]
44. Hyndman, R.; Koehler, A.B.; Ord, J.K.; Snyder, R.D. *Forecasting with Exponential Smoothing: The State Space Approach*; Springer Science & Business Media: Berlin, Germany, 2008.
45. Golyandina, N.; Osipov, E. The "Caterpillar"-SSA Method for Analysis of Time Series with Missing Values. *J. Stat. Plan. Inference* **2007**, *137*, 2642–2653. [CrossRef]
46. Golyandina, N.; Korobeynikov, A. Basic Singular Spectrum Analysis and Forecasting with R. *Comput. Stat. Data Anal.* **2014**, *71*, 934–954. [CrossRef]

47. Golyandina, N.; Zhigljavsky, A. *Singular Spectrum Analysis for Time Series*, 2nd ed.; Springer Briefs in Statistics; Springer: Berlin/Heidelberg, Germany, 2020. [CrossRef]
48. Golyandina, N.; Korobeynikov, A.; Zhigljavsky, A. *Singular Spectrum Analysis with R*; Springer: Berlin, Germany, 2018.
49. PetoLau/petolau.github.io. GitHub. Available online: https://github.com/PetoLau/petolau.github.io (accessed on 19 April 2021).
50. Hassani, H. Singular Spectrum Analysis: Methodology and Comparison. *J. Data Sci.* **2007**, *5*, 239–257. [CrossRef]
51. Hyndman, R.J.; Koehler, A.B. Another Look at Measures of Forecast Accuracy. *Int. J. Forecast.* **2006**, *22*, 679–688. [CrossRef]
52. Stone, R.J. Improved Statistical Procedure for the Evaluation of Solar Radiation Estimation Models. *Sol. Energy* **1993**, *51*, 289–291. [CrossRef]
53. Guimaraes, F.; Javedani Sadaei, H. Data for: Short-Term Load Forecasting by Using a Combined Method of Convolutional Neural Networks and Fuzzy Time Series. Mendeley Data 2019, Version 1. Available online: https://data.mendeley.com/datasets/f4fcrh4tn9/1 (accessed on 5 March 2021).
54. Sulandari, W.; Subanar; Lee, M.H.; Rodrigues, P.C. Indonesian Electricity Load Forecasting Using Singular Spectrum Analysis, Fuzzy Systems and Neural Networks. *Energy* **2020**, *190*, 116408. [CrossRef]
55. Montero-Manso, P.; Athanasopoulos, G.; Hyndman, R.J.; Talagala, T.S. FFORMA: Feature-Based Forecast Model Averaging. *Int. J. Forecast.* **2020**, *36*, 86–92. [CrossRef]
56. Smyl, S. A Hybrid Method of Exponential Smoothing and Recurrent Neural Networks for Time Series Forecasting. *Int. J. Forecast.* **2020**, *36*, 75–85. [CrossRef]

energies

Review

A Review of Auto-Regressive Methods Applications to Short-Term Demand Forecasting in Power Systems

Rafał Czapaj [1,*], Jacek Kamiński [2,*] and Maciej Sołtysik [3]

1 PSE Innowacje Sp.z o.o., Al. Jerozolimskie 132 a, 02-305 Warsaw, Poland
2 Mineral and Energy Economy Research Institute of the Polish Academy of Sciences, The Department of Policy and Strategic Research, The Division of Energy Economics, Wybickiego 7A, 31-261 Kraków, Poland
3 Faculty of Electrical Engineering, Częstochowa University of Technology, Armii Krajowej 17, 42-200 Częstochowa, Poland
* Correspondence: rafal.czapaj@wp.pl (R.C.); kaminski@min-pan.krakow.pl (J.K.)

Abstract: The paper conducts a literature review of applications of autoregressive methods to short-term forecasting of power demand. This need is dictated by the advancement of modern forecasting methods and their achievement in good forecasting efficiency in particular. The annual effectiveness of forecasting power demand for the Polish National Power Grid for the next day is approx. 1%; therefore, the main objective of the review is to verify whether it is possible to improve efficiency while maintaining the minimum financial outlays and time-consuming efforts. The methods that fulfil these conditions are autoregressive methods; therefore, the paper focuses on autoregressive methods, which are less time-consuming and, as a result, cheaper in development and applications. The prepared review ranks the forecasting models in terms of the forecasting effectiveness achieved in the literature on the subject, which enables the selection of models that may improve the currently achieved effectiveness of the transmission system operator. Due to the applied approach, a transparent set of forecasting methods and models was obtained, in addition to knowledge about their potential in the context of the needs for short-term forecasting of electricity demand in the national power system. The articles in which the MAPE error was used to assess the quality of short-term forecasts were analyzed. The investigation included 47 articles, several dozen forecasting methods, and 264 forecasting models. The articles date from 1997 and, apart from the autoregressive methods, also include the methods and models that use explanatory variables (non-autoregressive ones). The input data used come from the period 1998–2014. The analysis included 25 power systems located on four continents (Asia, Europe, North America, and Australia) that were published by 44 different research teams. The results of the review show that in the autoregressive methods applied to forecasting short-term power demand, there is a potential to improve forecasting effectiveness in power systems. The most promising prognostic models using the autoregressive approach, based on the review, include Fuzzy Logic, Artificial Neural Networks, Wavelet Artificial Neural Networks, Adaptive Neurofuse Inference Systems, Genetic Algorithms, Fuzzy Regression, and Data Envelope Analysis. These methods make it possible to achieve the efficiency of short-term forecasting of electricity demand with hourly resolution at the level below 1%, which confirms the assumption made by the authors about the potential of autoregressive methods. Other forecasting models, the effectiveness of which is high, may also prove useful in forecasting by electricity system operators. The paper also discusses the classical methods of Artificial Intelligence, Data Mining, Big Data, and the state of research in short-term power demand forecasting in power systems using autoregressive and non-autoregressive methods and models.

Keywords: short-term forecasting; electrical power demand; power systems; autoregressive forecasting methods; classical forecasting methods; artificial intelligence methods; Big Data; machine learning; Data Mining

Citation: Czapaj, R.; Kamiński, J.; Sołtysik, M. A Review of Auto-Regressive Methods Applications to Short-Term Demand Forecasting in Power Systems. *Energies* **2022**, *15*, 6729. https://doi.org/10.3390/en15186729

Academic Editors: Paweł Piotrowski, Grzegorz Dudek and Dariusz Baczyński

Received: 10 August 2022
Accepted: 5 September 2022
Published: 14 September 2022

1. Introduction

1.1. Overview

The economic development of countries is inextricably linked with the functioning of their power systems. Due to the development of power grids and the growing access to them, electricity is now indispensable for the proper functioning of the economy and the population, and the demand for it is systematically growing. Rising electricity prices in recent years and their fluctuations, in addition to insufficient development of the manufacturing sector, make it difficult to optimally meet the growing demand for electricity. Unfortunately, storage of electricity on a large scale and in the long term is a complex and very expensive issue. Thus, at any time in the operation of power systems, it is necessary to maintain a balance between the generation of electricity and its consumption, taking into account the technical limitations of electricity networks, in order to maintain continuity and security of power and electricity supplies while maintaining the optimal operating costs of the power system. In this context, forecasting the load of power systems is an essential element of planning their work in the short, medium, and long term, and is one of the greatest challenges faced by the power industry in every country. Electricity demand forecasting is a basic element of planning electricity generation, participation in electricity markets, and the development of the power grid. Short-term forecasting of the power system load, performed, inter alia, by operators of power systems, requires ensuring the highest possible accuracy for each hour of the day while maintaining the lowest computational cost at an appropriate time. Forecasting the load on systems with the use of prognostic models using explanatory variables is costly and time-consuming, in contrast to autoregressive methods which use only information about the earlier development of the analyzed parameter in the forecasting process. Thus, along with the observed trend indicating the reduction in forecast horizons from hours to minutes, and even seconds, it is necessary to search for cheap and quick forecasting methods that will allow the current forecasting effectiveness to be maintained at lower costs of their development and with a comparable or shorter development time.

1.2. Literature Survey

In short-term electrical power demand forecasting, both autoregressive methods using the properties of moving averages and exponential smoothing, and methods using machine learning [1–6]. Support Vector Machines and Particle Swarms, and artificial intelligence [7], including Artificial Neural Networks, have been used for years. Many research centers worldwide have developed more accurate forecasting methods and models, especially for short-term forecasting. Several teams have conducted research at the academic level, perfecting the methods and models they have developed. For the conducted analyses and simulations, usually, STATISTICA®, SAS/ETS, and SPSS environments [8], GRETL [9,10], and the R and Python programming languages are used, among others.

The demand for electrical power is characterized by large fluctuations [11]. In this case, the key factors exhibit daily, weekly, annual, and multi-year variability [12]. Moreover, the seasonal variability (which results in annual variability), quarterly variability (seasons), and monthly variability (part of the seasons) are distinguished. Continuity of power demand and the still "insufficient" (in the sense of high power/capacity) development of energy storage results in the inability to store it in large quantities, which makes it necessary to cover the demand for power at the time of the occurrence of this demand [13].

Other factors, apart from the passage of time (consecutive days, weeks, etc.), that influence the variability in the power system load [14,15] are the variability in weather conditions and the resulting variability in the ambient temperature, in addition to the transition from winter to summer time [16,17] and from summer to winter time (introduced to flatten the evening peak of power demand in the summer half of the year) [12]). Other weather factors influencing the level of demand in the power system include, among others, cloudiness, air humidity, and wind speed [12]. The ambient temperature significantly affects the load in the power system. The change in weather conditions directly impacts consumer behavior (municipal and industrial), consisting of increasing power consumption from lighting and heating devices (convector heating and electric heating).

1.3. Motivation and Incitement

Individual areas of the Polish Power System have a different share in shaping the domestic demand for electrical power. Naturally, areas with significant industrialization and, therefore, a significant population in Poland, translate into greater demand for electrical power (and, consequently, electrical power consumption), and thus, to a greater extent, changes in the weather (atmospheric conditions) affect these areas. The yearly demand forecasting error for the Polish National Power System is approximately 1%, which shows a high level of accuracy; thus, there is a need to search for the potential in well-known methods and models, including autoregressive models, to reduce the error below this level. In this context, this paper aims to review auto-regressive methods applied to short-term power demand forecasting in power systems.

1.4. Research Gaps

The conducted review of articles describing the methods and forecasting models used in short-term forecasting of electric power demand shows a great variety. Autoregressive methods are still an attractive and effective tool for forecasting. Their unquestionable advantage is low financial outlay and quickly obtaining forecast results. The current observation of scientific reports in the form of literature reviews is time-consuming. Therefore, it is important to develop rankings of forecasting models, taking into account their forecasting effectiveness. While preparing this review, the authors identified a gap in presenting the results of valuable research in this aspect, and thus attempted to develop such a ranking. The Mean Average Percentage Error was adopted as a measure for assessing the quality of forecasts developed with autoregressive methods. From the prepared ranking of 264 autoregressive models, a set of Top 10 models was distinguished, which can be a significant aid for researchers and scientists dealing with short-term forecasting of electricity demand in power systems.

1.5. Major Contributions

The main contribution of the authors is to present an overview of methods in the field of artificial intelligence, Data Mining (now often associated with Big Data issues), and Big Data. In addition, the state of research in short-term power demand forecasting for power systems using autoregressive and non-autoregressive methods and models is presented, along with a detailed table that describes the results of the review of 47 articles describing 264 forecasting models (Table 1, where MAPE is an ex post, and MAPE(ea) is an ex ante approach). Additionally, the authors present a new way to develop literature reviews in the context of selecting the most prospective prognostic models. In the proposed new approach (explained in the flowchart—Figure 1), ranking of forecasting models (Tables 2 and 3 and Figure 2) was used due to the selected measure of forecast quality (Mean Average Percentage Error). The applied new approach to the development of the results of literature reviews is an excellent source of knowledge for scientists, experts, and analysts, supporting the preparation of forecasts for power system operators, with particular emphasis on transmission system operators.

Table 1. The publications' preview results in short-term power demand forecasting methods and models used for power systems.

No.	Authors/Title/Publishing House	Year	Analysis Scope	Country	Method, Model	Effectiveness			Model No.
		-	Years	-	-	Error, %			-
1.	Al-Fuhaid A.S. et al. *Neuro-Short-Term Load Forecast of the Power System in Kuwait* Elsevier (21:215-219) [18]	1997	1994	Kuwait	ANN(ea)—*Artificial neural network*	MAPE(ea)	1.84	4.84	1
2.	Almeshaiei E., Soltan H. *A Methodology for Electric Power Load Forecasting* Alexandria Engineering Journal (50) [19]	2011	2006–2008	Kuwait	MA(ea7,30)—*Moving Average (7, 30)*	MAPE(ea)	3.84		2
3.	Al-Shobaki S., Mohsen M. *Modeling and Forecasting of Electrical Power Demands for Capacity Planning* Elsevier Energy Conversion and Management (49) [20]	2008	2002–2007	Jordan	ARIMA(ea)	MAPE(ea)	5.25		3
4.	Badran S.M., Abouelatta O.B. *Forecasting Electrical Load using ANN Combined with Multiple Regression Method* The Research Bulletin of Jordan ACM II(II) [21]	2013	1988–2006	Saudi Arabia	MR—*Multiple Regression*	MAPE	11.58	14.35	4
					ANN—*Multiple Regression*	MAPE	2.44	8.04	5
5.	Brodowski S. et al. *A Hybrid System for Forecasting 24-h Power Load Profile for Polish Electric Grid* Elsevier B.V. Applied Soft Computing (58) [22]	2017	2013, 2015	Poland (NPS)	HA + MR—*Hierarchical Approximator + Multiple Regression*	MAPE	1.08	2.26	6
6.	Buitrago J., Asfour S. *Short-Term Forecasting of Electric Loads Using Nonlinear Autoregressive Artificial Neural Networks with Exogenous Vector Inputs* Energies 10(40) [23]	2017	2005–2015	USA (New England)	ANN—*Artificial Neural Network*	MAPE	0.85		7
7.	Ceperic E., Ceperic V. *A Strategy for Short-Term Load Forecasting by Support Vector Regression Machines* IEEE Transactions on Power Systems (1) [24]	2013	2006	USA	ANN—*Artificial Neural Network*	MAPE	1.50	3.72	8
					SDBWNN—*Similar Day—Based Wavelet Neural Network*	MAPE	1.26	2.70	9
					SASVR—*Seasonality—Adjusted, Support Vector Regression*	MAPE	0.93	1.86	10
8.	Chahkoutahi F., Khashei M. *A Seasonal Direct Optimal Hybrid Model of Computational Intelligence and Soft Computing Techniques for Electricity Load Forecasting* Energy (140) [25]	2017	2011.05.02–2011.07.03	Australia	ARIMA	MAPE	0.76	1.07	11
					ANN(MLP)—*Artificial Neural Network (Multilayer Perceptron)*	MAPE	0.72	1.23	12
					ANFIS—*Adaptive Neuro Fuzzy Inference System*	MAPE	0.83	0.95	13
					DOPH—*Direct Optimum Parallel Hybrid*	MAPE	0.58	0.71	14
9.	Chapagain K., Kittipiyakul S. *Short-Term Electricity Load Forecasting Model and Bayesian Estimation for Thailand Data* MATEC Web of Conferences (55) [26]	2016	2013–2015	Thailand	MR—*Multiple Regression*	MAPE	1.75	33.45	15
					MR(Gibbs)—*Multiple Regression (Gibbs Sampling)*	MAPE	0.85	23.06	16
10.	Chen H. et al. *ANN-Based Short-Term Load Forecasting in Electricity Markets* University of Waterloo, Department of Electrical & Computer Engineering [27]	2001	1999	Canada	ANN—*Artificial Neural Network*	MAPE	0.48	3.00	17
11.	Chheepa T.K., Manglani T. *A Critical Review on Employed Techniques for Short Term Load Forecasting* IRJET 04(06) [28]	2017	1996–1997, 2000	Iran	ARIMA	MAPE	1.48	1.99	18
12.	Clements A.A. et al. *Forecasting Day-Ahead Electricity Load Using a Multiple Equation Time Series Approach* NCER Working Paper Series 103(5) [29]	2015	1999.07.12–2013.11.27	Australia	ARIMA	MAPE	1.36	2.89	19

Table 1. *Cont.*

No.	Authors/Title/Publishing House	Year	Analysis Scope	Country	Method, Model	Effectiveness			Model No.
		-	Years	-	-	Error, %			-
13.	Czapaj R. *Typowanie zmiennych objaśniających przy wykorzystaniu zautomatyzowanych metod statystycznych jako sposób optymalizacji wyboru metody estymacji szczytowego dobowego obciążenia KSE* Przegląd Elektrotechniczny 4(93) [30]	2016	2010–2014	Poland (NPS)	MARSplines	MAPE	1.86	6.99	20
					C&RT—*Classification and Regression Trees*	MAPE	2.57	7.18	21
					C&RT—*Classification and Regression Trees*	MAPE	2.56	6.77	22
					Chi^2—*Automatic interaction detector using Chi^2*	MAPE	4.06	7.33	23
					CHAID—*Chi^2 Automatic Interaction Detector*	MAPE	3.69	9.40	24
14.	Czapaj R., Kamiński J., Benalcazar P. *Dobór zmiennych objaśniających z wykorzystaniem metody MARSplines* Politechnika Częstochowska, XIV Konferencja PE [31]	2018	2009–2014	Poland (NPS)	MARSplines	MAPE	1.86	6.99	25
15.	Czapaj R., Kamiński J., Benalcazar P. *Prognozowanie krótkoterminowe z wykorzystaniem metody MARSplines* Politechnika Częstochowska, XIV Konferencja PE [32]	2018	2009–2014	Poland (NPS)	MARSplines	MAPE	3.36	6.04	26
					MARSplines(ea)	MAPE(ea)	6.57		27
16.	Dąsal K. *Dobór zmiennych wejściowych do Modelu Rozkładu Kanonicznego* Politechnika Częstochowska, VI Konferencja PE [33]	2002	1993–1995	Poland (NPS)	MRK(Mo-Fr) —*Canonical Vector Decomposition Method from Monday till Friday*	MAPE(Mo-Fr)	0.64	9.79	28
17.	Dudek G. *Short-Term Load Forecasting Based on Kernel Conditional Density Estimation* Przegląd Elektrotechniczny 8(86) [34]	2010	2002–2006	Poland (NPS)	SFS(5 years)—*Sequential Forward Selection Methods*	MAPE	1.84		29
					SBS(5 years)—*Sequential Backward Selection Methods*	MAPE	1.77		30
					NS(5 years)—*Nearest Neighbors*	MAPE	1.94		31
					ANN(5 years)—*Artificial Neural Network*	MAPE	2.02		32
					FE(5 years)—*Fuzzy Estimators*	MAPE	1.76		33
			1997–2000		SFS(4 years)—*Sequential Forward Selection Methods*	MAPE	2.19		34
					SBS(4 years)—*Sequential Backward Selection Methods*	MAPE	2.06		35
					NS(4 years)—*Nearest Neighbors*	MAPE	2.55		36
					ANN(4 years)—*Artificial Neural Network*	MAPE	2.24		37
					FE(4 years)—*Fuzzy Estimators*	MAPE	2.14		38
18.	Dudek G., Janicki M. *Nearest Neighbor Model with Weather Inputs for Pattern-based Electricity Demand Forecasting* Przegląd Elektrotechniczny 3(93) [35]	2017	2011–2014	Poland (NPS)	NNWISA(*working days*) —*Nearest Neighbors with Weather Inputs for Similarity Analysis*	MAPE (*working days*)	1.55	1.67	39
				Belgium		MAPE (*working days*)	2.82	2.88	40
				New England		MAPE (*working days*)	2.41	3.26	41
				USA		MAPE (*working days*)	3.43	4.82	42

Table 1. *Cont.*

No.	Authors/Title/Publishing House	Year	Analysis Scope	Country	Method, Model	Effectiveness			Model No.
		-	Years	-	-	Error, %			-
18.	Dudek G., Janicki M. *Nearest Neighbor Model with Weather Inputs for Pattern-based Electricity Demand Forecasting* Przegląd Elektrotechniczny 3(93) [35]	2017	2011–2014	Poland (NPS)	NNWISA(*weekends*)—*Nearest Neighbors with Weather Inputs for Similarity Analysis*	MAPE(*weekends*)	1.75	1.76	43
				Belgium		MAPE(*weekends*)	3.02	3.12	44
				New England		MAPE(*weekends*)	2.92	3.16	45
				USA		MAPE(*weekends*)	4.31	4.99	46
			2011–2014	Poland (NPS)	NNWISA(*Holidays*)—*Nearest Neighbors with Weather Inputs for Similarity Analysis*	MAPE(*Holidays*)	4.36	16.1[illegible]	47
				Belgium		MAPE(*Holidays*)	4.05	12.6[illegible]	48
				New England		MAPE(*Holidays*)	6.35	7.03	49
				USA		MAPE(*Holidays*)	6.05	7.62	50
19.	Dudek G. *Pattern-Based Local Linear Regression Models for Short-Term Load Forecasting* Elsevier, Electric Power System Research (130) [36]	2016	2002–2004	Poland (NPS)	MR(January)—*Multiple Regression*	MAPE(January)	2.37		51
					SR(January)—*Stepwise Regression*	MAPE(January)	1.52		52
					RR(January)—*Ridge Regression*	MAPE(January)	1.59		53
					Lasso(January)—*Least Absolute Selection Regression and the Constriction Operator*	MAPE(January)	1.51		54
					PCR(January)—*Principal Component Regression*	MAPE(January)	1.36		55
					PLSR(January)—*Partial Least Squares Regression*	MAPE(January)	1.18		56
					MR(July)—*Multiple Regression*	MAPE(July)	2.63		57
					SR(July)—*Stepwise Regression*	MAPE(July)	1.14		58
					RR(July)—*Ridge Regression*	MAPE(July)	1.23		59
					Lasso(July)—*Least Absolute Selection Regression and the Constriction Operator*	MAPE(July)	1.06		60
					PCR(July)—*Principal Component Regression*	MAP(July)	0.94		61
					PLSR(July)—*Partial Least Squares Regression*	MAPE(July)	1.00		62
			2002–2004	Poland (NPS)	PCR—*Principal Component Regression*	MAPE	1.35		63
					PLSR—*Partial Least Squares Regression*	MAPE	1.34		64
					ARIMA	MAPE	1.82		65
					ES—*Exponential Smoothing*	MAPE	1.66		66
					ANN(MLP)—*Artificial Neural Network (Multilayer Perceptron)*	MAPE	1.44		67
					NWE—*Nadaraya—Watson Estimator*	MAPE	1.30		38
					NM—*Naive Method*	MAPE	3.43		39

Table 1. *Cont.*

No.	Authors/Title/Publishing House	Year	Analysis Scope	Country	Method, Model	Effectiveness		Model No.
		-	Years	-	-	Error, %		-
19.	Dudek G. *Pattern-Based Local Linear Regression Models for Short-Term Load Forecasting* Elsevier, Electric Power System Research (130) [36]	2016	2007–2009	France	PCR—*Principal Component Regression*	MAPE	1.71	70
					PLSR—*Partial Least Squares Regression*	MAPE	1.57	71
					ARIMA	MAPE	2.32	72
					ES—*Exponential Smoothing*	MAPE	2.10	73
					ANN(MLP)—*Artificial Neural Network (Multilayer Perceptron)*	MAPE	1.64	74
					NWE—*Nadaraya—Watson Estimator*	MAPE	1.66	75
					NM—*Naive Method*	MAPE	5.05	76
			2007–2009	Great Britain	PCR—*Principal Component Regression*	MAPE	1.60	77
					PLSR—*Partial Least Squares Regression*	MAPE	1.54	78
					ARIMA	MAPE	2.02	79
					ES—*Exponential Smoothing*	MAPE	1.85	80
					ANN(MLP)—*Artificial Neural Network (Multilayer Perceptron)*	MAPE	1.65	81
					NWE—*Nadaraya—Watson Estimator*	MAPE	1.55	82
					NM—*Naive Method*	MAPE	3.52	83
			2006–2008	Australia	PCR—*Principal Component Regression*	MAPE	3.00	84
					PLSR—*Partial Least Squares Regression*	MAPE	2.83	85
					ARIMA	MAPE	3.67	86
					ES—*Exponential Smoothing*	MAPE	3.52	87
					ANN(MLP)—*Artificial Neural Network (Multilayer Perceptron)*	MAPE	2.92	88
					NWE—*Nadaraya–Watson Estimator*	MAPE	2.82	89
					NM—*Naive Method*	MAPE	4.88	90
20.	Dudek G. *Drzewa regresyjne i lasy losowe jako narzędzia predykcji szeregów czasowych z wahaniami sezonowymi* Politechnika Częstochowska [37]	2016	2002–2004	Poland (NPS)	RF(January)—*Random Forest*	MAPE(January)	1.42	91
					C&RT(January)—*Classification and Regression Trees*	MAPE(January)	1.70	92
					C&RTR(January)—*Fuzzy Classification and Regression Trees*	MAPE(January)	1.62	93
					ARIMA(January)	MAPE(January)	2.64	94
					ES(January)—*Exponential Smoothing*	MAPE(January)	2.35	95
					ANN(January)—*Artificial Neural Network*	MAPE(January)	1.32	96
					NM(January)—*Naive Method*	MAPE(January)	6.37	97
					RF(July)—*Random Forest*	MAPE(July)	0.92	98
					C&RT(July)—*Classification and Regression Trees*	MAPE(July)	1.16	99

Table 1. *Cont.*

No.	Authors/Title/Publishing House	Year	Analysis Scope	Country	Method, Model	Effectiveness			Model No.
		-	Years	-	-	Error, %			-
20.	Dudek G. *Drzewa regresyjne i lasy losowe jako narzędzia predykcji szeregów czasowych z wahaniami sezonowymi* Politechnika Częstochowska [37]	2016	2002–2004	Poland (NPS)	C&RTR(July)—*Fuzzy Classification and Regression Trees*	MAPE(July)	1.13		100
					ARIMA(July)	MAPE(July)	1.21		101
					ES(July)—*Exponential Smoothing*	MAPE(July)	1.19		102
					ANN(July)—*Artificial Neural Network*	MAPE(July)	0.97		103
					NM(July)—*Naive Method*	MAPE(July)	1.29		104
21.	Esener I.I., Yuskel T., Kurban M. *Short-Term Load Forecasting Without Meteorological Data Using AI-Based Structures* Turkish Journal of Electrical Engineering & Computer Sciences (23) [38]	2015	2009	Turkey	ANN—*Artificial Neural Network*	MAPE	3.67		105
					WM+ANN—WM—Wavelet Method + ANN—*Artificial Neural Network*	MAPE	3.73		106
					WM+ANN(RBF)—WM—Wavelet Method + ANN—*Artificial Neural Network (Radial Basis Functions)*	MAPE	2.89		107
					ED—*Empirical Decomposition*	MAPE	3.52		108
			2010		ANN—*Artificial Neural Network*	MAPE	3.81		109
					WM+ANN—WM—Wavelet Method + ANN—*Artificial Neural Network*	MAPE	4.18		110
					WM+ANN(RBF)—WM—Wavelet Method + ANN—*Artificial Neural Network (Radial Basis Functions)*	MAPE	2.99		111
					ED—*Empirical Decomposition*	MAPE	3.63		112
22.	Fan S. *Short-Term Load Forecasting Based on a Semi-Parametric Additive Model* IEEE Transactions on Power Systems [39]	2010	1997–2009 (training) 2009.01.01–2009.01.31 (test)	Australia	SPAM—*Semi-Parametric Additive Model*	MAPE	1.41	2.37	113
					ANN—*Artificial Neural Network*	MAPE	1.82	3.90	114
					SPAM+ANN—Hybrid Model (*Semi-Parametric Additive Model + Artificial Neural Network*)	MAPE	1.58	2.79	115
23.	Farahat M.A. *Short Term Load Forecasting Using Neural Networks and Particle Swarm Optimization* Journal of Electrical Engineering [40]	2018	2011.07.01–2011.08.10 (training) 2011.08.11–2011.08.17 (test)	Egypt	ANN(BP)—*Artificial Neural Network (Back Propagation Training)*	MAPE	4.60		116
					ANN(BP)+PSO -ANN(BP) —*Artificial Neural Network (Back Propagation Training)* + PSO—*Particle Swarm Optimization*	MAPE	1.90		117
24.	Gorwar M. *Short Term Load Forecasting Using Time Series Analysis:* *A Case Study for Karnataka, India* ResearchGate, IJESIT Conference [41]	2012	2011–2012	India	AR(ea)—*Autoregression*	MAPE	13.03		118
					ARMA(ea)	MAPE	11.73		119
					ARIMA(ea)	MAPE	6.15		120

Table 1. *Cont.*

No.	Authors/Title/Publishing House	Year	Analysis Scope	Country	Method, Model	Effectiveness			Model No.
		-	Years	-	-	Error, %			-
25.	Hassan S., Khosravi A., Jaafar J. *Examining Performance of Aggregation Algorithms for Neural Network-Based Electricity Demand Forecasting* ScienceDirect, Electrical Power and Energy Systems (64) [42]	2015	2011 (30-min Intervals)	Malaysia, Australia, Pakistan	ANN(I)—*Artificial Neural Network (Integration)*	MAPE(I 30 min.)	7.16		121
					ANN(TI)—Sztuczna sieć neuronowa (*Trimmed Integration*)	MAPE(I 30 min.)	10.13		122
					ANN(BA)—*Artificial Neural Network (Bayesian Averaging)*	MAPE(I 30 min.)	4.34		123
					NM—Metoda naiwna	MAPE(I 30 min.)	6.41		124
26.	He W. *Deep Neural Network Based Load Forecast* Computer Modelling & New Technologies 18(3) [43]	2014	2000.02.10–2012.11.30	China	ANN—*Artificial Neural Network*	MAPE	1.90	2.08	125
27.	Hong T., Wang P. *Fuzzy Interaction Regression for Short Term Load Forecasting* University of North Carolina at Charlotte 13(1) [44]	2014	2005–2007	USA	FRI(ea)—*Fuzzy Regression without Interaction*	MAPE(ea)	14.21		126
					FRICV(ea)—*Fuzzy Regression without Interaction with Categorical Variables*	MAPE(ea)	5.16		127
					FRI(ea) + MR—FRI(ea)—*Fuzzy Regression without Interaction* + MR—*Multiple Regression*	MAPE(ea)	4.63		128
					FRI(ea)+TV—FRI(ea)—*Fuzzy Regression without Interaction* + TV—*Temperature Variables*	MAPE(ea)	3.68		129
28.	Janicki M. *Temperature Correction Method for Pattern Similarity-Based Short-term Electricity Demand Forecasting Models* Przegląd Elektrotechniczny 3(93) [45]	2017	2013–2014	USA	IS+TC(USA 2013)—IS—*Image Similarities* + TC—*Temperature Correction* (USA 2013)	MAPE	4.50		130
				USA	NM(USA 2013)—*Naive Method* (USA 2013)	MAPE	10.78		131
				USA	IS+TC(USA 2014)—IS—*Image Similarities* + TC—*Temperature Correction* (USA 2014)	MAPE	4.86		132
				USA	NM(USA 2014)—*Naive Method* (USA 2014)	MAPE	10.94		133
				Belgium	IS+TC(BEL 2013)—IS—*Image Similarities* + TC—*Temperature Correction* (Belgium 2013)	MAPE	3.80		134
				Belgium	NM (BEL 2013)—*Naive Method* (Belgium 2013)	MAPE	8.54		135
				Belgium	IS+TC(BEL 2014)—IS—*Image Similarities* + TC—*Temperature Correction* (Belgium 2014)	MAPE	3.66		136
				Belgium	NM(BEL 2014)—*Naive Method* (Belgium 2014)	MAPE	9.47		137

Table 1. *Cont.*

No.	Authors/Title/Publishing House	Year	Analysis Scope	Country	Method, Model	Effectiveness		Model No.
		-	Years	-	-	Error, %		-
29.	Kheirkhah A. et al. *Improved Estimation of Electricity Demand Function by Using of Artificial Neural Network, Principal Component Analysis and Data Envelopment Analysis* Elsevier Ltd. Computers & Industrial Engineering (64) [46]	2013	1992.04.01–2003.02.28	Iran, Ireland Turkey	GA—Genetic Algorithm	MAPE	0.14	138
					FR—*Fuzzy Regression*	MAPE	0.08	139
					ANN—*Artificial Neural Network*	MAPE	0.16	140
					ANFIS—*Adaptive Neuro Fuzzy Inference System*	MAPE	0.15	141
					DEA—*Data Envelopment Analysis*	MAPE	0.01	142
30.	Kolcun M., Holka L. *Daily Load Diagram Prediction of Eastern Slovakia* Politechnika Częstochowska, VI Konferencja PE [47]	2002	1997–1998	Slovakia	ANN(Koh)—Kohonen's Artificial Neural Network	MAPE	3.50	143
31.	Lin Y. *An Ensemble Model Based on Machine Learning Methods and Data Preprocessing for Short-Term Electric Load Forecasting* Energies 10(1186) [48]	2017	2010.08.01–2011.07.31	Australia	EML—*Extreme Machine Learning*	MAPE	0.83	144
					EMLDE—*Extreme Machine Learning (optimized by) Differential Evolution*	MAPE	0.77	145
					ARIMA	MAPE	0.73	146
					WTWTMABC—Wavelet Transform—*Wavelet Transform—Modified Artificial Bee Colony—Extreme Machine Learning*	MAPE	0.59	147
					EMDDEEML—*Empirical Mode Decomposition—Differential Evolution – Extreme Machine Learning*	MAPE	0.39	148
					VMD—*Variational Mode Decomposition*	MAPE	0.30	149
32.	Liu N., Babushkin V., Afshari A. *Short-Term Forecasting of Temperature Driven Electricity Load Using Time Series and Neural Network Model* Journal of Clean Energy Technologies 2(4) [49]	2014	2010.01.01–2011.06.30	United Arab Emirates	SARIMAX	MAPE	1.58	150
					ANN—*Artificial Neural Network*	MAPE	2.29	151
33.	Magnano L., Boland J.W. *Generation of Synthetic Sequences of Electricity Demand: Application in South Australia* Elsevier Ltd. Energy (32) [50]	2006	2000–2001 (*Summer Time*)	Australia	ARMA(*Summer Time*)	MAPE (*Summer Time*)	2.40	152
34.	Nadtoka I.I., Al-Zihery B.M. *Mathematical Modelling and Short-Term Forecasting of Electricity Consumption of the Power System, with Due Account of Air Temperature and Natural Illumination, Based on Support Vector machine and Particle Swarm* Elsevier Ltd. Procedia Engineering (129) [51]	2015	2009–2012	Iraq	SVM+PSO—SVM *– Support Vector Machines* + PSO *– Particle Swarm Optimization* (including UV) 2011.05.11.	MAPE(UV; May 2011)	2.65	153
					2011.08.31.	MAPE(UV; August 2011)	1.23	154
					2011.11.30.	MAPE(UV; November 2011)	2.13	155
					2012.01.26.	MAPE(UV; January 2012)	1.73	156
					SVM+PSO—SVM *– Support Vector Machines* + PSO *– Particle Swarm Optimization* (including temperature) 2011.05.11.	MAPE(Temp.; May 2011)	2.60	157
					2011.08.31.	MAPE(Temp.; August 2011)	1.37	158
					2011.11.30.	MAPE(Temp.; November 2011)	1.94	159
					2012.01.26.	MAPE(Temp.; January 2012)	1.90	160

Table 1. *Cont.*

No.	Authors/Title/Publishing House	Year	Analysis Scope	Country	Method, Model		Effectiveness			Model No.
		-	Years	-	-		Error, %			-
34.	Nadtoka I.I., Al-Zihery B.M. *Mathematical Modelling and Short-Term Forecasting of Electricity Consumption of the Power System, with Due Account of Air Temperature and Natural Illumination, Based on Support Vector machine and Particle Swarm* Elsevier Ltd. Procedia Engineering (129) [51]	2015	2009–2012	Iraq	SVM+PSO—SVM – *Support Vector Machines* + PSO – *Particle Swarm Optimization* (including UV & Temperature)	2011.05.11.	MAPE(UV; Temp.; May 2011)	2.26		161
						2011.08.31.	MAPE(UV; Temp.; August 2011)	1.41		162
						2011.11.30.	MAPE(UV; Temp.; November 2011)	1.61		163
						2012.01.26.	MAPE(UV; Temp.; January 2012)	1.58		164
35.	Narayan A. *Long Short Term Memory Networks for Short-Term Electric Load Forecasting* IEEE International Conference on Systems, Man, and Cybernetics [52]	2017	2006–2016	Canada	ANN(January)—*Artificial Neural Network*		MAPE (January)	4.60		165
					ARIMA(May)		MAPE (May)	5.70		166
					ANN-LSTM-RNN(September)—*Long—Short—Term Memories—Recurrent Neural Network*		MAPE (September)	4.40		167
					ANN(sty.)—*Artificial Neural Network*		MAPE (January)	6.30		168
					ARIMA(May)		MAPE (May)	8.20		169
					ANN—LSTM—RNN(September)—*Long—Short—Term Memories—Recurrent Neural Network*		MAPE (September)	5.90		170
					ANN(January)—Sztuczna sieć neuronowa		MAPE (January)	3.80		171
					ARIMA(May)		MAPE (May)	3.90		172
					ANN-LSTM-RNN(September)—*Long—Short—Term Memories—Recurrent Neural Network*		MAPE (September)	3.80		173
36.	Nowicka-Zagrajek J., Weron R. *Modeling Electricity Loads in California: ARMA Models with Hyperbolic Noise* Hugo Steinhaus Center Wrocław University of Technology, KBN [53]	2002	1999.01.01–2000.12.31	USA	ARMA(1,6) (January 1.—February 28.)		MAPE	1.66		174
					ARMA *Adaptive* (January 3.—February 28.)		MAPE	1.66		175
					ARMA(1,6) (January 1.—February 28.)		MAPE	1.24		176
					ARMA Adaptive (January 3.—February 28.)		MAPE	1.23		177
37.	Nowotarski J. et al. *Improving Short Term Load Forecast Accuracy* via *Combining Sister Forecasts* Hugo Steinhaus Center Wrocław University of Technology, University of North Carolina at Charlotte [54]	2015	2007.01.01–2011.12.31	USA	SA—Simple Averaging(ea)		MAPE(ea)	2.10	2.82	178
					AT(PU—ea) (*Average Trimming*)		MAPE(ea)	2.10	2.82	179
					WA(UW—ea) (*Winsor's Averaging*)		MAPE(ea)	2.10	2.83	180
					OLS(MNKea) (*Ordinary Least Squares*)		MAPE(ea)	2.14	2.82	181
					RMAD(ea) (*Regression of the Minimum Absolute Deviation*)		MAPE(ea)	2.14	2.83	182
					LSLPW(ea) (*Least Squares Limited —Positive Weights*)		MAPE(ea)	2.12	2.81	183

Table 1. *Cont.*

No.	Authors/Title/Publishing House	Year	Analysis Scope	Country	Method, Model		Effectiveness			Model No.
-		-	Years	-	-		Error, %			-
37.	Nowotarski J. et al. *Improving Short Term Load Forecast Accuracy* via *Combining Sister Forecasts* Hugo Steinhaus Center Wrocław University of Technology, University of North Carolina at Charlotte [54]	2015	2007.01.01–2011.12.31	USA	LSL(ea) (*Least Squares Limited*)		MAPE(ea)	2.11	2.83	184
					IRMSEA(ea) (*IRMSE Averaging*)		MAPE(ea)	2.10	2.82	185
					BI—C(ea) (*The Best Individual Calibration Window*)		MAPE(ea)	2.25	2.93	186
					SM—Sister Model 1(ea)		MAPE(ea)	2.29	3.09	187
					SM—Sister Model 2(ea)		MAPE(ea)	2.24	3.15	188
					SM—Sister Model 3(ea)		MAPE(ea)	2.34	3.01	189
					SM—Sister Model 4(ea)		MAPE(ea)	2.32	3.17	190
					SM—Sister Model 5(ea)		MAPE(ea)	2.28	3.11	191
					SM—Sister Model 6(ea)		MAPE(ea)	2.30	3.18	192
					SM—Sister Model 7(ea)		MAPE(ea)	2.37	3.07	193
					SM—Sister Model 8(ea)		MAPE(ea)	2.31	3.21	194
38.	Hsiao-Ten P. *Forecast of Electricity Consumption and Economic Growth in Taiwan by State Space Modeling* Elsevier Ltd. Energy (34) [55]	2009	2002–2007	Taiwan	ECSTSP *—Error—Correction State Space Model*	2002–2007	MAPE	3.90		195
						2003–2007	MAPE	2.57		196
						2004–2007	MAPE	2.38		197
						2005–2007	MAPE	1.52		198
						2006–2007	MAPE	2.57		199
						2007	MAPE	2.04		200
					STSP *—State Space Model*	2002–2007	MAPE	4.04		201
						2003–2007	MAPE	2.62		202
						2004–2007	MAPE	2.43		203
						2005–2007	MAPE	1.75		204
						2006–2007	MAPE	2.34		205
						2007	MAPE	2.39		206
					SARIMA	2002–2007	MAPE	5.32		207
						2003–2007	MAPE	3.79		208
						2004–2007	MAPE	3.01		209
						2005–2007	MAPE	2.87		210
						2006–2007	MAPE	2.18		211
						2007	MAPE	1.20		212
39.	Rana M, Koprinska I. *Forecasting Electricity Load with Advanced Wavelet Neural Networks* Elsevier B.V. Neurocomputing (182) [56]	2016	2006–2007	Australia	WANN(F—Aus.)—*Wavelet Artificial Neural Network*		MAPE	0.27		213
				Australia	ANN(Aus.)—*Artificial Neural Network*		MAPE	0.28		214
				Australia	FL(Aus.)—*Fuzzy Logic*		MAPE	0.29		215
				Australia	MTR(Aus.)—*Model Tree Rules*		MAPE	0.35		216
				Australia	ESDS(n-1; Aus.)—*Exponential Smoothing—Daily Seasonality*		MAPE	0.30		217
				Australia	ESWS(n -7; Aus.)—*Exponential Smoothing—Weekly Seasonality*		MAPE	0.32		218
				Australia	ESDWS(n-1 i n -7; Aus.)—*Exponential Smoothing—Daily and Weekly Seasonality*		MAPE	0.30		219
				Australia	ARIMA(n-1; Aus.) *Daily*		MAPE	0.30		220
				Australia	ARIMA(n -7; Aus.) *Weekly*		MAPE	0.32		221

Table 1. *Cont.*

No.	Authors/Title/Publishing House	Year	Analysis Scope	Country	Method, Model	Effectiveness		Model No.
		-	Years	-	-	Error, %		-
39.	Rana M, Koprinska I. *Forecasting Electricity Load with Advanced Wavelet Neural Networks* Elsevier B.V. Neurocomputing (182) [56]	2016	2006–2007	Australia	ARIMA(n-1 i n -7; Aus.) *Daily & Weekly*	MAPE	0.30	222
				Australia	IM(Aus.)—*Industrial Model*	MAPE	0.31	223
				Australia	NAM(Aus.)—*Naive Averaged Method*	MAPE	13.48	224
				Australia	NDM(Aus.)—*Naive Delayed Method*	MAPE	0.47	225
				Australia	NM(n-1; Aus.)—*Naive Method (Previous Day)*	MAPE	5.05	226
				Australia	NM(n -7; Aus.)—*Naive Method (Previous Week)*	MAPE	4.94	227
			2010–2011	Spain	ANN(F—Esp)—*Wavelet Artificial Neural Network*	MAPE	1.72	228
				Spain	ANN(Esp)—*Artificial Neural Network*	MAPE	2.12	229
				Spain	FL(Esp)—*Fuzzy Logic*	MAPE	2.25	230
				Spain	MTR(Esp)—*Model Tree Rules*	MAPE	2.24	231
				Spain	ESDS(n-1; Esp)—*Exponential Smoothing—Daily Seasonality*	MAPE	2.54	232
				Spain	ESWS(n -7; Esp)—*Exponential Smoothing—Weekly Seasonality*	MAPE	2.01	233
				Spain	ESDWS(n-1 i n -7; Esp)—*Exponential Smoothing—Daily and Weekly Seasonality*	MAPE	1.95	234
				Spain	ARIMA(n-1; Esp) *Daily Seasonality*	MAPE	2.45	235
				Spain	ARIMA(n -7; Esp) *Weekly Seasonality*	MAPE	2.00	236
				Spain	ARIMA(n-1 i n -7; Esp) *Daily & Weekly Seasonality*	MAPE	1.89	237
				Spain	IM(Esp)—*Industrial Model*	MAPE	0.31	238
				Spain	NAM(Esp)—*Naive Averaged Method*	MAPE	21.18	239
				Spain	NDM(Esp)—*Naive Delayed Method*	MAPE	5.05	240
				Spain	NMPD(n-1; Esp)—*Naïve Method (Previous Day)*	MAPE	9.45	241
				Spain	NMPW(D-7; Esp)—Naive Method *(Previous Week)*	MAPE	7.42	242
40.	Siwek K., Osowski S. *Prognozowanie obciążeń 24-godzinnych w systemie elektroenergetycznym z użyciem zespołu sieci neuronowych* Przegląd Elektrotechniczny 8(85) [57]	2009	2006–2008	Poland (NPS)	ANN(MLP)—*Artificial Neural Network (Multilayer Perceptron)*	MAPE	2.07	243
					ANN(SVM)—*Artificial Neural Network* (Support Vector Machines)	MAPE	2.24	244
					ANN(Elman)—*Artificial Neural Network* (Elman)	MAPE	2.26	245
					ANN(Koh)—Kohonen's *Artificial Neural Network*	MAPE	2.37	246
					ANN(MLPZ1)—*Artificial Neural Network (Committee—Multilayer Perceptron 1)*	MAPE	1.48	247

Table 1. *Cont.*

No.	Authors/Title/Publishing House	Year	Analysis Scope	Country	Method, Model	Effectiveness			Model No.
		-	Years	-	-	Error, %			-
40.	Siwek K., Osowski S. *Prognozowanie obciążeń 24-godzinnych w systemie elektroenergetycznym z użyciem zespołu sieci neuronowych* Przegląd Elektrotechniczny 8(85) [57]	2009	2006–2008	Poland (NPS)	ANN(SVMZ)—*Artificial Neural Network (Committee—Support Vector Machines)*	MAPE	1.35		248
					ANN(BSSZ)—*Artificial Neural Network (Committee—*BSS)	MAPE	1.71		249
41.	Selivan R.A., Rajagopal R. *A Model For The Effect of Aggregation on Short Term Load Forecasting* IEEE, Stanford University [58]	2014	-	USA	ARMA	MAPE	2.00		250
					SVR (*Support Vector Regression*)	MAPE	4.00		251
					SSN(FF)—*Artificial Neural Network (Fast Forward Training)*	MAPE	2.40		252
42.	Sousa J.C., Neves LP., Jorge H.M. *Assessing the Relevance of Load Profiling Information in Electrical Load Forecasting Based on Neural Network Models* Elsevier Ltd. Electrical Power and Energy Systems (40) [59]	2012	2006.12.15–2009.11.30	Portugal	SSN(OK)—*Artificial Neural Network (Municipal Users)*	MAPE	6.13	22.39	253
					ANN(TSDSO)—*Artificial Neural Network (Transformer Station of a Distribution System Operator)*	MAPE	5.14	5.35	254
43.	Wang P., Liu B., Hong T. *Electric Load Forecasting with Recency Effect: A Big Data Approach* Hugo Steinhaus Center Wrocław University of Technology [60]	2015	2007	USA	REM—*Recent Effect Method (Forecasts for each day with a year in advance)*	MAPE	4.27	4.38	255
44.	Wang Y., Bielecki J.M. *Acclimation and the Response of Hourly Electricity Loads to Meteorological Variables* Elsevier Ltd. Energy (142) [61]	2018	1999.07.28–2007.12.31. (*Calibration Set*)	USA	GRM(Temp.)—*General Regression Model* (Temperature)	MAPE (Temp.)	~0.10	~4.10	256
					FGRM(ea; *Temperature; Wind)—Full General Regression Model* (hourly delays of thermosensitivity, binary variables of historical temperatures in months, wind speed)	MAPE (ea; Temp.; Wind)	~0.20	~4.30	257
			2008.01.01–2014.12.31.		2SGRM(*Temperature; Wind; Humidity)—2-Step General Regression Model (1.—Fit to the full model; 2.—Adjustment to the model of the influence of humidity on demand)*	MAPE (Temp.; Wind; Humid.)	1.00	2.00	258
45.	Wyrozumski T. *Prognozowanie neuronowe w energetyce* Politechnika Lubelska, Konferencja REE [62]	2005	-	Poland	ANN(ea)—*Artificial Neural Network*	MAPE (ea)	1.31	4.87	259
46.	Yang J. *Power System Short-term Load Forecasting* TU Darmstadt, *Doctoral Thesis* [63]	2006	2002	China	C&RT—*Classification and Regression Trees*	MAPE	2.63	11.64	260
					ANN—*Artificial Neural Network*	MAPE	1.51	4.13	261
					SVM—*Support Vector Machines*	MAPE	1.51	3.87	262
47.	Yu X., Ji H. *A PSO-SVM-Based 24 h Power Load Forecasting Model* MATEC Web of Conferences (25) [64]	2015	2014	China	ANN(BP)—*Artificial Neural Network (Back Propagation Training)*	MAPE	3.28	4.13	263
					SVM+PSO—SVM—*Support Vector Machines + PSO—Particle Swarm Optimization*	MAPE	2.58	2.68	264

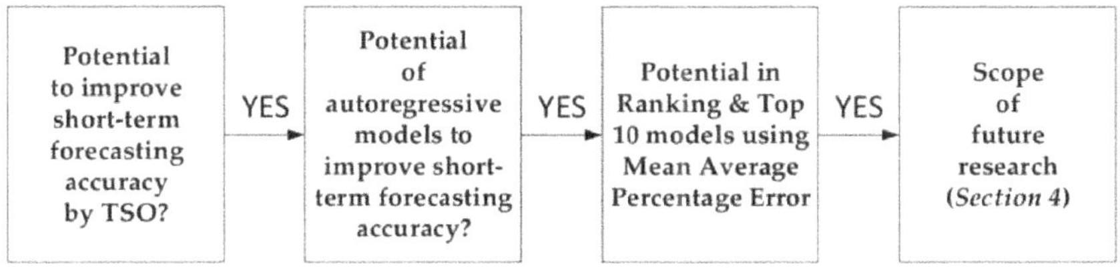

Figure 1. The design of the survey.

Table 2. Forecasting model ranking for the position from 1 to 132.

Ranking (1–33)	Model No.	Ranking (34–66)	Model No.	Ranking (67–99)	Model No.	Ranking (100–132)	Model No.
1	142	34	16	67	91	100	43
2	139	35	98	68	67	101	204
3	256	36	10	69	18	102	33
4	138	37	61	70	247	103	30
5	141	38	103	71	8	104	65
6	140	39	62	72	54	105	114
7	257	40	258	73	261	106	1
8	213	41	60	74	262	107	29
9	214	42	6	75	52	108	80
10	215	43	100	76	198	109	20
11	149	44	58	77	78	110	25
12	217	45	99	78	39	111	237
13	219	46	56	79	82	112	117
14	220	47	102	80	71	113	125
15	222	48	212	81	115	114	160
16	223	49	101	82	150	115	31
17	238	50	59	83	164	116	159
18	218	51	154	84	53	117	234
19	221	52	177	85	77	118	236
20	216	53	176	86	163	119	250
21	148	54	9	87	93	120	233
22	225	55	104	88	74	121	32
23	17	56	68	89	81	122	79
24	14	57	259	90	66	123	200
25	147	58	96	91	75	124	35
26	28	59	64	92	174	125	243
27	12	60	63	93	175	126	73
28	146	61	248	94	92	127	178
29	11	62	19	95	70	128	179
30	145	63	55	96	249	129	180

Table 2. *Cont.*

Ranking (1–33)	Model No.	Ranking (34–66)	Model No.	Ranking (67–99)	Model No.	Ranking (100–132)	Model No.
31	13	64	158	97	228	130	185
32	144	65	113	98	156	131	184
33	7	66	162	99	15	132	183

Table 3. Forecasting model ranking for the position from 133 to 264.

Ranking (133–165)	Model No.	Ranking (166–198)	Model No.	Ranking (199–231)	Model No.	Ranking (232–264)	Model No.
133	229	166	203	199	87	232	227
134	155	167	5	200	108	233	76
135	38	168	235	201	112	234	226
136	181	169	232	202	136	235	240
137	182	170	36	203	86	236	254
138	211	171	22	204	105	237	127
139	34	172	21	205	129	238	3
140	37	173	196	206	24	239	207
141	188	174	199	207	106	240	166
142	231	175	264	208	208	241	170
143	244	176	157	209	134	242	50
144	186	177	202	210	171	243	253
145	230	178	57	211	173	244	120
146	161	179	260	212	109	245	168
147	245	180	94	213	2	246	49
148	191	181	153	214	172	247	97
149	151	182	40	215	195	248	124
150	187	183	89	216	251	249	27
151	192	184	85	217	201	250	121
152	194	185	210	218	48	251	242
153	72	186	107	219	23	252	169
154	190	187	45	220	110	253	135
155	189	188	88	221	255	254	241
156	205	189	111	222	46	255	137
157	95	190	84	223	123	256	122
158	51	191	209	224	47	257	131
159	193	192	44	225	167	258	133
160	246	193	263	226	130	259	4
161	197	194	26	227	116	260	119
162	206	195	42	228	165	261	118
163	152	196	69	229	128	262	224
164	252	197	143	230	132	263	126
165	41	198	83	231	90	264	239

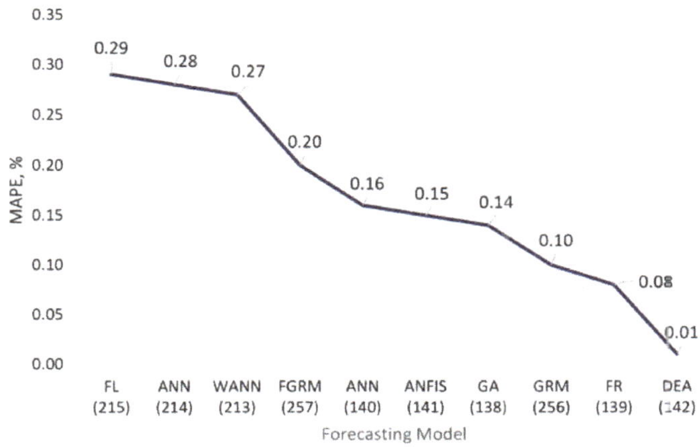

Figure 2. The effectiveness of forecasting models in the Top 10 set from a group of 264 models.

2. Short-Term Forecasting Methods and Models Used for Power Systems

2.1. Classical Methods of Artificial Intelligence

There main methods are successfully used in forecasting, optimization, diagnostics, detection, and design in the power industry: artificial neural networks, evolutionary algorithms, and expert systems. Neural networks are used, among others, in optimization of tap changer settings in transformers, optimization of capacitor bank settings, and forecasting of the peak load of the power system and its daily loads using Artificial Neural Networks [13,23,40,42,43,46,49,57,62,65–78], in addition to using Deep Neural Networks [43], and autoregressive models [79], Big Data [1,80], short-circuit analyses, and transformer damage detection. Artificial Neural Networks are the most commonly used artificial intelligence methods [81] in forecasting the operating parameters of power systems and networks. Artificial Neural Networks [82,83] are an effective tool for forecasting in the power industry (not only the loads mentioned above in the power grid [84–87], but also electricity prices [88], especially in short-term forecasting [72]. In practical applications, Artificial Neural Networks are also supported by the techniques of Fuzzy Logic functions [89] and the Neuro-Fuzzy Approach [90–93].

The indication of the greater effectiveness of Artificial Neural Networks over the improvement of traditional methods in short-term forecasting of power system loads, presented in [72], does not always translate into short-term forecasting of energy prices on Polish and foreign electricity trading floors [94]. In this context, it is possible to obtain an inverse relationship. For example, the multiple regression method gives significantly greater forecasting efficiency when compared to the models of Artificial Neural Networks [95]. Artificial Neural Networks are highly effective not only in the short term, but also in long-term forecasting [96,97].

Evolutionary algorithms are used, among others in [84]: forecasting daily loads of electric power systems [46,67], optimizing the configuration of power grids, optimizing voltage levels in power grids, designing power grids, planning power plant operation, creating an economical distribution of loads, planning power grid development, supporting regulatory activities in power systems, and protection automatics [83,98]. Expert systems are used, among other things, in [99]: designing power grids and stations and reconstruction of power systems in post-emergency states [100,101].

Additional information on the application of artificial intelligence methods, taking into account the studied subject of the variability in power system loads and their forecasting, can be found in [81,84,85,102,103].

2.2. Data Mining Methods

In the literature focusing on the analysis of large data sets and forecasting using Data Mining methods, there are many definitions of these methods and ideas [104].

The main definitions of Data Mining are:

1. An interdisciplinary approach using techniques from machine learning, image recognition, statistics, databases and visualization to extract information from large databases [42,105,106];
2. An analysis of large, previously collected data sets to discover new regularities and describe the data in a new way that is understandable and useful for the data owner [107].

The first definition comes from 1998, while the second comes from 2001; thus their evolution is noticeable.

Further definitions of Data Mining methods are:

3. The process of searching for valuable information (knowledge) when the researcher is dealing with a large amount of data [108–111];
4. The process of examining and analyzing large amounts of data by automated or semi-automatic methods to discover meaningful patterns and rules [112,113];
5. Methods of broadly understood data analysis aimed at identifying previously unknown regularities occurring in large data sets, from which the results are in a form that is easy to interpret by the researcher [109].

At the beginning of their development, Data Mining methods were accused of being unscientific, assuming no theory, having no elegance or formal evidence, and being primitive and for application only [114].

The classical approach to data analysis uses the scheme [115,116] from defining the problem through creating a mathematical model, preparing the input data, and analyzing the problem, to interpreting the obtained results. The Data Mining approach uses a scheme from problem definition through preparing input data, problem analysis, and creating a mathematical model, to interpreting the obtained results. The algorithms used in the field of Data Mining are divided into supervised learning and non-supervised learning [104]. In the supervised learning methods, the main goal is to recreate the value of the examined parameter. In the non-supervised learning methods, the aim is to detect structures or hidden patterns in the analyzed data due to the lack of distinguishing a single feature. Teaching forecasting models using a supervised learning approach can be conducted as an implementation of a classification or regression problem. In classification problems, the analyzed parameter is qualitative, and in regression problems, this parameter is quantitative.

The knowledge derived from empirical research is proven, and due to the collection of larger and larger sets of data, it is beneficial for further research, both empirical and forecasting (in a certain sense speculative); it is useful to analyze these sets and draw additional conclusions. Additional research, including experimental studies, may result in obtaining a greater number of answers than the questions posed by the researcher [117–119]. The classification indicated in [118] of problem types and their respective Data Mining methods concerning time series analysis notes the inclusion of MultiLayer Perceptron (MLP) and Radial Basis Function (RBF) Artificial Neural Networks in this method. It must be concluded that the classifications of methods overlap and do not function as hermetic.

The group of Data Mining methods and models also includes forecasting problems, which are divided into two groups. The first group includes regression and classification trees, and the second group includes advanced machine learning methods. Classification and regression trees include Classification and Regression Trees (C&RT) and Chi-Square Automatic Interaction Detection (CHAID) trees [96,120]. The advanced machine learning group consists of the methods Multivariate Adaptive Regression Splines (MARSplines), Support Vector Machines (SVMs), k Nearest Neighbors [121,122], k—Means [123–124], Naive Bayes Classifier (only applicable to classification problems), Random Forest [125], and Boosted Trees [96]. The use of Data Mining methods in forecasting regression problems consists of evaluating many models, comparing their effectiveness results, and creating hybrid systems, due to which it is possible to maintain the smallest deviations in the forecasted values from the realized values of the analyzed parameters. The distinguishing feature of Data Mining methods is the speed of their creation. The MARSplines and Boosted Tree methods are among the most effective predictive models from the group of Data Mining methods for forecasting power demand in power systems.

The MARSplines method is in the niche of practical applications in forecasting problems in large-scale power engineering. In the MARSplines method, a non-parametric type belonging to the group of supervised learning methods, the co-variability in features is used to predict the value of a selected feature, and in classification problems [126,127]. The indicated convenience excludes from research activities the necessity to analyze the correlation between the independent variables, which in many cases may correlate with the predicted variable, but do not affect it.

The Multivariate Adaptive Regression Splines (MARSplines) method [128–130] uses the method of recursive division of the feature space to build a regression model in the form of spline curves [131–133] and is an extension of the methods of regression trees and multiple regression [105]. Due to the above properties, the MARSplines [131–133] is an effective tool for Big Data applications [134,135].

The MARSplines method also enables the automatic selection of explanatory variables for forecasting models. The efficiency of this selection is in many cases greater than that for classical methods of selecting variables [30,31,136–138]. Thus, the method can be successfully used, in addition to the multiple regression method, in selecting input variables for forecasting models and short-term forecasting of time series, including power demand in power systems. [31,32,139].

The principal components method is an alternative to those analyzing the correlations between the explanatory variables in the forecasting process. It not only allows the removal of variables that are overly correlated with each other, but also the acquisition of uncorrelated variables that are responsible for part of the variability in groups of variables or even for the variability in entire groups of variables [140]. The application of the method creates new variables, which are linear combinations of the original variables, and the following components capture as much information contained in the original data as possible. The disadvantage of the method is the difficulty in interpreting the meaning of principal components [140].

2.3. Big Data

Big Data is a term that describes, on a very general level, exceptionally large data sets. These collections are characterized by a diversified structure of high complexity. The main difficulties are data storage, real-time analysis, and data visualization and analysis results [141,142]. The process of examining massive amounts of data to reveal hidden patterns and secret correlations is called Big Data analysis. In the 1990s and the first decade of the 21st century, Big Data analysis was understood as Data Mining. Big Data sets are characterized by: high volume (Volume) [98,141,143,144], high growth rate (Velocity) [98,141,143,144], reliability and accuracy (Veracity) [141,142], great variety (Variety) [98,144], and value for decision making processes (Value) [98,141,144,145].

The use of Big Data analysis for the needs of data sets containing electrical measurements, including the load size of power systems, includes practical applications, e.g., techniques, i.e., correlation analysis and machine learning techniques (including deep learning: Multilevel Deep Learning [146], Pooling Deep Recurrent Neural Network [147], Convolutional Neural Network Based Bagging Learning Approach [148], TensorFlow Deep Learning Framework and Clustering-regression [149], Long Short—Term Memory Neural Network [150], using Scikit-Learn and TensorFlow [151], with the Keras library [152], Deep Neural Networks [43,153], and introducing Multilevel Deep Learning Methods for Big Data Analysis [146] and databases [114]). Processing of electrical measurement data includes distributed processing (data storage and processing—Distributed Computing), memory processing (data reading and processing—Memory Computing) and stream processing (real-time data processing—Stream Processing) [141,154].

The use of Big Data techniques in the energy system in the energy sector [155–157] and in the field of Smart Grids [1,80,154,158] includes the use of RBF Artificial Neural Networks [159] using a Convolutional Neural Network Based Bagging Learning Approach [148]. This also encompasses compatibility of aid for technical measures concerning the integration of the generating sources [160], with special regard to renewable sources [161,162] and in creating backup data sets that can be used in situations of information and communication disruptions [163].

The use of sets, techniques, and processes concerning Big Data for the power industry is inextricably linked with the security of the stored data. The security of this type of data can be increased through its location dispersion (e.g., SCOOP system) [144].

Data streams supplying Big Data sets in transmission and distribution power systems come from [164–166]: Supervisory Control And Data Acquisition (SCADA) systems [167], phasor measurement systems in Wide Area Management System (WAMS) technology [168], Intelligent Electronic Devices (IEDs), network asset management systems, conventional and smart meters [147,169–171], and information exchange systems with electricity market participants, from seismic and meteorological institutes, Global Positioning System (GPS) systems, and Geographic Information System (GIS) systems. The practical method of the similarity of days [172–176] allows the quality of forecasting power demand to be below 3.00% per day and the efficiency achieved by the Polish Transmission System Operator (PSE S.A.) to be approx. 1.00%. Similar days are selected based on the most recent demand factor forecasts in the first step. In the second step, the weighted average is calculated for each hour of the day, considering the historical values. In the classical approach, there is a slight variation in the values of individual weights. Due to weighting of the most similar days, it is possible to obtain minimum, maximum, and average errors for the entire day below 2% [176]. The method of self-adaptive weighing is successfully used in forecasting the demand for electric power in microgrids. Compared to the standard methods of dynamic demand profiles, multiple regression, and Artificial Neural Networks, it almost doubles forecasting effectiveness (approx. 3.5%) [177]. A similar level of effectiveness (3.99%) using the multiple regression method for the power system shows that despite the longer computation time (for a seven-day horizon), its classical version [178], using as input data (explanatory variables) forecasts of weather parameters, gives a similar quality. The use of Artificial Neural Networks in short-term forecasting of electrical power demand in power systems does not always give exceptionally effective forecasting results compared to other methods. Artificial Neural Networks require significant research experience, and the results, even using efficient network learning methods [147], rarely give effectiveness below 1.00% per day. Often, advanced Artificial Neural Networks provide forecasting efficiency expressed by the values of Mean Average Percentage Error (MAPE) from approx. 3.00% to even approx. 13.00% (in the 20-day horizon) [5]. The knowledge of electrical power quality parameters is one of the key elements of entities operating in the electricity market [179]. Cyclical measurements of these parameters (including the assessment of the condition of electrical apparatus and devices [180]), and their transmission and collection, in addition to the conducted analyses, may affect the medium-term planning of outages of individual elements of the transmission network and, thus, indirectly, short-term forecasting of power demand.

3. The State of Research in Short-Term Power Demand Forecasting for Power Systems Using Autoregressive and Non-Autoregressive Methods and Models

The study (Figure 1) was planned in such a way as to answer the question of whether the use of autoregressive methods in short-term forecasting of electricity demand in power systems can be even more effective and, at the same time, inexpensive and quick to implement. In order to answer this question, scientific articles presenting the effectiveness of autoregressive forecasting models determined by the MAPE were analyzed. The result of the review is Table 1 and a ranking of forecasting models (Tables 2 and 3), and the Top 10 collection of the ten most effective forecasting models. As a result of the review and development of the ranking of forecasting models, it was confirmed that the use of autoregressive models may support the transmission system operator to achieve better forecasting efficiency.

The literature review (Table 1) included 47 unique items and titles, several dozen forecasting methods, and 264 forecasting models (Table 1). Scientific papers were published in the period from 1997 and concerned short-term forecasting of power demand. The source data used by the authors of the analyzed publications, constituting the input for the forecasting models, covered the period from 1998 to 2014. Diverse and international teams of authors conducted their research based on data on the functioning of power systems in 25 countries located on four continents—in the countries of the Near and the Far East, Western Europe (including the British Isles), Central Europe (including Poland), North America (USA), and Australia. The publications indicated were compiled by 44 different authors' teams and published in 23 publishing houses. The analysis concerning the nomenclature of forecasting models covers a set of 185 unique items. Diversifying the observed relationships in individual forecasting models results in identifying 197 unique abbreviations assigned to forecasting models. The MAPE(ea) in Table 1 means that the accuracy results are measured in ex ante mode.

All the reviewed references describe the effectiveness of the presented forecasting models, in terms of the MAPE measure, to assess the accuracy of the forecasts. To analyze the collected forecasting results, 27 unique names of MAPE errors were distinguished for this analysis, reflecting the forecasting models used in the analysis. Some of the forecast results described by the MAPE index, contained in selected publications, are presented from the lowest value (MAPE min) to the highest value (MAPE max). In contrast, the remaining part of the results is described by one value.

The analysis of monovalent results was decomposed into minimum and maximum values to standardize the dominant approach used in selected publications. The lowest values of MAPE min are recorded in the range from 0.01% to 21.18%, while in the MAPE max category, the corresponding range of variability in the MAPE ranges from 0.01% to 33.45%. The MAPE min category includes 196 unique items from a set of 264 models, while the MAPE max category includes 212 unique items from the same set.

Further analysis of the results of the effectiveness of the forecasts obtained, described by the forecasting quality measure using the MAPE, concerns the MAPE category, min. A set of the ten smallest results expressed as percentages was selected in this category (Figure 2). This collection was called Top 10. The smallest values of MAPE errors min, in ascending order, in the Top 10 set (Figure 2) are obtained for the following models: Data Envelopment Analysis (DEA), Fuzzy Regression (FR), General Regression Model (GRM), Genetic Algorithm (GA), Adaptive Neuro Fuzzy Inference System (ANFIS [181,182]), Artificial Neural Network (ANN), Full General Regression Model (FGRM), Wavelet Artificial Neural Network (WANN), Artificial Neural Network (ANN), and Fuzzy Logic (FL). The values of MAPE min were: 0.01%; 0.08%; 0.10%; 0.14%; 0.15%; 0.16%, 0.20%; 0.27%; 0.28%; and 0.29%. The summary of the abbreviations used for the forecasting methods and models in the Top 10 set is as follows: DEA; FR; GRM; GA; ANFIS; ANN; FGRM; WANN; ANN; and FL.

Only analytical studies on the GRM forecasting model in the Top 10 set are performed ex ante (ea). In the case of this model, the efficiency obtained in the third position should be considered very high. The GRM model uses information about the shaping of the ambient temperature as an input variable. The second model that uses the input variables is the FGRM model, which considers both the variability in the ambient temperature and the wind speed. The FGRM model ranks seventh in the Top 10 ranking in the MAPE category, min.

The forecasting effectiveness described by the lowest value of the MAPE min has an ambiguous effect on high forecasting efficiency. The power systems subject to forecast analysis in the Top 10 list are (in ascending order) the systems of Iran (two items), USA (one item), Iran (three items), USA (one item), and Australia (three items).

The length of the analyzed period significantly affects the quality of forecasting obtained. Along with the extension of the analysis period, including the natural impact of non-working days and holidays, both cyclical and non-cyclical, there is a decline in the effectiveness of the obtained forecasts of the load on power systems. The full forecasting model ranking is presented in Tables 2 and 3, where the column Model No. represents the model number from Table 1 (the last column on the right), and the column Ranking shows the position in the model ranking (1 equals the first position and 264 equals the last position). Table 2 consists of the models from Table 1 from 1 to 132 (in four pairs of Ranking and Model Number), and Table 3 shows the same scheme for the models from 133 to 264. Tables 2 and 3 present four sets of Ranking and Model Number. Articles [183–185] from 2019 to 2021 indicate that analysis and research are being continued, including with the use of some of the analyzed methods.

4. Conclusions

The 47 publications describing 264 models published from 1997 to 2018 were analyzed in detail by applying methods that use explanatory variables to broaden the background of analyses. Some relevant publications from 2019 to 2021 were also included to determine if autoregressive methods are still of interest. The results of the review confirm the significant potential of the autoregressive approach to power demand forecasting. The analyzed methods enable very high accuracy to be achieved in short-term forecasting with the resolution of one hour (accuracy measured in terms of MAPE is below 1%). The methods whose effectiveness were classified in the top ten sets are Fuzzy Logic (LR), Artificial Neural Network (ANN), Wavelet Artificial Neural Network (WANN), Full General Regression Model (FGRM), Artificial Neural Network (ANN), Adaptive Neurofuse Inference System (ANFIS), Genetic Algorithm (GA), General Regression Model (GRM), Fuzzy Regression (FR), and Data Envelope Analysis (DEA). These methods allowed them to achieve MAPE-determined values of: 0.29%; 0.28%; 0.27%; 0.20%; 0.16%; 0.15%; 0.14%; 0.10%; 0.08%; and 0.01%. All of the Top 10 models achieved high accuracy; however, the DEA model reached the accuracy of 0.01% MAPE. Models No. 257 (FGRM) and No. 256 (GRM) of the Top 10 set use the explanatory variables, and the other eight models were autoregressive (models No.: 215—FL, 214—ANN, 213—WANN, 140—ANN, 141—ANFIS, 138—GA, 139—FR, and 142—DEA). This shows the potential of the autoregressive prediction approach used in the models for short-term power demand forecasting in power systems.

5. Critical Discussion, Major Findings and Future Scope of Research

The results of the review show that the use of short-term forecasting of electric power demand with hourly resolution enables efficiency of below 1% to be achieved. It should be borne in mind that such effectiveness should apply to the entire calendar year. In the analyzed collection of 47 articles from all over the world, the analysis period ranges from several months to several years, which indicates that the research covers significant periods of time, and the analyzed models are stable and resistant to changes in external conditions (economic and climatic conditions). The group of the most effective prognostic models includes models using artificial intelligence techniques (e.g., Artificial Neural Networks, Fuzzy Logic, and Genetic Algorithms). The effective methods also include classic forecasting methods (e.g., ARIMA, Multiple Regression, Exponential Smoothing) and methods from the Data Mining group (e.g., Support Vector Machines, Nearest Neighbors, Random Forest).

The article confirms the authors' thesis about the enormous potential inherent in the use of the autoregressive approach for short-term forecasting of electricity demand. The results of the review (the prepared ranking of prognostic models and the knowledge from the analyzed articles) constitute an excellent starting point for further tests and pave the way for future research in this area.

The future research of the authors will focus on the first step of testing the prognostic models from the Top 10 set. The tests will take into account both the achieved effectiveness and the necessary financial costs and time consumption of the process. In the next step, the most effective prognostic methods selected in the first step will be tested, including individual testing in off-line mode. In the third step of further research, prognostic model committees will be established. The developed committees will assign weights to the participation of individual models (step 1) and test the suitability of individual models for forecasting individual hours of the day or periods of the day (step 2). The MAPE selected by the authors for the review analysis, despite the undoubted advantage of being able to be used to easily compare the effectiveness between forecasting models, has a tendency to average forecasts. Therefore, in future studies, the authors will also use other measures to assess the quality of forecasts, such as Mean Absolute Error, Mean Absolute Scaled Error, and Root Mean Square Error, and others as needed. The usefulness of the tested forecasting models will be assessed, taking into account the seasonality, periodicity, and ranges of hours during the day. The developed review encompasses an excellent range of forecasting methods and models that can be used at any time, and the usefulness of each of them may prove invaluable from the point of view of the needs of the Polish Transmission System Operator.

Author Contributions: Conceptualization, R.C., J.K. and M.S.; methodology, R.C.; data curation and data analysis, R.C.; supervision, J.K.; resources, M.S.; writing—original draft preparation, R.C., J.K. and M.S.; writing—review and editing, J.K. and M.S.; project administration, J.K.; funding acquisition, J.K. All authors have read and agreed to the published version of the manuscript.

Funding: This research received no external funding.

Institutional Review Board Statement: Not applicable.

Informed Consent Statement: Not applicable.

Data Availability Statement: The data are available within this document.

Conflicts of Interest: The authors declare no conflict of interest. The funders had no role in the design of the study; in the collection, analyses, or interpretation of data; in the writing of the manuscript, or in the decision to publish the results.

Abbreviations

The following abbreviations are used in this manuscript:

AG	Genetic Algorithm (GA)
ANFIS	Adaptive Neuro Fuzzy Inference System (ANFIS)
ANN	Artificial Neural Network (ANN)
ARIMA	Autoregressive Integrated Moving Average
C&RT	Classification And Regression Trees
CHAID	Chi-Square Automatic Interaction Detection
DEA	Data Envelopment Analysis
ea	ex ante
GIS	Geographic Information System
GPS	Global Positioning System
IED	Intelligent Electronic Device
FL	Fuzzy Logic
MAPE	Mean Average Percentage Error
MARSplines	Multivariate Adaptive Regression Splines
MLP	Multilayer Perceptron
GRM	General Regression Model
FGRM	Full General Regression Model
NPS	National Power System
PSE S.A.	Polskie Sieci Elektroenergetyczne S.A. (The Transmission System Operator in Poland)

RBF	Radial Basis Function
FR	Fuzzy Regression
SARIMAX	Seasonal Auto-Regressive Integrated Moving Average with eXogenous Factors
SCADA	Supervisory Control and Data Acquisition
WANN	Wavelet Artificial Neural Network (AWNN)
SVM	Support Vector Machines
WAMS	Wide Area Management System

References

1. Aurangzeb, K. Short Term Power Load Forecasting Using Machine Learning Models for Energy Management in a Smart Community. In Proceedings of the 2019 International Conference on Computer and Information Sciences (ICCIS), Sakaka, Saudi Arabia, 3–4 April 2019.
2. Conway, D.; Myles, J.; White, J.M. *Uczenie Maszynowe Dla Programistów*; Wydawnictwo Helion/O'Reilly: Gliwice, Poland, 2015.
3. Grus, J. *Data Science od Podstaw—Analiza Danych w Pythonie*; Wydawnictwo Helion/O'Reilly: Gliwice, Poland, 2018.
4. Kasperski, M.J. *Sztuczna Inteligencja—Droga Do Myślących Maszyn*; Wydawnictwo Helion: Gliwice, Poland, 2003.
5. Li, J.; Song, Z.; Zhong, Y.; Zhang, Z. Short-Term Load Forecasting Based on Improved Extreme Learning Machine. In Proceedings of the 2017 IEEE 2nd International Conference on Big Data Analysis, Beijing, China, 10–12 March 2017.
6. Theodoridis, S. *Probability and Stochastic Processes. Machine Learning: A Bayesian and Optimization Perspective. Chapter 1*; Academic Press: London, UK, 2015; pp. 9–51.
7. Tadeusiewicz, R. *Podstawy Biocybernetyki*; PWN: Warszawa, Poland, 1991.
8. Bian, H.; SPSS Discriminant Function Analysis. Office for Faculty Excellence. 2012. Available online: http://kharazmi-statistics.ir/Uploads/Public/MY%20article/SPSS%20Discriminant%20Function%20Analysis.pdf (accessed on 15 March 2016).
9. Cottrell, A.; Lucchetti, R.; Gretl User's Guide—Gnu Regression, Econometrics and Time—Series Library. (Gretl-Guide). 2015. Available online: http://gretl.sourceforge.net/gretl-help/gretl-guide.pdf (accessed on 15 March 2016).
10. Kufel, T. *Rozwiązywanie Problemów z Wykorzystaniem Programu GRETL*; Wydawnictwo Naukowe PWN: Warszawa, Poland, 2015.
11. Kamiński, J.; Kaszyński, P. Wybrane problemy implementacji zapotrzebowania na moc w matematycznych modelach systemów energetycznych. *Polityka Energetyczna* **2011**, *14*, 155–166.
12. Paska, J. *Ekonomika w Elektroenergetyce*; Wydawnictwo OWPW: Warszawa, Poland, 2007.
13. Dwijayanti, S.; Hagan, M. Short Term Load Forecasting Using a Neural Network Based Time Series Approach. In Proceedings of the 1st International Conference on Artificial Intelligence, Modelling and Simulation, Kota Kinabalu, Malaysia, 3–5 December 2013.
14. Dobrzańska, I. (Ed.) Zmienność obciążeń systemów energetycznych ich analiza i metoda prognozy na przykładzie zespołu systemów polskich. In *Analiza i Prognoza Obciążeń Elektroenergetycznych*; Wydawnictwa Naukowo-Techniczne: Warszawa, Poland, 1971.
15. Dobrzańska, I. (Ed.) Analiza zmienności obciążeń polskiego systemu energetycznego. In *Analiza i Prognoza Obciążeń Elektroenergetycznych*; Wydawnictwa Naukowo—Techniczne: Warszawa, Poland, 1971; Available online: http://elektroenergetyka.pl/upload/file/2005/3/elektroenergetyka_nr_05_03_2.pdf (accessed on 16 March 2016).
16. Dobrzańska, I. (Ed.) Wpływ czasu letniego na wykres obciążenia dobowego systemów energetycznych w Polsce. In *Analiza i Prognoza Obciążeń Elektroenergetycznych*; Wydawnictwa Naukowo—Techniczne: Warszawa, Poland, 1971.
17. Dyka, E.; Mróz-Radłowsk. Ekonomia w Energetyce. *Wydawnictwo Politechniki Łódzkiej, Łódź.* 2016. Available online: https://cybra.lodz.pl/Content/12517/EDyka_EkonomikaEnergetyka_2014_1.pdf (accessed on 28 December 2016).
18. Al-Fuhaid, A.S.; El-Sayed, M.A.; Mahmoud, M.S. Neuro-Short-Term Load Forecast of the Power System in Kuwait. *Appl. Math. Model.* **1997**, *21*, 215–219. [CrossRef]
19. Almeshaiei, E.; Soltan, H. A Methodology for Electric Power Load Forecasting. *Alex. Eng. J.* **2011**, *50*, 137–144. [CrossRef]
20. Al-Shobaki, S.; Mohsen, M. Modeling and Forecasting of Electrical Power Demands for Capacity Planning. *Energy Convers. Manag.* **2008**, *49*, 3367–3375. [CrossRef]
21. Badran, S.M.; Abouelatta, O.B. Forecasting Electrical Load Using ANN Combined with Multiple Regression Method. *Res. Bull. Jordan ACM* **2012**, *2*.
22. Brodowski, S.; Bielecki, A.; Filocha, M. A Hybrid System for Forecasting 24-h Power Load Profile for Polish Electric Grid. *Appl. Soft Comput.* **2017**, *58*, 527–539. [CrossRef]
23. Buitrago, J.; Asfour, S. Short-Term Forecasting of Electric Loads Using Nonlinear Autoregressive Artificial Neural Networks with Exogenous Vector Inputs. *Energies* **2017**, *10*, 40. [CrossRef]
24. Ceperic, E.; Ceperic, V.; Baric, A. A Strategy for Short-Term Load Forecasting by Support Vector Regression Machines. *IEEE Trans. Power Syst.* **2013**, *28*, 4356–4364. [CrossRef]
25. Chahkoutahi, F.; Khashei, M. A Seasonal Direct Optimal Hybrid Model of Computational Intelligence and Soft Computing Techniques for Electricity Load Forecasting. *Energy* **2017**, *140*, 988–1004. [CrossRef]
26. Chapagain, K.; Kittipiyakul, S. Short-Term Electricity Load Forecasting Model and Bayesian Estimation for Thailand Data. *MATEC Web Conf.* **2016**, *55*, 6003. [CrossRef]

27. Chen, H.; Canizares, C.A.; Singh, A. *ANN-Based Short-Term Load Forecasting in Electricity Markets*; University of Waterloo Department of Electrical & Computer Engineering: Waterloo, ON, Canada, 1999.
28. Chheepa, T.K.; Manglani, T. A Critical Review on Employed Techniques for Short Term Load Forecasting. *Int. Res. J. Eng. Technol. (IRJET)* **2017**, *4*, 2453–2460.
29. Clements, A.E.; Hurn, A.S.; Li, Z. *Forecasting Day-Ahead Electricity Load Using a Multiple Equation Time Series Approach*; NCER Working Paper Series; Working Paper #103; School of Economics and Finance; Queensland University of Technology: Brisbane, Australia, 2015.
30. Czapaj, R. Typowanie zmiennych objaśniających przy wykorzystaniu zautomatyzowanych metod statystycznych jako sposób optymalizacji wyboru metody estymacji szczytowego dobowego obciążenia KSE. *Przegląd Elektrotech.* **2018**, *93*, 58–61.
31. Czapaj, R.; Kamiński, J.; Benalcazar, P. Dobór zmiennych objaśniających z wykorzystaniem metody MARSplines. In Proceedings of the XIV Międzynarodowa Konferencja Naukowa Prognozowanie w Elektroenergetyce PE'18, Podlesice, Poland, 26–28 September 2018.
32. Czapaj, R.; Kamiński, J.; Benalcazar, P. Prognozowanie krótkoterminowe zapotrzebowania na energię elektryczną w KSE z wykorzystaniem metody MARSplines. In Proceedings of the XIV Międzynarodowa Konferencja Naukowa Prognozowanie w Elektroenergetyce PE'18, Podlesice, Poland, 26–28 September 2018.
33. Dąsal, K. Dobór zmiennych wejściowych do Modelu Rozkładu Kanonicznego. In Proceedings of the VI Konferencja Naukowa Prognozowanie w Elektroenergetyce, Prognozowanie, Efektywność Energetyczna, Częstochowa, Poland, 18–20 September 2002.
34. Dudek, G. *Short-Term Load Forecasting Based on Kernel Conditional Density Estimation*; Przegląd Elektrotechniczny (Electrical Review); Czestochowa University of Technology: Częstochowa, Poland, 2010.
35. Dudek, G.; Janicki, M. *Nearest Neighbour Model with Weather Inputs for Pattern-Based Electricity Demand Forecasting*; Department of Electrical Engineering, Czestochowa University of Technology: Częstochowa, Poland, 2017.
36. Dudek, G. *Pattern-Based Local Linear Regression Models for Short-Term Load Forecasting*; Department of Electrical Engineering, Czestochowa University of Technology: Częstochowa, Poland, 2016; Volume 130, pp. 139–147.
37. Dudek, G. Drzewa regresyjne i lasy losowe jako narzędzia predykcji szeregów czasowych z wahaniami sezonowymi. In Proceedings of the VI spotkanie Polskiej Grupy Badawczej Systemów Uczących Się, Częstochowa, Poland, 14 April 2016.
38. Esener, I.I.; Yuksel, T.; Kurban, M. Short-Term Load Forecasting without Meteorological Data Using AI-based Structures. *Turk. J. Electr. Eng. Comput. Sci.* **2015**, *23*, 370–380. [CrossRef]
39. Fan, S.; Hyndman, R.J. Short-Term Load Forecasting Based on a Semi-Parametric Additive Model. *IEEE Trans. Power Syst.* **2012**, *27*, 134–141. [CrossRef]
40. Farahat, M.A.; Elgawed, A.F.A.; Ibrahim, A.; Mustafa, H.M.M. Short Term Load Forecasting Using Neural Networks and Particle Swarm Optimization. *J. Electr. Eng.* **2014**, 1–10.
41. Gorwar, M.; Harsha, J.S. Short Term Load Forecasting using Time Series Analysis: A case study for Karnataka, India. *Int. J. Eng. Sci. Innov. Technol. (IJESIT)* **2012**, *1*.
42. Hassan, S.; Khosravi, A.; Jaafar, J. Examining performance of aggregation algorithms for neural network-based electricity demand forecasting. *Electr. Power Energy Syst.* **2015**, *64*, 1098–1105. [CrossRef]
43. He, W. Deep Neural Network Based Load Forecast. *Comput. Model. New Technol.* **2014**, *18*, 258–262.
44. Hong, T.; Wang, P. Fuzzy Interaction Regression for Short Term Load Forecasting. *Fuzzy Optim. Decis. Mak.* **2014**, *13*, 91–103. [CrossRef]
45. Janicki, M. *Temperature Correction Method for Pattern Similarity-Based Short-Term Electricity Demand Forecasting Models*; Faculty of Electrical Engineering, Czestochowa University of Technology: Czestochowa, Poland, 2017.
46. Kheirkhah, A.; Azadeh, A.; Saberi, M.; Azaron, A.; Shakouri, H. Improved Estimation of Electricity Demand Function by Using of Artificial Neural Network, Principal Component Analysis and Data Envelopment Analysis. *Comput. Ind. Eng.* **2013**, *64*, 425–441. [CrossRef]
47. Kolcun, M.; Holka, L. Daily Load Diagram Prediction of Eastern Slovakia. In Proceedings of the VI Konferencja Naukowa Prognozowanie w Elektroenergetyce, Prognozowanie, Efektywność Energetyczna, Częstochowa, Poland, 18–20 September 2002.
48. Lin, Y.; Luo, H.; Wang, D.; Guo, H.; Zhu, K. An Ensemble Model Based on Machine Learning Methods and Data Preprocessing for Short-Term Electric Load Forecasting. *Energies* **2017**, *10*, 1186. [CrossRef]
49. Liu, N.; Babushkin, V.; Afshari, A. Short-Term Forecasting of Temperature Driven Electricity Load Using Time Series and Neural Network Model. *J. Clean Energy Technol.* **2014**, *2*, 327–331. [CrossRef]
50. Magnano, L.; Boland, J.W. Generation of Synthetic Sequences of Electricity Demand: Application in South Australia. *Energy* **2007**, *32*, 2230–2243. [CrossRef]
51. Nadtoka, I.I.; Al-Zihery Balasim, M. Forecasting of Electricity Consumption of the Power System, with Due Account of Air Temperature and Natural Illumination, Based on Support Vector Machine and Particle Swarm *Procedia Eng.* **2015**, *129*, 657–663. [CrossRef]
52. Narayan, A.; Hipel, K.W. Long Short Term Memory Networks for Short-Term Electric Load Forecasting. In Proceedings of the IEEE International Conference on Systems, Man, and Cybernetics (SMC), Banff, AB, Canada, 5–8 October 2017.
53. Nowicka-Zagrajek, J.; Weron, R. Modeling Electricity Loads in California: ARMA Models with Hyperbolic Noise. *Signal Processing* **2002**, *82*, 1903–1915. [CrossRef]

54. Nowotarski, J.; Liu, B.; Weron, R.; Hong, T. *Improving Short Term Load Forecast Accuracy via Combining Sister Forecasts*; Hugo Steinhaus Center Wrocław University of Technology: Wrocław, Poland, 2015.
55. Hsiao-Tien, P. Forecast of Electricity Consumption and Economic Growth in Taiwan by State Space Modeling. *Energy* **2009**, *34*, 1779–1791.
56. Rana, M.; Koprinska, I. Forecasting Electricity Load with Advanced Wavelet Neural Networks. *Neurocomputing* **2016**, *182*, 118–132. [CrossRef]
57. Siwek, K.; Osowski, S. Prognozowanie obciążeń 24—godzinnych w systemie elektroenergetycznym z użyciem zespołu sieci neuronowych. *Przegląd Elektrotech.* **2009**, *85*, 540–544.
58. Sevlian, R.A.; Rajagopal, R. *A Model For The Effect of Aggregation on Short Term Load Forecasting*; Department of Electrical Engineering, Stanford University: Stanford, CA, USA, 2014.
59. Sousa, J.C.; Neves, L.P.; Jorge, H.M. Assessing the relevance of load profiling information in electrical load forecasting based on neural network Models. *Electr. Power Energy Syst.* **2012**, *40*, 85–93. [CrossRef]
60. Wang, P.; Liu, B.; Hong, T. Electric Load Forecasting with Recency Effect: A Big Data Approach. *Int. J. Forecast.* **2016**, *32*, 585–597. [CrossRef]
61. Wang, Y.; Bielicki, J.M. Acclimation and the Response of Hourly Electricity Loads to Meteorological Variables. *Energy* **2017**, *142*, 473–485. [CrossRef]
62. Wyrozumski, T. Prognozowanie neuronowe w energetyce. In Proceedings of the Konferencja Naukowa Rynek Energii Elektrycznej-Materiały Konferencyjne, Kazimierz Dolny, Poland, 25–27 April 2005.
63. Yang, J. Power System Short-Term Load Forecasting. Ph.D. Thesis, TU Darmstadt, Darmstadt, Germany, 2006.
64. Yu, X.; Ji, H. A PSO–SVM-based 24 Hours Power Load Forecasting Model. *MATEC Web Conf.* **2015**, *25*, 01008. [CrossRef]
65. Bartkiewicz, W. Metody określania niepewności prognoz krótkoterminowego obciążenia sieci dla modeli neuronowych i neuronowo—rozmytych. *Rynek Energii* **2011**, *1*, 41–46.
66. Bartkiewicz, W. Modelowanie Krótkoterminowego Popytu na Energie Elektryczną z Wykorzystaniem Sieci Neuronowych i Neuronowo—Rozmytych. Katedra Informatyki, Wydział Zarządzania, Uniwersytet Łódzki. Wydawnictwo Uniwersytetu Łódzkiego. 2013. Available online: www.wydawnictwo.uni.lodz.pl (accessed on 15 March 2016).
67. Bartkiewicz, W.; Matusiak, B. Sieci Neuronowe i Algorytmy Genetyczne a Krótkookresowe Prognozowanie Zużycia na Rynku Energii. w: J.S. Zieliński—50 Lat Pracy Naukowej. Wydawnictwo Biblioteka, 2004. Str. 345–353. Available online: https://docplayer.pl/10286641-Prognozowanie-w-elektroenergetyce-w-roznych-horyzontach-czasowych.html (accessed on 21 September 2017).
68. Filik, U.B.; Kurban, M.A. New Approach for the Short-Term Load Forecasting with Autoregressive and Artificial Neural Network Models. *Int. J. Comput. Intell. Res.* **2007**, *3*, 66–71.
69. Ishida, T.; Tamura, S. Daily Electric Load Forecasting Using Artificial Neural Network. *Electr. Eng. Jpn.* **1994**, *114*, 1109–1115. [CrossRef]
70. Jetcheva, J.G.; Majidpour, M.; Chen, W.P. Neural Network Model Ensembles for Building—Level Electricity Load Forecasts. *Energy Build.* **2014**, *84*, 214–223. [CrossRef]
71. Makino, K.; Shimada, T. Daily Peak Electric Load Forecasting Using an Artificial Neural Network and an Improvement Method for Reducing the Forecasting Errors. *Electr. Eng. Jpn.* **1995**, *114*, 27–42. [CrossRef]
72. Malko, J. Sztuczne sieci neuronowe—Nowe narzędzie prognozowania elektroenergetycznego. In Proceedings of the Materiały Konferencyjne Aktualne Problemy w Elektroenergetyce APE'97, Gdańsk, Poland, 11–13 June 1997.
73. Masters, T. *Sieci Neuronowe w Praktyce (Programowanie w Języku C++)*; WNT: Warszawa, Poland, 1996.
74. Osowski, S. *Sieci Neuronowe*; Oficyna Wydawnicza Politechniki Warszawskiej: Warszawa, Poland, 1996.
75. Piotrowski, P. Prognozowanie krótkoterminowe godzinowych obciążeń w spółce dystrybucyjnej z wykorzystaniem sieci neuronowych—Analiza wpływu doboru i przetworzenia danych na jakość prognoz. *Przegląd Elektrotech.* **2007**, *83*, 40–43.
76. Piotrowski, P. Analiza skuteczności wybranych algorytmów uczących sieci neuronowych typu MLP w zadaniu prognozowania krótkoterminowego zapotrzebowania na energię elektryczną na potrzeby spółek dystrybucyjnych. *Śląskie Wiadomości Elektryczne* **2011**, *1*, 18–24. [CrossRef]
77. Piotrowski, P. Prognozowanie w Elektroenergetyce w Różnych Horyzontach Czasowych. Rozprawa Habilitacyjna. Politechnika Warszawska, Instytut Elektroenergetyki, Prace Naukowe Elektryka z. 144. Warszawa 2013. Oficyna Wydawnicza Politechniki Warszawskiej. 2013. Available online: http://www.nazarko.pl/public/data/resource/upload/00003/2568/file/prognozowanie_w_elektroenergetyce_w_roznych_horyzontach_czasowych.pdf (accessed on 28 June 2017).
78. Piotrowski, P.; Górski, M. Ekonomiczne aspekty krótkoterminowego prognozowania zużycia energii elektrycznej wybranych grup odbiorców przy wykorzystaniu techniki sieci neuronowych. In Proceedings of the VIII Międzynarodowa Konferencja Naukowa pt. Aktualne problemy w Elektroenergetyce (APE-99), Jurata, Poland, 9–11 June 1999.
79. Kvhashei, M.; Bijari, M.; Ardali, G.A.R. Improvement of Auto-Regressive Integrated Moving Average Models using Fuzzy Logic and Artificial Neural Networks (ANNs). *Neurocomputing* **2009**, *72*, 956–967. [CrossRef]
80. Lang, J.; Pascoe, S.; Thomson, J.; Woyak, J.; Rahimi, K.; Broadwater, R. Smart Grid Big Data: Automating Analysis of Distribution Systems. In Proceedings of the 2016 IEEE Rural Electric Power Conference (REPC), Westminster, CO, USA, 15–18 May 2016.

81. Piotrowski, P. Krótkoterminowe prognozy zapotrzebowania na energię elektryczną dla spółek dystrybucyjnych w warunkach rynku energii przy wykorzystaniu technik sztucznej inteligencji. In Proceedings of the X Międzynarodowa Konferencja Naukowa pt. Aktualne problemy w Elektroenergetyce (APE-01), Jurata, Poland, 6–8 June 2001; pp. 117–124.
82. Tadeusiewicz, R. *Elementarne Wprowadzenie do Techniki Sieci Neuronowych z Przykładowymi Programami*; Akademicka Oficyna Wydawnicza PLJ: Warszawa, Poland, 1998.
83. Witkowska, D. *Sztuczne Sieci Neuronowe i Metody Statystyczne. Wybrane Zagadnienia Finansowe*; C.H. Beck: Warszawa, Poland, 2002.
84. Helt, P.; Parol, M.; Piotrowski, P. *Metody Sztucznej Inteligencji w Elektroenergetyce*; Oficyna Wydawnicza Politechniki Warszawskiej: Warszawa, Poland, 2000.
85. Bazzan, A.L.C.; Labidi, S. (Eds.) *Advances in Artificial Intelligence. 17th Brazilian Symposium on Artificial Intelligence, São Luis, Maranhão, Brazil, September 29–October 1, 2004, Proceedings*; Springer: Berlin/Heidelberg, Germany; New York, NY, USA, 2004; pp. 317–325.
86. Akademia Podlaska. *Sztuczna inteligencja—Materiały V Konferencji Naukowej*; Instytut Informatyki Akademii Podlaskiej, Instytut Podstaw Informatyki Polskiej Akademii Nauk i Wydział Cybernetyki Wojskowej Akademii Technicznej, Wydawnictwo Akademii Podlaskiej: Siedlce, Poland, 2002.
87. Xu, Z.; Dong, Z.Y.; Liu, W.Q. Neural Network Models For Electricity Market Forecasting. *Neural Netw. Appl. Inf. Technol. Web Eng.* **2005**, *1*, 233–245.
88. Czapaj, R. O możliwości krótkoterminowego prognozowania cen energii elektrycznej na polskich parkietach obrotu z uwzględnieniem indeksu niemieckiej giełdy EEX AG. *Przegląd Elektrotech.* **2009**, *85*, 140–143.
89. Dudek, G. Krótkoterminowe prognozowanie zapotrzebowania na moc elektryczną metodą klasteryzacji rozmytej. *Przegląd Elektrotech.* **2006**, *8*, 26–28.
90. Dudek, G. Neuro—Fuzzy approach to the next Day Load Curve Forecasting. *Przegląd Elektrotech.* **2011**, *87*, 61–64.
91. Rutkowska, D. *Neuro—Fuzzy Architectures and Hybrid Learning. Studies in Fuzziness and Soft Computing*; Springer: Berlin/Heidlberg, Germany, 2002.
92. Rutkowski, L. *Flexible Neuro-Fuzzy Systems—Structures, Learning and Performance Evaluation*; The Kluwer International Series in Engineering and Computer Science; Kluwer Academic Publishers: New York, NY, USA; Boston, MA, USA; Dordrecht, The Netherlands; London, UK; Moscow, Russia, 2004.
93. Wilczyński, A. Referat generalny do referatów podsekcji III.1. prognozowanie, planowanie i problemy rozwoju. In Proceedings of the Politechnika Wrocławska. Konferencja Naukowa Aktualne Problemy w Elektroenergetyce, Jurata, Poland, 8–10 June 2011.
94. Czapaj, R.; Gwóźdź, R.; Przygrodzki, M. Analiza techniczna i fundamentalna a dokładność prognozy cen energii elektrycznej na rynku giełdowym. *Przegląd Elektrotechn.* **2006**, *82*, 32–34.
95. Czapaj, R. Krótkoterminowe Prognozowanie Cen Energii Elektrycznej na Wybranych Polskich Parkietach Obrotu Energią Elektryczną z Uwzględnieniem Walorów Wybranych Europejskich Parkietów Obrotu Energią Elektryczną przy Wykorzystaniu Sieci Neuronowych. Praca Dyplomowa—Studia Podyplomowe, Zaoczne, 1—Roczne. Uniwersytet Ekonomiczny w Krakowie i StatSoft Polska—Czerwiec 2009. 2009. Available online: http://yadda.icm.edu.pl/yadda/element/bwmeta1.element.baztech-article-BPOG-0058-0032 (accessed on 16 March 2015).
96. Łapczyński, M.; Demski, T. *Data Mining—Metody Predykcyjne. Materiały Kursowe StatSoft Polska*; Wydawnictwo StatSoft Polska Sp. z o.o.: Kraków, Poland, 2018.
97. Hayashi, Y.; Iwamoto, S. Long Term Load Forecasting using Improved Recurrent Neural Network. *IEEJ Trans. Power Energy* **1993**, *113*, 1203–1212. [CrossRef]
98. Qing, L.; Boyu, Z.; Li, Q. Impact of Big Data on Electric-power Industry. In Proceedings of the 2017 IEEE 2nd International Conference on Big Data Analysis, Beijing, China, 10–12 March 2017.
99. Helt, P.; Parol, M.; Piotrowski, P. *Metody Sztucznej Inteligencji—Przykłady Zastosowań w Elektroenergetyce*; Oficyna Wydawnicza Politechniki Warszawskiej: Warszawa, Poland, 2012.
100. Poradnik Inżyniera Elektryka Tom 3. Wydanie Drugie Częściowo Zmienione. Warszawa, 1997. Wydawnictwa Naukowo-Techniczne. 1997. Available online: https://www.ksiegarniatechniczna.com.pl/poradnik-inzyniera-elektryka-komplet-3-tomow.html (accessed on 13 November 2018).
101. Winkler, W.; Wiszniewski, A. *Automatyka Zabezpieczeniowa w Systemach Elektroenergetycznych*; Wydawnictwo Naukowo—Techniczne: Warszawa, Poland, 2004.
102. Vouros, G.A.; Panayiotopoulos, T. (Eds.) Methods and Applications of Artificial Intelligence. In *Third Hellenic Conference on AI, SETN 2004 Samos, Greece, May 5–8, 2004, Proceedings*; Springer: Berlin/Heidelberg, Germany, 2004.
103. Zieliński, J.S. Artificial Intelligence Contemporary Applications in Power Systems. In Proceedings of the IV Sympozjum Metody Matematyczne w Elektroenergetyce, Zakopane, Poland, 22–24 October 1998.
104. Żołneczko, M.; Wilk, A. *Data Mining—Metody bez Nauczyciela*; Materiały kursowe StatSoft Polska; Wydawnictwo StatSoft Polska Sp. z o.o.: Kraków, Poland, 2018.
105. Cabena, P.; Hadjinian, P.; Stadler, R.; Verhees, J.; Zanasi, A. *Discovering Data Mining. From Concept to Implementation*; Prentice-Hall: Hoboken, NJ, USA, 1998.
106. StatSoft Polska. *Zastosowania Statystyki i Data Mining w Badaniach Naukowych i Innowacyjnych*; Wydawnictwo StatSoft Polska Sp. Z.o.o.: Kraków, Poland, 2014.

107. Gwóźdź, R.; Czapaj, R. *Wpływ Rozwoju PV na KSE*; niepubl.—na zlecenie PSE S.A.; PSE Innowacje Sp. z o.o.: Warszawa, Poland, 2020.
108. Kandel, A.; Last, M.; Bunke, H. *Data Mining and Computational Intelligence. Studies in Fuzziness and Soft Computing*; Springer: Berlin/Heidlberg, Germany, 2001.
109. Lula, P. *Sztuczne Sieci Neuronowe Jako Narzędzie Analiz Typu Data Mining*; Wydawnictwo StatSoft Polska Sp. z o.o.: Kraków, Poland, 2002.
110. Wątroba, J. *Przykład Rozwiązania Zagadnienia Predykcyjnego za Pomocą Technik Data Mining*; Wydawnictwo StatSoft Polska Sp. z o.o.: Kraków, Poland, 2002.
111. Weiss, S.M.; Indurkhya, N. *Predictive Data Mining—A Practical Guide*; Morgan Kaufmann Publishers: Burlington, MA, USA, 1998.
112. Demski, T. *Przykład Prognozowania z Wykorzystaniem Metod Data Mining (Materiał: STATISTICA 8 Bliżej CIEBIE)*; Wydawnictwo StatSoft Polska: Kraków, Poland, 2006.
113. StatSoft Polska. *Zastosowania Statystyki i Data Mining w Badaniach Naukowych*; Wydawnictwo StatSoft Polska Sp. Z.o.o.: Kraków, Poland, 2017.
114. Sokołowski, A. Data Mining—Automat czy Metoda Naukowa? StatSoft Polska, Akademia Ekonomiczna w Krakowie, Katedra Statystyki, Kraków. 2005. Available online: https://media.statsoft.pl/_old_dnn/downloads/data_mining_automat_czy_metoda.pdf (accessed on 30 July 2012).
115. Ratner, B. *Statistical Modeling and Analysis for Database Marketing: Effective Techniques for Mining Big Data*; Wydawnictwo CRC Press: Boca Raton, FL, USA, 2003.
116. StatSoft Polska. *Zastosowania Statystyki i Data Mining w Badaniach Naukowych*; Wydawnictwo StatSoft Polska Sp. Z.o.o.: Kraków, Poland, 2016.
117. StatSoft Polska. *Zastosowania Metod Statystycznych w Badaniach Naukowych I*; Wydawnictwo StatSoft Polska Sp. Z.o.o: Kraków, Poland, 2000.
118. StatSoft Polska. *Statystyka i Data Mining w Badaniach Naukowych*; Wydawnictwo StatSoft Polska Sp. Z.o.o.: Kraków, Poland, 2006.
119. Tadeusiewicz, R. Data Mining Jako Szansa na Relatywnie Tanie Dokonywanie Odkryć Naukowych Poprzez Przekopywanie Pozornie Całkowicie Wyeksploatowanych Danych Empirycznych. StatSoft Polska, Akademia Górniczo—Hutnicza, Laboratorium Biocybernetyki, Kraków. 2006. Available online: https://media.statsoft.pl/_old_dnn/downloads/dm_jako_szansa.pdf (accessed on 1 May 2014).
120. StatSoft Polska. *Zastosowania Metod Statystycznych w Badaniach Naukowych IV*; Wydawnictwo StatSoft Polska Sp. Z.o.o: Kraków, Poland, 2012.
121. Dudek, G. Prognozowanie Krótkoterminowe Obciążeń Systemów Elektroenergetycznych z Wykorzystaniem Estymatorów Najbliższego Sąsiedztwa. Konferencja Naukowa Aktualne Problemy w Elektroenergetyce—Materiały Konferencyjne, 3–5 czerwca 2009, Jurata. Available online: https://gdudek.el.pcz.pl/files/FuzzyTree_14.pdf (accessed on 15 March 2016).
122. StatSoft Polska. *Zastosowania Metod Statystycznych w Badaniach Naukowych V*; Wydawnictwo StatSoft Polska Sp. Z.o.o. Kraków, Poland, 2016.
123. StatSoft Polska. *Zastosowania Metod Statystycznych w Badaniach Naukowych II*; Wydawnictwo StatSoft Polska: Kraków, Poland, 2003.
124. Wang, Y.; Liu, L.; Pang, X.; Fan, E. Load Forecasting Based on Improved K-means Clustering Algorithm. In Proceedings of the 2018 China International Conference on Electricity Distribution, Tianjin, China, 17–19 September 2018.
125. Johannesen, N.J.; Kolhe, M.; Goodwin, M. Comparison of Regression Tools for Regional Electric Load Forecasting. In Proceedings of the 2018 3rd International Conference on Smart and Sustainable Technologies (SpliTech), Split, Croatia, 26–29 June 2018.
126. Hastie, T.; Tibshirani, R.; Friedman, J. *Data Mining, Inference, and Prediction. Stanford 2008*; Springer Series in Statistics. The Elements of Statistical Learning, 2nd ed. Springer: Berlin/Heidelberg, Germany, 2008.
127. Łapczyński, M.; Lula, P.; Demski, T. *Data Mining IIb—Modele i Metody. Materiały Kursowe StatSoft Polska*; Wydawnictwo StatSoft Polska Sp. z o.o.: Kraków, Poland, 2006.
128. Barron, A.R.; Xiao, X. Multivariate Adaptive Regression Splines. *Ann. Stat.* **1991**, *19*, 67–82. [CrossRef]
129. Pasztyła, A. Statystyczne Metody Prognozowania Popytu na Rynkach Energii Elektrycznej. Ph.D. Thesis, Uniwersytet Ekonomiczny w Krakowie, Wydział Zarządzania, Katedra Statystyki, Kraków, Poland, 2008.
130. StatSoft Polska. *Praktyczne Wykorzystanie Analizy Danych i Data Mining*; Wydawnictwo StatSoft Polska Sp. Z.o.o.: Kraków, Poland, 2008.
131. Friedman, J.H. Multivariate Adaptive Regression Splines. *Ann. Stat.* **1991**, *19*, 1–67. [CrossRef]
132. Friedman, J.H.; Fast MARS. *Technical Report No. 110*; Laboratory for Computational Statistics, Department of Statistics, Stanford Linear Accelerator Center, Stanford University: Stanford, CA, USA, 1993.
133. Friedman, J.H. *Multivariate Adaptive Regression Splines*; SLAC Pub-460, Rev; Tech Report 102 Rev; Laboratory for Computational Statistics, Department of Statistics. Stanford Linear Accelerator Center, Stanford University: Stanford, CA, USA, 1990.
134. Bansal, P.; Salling, J. Multivariate Adaptive Regression Splines (MARS). UT ECE. 16-Feb-2013. 2013. Available online: https://uc-r.github.io/mars (accessed on 15 October 2016).
135. Demski, T. *Data Mining I—Kurs Podstawowy. Materiały Kursowe StatSoft Polska*; Wydawnictwo StatSoft Polska: Kraków, Poland, 2006.

136. Czapaj, R.; Kamiński, J.; Benalcazar, P. Dobór zmiennych objaśniających z wykorzystaniem metody MARSplines na przykładzie prognozowania dobowego zapotrzebowania na moc szczytową 15-minutową w KSE. In Proceedings of the XIV Międzynarodowa Konferencja Naukowa Prognozowanie w Elektroenergetyce PE'18, Podlesice, Poland, 26–28 September 2018.
137. Czapaj, R.; Kamiński, J.; Benalcazar, P. Dobór zmiennych objaśniających z wykorzystaniem metody MARSplines na przykładzie prognozowania dobowego zapotrzebowania na moc szczytową 15-minutową w KSE. *Przegląd Elektrotech.* **2019**, *95*, 9–12. [CrossRef]
138. Czapaj, R.; Ponikowska, T.; Kozakiewicz, M. *Optymalizacja Metodyki Prognozowania oraz Interfejsu Użytkownika w Narzędziu Informatycznym Służącym do Prognozowania Strat Sieciowych (Zadanie w Ramach Projektu Poszukiwanie i Proponowanie Rozwiązań Innowacyjnych Zmierzających do Zapewnienia Bezpiecznej i Ekonomicznej Pracy KSE z Uwzględnieniem Bieżących Potrzeb PSE S.A.)*; niepubl.—na zlecenie PSE S.A.; PSE Innowacje Sp. z o.o.: Warszawa, Poland, 2020.
139. Czapaj, R.; Kamiński, J.; Benalcazar, P. Prognozowanie krótkoterminowe zapotrzebowania na energię elektryczną w KSE z wykorzystaniem metody MARSplines. *Przegląd Elektrotech.* **2019**, *95*, 133–136. [CrossRef]
140. Pasztyła, A. *Data Mining IIa—Metody bez Nauczyciela. Materiały Kursowe StatSoft Polska*; Wydawnictwo StatSoft Polska Sp. z o.o.: Kraków, Poland, 2006.
141. Aman, S.; Chelmis, C.; Prasanna, V. Addressing Data Veracity in Big Data Applications. In Proceedings of the 2014 IEEE International Conference on Big Data (Big Data), Washington, DC, USA, 27–30 October 2014.
142. Sagiroglu, S.; Sinanc, D. Big Data: A Review. In Proceedings of the 2013 International Conference on Collaboration Technologies and Systems (CTS), San Diego, CA, USA, 20–24 May 2013.
143. Williams, J.W.; Aggour, K.S.; Interrante, J.; McHugh, J.; Pool, E. Bridging High Velocity and High Volume Industrial Big Data Through Distributed In—Memory Storage & Analytics. In Proceedings of the 2014 IEEE International Conference on Big Data, Washington, DC, USA, 27–30 October 2014.
144. Yindong, Y.; Zhongqin, B. Advances and Future Challenges in Electric Power Big Data. In Proceedings of the 2014 Second International Conference on Advanced Cloud and Big Data, Huangshan, China, 20–22 November 2014.
145. Wang, Y.F.; Deng, M.H.; Bao, Y.K.; Zhang, H.; Chen, J.Y.; Qian, J.; Guo, C.X. Power System Disaster-Mitigating Dispatch Platform Based on Big Data. In Proceedings of the 2014 International Conference on Power System Technology, Chengdu, China, 20–22 October 2014.
146. Wang, X.Z.; Zhou, J.; Huang, J.L.; Bi, X.L.; Ge, Z.Q. A Multilevel Deep Learning Method for Big Data Analysis and Emergency Management of Power System. In Proceedings of the 2016 IEEE International Conference on Big Data Analysis (ICBDA), Hangzhou, China, 12–14 March 2016.
147. Shi, H.; Xu, M.; Li, R. Deep Learning for Household Load Forecasting—A Novel Pooling Deep RNN. *IEEE Trans. Smart Grid* **2017**, *9*, 5271–5280. [CrossRef]
148. Dong, X.; Qian, L.; Huang, L. A CNN Based Bagging Learning Approach to Short-Term Load Forecasting in Smart Grid. In Proceedings of the 2017 IEEE SmartWorld, Ubiquitous Intelligence & Computing, Advanced & Trusted Computed, Scalable Computing & Communications, Cloud & Big Data Computing, Internet of People and Smart City Innovation (SmartWorld/SCALCOM/UIC/ATC/CBDCom/IOP/SCI), San Francisco, CA, USA, 4–8 August 2017.
149. Hao, J.; Gu, H.; Wang, R.; Deng, M.; Liu, Y.; Yang, D.; Lei, J.; Huang, Z. Short-term Power Load Forecasting for Larger Consumer based on TensorFlow Deep Learning Framework and Clustering-regression Model. In Proceedings of the 2018 2nd IEEE Conference on Energy Internet and Energy System Integration (EI2), Beijing, China, 20–22 October 2018.
150. Wang, X.; Zhao, T.; Liu, H.; He, R. Power Consumption Predicting and Anomaly Detection Based on Long Short—Term Memory Neural Network. In Proceedings of the 2019 IEEE 4th International Conference on Cloud Computing and Big Data Analytics, Chengdu, China, 12–15 April 2019.
151. Geron, A. *Uczenie Maszynowe z Użyciem Scikit-Learn i Tensorflow—Pojęcia, Techniki i Narzędzia Służące Do Tworzenia Inteligentnych Systemów*; Wydawnictwo Helion/O'Reilly: Gliwice, Poland, 2018.
152. Chollet, F.; Allaire, J.J. *Deep Learning—Praca z Językiem R i Biblioteką Keras*; Wydawnictwo Helion: Gliwice, Poland, 2019.
153. Kollia, L.; Kollias, S. A Deep Learning Approach for Load Demand Forecasting of Power Systems. In Proceedings of the 2018 IEEE Symposium Series on Computational Intelligence (SSCI), Bangalore, India, 18–21 November 2018.
154. Zhan, J.; Huang, J.; Niu, L.; Peng, X.; Deng, D.; Cheng, S. Study of the Key Technologies of Electric Power Big Data and its Application Prospects in Smart Grid. In Proceedings of the 2014 IEEE PES Asia-Pacific Power and Energy Engineering Conference (APPEEC), Hong Kong, China, 7–10 December 2014.
155. Hu, L.; Liu, K.Y.; Lin, Z.; Diao, Y.; Sheng, W. An Abnormal State Detection Method for Power Distribution Network Based on Big Data Technology. In Proceedings of the 2018 International Conference on Cyber—Enabled Distributed Computing and Knowledge Discovery, China, Beijing, 18–20 October 2018.
156. Kamiński, J. Metody modelowania równowagi rynków energii elektrycznej. *Rynek Energii. Str.* **2010**, *3*, 127–136.
157. Kamiński, J. Założenia metodyczne do budowy modelu równowagi rynku energii elektrycznej dla warunków Polski. *Energetyka. Str.* **2011**, *1*, 10–16.
158. Correa, E.; Inga, E.; Inga, J.; Hincapié, R. Electrical Consumption Pattern base on Meter Data Management System using Big Data Techniques. In Proceedings of the 2017 International Conference on Information Systems and Computer Science, Quito, Ecuador, 23–25 November 2017.

159. Lu, Y.; Zhang, T.; Zeng, Z.; Loo, J. An Improved RBF Neural Network for Short-Term Load Forecast in Smart Grids. In Proceedings of the 2016 IEEE International Conference on Communication Systems (ICCS), Shenzhen, China, 14–16 December 2016.
160. Song, M.; Jia, L. Big Data Mining Method of Thermal Power Based on Spark and Optimization Guidance. In Proceedings of the 2018 IEEE 7th Data Driven Control and Learning Systems Conference, Enshi, China, 25–27 May 2018.
161. Ku, T.I.; Park, W.K.; Choi, H. Energy Peak Reduction Mechanism with Prediction of Demand and PV Generation on Big Data. In Proceedings of the 2018 IEEE International Conference on Big Data and Smart Computing, Shanghai, China, 15–17 January 2018.
162. Yuan, Q.; Zhou, K.; Lu, W.; Yao, J. Big Data Driven Optimal Sizing of Stand-alone Photovoltaic Energy Systems. In Proceedings of the 2018 13th IEEE Conference on Industrial Electronics and Applications (ICIEA), Wuhan, China, 31 May–2 June 2018.
163. Liu, N.; Ma, L.; Liu, X.; Yu, H.; Tan, H.; Zhao, Y.; Guodong, L. Research on Key Issues of Data Integration Technology in Electric Power System in Big Data Environment. In Proceedings of the 2017 9th IEEE International Conference on Communication Software and Networks, Guangzhou, China, 6–8 May 2017.
164. Liu, Z.; Zou, Y. Research on Distribution Network Operation and Control Technology Based on Big Data Analysis. In Proceedings of the 2018 China International Conference on Electricity Distribution, Tianjin, China, 17–19 September 2018.
165. Pei, Z.; Wu, X.; Wang, X.; Bi, S. Short—Term Load Forecasting Based on Big Data Technologies. *CSEE J. Power Energy Syst.* **2015**, *1*, 59–67.
166. Shady, S.R.; Haitham, A.-R. Amira Big Data, Better Energy Management and Control Decisions for Distribution Systems in Smart Grid. In Proceedings of the 2016 IEEE International Conference on Big Data (Big Data), Washington, DC, USA, 5–8 December 2016.
167. Huang, X.; Huang, A.; Huang, H.; Ouyang, W.; Zhu, Y. Economic Operation Platform for Power Grid Company. In Proceedings of the 2019 IEEE 4th International Conference on Cloud Computing and Big Data Analytics, Chengdu, China, 12–15 April 2019.
168. Jin, L.; Li, H.; Xu, Z.; Ting, W.; Jia, W.; Su, D. Research on Wide-Area Distributed Power Quality Data Fusion Technology of Power Grid. In Proceedings of the 2019 IEEE 4th International Conference on Cloud Computing and Big Data Analytics, Chengdu, China, 12–15 April 2019.
169. Gu, X.; Xiong, B. Analysis of Power User Behavior Based on Big Data Technology: A Review. In Proceedings of the 2018 2nd IEEE Conference on Energy Internet and Energy System Integration (EI2), Beijing, China, 20–22 October 2018.
170. Vijaya, T.; Vasirani, M.; Humeau, S.; Aberer, K. Cluster-Based Aggregate Forecasting for Residential Electricity Demand using Smart Meter Data. In Proceedings of the 2015 IEEE International Conference on Big Data (Big Data), Santa Clara, CA, USA, 29 October–1 November 2015.
171. Wu, P.; Tan, J. The Design of Distributed Power Big Data Analysis Framework and its Application in Residential Electricity Analysis. In Proceedings of the 2018 Sixth International Conference on Advanced Cloud and Big Data, Lanzhou, China, 12–15 August 2018.
172. Dudek, G. Analiza podobieństwa obrazów sekwencji szeregów czasowych obciążeń elektroenergetycznych. In Proceedings of the IX Międzynarodowa Konferencja Naukowa Prognozowanie w Elektroenergetyce, Wisła, Poland, 17–19 September 2008.
173. Dudek, G. Analiza podobieństwa obrazów sekwencji szeregów czasowych obciążeń elektroenergetycznych. *Przegląd Elektrotech.* **2009**, *85*, 149–152.
174. Dudek, G. Similarity-Based Approaches to Short-Term Load Forecasting. Electrical Department, Czestochowa University of Technology, Poland. Applied Soft Computing 37:277–287. December. 2015. Available online: https://gdudek.el.pcz.pl/files/STLF_similarity10.pdf (accessed on 16 March 2016).
175. Liao, X.; Cao, N.; Ma, L.; Kang, X. Research on Short-term Load Forecasting Using XGBoost Based on Similar Days. In Proceedings of the 2019 International Conference on Intelligent Transportation, Big Data & Smart City (ICITBS), Changsha, China, 12–13 January 2019.
176. Mu, Q.; Wu, Y.; Pan, X.; Huang, L. Short-term Load Forecasting Using Improved Similar Days Method. In Proceedings of the 2010 Asia-Pacific Power and Energy Engineering Conference, Chengdu, China, 28–31 March 2010.
177. Keitsch, K.; Bruckner, T. SAWing on Short Term Load Forecasting Errors: Increasing the Accuracy with Self AdaptiveWeighting. In Proceedings of the 2016 IEEE Innovative Smart Grid Technologies—Asia (ISGT-Asia), Melbourne, Australia, 28 November–28 December 2016.
178. Saber, A.Y.; Alam, R. Short Term Load Forecasting using Multiple Linear Regression for Big Data. In Proceedings of the 2017 IEEE Symposium Series on Computational Intelligence (SSCI), Honolulu, HI, USA, 27 November–1 December 2017.
179. Vatra, D.F.; Poida, D.A.; Stanescu, D.C. Data System for The Monitoring of Power Quality in The Transmission Substations Supplying Big Consumers. In Proceedings of the 2007 9th International Conference on Electrical Power Quality and Utilisation, Barcelona, Spain, 9–11 October 2007.
180. Huang, J.; Niu, L.; Zhan, J.; Peng, X.; Bai, J.; Cheng, S. Technical Aspects and Case Study of Big Data based Condition Monitoring of Power Apparatuses. In Proceedings of the 2014 IEEE PES Asia-Pacific Power and Energy Engineering Conference (APPEEC), Hong Kong, China, 7–10 December 2014.
181. Sowiński, J. Temperatura powietrza jako dodatkowa zmienna egzogeniczna w krótkoterminowym prognozowaniu modelem ANFIS obciążeń w podsystemie elektroenergetycznym. *Przegląd Elektrotech.* **2019**, *97*, 178–181. [CrossRef]
182. Sowiński, J.; Szydłowski, M. Forecast of Electricity Supply Using Adaptive Neuro—Fuzzy Inference System. In Proceedings of the 18th International Scientific Conference on Electric Power Engineering, Kouty nad Desnou, Czech Republic, 17–19 May 2017.
183. Gabaldón, A.; Ruiz-Abellón, M.C.; Fernández-Jiménez, L.A. *Short-Term Load Forecasting 2019*; MDPI: Basel, Switzerland, 2020.

184. Upadhaya, D.; Thakur, R.; Singh, N.K. A Systematic review on the Methods of Short Term Load Forecasting. In Proceedings of the 2019 2nd International Conference on Power Energy, Environment and Intelligent Control (PEEIC), Greater Noida, India, 18–19 October 2019.
185. Harrison, G. *NoSQL, NewSQL i BigData—Bazy Danych Następnej Generacji*; Wydawnictwo Helion: Gliwice, Poland, 2019.

energies

MDPI

Article

Offshore Wind Power Forecasting—A New Hyperparameter Optimisation Algorithm for Deep Learning Models

Shahram Hanifi [1,*], Saeid Lotfian [2,*], Hossein Zare-Behtash [1] and Andrea Cammarano [1]

[1] James Watt School of Engineering, University of Glasgow, Glasgow G12 8QQ, UK
[2] Department of Naval Architecture, Ocean and Marine Engineering, University of Strathclyde, Glasgow G4 0LZ, UK
* Correspondence: shahram.hanifi@glasgow.ac.uk (S.H.); saeid.lotfian@strath.ac.uk (S.L.); Tel.: +44-(0)-141-548-4371 (S.L.)

Abstract: The main obstacle against the penetration of wind power into the power grid is its high variability in terms of wind speed fluctuations. Accurate power forecasting, while making maintenance more efficient, leads to the profit maximisation of power traders, whether for a wind turbine or a wind farm. Machine learning (ML) models are recognised as an accurate and fast method of wind power prediction, but their accuracy depends on the selection of the correct hyperparameters. The incorrect choice of hyperparameters will make it impossible to extract the maximum performance of the ML models, which is attributed to the weakness of the forecasting models. This paper uses a novel optimisation algorithm to tune the long short-term memory (LSTM) model for short-term wind power forecasting. The proposed method improves the power prediction accuracy and accelerates the optimisation process. Historical power data of an offshore wind turbine in Scotland is utilised to validate the proposed method and compare its outcome with regular ML models tuned by grid search. The results revealed the significant effect of the optimisation algorithm on the forecasting models' performance, with improvements of the RMSE of 7.89, 5.9, and 2.65 percent, compared to the persistence and conventional grid search-tuned Auto-Regressive Integrated Moving Average (ARIMA) and LSTM models.

Keywords: auto-regressive integrated moving average (ARIMA); long short-term memory (LSTM); Optuna; isolation forest (IF); elliptic envelope (EE); one-class support vector machine (OCSVM)

Citation: Hanifi, S.; Lotfian, S.; Zare-Behtash, H.; Cammarano, A. Offshore Wind Power Forecasting—A New Hyperparameter Optimisation Algorithm for Deep Learning Models. *Energies* **2022**, *15*, 6919. https://doi.org/10.3390/en15196919

Academic Editors: Paweł Piotrowski, Grzegorz Dudek and Dariusz Baczyński

Received: 2 September 2022
Accepted: 16 September 2022
Published: 21 September 2022

Publisher's Note: MDPI stays neutral with regard to jurisdictional claims in published maps and institutional affiliations.

1. Introduction

Undoubtedly, to accelerate economic growth, power production through renewable energy sources needs to increase because conventional methods such as using fossil fuels have irreparable consequences, including pollution, climate change, and the depletion of the ozone layer [1].

In recent decades, various renewable energies, such as wind, solar, waves, etc., have received increasing attention. Among all these energies, wind power has played the most important role in replacing fossil fuels [2]. As reported by the World Wind Energy Council, the installed global capacity of wind energy in the world in 2021 has reached 837 GW, with an increase of 92 GW compared to 2020 [3]. Figure 1 shows the global wind power installed capacity increment over the past 21 years [3]. In this figure, the blue columns represent the capacity of installed wind power on land, while the red columns represent the offshore installed wind energy.

One main obstacle hindering the increase of wind power penetration into the power grid is the production uncertainty due to fluctuations in wind speed [1]. Therefore, adequate planning in electricity distribution to meet consumers' demand, determining the best time for operation and maintenance, and the fairest pricing on the market requires accurate wind power forecasting in the upcoming time steps.

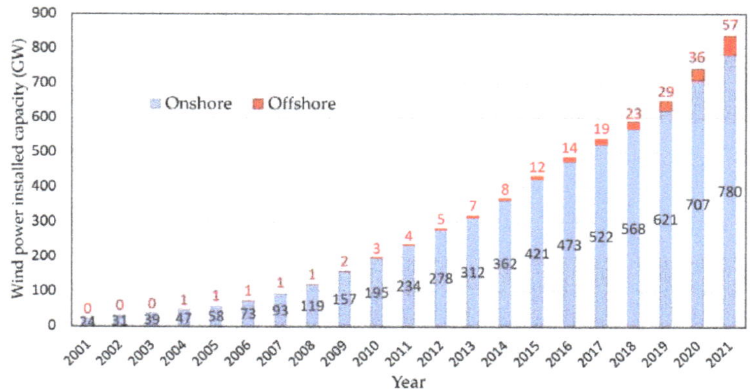

Figure 1. Global wind power installed capacity increment during the last 21 years [3].

Hanifi et al. [1] categorised wind power forecasting into three main methods, including physical, statistical, and hybrid approaches. Physical methods utilise numerical weather prediction (NWP) data, wind turbine geographic descriptions, and weather information to predict wind power [1]. These methods are computationally complex and very sensitive to initial information [2]. On the other hand, statistical methods work based on building an accurate mapping between input variables (such as NWP data, historical data, etc.) and target variables (wind speed or wind power). These methods include two main approaches: time-series-based methods and machine learning (ML) approaches [1]. Time-series-based methods can predict wind speed or wind power based on the history of the predicted variable itself. They can recognise the concealed random features of wind speed and are used for very short-term (minutes to a few hours) forecasting. The Auto-Regressive Integrated Moving Average (ARIMA) model proposed by Box–Jenkins [4] is one of the common statistical methods which is used in various research. For example, in Western Australia, Yatiyana et al. [5] applied the ARIMA model for wind speed and direction forecasting. They proved that their proposed model could predict wind speed and direction with a maximum of 5% and 16% error, respectively. Firat et al. [6] proposed an autoregressive (AR) wind speed prediction model for a wind farm in the Netherlands. They used six years of hourly wind speed and achieved a high accuracy for 2–14 h ahead. In another study, De Felice et al. [7] applied 14 months of temperature readings in Italy to train an ARIMA model for electricity demand prediction. Their proposed method demonstrated higher accuracy, particularly in hot locations, compared with persistence methods. Duran et al. [8] proposed a method to combine AR and exogenous variable (ARX) models to predict the wind power generation in a wind farm located in Spain up to one day in advance. They used different model orders and training periods to prove that the application of the AR models presents lower errors than a persistent model. Kavasseri et al. [9] examined the application of fractional ARIMA models to predict wind farm hourly average wind speed for one- and two-day-ahead time horizons. The results of the predictions showed a 42% improvement compared to persistent methods. Later, the predicted wind speeds were applied to the power curve of an operating wind turbine to predict the relevant wind powers. In another study, Torres et al. [10] used the ARMA and the persistence model for hourly average wind speed forecasting up to 10 h ahead. The ARMA model demonstrated a better performance compared to the persistence method, with a 12% to 20% lower root mean square error (RMSE) when forecasting 10 h in advance.

ML methods such as neural networks (NNs) can establish deductive models by learning dependencies between input and output variables. These methods are easy to create, do not require further geographic information, and can predict over longer timeframes. One of the common ML methods is the LSTM model, which can address the long-term dependency issues [11], which is important in forecasting time-series with long input sequences [12].

LSTM is variously used in research for wind power prediction. For instance, Zhang et al. [13] proposed an LSTM wind power forecasting model for three wind turbines of a wind farm in China. They utilised three months of wind speed and historical power data and achieved the highest forecasting accuracy in a one-to-five time-steps ahead compared to the radial basis function (RBF) and deep belief network (DBN). Fu et al. [14] demonstrated LSTM and gated recurrent unit (GRU) for a one-to-four step-ahead forecasting of a 3 MW wind turbine in China, based on the first three-month dataset of 2014, with a resolution of 15 min. The comparison with ARIMA and support vector machine (SVM) methods showed the superiority of their proposed methods. Cali and Sharma [15] proposed an LSTM-based model with one hidden layer for 1 to 24 h ahead of wind power forecasting. The model was trained with 9-month data and evaluated in the last three months of 2016. They used nine combinations of input data, including wind speed at various levels, wind direction, temperature, and surface pressure. They demonstrated that temperature, wind speed, and direction positively impacted model performance; however, adding surface pressure to the input features led to worse performance.

As well as the training data, ML models' accuracy strongly depends on the adequate selection of their parameters and hyperparameters. The parameters of ML models (e.g., the weights of each neuron) are determined during the training process of the algorithm. In contrast, hyperparameters are not directly learnt by the learning algorithm and need to be specified outside the training process. The main role of the hyperparameters is to control the capacity of the models in learning dependencies. They also prevent overfitting and improve the generalisation of the algorithm. Hyperparameter optimisation or tuning improves forecasting accuracy and reduces models' complexity [16].

The literature's most common hyperparameter tuning methods are the grid search and random search. Grid search can be used for simple models with a few parameters. The calculation will be extremely time-consuming by increasing the number of parameters and expanding the space of the possible configurations [17]. Therefore, researchers usually consider a narrow range of hyperparameters during the grid search [16]. On the other hand, a random search algorithm looks randomly for a set of combinations rather than searching for better results.

Both these search methods generate all candidate combinations of hyperparameters upfront and then evaluate them in parallel. Based on the evaluation of all combinations, the best hyperparameters can be selected. Trying all possible combinations is very costly; as a result, it is vital to develop advanced techniques to intelligently select which hyperparameters to assess and then decide where to sample next after evaluating their quality.

The advanced optimisation of the ML-based time-series forecasting models for wind turbine-related predictions remained untouched. However, a few studies have proposed methods for optimising ARIMA and LSTM models within other applications than wind power forecasting. For example, Al-Douri et al. [18] designed a genetic algorithm (GA) to find the best parameters of an ARIMA model for the better cost prediction of used fans in Swedish road tunnels, and provided results which proved a significant improvement in data forecasting. In another study, F. Shahid et al. [19] employed GA to optimise the window size and neuron numbers of LSTM layers. This approach improved the power prediction accuracy of wind farms in Europe by up to about 30% compared to existing methods such as support vector regressors.

As the review of the literature indicates, several examples use linear and nonlinear regression models for challenges related to predicting wind power. Each study provides the use of one model type or a comparison of various model types Nevertheless, without the tuning and selection of the hyperparameters, it is not possible to obtain their maximum benefit [16]. This advanced tuning method plays an important role when the hyperparameter search space grows exponentially, and the use of exhaustive grid search becomes extremely time-consuming.

This paper proposes a framework for developing accurate and robust ML models for wind power forecasting. The framework outlines the model development procedure from data engineering to precision evaluation and fine-tuning. Furthermore, an advanced algorithm is utilised to optimise wind power forecasting models to reduce time calculation costs, as well as to improve accuracy. For the case study, two ML models were selected: the LSTM model, which is proven to have remarkable prediction performance on time series-based models, and ARIMA, a traditional model, for the purpose of benchmarking.

The novelty of this work lies in developing a short-term wind power forecasting model through an intelligent application of the long short-term memory (LSTM) model, while a new optimisation algorithm tunes its main hyperparameters. In addition, the distinguished aspects of the methodology are summarised, based on importance, as follows:

- LSTM is used on a wind power dataset to take advantage of its ability to learn nonstatic features from nonlinear sequential data automatically.
- The ARIMA model is applied as a forecasting model because of its short response time and ability to capture the correlations in time series.
- Instead of the trial-and-error method to select the best hyperparameters of the ARIMA and LSTM forecasting models, which require a great deal of time, grid search is used to tune both these models.
- The new Optuna optimisation framework is employed to optimise the hyperparameters of the LSTM model, including the number of lag observations, the quantity of LSTM units for the hidden layer, the exposure frequency, the number of samples inside an epoch, and the used difference order for making a nonstationary dataset stationary.
- Unlike most previous studies, which is for onshore wind turbines, forecasting assessments have been done for an offshore wind turbine in this study.
- How to deal with the negative values of wind power (which are normally found in active power observations), in terms of removal or replacement, has been thoroughly investigated in this study and the results have been discussed.
- After a detailed discussion about the reasons for having outliers, three different methods, including isolation forest (IF), elliptic envelope (EE), and the one-class support vector machine (OCSVM), are used to detect and treat them. A comparison of the results will help researchers to choose the best outlier detection method for future studies.
- The proposed Optuna–LSTM model is assessed by the comparison of its forecasted power with actual values and predictions by persistence and ARIMA based on the RMSE statistical error measure.

The rest of this paper is organised as follows: Section 2 discusses the optimisation process, the forecasting models, and the studied supervisory control and data acquisition (SCADA) data. This section includes the steps taken for preprocessing, resampling, and outlier treatment. Section 3 presents the results of the trained, optimised LSTM model in terms of model accuracy and the time cost compared to other prediction methods. Finally, Section 4 summarises the paper's contributions.

2. Methodology

The proposed procedure of this study is illustrated in Figure 2. At the beginning of this study, three required features, including the time stamps, wind speeds, and active wind powers, are selected to improve the computational time. At the next step, negative power values are removed or replaced. This data preprocessing is followed by resampling the dataset and removing outliers in three different ways. After finishing the data preprocessing and providing proper data for forecasting, data predictability and stationarity are assessed as two important specifications for accurate power forecasting. Afterwards, three different approaches are employed for forecasting, and their best performance is gained by the selection of their most appropriate hyperparameters.

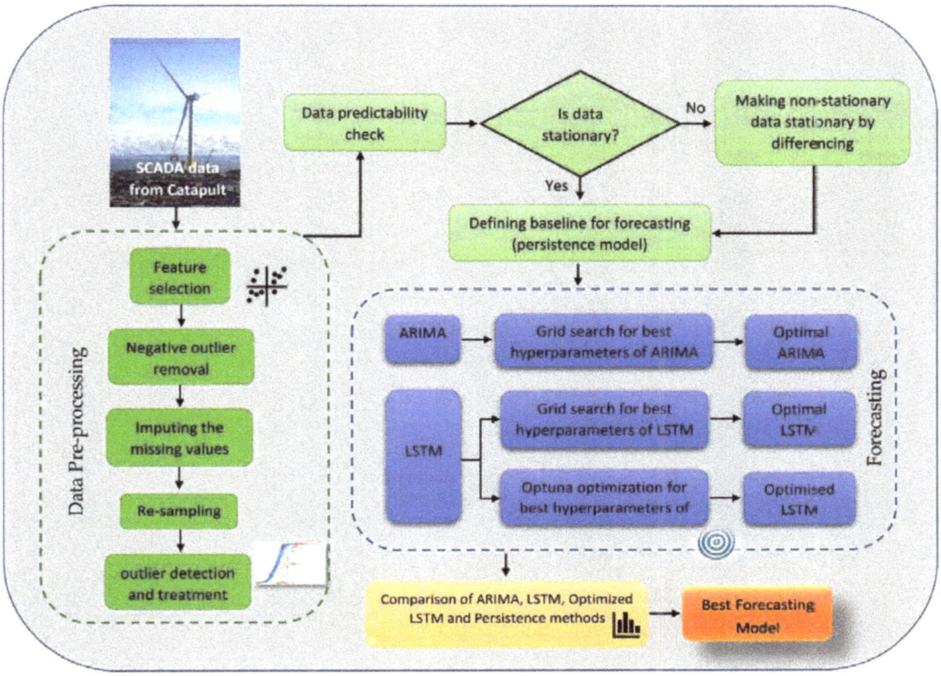

Figure 2. Diagram of applied methodology.

2.1. ARIMA Model

In this study, the standard approach of the Box–Jenkins method [20] was traced for the ARIMA model development. The ARIMA model is a widely used set of statistical models for analysing and predicting time-series data [21]. This model can be expressed as [22]:

$$X_t = \phi_1 X_{t-1} + \phi_2 X_{t-2} + \cdots + \phi_p X_{t-p} + e_t - \theta_1 e_{t-1} - \theta_2 e_{t-2} - \cdots - \theta_q e_{t-q} \quad (1)$$

While ϕ_t and θ_t are coefficients, p, q, and d are the lag number of observations in the model, the order of moving average, and the degree of difference, respectively. Degree of difference (d) values greater than 0 imply that the data has been nonstationary but has become stationary after some degree of difference.

The ARIMA model combines the AR, moving average (MA), and the Integrated (I) components, which denotes the data substitution with the value of the difference between its values and the preceding values [23]. The forecasting accuracy of the ARIMA model depends on selecting the most appropriate combination of p, d, and q. Normally, for small data sets, the autocorrelation function (ACF) and partial autocorrelation function (PACF) can be used to determine which AR or MA component should be selected in the ARIMA model [24].

These two factors, which can be graphically plotted, are widely used elements in analysing and predicting time-series. They highlight the relationship between an observation and the observations' value at prior time steps. The difference between ACF and PACF is that, in PACF, while assessing the relationship between observation of two time steps, the relationships of the intervening observations are removed. Figure 3a,b show the observations' ACF and PACF plots. An appropriate ARIMA model can be selected based on the simple explanations in Table 1 [9], and the value of d (degree of difference) depends on the number of differencing until the data is stationary.

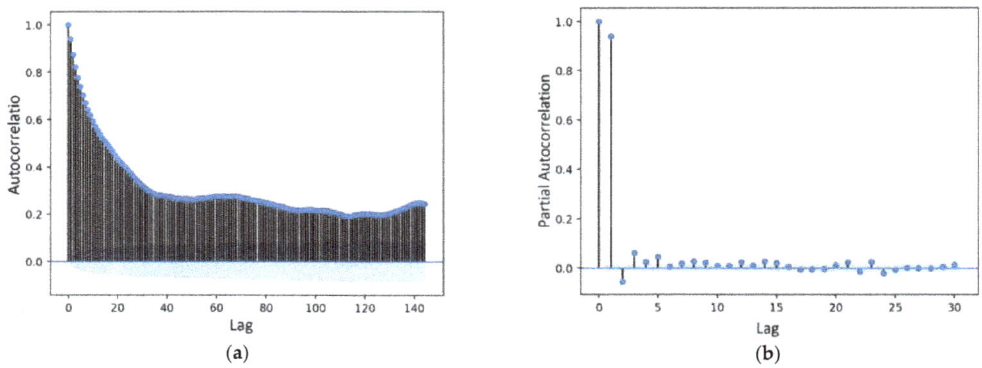

Figure 3. ACF (**a**) and PACF (**b**) plots for generated power of LDT. The blue points represent the value of autocorrelation and partial autocorrelation of different time lags.

Table 1. ACF and PACF application for statistical model selection.

Model	Autocorrelation	Partial Autocorrelation
AR (p)	Tails off gradually	Cuts off after p lags
MA(q)	Cuts off after q lags	Tails off gradually
ARMA (p, q)	Tails off gradually	Tails off gradually

The ARIMA model forecasting steps after resampling and outlier treatment can be seen in Figure 4. The first step is assessing the stationarity of the time-series. Stationary is one of the assumptions during time-series modelling, which shows the consistency of the summary statistics of the observations.

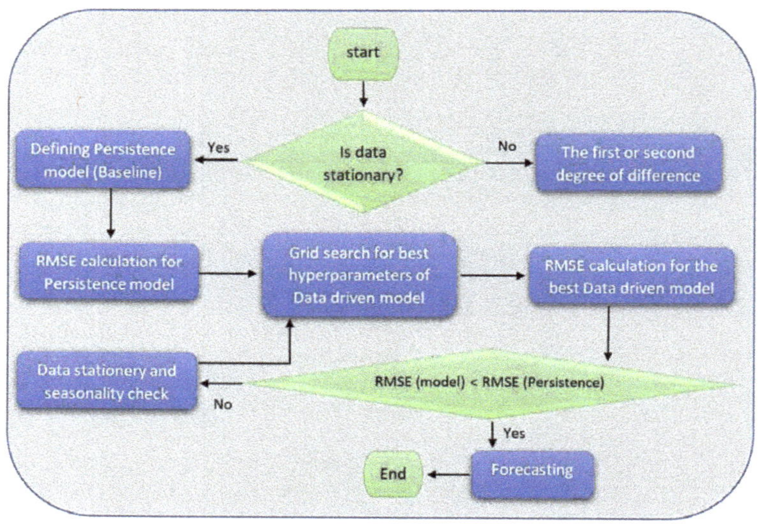

Figure 4. Flowchart of ARIMA and LSTM wind power forecasting models.

When a time-series is stationary, it means that the statistical properties of the time-series (such as mean, variance, and autocorrelation) do not change over time. This property can be violated by having any trend, seasonality, and other time-dependent structures.

There are two main methods for the stationarity assessment of time-series, the visualisation approach and the augmented Dickey–Fuller (ADF) test. The visualisation method uses graphs to show whether the standard deviation changes over time. On the other hand, the ADF method is a statistical significance test that compares the p-value with the critical values and does hypothesis testing. This test makes the stationarity of data clear at different levels of confidence.

Regarding the data used in this study, due to the high number of observations and wide dispersion, it is not possible to check stationarity through the visualisation method. Therefore, in this study, the ADF method was used.

The ADF test's execution provides a p-value which, by comparing it with a threshold (such as 5% or 1%), can identify the stationarity of the data. Nonstationary data in this step need to be changed to stationary by methods such as differencing. After ensuring the time-series is stationary, a persistence method as a baseline is created. Then, through a detailed grid search, the best hyperparameters for the ARIMA forecasting for each preprocessed data were found. The last step is ARIMA forecasting and comparing its error with the error of the persistence method.

2.2. LSTM Model

The recurrent neural network (RNN) is a model in which the connection of its units creates cycles. RNN has a high ability to represent all dynamics. However, its effectiveness is affected by the limitations of the learning process. The main limitation of gradient-based methods that use back propagation is their path integral time-dependence on assigned weight [13]. When the time lag between the input signal and the target signal increases to more than 5–10 time-steps, the normal RNN loses the learning ability, and the back-propagation error either vanishes or explodes. This error elimination raises the question of whether normal RNNs can show practical benefits for feed-forward networks. To address this problem, the LSTM has been developed based on memory cells. The LSTM consists of a recurrently attached linear unit known as the constant error carousel (CEC). CECs, by keeping the local error backflow constant, mitigate the gradient's vanishing problem [25]. They can be trained by adjusting both the back propagation over time and the real-time recurrent learning algorithm [26]. Figure 5 shows the typical structure of the LSTM.

As can be seen, there are three gate units in a basic LSTM cell, including the input, output, and forget gates. The gate activation vectors of i_t, o_t and f_t for input, output, and forget gates, respectively, are calculated in Equations (2)–(4).

$$i_t = \sigma_l(W_i x_t + U_i h_{t-1} + b_i) \tag{2}$$

$$o_t = \sigma_l(W_o x_t + U_o h_{t-1} + b_o) \tag{3}$$

$$f_t = \sigma_l\left(W_f x_t + U_f h_{t-1} + b_f\right) \tag{4}$$

In these equations, W_i, W_o, W_f U_i, U_o, and U_f represent the assigned weights, and b_i, b_o, and b_f represent the biases in conjunction with relevant activation functions σ_l. In addition, x_t is the neuron input at time step t, and the cell state vector at time step t − 1 is h_{t-1}. As shown in Equation (5), the next evaluated value of the state $\widetilde{S}_t$ can be calculated based on the relevant activation function σ_s.

$$\widetilde{S}_t = \sigma_s(W_s x_t + U_s h_{t-1} + b_s) \tag{5}$$

In Equation (7), the newly assessed value of $\widetilde{S}_t$ and the prior cell state S_{t-1} are used to calculate cell state S_t, which by itself will be used with the output gate control signal o_t and the activation function σ_{lh} to obtain the overall output h_t according to Equation (8).

$$S_t = f_t S_{t-1} + i_t \widetilde{S}_t \tag{6}$$

$$h_t = o_t \sigma_{lh}\,(S_t) \tag{7}$$

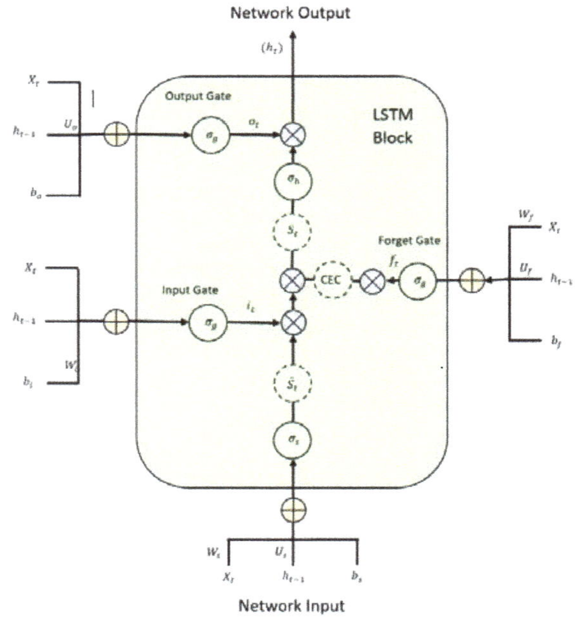

Figure 5. Typical structure of the LSTM.

As can be seen in Equations (6) and (7), the output h_t is dependent on the state S_t of the LSTM cell and the activation function σ_{lh} that is usually tanh (x). The state S_t depends on the state of the prior step S_{t-1} as well as the new value of the state $\widetilde{S}_t$.

In accordance with all the relations mentioned above, the function of the LSTM model can be concluded as:

- Input gate (i_t) controls the extent to which $\widetilde{S}_t$ flows into the memory.
- Output gate (o_t) regulates the extent to which S_t gives to the output (ht).
- Forget gate (f_t) controls the extent to which S_{t-1} (i.e., previous state) is kept in the memory.

Specifying the best LSTM model for wind power forecasting requires the determination of the neural network's best combination of hyperparameters. LSTMs have five main hyperparameters, including the number of lag observations as inputs of the model, the quantity of LSTM units for the hidden layer, the model exposure frequency to the whole training dataset, the number of samples inside an epoch in each weight updating, and finally, the used difference order for making nonstationary data stationary.

2.3. *Grid Search for ARIMA and LSTM Models*

ARIMA model factors (i.e., *p*, *d*, and *q*) can be estimated through iterative trial and error by revising the ACF and PACF plot. This part of defining the ARIMA forecasting model can be very challenging and time-consuming, leading to prediction errors. As a result, researchers attempt to find these hyperparameters using an automatic grid search approach. Similar to the ARIMA model, specifying the best LSTM model for wind power forecasting requires the determination of the best combination of hyperparameters in this neural network. This study also specified a grid of the LSTM parameters to iterate. An LSTM model is created based on each combination, and its forecasting accuracy is assessed by calculating its RMSE.

2.4. Persistence Method

It is vital to create a baseline for any time-series prediction approach. As a reference, for comparing all modelling approaches, this baseline can show how well a model makes predictions. Models which perform worse than the performance level of the baseline can be ignored.

Benchmarks for forecasting problems need to be very simple to train, fast to implement, and repeatable. The persistence model is one of the most commonly used references for wind speed and power prediction (short-term forecasting methods in particular). Based on the definition of this method, wind power in the future will be equivalent to the generated power in the present [27], as given by Equation (8):

$$\hat{P}_{t+k/t} = P_t \tag{8}$$

where P_t is the measured wind power at time t and $\hat{P}_{t+k/t}$ is the predicted wind power for the future time k. This model performs better than most short-term physical and statistical forecasting methods. Therefore, it is still widely used in very short-term prediction [28]. This research uses the persistence model to compare the performance of the ARIMA and LSTM models for different datasets.

2.5. Hyperparameter Optimisation with Optuna

This study uses the Optuna optimisation method to optimise the forecasting models. Optuna is an open-source optimisation software with several advantages over the other optimisation frameworks [29]. Other optimisation tools usually differ depending on the algorithm used to select the parameters. For example, GPyOpt and Spearmint [30] apply Gaussian processes, SMAC [31] employs random forests, and Hyperopt [32] uses a tree-structured Parzen estimator (TPE). These methods have three main drawbacks. Firstly, they need the parameter search space to be statically defined by the user, a process that is extremely hard for large-scale experiments with many possible parameters. Furthermore, they do not have an efficient pruning strategy for high-performance optimisation when accessing limited resources. In addition, they cannot handle large-scale experiments with minimal setup requirements. On the other hand, Optuna, with a define-by-run design, enables the user to create the search space dynamically. This optimisation framework is an open-source, easy-to-set-up package that benefits effective sampling and pruning algorithms [29]. Optuna optimises the model through minimising/maximising an objective function (here, the RMSE of the forecasted wind power rather than the real generated values) that assumes a group of hyperparameters as input and returns its validation core. The optimisation process is called a study, and each objective function's evaluation is called a trial [29].

At the beginning of the optimisation, the user is asked to provide the search space for the dynamic generation of the hyperparameters for each trial. Then, the model builds the objective function by interacting with the trial object. After this step, the next hyperparameter selection is based on the history of previously evaluated trials. This algorithm optimises ML models in two steps. First, a search strategy determines a set of parameters to be examined, and second, a performance assessment strategy known as a pruning algorithm excludes the improper parameters based on the estimation of the value of the currently investigated parameters [29].

Since the initial prediction accuracy assessment of the ARIMA and LSTM models (both tuned by grid search) highlighted the better performance of the LSTM model compared to ARIMA, it was decided to apply the optimisation framework only to the LSTM model.

In this way, the hyperparameter ranges of the LSTM model increased from what was examined in its grid search to wider ranges, as shown in Table 2. In other words, the hyperparameter combinations increased from 48 combinations to more than a million combinations.

Table 2. Hyperparameter ranges and their total combinations in LSTM–grid search and LSTM–Optuna methods.

Parameters	LSTM–Grid Search		LSTM–Optuna	
	Values	Possible Combinations	Values	Possible Combinations
No. of lag observations	(3, 4, 6)	48	(1, 3, 4, . . . , 10)	More than 1 M
No. of LSTM units	(100, 150)		(50, 60, 70, . . . , 300)	
Exposure frequency	(100, 150)		(50, 60, 70, . . . , 300)	
No. of samples inside an epoch	(100, 150)		(50, 60, 70, . . . , 300)	
Difference order	(0, 1)		(0, 1, 2, . . . , 5)	

2.6. Wind Power Dataset

The source SCADA data are measured at a 1 Hz frequency from the Levenmouth Demonstration Turbine (LDT), an offshore wind turbine which is located just 50 m from the coast at Leven, a seaside town in Fife, Scotland [33]. This wind turbine was acquired by the Offshore Renewable Energy (ORE) Catapult in 2015, while its construction was completed by Samsung in October 2013 [34].

ORE Catapult's wind turbine is a three-bladed upwind turbine installed on a jacket structure [25]. The turbine is ranked to work at 7 MW, but to decrease the noise, it is limited to operating at the highest power of 6.5 MW [33]. This turbine's rotor diameter is 171.2 m, and its hub height is 110.6 m. Each blade of this turbine measures 83.5 m and weighs 30 tons. The defined cut-in speed for this turbine is 3.5 m/s, which means its electricity generation will start when wind speeds reach this speed. It will shut down if the wind is blowing too hard (roughly 25 m/s) so to prevent equipment damage. Its operating temperature is between −10 °C to +25 °C, and it has been designed to work for 25 years [35]. Figure 6 shows the configuration and main parameters of the LDT.

Features	Values
Wind Class	IEC Class IA/SB
Capacity	7 MW
Rotor diameter	171.2 m
Blade length	83.5 m
Hub height	110.6 m
Total height	196 m blade tip to sea level
Generator	Medium (3.3 kV), PMG
Converter	Full power conversion
Dive train	Medium speed (400 rpm)
Rated frequency	50 Hz
Rotor speed	5.9–10.6 rpm
Wind speed	3.5–25 m/s
Design life	25 years

Figure 6. Main parameters and schematic of Levenmouth wind turbine [35].

2.7. Feature Selection

This study recorded the SCADA datasets for five months, from 1 January 2019 to 31 May 2019, at a 1 Hz frequency (with one-second intervals). Each timestamp in this time-series data includes 574 different observations, including the generated power, wind speed at different levels, blade pitch angle, nacelle orientation, etc. At the beginning of the data processing, a feature selection was carried out to decrease the size of the dataset to reduce the computation time by excluding unnecessary variables. This process was vital to making this study possible. All variables except the time stamp, wind speed, and active power were removed at this stage, which was useless in the ARIMA and univariate LSTM forecasting methods. Keeping the wind speed variable was vital in this project, as it verified the accuracy of generated power. For example, failure to generate power when high wind speeds were recorded was recognised as a stop in power generation due to reasons such as maintenance. After removing the redundant information, observations of wind speed and active power were plotted as shown in Figure 7a,b.

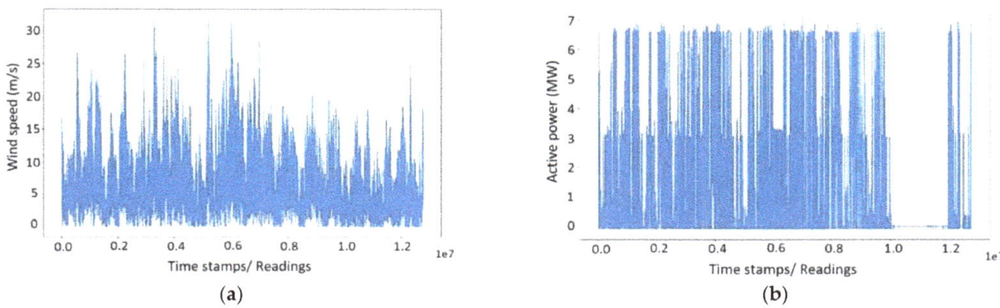

Figure 7. Wind speed observations (**a**), wind active power observations (**b**).

The histograms of this dataset for wind speed and active power are presented in Figure 8a,b, and Table 3 shows their statistical descriptions.

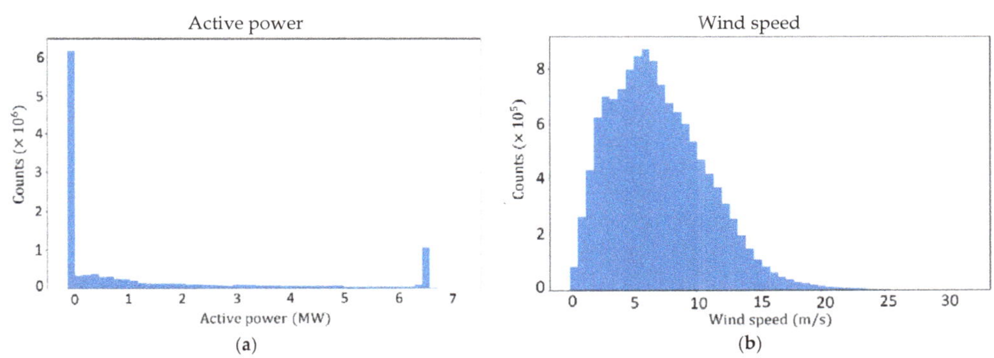

Figure 8. Histogram of active power (**a**) and wind speed (**b**).

Table 3. Statistical descriptions of the SCADA datasets.

	Active Power, kW	Wind Speed, m/s
Count	1.0×10^7	1.0×10^7
Mean	1.8×10^3	7.6
Standard deviation	2.3×10^3	3.9
Minimum	-1.2×10^2	-3.3×10^{-2}
25%	-6.0×10	4.7
Medium	5.9×10^2	7.1
75%	3.2×10^3	1.0×10
Maximum	7.2×10^3	3.2×10

2.8. Obvious Outlier Removal

An initial assessment of Figure 7b specified that a large part of the recorded generated power at the end of this time-series (May 2019) equals zero. Usually, the generated power of a turbine can be zero when no wind is blown. However, the evaluation of Figure 7a shows a continuous wind blowing with fluctuations similar to previous months. Therefore, it is speculated that the turbine was out of production during this period. Based on this assumption, it was decided that this month (May 2019) should be removed entirely from the dataset. The time-series after this omission was reduced to four months, from 1 January 2019 to 30 April 2019. A closer look at the active power, as shown in Figure 9, revealed another obvious error in the SCADA data, the existence of negative values. Negative values are values of which there is no practical meaning in wind power generation. Shen et al. [36] believe that these values represent time stamps when turbine blades do not rotate, but the turbine's control system needs electricity [36]. These values need to be eliminated along with the corresponding parameters of the same timestamp for better forecasting results [25]. Since the elimination of these negative values disrupts the time continuity of the time-series, and can possibly lead to errors in wind power prediction, at this stage it was decided to create and assess three types of datasets based on different actions against negative values. Assessment of the impact of these actions on forecasting accuracy became another goal of this study.

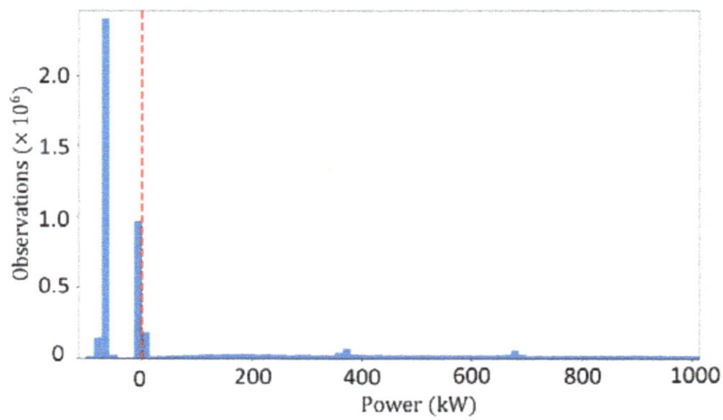

Figure 9. Wind power observations (only power values under 1000 kW are shown). The dotted red line indicates the power value of zero (the boundary of negative/positive values).

These three preprocessing methods against the negative values are:

- Total elimination of negative values without any substitution;
- Replacement of negative values with the average amount of power in the whole 4-month period;
- Replacement of negative values with positive values of power at the nearest timestamp.

2.9. Resampling

The effect of wind turbulence as one of the obstacles to increasing the wind energy penetration in energy markets is more significant in horizontal axis wind turbines. This is because the wind speed and direction change rapidly after hitting swept blade rotors. Therefore, the amount of wind speed measurements by installed anemometers are not equal to the speed of the wind flow hitting turbine blades [25]. These differences, which lead to a decrease in the correlation between the measured wind speed and the output power, and then scattering of the power curve, can be resolved by averaging the samples in a reasonable average period [25]. The SCADA data for this study was recorded with a 1 Hz frequency; as a result, it was possible to create multiple averaged sets for removing the mentioned obstacle. According to a review conducted by Hanifi et al. [1], the maximum sampling rate used for wind speed and power forecasting in the previous research is 10 min. This is equivalent to an average time that the international standard for power performance measurements of electricity-producing wind turbines (IEC 61400-12-1) establishes for large wind turbines [37]. Based on the IEC 61400-12-1 and reviewed literature, the data presented here was averaged for each 10 min of data collection. Figure 10a,b show the wind power curves for the original and 10 min resampled data.

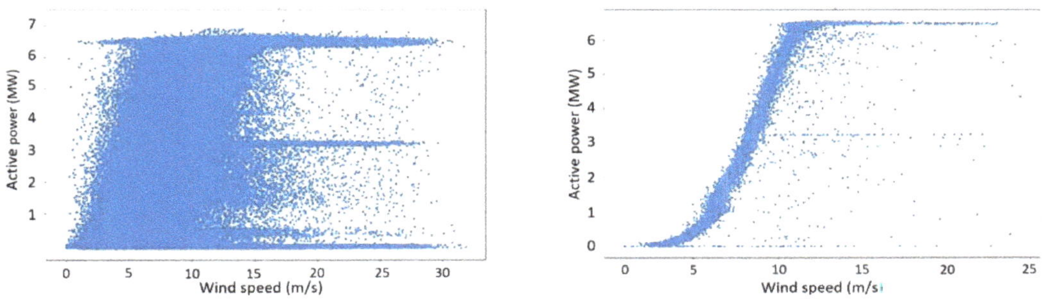

Figure 10. Wind power curves. (**a**) Original 1s data (**b**) and 10 min resampled data.

2.10. Anomalies Detection and Treatment

Outliers in a dataset are specific data points that are different or far from most other regular data points [38]. Undetected or improperly treated anomalies can adversely affect wind power forecasting applications. They may be biased with high prediction errors [38].

There are various reasons for having outliers among wind turbine and wind farm measurements, including wind turbine downtime [36], data transmission, processing or management failure [39], data acquisition failure [40], electromagnetic disturbance [36], wind turbine control system fault (such as the pitch control system fault) [41], damage of the blades or the existence of ice or dust [42], shading effect of neighbouring turbines, fluctuation of air density [43], etc.

Figure 11 shows four different types of anomalies in the current SCADA data. Category A points have negative, zero, or low values of generated power during speeds larger than the cut-in speed [25]. The leading causes of these outliers are wrong wind power measurements, wind turbine failure, and unexpected maintenance. Wind speed sensors and communication errors cause category B outliers. The mid-curve outliers (category C) represent power values lower than ideal—this is caused by the down-rating of the wind turbines and data acquisition. Outliers in category D are scattered irregular points due to faulty sensors exacerbated during harsh weather circumstances [36].

There are different methods for anomaly detection in machine learning, such as Density-Based Spatial Clustering of Applications with Noise (DBSCAN), IF, local outlier factor, and EE. In this study, three common methods for wind power forecasting are investigated. EE is used based on the assumptions described in [44]. IF, which is an unsupervised learning algorithm, recognises anomalies by isolating them in the data. This algorithm works based on two main features of anomalies, that they are few and different. The one-class support vector machine (OCSVM) is a common unsupervised learning algorithm for outlier detection, assuming rare anomalies create a boundary for most data, and considering data points out of the boundary as outliers [45]. This method of outlier detection and treatment chose the third method.

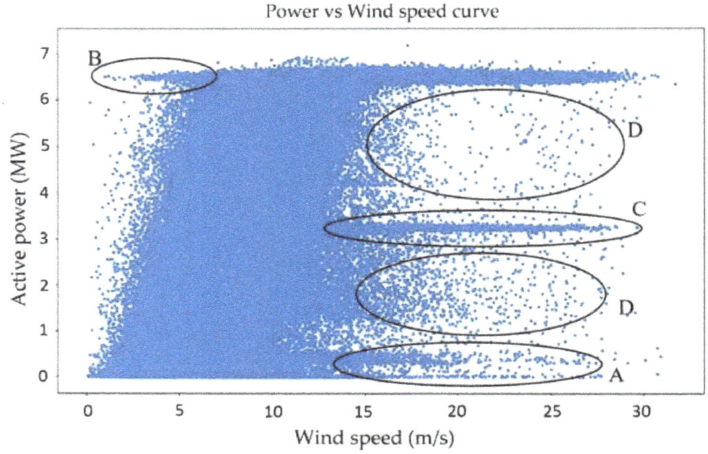

Figure 11. Observed anomalies coupled with the power curve of the 1 Hz original data. (A) Low power output in high wind speeds in turbine failure cases; (B) Outliers due to the wind speed sensor and communication errors; (C) Power outputs less than the rated power as a result of the turbine's down-rating; (D) Scattered outliers caused by sensor malfunctions or noise in signal processing.

3. Experimental Results and Discussion

This research employs packages and subroutines written in Python to implement the proposed algorithms. A PC with an Intel Core i5–7300 32.6 GHz CPU and 8 GB RAM (without any GPU processing) was used to run the experiments. Three outlier detection methods, which were described in Section 2.8, were used to detect and remove the outliers of the resampled dataset. The results of these treatments can be seen in Figures 12–14:

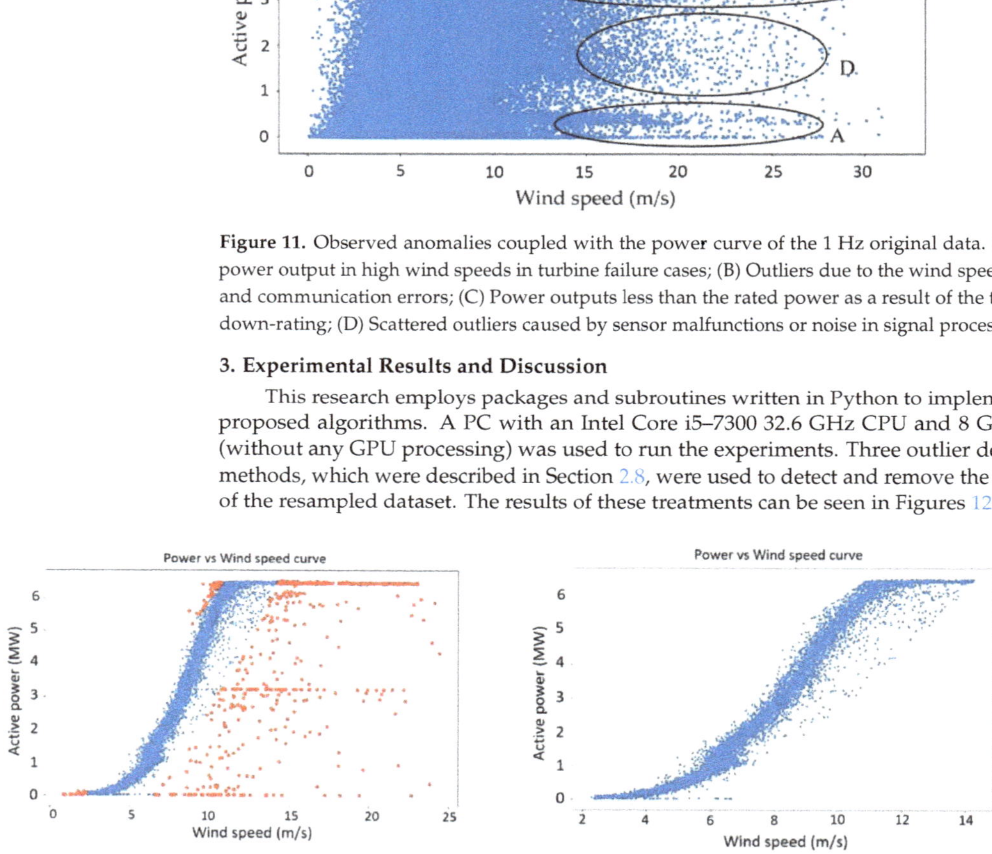

Figure 12. Elliptic envelope application for outlier detection and treatment. The blue points represent the normal data, and the red represents the detected anomalies.

This study considers six different preprocessing methods based on applying three different outlier detection methods and three approaches against the negative power values (Table 4). Different cases of preprocessed data are fed to the ARIMA and LSTM forecasting models. The grid search method is applied for the initial hyperparameter tuning; Table 4 shows the selected hyperparameters for the ARIMA and LSTM models. As expected, the values of the hyperparameters vary depending on the different employed preprocessing methods (Table 4).

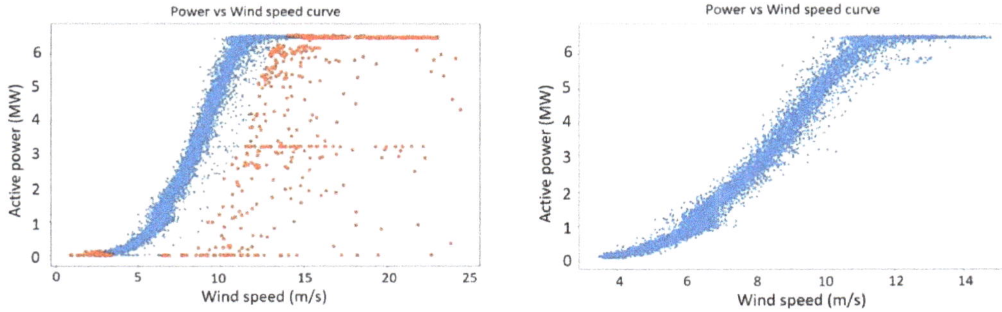

Figure 13. Isolation forest application for outlier detection and treatment. The blue points represent the normal data, and the red represents the detected anomalies.

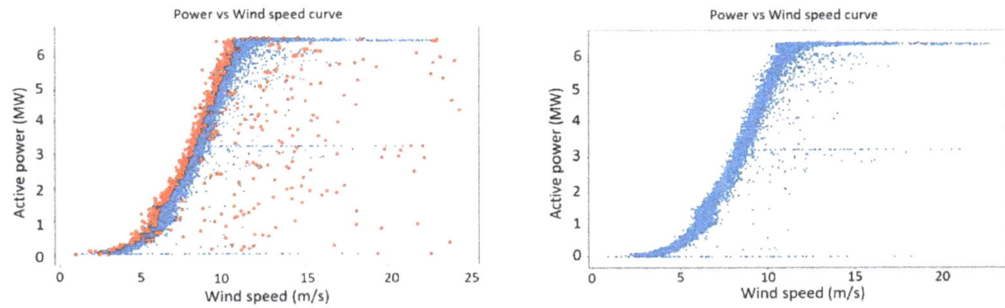

Figure 14. OCSVM application for outlier detection and treatment. The blue points represent the normal data, and the red represents the detected anomalies.

Table 4. Best ARIMA and LSTM hyperparameters resulting from the grid search.

Case Data	Preprocessing Approach	ARIMA Hyperparameters	LSTM Hyperparameters
Case 1	Negatives replaced by mean *; Outliers not removed	(2, 0, 1)	(3, 100, 100, 150, 0)
Case 2	Negatives replaced by nearest positive value *; Outliers not removed	(1, 1, 1)	(6, 100, 150, 150, 0)
Case 3	Negatives removed; Outliers not removed	(1, 0, 2)	(3, 100, 150, 150, 0)
Case 4	Negatives removed; Outliers removed by EE method	(1, 1, 3)	(3, 100, 100, 150, 0)
Case 5	Negatives removed; Outliers removed by IF method	(3, 1, 1)	(3, 150, 150, 150, 0)
Case 6	Negatives removed; Outliers removed by OCSVM method	(1, 1, 3)	(3, 150, 150, 150, 0)

*: Mean value has been calculated after removing negative values.

After selecting the best ARIMA and LSTM prediction methods, both models were trained by the first 95% part of the dataset (as training data) to make predictions for the last 5% of the dataset. The predicted values were compared with the measured values to determine the RMSE of each forecasting process. Table 5 provides the RMSE values of the ARIMA, LSTM, and persistence methods.

Table 5. RMSE values of persistence, ARIMA, and LSTM models for six different treated case data.

	Data 1	**Data 2**	**Data 3**	**Data 4**	**Data 5**	**Data 6**
Negative values	Mean [1]	Nearest Positive 2	Removed	Removed	Removed	Removed
Outliers	Not removed	Not removed	Not removed	EE removed	IF removed	OCSVM removed
Persistence	636.3	720.5	830.5	512.7	509	566
ARIMA	622.8	713	813	505.3	503.3	559
LSTM	626	695.8	785	501.5	497.5	550.5

[1] Replaced by mean value calculated after removing negative values. [2] Replaced by nearest positive value.

Comparing the RMSE values of all three models (Table 5) for case data 1, 2, and 3 clarifies that the complete elimination of the negative values (without any replacement) will lead to worse forecasting. The highest RMSE value of case 3 means that removing the negative values will decrease the forecasting accuracy. One of the reasons for this performance drop can be the creation of discontinuity in the dataset.

Regarding the best specific value to be considered instead of negatives, a comparison of case data 1 and 2 proves that replacing the negative values with the average wind power values has a better impact than replacing them with the nearest (neighbour) positive value. Replacing the negative values with the average values can lead to about a 15% forecasting improvement for ARIMA and 11% for the LSTM models.

The results also highlight the importance of dealing with outliers in wind power forecasting. Cases 4, 5, and 6, representing the outlier removed data, show a significant enhancement of the accuracy rather than the other cases, without any action against the anomalies. Comparing the error levels of case data 3 with cases 4, 5, and 6 (for both ARIMA and LSTM models) shows a 30% to 38% forecasting improvement by the elimination of the outliers, either by isolation forest, elliptic envelope, or the one-class SVM outlier detection methods.

The assessment of the RMSE values of cases 4, 5, and 6 show that the IF and EE outlier detection methods overcome the OCSVM method. An elliptic envelope can improve forecasting performance up to 9.61% and 8.92% rather than OCSVM for ARIMA and LSTM methods. This performance enhancement can reach 9.96% and 9.64% for ARIMA and LSTM, respectively, by applying the isolation forest.

As shown in Table 5, the ARIMA and LSTM methods for all the treated case data have better performances than the persistence methods. This is understandable if one remembers that, in the persistence method, only one preceding step data is used for forecasting, whilst the ARIMA and LSTM models consider a more extensive range of prior data.

It is also clear that the LSTM performs better than the ARIMA almost for all approaches against the negative values and outliers. This is probably due to the fact that LSTMs are better equipped to learn long-term correlation. In addition, the LSTM can better capture the nonlinear dependencies between the features.

In this study, because of the better prediction performance of the LSTM model compared to the ARIMA model, the proposed optimisation algorithm is applied to the LSTM model to tune its hyperparameters even more. As discussed in Section 2.5, the hyperparameter ranges of the LSTM model are increased from what was examined in its grid search to the wider ranges shown in Table 2.

The six preprocessed case data are again divided into the first 95% as the training dataset and the rest 5% as the test data. These divisions were developed to establish the same conditions and logically compare the new and previous methods. The developed optimisation algorithm, with the two described strategies, including search and pruning, started the selection of different combinations to minimise the RMSE value. Table 6 shows the new hyperparameters found by the Optuna optimisation algorithm, and Figure 15 shows the measured power values of the turbine and prediction results of all the forecasting methods, including ARIMA, LSTM–grid, and LSTM–Optuna, for one of the datasets (data 4—removed negative values and removed outliers with the EE method).

Table 6. Best LSTM hyperparameters resulted from Optuna optimisation.

Case Data	Preprocessing Approach	LSTM Hyperparameters
Case 1	Negatives replaced by mean *, Outliers not removed	(9, 190, 230, 130, 0)
Case 2	Negatives replaced by nearest positive value *, Outliers not removed	(8, 60, 270, 170, 0)
Case 3	Negatives removed; Outliers not removed	(3, 120, 120, 160, 0)
Case 4	Negatives removed; Outliers removed by EE method	(6, 180, 180, 180, 0)
Case 5	Negatives removed; Outliers removed by IF method	(4, 260, 120, 280, 0)
Case 6	Negatives removed; Outliers removed by OCSVM method	(2, 130, 280, 90, 1)

*: Mean value has been calculated after removing negative values.

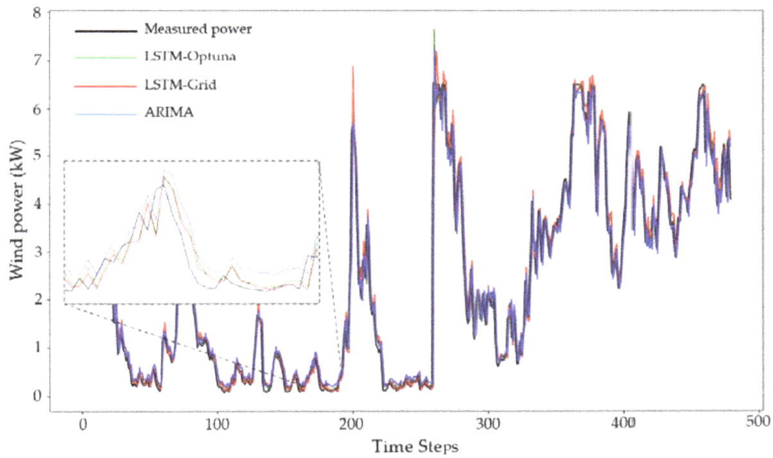

Figure 15. Comparison of measured wind power and forecasted values by ARIMA, LSTM–grid, and LSTME–Optuna models for data 4 (removed negative values and removed outliers with EE method).

As can be seen in Figure 15, the LSTM model optimised by Optuna can predict more accurately by better learning the wind power's short-term and long-term dependencies. The diagram illustrated in Figure 16 is plotted to better compare the error levels of the different wind power forecasting methods. It can be recognised from this diagram that the LSTM–Optuna approach follows rules similar to the ARIMA and LSTM–grid models. To achieve a higher prediction accuracy, it is essential to eliminate the outliers and replace the negative power values with the average wind power value.

Building the LSTM models based on the new values of the hyperparameters, as shown in Table 6, improves the prediction accuracy of the LSTM model in a range from 1.22% to 2.65% for different cases of preprocessed data. These accuracy improvements can be seen in Table 7.

The results show that the highest accuracy improvement is related to case 5, a case in which negative values were replaced with the mean power value and the outliers were removed through the IF method. A comparison of the required search times to find the best combination of the hyperparameters in LSTM–grid and LSTM–Optuna proves the faster performance of the proposed method, as it spends from 13.79% to 20.59% less time adjusting the model for the most accurate prediction (Table 8).

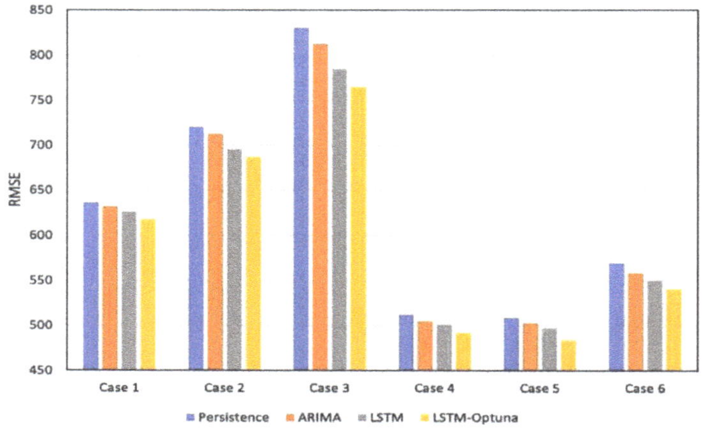

Figure 16. Error comparison of persistence, ARIMA, LSTM, and LSTM optimised by Optuna forecasting methods.

Table 7. A comparison of RMSE, the LSTM–grid search, and LSTM–Optuna methods.

Predictive Model	Case 1	Case 2	Case 3	Case 4	Case 5	Case 6
LSTM tuned by grid search	626	695.8	785	501.5	497.5	550.5
LSTM optimised by Optuna	617.8	687.3	765	492.4	484.3	540.3
Accuracy improvement	1.31%	1.22%	2.55%	1.81%	2.65%	1.78%

Table 8. A comparison of the required tuning time of the LSTM–grid search and LSTM–Optuna methods.

Predictive Model	Case 1	Case 2	Case 3	Case 4	Case 5	Case 6
LSTM tuned by grid search	680	660	290	396	380	340
LSTM optimised by Optuna	540	530	250	340	320	280
Spent time improvement	20.59%	19.70%	13.79%	14.14%	15.79%	17.65%

4. Conclusions

This study addresses issues regarding inaccurate wind power prediction using ML approaches. As discussed in the reviewed literature, most previous research applied ML without advanced model optimisation. At the same time, in this paper, a novel concept of Optuna–LSTM is reported to expedite the process of selecting the hyperparameters and tuning the wind power forecasting models. This model not only reduces the time complexity of creating reliable models, but also improves the accuracy of the predictions.

To accurately evaluate the proposed model, SCADA data of an offshore wind turbine was preprocessed by eliminating its negative values and outliers to help find the best preprocessing method. The performance of the proposed forecasting was demonstrated through comparisons with the persistence, ARIMA, and LSTM models, which were already tuned by grid search. This comparison proved the better performance of the proposed model, with a range up to 7.89, 5.9, and 2.65 percent compared to the persistence and conventional grid-search-tuned ARIMA and LSTM models.

This study also highlights the importance of eliminating negative values in the power recordings. The results of this study confirmed that replacing the negative values with the average power value has the most positive effect on the forecasting accuracy. In addition, comparisons between several data cases showed the significant impact of the outlier treatment methods on the forecasting performance. The results proved that removing the outliers by the isolation forest method improves the forecast accuracy compared to the elliptic envelope and OCSVM methods. This novel forecasting method combining the capacity of the LSTM model in the prediction of nonlinearities and the optimisation tool for better tuning the hyperparameters can be used for different time-series-based predictions.

Author Contributions: Conceptualisation, S.H. and S.L.; methodology, S.H. and S.L.; investigation, S.H. and S.L.; writing—original draft preparation, S.H.; writing—review and editing, S.L., H.Z.-B., A.C. and S.H.; supervision, S.L.; data curation, S.H. All authors have read and agreed to the published version of the manuscript.

Funding: This research was funded by the EPSRC Doctoral Training Partnership (EP/R513222/1).

Institutional Review Board Statement: Not applicable.

Informed Consent Statement: Not applicable.

Data Availability Statement: The data presented in this study are available on request from the corresponding author. The data are not publicly available because it also forms part of an ongoing study.

Acknowledgments: The authors are grateful to Xiaolei Liu for his assistance and contribution to this research.

Conflicts of Interest: The authors declare no conflict of interest.

Nomenclature

Latin symbols	
b_i, b_o, b_f	LSTM model biases
d	degree of ARIMA differencing
f_t	LSTM forget gate
h_t	LSTM overall output
i_t	LSTM input gate
o_t	LSTM output gate
p	order of autoregressive
P_t	measured wind power at the time t
$\hat{P}_{t+k/t}$	predicted wind power for the future time k
q	order of moving average model
U_i, U_o, U_f	LSTM assigned weights
W_i, W_o, W_f	LSTM assigned weights
h_{t-1}	cell state vector at time step $t-1$
x_t	neuron input at time step t
X_t	forecasted wind power
Greek symbols	
σ_l	activation function
σ_s	activation function
ϕ_t	ARIMA model coefficient
θ_t	ARIMA model coefficient
Abbreviation	
ACF	autocorrelation function
ADF	augmented Dickey–Fuller
ANN	artificial neural network
AR	auto-regressive
ARMA	Auto-Regressive Moving Average Model
ARIMA	Auto-Regressive Integrated Moving Average
ARX	Auto-Regressive with Exogenous variable

BP	back propagation
BPNN	back propagation neural network
CEC	constant error carousel
DBN	deep belief network
DGF	double Gaussian function
EE	elliptic envelope
FFNN	feed-forward neural network
IF	isolation forest
LDT	Levenmouth Demonstration Turbine
LSTM	long short-term memory
MA	moving average
MAE	mean absolute error
MSE	mean square error
NN	neural network
NWP	numerical weather prediction
ORE	offshore renewable energy
PACF	partial autocorrelation function
RBF	radial basis function
RMSE	root mean square error
RNN	recurrent neural network
SCADA	Supervisory Control and Data Acquisition

References

1. Hanifi, S.; Liu, X.; Lin, Z.; Lotfian, S. A critical review of wind power forecasting methods-past, present and future. *Energies* **2020**, *13*, 3764. [CrossRef]
2. Yang, B.; Zhong, L.; Wang, J.; Shu, H.; Zhang, X.; Yu, T.; Sun, L. State-of-the-art one-stop handbook on wind forecasting technologies: An overview of classifications, methodologies, and analysis. *J. Clean. Prod.* **2021**, *283*, 124628. [CrossRef]
3. Global Wind Energy Council. *Global Wind Report 2022*; Global Wind Energy Council: Brussels, Belgium, 2022.
4. Yu, C.; Li, Y.; Xiang, H.; Zhang, M. Data mining-assisted short-term wind speed forecasting by wavelet packet decomposition and Elman neural network. *J. Wind Eng. Ind. Aerodyn.* **2018**, *175*, 136–143. [CrossRef]
5. Yatiyana, E.; Rajakaruna, S.; Ghosh, A. Wind speed and direction forecasting for wind power generation using ARIMA model. In Proceedings of the 2017 Australasian Universities Power Engineering Conference, AUPEC 2017, Melbourne, Australia, 19–22 November 2017; pp. 1–6.
6. Firat, U.; Engin, S.N.; Sarcalar, M.; Ertuzum, A.B. Wind speed forecasting based on second order blind identification and autoregressive model. In Proceedings of the 2010 Ninth International Conference on Machine Learning and Applications, Washington, DC, USA, 12–14 December 2010; pp. 686–691.
7. de Felice, M.; Alessandri, A.; Ruti, P.M. Electricity demand forecasting over Italy: Potential benefits using numerical weather prediction models. *Electr. Power Syst. Res.* **2013**, *104*, 71–79. [CrossRef]
8. Durán, M.J.; Cros, D.; Riquelme, J. Short-term wind power forecast based on ARX models. *J. Energy Eng.* **2007**, *133*, 172–180. [CrossRef]
9. Kavasseri, R.G.; Seetharaman, K. Day-ahead wind speed forecasting using f-ARIMA models. *Renew. Energy* **2009**, *34*, 1388–1393. [CrossRef]
10. Torres, J.L.; García, A.; de Blas, M.; de Francisco, A. Forecast of hourly average wind speed with ARMA models in Navarre (Spain). *Sol. Energy* **2005**, *79*, 65–77. [CrossRef]
11. Mujeeb, S.; Alghamdi, T.A.; Ullah, S.; Fatima, A.; Javaid, N.; Saba, T. Exploiting deep learning for wind power forecasting based on big data analytics. *Appl. Sci.* **2019**, *9*, 4417. [CrossRef]
12. Azimi, R.; Ghofrani, M.; Ghayekhloo, M. A hybrid wind power forecasting model based on data mining and wavelets analysis. *Energy Convers. Manag.* **2016**, *127*, 208–225. [CrossRef]
13. Zhang, J.; Yan, J.; Infield, D.; Liu, Y.; Lien, F.-S. Short-term forecasting and uncertainty analysis of wind turbine power based on long short-term memory network and Gaussian mixture model. *Appl. Energy* **2019**, *241*, 229–244. [CrossRef]
14. Fu, Y.; Hu, W.; Tang, M.; Yu, R.; Liu, B. Multi-step ahead wind power forecasting based on recurrent neural networks. In Proceedings of the 2018 IEEE PES Asia-Pacific Power and Energy Engineering Conference (APPEEC), Kota Kinabalu, Malaysia, 7–10 October 2018; pp. 217–222.
15. Cali, U.; Sharma, V. Short-term wind power forecasting using long-short term memory based recurrent neural network model and variable selection. *Int. J. Smart Grid Clean Energy* **2019**, *8*, 103–110. [CrossRef]
16. Seyedzadeh, S.; Rahimian, F.P.; Oliver, S.; Glesk, I.; Kumar, B. Data driven model improved by multi-objective optimisation for prediction of building energy loads. *Autom. Constr.* **2020**, *116*, 103188. [CrossRef]
17. Seyedzadeh, S.; Rahimian, F.P.; Rastogi, P.; Glesk, I. Tuning machine learning models for prediction of building energy loads. *Sustain. Cities Soc.* **2019**, *47*, 101484. [CrossRef]

18. Al-Douri, Y.K.; Hamodi, H.; Lundberg, J. Time series forecasting using a two-level multi-objective genetic algorithm: A case study of maintenance cost data for tunnel fans. *Algorithms* **2018**, *11*, 123. [CrossRef]
19. Shahid, F.; Zameer, A.; Muneeb, M. A novel genetic LSTM model for wind power forecast. *Energy* **2021**, *223*, 120069. [CrossRef]
20. Box, G.; Jenkins, G.; Reinsel, G.; Ljung, G. *Time Series Analysis: Forecasting and Control*; John Wiley & Sons: Somerset, NJ, USA, 2015.
21. Erdem, E.; Shi, J. ARMA based approaches for forecasting the tuple of wind speed and direction. *Appl. Energy* **2011**, *88*, 1405–1414. [CrossRef]
22. Ho, S.L.; Xie, M.; Goh, T.N. A comparative study of neural network and Box-Jenkins ARIMA modeling in time series prediction. *Comput. Ind. Eng.* **2002**, *42*, 371–375. [CrossRef]
23. Nair, K.R.; Vanitha, V.; Jisma, M. Forecasting of wind speed using ANN, ARIMA and hybrid models. In Proceedings of the 2017 International Conference on Intelligent Computing, Instrumentation and Control Technologies (ICICICT), Kerala, India, 6–7 July 2017; pp. 170–175.
24. Liu, H.; Tian, H.-Q.; Li, Y.-F. Comparison of two new ARIMA-ANN and ARIMA-Kalman hybrid methods for wind speed prediction. *Appl. Energy* **2012**, *98*, 415–424. [CrossRef]
25. Kisvari, A.; Lin, Z.; Liu, X. Wind power forecasting—A data-driven method along with gated recurrent neural network. *Renew. Energy* **2021**, *163*, 1895–1909. [CrossRef]
26. López, E.; Valle, C.; Allende, H.; Gil, E.; Madsen, H. Wind power forecasting based on echo state networks and long short-term memory. *Energies* **2018**, *11*, 526. [CrossRef]
27. Jung, J.; Broadwater, R.P. Current status and future advances for wind speed and power forecasting. *Renew. Sustain. Energy Rev.* **2014**, *31*, 762–777. [CrossRef]
28. Soman, S.S.; Zareipour, H.; Malik, O.; Mandal, P. A review of wind power and wind speed forecasting methods with different time horizons. In Proceedings of the North American Power Symposium 2010, Arlington, TX, USA, 26–28 September 2010; pp. 1–8.
29. Akiba, T.; Sano, S.; Yanase, T.; Ohta, T.; Koyama, M. Optuna: A next-generation hyperparameter optimization framework. In Proceedings of the KDD '19: The 25th ACM SIGKDD Conference on Knowledge Discovery and Data Mining, Anchorage, AK, USA, 4–8 August 2019; pp. 2623–2631.
30. Snoek, J.; Larochelle, H.; Adams, R.P. Practical bayesian optimisation of machine learning algorithms. In Proceedings of the 25th International Conference on Neural Information Processing Systems, Siem Reap, Cambodia, 13–16 December 2018; pp. 159–170.
31. Hutter, F.; Hoos, H.; Leyton-Brown, K. Sequential model-based optimisation for general algorithm configuration lecture notes in computer science. In Proceedings of the International Conference on Learning and Intelligent Optimization, Rome, Italy, 17–21 January 2011; pp. 507–523.
32. Bergstra, J.; Komer, B.; Eliasmith, C.; Yamins, D.; Cox, D.D. Hyperopt: A Python library for model selection and hyperparameter optimisation. *Comput. Sci. Discov.* **2015**, *8*, 14008. [CrossRef]
33. Chatterjee, J.; Dethlefs, N. Deep learning with knowledge transfer for explainable anomaly prediction in wind turbines. *Wind Energy* **2020**, *23*, 1693–1710. [CrossRef]
34. Anon Platform for Operational Data (POD). Available online: https://pod.ore.catapult.org.uk/ (accessed on 31 August 2022).
35. Serret, J.; Rodriguez, C.; Tezdogan, T.; Stratford, T.; Thies, P. Code comparison of a NREL-fast model of the levenmouth wind turbine with the GH bladed commissioning results. In Proceedings of the ASME 2018 37th International Conference on Ocean, Offshore and Arctic Engineering, Madrid, Spain, 17–22 June 2018.
36. Shen, X.; Fu, X.; Zhou, C. A combined algorithm for cleaning abnormal data of wind turbine power curve based on change point grouping algorithm and quartile algorithm. *IEEE Trans. Sustain. Energy* **2019**, *10*, 46–54. [CrossRef]
37. Jafarian, M.; Ranjbar, A.M. Fuzzy modeling techniques and artificial neural networks to estimate annual energy output of a wind turbine. *Renew. Energy* **2010**, *35*, 2008–2014. [CrossRef]
38. Zou, M.; Djokic, S.Z. A review of approaches for the detection and treatment of outliers in processing wind turbine and wind farm measurements. *Energies* **2020**, *13*, 4228. [CrossRef]
39. Sainz, E.; Llombart, A.; Guerrero, J.J. Robust filtering for the characterisation of wind turbines: Improving its operation and maintenance. *Energy Convers. Manag.* **2009**, *50*, 2136–2147. [CrossRef]
40. Villanueva, D.; Feijóo, A. Normal-based model for true power curves of wind turbines. *IEEE Trans. Sustain. Energy* **2016**, *7*, 1005–1011. [CrossRef]
41. Taslimi-Renani, E.; Modiri-Delshad, M.; Elias, M.F.M.; Rahim, N.A. Development of an enhanced parametric model for wind turbine power curve. *Appl. Energy* **2016**, *177*, 544–552. [CrossRef]
42. Zhao, Y.; Ye, L.; Wang, W.; Sun, H.; Ju, Y.; Tang, Y. Data-driven correction approach to refine power curve of wind farm under wind curtailment. *IEEE Trans. Sustain. Energy* **2018**, *9*, 95–105. [CrossRef]
43. Schlechtingen, M.; Santos, I.F.; Achiche, S. Using data-mining approaches for wind turbine power curve monitoring: A comparative study. *IEEE Trans. Sustain. Energy* **2013**, *4*, 671–679. [CrossRef]
44. Lin, Z.; Liu, X.; Collu, M. Electrical power and energy systems wind power prediction based on high-frequency SCADA data along with isolation forest and deep learning neural networks. *Electr. Power Energy Syst.* **2020**, *118*, 105835. [CrossRef]
45. Tran, L.; Fan, L.; Shahabi, C. Outlier detection in non-stationary data streams. In Proceedings of the 31st International Conference on Scientific and Statistical Database Management, Santa Cruz, CA, USA, 23–25 July 2019; pp. 25–36.

energies

MDPI

Article

A Spiking Neural Network Based Wind Power Forecasting Model for Neuromorphic Devices

Juan Manuel González Sopeña [1,*], Vikram Pakrashi [2,3,4] and Bidisha Ghosh [1,5]

1 QUANT Group, Department of Civil, Structural and Environmental Engineering, Trinity College Dublin, D02 PN40 Dublin, Ireland
2 UCD Centre for Mechanics, Dynamical Systems and Risk Laboratory, School of Mechanical & Materials Engineering, University College Dublin, D04 V1W8 Dublin, Ireland
3 SFI MaREI Centre, University College Dublin, D04 V1W8 Dublin, Ireland
4 The Energy Institute, University College Dublin, D04 V1W8 Dublin, Ireland
5 CONNECT: SFI Research Centre for Future Networks & Communications, Trinity College Dublin, D02 PN40 Dublin, Ireland
* Correspondence: gonzlezj@tcd.ie

Abstract: Many authors have reported the use of deep learning techniques to model wind power forecasts. For shorter-term prediction horizons, the training and deployment of such models is hindered by their computational cost. Neuromorphic computing provides a new paradigm to overcome this barrier through the development of devices suited for applications where latency and low-energy consumption play a key role, as is the case in real-time short-term wind power forecasting. The use of biologically inspired algorithms adapted to the architecture of neuromorphic devices, such as spiking neural networks, is essential to maximize their potential. In this paper, we propose a short-term wind power forecasting model based on spiking neural networks adapted to the computational abilities of Loihi, a neuromorphic device developed by Intel. A case study is presented with real wind power generation data from Ireland to evaluate the ability of the proposed approach, reaching a normalised mean absolute error of 2.84 percent for one-step-ahead wind power forecasts. The study illustrates the plausibility of the development of neuromorphic devices aligned with the specific demands of the wind energy sector.

Keywords: neuromorphic computing; spiking neural network; short-term wind power forecasting

Citation: González Sopeña, J.M.; Pakrashi, V.; Ghosh, B. A Spiking Neural Network Based Wind Power Forecasting Model for Neuromorphic Devices. *Energies* **2022**, *15*, 7256. https://doi.org/10.3390/en15197256

Academic Editors: Paweł Piotrowski, Grzegorz Dudek and Dariusz Baczyński

Received: 5 September 2022
Accepted: 28 September 2022
Published: 2 October 2022

1. Introduction

A large number of machine learning (ML) and deep learning (DL) models have been developed and applied to time series data of a varied nature for tasks such as forecasting [1], classification [2], and clustering [3]. This trend has been also been observed in the field of wind power forecasting (WPF) [4], particularly the use of artificial neural networks (ANNs) [5], which are usually trained with the backpropagation algorithm [6]. Recurrent neural networks, such as the gated recurrent unit (GRUs) [7] and long short-term memory (LSTM) neurons [8], can learn temporal features on wind data, whereas convolutional neural networks (CNNs) capture spatial ones [9]. Other ML algorithms that have been applied in the literature are support-vector machines [10], random forests [11], gradient boosting machines [12], and neuro-fuzzy models [13,14]. DL methods such as deep neural networks are built by stacking multiple layers between the input and output layers to extract higher-level features from the data [15]. Deep neural architectures such as deep belief networks [16], deep convolutional networks [17] and N-BEATS [18] have been applied in the WPF literature. Furthermore, the abilities of ML/DL as a modelling tool have proven valuable for solar power forecasting [19] and renewable energy systems [20].

Accurate WPFs can be estimated using ML/DL architectures considering different types of data collected at a wind farm [21,22]. However, such models may be associated with a high computational cost, a critical factor for edge computing [23], including those applications for renewable energy [24]. For instance, the low latency inherent in neuromorphic devices can be critical for transmission system operators to manage the grid in real time and the decision-making process of traders participating in electricity markets, specifically to correct their positions in intraday markets. Thus, neuromorphic computing provides an alternative to the computational complexity of ML/DL models [25] with the development of devices inspired by the energy-efficient nature of biological systems such as the Intel's Loihi chip [26]. The architecture of spiking neural networks (SNNs) [27] resemble more closely biological neurons, and are thus adequate for implementation in neuromorphic devices to unleash their potential in terms of low latency and lower energy consumption. However, training SNNs remains a challenge, as the well-known backpropagation algorithm cannot be applied, due to the non-differentiable nature of spikes. The current approaches to train spiking DL algorithms can be broadly divided into *online* and *offline* approaches [28]. Online approaches first implement an SNN in neuromorphic hardware, leveraging on-chip plasticity to train the spiking network and evolve its parameters with the arrival of new data [29]. This approach includes online approximations of the backpropagation algorithm [30,31] and evolving SNNs [32]. On the other hand, the SNN is trained before deploying the model for offline approaches. These can be further divided into two categories, considering how the training stage is performed. One possibility is to train a conventional ANN using the backpropagation algorithm, and later map the parameters into an equivalent SNN model [33]. This approach is known as *ANN-to-SNN conversion*. Alternatively, a direct training approach uses a variation of error backpropagation to optimize directly the parameters of an SNN [34].

In addition, the research community has been developing specific software platforms to implement applications based on SNNs. For instance, Nengo [35] is a software based on the principles of the Neural Engineering Framework (NEF), a theoretical framework to implement large-scale neural models with cognitive abilities [36]. This same software was later extended with the sister library NengoDL [37], aiming to combine the principles of neuromorphic modelling with the well-known deep learning framework TensorFlow [38] to build deep spiking neural models by ANN-to-SNN conversion. Alternatively, other frameworks can directly train SNNs, such as the Spike Layer Error Reassignment (SLAYER) algorithm proposed by Shrestha and Orchard [39]. Recently, in October 2021, Intel's Neuromorphic Computing Lab released the first version of Lava [40], an open-source software framework, to implement neuromorphic applications for the Intel Loihi architecture [41].

The SNN features are maximized within the framework provided by neuromorphic computing. However, no study has realistically attempted to model short-term WPFs using SNNs while considering the current computational abilities of neuromorphic devices to date. We have to remember that neuromorphic computing is still in its infancy, so the goal is to reach an acceptable level of performance to build up our knowledge regarding the implementation of spiking-based models in WPF, and not to outperform the current well-established neural network models [28]. Therefore, we propose a SNN model for short-term WPF, adapted to the hardware capacity of the current state-of-the-art neuromorphic devices, particularly the neuromorphic chip Loihi developed by Intel. The aim of this study is not solely constrained to achieving highly accurate WPFs, but also to efficiently design WPF models that leverage the neuromorphic processors' power efficiency. The proposed forecasting approach was designed by applying the modelling framework provided by NengoDL to build spiking neuron models, and NengoLoihi, a complementary library, to implement such models on Loihi hardware.

The rest of this paper is structured as follows. Section 2 describes the ANN-to-SNN conversion method used to train spiking neural networks, as well as the spiking model architecture tailored to Loihi hardware. Section 3 presents a case study using this methodology for short-term WPF, using real data from an Irish wind farm. Section 4 contains the concluding remarks and the scope for future research work.

2. Methodology

NengoDL [37] is a modelling framework that includes tools to design biological neuronal models and the optimization methods used to train ML/DL models. Such optimization methods are usually incompatible with SNNs, as spikes are not differentiable. NengoDL links SNNs and these optimization methods by performing the necessary transformations to apply the ANN-to-SNN conversion method proposed by Hunsberger and Eliasmith [42], which allows for the use of a rate-based version of the spiking model in the training stage and the SNN for inference. The design of this rate-based approximation is key to successfully mapping its parameters into a spiking network, so the parameters of the network must be carefully tuned to ensure a minimal loss of performance during the conversion, and the architecture of the model must be tailored to subsequently build the network on Loihi hardware. Typically, six steps were followed to build and evaluate the performance of our proposed SNN model within the framework provided by NengoDL, as follows:

1. Build the non-spiking neural model as usual. The network must be designed considering the specific requirements for its implementation on Loihi hardware, such as the communication with the chip.
2. Train the equivalent rate-based network with the methodology described by Hunsberger and Eliasmith [42], the default method implemented in NengoDL to train SNNs.
3. Replace the activation functions with their spiking counterparts. We used spiking Rectified Linear Unit (ReLU) activations for the inference process. The activation profile of this function is restricted by the discretization required for the Loihi chip [43], leading to discrepancies compared to the theoretical spiking ReLU activation (Figure 1). Such discrepancies increase for higher firing rates due to this discretization. Furthermore, the Loihi chip can only fire a spike once per timestep, limiting its firing rate at a maximum of 1000 Hz. This constraint does not exist otherwise, and multiple spikes could, in theory, be fired simultaneously and exceed that value [44].
4. Run the network using the NengoDL framework, setting parameters such as the number of timesteps that each input will present to the spiking model, allowing for the network to settle and spike in the given timeframe, and the firing rate scale, letting the network spike at a higher rate. These preliminary results will help us monitor the neural activities and tune the parameters of the SNN.
5. Once an acceptable model performance is reached, we need to configure some additional parameters to set up the SNN for Loihi and simulate it for either Loihi hardware or the emulator [45] to replicate the chip's behavior. This is achieved with the extra functionalities provided by the library NengoLoihi.
6. Collect the results to evaluate them. One-step ahead point predictions are calculated, and normalized mean absolute error (NMAE) [46] is the metric used to measure the accuracy of these forecasts.

In the remainder of this section, we introduce how the ANN-to-SNN conversion is performed and the model architecture chosen to forecast wind power.

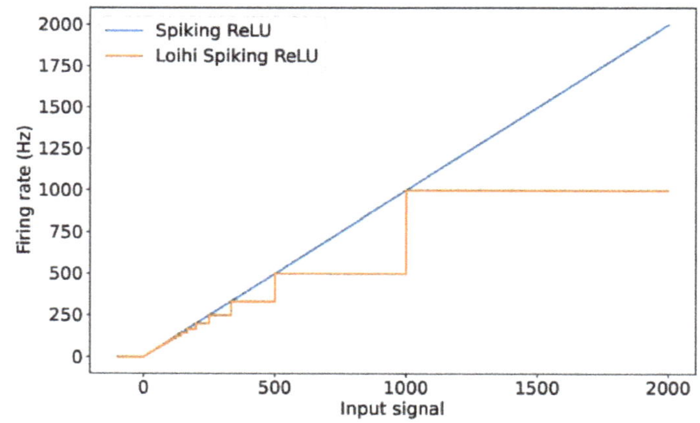

Figure 1. Spiking ReLU activation profile (based on DeWolf et al. [43]).

2.1. ANN-to-SNN Conversion

The non-differentiable nature of spikes impedes the use of the backpropagation algorithm to train spiking neurons [47]. ANN-to-SNN conversion sorts this out by mapping the parameters of a trained ANN to an equivalent SNN. Thus, the main challenge is how to train the non-spiking model so that there is only a small loss of performance in the conversion process. The first point is choosing an adequate spiking activation function. Cao et al. [48] established an equivalence between the ReLU activation function [49] and the spiking neuron's firing rate. The ANN-to-SNN conversion method implemented in NengoDL was proposed by Hunsberger and Eliasmith [42]. This method is valid for both linear (such as ReLU) and non-linear activation functions such as leaky integrate-and-fire (LIF) by smoothing the equivalent rate equation employed to train the ANN. To understand this, let us look at the equation governing the dynamics of an LIF neuron:

$$\tau_{RC}\frac{dv(t)}{dt} = -v(t) + I(t) \tag{1}$$

where τ_{RC} is the membrane time constant, $v(t)$ is the membrane voltage, and $I(\cdot)$ is the input current. The neuron will fire a spike if it reaches a certain threshold $\mathcal{V}$ and after the potential is reset during a certain period of time (known as refractory period τ_{ref}). The dynamics of the neuron are recovered after the refractory period τ_{ref} is ended. If a constant input current is given to the neuron, the steady-state firing rate (i.e., the time that it takes to the neuron to reach the threshold to fire a spike) can be determined as:

$$r(j) = \left[\tau_{ref} + \tau_{RC}\log(1 + \frac{\mathcal{V}}{\rho(j-\mathcal{V})})\right]^{-1} \tag{2}$$

where $\rho(x) = max(x, 0)$. However, this function is not completely differentiable, so the LIF rate equation is softened to address this problem and allow for use of the backpropagation algorithm [42]. The hard maximum ρ is replaced by a soft maximum ρ' defined as:

$$\rho'(x) = \gamma\log(1 + e^{x/\gamma}) \tag{3}$$

After training the conventional ANN, the parameters of the SNN are identical to its non-spiking counterpart, with only the neurons themselves changing. The performance of the spiking network can be further enhanced by tuning additional parameters. For instance, if using a linear activation function for the spiking forecasting model, the spiking firing rate can easily be increased after training by applying a scale to the input weights of the neurons to make them spike at a faster rate. The output of the network is divided for the same scale to not affect the behavior of the trained network. This way of proceeding is not optimal for non-linear activation functions. Instead, the firing rates can be optimized during training with regularization, so the firing rates are encouraged to spike at a certain firing rate [43]. Furthermore, a synaptic filter can be applied to reduce any possible noise found in the output of the spiking network.

2.2. Spiking Model Architecture

The model architecture (Figure 2) is slightly different to conventional ANNs, as it has to be adapted to the requirements of the Loihi hardware. The first distinctive feature of this network is the *off-chip* layer. This layer is a prerequisite to transmitting any information with the hardware, as it only communicates with spikes. Thus, this initial layer is run off-chip and converts the input into spikes [44]. The rest of the network is run on the hardware. A convolutional (*conv-layer*) and a regular fully connected layer (*dense-layer*) are used to process the data and generate the forecast. The convolutional layer is constituted by filters in the form of convolutions:

$$(f * g)(t) = \int_{-\infty}^{\infty} f(\tau)g(t-\tau)d\tau \quad (4)$$

where $(f * g)$ indicates the convolution between the functions f and g, in which the function f can be considered as a *filter* or *kernel* and g as the input data. On the right hand side, $g(t-\tau)$ indicates that the input data g are reversed and shifted to a certain time t. It is important to notice that not all types of neural networks are currently available in this ANN-to-SNN conversion framework (e.g., LSTM neurons are not supported). The activation function of all these three layers will be a spiking ReLU activation for inference. The equivalent ANN used during training follows the same architecture, including the *off-chip* layer, although it only behaves as a regular convolutional layer in this case. ReLU activation functions are used instead during the training stage.

Following up on our previous work [46,50], this model architecture is applied to 10-min resolution wind power data. The data used in this paper were collected for a wind turbine of a site located in Ireland (the exact location cannot be disclosed due to confidentiality reasons) for a two-year-and-a-half period (from January 2017 to June 2019). As input, the model uses previous wind power observations to provide the one-step-ahead forecasts as the output.

The wind power data were preprocessed using the variational mode decomposition (VMD) algorithm [51]. In particular, the data were decomposed into 8 subseries (known as modes) with different levels of complexity [52], giving us the opportunity to examine and adapt the SNN architecture under varied conditions. The forecasts of each mode were later aggregated to subsequently estimate the WPF [53].

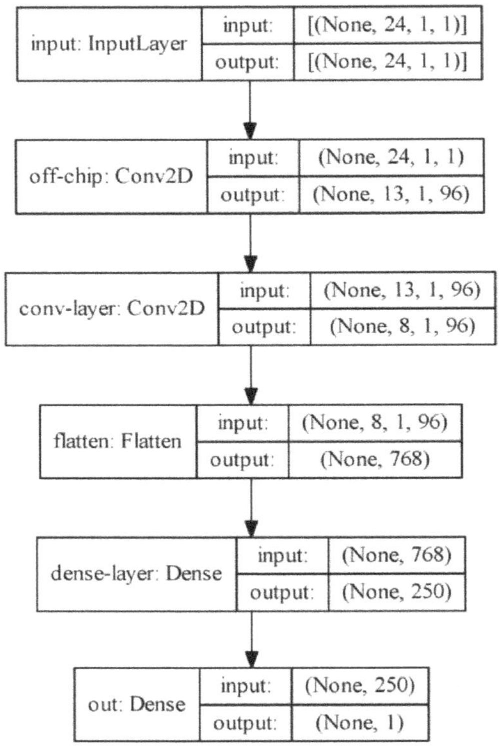

Figure 2. SNN model architecture.

3. Results

First, examples using a synthetic sine wave signal and load data are given to clarify some details of the steps that need to be followed to successfully convert an ANN model into a spiking one. Later, a case study is presented using data from an Irish wind farm.

3.1. Synthetic Signal Forecasting

Before applying the methodology to wind power data, let us present an example with a more simple signal (a synthetic sine wave) to clarify and further explain the details of tuning the parameters to achieve a good performance with the spiking network. For simplicity, the example using this signal was run within the NengoDL framework, so any additional parameters used to implement the model on Loihi hardware can be dismissed (such as the off-chip layer); therefore, a basic feedforward neural network (FFNN) model was used instead of the previously described model architecture, which suffices to accurately predict such a basic signal.

During the initial evaluation of the spiking network model (Steps 3 and 4), considering the discretization of the activation function required for Loihi hardware is of importance to posteriorly transfer our model without a significant drop in performance. Therefore, we must be particularly careful when scaling the firing rate of the spikes, as very high rates will not work on Loihi hardware. Let us examine the implications of disregarding this point with the example shown in Figure 3: we build the FFNN model (Step 1) and train it with a rate-based (i.e., non-spiking) ReLU activation (Step 2). Then, we replace the activation for its spiking counterpart, scaling the firing rate with a high enough value (Step 3). The neural activities of three neurons when presenting an input are shown in

Figure 3a,b, having replaced the ReLU activation function with the theoretical spiking ReLU and the discretized version with Loihi, respectively. Two of these neurons (shown in green and yellow) fire very fast in the first case, but their behavior is diminished in the second one due to the activation profile, impacting the performance of the model when all the input vectors conforming with the testing set are presented to the network (Step 4), as displayed in Figure 3c). Thus, the firing rate of this network should be lowered to satisfy the hardware specifications required in the following steps to implement the model on neuromorphic devices.

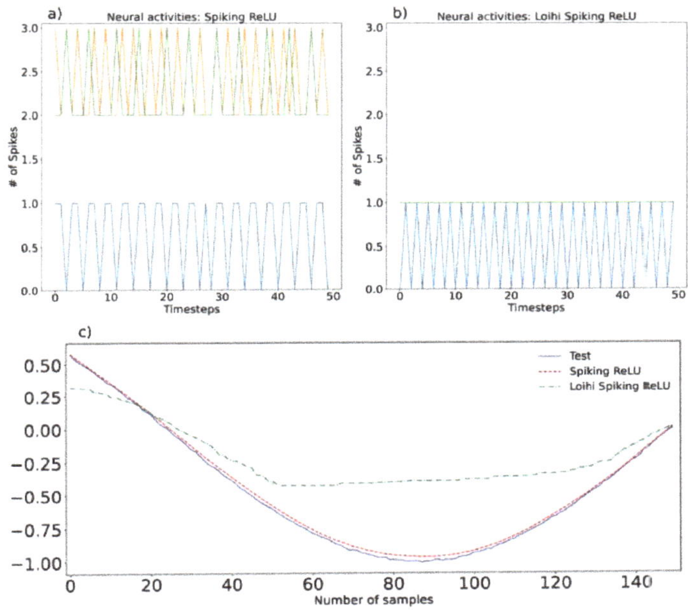

Figure 3. (**a**) Neural activities using a spiking ReLU activation for inference (one input vector is shown to the network during 50 timesteps), (**b**) neural activities using the discretized version of the spiking ReLU activation, and (**c**) predictions over the testing set.

The tuning of the firing rate scale, as well as the amplitude of the spikes, are essential to achieve a good forecasting accuracy while simultaneously trying to find a balance between the firing rates (enough spikes must be generated to transmit the information to the network) and the sparsity of spiking networks (leveraging the promise of low energy consumption by neuromorphic devices). Following the same example, let us a fix a certain spiking amplitude and experiment with different firing rate scales to find this trade-off, considering the Loihi-tailored spiking ReLU activation. The neural activities of the same three neurons are shown in Figure 4a for a scale of 1 (i.e., keeping the same input weights as the original SNN), in Figure 4b for a scale of 5 (a linear scale of 5 is applied to the inputs of the neurons), and in Figure 4c using a scale of 50. As expected, the spikes fire much faster when increasing this parameter, with the spikes being almost indistinguishable in the latter case, thus reducing the sparsity of this network. Between the neural activities shown in Figure 4a,b, the mean firing rates are low (6 and 30.9 Hz) and show a more sparse firing rate, meaning that both are, in principle, better-suited to this application. The preliminary results computed within NengoDL (Figure 4d) indicate that a scale of 5 provides a slightly better performance, being the most adequate value for this parameter. Naturally, tuning these parameters is a harder task when dealing with more complex data and more complex spiking architectures, as we will see in the following section.

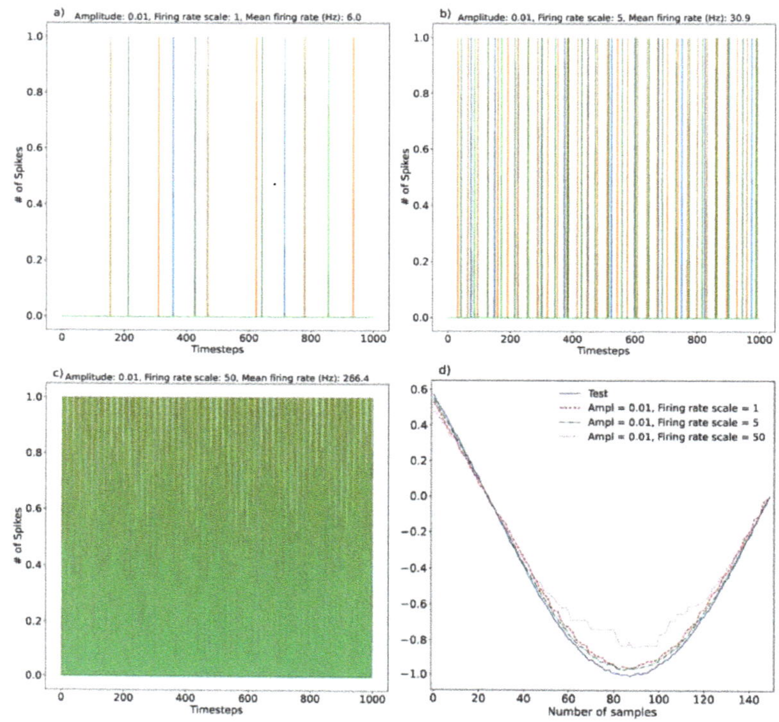

Figure 4. (**a**) Neural activities setting an amplitude = 0.01 and a firing rate scale = 1, (**b**) Neural activities setting an amplitude = 0.01 and a firing rate scale = 5, (**c**) Neural activities setting an amplitude = 0.01 and a firing rate scale = 50, and (**d**) predictions over the testing set.

3.2. Load Forecasting

Let us set another example using real data to calculate one-step ahead forecasts. In particular, short-term load forecasting is of interest due to its close relation with WPF, as both are necessary to operate and maintain the stability of the electrical grid [54]. Furthermore, load demand data show regular daily and weekly patterns, which are not observed in wind power data [55], so a model architecture formed of CNNs is a good candidate to extract such features [56]. Records of aggregated hourly demand data from Ireland can be found on the European Network of Transmission System Operators for Electricity (ENTSO-E) website [57]. The available measurements were recorded between 2016 and 2018.

As usual, we built and trained the rate-based equivalent of the model, and subsequently the activation functions were replaced. Then, the spike parameters were tuned without specifying any hardware requirements, and we monitored the initial results to choose the best values for these parameters. Some of these initial forecasts are shown in Figure 5. The existing patterns in load data were captured by the model, and adjusting the spikes' parameters is fairly straightforward. The dashed red line (obtained using an amplitude of 0.05 and a firing rate scale of 50) more closely matches the test data than the rest, so these values were chosen for its implementation on Loihi's emulator (or the hardware itself, if available).

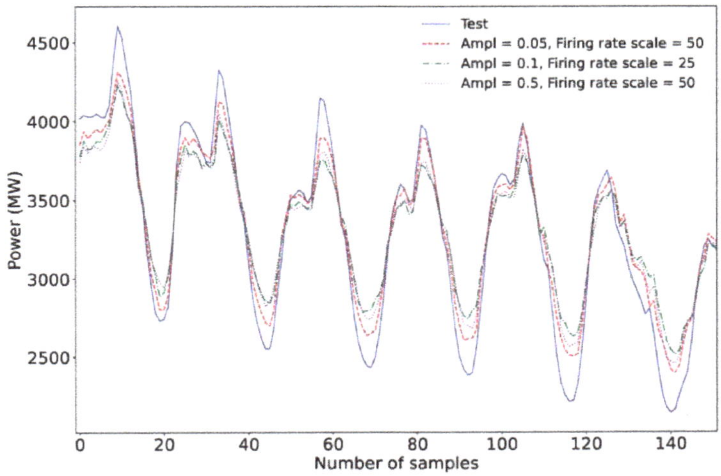

Figure 5. Preliminary one-step-ahead load forecasts, setting different spike amplitudes and firing rates.

As indicated in Step 5, the network must be further adjusted to be run on Loihi. In our particular case, we must indicate what layers are run on- and off-chip, but other adjustments might be needed for more complex networks, such as distributing the connections of the network over multiple cores on Loihi [44]. Figure 6a shows that neurons are effectively firing in each layer, whereas Figure 6b compares the initial forecasts thaat were obtained previously while tuning the spike parameters (the red dashed line) and the load forecasts emulating the Loihi chip (dash-dot green line). It can be observed that the model architecture translates well to the emulator after fine-tuning those hardware specifications, resulting in similar load forecasts with respect to the initial evaluation of Step 4.

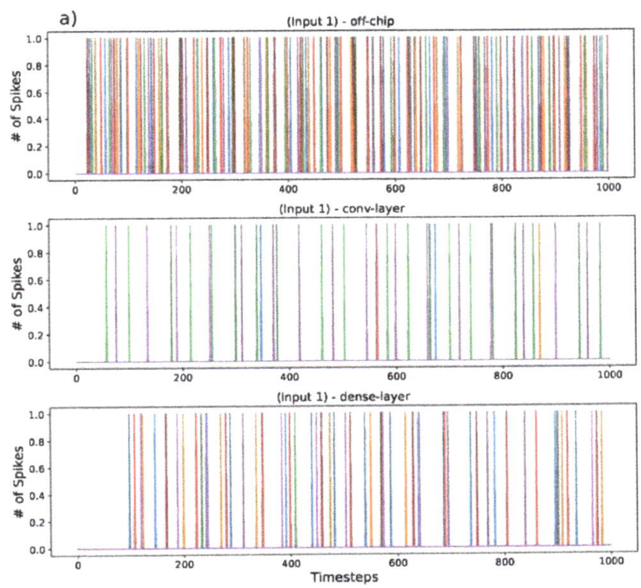

Figure 6. *Cont.*

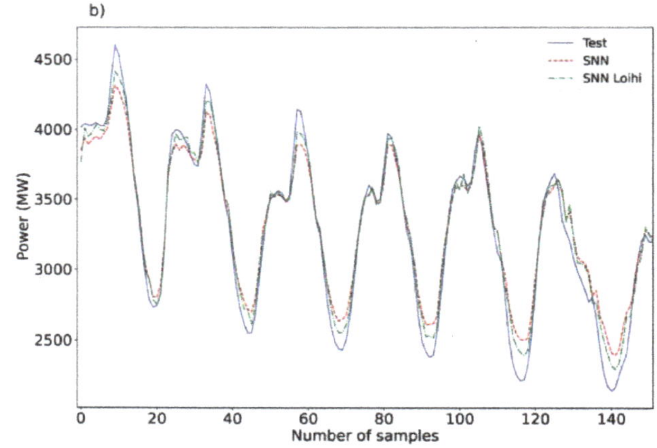

Figure 6. Results for one-step ahead load forecasts: (**a**) Neural activities of 5 neurons of each layer. One input vector is shown over 1000 timesteps. (**b**) Predictions over the testing set with the SNN architecture (dashed red line) and running the SNN on the Loihi emulator (dash-dot green line)

3.3. *Wind Power Forecasting*

Noting the computational abilities of current neuromorphic devices, let us apply the proposed model architecture to build the spiking forecasting models for each mode extracted from Irish wind-power data after preprocessing the data with the VMD algorithm. The exact location of the wind farm is not disclosed for confidentiality reasons. The library Nengo [35] was used to simulate neuromorphic algorithms, together with the extensions NengoDL [37] for deep learning and NengoLoihi to emulate the behavior of Loihi hardware. The original neural-network-based models were implemented using Keras with Tensorflow backend for the rest of the models [38,58].

Following the proposed methodology, we first built the model (Step 1), and trained the rate-based neural network model to set its network parameters (Step 2). Then, we transformed it into an SNN by switching the activation functions to spiking ones (Step 3). In Step 4, we set empiric values for the amplitude and firing rate of the spikes (Table 1) within the NengoDL framework until we obtained a reasonable performance from the spiking model. The spiking amplitude modulates the amount of information transmitted to the subsequent layers of the network, whereas the firing rate adjusts how fast the spikes are being fired. If the firing rate is high, the behavior will be closer to the non-spiking model, and thus the performance will increase, but at the cost of losing the characteristic temporal sparsity provided by the spikes [59]. In addition, a high firing rate will lead to detrimental results on Loihi because of the discrepancy resulting from discretizing the spiking activation function (as shown in Figure 1). The low mean firing rates of these preliminary results (Table 2) suggest that the selected parameters are potentially good for implementation on Loihi. Afterwards, we configured some additional parameters to run the model on the Loihi emulator (Step 5). In particular, we must indicate what part of the model is run off-chip (in this case, the *off-chip* layer we use to communicate with the chip) and how long each input vector is presented to the network (in our case, we show each one for 0.4 s).

Table 1. Main spiking network parameters.

	Neuron Type	Spiking Amplitude	Firing Rate Scale
Mode 1	Spiking ReLU	0.1	25
Mode 2	Spiking ReLU	0.05	40
Mode 3	Spiking ReLU	0.01	70
Mode 4	Spiking ReLU	0.1	90
Mode 5	Spiking ReLU	0.3	200
Mode 6	Spiking ReLU	0.3	200
Mode 7	Spiking ReLU	0.5	400
Mode 8	Spiking ReLU	1.5	500

Table 2. Mean firing rates (Hz) for each layer.

	Off-Chip Layer	Conv Layer	Dense Layer
Mode 1	8.1	8.3	12.0
Mode 2	1.9	1.8	2.1
Mode 3	7.2	4.9	2.5
Mode 4	1.6	1.2	1.0
Mode 5	1.2	1.3	1.0
Mode 6	1.1	1.0	1.0
Mode 7	1.4	1.1	1.0
Mode 8	3.7	9.6	11.5

The information recorded in Steps 4 and 5 is shown in Figures 7–14 for modes 1–8, respectively. Part a of these figures shows the neural activities of each layer (limited to 5 neurons for illustrative purposes). These neural activities correspond to the first input vector fed to the model, and produce the first point forecast, shown on part b. This constitutes one of the main differences in comparison to ANNs. Neurons in regular ANNs are static entities, which are always activated every time a new input arrives to the model, whereas neurons of SNN models will only be activated if certain dynamic conditions are met. From an user perspective, the neural activities help us visualize the mean firing rates shown in Table 2: modes 1, 3, and 8 exhibit higher firing rates, which translates into a large number of spikes being generated during this timeframe, whereas the rest of the modes present a more sparse behavior, resulting in a lower generation of spikes. In some cases, such as mode 4 (Figure 10) and mode 5 (Figure 11), the neurons of the *off-chip* layer need a long time to settle and thus start to spike, delaying the neural response of subsequent layers. Even if temporal sparsity is a desirable feature in a spiking model, in the sense that a smaller number of spikes means a lower consumption of energy (as a non-activated neuron will consume no energy), it might occasionally be advisable to finetune the firing rate of the *off-chip* layer to propagate the information faster to the rest of the network, as delays in the neural responses could degrade model performance. A quote on exact power consumption on the chip is erroneous in the current context, since the entire hardware is active during implementation, while only a minute fraction is actually used for the proposed problem. Under such circumstances, power quotes become relevant with sector-customised chips and a better handling of SNN architecture for DL, a direction in which the industry and current research is quickly moving.

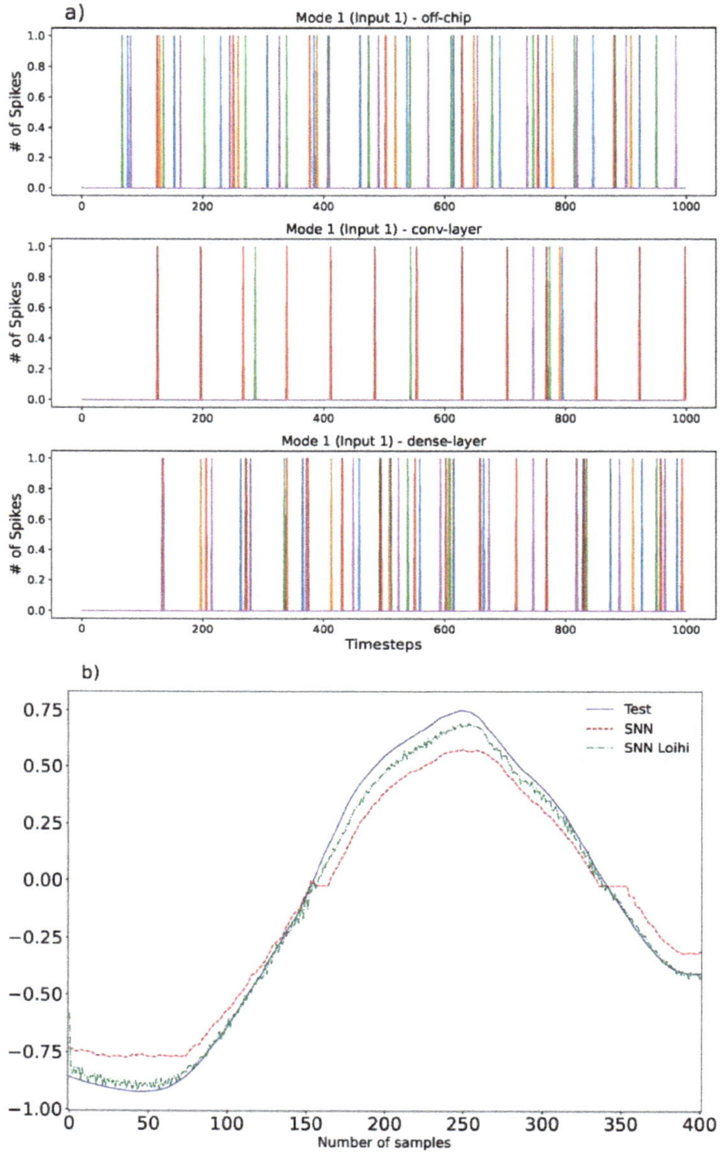

Figure 7. Results for mode 1: (**a**) Neural activities of 5 neurons of each layer. One input vector is shown over 1000 timesteps. (**b**) Predictions over the testing set with the SNN architecture (dashed red line) and running the SNN on the Loihi emulator (dash-dot green line).

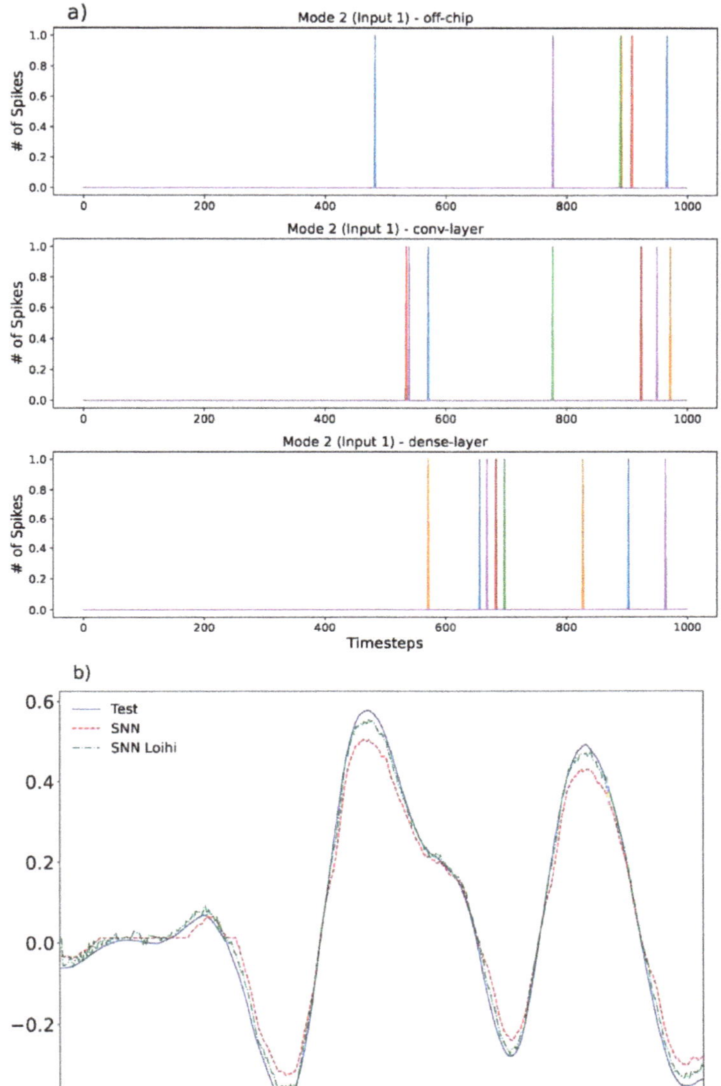

Figure 8. Results for mode 2: (**a**) Neural activities of 5 neurons of each layer. One input vector is shown over 1000 timesteps. (**b**) Predictions over the testing set with the SNN architecture (dashed red line) and running the SNN on the Loihi emulator (dash-dot green line).

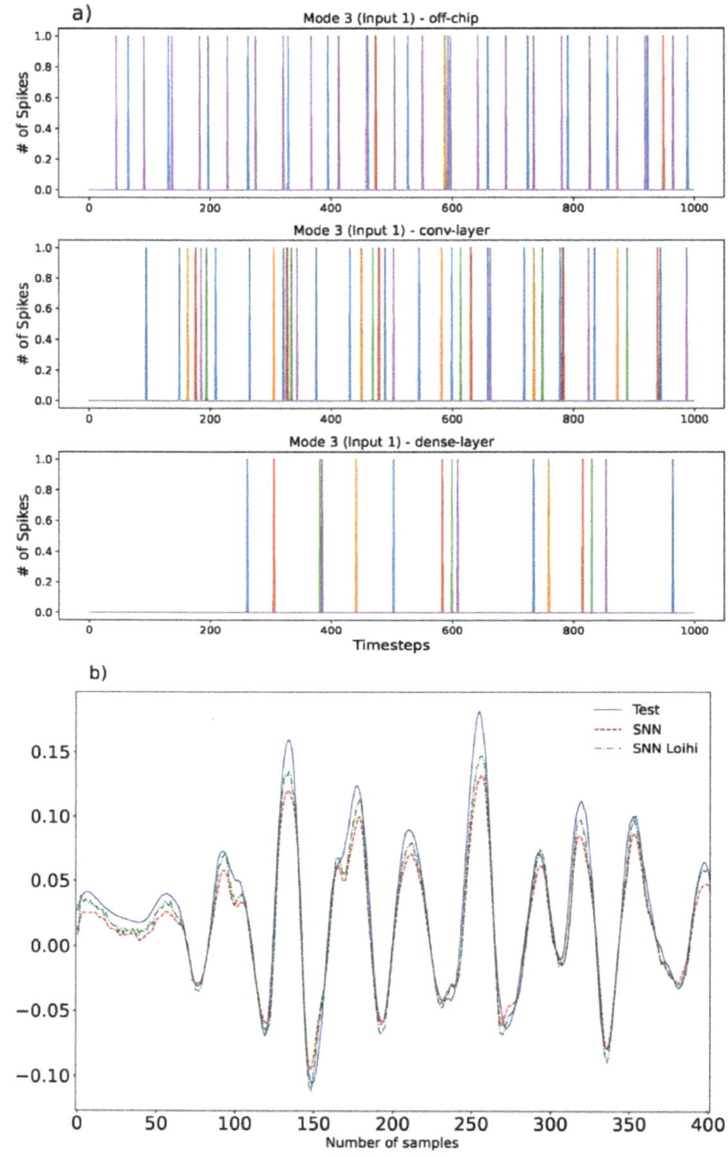

Figure 9. Results for mode 3: (**a**) Neural activities of 5 neurons of each layer. One input vector is shown over 1000 timesteps. (**b**) Predictions over the testing set with the SNN architecture (dashed red line) and running the SNN on the Loihi emulator (dash-dot green line).

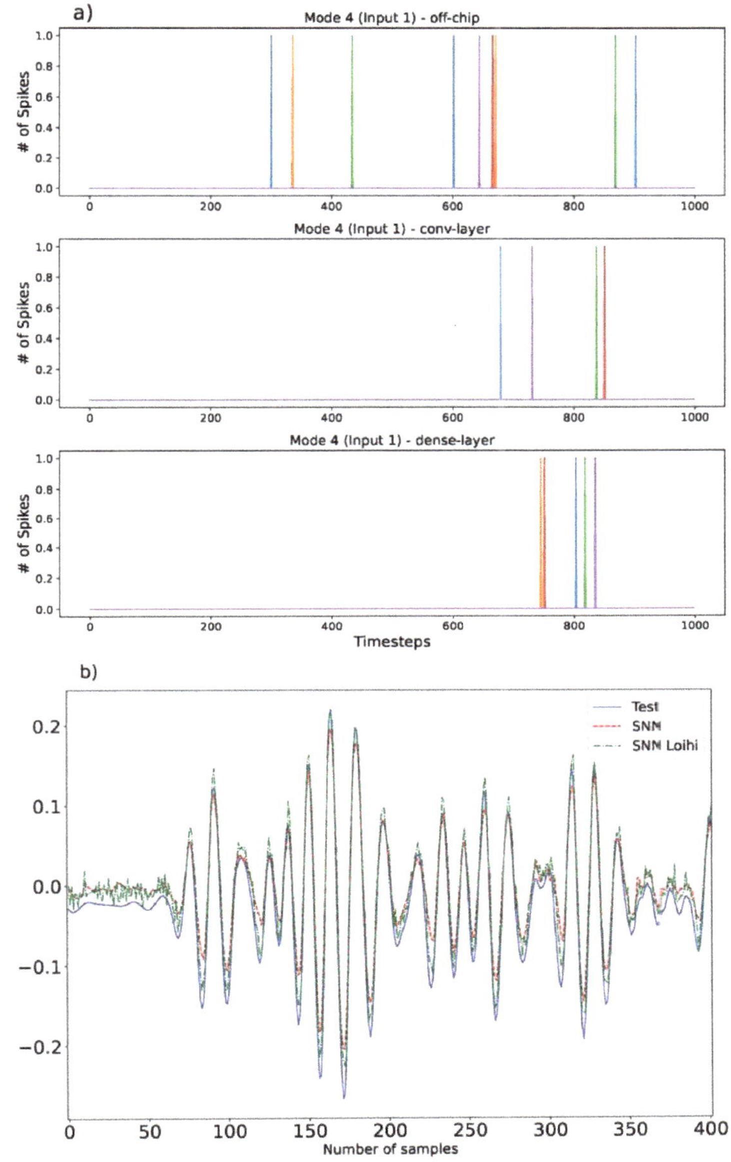

Figure 10. Results for mode 4: (**a**) Neural activities of 5 neurons of each layer. One input vector is shown over 1000 timesteps. (**b**) Predictions over the testing set with the SNN architecture (dashed red line) and running the SNN on the Loihi emulator (dash-dot green line).

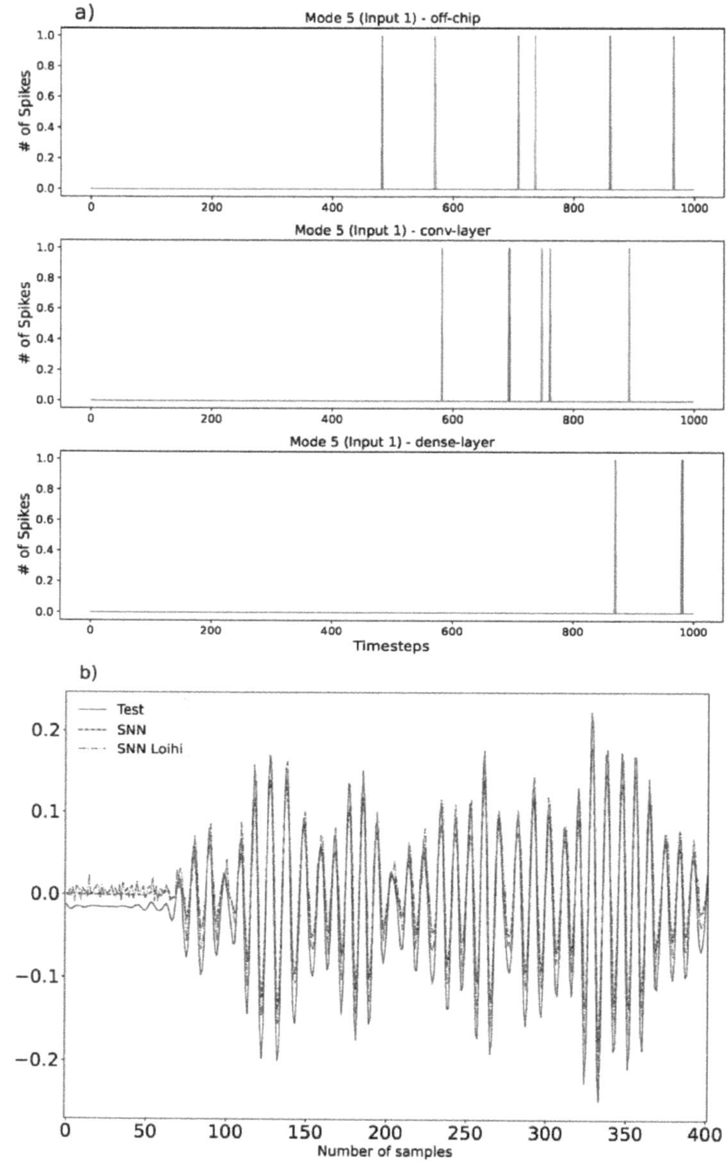

Figure 11. Results for mode 5: (**a**) Neural activities of 5 neurons of each layer. One input vector is shown over 1000 timesteps. (**b**) Predictions over the testing set with the SNN architecture (dashed red line) and running the SNN on the Loihi emulator (dash-dot green line).

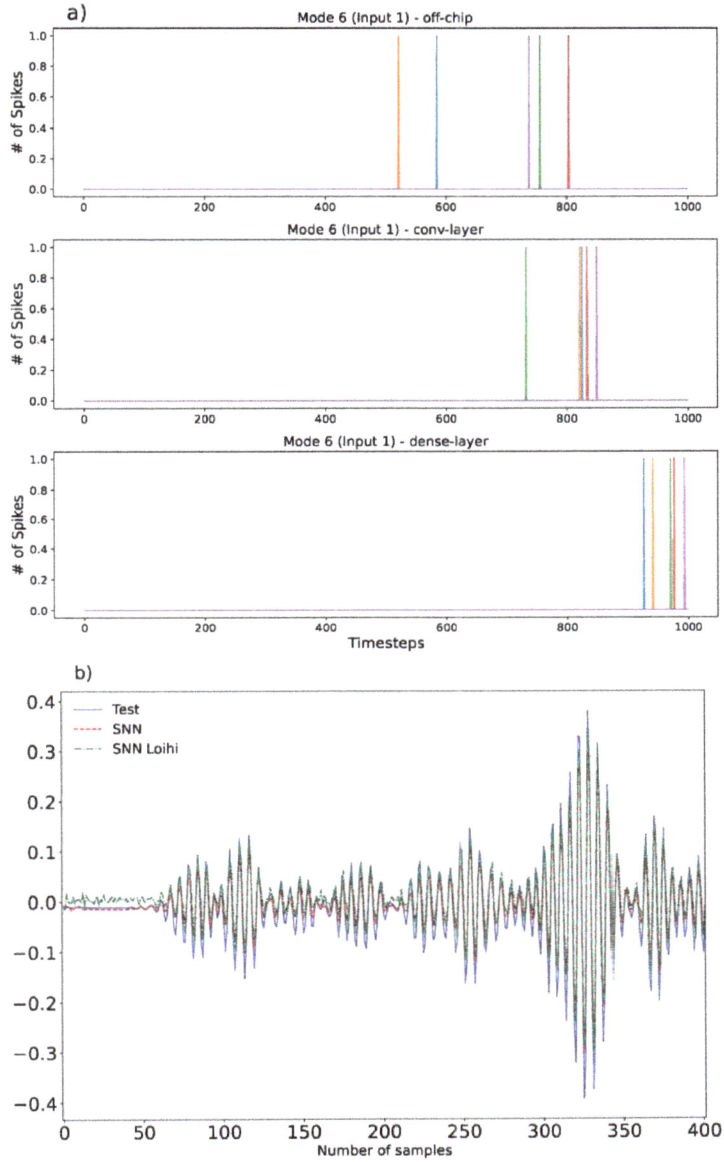

Figure 12. Results for mode 6: (**a**) Neural activities of 5 neurons of each layer. One input vector is shown over 1000 timesteps. (**b**) Predictions over the testing set with the SNN architecture (dashed red line) and running the SNN on the Loihi emulator (dash-dot green line).

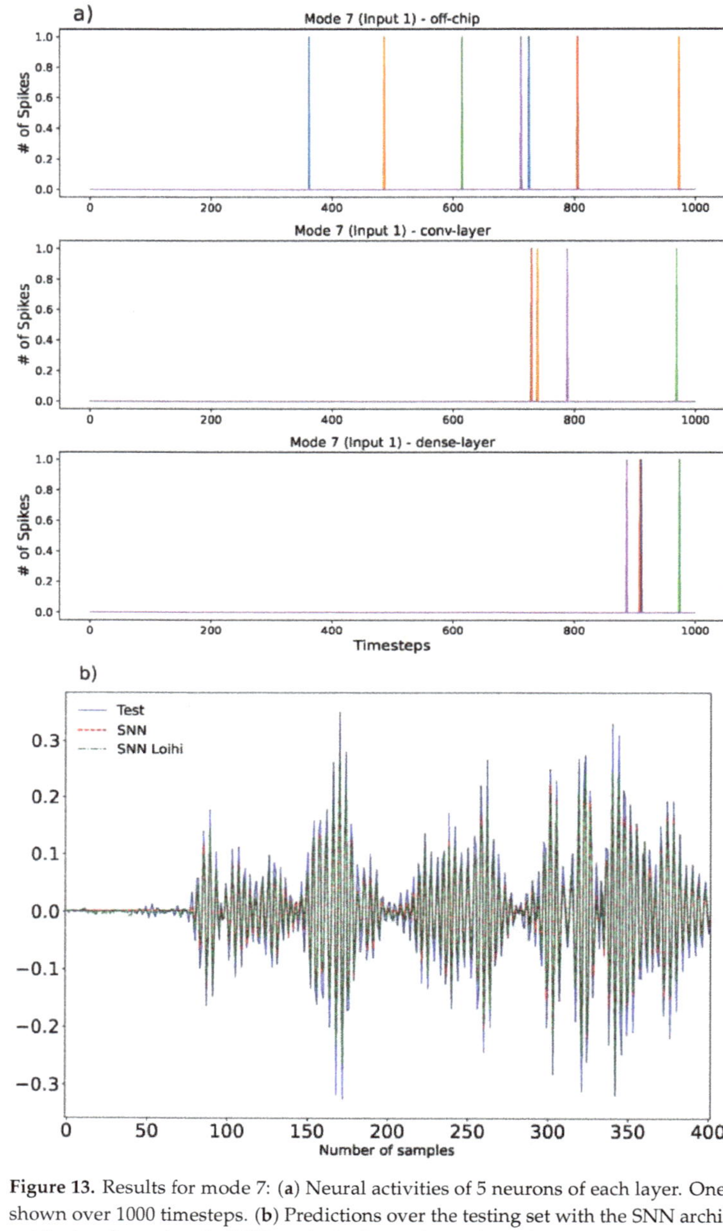

Figure 13. Results for mode 7: (**a**) Neural activities of 5 neurons of each layer. One input vector is shown over 1000 timesteps. (**b**) Predictions over the testing set with the SNN architecture (dashed red line) and running the SNN on the Loihi emulator (dash-dot green line).

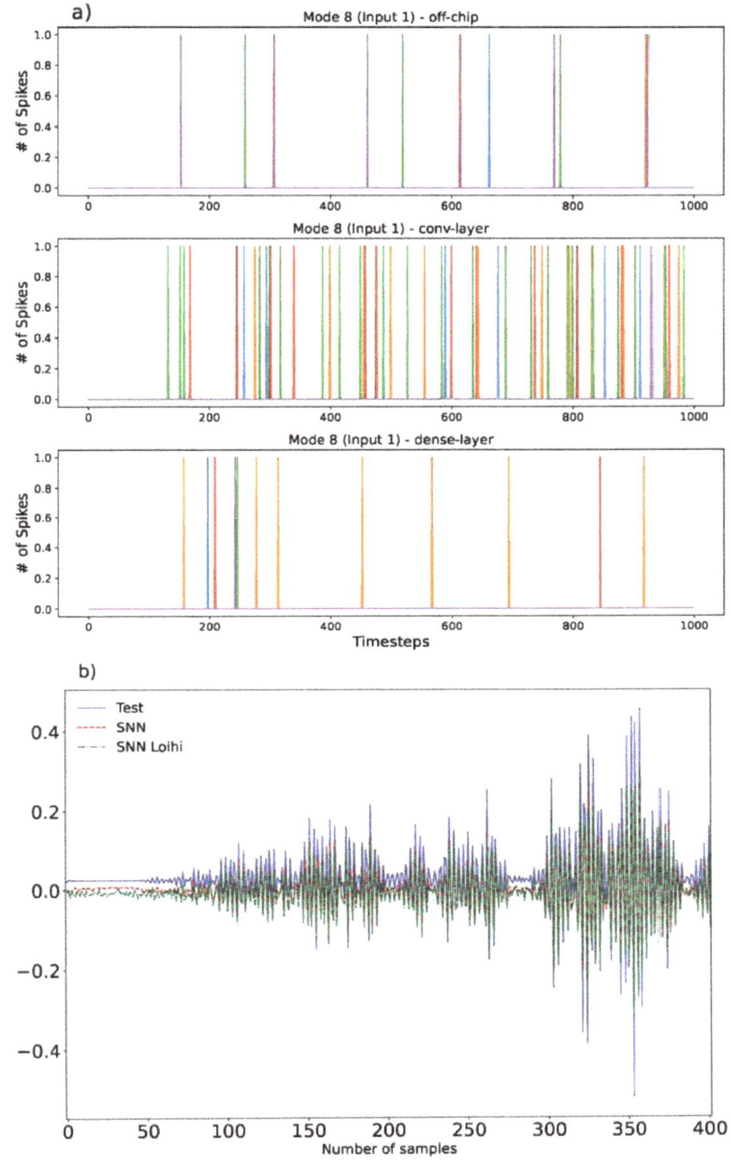

Figure 14. Results for mode 8: (**a**) Neural activities of 5 neurons of each layer. One input vector is shown over 1000 timesteps. (**b**) Predictions over the testing set with the SNN architecture (dashed red line) and running the SNN on the Loihi emulator (dash-dot green line).

At this stage, the performance of our models can finally be examined (Step 6). The model is designed to provide one-step-ahead point forecasts (Figure 15). The dashed-red lines show the forecasts obtained while tuning the model using the NengoDL framework in Step 4. While this preliminary model is able to forecast increasing/decreasing trends of power generation, it is not as accurate for high or low power-generation scenarios. Nonetheless, this initial assessment allows for us to prepare our model for Loihi (dash-dot green line), which demonstrates the same skill in detecting increasing/decreasing trends of power generation as the preliminary model, while showing a better ability to forecast high/low power-generation values. This difference in performance also arises from the model architecture itself. When the model is initially evaluated outside the Loihi framework, it cannot discern that the first layer is only set to start to generate spikes. Such nuance is captured when the model is configured for implementation on Loihi. Additionally, we observe that the forecasts are not as accurate compared to a non-spiking VMD-GRU model that we used in a previous study using the same data [50], and the outputs are generally noisier. However, this is an expected outcome due to the current limitations of neuromorphic hardware.

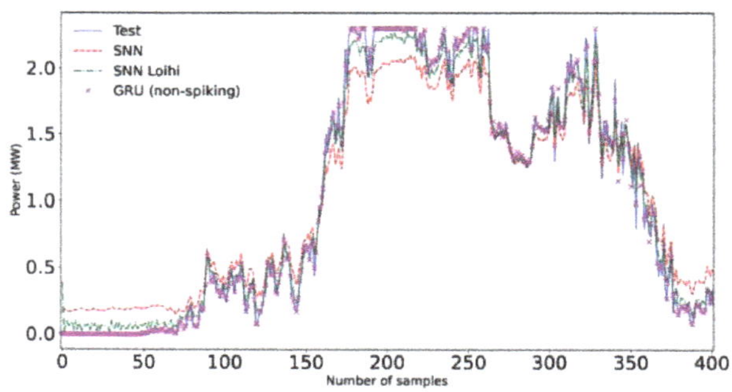

Figure 15. One-step ahead WPFs with the SNN architecture (dashed red line), running the SNN on the Loihi emulator (dash-dot green line) and a non-spiking VMD-GRU model (purple crosses) over the testing set.

In conclusion, we have successfully transformed a non-spiking neural model into a spiking one with a reasonably good performance, having achieved a 2.84% NMAE for one-step ahead forecasts with the model being adapted to neuromorphic hardware. This type of proof is not only helpful to prove that industrial applications such as WPF modelling can be transferred to non-von Neumann architectures such as neuromorphic computing, but to provide guidelines to the manufacturers of such hardware (e.g., Intel and the development of devices as Loihi) to cater to the industry's needs.

4. Conclusions

Neuromorphic computing provides a new paradigm to build energy-efficient, low-latency algorithms in contrast to the current state-of-the-art ML/DL strategies, thus potentially reducing the computational cost of training and deploying artificial-intelligence-based models. In particular, SNNs aim to learn in a more biologically plausible manner [60] by more closely mimicking the spike-based transmission of information that occurs in the brain [61]. At present, the two major challenges for the use and implementation of SNNs are (1) the training of such models, as the well-established training strategies based on the backpropagation algorithm applied to ML/DL cannot be directly used, as spikes are not differentiable, and (2) the implementation of SNNs on neuromorphic hardware, as SNNs must be tailored to cater to the specific requirements of the hardware. The first challenge has been addressed with different approaches to date, such as ANN-to-SNN conversion, and using variations in error backpropagation to directly train SNNs. The second challenge is hardware-dependent, and should be addressed according to the requisites of the hardware used to implement the SNN. Additionally, there is currently a lack of studies applying neuromorphic computing for practical cases that are useful for both research and industrial practices, such as the design of WPF models [62].

In this paper, we adopt an ANN-to-SNN conversion approach to forecast wind power, and obtain these WPFs emulating or running the spiking model using the neuromorphic hardware Loihi [41]. SNNs are designed using the framework provided by the software Nengo [35,37]. First, we build and train the non-spiking neural network. After training, we map the parameters and replace the activation functions for their spiking counterparts, which will be used during the prediction stage. Then, without considering hardware specific constraints, some preliminary results are evaluated to tune some spike-related parameters such as the firing rate or the amplitude of the spikes. Finally, the SNN is further adjusted to be run on the hardware emulator (or actually running the model on Loihi if available) to obtain the WPFs. Following all these steps, we managed to reach our goal of achieving a good level of performance with the proposed spiking architecture, obtaining a NMAE of 2.84% for one-step ahead forecasts when the model is emulated on Loihi.

As neuromorphic computing is not a well-established technology at present, there is room for future research. First, the proposed ANN-to-SNN conversion approach for short-term WPF can be further refined by tuning the firing rates of each layer individually and considering the use of synaptic filters to smooth the output. Second, the modelling of spiking neural networks can be improved by (1) directly training the network to increase the efficiency of neuromorphic devices from the energy point of view and (2) using online approximations of the backpropagation algorithm to adjust network parameters with the arrival of new data. Third, it will be possible to implement more complex biological inspired neural networks in the near future, as the computational capability of neuromorphic computing continues to increase with the development of new devices, such as Loihi 2 [63]. As reducing the computational cost is one of the main reasons to use neuromorphic devices, any future line of research must address model performance in terms of energy consumption to verify that it achieves a significant reduction in consumption compared to conventional computer architectures. To achieve that, the neural spiking models must not only be emulated but run on a real neuromorphic device to realistically measure this feature.

Author Contributions: Conceptualization, J.M.G.S., V.P. and B.G.; methodology, J.M.G.S. and B.G.; software, J.M.G.S.; validation, J.M.G.S., V.P. and B.G.; formal analysis, J.M.G.S.; investigation, J.M.G.S., V.P. and B.G.; resources, J.M.G.S., V.P. and B.G.; data curation, J.M.G.S., V.P. and B.G.; writing—original draft preparation, J.M.G.S. and B.G.; writing—review and editing, J.M.G.S., V.P. and B.G.; visualization, J.M.G.S.; supervision, V.P. and B.G.; project administration, V.P. and B.G.; funding acquisition, V.P. and B.G. All authors have read and agreed to the published version of the manuscript.

Funding: The authors acknowledge the funding of SEAI WindPearl Project 18/RDD/263.

Data Availability Statement: Relevant computed data are available from authors upon reasonable request to the corresponding author.

Acknowledgments: The authors like to thank George Vathakkattil Joseph and Aasifa Rounak for their technical support. Bidisha Ghosh would like to acknowledge the support of ENABLE (Grant number 16/SP/3804) and Connect Center (Grant number 13/RC/2077_P2). Vikram Pakrashi would like to acknowledge the support of SFI MaREI centre (Grant number RC2302_2), the resources and support of Intel Neuromorphic Research Community, Accenture NeuroSHM project with UCD, Science Foundation Ireland NexSys 21/SPP/3756, and UCD Energy Institute.

Conflicts of Interest: The authors declare no conflict of interest.

Abbreviations

The following abbreviations are used in this manuscript:

ANN	Artificial Neural Network
CNN	Convolutional Neural Network
DL	Deep Learning
ENTSO-E	European Network of Transmission System Operators for Electricity
FFNN	Feedforward Neural Network
GRU	Gated Recurrent Unit
LIF	Leaky Integrate-and-Fire
LSTM	Long Short-Term Memory
ML	Machine Learning
NEF	Neural Engineering Framework
NMAE	Normalized Mean Absolute Error
SLAYER	Spike Layer Error Reassignment
ReLU	Rectified Linear Unit
SNN	Spiking Neural Network
VMD	Variational Mode Decomposition
WPF	Wind Power Forecasting

References

1. Lim, B.; Zohren, S. Time-series forecasting with deep learning: A survey. *Philos. Trans. R. Soc. A* **2021**, *379*, 20200209.
2. Fawaz, H.I.; Forestier, G.; Weber, J.; Idoumghar, L.; Muller, P.A. Deep learning for time series classification: A review. *Data Min. Knowl. Discov.* **2019**, *33*, 917–963.
3. Ma, Q.; Zheng, J.; Li, S.; Cottrell, G.W. Learning representations for time series clustering. *Adv. Neural Inf. Process. Syst.* **2019**, *32*, 3781–3791.
4. Wang, Y.; Zou, R.; Liu, F.; Zhang, L.; Liu, Q. A review of wind speed and wind power forecasting with deep neural networks. *Appl. Energy* **2021**, *304*, 117766.
5. Marugán, A.P.; Márquez, F.P.G.; Perez, J.M.P.; Ruiz-Hernández, D. A survey of artificial neural network in wind energy systems. *Appl. Energy* **2018**, *228*, 1822–1836.
6. Rumelhart, D.E.; Durbin, R.; Golden, R.; Chauvin, Y. Backpropagation: The basic theory. In *Backpropagation: Theory, Architectures and Applications*; Psychology Press: London, UK, 1995; pp. 1–34.
7. Wang, R.; Li, C.; Fu, W.; Tang, G. Deep learning method based on gated recurrent unit and variational mode decomposition for short-term wind power interval prediction. *IEEE Trans. Neural Netw. Learn. Syst.* **2019**, *31*, 3814–3827.
8. Li, C.; Tang, G.; Xue, X.; Chen, X.; Wang, R.; Zhang, C. The short-term interval prediction of wind power using the deep learning model with gradient descend optimization. *Renew. Energy* **2020**, *155*, 197–211.
9. Yildiz, C.; Acikgoz, H.; Korkmaz, D.; Budak, U. An improved residual-based convolutional neural network for very short-term wind power forecasting. *Energy Convers. Manag.* **2021**, *228*, 113731.
10. He, Y.; Li, H.; Wang, S.; Yao, X. Uncertainty analysis of wind power probability density forecasting based on cubic spline interpolation and support vector quantile regression. *Neurocomputing* **2021**, *430*, 121–137.
11. Lahouar, A.; Slama, J.B.H. Hour-ahead wind power forecast based on random forests. *Renew. Energy* **2017**, *109*, 529–541.
12. Landry, M.; Erlinger, T.P.; Patschke, D.; Varrichio, C. Probabilistic gradient boosting machines for GEFCom2014 wind forecasting. *Int. J. Forecast.* **2016**, *32*, 1061–1066.
13. Mohammadzaheri, M.; Mirsepahi, A.; Asef-afshar, O.; Koohi, H. Neuro-fuzzy modeling of superheating system of a steam power plant. *Appl. Math. Sci* **2007**, *1*, 2091–2099.
14. Mohammadzaheri, M.; AlQallaf, A.; Ghodsi, M.; Ziaiefar, H. Development of a fuzzy model to estimate the head of gaseous petroleum fluids driven by electrical submersible pumps. *Fuzzy Inf. Eng.* **2018**, *10*, 99–106.
15. Bengio, Y. *Learning Deep Architectures for AI*; Now Publishers Inc.: Hannover, MA, USA 2009.
16. Wang, K.; Qi, X.; Liu, H.; Song, J. Deep belief network based k-means cluster approach for short-term wind power forecasting. *Energy* **2018**, *165*, 840–852.
17. Hong, Y.Y.; Rioflorido, C.L.P.P. A hybrid deep learning-based neural network for 24-h ahead wind power forecasting. *Appl. Energy* **2019**, *250*, 530–539.
18. Putz, D.; Gumhalter, M.; Auer, H. A novel approach to multi-horizon wind power forecasting based on deep neural architecture. *Renew. Energy* **2021**, *178*, 494–505.
19. Munawar, U.; Wang, Z. A framework of using machine learning approaches for short-term solar power forecasting. *J. Electr. Eng. Technol.* **2020**, *15*, 561–569.
20. Nam, K.; Hwangbo, S.; Yoo, C. A deep learning-based forecasting model for renewable energy scenarios to guide sustainable energy policy: A case study of Korea. *Renew. Sustain. Energy Rev.* **2020**, *122*, 109725.
21. González Sopeña, J.; Pakrashi, V.; Ghosh, B. Can we improve short-term wind power forecasts using turbine-level data? A case study in Ireland. In Proceedings of the 2021 IEEE Madrid PowerTech, Madrid, Spain, 28 June–July 2021; pp. 1–6.

22. González Sopeña, J.; Maury, C.; Pakrashi, V.; Ghosh, B. Turbine-level clustering for improved short-term wind power forecasting. In *Proceedings of the Journal of Physics: Conference Series;* IOP Publishing: Bristol, UK, 2022; Volume 2265, p. 022052.
23. Wang, X.; Han, Y.; Leung, V.C.; Niyato, D.; Yan, X.; Chen, X. Convergence of edge computing and deep learning: A comprehensive survey. *IEEE Commun. Surv. Tutor.* **2020**, *22*, 869–904.
24. Li, W.; Yang, T.; Delicato, F.C.; Pires, P.F.; Tari, Z.; Khan, S.U.; Zomaya, A.Y. On enabling sustainable edge computing with renewable energy resources. *IEEE Commun. Mag.* **2018**, *56*, 94–101.
25. Justus, D.; Brennan, J.; Bonner, S.; McGough, A.S. Predicting the computational cost of deep learning models. In Proceedings of the 2018 IEEE International Conference on Big Data (Big Data), Seattle, WA, USA, 10–13 December 2018; pp. 3873–3882.
26. Lin, C.K.; Wild, A.; Chinya, G.N.; Cao, Y.; Davies, M.; Lavery, D.M.; Wang, H. Programming spiking neural networks on Intel's Loihi. *Computer* **2018**, *51*, 52–61.
27. Maass, W. Networks of spiking neurons: The third generation of neural network models. *Neural Netw.* **1997**, *10*, 1659–1671.
28. Davies, M.; Wild, A.; Orchard, G.; Sandamirskaya, Y.; Guerra, G.A.F.; Joshi, P.; Plank, P.; Risbud, S.R. Advancing neuromorphic computing with Loihi: A survey of results and outlook. *Proc. IEEE* **2021**, *109*, 911–934.
29. Stewart, K.; Orchard, G.; Shrestha, S.B.; Neftci, E. On-chip few-shot learning with surrogate gradient descent on a neuromorphic processor. In Proceedings of the 2020 2nd IEEE International Conference on Artificial Intelligence Circuits and Systems (AICAS), Genova, Italy, 31 August–2 September 2020; pp. 223–227.
30. Tavanaei, A.; Maida, A. BP-STDP: Approximating backpropagation using spike timing dependent plasticity. *Neurocomputing* **2019**, *330*, 39–47.
31. Bellec, G.; Scherr, F.; Subramoney, A.; Hajek, E.; Salaj, D.; Legenstein, R.; Maass, W. A solution to the learning dilemma for recurrent networks of spiking neurons. *Nat. Commun.* **2020**, *11*, 3625.
32. Kasabov, N.; Scott, N.M.; Tu, E.; Marks, S.; Sengupta, N.; Capecci, E.; Othman, M.; Doborjeh, M.G.; Murli, N.; Hartono, R.; et al. Evolving spatio-temporal data machines based on the NeuCube neuromorphic framework: Design methodology and selected applications. *Neural Netw.* **2016**, *78*, 1–14.
33. Diehl, P.U.; Neil, D.; Binas, J.; Cook, M.; Liu, S.C.; Pfeiffer, M. Fast-classifying, high-accuracy spiking deep networks through weight and threshold balancing. In Proceedings of the 2015 International Joint Conference on Neural Networks (IJCNN), Killarney, Ireland, 12–17 July 2015; pp. 1–8.
34. Taherkhani, A.; Belatreche, A.; Li, Y.; Maguire, L.P. DL-ReSuMe: A delay learning-based remote supervised method for spiking neurons. *IEEE Trans. Neural Netw. Learn. Syst.* **2015**, *26*, 3137–3149.
35. Bekolay, T.; Bergstra, J.; Hunsberger, E.; DeWolf, T.; Stewart, T.C.; Rasmussen, D.; Choo, X.; Voelker, A.; Eliasmith, C. Nengo: A Python tool for building large-scale functional brain models. *Front. Neuroinform.* **2014**, *7*, 48.
36. Eliasmith, C.; Anderson, C.H. *Neural Engineering: Computation, Representation, and Dynamics in Neurobiological Systems;* MIT Press: Cambridge, MA, USA, 2003.
37. Rasmussen, D. NengoDL: Combining deep learning and neuromorphic modelling methods. *Neuroinformatics* **2019**, *17*, 611–628.
38. Abadi, M.; Agarwal, A.; Barham, P.; Brevdo, E.; Chen, Z.; Citro, C.; Corrado, G.S.; Davis, A.; Dean, J.; Devin, M.; et al. TensorFlow: Large-Scale Machine Learning on Heterogeneous Systems, 2015. Available online: http://tensorflow.org (accessed on 30 September 2022).
39. Shrestha, S.B.; Orchard, G. Slayer: Spike layer error reassignment in time. In Proceedings of the Advances in Neural Information Processing Systems, Montreal, QC, Canada, 3–8 December 2018; pp. 1412–1421.
40. Intel's Neuromorphic Computing Lab. Lava: A Software Framework for Neuromorphic Computing. 2021. Available online: https://github.com/lava-nc/lava (accessed on 25 March 2022).
41. Davies, M.; Srinivasa, N.; Lin, T.H.; Chinya, G.; Cao, Y.; Choday, S.H.; Dimou, G.; Joshi, P.; Imam, N.; Jain, S.; et al. Loihi: A neuromorphic manycore processor with on-chip learning. *IEEE Micro* **2018**, *38*, 82–99.
42. Hunsberger, E.; Eliasmith, C. Training spiking deep networks for neuromorphic hardware. *arXiv* **2016**, arXiv:1611.05141.
43. DeWolf, T.; Jaworski, P.; Eliasmith, C. Nengo and low-power AI hardware for robust, embedded neurorobotics. *Front. Neurorobot.* **2020**, *14*, 568359.
44. Applied Brain Research. Converting a Keras Model to an SNN on Loihi. 2021. Available online: https://www.nengo.ai/nengo-loihi/v1.0.0/examples/keras-to-loihi.html (accessed on 20 April 2022).
45. Voelker, A.R.; Eliasmith, C. Programming neuromorphics using the Neural Engineering Framework. In *Handbook of Neuroengineering;* Springer Nature: Singapore, 2020; pp. 1–43.
46. González Sopeña, J.; Pakrashi, V.; Ghosh, B. An overview of performance evaluation metrics for short-term statistical wind power forecasting. *Renew. Sustain. Energy Rev.* **2021**, *138*, 110515.
47. Tavanaei, A.; Ghodrati, M.; Kheradpisheh, S.R.; Masquelier, T.; Maida, A. Deep learning in spiking neural networks. *Neural Netw.* **2019**, *111*, 47–63.
48. Cao, Y.; Chen, Y.; Khosla, D. Spiking deep convolutional neural networks for energy-efficient object recognition. *Int. J. Comput. Vis.* **2015**, *113*, 54–66.
49. Schmidt-Hieber, J. Nonparametric regression using deep neural networks with ReLU activation function. *Ann. Stat.* **2020**, *48*, 1875–1897.
50. González Sopeña, J.M.; Pakrashi, V.; Ghosh, B. Decomposition-based hybrid models for very short-term wind power forecasting. *Eng. Proc.* 2021, *5*, 39.

51. Dragomiretskiy, K.; Zosso, D. Variational mode decomposition. *IEEE Trans. Signal Process.* **2013**, *62*, 531–544.
52. Tang, L.; Lv, H.; Yang, F.; Yu, L. Complexity testing techniques for time series data: A comprehensive literature review. *Chaos Solitons Fractals* **2015**, *81*, 117–135.
53. Ren, Y.; Suganthan, P.; Srikanth, N. A comparative study of empirical mode decomposition-based short-term wind speed forecasting methods. *IEEE Trans. Sustain. Energy* **2014**, *6*, 236–244.
54. Hong, T.; Fan, S. Probabilistic electric load forecasting: A tutorial review. *Int. J. Forecast.* **2016**, *32*, 914–938.
55. Quan, H.; Srinivasan, D.; Khosravi, A. Short-term load and wind power forecasting using neural network-based prediction intervals. *IEEE Trans. Neural Netw. Learn. Syst.* **2013**, *25*, 303–315.
56. Sadaei, H.J.; e Silva, P.C.d.L.; Guimarães, F.G.; Lee, M.H. Short-term load forecasting by using a combined method of convolutional neural networks and fuzzy time series. *Energy* **2019**, *175*, 365–377.
57. ENTSO-E. Hourly Load Demand Data. 2021. Available online: https://www.entsoe.eu/data/power-stats/ (accessed on 20 April 2022).
58. Chollet, F. Keras. 2015. Available online: https://github.com/fchollet/keras (accessed on 20 April 2022).
59. Patel, K.; Hunsberger, E.; Batir, S.; Eliasmith, C. A spiking neural network for image segmentation. *arXiv* **2021**, arXiv:2106.08921.
60. Tan, C.; Šarlija, M.; Kasabov, N. Spiking neural networks: Background, recent development and the NeuCube architecture. *Neural Process. Lett.* **2020**, *52*, 1675–1701.
61. Kasabov, N.K. *Time-Space, Spiking Neural Networks and Brain-Inspired Artificial Intelligence*; Springer: Berlin, Germany, 2019.
62. Davies, M. Benchmarks for progress in neuromorphic computing. *Nature Mach. Intell.* **2019**, *1*, 386–388.
63. Orchard, G.; Frady, E.P.; Rubin, D.B.D.; Sanborn, S.; Shrestha, S.B.; Sommer, F.T.; Davies, M. Efficient Neuromorphic Signal Processing with Loihi 2. In Proceedings of the 2021 IEEE Workshop on Signal Processing Systems (SiPS), Coimbra, Portugal, 19–21 October 2021; pp. 254–259.

energies MDPI

Article

A Comprehensive Study of Random Forest for Short-Term Load Forecasting

Grzegorz Dudek

Department of Electrical Engineering, Częstochowa University of Technology, 42-200 Częstochowa, Poland; grzegorz.dudek@pcz.pl

Abstract: Random forest (RF) is one of the most popular machine learning (ML) models used for both classification and regression problems. As an ensemble model, it demonstrates high predictive accuracy and low variance, while being easy to learn and optimize. In this study, we use RF for short-term load forecasting (STLF), focusing on data representation and training modes. We consider seven methods of defining input patterns and three training modes: local, global and extended global. We also investigate key RF hyperparameters to learn about their optimal settings. The experimental part of the work demonstrates on four STLF problems that our model, in its optimal variant, can outperform both statistical and ML models, providing the most accurate forecasts.

Keywords: random forest; regression tree; pattern representation of time series; short-term load forecasting

Citation: Dudek, G. A Comprehensive Study of Random Forest for Short-Term Load Forecasting. *Energies* **2022**, *15*, 7547. https://doi.org/10.3390/en15207547

Academic Editor: Andrzej Bielecki

Received: 15 September 2022
Accepted: 10 October 2022
Published: 13 October 2022

1. Introduction

Electricity demand forecasting is extremely important for energy providers to ensure the secure, effective and economic operation of the power system. Short-term load forecasting (STLF) covers a forecast horizon of a few hours to a few days. STLF is necessary for generation resource planning to meet electricity demands and optimize the power flow on the transmission grid to avoid overloads. As electricity demand is a major driver of electricity prices, load forecasting plays a key role in competitive energy markets. The STLF accuracy directly affects the financial performance of energy market participants.

The importance of accurate electricity demand forecasts for the safe, reliable and effective operation of power systems is behind the great interest of researchers in this area. STLF problems are complex because electricity demand time series express a nonlinear trend, multiple seasonality, variable variance, significant random disruptions and changing daily profile. These challenging factors place high demands on STLF models.

1.1. Related Work

Roughly, STLF methods can be divided into statistical and ML methods. The most popular representatives of the first group are: auto-regressive integrated moving average (ARIMA) [1], exponential smoothing (ETS) [2], linear regression [3], and Kalman filtering [4]. The main drawbacks of the statistical methods are their linear character, limited adaptability, limited ability to deal with complex seasonal patterns, and problems with capturing long-term dependencies in time series and introducing exogenous variables into the model [5].

ML models provide more flexibility in modeling nonlinear functions. Unlike statistical methods, they do not require strong assumptions about the mapping function, and they learn relationships between predictors and targets directly from historical data. Among ML methods for forecasting, neural networks (NNs) have gained the most popularity in recent years [6]. The multitude of architectural solutions and mechanisms to improve performance encourage the use of NNs to solve complex forecasting problems such as STLF. Classical NNs were investigated for suitability for STLF in [7]. To deal with triple seasonality in time series, patterns of the daily profiles were introduced, which filter out the trend and the weekly and yearly seasonality (a similar approach was used in this study, where different definitions of patterns are examined). Among the considered NN architectures, which included Multilayer Perceptron (MLP), Radial Basis Function NN, General Regression NN (GRNN), Fuzzy Counterpropagation NN, and Self-Organizing Maps, GRNN and MLP stand out as performing with the highest accuracy.

In recent years, a development in the NN field has been the move towards deep and, especially useful in forecasting, recurrent architectures [6]. Deep NNs (DNNs) are especially beneficial in learning the most useful data representation for modeling a given target function, while Recurrent NNs (RNNs) are beneficial in modeling complex, short- and long-term temporal relationships in data. New mechanisms and procedures introduced to RNNs such as delayed connections, attention, hybrid architecture, dynamic training sets, residual connections and flexible loss functions improve their learning capabilities and expressive power to solve forecasting problems [5]. Some examples of using DNNs and RNNs for STLF are [8], where Convolutional NNs (CNNs) are utilized to extract load and temperature features, which are fed as inputs into the bidirectional propagating RNN to perform hourly electrical load forecasting [9], where a recurrent inception CNN is proposed for STLF that combines RNN and 1-dimensional CNN [10], where RNN with attention significantly reduced forecasting errors as compared to the current state-of-the-art results; and [11], where deep residual networks integrate domain knowledge and researchers' understanding of the problem and enables probabilistic load forecasting using Monte Carlo dropout.

An effective way to increase the forecast accuracy and robustness of both statistical and ML models is ensembling. This combines multiple models for a common response to improve both the accuracy and stability of the final solution compared to a single model [12]. The theoretical properties of forecast combination investigated in [13] answer the question why a simple average of forecasts often outperforms forecasts from single models. They also prove that simple averages in many cases perform better than more complicated weighting schemes. The beneficial effects of the forecast aggregation on STLF accuracy are shown in many papers: in [14], several ML methods are aggregated in ensembles for one-day-ahead wind power forecasting; in [15], to forecast an interval-valued load, ensemble of RNNs is applied, which learns on the components of the bivariate empirical mode decomposition; in [5], ensembling of a hybrid model, which combines ETS and RNN, leads to a significant reduction in the forecast error; in [16], an ensemble of randomized DNN combined with a walk-forward decomposition is proposed; in [17], a stacking ensemble approach is used to combine DNNs, and in [18], several methods of aggregating base models (MLPs) are considered. Stacking, used in the last two papers, is a way of combining base models via meta-learning, i.e., a meta-model is trained on the predictions of the base models.

Alternatives to stacking are boosting and bagging. Popular representatives of these are gradient-boosted trees and random forest (RF), respectively. Both, used as forecasting models, are based on regression trees. It was shown in [19] that RF can compete with both classical models and NNs in STLF. It can deal with complex time series using appropriate data preprocessing, which produces normalized patterns of the daily profiles. Based on this research, in this study, to improve RF performance, we extend pattern definitions and introduce additional predictors. To enrich input information, in [20], the input patterns are extracted from electrical, meteorological and calendar data by temporal CNN. Fed with these patterns, a Light Gradient Boosting Machine, a type of gradient-boosted trees algorithm, was able to forecast very volatile industrial customer loads. A novel tree-based ensemble method called Warm-start Gradient Tree Boosting (WGTB) was proposed in [21]. It combines four different inference models and aggregates their outputs by a warm-start bagging and boosting, which at the same time reduces bias and variance. The result proves the efficiency of the proposed strategy and shows an improvement in STLF accuracy over baseline models. Another type of tree-based ensemble, eXtreme Gradient Boosting (XGBoost), was used in [22] for forecasting electricity consumption by industrial customers.

To deal with multiple seasonality, the time series were first decomposed using variational mode decomposition. Then, a linear regression model was applied for the trend series and a XGBoost regression model was applied for each fluctuation sub-series.

To close this section, we note that, according to some studies, tree-based ensembles are not inferior to NNs in terms of forecast accuracy. In [23], the authors have examined and reproduced a number of state-of-the-art DNNs for time series forecasting. DNNs were compared on different datasets to a Gradient Boosting Regression Tree (GBRT). The experimental results show that a conceptually simpler model such as GBRT can compete and sometimes outperform modern DNNs by efficiently feature-engineering the input and output structures of GBRT.

1.2. Motivation and Contribution

Tree-based methods are widely used as prediction models as they have very attractive properties such as a capacity for flexible nonlinear regression, which can capture complex interactions between variables and effectively handle multiple predictors (including exogenous ones) of various types (numeric, binary, and categorical). Moreover, they are robust against over-fitting of the training data, they are relatively simple to tune, and they are easy to implement with the available software. Their effectiveness has been confirmed in many forecasting competitions, for example those carried out on the Kaggle platform [24].

The excellent performance of tree-based approaches was demonstrated in the 2020 M5 forecasting competition. The top places in this competition, in terms of both accuracy and uncertainty, were dominated by entries that used tree-based ML methods such as gradient-boosted trees [25]. Four out of the five winning models used a variant of the tree-based method and most of the other top 50 best-performing models adopted similar approaches to the winning submission by training recursive and non-recursive tree-based models [26]. Thus, tree-based forecasting models appear to be strong competitors to NNs, which in the form of deep learning-based models dominate the recent literature on forecasting methods [6].

In this study, motivated by the excellent results of tree-based models in forecasting competitions, we apply RF to the challenging problem of STLF. RF gives similar results to boosting, but is easier to train and tune [27]. The main contribution of this work is to examine RF models using a variety of time series preprocessing methods and training modes. In the local mode, RF learns on samples similar to the query sample, which enables the model to focus on the local features of the target function around the query pattern and improve accuracy in this region. In the global mode, to achieve the same goal, i.e., focusing on the proper region of the target function, we introduce additional calendar variables. By examining different methods of time series preprocessing, we find the most useful data representation for achieving the highest accuracy of the model. We empirically demonstrate that the proposed approach outperforms in terms of accuracy both standard statistical models as well as more sophisticated ML approaches.

The novelty of this work in relation to our previous work [19] is twofold. First, we extend pattern definition by introducing seven types of patterns based on the historical data. They incorporate daily and/or weekly seasonality, while in [19], the patterns captured only daily seasonality. Second, we introduce a global mode of training with additional predictors representing calendar data. In [19], only a local training was considered without calendar inputs.

The rest of the paper is organized as follows. In Section 2, we propose several data prereprocessing methods for electricity demand times series. Section 3 defines the forecasting problem and RF training modes. Section 4 describes the RF algorithm in application to STLF. The experimental framework used to evaluate the performance of the proposed model and compare it with baseline models is described in Section 5. Finally, Section 6 concludes the work.

2. Data Preprocessing

The power system load or electricity demand time series express a trend, triple seasonality (annual, weekly and daily) and random fluctuations. These components are dependent on the system size, size of the economy served, customer structure, as well as weather and climatic conditions. The daily load profiles that we focus on in STLF vary throughout the year and depend on the day of the week [5].

To forecast future demand with the least possible error, the forecasting model should be fed by the most relevant predictors. In our univariate STLF model, which produces forecasts for the next day, the predictors are selected from recent history and are preprocessed accordingly. The forecasting model based on RF is fed by input patterns $\mathbf{x}_i$ and produces encoded forecasts for hour t of day i, $y_{i,t}$ (MISO model).

Let $\{z_\tau\}_{\tau=1}^{M}$ be an electricity demand time series with hourly resolution and vector $\mathbf{z}_i = [z_{i,1}, ..., z_{i,24}]$ represents its 24-hour-long sequence for day i. To capture the characteristic properties of the series, remove the trend and unify data, we define the input patterns as follows:

r1 The input patterns are defined based on the weekly sequence which precedes forecasted day i:

$$\mathbf{x}_i = \frac{\mathbf{s}_i - \bar{s}_i}{\|\mathbf{s}_i - \bar{s}_i\|} \tag{1}$$

where $\mathbf{x}_i \in \mathbb{R}^{168}$ is the input pattern, $\mathbf{s}_i = [\mathbf{z}_{i-7}, ..., \mathbf{z}_{i-1}]$ is the demand sequence of the week preceding the forecasted day i and $\bar{s}_i$ is the mean of this sequence.
Input vectors (1), which for successive i represent overlapping weekly sequences shifted by one day, are normalized versions of centered vectors $\mathbf{s}_i$. They all have zero mean, the same variance and the same unity length. However, they differ in shape. Thus, we assume that the weekly shape carries the information about the forecasted demand of the day following this week.

r2 The input patterns are defined based on the daily sequence which precedes the forecasted day i. The encoding equation is (1), where $\mathbf{x}_i \in \mathbb{R}^{24}$ and $\mathbf{s}_i = \mathbf{z}_{i-1}$ is the demand sequence of the day preceding the forecasted day i. In case of r2, a carrier of the information about the forecasted value is the shape of the preceding day.

r3 The input patterns are defined based on the sequence composed of the demands at hour t of seven consecutive days preceding the forecasted day i. In (1), $\mathbf{x}_i \in \mathbb{R}^7$ and $\mathbf{s}_i = [z_{i-7,t}, ..., z_{i-1,t}]$, where t is the forecasted hour of day i.

r4 The input patterns are defined based on the sequence composed of the demands at hour t of 21 consecutive days preceding the forecasted day i. In (1), $\mathbf{x}_i \in \mathbb{R}^{21}$ and $\mathbf{s}_i = [z_{i-21,t}, ..., z_{i-1,t}]$, where t is the forecasted hour of day i.

r5 The input patterns are defined based on the sequence composed of the demands at hour t of seven days preceding forecasted day i and representing the same day of the week as the forecasted day. For example, when the model predicts demand at hour t on Monday, the input pattern is composed of the demands at hour t of seven preceding Mondays. In (1), $\mathbf{x}_i \in \mathbb{R}^7$ and $\mathbf{s}_i = [z_{i-49,t}, z_{i-42,t}, ..., z_{i-7,t}]$, where t is the forecasted hour of day i.

r6 Cross-pattern combining r2 and r3—the input patterns are defined based on both the daily sequence and the sequence composed of the demands at hour t of seven consecutive days preceding the forecasted day i. In (1), $\mathbf{x}_i \in \mathbb{R}^{30}$ and $\mathbf{s}_i = [\mathbf{z}_{i-1}, z_{i-7,t}, ..., z_{i-2,t}]$, where t is the forecasted hour of day i.

r7 Cross-pattern combining r2 and r4—the input patterns are defined based on both the daily sequence and the sequence composed of the demands at hour t of 21 consecutive days preceding the forecasted day i. In (1), $\mathbf{x}_i \in \mathbb{R}^{44}$ and $\mathbf{s}_i = [\mathbf{z}_{i-1}, z_{i-21,t}, ..., z_{i-2,t}]$, where t is the forecasted hour of day i.

Figure 1 shows the sequences which are used for x-patterns construction and Figure 2 shows data used for construction patterns r6 and r7. Depending on the definition, x-patterns introduce different input information to the model. Pattern r1 introduces detailed information about the weekly sequence which precedes the forecasted day. Note that r1 expresses both daily and weekly seasonality unlike r2, which carries information only about the daily seasonality. To deal with weekly seasonality when r2-patterns are used, the model can be trained in the local mode, i.e., on the subset of x-patterns corresponding to the forecasting task (see Section 3).

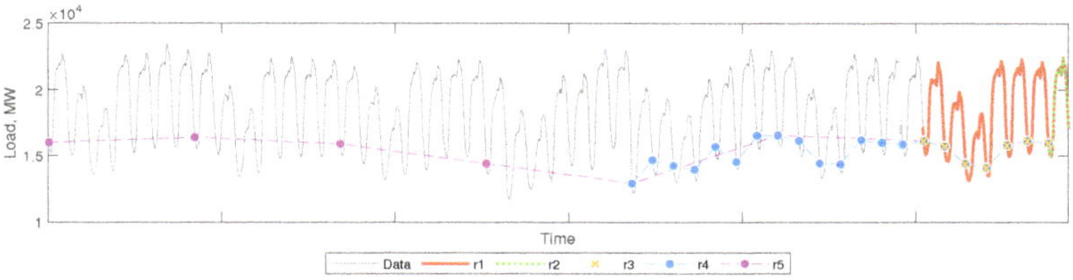

Figure 1. Load time series points used for input patterns construction.

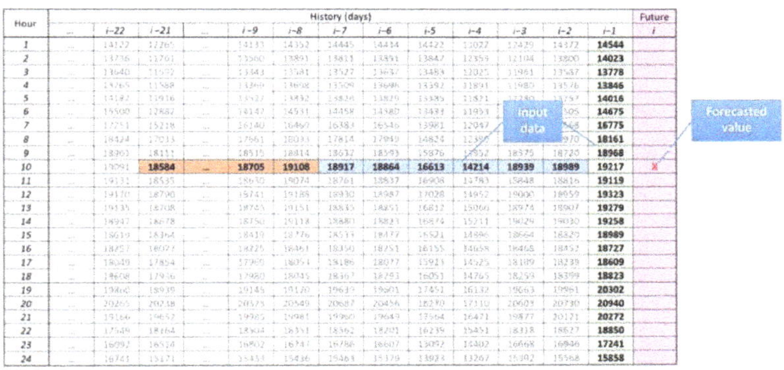

Figure 2. Cross-patterns: r6 (green + blue) and r7 (green + orange + blue).

Pattern r3 introduces information on the demand in the previous seven days at the same hour as the forecasted one. It expresses only weekly seasonality. Information about the daily seasonality is not included, so the local mode of training, i.e., training on the selected r3-patterns corresponding to the forecasting task (see Section 3), can help with dealing with daily seasonality. Similar information as in r3 is contained in pattern r4 but from a longer 3-week period. Pattern r5 shows neither daily nor weekly seasonality. It carries information about the demand at the same hours as forecasted in previous days of the same type as the forecasted day.

Cross-patterns r6 and r7 express both daily and weekly seasonalities as r1, but in a more sparing form, using respectively 30 or 44 instead of 168 components. In [28], we showed that STLF based on both daily and weekly patterns gives better results than forecasting based on separate daily or weekly patterns. In [28], we aggregated forecasts generated by two neural models: a daily pattern-based model and a weekly pattern-based model, while in this study, we combine daily and weekly patterns into one pattern and use only one model.

Examples of input patterns r1–r7 are depicted in Figure 3. Note different shapes of patterns, carrying different input information.

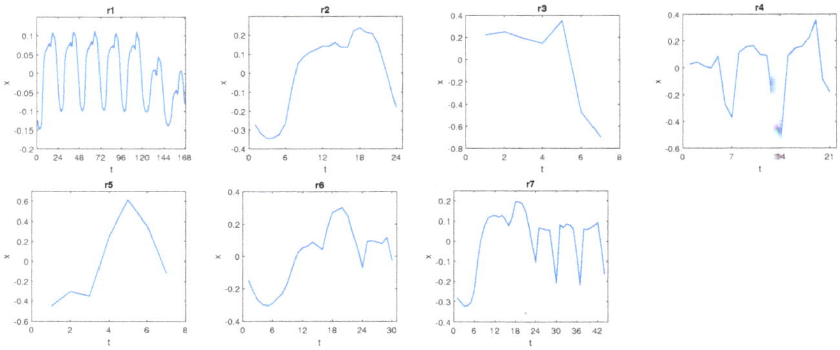

Figure 3. Examples of input patterns r1–r7.

The output data, i.e., the electricity demand at hour t of day i, is encoded as follows:

$$y_{i,t} = \frac{z_{i,t} - \bar{s}_i}{\|\mathbf{s}_i - \bar{s}_i\|} \quad (2)$$

where $\mathbf{s}_i$ is the demand sequence preceding forecasted day i, defined depending on the x-pattern type r1–r7 (this is the same sequence based on which pattern $\mathbf{x}_i$ was defined) and $\bar{s}_i$ is the mean of this sequence.

The output data are encoded similarly to the input data, using the same coding variables: $\bar{s}_i$ and $\|\mathbf{s}_i - \bar{s}_i\|$. Thus, Equation (2) like Equation (1) filters the data by removing the local trend ($\bar{s}_i$) and unifying the variance (this is a function of the denominator of these equations, which can be thought of as a measure of diversity of the input sequence). Such filtered and unified data are predicted by the forecasting model (RF). Then, the real forecast is determined from transformed Equation (2):

$$\hat{z}_{i,t} = \hat{y}_{i,t}\|\mathbf{s}_i - \bar{s}_i\| + \bar{s}_i \quad (3)$$

where $\hat{y}_{i,t}$ is the model prediction and $\hat{z}_{i,t}$ is the real forecast.

Note that (3) brings back the local current properties of the time series (level and dispersion), which were removed by (1) and (2) to simplify the relationships between input and output data. We have successfully used this kind of preprocessing of input and output data in our previous load forecasting models to deal with multiple seasonality, simplify the model and speed up training, see, e.g., [7,19,28–32].

3. Forecasting Problem and Training Modes

The forecasting task is defined as follows: predict electricity demand for hour t^* (1, ..., 24) of day i^* based on historical data. Day i^* represents day of the week d^* (*Monday*, ..., *Sunday*). To maximize the forecasting performance and to make the most of all available training data up to day i^* (forecasted day), the forecasting model is trained individually for each forecasting task and it performs only one prediction: $\hat{y}_{i^*,t^*}$. Note that the "global" generalization property of the model is not important because it is built to make only one prediction. What is important is the "local" performance in the neighborhood of pattern $\mathbf{x}_{i^*}$. To increase this property, we use two approaches. In the first one, we train the model in the local mode and in the second one, we extend input patterns with calendar variables when we use global learning. For comparison we also train the model in the standard global mode.

The full training set determined on the historical data is $\Psi = \{(\mathbf{x}_i, y_{i,t})\}$, where $i = 1, \ldots, i^* - 1$, $t = 1, \ldots, 24$ and pair $(\mathbf{x}_i, y_{i,t})$ includes the input pattern and target defined according to r1–r7. The three training modes are as follows:

Local The model is trained on the subset of Ψ containing pairs $(\mathbf{x}_i, y_{i,t})$ which correspond to the forecasting task, i.e., the pairs which include targets representing the same day type as forecasted, $d(y_{i,t}) = d^*$, and the same hour as forecasted, $t = t^*$.

Global The model is trained on full training set Ψ.

Global extended The input data are extended with calendar information:

- season of the year encoded as follows [29]:

$$\mathbf{p}_i = \left[\sin\frac{2\pi\#i}{366}, \cos\frac{2\pi\#i}{366}\right] \tag{4}$$

  where $\#i$ is the number of day i in the year,
- day of the week, $d = Monday, ..., Sunday$ (categorical variable), and
- hour of the day, $t = 1, ..., 24$ (categorical variable).

The training set in the global extended mode is of the form: $\Psi = \{(\langle \mathbf{x}_i, \mathbf{p}_i, d_i, t\rangle, y_{i,t})\}$, $i = 1, ..., i^* - 1$, $t = 1, ..., 24$.

In the local training mode, the model solves the forecasting task by learning on the samples expressing similar properties and relationships between input and output data as those expressed by input pattern $\mathbf{x}_{i^*}$ and forecasted value y_{i^*,t^*}. That is, the training input patterns are limited to those that are similar in shape to pattern $\mathbf{x}_{i^*}$ (further limitations in this regard can be made by selecting training data from the same period of the year as day i^* or by selecting training data based on similarity to pattern $\mathbf{x}_{i^*}$ [7], but we have not employed these approaches in this study). The relationships between input and output data expressed in the local training set are limited to those corresponding to day type d^* and hour t^*, so we expect the model to more accurately approximate the relationship between $\mathbf{x}_{i^*}$ and y_{i^*,t^*} than in case of global training on the full training set expressing the relationship for all day types and hours.

In the global training mode with extended inputs, the model has additional input information to more accurately solve the forecasting task, i.e., the calendar variables. Although the model is global, we expect that the calendar data will help to increase its local accuracy around pattern $\mathbf{x}_{i^*}$. A regression tree as a base forecasting model can more appropriately divide the input space using the calendar variables than without these variables, and therefore approximate locally the target function with greater accuracy. Although this will lead to a more complex model than in the other two training modes.

4. Random Forest for STLF

RF is a ensemble learning algorithm based on decision trees (CART [33]) as the base models [34]. It is suitable for either regression or classification problems. In this study, for forecasting problems, we focus on the regression RF based on regression trees.

RF is devoid of the well-known drawbacks of single trees such as unstable splits and a lack of smoothness [27]. It combines bagging [35] with a random subspace method [36]. The key idea in bagging is to average multiple noisy but approximately unbiased base models and thus reduce the variance. Trees as noisy and low biased models if they have grown sufficiently deep, are great candidates for bagging. The main goal of the random subspace method is to increase diversity between trees by restricting them to work on different random subsets of the full predictor space (more specifically, at each node of the tree, a random predictor subset is selected). Each tree in the forest is built from a bootstrap sample of the original dataset, which is an additional source of diversity. Random predictors selected in the nodes of bagged trees help to decorrelate the trees and improve prediction accuracy as well as reduce the model variance.

The RF algorithm draws a bootstrap sample Ψ_k of size N from training set Ψ for each of K trees, $k = 1, \ldots, K$. For each bootstrapped sample, a tree T is grown by recursive partitioning the input space in each node until a minimum leaf size is reached. At each node, data splits based on p out of n predictors chosen at random are considered. The best split is determined by maximizing the reduction in mean squared error (MSE) over all splitting candidates and cutpoints. After all K trees are grown in this fashion, the RF predictor is [27]:

$$\hat{f}_K(\mathbf{x}) = \frac{1}{K}\sum_{k=1}^{K} T(\mathbf{x}; \Theta_k) \quad (5)$$

where $\mathbf{x}$ is the input pattern and Θ_k characterizes the k-th tree in terms of split predictors, cutpoints, and terminal-node values.

RF has the following hyperparameters:

- number of trees K. Intuitively, the number of trees should be as large as possible because the model variance reduces with K. Usually, the forecast error stabilizes with the number of trees, and the most reasonable RF size is selected.
- minimum leaf size m. As deeper trees have low bias and large variance, they are strongly recommended as RF members. Thus, the minimum leaf size should be small. The inventors of the algorithm recommend $m = 5$ for regression forest.
- number of predictors to select at random for each split p. As p decreases, the correlation between trees reduces, and hence the variance of the average reduces. Typically, a value for p for regression RF is $n/3$, as the inventors recommend.

In practice the best values for the hyperparameters are dependent on the problem, and they should be treated as tuning parameters.

The standard method of selecting split predictors [33] has two drawbacks. Firstly, it tends to miss important interactions between pairs of predictors and the response. Secondly, it tends to select continuous predictors that have many levels, which masks more important predictors that have fewer levels, such as categorical predictors. To mitigate selection bias and increase detection of important interactions, curvature or interaction tests can be applied [37,38]. Therefore, in this study we consider three methods of selecting the split predictors:

s1 Standard CART method. This selects the split predictor that maximizes the split-criterion gain over all possible splits of all predictors.

s2 Curvature test. This selects the split predictor that minimizes the p-value of chi-square tests of independence between each predictor and the response.

s3 Interaction test. This performs the curvature test extended by the minimization of the p-value of a chi-square test of independence between each pair of predictors and the response.

The algorithm of RF construction for STLF is shown in Algorithm 1. It produces a set of K trees, $\{T_k\}_{k=1}^K$. Based on them, to make a prediction for new point $\mathbf{x}$, we use (5). Then, the real forecast is calculated from (3). Note that training set Ψ is prepared for the selected training mode and input pattern type.

Algorithm 1 Random forest construction for STLF

Input: training set Ψ containing N samples, number of trees K, minimum leaf size m, number of predictors to select at random for each split p, split predictor selection method s
Output: set of trees $\{T_k\}_{k=1}^K$
Procedure:
for $k = 1$ **to** K **do**
  Draw a bootstrap sample Ψ_k of size N from Ψ
  Grow tree T_k to Ψ_k by recursively repeating the following steps for each terminal node, until the minimum node size m is reached:
  – Select p predictors from the n predictors
  – Pick the best predictor/cutpoint among the p using predictor selection method s
  – Split the node into two daughter nodes
end for

In the experimental study (Section 5), we use RF specified in Algorithm 1 in several variants depending on the data preprocessing method (r1–r7) and training modes, i.e., local, global and global extended. In the global extended mode, the predictor vector is composed of $\langle \mathbf{x}_i, \mathbf{p}_i, d_i, t \rangle$. In the other training modes, the predictor vector is the same as input pattern $\mathbf{x}$ (1).

5. Simulation Study

In this section, we investigate RF variants with different data preprocessing methods and training modes. We compare RF performance with that of other models based on classical statistical methods and ML methods.

STLF for four countries is performed: Poland (PL), Great Britain (GB), France (FR) and Germany (DE). The real-world data was collected from ENTSO-E repository (www.entsoe.eu/data/power-stats; accessed on 6 April 2016). It details the hourly power system load in the period from 2012 to 2015. The last year of the data (2015) is treated as a test period. We predict the daily load profiles for each day of this period, excluding atypical days such as public holidays (between 10 and 20 days a year). RF models were optimized on the data from 2012 to 2014, with validation data composed of 100 patterns selected randomly from 2014 and training data preceding the validation pattern.

5.1. Results for Different Preprocessing Methods and Training Modes

Tables 1 and 2 show mean absolute percentage error (MAPE) and root mean square error (RMSE), respectively, for input patterns r1–r7 and different training modes. Figures 4–6 show the boxplots of MAPE. The results can be summarised as follows:

- It is evident from these tables and figures that the global extended mode yields the lowest errors, when combined with patterns r4 (for PL, FR and DE) or r6 (for GB).
- The local training mode brings lower errors than the global one when patterns r1, r2, r6 and r7 are used, i.e., patterns which are composed of the daily curves. The local mode is usually better than the global extended one when patterns r1 and r2 are used, but it is worse than the global extended mode when cross-patterns are used, which also reflect a weekly seasonality.
- The highest errors for the global mode were observed when patterns r1 and r2 were used. In these cases, the errors were up to nine times greater than in the alternative training modes. Pattern r4 is recommended for the global mode, which for all countries provided the lowest errors. Modifying the global mode by extending the input patterns with calendar variables has always resulted in a reduction of errors.

Table 1. Validation MAPE for different input patterns r1–r7 and training modes (lowest errors in bold, second lowest errors in italics).

Data	Training Mode	r1	r2	r3	r4	r5	r6	r7
PL	Local	1.54	1.55	1.89	2.17	3.66	1.45	1.63
	Global	11.55	11.61	2.11	*1.30*	2.94	1.82	1.82
	Global ext.	1.92	1.79	1.40	**1.24**	1.96	1.50	1.55
GB	Local	1.92	1.91	2.29	2.21	3.68	1.76	1.75
	Global	14.88	15.01	2.21	1.75	2.92	1.84	1.83
	Global ext.	2.22	2.09	1.79	1.59	2.12	**1.47**	*1.52*
FR	Local	1.88	1.77	2.12	2.43	5.10	1.74	1.86
	Global	8.50	8.54	2.09	*1.53*	2.90	2.11	1.92
	Global ext.	1.98	1.71	1.59	**1.37**	1.71	1.63	1.64
DE	Local	1.34	1.31	1.67	1.93	3.47	1.21	1.45
	Global	12.33	12.34	1.62	*1.07*	2.61	1.79	1.54
	Global ext.	1.55	1.57	1.17	**0.99**	1.65	1.17	1.19

Table 2. Validation RMSE for different input patterns r1–r7 and training modes (lowest errors in bold, second lowest errors in italics).

Data	Training Mode	r1	r2	r3	r4	r5	r6	r7
PL	Local	436	522	489	540	1014	412	473
	Global	2420	2427	577	*347*	741	454	458
	Global ext.	468	448	364	**337**	509	406	410
GB	Local	918	825	1100	1118	1816	811	841
	Global	6309	6358	1035	860	1445	811	824
	Global ext.	925	937	784	757	985	**669**	*690*
FR	Local	1630	1576	1851	1939	3893	1406	1626
	Global	5602	5635	1714	*1144*	2697	1538	1437
	Global ext.	1414	1280	1287	**1048**	1450	1232	1272
DE	Local	1590	1190	1748	2099	3046	1168	1641
	Global	8927	8926	1410	878	2094	1411	1131
	Global ext.	1224	1229	*833*	**713**	1261	915	883

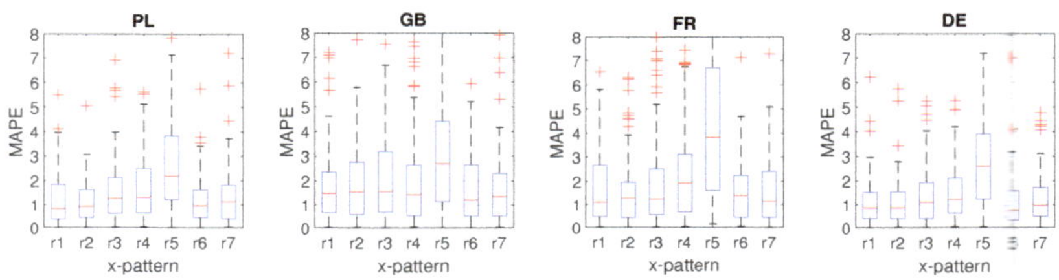

Figure 4. Local training mode: Boxplots of validation MAPE for different input patterns r1–r7.

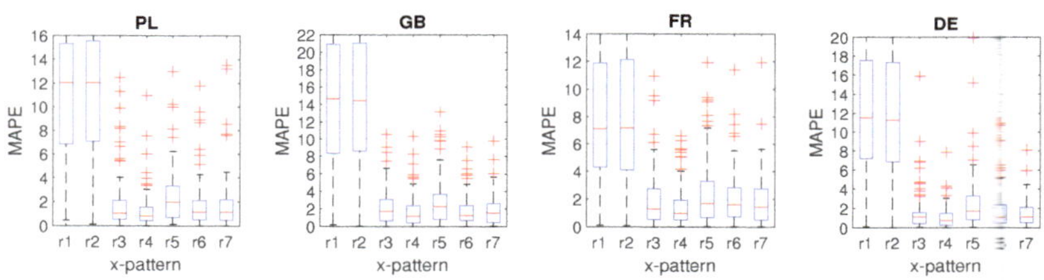

Figure 5. Global training mode: Boxplots of validation MAPE for different input patterns r1–r7.

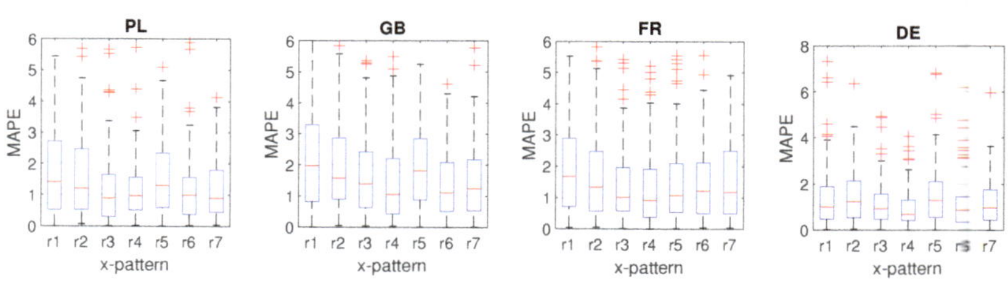

Figure 6. Global extended training mode: Boxplots of validation MAPE for different input patterns r1–r7.

Based on the results, the recommended training mode is global extended with r4 patterns for PL, FR and DE, and r6 patterns for GB. These variants of RF were used in the experiments described in the next sections.

5.2. Tuning Hyperparameters

In this experiment, we change the selected hyperparameter in the range shown in Figure 7, keeping the remaining hyperparameters at their constant values as follows: number of trees in the forest—$K = 100$, minimum number of leaf node observations—$m = 1$, and number of predictors to select at random for each decision split—$p = n/3$.

Figure 7 shows the impact of hyperparameters on the forecasting error (MAPE). As expected, the error decreases with the number of trees in the forest. The reduction in MAPE when the RF size changes from 1 do 300 trees was from 38.7% for PL to 50.0% for DE. At the same time, a significant reduction in the forecast variance was also observed from 63.8% for PL to 82.5% for DE. It can be seen from Figure 7b that an increase in the minimum leaf size leads to a deterioration in the results. Small values of m, close to 1, are preferred. This means that trees as deep as possible are the most beneficial in RF. The optimal number of predictors selected in the nodes to perform a split varies from country to country (see Figure 7c). For PL and DE it is 15, for GB it is 20, and for FR it is 6. These values differ from the recommended $p = n/3$, which are 8 for PL, FR and DE, and 11 for GB.

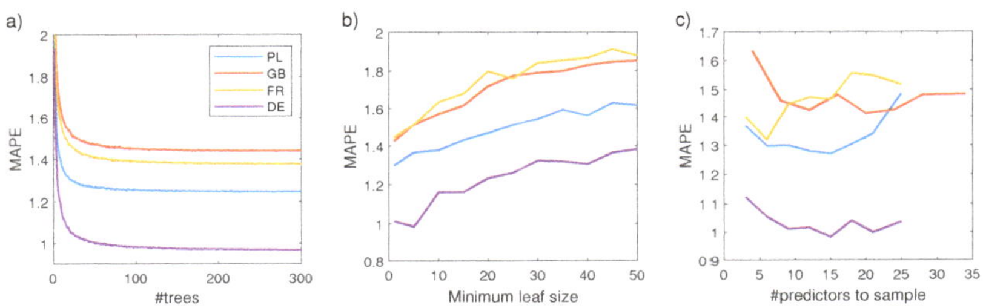

Figure 7. Validation MAPE depending on hyperparameters: number of trees in the forest (**a**), minimum number of leaf node observations (**b**) and number of predictors to select at random for each decision split (**c**).

Using optimal values of the hyperparameters for each country, we investigate the methods of split predictor selection s1–s3. Table 3 shows the results, validation MAPE and RMSE. Both accuracy measures show similar results for all methods of split predictor selection. Therefore, s1 is recommended as a simple, standard method, which does not cause any additional computational burden.

Table 3. Validation MAPE and RMSE for different methods of split predictor selection (training mode: global extended, #trees: 300, minimum leaf size: 1; lowest errors in bold).

Data	Variant	MAPE			RMSE		
		s1	s2	s3	s1	s2	s3
PL	r4, $p = 15$	**1.26**	1.30	1.30	**339**	345	345
GB	r6, $p = 20$	1.44	**1.42**	**1.42**	652	**649**	**649**
FR	r4, $p = 6$	**1.35**	1.39	1.38	**1046**	1071	1053
DE	r4, $p = 15$	**0.98**	1.00	1.00	**709**	714	714

Figure 8 shows the "importance" or "predictive strength" of the predictors estimated on the out-of-bag data (this is discussed further in Section 5.4). As can be seen from this figure, when r4 extended pattern is used (PL, FR and DE), the most important predictor is the last component of the r-pattern, i.e., the predictor expressing electricity demand at forecasted hour t of the day preceding the forecasted day, $z_{i-1,t}$. The importance of this predictor reaches 3.5 for FR and over 5 for PL and DE, while the importance of other demand predictors is usually below 2. Among the calendar predictors, the most important for r4 extended pattern are those coding the season of the year, especially for DE. For cross-pattern r6 (GB), the most important predictors are the calendar ones: day of the week and season of the year ($p1$). The next positions are occupied by predictors coding the demand for the last four hours of the day before the forecasted day ($z_{i-1,24}$ is clearly the most important of these) and predictor coding demand at forecasted hour t week ago $z_{i-7,t}$. Note the low importance of the other predictors representing demand at hour t of the preceding days, $z_{i-6,t}$–$z_{i-2,t}$.

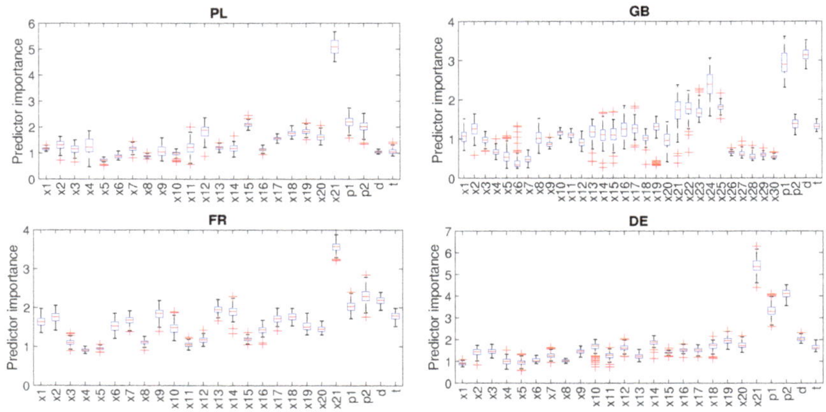

Figure 8. Importance of the predictors ($x1 - x30$—predictors expressing demand pattern r4 or r6, $p1$ and $p2$—components of vector **p** expressing season of the year, d—day of the week, and t—hour of the day).

Table 4 shows the forecasting results for the test set when using RF with the optimal values of hyperparameters. As performance metrics we use: MAPE, MdAPE (median of absolute percentage error), IqrAPE (interquartile range of APE), RMSE, MPE (mean percentage error), and StdPE (standard deviation of PE). MdAPE measures the mean error without the influence of outliers, while RMSE, as a square error, is especially sensitive to outliers.

Table 4. Results for test data (training mode: global extended, #trees: 300, minimum leaf size: 1, spl predictor selection method: s1).

Data	Variant	MAPE	MdAPE	IqrAPE	RMSE	MPE	StdPE
PL	r4, $p = 15$	1.05	0.78	1.06	259	0.03	1.52
GB	r6, $p = 20$	2.36	1.78	2.39	1058	−0.32	3.36
FR	r4, $p = 6$	1.67	1.18	1.67	1338	−0.22	2.42
DE	r4, $p = 15$	1.06	0.80	1.07	815	−0.04	1.48

The MAPE and MdAPE values in Table 4 indicate that the most accurate forecasts were obtained for PL and DE, while the least accurate were for GB. MPE allows us to assess the forecast bias. Positive values of MPE indicate underprediction, while its negative values indicate overprediction. Note that for PL and DE the forecast bias was significantly smaller than for GB and FR. The same can be said about the forecasts dispersion measured by IqrAPE and StdPE.

5.3. Results Comparison with Other Models

We compare the performances of RF with other models including statistical models and ML models. The comparative models are outlined below (see [30,39] for further description). Their hyperparameters were selected on the data from 2012–2014 in grid search procedures using a variant of cross-validation or selected by experimentation (this applies to models with a large number of hyperparameters, which are difficult to optimize using standard methods due to the huge search space).

- Naive—naive model: $\hat{\mathbf{z}}_{i*} = \mathbf{z}_{i*-7}$;
- ARIMA—Auto-Regressive Integrated Moving Average model;
- ETS—Exponential Smoothing model;
- Prophet—a modular additive regression model with nonlinear trend and seasonal components;
- MLP—perceptron with a single hidden layer and sigmoid nonlinearities;
- SVM—linear epsilon insensitive Support Vector Machine (ϵ-SVM);
- ANFIS—Adaptive Neuro-Fuzzy Inference System;
- LSTM—Long Short-Term Memory;
- FNM—Fuzzy Neighborhood Model;
- N-WE—Nadaraya–Watson Estimator;
- GRNN—General Regression NN;
- RandNN—ensemble of 100 Randomized NNs;
- MTGNN—Graph NN for multivariate time series forecasting;
- ES-adRNNe—ensemble of five hybrid and hierarchical models combining ETS and dilated RNN with attention mechanism.

We also compare our model with competitive tree-based ensembles: XGBoost [40] and LightGBM [41]. Their predictors include both calendar data (hour of the day, day of the week, quarter, month, year, day of the year, day of the month and week of the year) and historical demands (demands at hour t of 21 consecutive days preceding the forecasted day).

Table 5 compares MAPE for RF and the baseline models. From this table, you can clearly see the better performance of RF compared to the other models. RF outperformed all other models in terms of accuracy for PL, GB and DE. For FR it took third place after RandNN and SVM. To confirm the results, a pairwise one-sided Giacomini-White test was performed (GM test) [42]. Its results, p-values, are shown in Figure 9 (we used GW test implementation from [43]). Small p-values, below 0.05 (green color), indicate that the model on the X-axis significantly outperforms in terms of accuracy the model on the Y-axis.

Table 5. MAPE comparison between RF and baseline models (lowest errors in bold).

Model	PL	GB	FR	DE
RF (proposed)	**1.05**	**2.36**	1.67	**1.06**
Naive	2.96	4.80	5.53	3.13
ARIMA	2.31	3.50	3.00	2.31
ETS	2.14	3.19	2.79	2.10
Prophet	2.63	4.00	4.71	3.23
MLP	1.39	2.84	1.93	1.58
SVM	1.32	2.54	1.63	1.38
ANFIS	1.64	2.80	2.12	2.48
LSTM	1.57	2.92	1.81	1.57
FNM	1.21	3.02	1.84	1.30
N-WE	1.19	3.12	1.86	1.29
GRNN	1.22	3.01	1.81	1.30
RandNN	1.14	2.51	**1.57**	1.18
MTGNN	1.95	3.44	2.59	2.04
ES-adRNNe	1.22	2.45	1.73	1.15
XGBoost	1.67	3.36	2.51	1.74
LightGBM	1.67	3.39	2.54	1.79

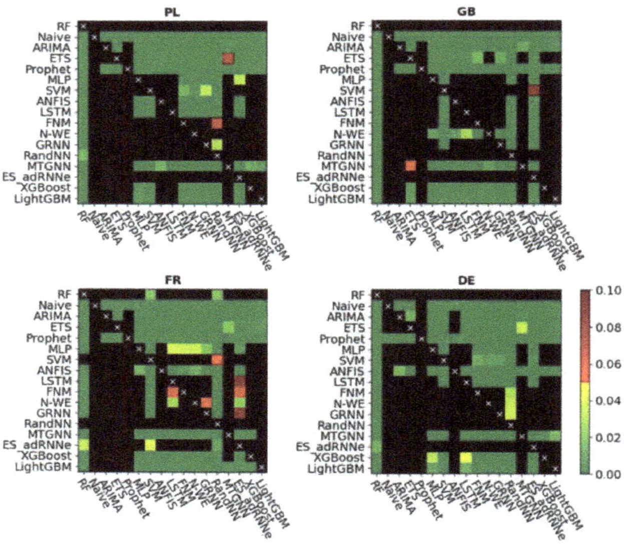

Figure 9. Results of the Giacomini-White test for the proposed and baseline models (black color is for p-values larger than 0.10).

5.4. Discussion

In our previous work [19], we used a local training mode with input patterns r2, which express daily profiles. Our current research revealed that input patterns incorporating weekly seasonality (r4) or both daily and weekly seasonality (r6) combined with global training with extended inputs improve the results (note that in Tables 1 and 2, patterns r4 and r6 provide lower errors than r2 for all countries). Calendar data used as additional input in the global extended training helps the trees to properly partition the input space and thus approximate the target function with greater accuracy. It does not take place without costs: the complexity of the model increases due to learning on all data, not just the selected data as in local training.

Table 5 and Figure 9 show that the proposed RF outperforms classical statistical models (ARIMA and ETS), modern statistical model (Prophet), classical ML models (MLP, SVM, ANFIS, GRNN), modern ML models (LSTM, RandNN), similarity-based models (FNM, N-WE) as well as state-of-the-art ML models (MTGNN, ES-adRNNe) and boosted regression trees (XGBoost, LightGBM). The last two models, as well as the proposed RF model, also used calendar variables, even in larger numbers. However, in contrast to these models, our model uses specific time series preprocessing, which may be a decisive advantage. Our model also outperforms ES-adRNNe, which is a very sophisticated and complex model developed especially for STLF [39]. To increase its predictive power, it is equipped with a new type of RNN cell with delayed connections and inherent attention, it processes time series adaptively, learning their representation and it learns in the cross-learning mode (i.e., it learns from many time series in the same time). It reveals its strength with a large amount of data, numerous and long time series. In our case, this condition was not met—there were only four, relatively short series available for training the models. In this case, the proposed RF model, which learns from individual series, generated more accurate forecasts than ES-adRNNe.

It is worth noting that RF has few hyperparameters to tune, which makes it easy to optimize (compare with DNNs with many hyperparameters). The results of our experiments confirmed that the number of trees in the forest should be as large as possible and the mimimum leaf size can be set to one. Therefore, the key hyperparameter remains the number of predictors to sample in the nodes. Its optimal values significantly differed from the recommended default values. Our attempt to increase the performance of the RF model through alternative methods of selecting predictors for split failed. Neither the curvature test nor the interaction test, which take into account the relationship between predictors and response when splitting data in nodes, improves the results significantly over the default CART method.

In our study, we used both continuous and categorical predictors. Such a mix causes many difficulties for other models such as ARIMA, ETS, NNs, SVM, LSTM and others. Categorical variables cannot be processed by these models directly. Such predictors must be converted into numerical data, so as to maintain the relationship between their values. The method of this conversion can be treated as an additional hyperparameter. RF has no problems with categorical variables, which is its big advantage. Moreover, RF can deal easily with any number of additional exogenous predictors and does not need to unify predictor ranges, which is often necessary for other models. RF can even deal with raw data because the predictors are not processed by the tree in any way, just selected in nodes, to construct a specific decision model (flowchart-like structure).

Regression tree provides fast one-pass training which does not need to repeatedly refer to the data. In contrast, NNs, which use a variant of the gradient descent optimization algorithm with multiple scanning of a dataset, are more time-consuming to train. Additionally due to the number of hyperparameters, they are also much more expensive to optimize in terms of time than RF. The training of a tree does not provide an optimal result because decisions about data split are made in nodes using a local rather than a global criterion, i.e., the split made may not be optimal from the point of view of the final result. However, the NN learning process also does not lead to optimal results due to sensitivity to the starting point and tendency to fall into the traps of the local minimum. Note that non-optimality of the trees is mitigated by their aggregation in the forest. Aggregation also smoothes out functions modeled by individual trees and reduces their variance. The learning process of RF can be easily paralleled because the individual trees learn independently.

One useful feature of RF is that it enables the generalization error to be estimated using out-of-bag (OOB) patterns, i.e., training patterns not selected for the bootstrap sample (approximately one third of the training patterns are left out in each bootstrap sample). Therefore, the time-consuming cross-validation that is widely used in other models for estimating the generalization error is not needed. Using OOB patterns, the generalization error can be estimated during one training session, along the way. Although for forecasting problems, where training patterns should precede validation to prevent data leakage, the OOB approach as well as standard cross-validation may be questionable. For this reason, we did not use the OOB approach in this study. Instead, we applied a different strategy. We chose a set of 100 validation patterns from 2014 and for each of them we trained RF on training patterns preceding the validation pattern.

A valuable feature of RF is its built-in mechanism for predictor selection. In each node, the predictor which improves the split-criterion the most is selected. The splitting criterion favors informative predictors over noisy ones, and can completely disregard irrelevant ones. Thus, in RFs an additional feature selection procedure is unnecessary. Based on the internal mechanism for selecting predictors, the predictor importance or strength can be estimated. The importance measure attributed to the splitting predictor is the accumulated improvement this predictor gives to the split-criterion at each split in each tree. RF also offers another method of estimating the predictor importance based on the OOB patterns [27]. When the tree is grown, the OOB patterns are passed down the tree, and the prediction error is recorded. Then the values for the given predictor are randomly permuted across the OOB samples, and the error is computed again. The importance measure is defined as the increase in error. This measure is computed for every tree, then averaged over the entire forest and divided by the standard deviation over the entire forest. Such a measure is presented in Figure 8. Note that information about the predictor importance is a key factor, which helps to improve the interpretability of the model and can be used for feature selection for other models.

Model interpretability is an emerging area in ML that aims to make the model more transparent and strengthen confidence in its results. This topic is also explored in electricity demand forecasting literature [44]. In [45], it was shown that the predictor importance is related to the model sensitivity to inputs and also to the method of importance estimation. An LSTM-based model, which is proposed in [45], is equipped with a built-in mechanism based on a mixture attention technique for temporal importance estimation of predictors. In the experimental study, this model demonstrated higher sensitivity to inputs than tree-based models (RF and XGBoost) which showed very low sensitivity on the predictors except one, which strongly dominated (the authors used built-in functions of scikit-learn to calculate the predictor importance for tree-based models). In our study, the predictor importance is more diverse (see Figure 8), which may result from the fact that our trees are very deep and thus involve a great number of predictors. Note that tree-based models enhance interpretability not only through built-in mechanisms of predictor importance estimation, which show predictive power of individual predictors, but also through their flowchart-like tree structure. They can be interpreted simply by plotting a tree and observing how the splits are made and what is the arrangement of the leaves. It should be noted, however, that while following the path that a single tree takes to make a decision is trivial and self-explanatory, following the paths of hundreds of trees in the ensemble is much more difficult. To facilitate this, in [46], model compression methods were proposed that transform a tree ensemble into a single tree that approximates the same decision function.

In this study we use a standard RF formulation which is a MISO model producing point forecasts. Thus for prediction of 24 values of the daily curve of electricity demand, we need to train 24 RF models. In [32], we proposed a multivariate regression tree for STLF, which produces a vector as an output, representing the 24 predicted values. Using such MIMO trees as ensemble members simplifies and speeds up the forecasting process. A promising extension of the RF in the direction of probabilistic forecasting can be achieved using a quantile regression forest, which can infer the full conditional distribution of the response variable for high-dimensional predictor variables [47].

6. Conclusions

ML ensemble models are state-of-the-art for forecasting problems. They dominate the most recent literature on forecasting. Among them, tree-based ensembles have a solid theoretical basis and have been thoroughly researched in a huge number of papers. Their predictive power has been confirmed in numerous forecasting competitions [24].

In this study, we propose a RF model for a challenging STLF problem with multiple seasonality, nonlinear trend, and varying variance in time series. Unlike DNNs, RF is simple and transparent, it does not require a complex, deep architecture, equipped with additional sophisticated mechanisms to deal with complex time series. The greatest advantages of RF as a forecasting model are: small number of tuning hyperparameters (we show that only one is key), fast training and optimization, ability to deal with multiple exogenous predictors of different types, and built-in mechanism for selecting predictors and estimating their importance.

As with any predictive model, the performance of RF depends significantly on data preprocessing and proper organization of the training process. In the simulation study, we show how the results of RF depend on the training method, definition of input variables and hyperparameters. Based on the results, we recommend the best method of predictor definition (r4 and r6) and training mode (global extended) for STLF. Comparing the performances of RF and baseline models including statistical and ML ones, we showed that RF can successfully compete with them, providing the most accurate forecasts.

In our future work, we plan to extend RF with random data projection (to further smooth the estimator and provide an additional source of diversity) and use RF for probabilistic forecasting. A quantile regression forest [47] is a promising tool for the latter task.

Funding: This research received no external funding.

Data Availability Statement: We use real-world data collected from www.entsoe.eu (accessed on 6 April 2016).

Conflicts of Interest: The author declares no conflict of interest.

Abbreviations

The following abbreviations are used in this manuscript:

ANFIS	Adaptive Neuro-Fuzzy Inference System
APE	Absolute Percentage Error
ARIMA	Auto-Regressive Integrated Moving Average
CART	Classification And Regression Trees
CNN	Convolutional Neural Network
DE	Germany
DNN	Deep Neural Network
ES-adRNNe	ensemble of five hybrid and hierarchical models combining ES and dilated RNN with attention mechanism
ETS	Exponential Smoothing
FR	France
FNM	Fuzzy Neighborhood Model
GB	Great Britain
GBRT	Gradient Boosting Regression Tree
GRNN	General Regression Neural Network
IqrAPE	Interquartile Range of Absolute Percentage Error
LightGBM	Light Gradient Boosting Machine
LSTM	Long Short Term Memory Neural Network
MAPE	Mean Absolute Percentage Error
MdAPE	Median of Absolute Percentage Error
MIMO	Multiple Input Multiple Output
MISO	Multiple Input Single Output
ML	Machine Learning
MLP	Multilayer Perceptron
MPE	Mean Percentage Error
MTGNN	Graph Neural Network for Multivariate Time series forecasting
N-WE	Nadaraya–Watson Estimator
NN	Neural Network
PE	Percentage Error
PL	Poland
RandNN	Randomized Neural Network
RF	Random Forest
RMSE	Root Mean Square Error
RNN	Recurrent Neural Network
StdPE	Standard Deviation of Percentage Error
SVM	Support Vector Machine
STLF	Short-Term Load Forecasting
WGTB	Warm-start Gradient Tree Boosting
XGBoost	eXtreme Gradient Boosting

References

1. Arora, A.; Taylor, J.W. Rule-based autoregressive moving average models for forecasting load on special days: A case study for France. *Eur. J. Oper. Res.* **2018**, *266*, 259–268. [CrossRef]
2. Taylor, J.W. Short-term load forecasting with exponentially weighted methods. *IEEE Trans. Power Syst.* **2012**, *27*, 458–464. [CrossRef]
3. Charlton, N.; Singleton, C. A refined parametric model for short term load forecasting. *Int. J. Forecast.* **2014**, *30*, 364–368. [CrossRef]
4. Takeda, H.; Tamura, Y.; Sato, S. Using the ensemble Kalman filter for electricity load forecasting and analysis. *Energy* **2016**, *104*, 184–198. [CrossRef]
5. Smyl, S.; Dudek, G.; Pełka, P. ES-dRNN: A hybrid exponential smoothing and dilated recurrent neural network model for short-term load forecasting. *arXiv* **2021**, arXiv:2112.02663.

6. Benidis, K.; Rangapuram, S.S.; Flunkert, V.; Wang, Y.; Maddix, D.; Turkmen, C.; Gasthaus, J.; Bohlke-Schneider, M.; Salinas, D.; Stella, L.; et al. Deep Learning for Time Series Forecasting: Tutorial and Literature Survey. *ACM Comput. Surv.* **2022**, *accepted*. [CrossRef]
7. Dudek, G. Neural networks for pattern-based short-term load forecasting: A comparative study. *Neurocomputing* **2016**, *205*, 64–74. [CrossRef]
8. Eskandari, H.; Imani, M.; Moghaddam, M.P. Convolutional and recurrent neural network based model for short-term load forecasting. *Electr. Power Syst. Res.* **2021**, *195*, 107173. [CrossRef]
9. Kim, J.; Moon, J.; Hwang, E.; Kang, P. Recurrent inception convolution neural network for multi short-term load forecasting. *Energy Build.* **2019**, *194*, 328–341. [CrossRef]
10. Chitalia, G.; Pipattanasomporn, M.; Garg, V.; Rahman, S. Robust short-term electrical load forecasting framework for commercial buildings using deep recurrent neural networks. *Appl. Energy* **2020**, *278*, 115410. [CrossRef]
11. Chen, K.; Chen, K.; Wang, Q.; He, Z.; Hu, J.; He, J. Short-term load forecasting with deep residual networks. *IEEE Trans. Smart Grid* **2019**, *10*, 3943–3952. [CrossRef]
12. Reeve, H.W.J.; Brown, G. Diversity and degrees of freedom in regression ensembles. *Neurocomputing* **2018**, *298*, 55–68. [CrossRef]
13. Chan, F.; Pauwels, L.L. Some theoretical results on forecast combinations. *Int. J. Forecast.* **2018**, *34*, 64–74. [CrossRef]
14. Piotrowski, P.; Baczyński, D.; Kopyt, M.; Gulczyński, T. Advanced ensemble methods using machine learning and deep learning for one-day-ahead forecasts of electric energy production in wind farms. *Energies* **2022**, *15*, 1252. [CrossRef]
15. Yang, D.; Guo, J.; Sun, S.; Han, J.; Wang, S. An interval decomposition-ensemble approach with data-characteristic-driven reconstruction for short-term load forecasting. *Appl. Energy* **2022**, *306A*, 117992. [CrossRef]
16. Gao, R.; Du, L.; Suganthan, P.N.; Zhou, Q.; Yuen, K.F. Random vector functional link neural network based ensemble deep learning for short-term load forecasting. *Expert Syst. Appl.* **2022**, *206*, 117784. [CrossRef]
17. Moon, J.; Jung, S.; Rew, J.; Rho, S.; Hwang, E. Combination of short-term load forecasting models based on a stacking ensemble approach. *Energy Build.* **2020**, *216*, 109921. [CrossRef]
18. Ribeiro, G.T.; Mariani, V.C.; Coelho, L.S. Enhanced ensemble structures using wavelet neural networks applied to short-term load forecasting. *Eng. Appl. Artif. Intell.* **2019**, *82*, 272–281. [CrossRef]
19. Dudek, G. Short-term load forecasting using random forests. In *Intelligent Systems'2014, Advances in Intelligent Systems and Computing*; Filev, D., Jabłkowski, J., Kacprzyk, J., Krawczak, M., Popchev, I., Rutkowski, L., Sgurev, V., Sotirova, E., Szynkarczyk, P., Zadrożny, S., Eds.; Springer: Cham, Switzerland, 2015; Volume 323, pp. 821–828.
20. Wang, Y.; Chen, J.; Chen, X.; Zeng, X.; Kong, Y.; Sun, S.; Guo, Y.; Liu, Y. Short-term load forecasting for industrial customers based on TCN-LightGBM. *IEEE Trans. Power Syst.* **2021**, *36*, 1984–1997. [CrossRef]
21. Zhang, Y.; Wang, J. Short-term load forecasting based on hybrid strategy using warm-start gradient tree boosting. *J. Renew. Sustain. Energy* **2020**, *12*, 066102. [CrossRef]
22. Wang, Y.; Sun, S.; Chen, X.; Zeng, X.; Kong, Y.; Chen, J.; Guo, Y.; Wang, T. Short-term load forecasting of industrial customers based on SVMD and XGBoost. *Int. J. Electr. Power Energy Syst.* **2021**, *129*, 106830. [CrossRef]
23. Elsayed, S.; Thyssens, D.; Rashed, A.; Jomaa, H.S.; Schmidt-Thieme, L. Do we really need deep learning models for time series forecasting? *arXiv* **2021**, arXiv:2101.02118.
24. Bojer, C.S.; Meldgaard, J.P. Kaggle forecasting competitions: An overlooked learning opportunity. *Int. J. Forecast.* **2021**, *37*, 587–603. [CrossRef]
25. Januschowski, T.; Wang, Y.; Torkkola, K.; Erkkilä, T.; Hasson, H.; Gasthaus, J. Forecasting with trees. *Int. J. Forecast.* **2021**, *38*, 1473–1481. [CrossRef]
26. Makridakis, S.; Spiliotis, E.; Assimakopoulos, V. M5 accuracy competition: Results, findings, and conclusions. *Int. J. Forecast.* **2022**, *38*, 1346–1364. [CrossRef]
27. Hastie, T.; Tibshirani, R.; Friedman, J. *The Elements of Statistical Learning. Data Mining, Inference, and Prediction*; Springer: New York, NY, USA, 2009.
28. Dudek, G. Short-term load cross-forecasting using pattern-based neural models. In Proceedings of the 2015 16th International Scientific Conference on Electric Power Engineering, EPE 2015, Kouty nad Desnou, Czech Republic, 20–22 May 2015; pp. 179–183.
29. Dudek, G. Multilayer perceptron for short-term load forecasting: From global to local approach. *Neural Comput. Appl.* **2019**, *32*, 3695–3707. [CrossRef]
30. Dudek, G.; Pełka, P. Pattern similarity-based machine learning methods for mid-term load forecasting: A comparative study. *Appl. Soft Comput.* **2021**, *104*, 107223. [CrossRef]
31. Pełka, P.; Dudek G. Pattern-based long short-term memory for mid-term electrical load forecasting. In Proceedings of the 2020 International Joint Conference on Neural Networks, IJCNN 2020, Glasgow, UK, 19–24 July 2020; pp. 1–8.
32. Dudek, G. Multivariate regression tree for pattern-based forecasting time series with multiple seasonal cycles. In *Information Systems Architecture and Technology, Proceedings of the 38th International Conference on Information Systems Architecture and Technology, ISAT 2017, Szklarska Poręba, Poland, 17–19 September 2017*; Świątek, J., Borzemski, L., Wilimowska, Z., Eds.; Springer: Cham, Switzerland, 2018; Volume 655, pp. 85–94.
33. Breiman, L.; Friedman, J.H.; Olshen, R.A.; Stone C.J. *Classification and Regression Trees*; Chapman and Hall: London, UK, 1984.
34. Breiman, L. Random forests. *Mach. Learn.* **2001**, *45*, 5–32. [CrossRef]
35. Breiman, L. Bagging predictors. *Mach. Learn.* **1996**, *24*, 123–140. [CrossRef]

36. Ho, T.K. The random subspace method for constructing decision forests. *IEEE Trans. Pattern Anal. Mach. Intell.* **1998**, *20*, 832–844.
37. Loh, W.Y.; Shih, Y.S. Split selection methods for classification trees. *Stat. Sin.* **1997**, *7*, 815–840.
38. Loh, W.Y. Regression trees with unbiased variable selection and interaction detection. *Stat. Sin.* **2002**, *12*, 361–386.
39. Smyl, S.; Dudek, G.; Pełka, P. ES-dRNN with dynamic attention for short-term load forecasting. In Proceedings of the 2022 International Joint Conference on Neural Networks IJCNN 2022, Padova, Italy, 18–23 July 2022.
40. Chen, T.; Guestrin, C. XGBoost: A scalable tree boosting system. In Proceedings of the 22nd International Conference on Knowledge Discovery and Data Mining, ACM SIGKDD 2016, San Francisco, CA, USA, 13–17 August 2016; pp. 785–794.
41. Ke, G.; Meng, Q.; Finley, T.; Wang, T.; Chen, W.; Ma, W.; Ye, Q.; Liu, T.-Y. LightGBM: A highly efficient gradient boosting decision tree. In Proceedings of the Conference on Advances in Neural Information Processing Systems 30, NIPS 2017, Long Beach, CA, USA, 4–9 December 2017; pp. 3149–3157.
42. Giacomini, R.; White, H. Tests of conditional predictive ability. *Econometrica* **2006**, *74*, 1545–1578. [CrossRef]
43. Lago, J.; Marcjasz, G.; De Schutter, B.; Weron, R. Forecasting day-ahead electricity prices: A review of state-of-the-art algorithms, best practices and an open-access benchmark. *Appl. Energy* **2021**, *293*, 116983. [CrossRef]
44. Xu, C.; Liao, Z.; Li, C.; Zhou, X.; Xie, R. Review on interpretable machine learning in smart grid. *Energies* **2022**, *15*, 4427. [CrossRef]
45. Li, C.; Dong, Z.; Ding, L.; Petersen, H.; Qiu, Z.; Chen, G.; Prasad, D. Interpretable memristive LSTM network design for probabilistic residential load forecasting. *IEEE Trans. Circuits Syst. I* **2022**, *69*, 2297–2310. [CrossRef]
46. Sagi, O.; Lior, R. Approximating XGBoost with an interpretable decision tree. *Inf. Sci.* **2021**, *572*, 522–542. [CrossRef]
47. Meinshausen, N. Quantile regression forests. *J. Mach. Learn. Res.* **2006**, *7*, 983–999.

energies

MDPI

Article

Multi-Task Autoencoders and Transfer Learning for Day-Ahead Wind and Photovoltaic Power Forecasts

Jens Schreiber * and Bernhard Sick

Intelligent Embedded System, University of Kassel, Wilhelmshöher Allee 71, 34121 Kassel, Germany; bsick@uni-kassel.de

* Correspondence: j.schreiber@uni-kassel.de

Abstract: Integrating new renewable energy resources requires robust and reliable forecasts to ensure a stable electrical grid and avoid blackouts. Sophisticated representation learning techniques, such as autoencoders, play an essential role, as they allow for the extraction of latent features to forecast the expected generated wind and photovoltaic power for the next seconds up to days. Thereby, autoencoders reduce the required training time and the time spent in manual feature engineering and often improve the forecast error. However, most current renewable energy forecasting research on autoencoders focuses on smaller forecast horizons for the following seconds and hours based on meteorological measurements. At the same time, larger forecast horizons, such as day-ahead power forecasts based on numerical weather predictions, are crucial for planning loads and demands within the electrical grid to prevent power failures. There is little evidence on the ability of autoencoders and their respective forecasting models to improve through multi-task learning and time series autoencoders for day-ahead power forecasts. We can close these gaps by proposing a multi-task learning autoencoder based on the recently introduced temporal convolution network. This approach reduces the number of trainable parameters by 38 for photovoltaic data and 202 for wind data while having the best reconstruction error compared to nine other representation learning techniques. At the same time, this model decreases the day-ahead forecast error up to 18.3% for photovoltaic parks and 1.5% for wind parks. We round off these results by analyzing the influences of the latent size and the number of layers to fine-tune the encoder for wind and photovoltaic power forecasts.

Keywords: transfer learning; wind power; photovolatic power; autoencoders; deep learning; time series

Citation: Schreiber, J.; Sick, B. Multi-Task Autoencoders and Transfer Learning for Day-Ahead Wind and Photovoltaic Power Forecasts. *Energies* **2022**, *15*, 8062. https://doi.org/10.3390/en15218062

Academic Editors: Paweł Piotrowski, Grzegorz Dudek and Dariusz Baczyński

Received: 30 September 2022
Accepted: 26 October 2022
Published: 30 October 2022

1. Introduction

Due to the increase in the amount of renewable energy in the electric grid, it is essential to find suitable compact representations that allow prediction of expected power generation. Such compact representations often improve the prediction quality and can be used to save computational resources and time [1]. The area of research that deals with finding a suitable representation, e.g., through autoencoders, is called representation learning [2]. An autoencoder, an artificial neural network architecture, consists of an encoder, a bottleneck layer, and a decoder. In the case of an undercomplete autoencoder, an encoder learns a transformation of the original features into a lower-dimensional feature space, e.g., through a bottleneck in the neural network [3]. The decoder utilizes this latent representation to reconstruct the original features. Although this area of research has existed in the literature on renewable energy for some time, it is mainly concerned with short-term forecasts, which use wind speed measurements to predict the expected wind power in the following minutes and hours [4–7].

At the same time, for planning and ensuring the stability of the electric grid, larger forecast horizons, such as day-ahead forecasts between 24 and 47 h into the future are inevitable. Such prediction horizons are possible with predicted features of numerical weather prediction (NWP) models.

Due to the weather's chaotic and non-linear behavior, these forecast horizons have more substantial forecast errors than smaller horizons. Therefore, finding suitable representations for those horizons is even more essential. The research questions to find an appropriate representation are:

Research Question 1. To what extent are multi-task learning (MTL) autoencoders beneficial for learning latent features of NWP for day-ahead forecasts?

Answering and developing methods for this question is critical, as in practice, an individual autoencoder is often trained for each wind or photovoltaic (PV) park. This training setting is also called single-task learning (STL), where we train one model for each park individually. At the same time, an MTL architecture reduces the required training time and often improves the forecast error. To our knowledge, MTL autoencoders, trained in a semi-supervised setting, have not been considered for day-ahead power forecasts.

Research Question 2. To what extent are time series-specific layers in autoencoders beneficial for learning latent features of NWP day-ahead features?

Answering this question allows us to consider seasonality within forecasts, e.g., given through the diurnal cycle, which influences the forecast error [8].

Research Question 3. To what extent is it necessary to fine-tune the encoder for day-ahead power forecasts?

In the literature, the encoder is often fine-tuned completely for predicting power forecasts. However, as we will see later on, this is not always beneficial and depends on, e.g., the architecture of the autoencoder.

To answer those research questions, we combine techniques from unsupervised learning and MTL. We utilize a discriminative STL autoencoder based on a multi-layer perceptron (MLP) and a temporal convolution network (TCN) for learning latent features of the weather, i.e., referred to as AEMLP and AETCN. The TCN model type is recently introduced for time series forecasts in fields such as speech processing, traffic estimation, short-term wind power predictions [9], and day-ahead wind and PV power forecasts [10]. We extend both models in an MTL setting and refer to them as AEMLP-MTL and AETCN-MTL. Our experiments consider a generative approach for all variants through a variational autoencoder (VAE). The generative approach learns to represent the (whole) distribution, while the discriminative autoencoders learn the most efficient data encoding to represent the data. Finally, we train the encoder with a forecasting model in a supervised fashion to forecast the expected power and evaluate the different amount of layers to fine-tune the encoder, leading to the following contributions:

- Through the proposed MTL autoencoders, we reduce the number of parameters for training by a factor of 38 for PV data and by 202 times for wind data.
- At the same time, we improve the reconstruction error by up to 300% for PV and 134 times for wind through the proposed AETCN-MTL architecture.
- During our study, to answer Research Question 3, we found that the number of layers to be fine-tuned depends on the model type, e.g., MLP or TCN and the model architecture, e.g., STL or MTL.
- Based on the encoder of the proposed AETCN-MTL model, we achieve one of the best results with improvements up to 18.3 and 1.5 percent for PV and wind day-ahead power forecasts, respectively.

The remaining article shows the necessity for a systematic analysis of autoencoders in day-ahead forecasts in Section 2 through the literature review. Afterward, we detail different autoencoder techniques in Section 3. The following section, Section 4, summarizes the datasets and challenges for day-ahead power forecasts. In Section 5, we detail our analysis to answer the identified research questions. Finally, in Section 6, we revisit our work and propose future work.

2. Related Work

The following section reviews the related work on wind and PV power forecasts using autoencoders. Within the review, we focus on work that utilizes autoencoder in a transfer learning (TL) setting, e.g., to reduce the required training time. Generally, we consider transfer learning as a knowledge transfer between two tasks. Nowadays, most researchers perform knowledge transfer through fine-tuning layers of a deep learning model. Additional work that applies autoencoders for predicting wind speed and solar radiation is summarized in [11].

The authors of [12] are one of the first that considered long-tem short memories (LSTMs) for day-ahead PV power forecasts. Therefore, they initially trained a vanilla autoencoder based on day-ahead NWP inputs from 21 parks. Vanilla autoencoder refers to an autoencoder based on an MLP architecture. Afterward, the authors trained an LSTM attached to the encoder of the autoencoder for renewable power forecasts.

The authors of [4] proposed the utilization of a stacked autoencoder for multi-step wind power prediction based on an MLP model. The authors trained a three-layer stacked autoencoder on a single wind park based on historical power measurements. Afterward, adding a final layer to the model and fine-tuning the whole network allowed for the creation of multi-step-ahead prediction. The authors evaluated the results on roughly 1.5 weeks of data for a one-hour-ahead prediction.

In [5], a stacked autoencoder was proposed for wind speed and wind power prediction for horizons up to four hours ahead. Compared to [4], a recurrent autoencoder allowed for learning of the relations in the time series in [5]. Again, the whole network was fine-tuned after the pre-training and evaluated on two wind speed prediction experiments and one for predicting the expected wind power generation in Belgium.

The third variant of a stacked autoencoder was presented in [6] for wind power predictions up to two-hours ahead. A particularly engaging aspect of their approach is that they learn the feature extraction of the stacked autoencoder with the power prediction layer jointly in an end-to-end fashion. At the same time, they considered an MTL approach for multi-output predictions, where the model predicts multiple horizons simultaneously. The results were evaluated on a single wind farm in the United States after fine-tuning the complete network.

While the former articles did not explicitly consider TL techniques, the following articles consider techniques from this field. One of the earliest works trained nine MLP-based autoencoders on a single wind park [7]. Those autoencoders were adapted to four other wind parks and an ensemble given by a deep belief network that combines each park's extracted features. The authors made a one-hour ahead prediction by utilizing the previous 24 h NWP data, where they reduced the training time through the fine-tuning process from the source park.

The same leading author randomly selected one out of five parks in [13]. This park was utilized for training a single MLP autoencoder from a randomly selected park. This autoencoder was adapted for every four months of available training data. Over time, there are various autoencoders acting as feature extractors for intra-day forecast horizons. Features extracted from those autoencoders were selected through mutual information and forecasts were combined through an ensemble to provide the final prediction.

The article closest to ours is [1]. This article compared traditional feature extraction methods with feature extraction techniques from deep learning. The authors showed that fine-tuning helps improve the forecast error for MLP-based models for day-ahead wind and PV power forecasts on a total of 21 PV and 55 wind parks. However, they did not consider autoencoders based on TCNs. Furthermore, an MTL approach was not considered, nor was a different number of layers to unfreeze in the model.

The authors of [14] trained a unified autoencoder considering data from wind turbines. This model allowed them to extract homogeneous features of all turbines in a base model. To reduce the training time and problem of vanishing gradient, they adopted a model fn a single turbine to extract heterogeneous features and fine-tune the model for the final prediction. Their work considered two experiments with forecast horizons between 10 and 60 s based on meteorological measurements to predict the expected power generation. In the first experiment, they considered 15 parks as the source and target task, whereas in the second experiment they considered 50 parks as the source and target. The unified approach within the article is similar to our proposed MTL approach. However, they only considered an MLP-based model for short forecast horizons, whereas we are interested in the horizon between 24 and 47 h into the future and TCN architecture.

Finally, the authors of [15] trained variational autoencoders on source parks and fine-tuned this on a target with limited data. They simultaneously utilized five variational datasets as source models to generate several different latent features. The proposed approach reduced the training time while having excellent forecast errors. However they only used two wind parks as the source domain and three wind parks as the target domain. At the same time, this article utilizes meteorological measurements in a regression task.

Overall we can summarize that only two articles have considered day-ahead power forecast and only one utilized an MTL approach. Even though various articles have considered fine-tuning an initial representation, to the best of our knowledge, none of the articles have evaluated the different number of layers to fine-tune. Furthermore, those techniques are often considered separately. We aim to close this gap with our article. Finally, the maximum number of considered wind parks in the literature are 55 and 21 PV, whereas we considered 445 wind and 117 PV parks.

3. Method

This section explains the deep learning-based techniques for latent feature extraction and concepts of transfer learning and MTL. Latent feature extraction from input features allows the derivation of valuable latent features for downstream tasks such as classification or regression.

We concentrate on undercomplete autoencoders (Figure 1), as they allow learning a representation $\mathbf{z} \in \mathbb{R}^{D_\mathbf{z}}$ of the input $\mathbf{x} \in \mathbb{R}^{D_\mathbf{x}}$, where the number of latent features $D_\mathbf{z} \in \mathbb{N}_{\geq 1}$ is less or equal than the input feature dimension $D_\mathbf{x} \in \mathbb{N}_{\geq D_\mathbf{z}}$. The concept of undercomplete autoencoders assures that after the training the computational effort is less than with the original input features. Moreover, undercomplete autoencoders are often preferred over overcomplete autoencoders for the same reasons, as they learn a higher dimensional representation of the input features and increase the computational effort in further processing. Within the context of undercomplete autoencoders, we analyze and explain three types of autoencoders:

1. We present the vanilla autoencoder that builds the foundation for the other architectures.
2. We introduce time series autoencoders. These consists of recurrent networks or a convolutional neural network (CNN)-based approach. Due to the recent success and advantages of CNNs over recurrent networks [16,17], we focus on the latter.
3. We detail variational autoencoders as an example of generative autoencoders. A generative approach has the advantage that we can impute missing values, reduce noise in data, or sample from the learned representation. The generative approach learns to represent the (whole) distribution, while a discriminative autoencoder learns the most efficient data encoding to represent the data. On the other hand, discriminative autoencoders have the advantage that they are easy to implement and often are sufficient for various tasks [1].

A combination of those concepts is applicable. For instance, we can train a variational time series autoencoder. Further, we must differentiate if an architecture is learned from single or multiple tasks simultaneously. In the former, we refer to it as an STL and, in the latter case, as an MTL autoencoder.

3.1. Vanilla Autoencoder

Autoencoders describe the concept of learning latent features through neural networks. Within this article, we refer to a vanilla autoencoder as architecture that utilizes an MLP architecture to extract those features. Additionally, we only consider undercomplete autoencoders for reducing the original input features from the NWP, as detailed in Section 4.1.

In such a vanilla autoencoder, as visualized in Figure 1, we have three main components: the encoder, the bottleneck, and the decoder. The encoder has the features $\mathbf{x}$ from the NWP as input, where the dimension is of size $D_{\mathbf{x}}$. In each successive layer, we reduce the number of features to the required number of latent features $D_{\mathbf{z}}$, which we refer to as the bottleneck. Afterward, the decoder aims to reconstruct the original features from the latent features $\mathbf{z}$ at the bottleneck by increasing the number of features in each successive layer. The difference between the original input and the reconstructed features is referred to as the reconstruction error. We evaluate the reconstruction error often through a squared error loss.

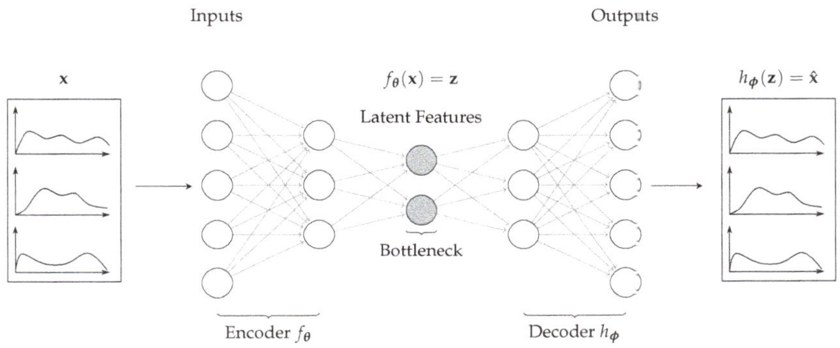

Figure 1. An example undercomplete autoencoder (AE) topology. The AE reduces the dimensionality in each layer of the encoder. The representation of the latent features at the bottleneck are the extracted hidden features sufficient to reconstruct the original input successively in each layer of the decoder.

One of the fundamental concepts of autoencoders is the bottleneck. We ensure that the network is not learning an identity mapping during training by having a lower dimension in the bottleneck than in the input. Other alternatives to avoid this problem are, e.g., denoising autoencoders, where we induce random noise on the input features [1]. However, we excluded those variants as the current results suggest that they are not beneficial over vanilla autoencoders for day-ahead power forecasting [1].

To introduce the loss function, consider that the latent features are given by the encoder with $\mathbf{z} = f_\theta(\mathbf{x})$, where f_θ is the encoding neural network with parameters $\boldsymbol{\theta}$. At the same time, the reconstructed features $\hat{\mathbf{x}} \in \mathbb{R}^{D_{\mathbf{x}}}$ and $D_{\mathbf{x}} \in \mathbb{N}_{\geq 1}$ are given by $\hat{\mathbf{x}} = h_\phi(\mathbf{z})$, where h_ϕ is the decoding function given by the neural network with parameters $\boldsymbol{\phi}$. In the case that we are interested in reconstruction of the same input features $\mathbf{x}$ as output features $\hat{\mathbf{x}}$, then we aim to approximate the input features with $\mathbf{x} \approx \hat{\mathbf{x}}$. We achieve this approximation through the following loss function:

$$\mathcal{L}_{AE} = L(\mathbf{x}, h_\phi(f_\theta(\mathbf{x}))), \tag{1}$$

where L is often a quadratic loss such as the mean squared error (MSE).

After the unsupervised training of an autoencoder through the above loss function, we typically remove the decoder and solely use the learned latent feature as input for supervised machine learning (ML) algorithms for regression or classification tasks. Often, we train this model through a gradient descent method such as the Adam optimizer, where we update the weights of the encoder and those of the forecasting model, e.g., based on a squared error loss.

For results of a squared error loss and a linear decoder, the latent space of the autoencoder lies in a similar sub-space to principal component analysis (PCA). Moreover, ref. [18] shows that PCA components can be estimated from latent features through singular value decomposition. This result motivates us to utilize PCA as a reference for later experiments.

To compare the predictions obtained with AEs to extended techniques, we extend the idea of AE to more complex structures and further exploit the potential of deep architectures for representation learning in the following.

3.2. Time Series Autoencoder

While a latent representation through an MLP has the advantage that it is easy and efficient to train, this representation neglects cyclic influences, e.g., caused by the diurnal cycle within a day [8]. Therefore, an autoencoder that learns correlations between timesteps is often beneficial. One choice would be an autoencoder based on a recurrent network such as LSTMs. However, due to the recent success of 1D CNNs for time series forecasts, their reduced training time, and often improved performance, see, e.g., [17], we focus on these over recurrent architectures. As the principled structure of a time series autoencoder is identical to the vanilla autoencoder detailed in the previous section, we initially focus on the general concept of a one-dimensional convolution autoencoder. Afterward, we detail the proposed TCN autoencoder to learn the latent features from NWP models for day-ahead forecasts.

Figure 2 visualizes an example of a 1D-CNN. Let us, therefore, assume we have a one-dimensional input time series with 24 timesteps t from $\{0, \ldots, 23\}$ similar to the time series length in day-ahead forecasts with an hourly resolution. Let us further assume the time series of a single input feature is given as an ordered set with $\{x_0, \ldots, x_t, \ldots, x_{23}\}$, also referred to as input channel, and a filter $\mathbf{a} \in \mathbb{R}^3$. Then, the result of the 1D-CNN at timestep t is simply given by the dot product between the filter $\mathbf{a}$ and $\{x_{t-1}, x_t, x_{t+1}\}$. By adding, e.g., a zero padding at the beginning and end of the time series, we ensure that we maintain the length of the time series, similar to a recurrent network [19], in the output channel. Padding the time series is also essential, as later on, we aim to forecast the expected power generation with an equivalent time series length.

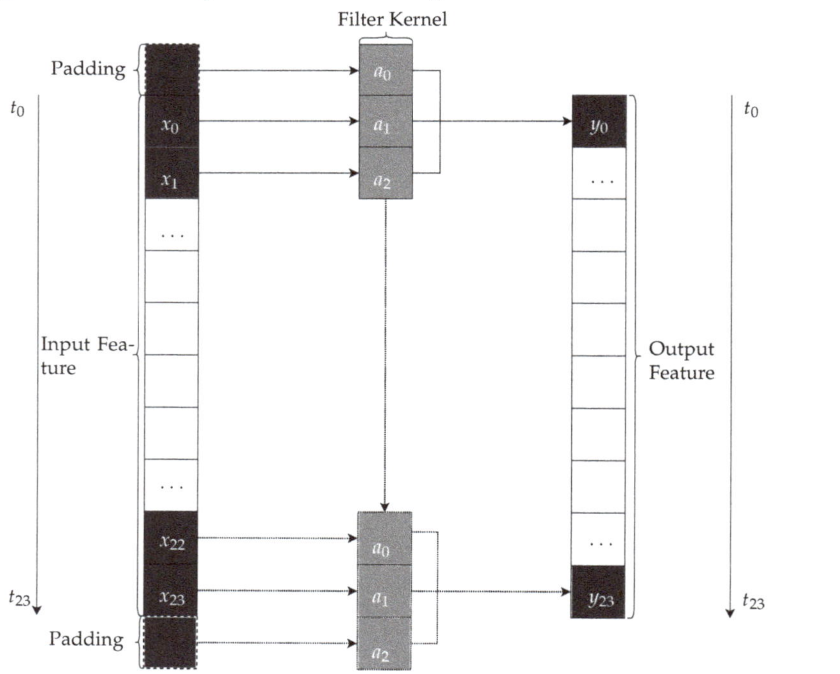

Figure 2. Example one-dimensional CNN with a filter of size 1×3. We keep the time series' dimension and extract relevant information by applying the filter to an input time series size of 1×24 with additional padding.

Figure 3 visualizes this concept for multiple inputs. We can observe that in each layer of the encoder, the length of the time series stays the same while the number of features reduces. In the decoder, we increase the number of features. The time series autoencoder now has a tensor $\mathbf{X} \in \mathbb{R}^{N \times D_{\mathbf{x}} \times K}$ with $N, D_{\mathbf{x}}, K \in \mathbb{N}_{\geq 1}$ as input. Here, N refers to the number of samples, $D_{\mathbf{x}}$ is the number of features, and K is the length of the time series. Again, in our case for day-ahead forecasts, the length of the time series is 24 considering an hourly resolution. The encoder now reduces the dimension to obtain the latent feature tensor $\mathbf{Z} \in \mathbb{R}^{N \times D_{\mathbf{z}} \times K}$. The decoder approximates the original input tensor with $\mathbf{X} \approx \hat{\mathbf{X}}$.

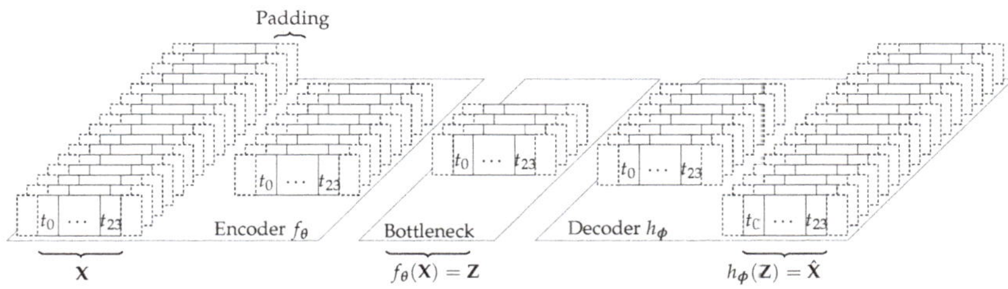

Figure 3. An example undercomplete time series AE topology. The AE reduces the dimensionality in each layer of the encoder. The latent features' representation at the bottleneck are the extracted hidden features sufficient to reconstruct the original input successively in each layer of the decoder.

As pointed out earlier, we use the TCN network as the building block for the time series autoencoder. The principle approach is inspired by [20]. We adapted their proposal to the needs for day-ahead power forecasts. We do not use upsampling or downsampling, as we use zero-padding in all layers and have short time series. As a result, the TCN autoencoder simplifies to a sequential concatenation of residual blocks, as visualized in Figure 4, of the original TCN network [21]. The concept of residual blocks is well known in computer vision [22]. The principle idea behind a residual block is to add a skip connection for the input from previous layers to reduce the risk of the vanishing gradient. Therefore, in each residual block, the input is processed twice in the following pattern: dilated convolution, weight norm, ReLU activation, and dropout for regularization (see Figure 4). Note that a dilated convolution is a particular convolutional layer that increases the receptive field, is computationally efficient, and requires less memory. The skip connection adds the original input to the output. An optional convolution matches the dimensions in the skip connection if a single layer's input and output dimensions are unequal.

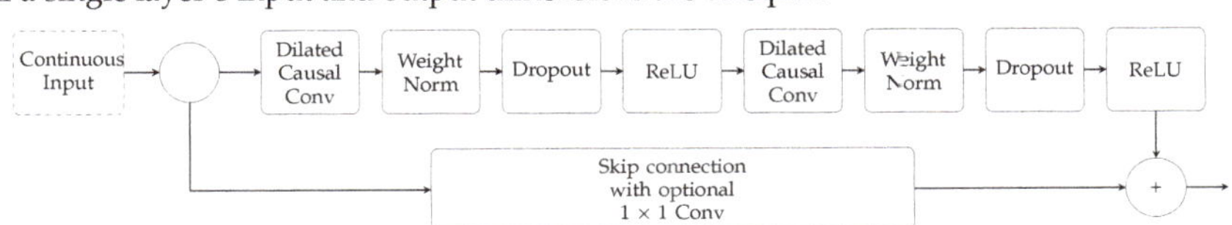

Figure 4. Residual block of the TCN.

3.3. Variational Autoencoder

A drawback of the discriminative architectures in the previous sections is that they cannot be used to reconstruct missing values or generate new samples. VAEs are a generative approach that extends the idea of a simple autoencoder by adding a constraint on the encoding site to generative properties.

The encoder f_θ, with parameters $\boldsymbol{\theta}$, is forced to learn the mean $\boldsymbol{\mu}$ and standard deviation $\boldsymbol{\sigma}$ of a Gaussian distribution. $\boldsymbol{\mu}$ and $\boldsymbol{\sigma}$ are used to create latent features $\mathbf{z}$ by sampling from a unit Gaussian translated and are scaled with the learned $\boldsymbol{\mu}$ and $\boldsymbol{\sigma}$ to obtain $q_\theta(\mathbf{z}|\mathbf{x})$. This is also called the reparameterization trick [23]. The scaled samples are used to reconstruct the original features $\mathbf{x}$ with the decoder h_ϕ, with parameters $\boldsymbol{\phi}$. More formally, this can be done using the loss function:

$$\mathcal{L}_{VAE} = \underbrace{-\mathbb{E}_{\mathbf{z}\sim q_{\theta}(\mathbf{z}|\mathbf{x})}\left[\log p_{\phi}(\mathbf{x}|\mathbf{z})\right]}_{\mathcal{L}_{AE}} + D_{KL}(q_{\theta}(\mathbf{z}|\mathbf{x}) \parallel p_{\phi}(\mathbf{z})), \quad (2)$$

where the first part, the likelihood function, is equal to the loss function $\mathcal{L}_{AE}$ of an autoencoder (see Equation (1)). The Kullback–Leibler Divergence D_{KL} penalizes the deviation between the learned distribution q_θ from a unit Gaussian with $p_{\phi}(\mathbf{z}) = \mathcal{N}(0, \mathbf{I})$.

By applying the reparameterization trick, it is possible to extend the original idea of an AE and achieve the following properties:

- Often, the combination of a generative network with an encoder forces the VAE to learn a representation in a much lower-dimensional space (see [3] (p. 699) and [23]).
- The decoder and the latent vector, which can be drawn from prior $p_{\phi}(\mathbf{z})$, provide a generative framework.

3.4. Transfer Learning

Transfer learning describes the concept of knowledge transfer from one task to another. Knowledge transfer from a source to a target task often has better generalization capabilities and improves the forecast error for problems with limited data. One sub-field of transfer learning is multi-task learning. The term MTL refers to two different concepts: Soft parameter sharing (SPS) and hard parameter sharing (HPS). The idea that shared knowledge should be used across tasks is included in both principles. However, HPS is advantageous when tasks are closely related and it is advantageous to share a lot of information [10]. When activities are only slightly related, SPS is helpful, and it is good to have primarily specific knowledge for each task [10]. When considering a neural network, most layers in the case of HPS are the same for all tasks, and only the final few layers differ. For SPS, we train a single network for each task and regularize the training to make the learned representations of each task similar. Within our work, we consider the extreme case of an HPS autoencoder, where layers of all tasks are the same, as we assume that there is a common latent representation of the weather. While in principle, this might be a simplified consideration, it is the first important step towards finding a common latent representation for forecasting tasks and we can extend the concept to task-specific representation later on, e.g., through task embeddings as proposed in [10].

Once we learn a joint representation of the weather, we must ensure that the knowledge is appropriate for the target task, in our case, the forecasting of the expected power generation. While an adaption is unnecessary for some latent representations and tasks, this is probably untrue for most. Therefore, we have to adapt knowledge for the task at hand. A common approach in deep learning is the fine-tuning of the final layers of a network. In our case, the encoder extracts the latent features and we adapt the layers of this network. Training an initial network on a source task and adapting it for a target task through fine-tuning is often referenced as sequential transfer learning [24].

4. Datasets and Challenges in Day-Ahead Power Forecasts

The following sections summarize the datasets and the challenges in day-ahead power forecasts.

4.1. Overall Process of Day-Ahead Forecasts and Challenges

The following section details the challenges associated with day-ahead power forecasts. We include a description of the overall process for generating power forecasts for a wind or PV park. Figure 5 summarizes this process. Due to the weather dependency of renewable power plants, we require weather predictions from so-called NWP models. The NWP model receives input from sensors that approximate the current weather situation. Based on the latest sensory data, a so-called model run is calculated. This model is ran, e.g., at 0.00 a.m. Due to the complex and manifold stochastic differential equations involved in predicting the weather, such a model run typically requires about six hours. Afterward, the NWP provides forecasts, e.g., up to 72 h into the future. In our case, we are interested in so-called day-ahead power forecasts based on weather forecasts between 24 and 47 h into the future. Based on these weather forecasts and historical power measurements, we can train the ML model to predict the expected power generation for day-ahead forecasting problems.

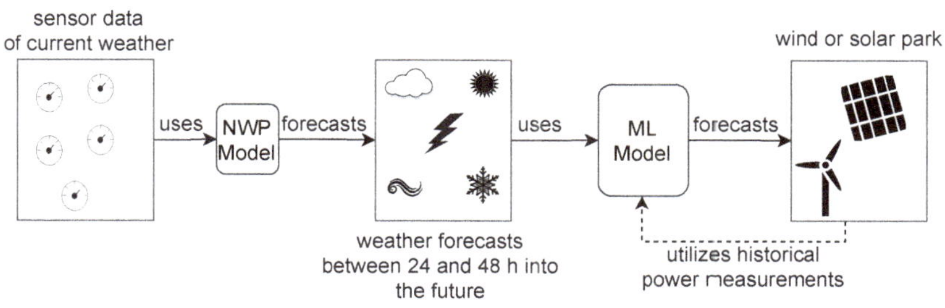

Figure 5. Overview of renewable power forecast process.

However, due to the dependency of renewable power forecasts on weather forecasts as input features, substantial uncertainty is associated with these forecasts making it a challenging problem. At the same time, weather forecasts are valid for larger grid sizes, e.g., three kilometers, and a mismatch between these grids and the location of a wind or PV causes additional uncertainty in the power forecasts [25]. These mismatches and the non-linearity of the forecasting problem are depicted in Figure 6. We can observe mismatches between the predicted wind speed (or radiation) and historical power measurements. For instance, we can observe (outliers) where a large amount of power is generated for low values of those features. These mismatches are also shown in the time series plots in Figure 7. This observation indicates that the weather forecasts were incorrect. While the scatter plot indicates a more linear and straightforward forecasting problem for PV, we can also observe a stronger correlation in the time series than for wind.

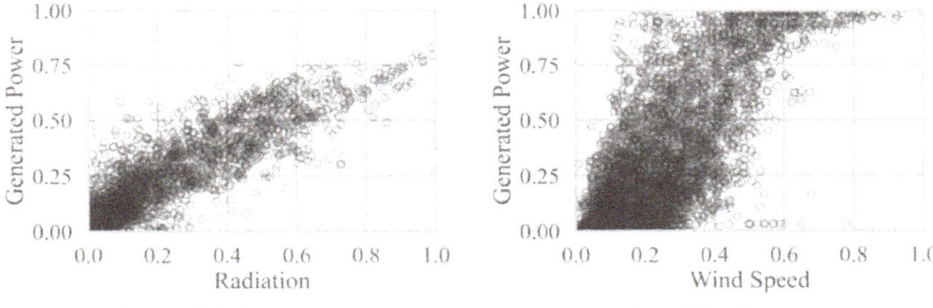

(**a**) PVREAL dataset. (**b**) WINDREAL dataset.

Figure 6. Scatter plots of the most relevant features for power forecasts from day-ahead weather forecasts and the historical power measurements. Large radiation or wind speed values and no power generation indicate an incorrect weather forecast or feed-in management. Large power generation values and a low value of wind speed or radiation indicate an incorrect weather forecast.

Examples are also present where we observe a considerable amount of wind speed or radiation, but no or little power is generated. An incorrect weather forecast can cause this problem. However, often it is associated with regular interventions. For instance, in some regions in Germany, wind turbines must limit the rotation speed at night. Furthermore, there is a large portion of feed-in management interventions in Germany. These interventions are used to stabilize the electrical grid. A typical pattern for those interventions is given in Figure 7b at time step 275. We can observe an initial large power production associated with large wind speed values. At this point, the power generation drops to zero while the wind speed remains high. Such a sudden drop is typically associated with feed-in management interventions. As those interventions depend on the power grid's state, we typically have no information about such drops, making the forecasting problem even more challenging.

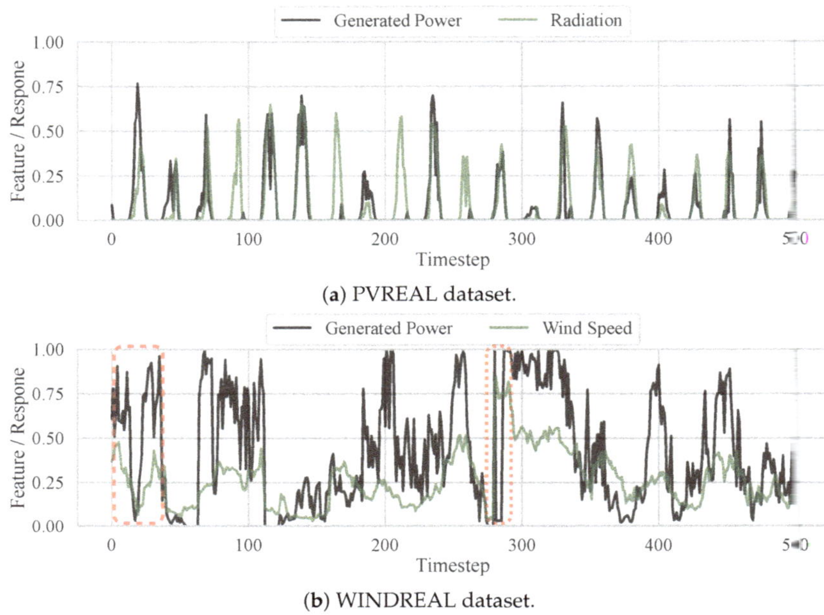

Figure 7. Time series plots of most relevant features for power forecasts from day-ahead weather forecasts and the historical power measurements. The radiation, as well as the historical power, shows a typical Gaussian-shaped behavior during the day. The dashed rectangle shows a potential bad weather forecast of the wind speed. The dotted rectangle for the WINDREAL dataset indicates a typical pattern of feed-in management interventions.

4.2. Summary of Datasets

We summarize the considered datasets for learning latent features and predicting the day-ahead power forecasts between 24 and 47 h into the future in Table 1. The table shows the diversity of the datasets. Each dataset has a different number of parks, input features, training, and test samples. For a better comparison between datasets, we linearly interpolated the PVOPEN datasets from a three-hour to an hourly resolution. All other datasets already had the respective resolution. The datasets also differ in the utilized NWP model. In those weather predictions, from either the European center for medium-range weather forecasts (ECMWF) [26] or the Icosahedral Nonhydrostatic-European Union (ICON-EU) [27] weather model, features such as wind speed, wind direction, air pressure, and direct and diffuse radiation are included. For the PVOPEN dataset, various manually engineered features are included, taking seasonal patterns of the sun into account. In the case that a dataset initially did not include seasonal features from the month, day of the year, and hour of the day, we incorporated those through a sine and cosine encoding [28].

All datasets except the PVREAL and WINDREAL datasets are openly accessible. At the same time, those two datasets are the most recent and diverse. For instance, the WINDREAL dataset includes 13 turbine manufacturers, six hub heights, and 99 different nominal capacities. Furthermore, the PVREAL dataset has various distinct physical characteristics, including ten tilt orientations, 31 different nominal capacities, and nine azimuth orientations. Both datasets include parks located in and around Germany.

Table 1. Overview of the evaluated datasets. All datasets except PVREAL and WINDREAL are openly accessible.

Dataset	#Parks	#Features	Mean Train Samples	Mean Test Samples	Resolution	NWP	Ref.
PVOPEN	21	43	6336	8424 (57%)	hourly	ECMWF	[29]
PVSYN	114	16	7596	3730 (32%)	hourly	ICON-EU	[30]
PVREAL	38	25	14,513	4836 (25%)	hourly	ICON-EU	[31]
WINDOPEN	45	9	6931	6659 (49%)	hourly	ECMWF	[29]
WINDSYN	260	25	8428	4169 (33%)	hourly	ICON-EU	[30]
WINDREAL	185	29	9032	3023 (25%)	hourly	ICON-EU	[31]

The length of these two datasets varies dramatically between parks. Therefore, we considered 25% randomly sampled days as test data such that we have an equal fraction for training and testing in each park. Note that each day is based on independent day ahead NWP forecasts, so no information is leaked from the future to the past [25]. This splitting allows for a split without generalization of the test error. For the WINDSYN and PVSYN dataset, on the other hand, the predefined test set is utilized [30]. Similar to [29], for the WINDOPEN and PVOPEN, we use the first year's data as training data and the remaining data as test data. We use 25% randomly sampled for validation from the training data for all datasets.

5. Experiments

The following sections summarize our results to answer our research questions for the PV and WIND parks from the six datasets. PV parks contain all parks from the datasets PVREAL, PVSYN, and PVOPEN, and WIND refers to all parks from the datasets WINDREAL, WINDSYN, and WINDOPEN.

We first provide details on the overall experimental setup. Afterward, we describe one experiment to evaluate the reconstruction error for the different representation learning techniques for the evaluated latent sizes 2, 4, 6, 8, and 10. The trained autoencoder from this experiment extracting the latent features of the weather (Figure 8) is then reused for forecasting the expected power in the second experiment described afterward.

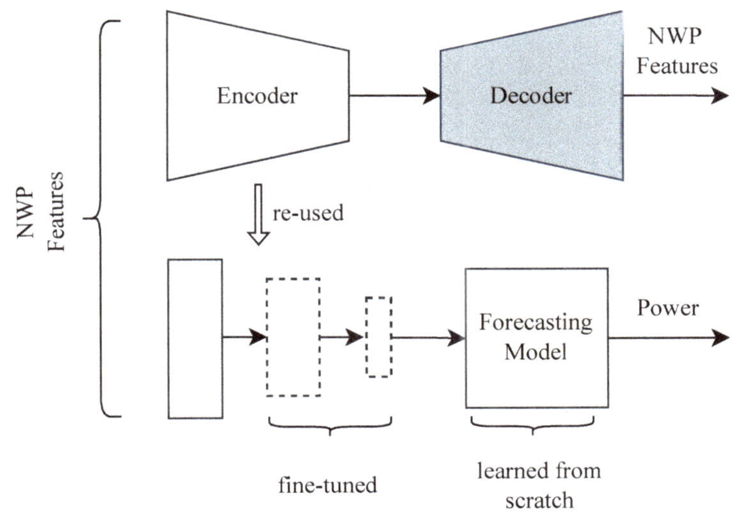

Figure 8. In our experiments, we initially trained an autoencoder to reconstruct the NWP features and learn the latent features of the weather. Afterward, the encoder was fine-tuned with a forecasting model to forecast the day-ahead power.

5.1. Overall Experimental Setup

For evaluation, we considered the normalized root mean squared error (nRMSE) given by Equation (4) based on the root mean squared error (RMSE) in Equation (3). We use Equation (4) in two ways. First, we utilize it in calculating the reconstruction error to quantify how well a representation learning technique is capable of reconstructing the original features from a latent representation. Secondly, it measures the forecast error for day-ahead power forecasts.

$$\text{RMSE} = \sqrt{\frac{1}{N}\sum_{i=1}^{i=N}(y_i - \hat{y}_i)^2} \quad (3)$$

$$\text{nRMSE} = \frac{\text{RMSE} - y_{\text{min}}}{y_{\text{max}} - y_{\text{min}}} \quad (4)$$

In Equations (3) and (4), y_i is the i-th value of the response, $\hat{y}_i$ is the prediction from a model, $N \in \mathbb{R}_{\geq 1}$ is the number of samples, and y_{max} and y_{min} are the maximum and minimum values of the response, respectively, a feature. Note that for normalization of the power in all datasets, y_{max} is given by the nominal power, whereas for features from the NWP, we consider the empirical minimum and maximum values from the training set.

We considered between two and ten latent features for each experiment. All experiments were conducted on a Slurm cluster and all processes run on one out of four computing nodes, each with 256 AMD EPYC 7742 CPUs and 1008 GB Ram. To answer Research Questions 1 and 2, in the first experiment we evaluated the reconstruction error of the autoencoders. In the second experiment, we evaluated how the encoders from the first experiments can be utilized for power forecasts to answer Research Question 3. In both experiments, we differentiate between WIND and PV parks due to their differences in the expected forecast errors [8]. The former park type has a total of 490 parks, whereas the latter has 173 parks. For a given park type and the number of latent features, we calculated the mean performance rank based on the nRMSE for both park types. We test for a significant difference compared to the baseline by the Wilcoxon test ($\alpha = 0.01$).

5.2. Experiment on Representation Learning for Dimension Reduction

In this section, we answer Research Questions 1 and 2 to evaluate the different representation learning techniques for day-ahead weather forecasts.

5.2.1. Experimental Setup

As a traditional dimension reduction technique we considered a PCA and a kernel PCA with a cosine kernel, referred to as PCA-COSINE. The latter had one of the best results in representation learning for day-ahead power forecasts in [1]. We compared those two techniques with eight variants of autoencoders as summarized in Table 2.

Table 2. Overview of autoencoder-based representation learning techniques. The model type is abbreviated as MLP for the multi-layer perceptron and TCN for temporal convolutional networks. MTL is an abbreviation for multi-task learning, where all parks of a datatype are trained in a unified autoencoder model. VAE is an abbreviation for a variational autoencoder as a generative approach for dimension reduction.

Model Type	Architecture (STL or MTL)	Discriminative or Generative	Abbreviation
MLP/TCN	STL	Discriminative	AEMLP/AETCN
MLP/TCN	MTL	Discriminative	AEMLP-MTL/AETCN-MTL
MLP/TCN	STL	Generative	VAEMLP/VAETCN
MLP/TCN	MTL	Generative	VAEMLP-MTL/VAETCN-MTL

Within those autoencoders we either considered an MLP or a TCN model. The former consists of a linear layer followed by the rectified linear unit (ReLU) activation and batchnorm for the input and hidden layers. For the output layers of the encoder and decoder, the activation and batchnorm were not included. The TCN model allows us to take the diurnal behavior of the weather into account and has the structure and non-linear activation functions as described in Section 3.2.

For each model type, we considered an STL and an MTL approach. For the STL architecture for each park and latent size, we trained an autoencoder. In the case of the MTL architecture, we combined the data of all parks to train a single autoencoder. As all data were combined, we considered this to be a unified AE, where we learn a unified latent space across all parks. It is important to consider here that the MTL architecture has the same number of parameters as the STL approach for a single park. This way, we can evaluate how we can compress features robustly from various parks through a unifying autoencoder.

Within this experiment, the baseline refers to the AEMLP dimension reduction technique. Choosing the AEMLP as baseline allows to compare how MTL architectures improve the reconstruction error for Research Question 1 over STL architectures. Furthermore, it gives insights into how a time series autoencoder improves upon an MLP-based autoencoder for Research Question 2.

The number of input features for autoencoders is equal to that described in Table 1. Each dataset includes six seasonal features, such as the day of the year, as described in Section 4.2. These manually engineered features were not included in the output as they are often challenging to learn for autoencoders and we can add those manually. At the same time, including them in the input is essential so that the autoencoder can learn latent features that depend on the seasons. For training, we reduced the number of features in each successive layer by 70%, up to a minimum of the latent size plus one for all encoder models. The final number of features in the encoder was then equal to the latent size. The decoder had the reverse structure and an additional output layer to map it to the number of features within a park without the seasonal features. For the variational autoencoders, we added one additional layer in the encoder, transforming the latent features into μ and σ (see Section 3.3).

To find the best hyperparameters, we conducted a grid search for the learning rate and the number of epochs based on the reconstruction error on the validation dataset. For an STL AE, we selected the best number of epochs from the set $\{50, 100, 200\}$ along with a learning rate from the set $\{10^{-2}, 10^{-3}, 10^{-4}\}$. We initiated five parallel jobs for STL models to train individual parks, where each process had 20 CPUs. While the learning rate was the same, due to the additional data the number of epochs was selected from the set $\{50, 100, 200, 300, 400\}$ for MTL architectures. For MTL, we trained a model for all parks simultaneously with a single process with 50 CPUs. We trained all MTL deep learning models through the Adam optimizer.

5.2.2. Findings

To show how the required number of parameters is reduced through MTL architectures, Table 3 provides an overview of the number of parameters for the different models for an example latent size of two. Other latent sizes solely differ slightly in the additional parameters required for the latent features. For both model types, MLP and TCN, when comparing the STL with the MTL architecture, the required number of parameters is reduced 38 times for the PV datasets and 202 times for the wind datasets. The TCN model type, on the other hand, increases the required number of parameters five times over an MLP-based model due to additional parameters in the residual block of the TCN.

We summarize the time consumption of these models in Table 4. Even though we trained STL models with 100 CPUs through parallelization for each data and model type and the MTL models were trained only with 50 CPUs, we can observe a substantial reduction in the training time. Often the STL models require at least five times more computation time, even though we trained them with additional CPUs. For a few cases, the computation time of MTL is even ten times less than that of an STL architecture.

Table 3. The number of parameters for all autoencoder architectures for an example latent size of two. Other latent sizes solely differ slightly in the additional parameters required for the latent features. The number of parameters for the three wind and three PV datasets are summed for readability. The abbreviations here are the same as those in Table 2.

Model Type Data Type	AEMLP-MTL	AEMLP-MTL	AETCN	AETCN-MTL	VAE	VAE-MTL	VAETCN	VAETCN-MTL
PV	118,707	3080	624,762	16,514	120,831	3116	629,718	16,[illegible]98
WIND	421,870	2076	2,243,970	10,986	427,750	2112	2,257,690	11,[illegible]70

Table 4. Duration to train the different autoencoders in minutes. For the STL models, we trained with five parallel jobs. Each job utilized 20 CPUs and trained a single park. An MTL trained all parks simultaneously through a single job with 50 CPUs.

Model Type	Data Type	Duration	
		STL	MTL
AE	PV	1473	1295
	WIND	5723	1126
AETCN	PV	5043	940
	WIND	13,174	1289
VAE	PV	1592	1309
	WIND	7456	1364
VAETCN	PV	5495	1009
	WIND	14,249	1342

The mean performance rank results for the reconstruction error of all models are depicted in Table 5. We summarize the median reconstruction errors based on Equation (4) in Table 6. These tables show that the variational autoencoders have the worst results, regardless of the architecture and model type used. The variational autoencoders are significantly worse than the baseline for all latent sizes and the two data types. For instance, for a latent size of ten, the median nRMSE of the baseline is 74% lower than the best variational autoencoder for the WIND parks. Similarly, for PV parks, the baseline is at least 78% better than the variational autoencoder. This observation likely occurs due to the limitation of the variational distribution that restricts the latent space through the normal distribution. This effect can potentially be reduced by scaling the Kullback–Leibler divergence in Equation (2).

Another interesting observation is that the mean performance rank in all cases is larger for variational MTL autoencoders than for the same STL architecture. At the same time, the difference in the median reconstruction error is only about 1%. Again, we can explain this result with the variational distribution. In the case of the MTL architecture, the same variational distribution must express multiple parks. Intuitively enough, as the STL architecture already has difficulties compressing the information in the latent space, it is even more challenging in an MTL setting. As all models share the same number of layers, it might be beneficial for the variational autoencoder to utilize a deeper network or a larger latent size to ease the training. Again, another option would be to reduce the constraint given through the Kullback–Leibler divergence. With the evaluated hyperparameters, we can summarize that the MTL architecture is not beneficial for variational autoencoders.

Table 5. Rank summary of the reconstruction error for all models and latent sizes for the autoencoder model (see Table 2), and the PCA-based models. AEMLP is the baseline and all models were tested to determine whether the reconstruction error is significantly better (∨), worse (∧), or not significantly different (⋄) compared to the baseline. We test for a significant difference with the Wilcoxon test ($\alpha = 0.01$). The colors denote the respective rank. Blue indicates a smaller (better) rank and red a higher (worse) rank. The best latent size and data type model is highlighted in bold.

Data Type	Latent Size	Baseline	AEMLP-MTL	AETCN	AETCN-MTL	PCA	PCA-COSINE	VAE	VAEMLP-MTL	VAETCN	VAETCN-MTL
PV	2	3.954	2.387∨	2.757∨	**1.006**∨	4.896∧	8.249∧	8.844∧	9.156∧	6.659∧	7.092∧
PV	4	3.682	2.971∨	2.410∨	**1.087**∨	4.850∧	6.173∧	9.069∧	9.474∧	7.451∧	7.832∧
PV	6	3.618	2.301∨	3.075∨	**1.393**∨	4.613∧	6.017∧	9.092∧	9.468∧	7.497∧	7.925∧
PV	8	3.254	1.983∨	3.983∧	**1.260**∨	4.520∧	6.006∧	9.017∧	9.538∧	7.486∧	7.954∧
PV	10	4.017	1.925∨	4.520∧	**1.179**∨	3.358∨	6.000∧	9.104∧	9.468∧	7.526∧	7.902∧
WIND	2	3.820	2.332∨	2.992∨	**1.012**∨	4.853∧	8.120∧	6.925∧	6.641∧	8.888∧	9.417∧
WIND	4	4.447	2.393∨	4.285∨	**1.249**∨	2.636∨	6.838∧	7.299∧	7.318∧	8.99∧	9.545∧
WIND	6	4.008	2.162∨	3.919⋄	**1.110**∨	3.813∨	6.401∧	7.376∧	7.553∧	9.081∧	9.576∧
WIND	8	4.080	2.037∨	4.439∧	**1.000**∨	3.458∨	6.165∧	7.256∧	7.570∧	9.222∧	9.773∧
WIND	10	4.014	2.199∨	4.828∧	**1.002**∨	2.970∨	6.103∧	7.387∧	7.501∧	9.231∧	9.764∧

Table 6. Median nRMSE of the reconstruction error for all models and latent sizes for the autoencoder models (see Table 2), and the PCA-based models. AEMLP is the baseline and all models were tested to determine whether the reconstruction error is significantly better (∨), worse (∧), or not significantly different (⋄) compared to the baseline. We test for a significant difference with Wilcoxon test ($\alpha = 0.01$). The colors denote the respective rank. Blue indicates a smaller (better) rank and red a higher (worse) rank. The best latent size and data type model is highlighted in bold.

Data Type	Latent Size	Baseline	AEMLP-MTL	AETCN	AETCN-MTL	PCA	PCA-COSINE	VAE	VAEMLP-MTL	VAETCN	VAETCN-MTL
PV	2	0.126	0.108∨	0.112∨	**0.075**∨	0.141∧	0.201∧	0.227∧	0.227∧	0.190∧	0.191∧
PV	4	0.081	0.068∨	0.063∨	**0.04**∨	0.100∧	0.180∧	0.226∧	0.227∧	0.190∧	0.191∧
PV	6	0.058	0.046∨	0.052∨	**0.028**∨	0.072∧	0.153∧	0.227∧	0.227∧	0.189∧	0.191∧
PV	8	0.044	0.027∨	0.046∧	**0.011**∨	0.048∧	0.135∧	0.226∧	0.227∧	0.190∧	0.191∧
PV	10	0.042	0.015∨	0.044∧	**0.014**∨	0.024∨	0.124∧	0.227∧	0.227∧	0.189∧	0.191∧
WIND	2	0.119	0.109∨	0.110∨	**0.081**∨	0.135∧	0.255∧	0.217∧	0.214∧	0.276∧	0.278∧
WIND	4	0.103	0.080∨	0.097∨	**0.062**∨	0.088∨	0.209∧	0.213∧	0.214∧	0.276∧	0.278∧
WIND	6	0.068	0.054∨	0.067⋄	**0.033**∨	0.070∨	0.191∧	0.208∧	0.210∧	0.276∧	0.278∧
WIND	8	0.061	0.046∨	0.064∧	**0.026**∨	0.059∨	0.186∧	0.209∧	0.213∧	0.278∧	0.279∧
WIND	10	0.054	0.035∨	0.060∧	**0.023**∨	0.048∨	0.181∧	0.208∧	0.209∧	0.278∧	0.279∧

On the other hand, regarding Research Question 2, the results are different. We can observe that regardless of the architecture, the mean performance rank is better for time series autoencoders for PV parks. For WIND parks, the time series autoencoders are worse than the MLP-based variational autoencoders. These differences in the data types are best explained by the diurnal cycle of a day that influences the forecast for PV parks more than for WIND parks (see [8]).

We can make similar conclusions for the PCA-COSINE representation technique than for the variational autoencoders. A transformation through a cosine kernel seems too restrictive or unsuitable for the data at hand. On the other hand, the PCA model is, at least for the WIND data type, better than the baseline up to a latent size of four. For PV parks, the PCA is significantly worse except for with a latent size of ten.

As the baseline substantially outperforms the PCA-based models and the variational autoencoders, we can safely assume that the AEMLP is a reliable baseline method. The MTL architecture of the baseline, the AEMLP-MTL model, significantly outperforms this reference in all cases. The improvements in the median nRMSE range from 9 to 54% for WIND parks and between 16 and 179% for PV parks. We also achieve improvements through the MTL architecture for time series autoencoders. The MTL time series autoencoder has the best median performance rank for all latent sizes and data types. Thereby, concerning Research Question 1, we can conclude that we substantially improve the reconstruction error for discriminative autoencoders through an MTL architecture. This result contrasts with the generative autoencoder results, where the MTL architecture worsens the results. However, the discriminative autoencoder is not limited by the variational distribution. Without this limitation, the autoencoder learns a better encoding for the weather features through the MTL approach. The additional training samples from all parks allow us to learn an encoding that generalizes better during test time. The improvements can be best explained by the fact that in an MTL setting, we train with weather conditions from other parks that would otherwise not be available in an STL training. During test time, this allows the network to use knowledge from all parks.

To answer Research Question 2 for discriminative autoencoders we need to differentiate between STL and MTL architectures. While the median nRMSE of the STL time series autoencoder is often of similar magnitude to the baseline, the results are only significantly better or equal in six cases. The critical improvements come in combination with the MTL architecture. As pointed out earlier, this model is the best for all tested latent sizes and data types. For PV parks, the improvements of the median error range from 68–300% and between 46 and 134% for WIND. Here, we can assume that the STL time series autoencoder differs from the MTL approach, as additional training samples are required for the additional parameters and learning the particular requirements in learning the diurnal cycle. Overall we can summarize the results as follows:

- Generative autoencoders need further investigation regarding influences of the latent size, the network structure, and the scaling of the Kullback–Leibler divergence to answer the research questions.
- MTL architectures improve the reconstruction error substantially for discriminative autoencoders.
- Time series autoencoders significantly improve when combined with an MTL architecture.

5.3. Experiment on Wind and PV Power Forecasts

Within this section, we answer Research Question 3 to evaluate the importance of fine-tuning the encoder for day-ahead power predictions.

5.3.1. Experimental Setup

In the experiment described in this section, we considered the models from Section 5.2 as models for feature extraction. These models were used to extract the latent features utilized to make predictions for wind and PV day-ahead power forecasts. As forecasting models, we considered an MLP and a TCN attached to the encoder with the same model type from the previous experiment. For the PCA and PCA-COSINE representation learning technique, we also considered an MLP for forecasting the expected power generation. We chose the hidden layers to be 200 and 100 neurons for the MLP. Due to the additional parameters in the residual block of the TCN, the hidden layers were of size 60 and 30. These two forecasting models were trained for ten epochs with cosine annealing with a maximum learning rate of 10^{-1} and afterward with a maximum learning rate of 10^{-2} for another ten epochs with cosine annealing through the Adam optimizer.

The batch size was selected to have ten iterations within each epoch. We evaluated whether it was beneficial to fine-tune the encoder models and we fine-tuned zero, one, or two layers of the encoder models. As in the previous experiment, we utilized the AEMLP for latent feature extraction as a baseline and attached an MLP for forecasting the power. In principle, we could have also attached other models to the encoder, such as a linear model or a gradient boosting regression tree. However, forecast errors for different forecasting models are often in a similar range [1] and it is difficult to fine-tune the encoder in an end-to-end fashion for those models.

5.3.2. Findings

The results that we use to answer Research Question 3 are summarized in Tables 7 and 8. We only consider those models that are within the two best ranked models for readability. The naming conventions are the same as in the previous section for the dimension reduction techniques. Additionally, we add the considered forecasting model. The results indicate the number of fine-tuned layers that are 1 or 2. For example, AEMLP-MTL-MLP1 depicts the AEMLP-MTL feature extraction technique coupled with the MLP forecasting model and the fine-tuning of the last layer of the encoder.

Table 7. Rank summary of the forecast error for best models and latent sizes for the autoencoder models (see Table 2), and the PCA-based models. AEMLP-MTL-MLP is the baseline and all models were tested to determine whether the mean performance rank was significantly better (∨), worse (∧), or not significantly different (⋄) compared to the baseline. We tested for a significant difference with the Wilcoxon test ($\alpha = 0.01$). The colors denote the respective rank. Blue indicates a smaller (better) rank and red a higher (worse) rank. The best model for a latent size and data type is highlighted in bold.

Data Type	Latent Size	Baseline	AEMLP-MLP1	AEMLP-MTL-MLP	AEMLP-MTL-MLP1	AEMLP-MTL-MLP2	AETCN-MTL-TCN1	AETCN-MTL-TCN2	AETCN-TCN2	PCA-COSINE-MLP	PCA-MLP
PV	2	6.821	6.682⋄	7.052⋄	5.665∨	5.780∨	2.503∨	**2.480**∨	4.671∨	6.775⋄	6.572⋄
PV	4	7.260	6.277∨	8.983∧	6.272∨	5.931∨	3.659∨	3.335∨	**3.006**∨	5.624∨	4.653∨
PV	6	5.514	5.445⋄	8.393∧	7.098∧	7.341∧	4.538∨	**3.746**∨	4.312∨	4.601∨	4.012∨
PV	8	6.936	5.746∨	8.220∧	6.110∨	6.491∨	4.965∨	4.046∨	5.121∨	4.035∨	**3.329**∨
PV	10	6.717	7.052⋄	6.023∨	6.422⋄	6.358⋄	5.225∨	4.133∨	5.017∨	**3.838**∨	4.214∨
WIND	2	4.707	4.440∨	6.606∧	**3.778**∨	3.830∨	5.349∧	4.641⋄	4.865⋄	9.683∧	7.102∧
WIND	4	6.312	6.341⋄	**3.653**∨	4.370∨	4.012∨	7.647∧	7.973∧	6.243⋄	4.366∨	4.083∨
WIND	6	5.212	**3.992**∨	7.574∧	5.805∧	4.954⋄	5.310⋄	4.239∨	4.031∨	7.160∧	6.723∧
WIND	8	4.879	4.989⋄	5.513∧	5.703∧	5.812∧	4.654⋄	**4.071**∨	4.527⋄	7.632∧	7.121∧
WIND	10	5.590	5.387⋄	6.130∧	4.346∨	**4.039**∨	5.007∨	4.121∨	4.320∨	8.018∧	8.041∧

Table 8. Median nRMSE of the forecast error for best models and latent sizes for the autoencoder models (see Table 2), and the PCA-based models. AEMLP-MTL-MLP is the baseline and all models were tested to determine whether the mean performance rank is significantly better (∨), worse (∧), or not significantly different (⋄) compared to the baseline. We tested for a significant difference with the Wilcoxon test ($\alpha = 0.01$). The colors denote the respective rank. Blue indicates a smaller (better) rank and red a higher (worse) rank. The best latent size and data type model is highlighted in bold.

Data Type	Latent Size	Baseline	AEMLP-MLP1	AEMLP-MTL-MLP	AEMLP-MTL-MLP1	AEMLP-MTL-MLP2	AETCN-MTL-TCN1	AETCN-MTL-TCN2	AETCN-TCN2	PCA-COSINE-MLP	PCA-MLP
PV	2	0.116	0.117⋄	0.116⋄	0.110∨	0.111∨	0.099∨	**0.098**∨	0.105∨	0.117⋄	0.116⋄
PV	4	0.099	0.095∨	0.106∧	0.092∨	0.093∨	**0.087**∨	**0.087**∨	**0.087**∨	0.093∨	0.092∨
PV	6	0.088	0.087⋄	0.094∧	0.090∧	0.09∧	0.086∨	**0.084**∨	0.085∨	0.087∨	0.085∨
PV	8	0.087	0.087∨	0.091∧	0.086∨	0.087∨	0.084∨	**0.083**∨	0.084∨	0.084∨	**0.083**∨
PV	10	0.087	0.087⋄	0.085∨	0.085⋄	0.087⋄	0.085∨	0.084∨	0.085∨	**0.083**∨	**0.083**∨
WIND	2	0.178	0.176∨	0.191∧	**0.171**∨	0.173∨	0.179∧	0.176⋄	0.178⋄	0.226∧	0.192∧
WIND	4	0.158	0.159⋄	**0.143**∨	0.146∨	0.145∨	0.174∧	0.175∧	0.163⋄	0.150∨	0.149∨
WIND	6	0.137	**0.135**∨	0.143∧	0.139∧	0.138⋄	0.136⋄	**0.135**∨	**0.135**∨	0.147∧	0.147∧
WIND	8	0.136	0.135⋄	0.135∧	0.135∧	0.135∧	0.134⋄	**0.133**∨	0.134⋄	0.145∧	0.146∧
WIND	10	0.135	0.135⋄	0.136∧	**0.133**∨	**0.133**∨	**0.133**∨	**0.133**∨	**0.133**∨	0.147∧	0.149∧

Compared to the previous experiment, there is no prominent best model. Not surprisingly, none of the variational autoencoder architectures are among the best models, as learned latent features are insufficient for reconstructing the original features. Unlike the previous experiment, PCA and PCA-COSINE now achieve the best results in two cases for PV parks. Except for a latent size of two, the PCA-MLP and the PCA-COSINE-MLP outperform the baseline significantly for this data type. For WIND parks, these models are significantly worse for most latent sizes.

For the AEMLP-MLP model, we can observe that fine-tuning the final layer is at least as good as the baseline (the same model without fine-tuning). Fine-tuning the second last layer of the encoder leads to worse results and is not within the best models. We can assume that due to the STL approach, the latent features are already close to a representation required for forecasting the expected power and only slight adjustments in the encoder are required.

For the STL time series encoder, we need to fine-tune two layers such that the model is among the best and at least as good as the baseline. For the PV parks, improvements up to 13% are present, whereas for WIND, these are at about 1.5%.

For the MTL variant of the baseline, the AEMLP-MTL-MLP model, we can observe no substantial difference between fine-tuning one or two layers. Nevertheless, the results of this model without fine-tuning are worse than with the adaption of the encoder. Furthermore, in three cases, AEMLP-MTL-MLP1 and AEMLP-MTL-MLP2 are significantly worse than the baseline. These results are engaging in a manifold way. In the previous experiment, we saw that the MTL architecture outperforms the STL approach for representation learning significantly in all cases for the MLP-based autoencoder. This relation is no longer the case for forecasting historical power. We can assume here that, on the one hand, the learned latent space needs to be adapted for forecasting the power in general. On the other hand, due to the MTL approach, the latent space is too broad and careful fine-tuning is needed to adapt the encoder model for the specific requirements of a single park. This problem was not present for the STL architecture, where the forecast errors were always as least as good as the baseline.

The AETCN-MTL-TCN model is consistently among the best for latent sizes up to six. When fine-tuning two layers of the encoder, it is only worse than the baseline in one case. We can also observe, except for this case, that fine-tuning two layers improves the forecast error compared to a single layer. For the PV parks, we achieved improvements between 3.5% and 18.3% for the median forecast error, whereas for WIND parks, we accomplished advances in the error ranging from 1.1 to 1.5%. In contrast to the MLP-based MTL autoencoder, we can assume that the fine-tuning of the encoder is more reliable for CNN layers than for MLP layers. Overall the results can be summarized as follows:

- For forecasting the day-ahead power generation based on an AE-MLP autoencoder and MLP forecasting model, fine-tuning only the last layer of the encoder for STL architectures leads to the most reliable results.
- When considering a time series autoencoder such as the proposed AE-TCN, an MTL approach and the fine-tuning of two layers of the encoder leads to the best and most consistent results regardless of the data type and latent size.

6. Conclusions and Future Work

This article studied autoencoders for day-ahead wind and solar power forecasts. By considering generative, discriminative, vanilla, and time series autoencoders, we considered a broad range of architectures to find the most applicable representation learning technique. We found that multi-task autoencoders improve reconstruction errors for discriminative autoencoders. A combination of multi-task and time series autoencoders led to an almost perfect ranking of the reconstruction error. By considering a multi-task approach, we reduced the trainable parameters by up to 203 times. Finally, we can conclude that the amount of layers to fine-tune depends on the architecture and the model. For single-task learning architectures and a multi-layer perceptron-based autoencoder, fine-tuning a single layer is sufficient. In contrast, for single-task time series, an autoencoder including additional layers is beneficial, whereas for multi-task architectures, it is always beneficial to include multiple layers during fine-tuning.

Due to the restrictions of the variational distribution of variational autoencoders we could not find a sufficiently good representation. In future work, we will need to extend the analysis for this model concerning the latent size, the number of layers in the network, and the scaling of the Kullback–Leibler divergence.

Author Contributions: Conceptualization, J.S.; methodology, J.S.; software, J.S.; validation, J.S.; formal analysis, J.S.; investigation, J.S.; resources, J.S.; data curation, J.S.; writing—original draft preparation, J.S.; writing—review and editing, J.S.; visualization, J.S.; supervision, B.S.; project administration, J.S.; funding acquisition, B.S. All authors have read and agreed to the published version of the manuscript.

Funding: This work results from the project TRANSFER (01IS20020B) funded by BMBF (German Federal Ministry of Education and Research). Enercast GmbH has provided the real-world datasets.

Data Availability Statement: PVOPEN and WINDOPEN are accessible at https://www.uni-kassel.de/eecs/en/sections/intelligent-embedded-systems/downloads, accessed on 28 October 2022. PVSYN and WINDSYN are available at http://dx.doi.org/10.48662/daks-11, accessed on 28 October 2022.

Acknowledgments: We thank Mohammad Wazed Ali, Diego Botache, Christian Gruhl, Stephan Vogt, and Alyssa Opland for their valuable input and feedback.

Conflicts of Interest: The authors declare no conflict of interest.

References

1. Henze, J.; Schreiber, J.; Sick, B. Representation Learning in Power Time Series Forecasting. In *Deep Learning: Algorithms and Applications*; Springer: Cham, Switzerland, 2020; pp. 67–101. [CrossRef]
2. Bengio, Y.; Courville, A.; Vincent, P. Representation Learning: A Review and New Perspectives. *arXiv* **2012**, arXiv:1206.5538.
3. Goodfellow, I.; Bengio, Y.; Courville, A. *Deep Learning*; MIT Press: Cambridge, MA, USA, 2016; p. 775.
4. Jiao, R.; Huang, X.; Ma, X.; Han, L.; Tian, W. A Model Combining Stacked Auto Encoder and Back Propagation Algorithm for Short-Term Wind Power Forecasting. *IEEE Access* **2018**, *6*, 17851–17858. [CrossRef]
5. Wang, L.; Tao, R.; Hu, H.; Zeng, Y.-R. Effective wind power prediction using novel deep learning network: Stacked independently recurrent autoencoder. *Renew. Energy* **2021**, *164*, 642–655. [CrossRef]
6. Chen, J.; Zhu, Q.; Li, H.; Zhu, L.; Shi, D.; Li, Y.; Duan, X.; Liu, Y. Learning Heterogeneous Features Jointly: A Deep End-to-End Framework for Multi-Step Short-Term Wind Power Prediction. *IEEE Trans. Sustain. Energy* **2020**, *11*, 1761–1772. [CrossRef]
7. Qureshi, A.S.; Khan, A.; Zameer, A.; Usman, A. Wind power prediction using deep neural network based meta regression and transfer learning. *Appl. Soft Comput.* **2017**, *58*, 742–755. [CrossRef]
8. Schreiber, J.; Buschin, A.; Sick, B. Influences in Forecast Errors for Wind and Photovoltaic Power: A Study on Machine Learning Models. In *INFORMATIK 2019*; GI e.V.: Bonn, Germany, 2019; pp. 585–598. [CrossRef]
9. Zhu, R.; Liao, W.; Wang, Y. Short-term prediction for wind power based on temporal convolutional network. *Energy Rep.* **2020**, *6*, 424–429. [CrossRef]
10. Schreiber, J.; Sick, B. Emerging Relation Network and Task Embedding for Multi-Task Regression Problems. In Proceedings of the ICPR 2020, Milan, Italy, 22–24 September 2020; pp. 2663–2670. [CrossRef]
11. Alkhayat, G.; Mehmood, R. A review and taxonomy of wind and solar energy forecasting methods based on deep learning. *Energy AI* **2021**, *4*, 100060. [CrossRef]
12. Gensler, A.; Henze, J.; Sick, B.; Raabe, N. Deep Learning for solar power forecasting—An approach using AutoEncoder and LSTM Neural Networks. In Proceedings of the 2016 IEEE International Conference on Systems, Man, and Cybernetics, Budapest, Hungary, 9–12 October 2016; pp. 2858–2865.
13. Qureshi, A.S.; Khan, A. Adaptive transfer learning in deep neural networks: Wind power prediction using knowledge transfer from region to region and between different task domains. *Comput. Intell.* **2019**, *35*, 1088–1112. [CrossRef]
14. Liu, X.; Cao, Z.; Zhang, Z. Short-term predictions of multiple wind turbine power outputs based on deep neural networks with transfer learning. *Energy* **2021**, *217*, 119356. [CrossRef]
15. Khan, M.; Naeem, M.R.; Al-Ammar, E.A.; Ko, W.; Vettikalladi, H.; Ahmad, I. Power Forecasting of Regional Wind Farms via Variational Auto-Encoder and Deep Hybrid Transfer Learning. *Electronics* **2022**, *11*, 206. [CrossRef]
16. Fawaz, H.I.; Forestier, G.; Weber, J.; Idoumghar, L.; Muller, P.-A. Transfer learning for time series classification. In Proceedings of the 2018 IEEE BigData, Seattle, WA, USA, 10–13 December 2018; pp. 1367–1376. [CrossRef]
17. Yan, J.; Mu, L.; Wang, L.; Ranjan, R.; Zomaya, A.Y. Temporal Convolutional Networks for the Advance Prediction of ENSO. *Sci. Rep.* **2020**, *10*, 8055. [CrossRef] [PubMed]
18. Plaut, E. From Principal Subspaces to Principal Components with Linear Autoencoders. *arXiv* **2018**, arXiv:1804.10253.
19. Yan, C.; Pan, Y.; Archer, C.L. A general method to estimate wind farm power using artificial neural networks. *Wind Energy* **2019**, *22*, 1421–1432. [CrossRef]
20. Thill, M.; Konen, W.; Bäck, T. Time Series Encodings with Temporal Convolutional Networks. In *BIOMA 2020: Bioinspired Optimization Methods and Their Applications*; Lecture Notes in Computer Science (including subseries Lecture Notes in Artificial Intelligence and Lecture Notes in Bioinformatics); Springer: Berlin/Heidelberg, Germany, 2020; Volume 12438, pp. 161–173. [CrossRef]
21. Lea, C.; Vidal, R.; Reiter, A.; Hager, G.D. Temporal convolutional networks: A unified approach to action segmentation. In *Computer Vision—ECCV 2016 Workshops*; Springer: Berlin/Heidelberg, Germany, 2016; pp. 47–54. [CrossRef]
22. He, K.; Zhang, X.; Ren, S.; Sun, J. Deep Residual Learning for Image Recognition. In Proceedings of the IEEE Conference on Computer Vision and Pattern Recognition (CVPR), Las Vegas, NV, USA, 27–30 June 2016; pp. 1–112. [CrossRef]
23. Kingma, D.P.; Welling, M. Auto-Encoding Variational Bayes. *arXiv* **2013**, arXiv:1312.6114.
24. Ruder, S. Neural Transfer Learning for Natural Language Processing. Ph.D. Thesis, National University of Ireland, Galway, Ireland, 2019.
25. Gensler, A. Wind Power Ensemble Forecasting. Ph.D. Thesis, University of Kassel, Kassel, Germany, 2018.

26. European Centre for Medium-Range Weather Forecasts. 2020. Available online: http://www.ecmwf.int/ (accessed on 24 August 2022).
27. Icosahedral Nonhydrostatic-European Union. 2022. Available online: https://www.dwd.de/SharedDocs/downloads/DE/modelldokumentationen/nwv/icon/icon_dbbeschr_aktuell.pdf? (accessed on 24 August 2022).
28. Lewinson, E. Three Approaches to Encoding Time Information as Features for ML Models. 2022. Available online: https://developer.nvidia.com/blog/three-approaches-to-encoding-time-information-as-features-for-ml-models/ (accessed on 8 July 2022).
29. Schreiber, J.; Vogt, S.; Sick, B. Task Embedding Temporal Convolution Networks for Transfer Learning Problems in Renewable Power Time-Series Forecast. In Proceedings of the ECML, Virtual, 13–17 September 2021; pp. 118–134. [CrossRef]
30. Vogt, S.; Schreiber, J. Synthetic Photovoltaic and Wind Power Forecasting Data. *arXiv* **2022**, arXiv:2204.00411.
31. Schreiber, J.; Sick, B. Model Selection, Adaptation, and Combination for Transfer Learning in Wind and Photovoltaic Power Forecasts. *arXiv* **2022**, arXiv:2204.13293.

energies

Review

A Selective Review on Recent Advancements in Long, Short and Ultra-Short-Term Wind Power Prediction

Manisha Sawant [1], Rupali Patil [2], Tanmay Shikhare [2], Shreyas Nagle [2], Sakshi Chavan [2], Shivang Negi [2] and Neeraj Dhanraj Bokde [3,4,*]

1 Department of Electronics and Communication Engineering, Indian Institute of Information Technology, Nagpur 441108, India
2 Department of Electronics and Telcommunication Engineering, K. J. Somaiya College of Engineering, Somaiya Vidyavihar University, Mumbai 400077, India
3 Center for Quantitative Genetics and Genomics, Aarhus University, 8000 Aarhus, Denmark
4 Interdisciplinary Centre for Climate Change, iCLIMATE Aarhus University , Foulum, 8830 Tjele, Denmark
* Correspondence: neerajdhanraj@qgg.au.dk; Tel.: +45-6090-1986

Abstract: With large penetration of wind power into power grids, the accurate prediction of wind power generation is becoming extremely important. Planning, scheduling, maintenance, trading and smooth operations all depend on the accuracy of the prediction. However due to the highly non-stationary and chaotic behaviour of wind, accurate forecasting of wind power for different intervals of time becomes more challenging. Forecasting of wind power generation over different time spans is essential for different applications of wind energy. Recent development in this research field displays a wide spectrum of wind power prediction methods covering different prediction horizons. A detailed review of recent research achievements, performance, and information about possible future scope is presented in this article. This paper systematically reviews long term, short term and ultra short term wind power prediction methods. Each category of forecasting methods is further classified into four subclasses and a comparative analysis is presented. This study also provides discussions of recent development trends, performance analysis and future recommendations.

Keywords: wind power prediction; machine learning; deep learning; hybrid methods; time series analysis

Citation: Sawant, M.; Patil, R.; Shikhare, T.; Nagle, S.; Chavan, S.; Negi, S.; Bokde, N.D. A Selective Review on Recent Advancements in Long, Short and Ultra-Short-Term Wind Power Prediction. *Energies* **2022**, *15*, 8107. https://doi.org/10.3390/en15218107

Academic Editors: Paweł Piotrowski, Grzegorz Dudek and Dariusz Baczyński

Received: 10 September 2022
Accepted: 24 October 2022
Published: 31 October 2022

1. Introduction

According to the global wind energy council report 2022, wind power capacity added in 2021 was 93.6 GW which was the second best year. However, in the same report it is mentioned that, to meet net zero, a four times increase in installation is required by the end of the decade. This implies that wind power is going to play a key role in future worldwide energy requirements. However, irregularities and randomness in wind power generation severely affect large-scale access of wind power to the grid [1,2]. This impacts dispatch operation, power quality and stable power system operations. Therefore, an accurate wind power prediction method is very important to reduce the burden on grid dispatching operations and to improve wind farm management [3,4]. However, the accurate prediction of wind power generation is a complex task owing to the stochastic nature of wind speed. The accurate prediction of wind power is challenging due to the nonlinear behaviour of wind speed, its random patterns and its dependence on atmospheric pressure and temperature [5,6]. Due to the stochastic nature of wind speed, the accurate prediction of wind power generation is a complex task.

Being a very active field of research, a large number of wind power prediction models have been developed. Few review articles [7,8] on this topic are available. Reference [7] presents a detailed review of past and present methods in WPP along with the future scope in this area. Reference [8] presented a review of hybrid models based on empirical mode decomposition. In this article we present recent developments in WPP and provide quick access to meaningful works. Based on the type of prediction models, existing wind power prediction (WPP) methods are categorised into physical models [9], statistical models [10,11], and hybrid models [12,13]. Detailed physical analyses and descriptions of wind farm layout and wind turbines, and physical descriptions of meteorological and geographical conditions are required to build physical models. Physical attributes of geographical location such as terrain and wind turbulence affect the accuracy of the physical models. The physical models need various environmental parameters such as wind speed, wind direction, and air pressure. These parameters are obtained from the numerical weather prediction data which is updated once every few hours. In areas where the numerical weather prediction (NWP) system is not available the physical models are not useful. Due to the low update frequency of NWP data, physical models are not suitable for a prediction horizon of more than 6hrs, i.e., short-term prediction [14,15].

Several statistical models have been developed in the literature for WPP. In this review we have classified them into time series methods, machine learning methods and deep learning methods. Time series methods consist of linear and nonlinear time series-based models such as auto-regressive (AR), auto-regressive moving average (ARMA) [16], moving average (MA) and auto-regressive integrated moving average (ARIMA) [17]. Various machine learning (ML) methods such as support vector machine (SVM) [18], support vector regression (SVR), Gaussian process regression (GPR) [19], random forest and k-means clustering artificial neural networks are used for WPP.

With the development of high computing power and advanced machine learning and deep learning methods, accurate and effective wind power prediction methods have been developed. Various combinations of physical, statistical and deep learning methods have also evolved to improve prediction accuracy. Furthermore, data cleaning, preprocessing and feature extraction methods combined with advanced learning algorithms lead to improved results.

In this paper, we have systematically investigated WPP methods based on different prediction horizons, algorithms and evaluation criteria. In this review, we present detailed documentation of various algorithms, their performance and discussions. Wind power forecasting can be categorised depending on prediction horizon or prediction methodology. In Table 1 prediction horizons and the corresponding time range are listed. According to the prediction horizon, they can be categorised as long-term, short-term or ultra-short-term methods. Prediction methodologies are classified as physical, statistical and hybrid methods. With the recent developments in computing power and ML techniques, the statistical methods are further classified as time series methods, machine learning methods and deep learning (DL) methods. Hybrid methods are a combination of different prediction methodologies; it can be a combination of time series and ML or ML and DL or a combination of all of them. In this article, we have followed the prediction horizon for the categorization of WPP methods and, for each category, related prediction methodologies are discussed. The statistics of the number of articles referred to in this article are shown in Figure 1. In this review, we have considered articles published on WPP after 2015. It is clear from Figure 1 that recent research in this field mainly focuses on short-term wind power prediction.

Table 1. Prediction horizons in WPP.

Prediction Horizon	Time Range
Long term	a day to 6 days ahead
short-term	an hour to a day ahead
Ultra short-term	5 min to 1 h

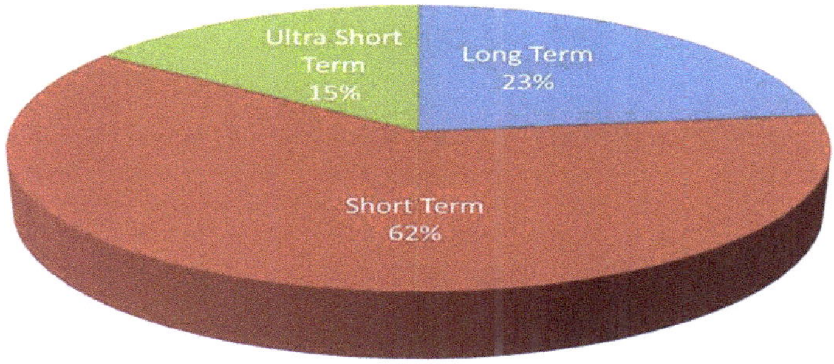

Figure 1. The distribution of articles reviewed into long-term, short-term and ultra-short-term predictions.

Depending on the amount of power generated in different time scales, the prediction methodologies are classified as long term, short-term and ultra short-term. For example, for turbine maintenance scheduling, optimization of operating cost and other management issues, a day ahead or 2 to 3 days ahead, i.e., long term predictions are required. From a day until several hours ahead, predictions are enough for planning related to load dispatch and for treading issues. Further, a shorter prediction horizon is required for turbine control, and real time grid operations. Several learning algorithms have been developed since the past decade that cover the wide range of forecasting horizons. Figure 2 shows recently published prediction methodologies (referred in this article) for long term, short-term and ultra short-term WPP. Clearly it can be seen that recently, researchers are focusing on short-term WPP. As far as prediction methodologies are concerned, a number of publications on hybrid methods are more for all the three prediction horizons. This implies that the hybrid models are more suitable and widely used for short-term prediction applications.

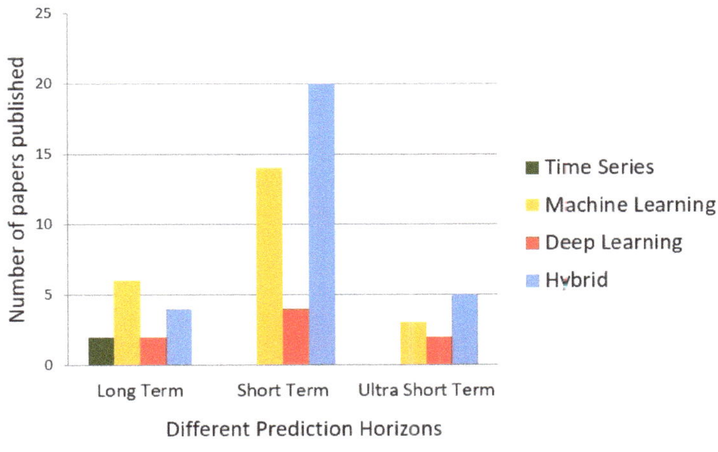

Figure 2. Distribution of different categories of articles reviewed.

In this paper, recent advancements in wind power forecasting approaches are reviewed. Performance evaluation metrics are presented in Section 2. Classification of WPP methods based on prediction algorithms is presented in Sections 3–5. Detailed review of long term forecasting methods is discussed in Section 3. Different prediction algorithms developed for short-term forecasting are presented in Section 4, Furthermore, Section 5 reviews ultra short-term WPP methods in detail. Section 6 is devoted to detailed discussions and future scope in this area. Finally, the conclusions are presented in Section 7.

2. Performance Evaluation Metrics

Performance evaluation metrics are measures to judge or quantify the goodness or usefulness of the prediction algorithm. These metrics generally estimate the distance between original output and the estimated output. Performance of the wind power prediction models is evaluated using several statistical metrics; the following are frequently used performance measures. Let Y_i be the ith actual load value, $\hat{Y}_i$ be the ith predicted load value, $\bar{Y}_i$ the mean of the actual load value and N the total number of predicted points. Different evaluation metric used as performance measures are listed below.

1. *Mean absolute error (MAE)* :
   MAE is average value of absolute different between predicted and actual value.

$$MAE = \frac{1}{N}\sum_{i=1}^{N}\left|Y_i - \hat{Y}_i\right|. \tag{1}$$

2. *Root Mean Square Error (RMSE)*:
   $RMSE$ computes the standard deviation of the residuals between predicted and actual values. Residuals defined the distance between regression line data points and $RMSE$ measures the spread of these residuals.

$$RMSE = \sqrt{\frac{\sum_{i=1}^{N}\left(\hat{Y}_i - Y_i\right)^2}{N}}. \tag{2}$$

3. *Mean Square Error (MSE)*:
   The mean squared error calculates the average of the squares of the error in the prediction.

$$MSE = \frac{1}{N}\sum_{i=1}^{N}\left(Y_i - \hat{Y}_i\right)^2. \tag{3}$$

4. *Mean Absolute Percentage Error (MAPE)*:
   The mean absolute percentage error ($MAPE$) is average of the absolute percentage error in the forecasts.

$$MAPE = \frac{1}{N}\sum_{i=1}^{N}\left|\frac{Y_i - \hat{Y}_i}{Y_i}\right|. \tag{4}$$

5. *Normalised RMSE (NRMSE)*:
   Normalization of the $RMSE$ value is useful for fair comparison of the model on different scales. The normalization can be performed with respect to mean or standard deviation. The following is the Mean $NRMSE$.

$$MeanNRMSE = \frac{RMSE}{\bar{Y}}. \tag{5}$$

6. *Normalized Mean Absolute Error (NMAE)*:
   $NMAE$ is used to compare the MAE of models with different scales. The $NMAE$ is a two-step process. The normalization can be performed with respect to mean, range or inter quartile range. The following is Range $NMAE$.

$$RangeNMAE = \frac{MAE}{range(Y)}. \tag{6}$$

7. *Root Mean Square Prediction Error (*RMSPE*)*:
   RMSPE calculates the root mean squared percentage error regression loss.

$$RMSPE = \sqrt{\frac{\sum_{i=1}^{N}(Predicted_i - Actual_i)^2}{N}} \times 100. \tag{7}$$

8. *R-Square* (R^2):
   R^2 is the coefficient of determination, it computes the variance of the prediction from the measured data. A negative value of R^2 implies worse prediction while it can reach a maximum value of 1.

$$R^2 = 1 - \frac{\sum_{i=1}^{n_{test}}(\hat{Y}_i - Y_i)^2}{\sum_{i=1}^{n_{test}}(Y_i - \bar{Y})^2}. \tag{8}$$

3. Long Term Prediction

Maintenance of wind turbines and other management issues are planned with the help of long term prediction. These activities do not require highly accurate prediction accuracy.

3.1. Time Series Analysis

Time series prediction models are mathematical models that estimate model parameters from the historic data. Time series prediction models can capture nature of system and generate predictions. Time series models with different orders generate different results.

A polynomial extension of the AR model, i.e., PAR is presented in [20]. A polynomial AR (PAR) model of degree 2 derived from Volterra series expansion (9) is used for wind power prediction. A comparative study of PAR with MLFF, MLP, ANN, AR and ANFIS is also presented in [20]. Compared to these nonlinear models PAR requires less parameters, is computationally efficient, and performs better for longer prediction horizons (more than 12 h). Experimental analysis was performed on the data published for Global Energy Forecasting Competition 2012 [21] and NRMSE, NMAPE and bias were used as error measures. In (9), μ is the intercept, excitation sequence $\epsilon(l)$ is n independent and identically distributed with distribution $\mathcal{N}(0,\sigma^2)$ and $a_i^{(1)}, a_{i,j}^{(2)}, \cdots, a_{i,\cdots}^{(p)}$ are coefficients for first, second and pth order polynomials, respectively, degree of the non-linearity is p and the AR order k.

$$x(l) = \mu + \sum_{i}^{k} a_i^{(1)} x(l-i) + \sum_{i}^{k}\sum_{j}^{k} a_{i,j}^{(2)} x(l-i)x(l-j) + \cdots + \sum_{i,\cdots}^{k,\cdots} a_{i,\cdots}^{(p)} x(l-i)\cdots + \epsilon(l). \tag{9}$$

Large fluctuations in wind power within a relatively short time interval caused by wind is defined as a wind power ramp event. These power ramps lead to a potential disaster and affect the stability and safety of the wind farms and power grids. In order to take preventive action before such disaster happens, the accurate prediction of power ramp events is most important. Wind power prediction and ramp event detection algorithm is presented in [22]. Two models have been proposed for wind power prediction. Long term trends in the data are captured using wind power curve model utilizing NWP. A correction model improves the local prediction accuracy using a multivariate prediction algorithm. For power ramp event detection a well-known swinging door algorithm [23] is used and a higher accuracy of ramp event prediction was reported. Table 2 list the time series methods for long term WPP and their respective performances.

3.2. Machine Learning

A wide range of machine learning models including extreme learning machine, support vector machine (SVM) [18] and the Gaussian process [19], backpropagation network [24] and radial basis function are applied for WPP. These methods learn a nonlinear regression function that fits the relationship between the input feature space and the output wind power from the data.

Table 2. Time series methods for long term WPP.

Sr. No.	Articles	Models	Evaluation Metric	Evaluation Value
1.	Karakus et al., 2017 [20]	Polynomial AR	NRMSE NMAPE	0.1199 6.8146
2	Ouyang et al., 2019 [22]	Residue correction	MAE RMSE Bias	4.9647 8.9453 0.0013

Back propagation neural network (BPNN) is a widely used nonlinear method for wind power prediction. In [25], the basic structure of BPNN is used along with the conjugate gradient method for weight optimization and termed the method as conjugate gradient neural network (CGNN). There are various meteorological factors such as air pressure, humidity, temp etc. influence the wind power, in [25] along with wind speed and wind cos above mentioned parameters are also taken as input to the CGNN. For experimental validation the data from wind farms in Mongolia and China is used. Accuracy as well as time taken by the proposed CGNN and existing Racial Basis Function Neural Network (RBFNN), Steepest Gradient Neural Network (SGNN), and Extreme Learning Machine (ELM) is reported. Due to conjugate gradient optimization, the training time as well as MAE of the CGNN are less than those of the other compared methods. Ref. [26] also used the swinging door algorithm for power ramp prediction. In [26], first, the data are divided into two data segments—a ramp window or a non-ramp window. The optimum window size for these two events is decided using a genetic algorithm. Once the optimum window size is decided then the power in the optimized window is predicted using SVM, which receives NWP data as input data. Depending on the predicted power, the swinging door algorithm detects whether it is a power ramp event or not. To validate the results, different window sizes were analysed and the accuracy and false positive rate of ramp detection were reported.

Support vector machine (SVM) [18] is a popular machine learning algorithm due to its generalization ability and high dimensional data handling capability. SVM is widely used for wind speed and wind power prediction. The accuracy of the SVM depends on various hyper-parameters of kernel and cost function. In [15], SVM with hybrid kernel function is proposed for wind power prediction. Two separate kernels polynomial and radial basis function (RBF) were used to build a hybrid kernel that can capture correlation in the local and distant data samples. The parameters of the hybrid kernel are estimated using an improved particle swarm optimization algorithm. Experimental analysis showed better accuracy of SVM with hybrid kernel in terms of RMSE, MAE and MAPE compared to ARMA, SVM with only RBF and the echo state network. Table 3 shows machine learning models developed for long term WPP.

Due to the no-nstationary behaviour of the wind energy, a single algorithm is not able to fit the data accurately. In this situation, ensemble learning methods are used to improve the accuracy. In ensemble learning, multiple base learning methods are combing. Improvement of the accuracy can be achieved by one of the following approaches—to perturb the training data, model parameters, attributes of the data and base models. The selection of appropriate base learners is also important to increase accuracy. Non-probabilistic learning methods provide point prediction outputs but do not provide an estimate of uncertainty. Gaussian process regression is a powerful nonparametric Bayesian method for supervised learning. Along with the probabilistic predictions, it also provides confidence intervals of predictions. Ensemble learning model in [27] utilizes Gaussian process regression as the base learners. In order to improve the accuracy and diversity of the learning methods, first the perturbations on training data and input attributes are combined. Next the Gaussian mixture model (GMM) clustering is applied to create different clusters of the data. Further, GPR is applied on each cluster separately to fit each cluster individually. This method is termed the selective ensemble of finite mixture Gaussian process regression models (SEFMGPR). Performance of the ensemble learning improves with the pruning method. In [27], a genetic algorithm-based pruning method has been adopted to select significant models. The pruning algorithm enhanced the performance as well as reduced the model complexity.

Table 3. Machine learning methods for long term WPP.

Sr. No.	Articles	Models	Evaluation Metric	Evaluation Value
1	Fang et al., 2016 [28]	Improved GPR	NRMSE	0.14564
2	Tian et al., 2016 [25]	CGNN	MSE	0.004
3	Tinghui et al., 2017 [26]	Swinging door+SVM	Precision, Recall Accuracy, Error	0.8059, 0.8390 [1] 0.8747, 0.1253
4	Zhongda et al., 2018 [15]	Modified SVM	RMSE MAE MAPE	59.313 50.344 0.038
5	Bogdan et al., 2021 [29]	Comparison of ML	RMSE MAPE	412 [2] 0.267
6	Huaiping et al., 2021 [27]	ensemble of mixture of GPR	RMSE R^2	1.7771 [3] 0.9057

[1] for 2 h time window. [2] indicate hourly prediction, daily prediction values are 13.9% 65.60. [3] indicate 4 step ahead prediction.

Another Gaussian process based approach in [28] proposed a composite covariance function (CF) for the GP. Performance of the GPR varies with the CF. The composite CF proposed in models the relation between wind features and auxiliary features. The composite CF is the multiplication of squared exponential CF that can integrate multiple NWP features into a single composite CF. The GP approach in [28] used the 2012 global energy forecasting competition wind power forecasting data, and outperformed all of the competitors on this data.

A comparative analysis of different machine learning techniques to forecast the production of wind energy not for a single wind farm but for an entire country Poland has been presented in [29]. They have presented the results of two decision tree based algorithms, i.e., random forest (RF) and Extreme Gradient Boosting (XGB) and two neural network based algorithms, i.e., artificial neural network (ANN) and deep neural network (DNN). For the experimental analysis, various interesting inferences were also presented in [29]. Although all four algorithms predicted wind power with high accuracy, XGB was better in terms of MAPE for hourly predictions and ANN for daily sums of produced energy. Performance analysis for different seasons was also presented and it was inferred that MAPE was the highest in June and the lowest in January. This is due to the fact that the windiest day occurs in January and the calmest day appears in August. The lowest variance in prediction was reported in the winter season and was highest in the summer season.

3.3. Deep Learning Models

Wind power data are characterised as highly nonlinear as well as high dimensional. Compared to shallow machine learning models, deep learning models are more suitable for such data. With high computing power and the ability to fit complex and nonlinear function deep learning methods are widely used for WPP.

In [30], instead of statistical features, stacked autoencoder (SAE) features are proposed for wind power prediction. Structural properties of the wind data are effectively extracted using an autoencoder. A two level autoencoder is designed for structural feature extraction. During the training stage, the input data are divided into small data segments and predictions are performed on those segments individually. Features are also extracted for each segment. For wind power, a cluster-based ensemble regression is proposed, where the data segments are first clustered and then a regression model is learned separately for each cluster. Compared to statistical features 12.63% improvement in the prediction accuracy was reported when SAE features are used.

A combinatory approach for feature generation, feature selection and power prediction is presented in [31]. They presented an improved wavelet neural network (WNN) that uses the Morlet wavelet as an activation function of the neural network for feature extraction. Next, relevant features are selected using a maximum dependence, maximum relevancy, and minimum redundancy (MDMRMR) feature selection algorithm. Later they trained a 2D CNN using these selected features as input and PSO based improved optimization algorithm. Shallow 2D CNN is build that consist of input layer, two convolutional layers, two pooling layers and one fully connected layer. Extensive experimentation was performed to validate the proposed method. This combinatory approach is evaluated for three different prediction horizons i.e., an hour ahead, day-ahead, and 48 h and two separate databases for this analysis. Accuracy of both the methods is reported with respect to different measures. Comparative analysis of deep learning methods in long term WPP is shown in Table 4.

Table 4. Deep learning methods for long term WPP.

Sr. No.	Articles	Models	Evaluation Metric	Evaluation Value
1	Abedinia et al., 2017 [30]	autoencoder features and SVR	PE	0.152 [1]
2	Abedinia et al., 2020 [31]	2D CNN	MAPE NRMSE	5.93% 9.40%

[1] for ensemble of SVR.

3.4. Hybrid Approach

Multiple prediction horizons are proposed in [32] using a hybrid approach for long term prediction and reinforcement learning for short-term prediction. For long term wind power prediction sigma point Kalman filter is modified using complementary ensemble empirical mode decomposition. Initially, sigma points are used to limit the boundary effects; next the historic data are decomposed into various intrinsic mode functions with steady-state features using the complementary ensemble empirical mode decomposition (CEEMD) method. For power prediction, each stable sequence is updated to reconstructed using sigma point based Kalman filter. For short-term prediction, a deep deterministic policy gradient (DDPG) method is proposed in [32]. Prediction results are compared with different state-of-the-art methods on the basis of MAE, MAPE, SDE and RMSE.

A hybrid approach to predicting wind power from the numerical weather prediction data and actual wind power data are presented in [33]. Daily similarities are observed in the wind power and based on these similarities the data can be easily clustered. Using spatial similarities in the NWP data, k-means clustering is used to split the data into different subsets. Next, the samples matching with the predicted day are used to train the generalized regression neural network (GRNN) model. Experimental analysis shows that GRNN can effectively model the nonlinear relationship between the wind data and the predicted output. The results also show the impact of clustering on the long term wind power prediction. In [34], a bagging neural network (BaNN) is also combined with k-means clustering for long term wind power prediction. Prediction accuracy is enhanced by fine tuning the BaNN parameters using an optimization method. They also used improved empirical mode decomposition (IEMD) to reduce the fluctuations during the forecasting process to improve the accuracy. The Experimental analysis was performed on the data collected from three different farms. Since CNN do not provide good prediction results, such hybrid approaches with improved clustering methods and advance neural networks need to be further explored for long term predictions.

An hourly forecast of day-ahead wind power method proposed in [35] combines variational mode decomposition (VMD) and LSTM. Compared to empirical mode decomposition (EMD), VMD provides less fluctuation and retains more adequate data information for forecasting. VMD decomposes the input wind series data into different modes, separate LSTM with three layers is trained for each mode. Experimental analysis of two different VMDs, recursive (R) VMD and direct (D) VMD with LSTM is presented in [35]. Performance of the VMD-LSTM method is compared with the BP, ELM, and SVM and results show that VMD- LSTM achieved significant results. RMSE, MAE and MAPE are reported for one-day, two day and three day-ahead prediction. Hybrid methods and their corresponding performances referred to in this article are listed in Table 5.

Table 5. Hybrid methods for long term WPP.

Sr. No.	Articles	Models	Evaluation Metric	Evaluation Value
1	Dong et al., 2016 [33]	NWP + GRNN	NMAE NRMSE	10.67 14.01%
2	Shi et al., 2018 [35]	LSTM + VMD	RMSE MAE MAPE	6.44 7.16 22%
3	Abedinia et al., 2020 [34]	BaNN + K-means clustering	MAPE NRMSE	2.71 19.78%
4	Zhang et al., 2021 [32]	E-EMD + RNN	MAE, SDE RMSE, MAPE	144.79, 60.32 81.77, 12.79%

4. Short-Term Prediction

Short-term wind power prediction assists with deciding power generation plans, regional dispatching, and maintenance plans.

4.1. Machine Learning

Instead of using a single ML algorithm, the combining of different ML models into an ensemble predictor provides improved results. The heterogeneous ensemble approach in [36] used decision trees (DT), k-NN, or support vector regressors as base algorithms. They also analysed the performance of individual as well as different combinations of the base predictors. Different combinations provide different accuracies and computational complexity. On the basis of these two parameters, the combination of DT and SVR provided improved results. Experiments were conducted using the power output data of the five wind parks. Ensemble of Boosted Trees, Random Forest, and Generalized Random Forest are presented in [37] for short-term wind power prediction. Correlation or time dependencies in the data are considered in ensemble learning, which improved the accuracy. Time lagged values are added as new features and a feature importance analysis is performed to decide the impact of features on the forecast. The proposed method is evaluated using data from five farms and compared with SVR and GPR in terms of R^2, RMSE and MAE.

A variant of Gaussian process regression model is proposed for short-term prediction in [38]. Computational complexity of the GPR increases with dimensionality of the data. In order to reduce computational complexity and to model the non-stationarities in the wind data, a new teaching learning based optimization (TLBO) is proposed. Optimal parameters of the Gaussian process are learned during the training process using the TLBO. It also helped to improve the learning rate and computational complexity. This method effectively forecasted the data from a single farm as well as from an entire Ireland. Performance of the GPR outperforms many other ML methods with proper selection of covariance function and optimization. However, even with the optimally tuned parameters, a drastic reduction in accuracy and confidence interval is observed if the missing data are encountered. In [39], data imputation approach is used to handle missing data and new datasets are generated. Missing data are reconstructed from the distribution of the data and the iterative learning algorithm. Next, the GPR model is built using the reconstructed data for wind power prediction. Performance of the proposed approach is compared with

SVM and MLP using data from three different wind farms and GPR reported better results in terms of RMSE and MAE.

Appropriate parameter selection usually helps to improve the accuracy of the learning algorithm. Out of multiple parameters and features available in the meteorological data, the work in [40], used most useful features of the wind data. Using the correlation and the importance measures, spatially averaged wind speed and wind direction are selected for the wind power prediction. Next, random forest is selected as a prediction algorithm due to its low computation complexity compared to other ML methods. Impact of selected features on the prediction accuracy is also analysed.

Short-term wind power prediction in [41] explores correlation between wind speed and wind power data. This method combines NN and PLS to form Nonlinear Partial Least Square (NPLS) method. The historic data are applied as input to the NN and the output of the NN is provided as input to the PLS which provides the final wind power predicted value. For experimental analysis two well known NNs i.e., BPNN and RBFNN are used and combined with PLS. The performance of BPE-PLS and RBFE-PLS and compared with SVM, BPNN, RBFNN and PLS. For experimentation, datasets with three different weather conditions are used. In terms of RMSEP measure, the RBFE-PLS algorithms outperformed all the other method on all the dataset.

Extreme learning machine is a feed forward neuron network with three layers, input layer, hidden layer and output layer. A few wind power prediction method used ELM as a regressor, but the training strategy (leave one out) is not suitable for high dimensional data. Kernel ELM is proposed in [42] adopts k-fold cross validation with its average MSE as error function. High nonlinearity of wind data are effectively captured by kernel ELM. Performance of learning machines depend on the parameter selection, in case of KELM, optimal values of regularization coefficient and kernel width improves the performance. In [42], KELM is trained using the wind power data and optimal parameters are learned using differential evolution (DE) optimizer and average MSE of k-fold cross validation. This approach improves the generalizability as well as stability of the model. Performance of the KLEM with cross validation and DE optimizer (DECVKLEM) is compared with KLEM with cross validation and GA as an optimizer. Compared to GACVKLEM, only 8.34% improvement has been observed in case of DECVKLEM, but the convergence speed of DE base CVKLEM is more. In [43] ELM is trained using PSO and combined with Adaboost for short-term wind power prediction. Performance of the Adaboost-PSO-ELM is compared with PSO-BP, GPR, PSO-SVM, PSO-ELM, GA-ELM, few tree based methods and Adaboost-PSO-BP and better performance of Adaboost-PSO-ELM is reported.

Accurate wind speed prediction is important for NWP based WPP, so to improve the accuracy of WPP, the NWP data from three different organizations is combined in [44] and used for prediction. Three forecasted wind speeds from NWP are fused using weighted naive Bayes (WNB) method and accurate wind speed is estimated. Next, wind power prediction is performed using BPNN. Ref. [45] modified BPNN and proposed a small-world BPNN (SWBP). Small-world networks ties to reduce the gap between artificial and biological neural networks [46] by modifying node type, connections between the nodes, and realization function. Input features for the SWBP are selected using modified mutual information (MI) and applied to the SWBP. The proposed model is compared with BPNN for 15 min-ahead power prediction and found better than BPNN in terms of training time, prediction accuracy and convergence.

Uncertainty and missing values in the wind data incurs difficulties in the accurate prediction. In such cases, grey models are found useful. In [47], grey model GM(1,1) with background value optimization is proposed for wind speed prediction. Two separate grey models are designed and combined to improve the wind speed prediction accuracy. Further, for WPP, SVR is designed. Various parameters of SVR such as cost function, precision and variance of the kernel function are estimated using PSO optimizer. Results of PSO-SVR are compared with ARIMA on the basis of MAE, MAPE and RMSE and nearly 30% improvement in speed prediction and 35% improvement in power prediction are reported.

Multimodality, nonstationarity, and skewness are characteristics of the wind power which make wind power prediction a challenging task. In [48], an infinite Markov switching autoregressive model is used for wind power forecast. Using a nonparametric Bayesian approach, a posterior predictive distribution is computed which is further used to predict wind power and uncertainty of the forecast. Probabilistic methods provide estimate of the value as well as uncertain in prediction. Compared to MSAR, TVQR and BELM model, the proposed nonparametric method performed better.

Pattern Sequence-based forecasting (PSF) [49–51] method has shown its potential in short-term wind speed forecasting accurately [52], but for the first time, it offers higher accuracy for wind power time series in [53]. This is a kind of its own approach, where the wind power time series was first smoothened down with the reference of the corresponding wind speed time series dataset and then a smoother wind power time series was forecasted with the PSF algorithm. This smoothening process comprised of generation of label sequences in the PSF method and a matching process with Naive Bayesian. The proposed approach was observed to be less chaotic for wind speed predictions than the existing ones.

An integrated approach is employed in [54] for short-term wind power prediction. Uncertainty, nonlinearity, missing data extended training time and computational complexity these are various factors that affects performance of the prediction system. Uncertainties and missing information in the data are modelled and the controlled fuzzy network, wavelet decomposition models the dynamic behaviour, and nonlinearities are modelled with NN. In [54], an integrated approach of these methods is presented. Similar to ANFIS, a fuzzy NN is proposed where a wavelet function is used as an activation function and the combined model is termed as Fuzzy WNN (FWNN). Optimization of this combined model is carried out using PSO and gradient descent. Performance of the FWNN is compared with 7 ML methods, RBF, SVR, ANN, ANN-GA, ANN-PSO, ANFIS, ANFIS-GA and ANFIS-PSO. Table 6 list the details of the machine learning methods in short-term WPP.

Table 6. Machine learning methods for short-term WPP.

Sr. No.	Articles	Models	Evaluation Metric	Evaluation Value
1	Heinermann et al., 2016 [36]	SVR	MSE	588.60
2	Yan et al., 2016 [38]	Variant of GPR	NRMSE MAE	0.05 0.1
3	Lahouar et al., 2017 [40]	random forests	MAE,RMSE NAME,MAPE,MASE	0.93, 1.76 1 1.72, 10.29, 0.4
4	Wang et al., 2018 [41]	Nonlinear PLS	RMSECV RMSEP	206.2611 225.5973
5	Liu et al., 2018 [39]	GPR with data imputation	RMSE MAE	0.9763 2 0.8321
6	Xie et al., 2018 [48]	Bayesian Framework	Skill score Coverage	−5.314 0.164
7	Wang et al., 2019 [55]	Small-world NN	eNRMSE eNMAE	7.5 4.9
8	Zhang et al., 2019 [47]	PSO-SVR	MAE RMSE MAPE	76.9 109.4 15.7%
9	Jianqi et al., 2020 [44]	Weighted Naive Bayes	MAE NRMSE	8.97 77.23%
10	Junho 2020 [37]	Boosted Trees + RF	MAE RMSE R^2	22.54 50.44 98.88

Table 6. *Cont.*

Sr. No.	Articles	Models	Evaluation Metric	Evaluation Value
11	Ning et al., 2020 [42]	Kernel ELM	MAE RMSE MSE R^2	7.30 10.22 104.59 0.53
12	Guoqing et al., 2021 [43]	Adaboost	MAE, MSE RMSE, MAPE R^2	23.38, 1023.4 31.99, 0.028 0.99
13	Ghoushchi et al., 2021 [54]	fuzzy Wavenet	RMSE MAPE	0.198 0.00123

[1] values not explicitly mentioned noted from graph for 6-step ahead. [2] for 20% missing rate.

4.2. Deep Learning Methods

In [56], DNN based ensemble learning is proposed where base-regressors and a meta-regressor both are built using DNN. First several autoencoders act as base methods and are trained using the training data and transfer learning. Transfer learning saves the time to train the system from scratch as well as provides suitable weight initialization for training. Due to abrupt changes in the meteorological conditions, the transients are observed in the predictions. These transients are smoothed with the help of a meta-regressor. In [56], Restricted Boltzmann Machines (RBMs) are stacked to Deep Belief Network (DBN) which acts as a meta learner. Once the base learner is trained, the test data features and predictions from the base learners on test data are applied as input to DBN for final prediction value. Data from five wind farms is uses for evaluation of the algorithm and the results in terms of RMSE, MAE and SDE are reported. Two step approach in [57] uses DBN and k-means clustering for wind power prediction. The noise in the NWP greatly affects the accuracy of the learning method, so the NWP data are divided into different clusters using k-means clustering. Next the clustered data (e.g., NWP wind speed, wind direction, humidity, temp etc.) is applied as input to the DBN. The DBF consists of five layers with three hidden layers. For prediction, the test data are divided into clusters and the clusters belonging to those data are fed to the trained model to obtain the wind power. In comparison to BPNN and WMNN, the performance of the proposed method improved by 44%.

A Gaussian mixture model combined with NN is termed the Gaussian Mixture Density Network (MDN). The conditional density function of the data is predicted using a trained MDN which is further used to predict the required uncertainty information. The parameters of the Gaussian mixture are computed using a feed forward NN. An improved deep MDN proposed in [58] uses beta distribution to solve density leakage associated with MDN and modified ReLU activation function to handle NaN issue associated with activation function. Data from seven wind farms is used and proposed method is compared with 8 existing methods and the improvement in the performance is recorded. Time and memory complexity analysis is also presented and the proposed method requires 10 min training time.

NWP provides various parameters such as wind speed, wind direction, temp. air pressure etc. of which wind speed is an important parameter for power prediction. A gated recurrent unit neural network (GRUNN) presented in [59] makes use of the variance of the NWP wind speed prediction error for wind power prediction. It utilizes both temporal as well as statistical characteristics of the time series data. Bidirectional GRUNN in [59] is a simple version of LSTM [60] with two gates in GRU. In the proposed method, first, local features are extracted from the NWP data and a weight time series is constructed using the NWP wind speed prediction error and extracted features. This weight time series is applied as input to GRUNN which corrects the NWP wind speed. Once correct wind speed is obtained then Power Forecasting is performed using the Wind Power Curve Model. This computationally efficient method is compared with SVN and ANN and the results in terms of RMSE and MAE are presented.

Data cleaning and feature reconfiguration approach using CNN is proposed in [61] for WPP. It has been observed that performance of the prediction system degrades in presence of outliers. In [61], outliers are identified using density based clustering method. After data cleaning the wind data are applied to CNN, since CNN requires images as an input; a feature reconfiguration is an essential step. Each sample of wind data has two features; wind direction and wind speed along with the label, i.e., wind power. Wind direction and wind speed sample along with the corresponding temporal information and label are arranged in a 2D matrix which is applied as an input to the CNN. The CNN architecture consists of one input layer, two convolutional layers, and one fully connected layer, ReLU is used as an activation function, no pooling layers are used and parameter tuning is performed by trial and error. MAE, MAPE and NRMSE are used as performance measures for the evaluation of the proposed scheme. This is the only method that reconfigures wind data as a 2D matrix and uses an image based deep learning approach. In our opinion, if parameter tuning is performed by an optimization method then accuracy can further increase. Table 7 shows the deep learning methods for short-term WPP and their performances.

Table 7. Deep learning methods for short-term WPP.

Sr. No.	Articles	Models	Evaluation Metric	Evaluation Value
1	Qureshi et al., 2017 [56]	DNN with transfer learning	MAE SDE RMSE	0.0658 0.0929 [1] 0.0939
2	Wang et al., 2018 [57]	Deep belief network	NMAE NRMSE	0.0236 0.0322
3	Ding et al., 2019 [59]	Gated RUNN	RMSE MAE	13.45 [2] 6.87
4	Zhang et al., 2020 [58]	Deep Mixture Density Network	NMAE NRMSE	0.108 0.147
5	Wang et al., 2021 [61]	Multidimensional data cleaning + feature reconfiguration	MAE, MAPE NRMSE	2.18, 4.36% 7.29%

[1] results of test data from farm 1. [2] % value of MAE and RMSE for day 1.

4.3. Hybrid Methods

LSTM and genetic algorithm (GA) are combined for wind power prediction in [62]. The performance of the LSTM algorithm largely depends on the window size. A smaller window size implies no information is forwarded and carried, whereas a larger window size implies noise in the past samples. A genetic algorithm is used to learn optimum window size. Experiments were performed on the dataset from seven wind farms in the European region. The data consisting of sixteen features measures a duration of 48 h and a 12 h interval is applied as an input to the Genetic LSTM (GLSTM) network. GA is used to train the network to find the optimum window size and number of neurons. Performance of GSTM is compared with the ARIMA, a few deep learning methods and SVR of three different kernel functions. To validate the effectiveness, six variants of GLSTM are applied on the seven datasets and improvement in the performance was reported. Closed to zero MSE, MAE and RMSE were reported with the proposed GLSTM network.

Wind power ramps events are predicted in [63] using different ML algorithms. A comparative analysis of ML methods to predict ramp events has been presented. In this hybrid approach the data from numerical-physical models is applied as input the various ML algorithms. The effectiveness of SVM, GPR, ELM and MLP for ramp event prediction is experimentally verified in [63]. RMSE, MAE and sensitivity are used as performance measures for evaluation of these methods on three wind farm dataset. The performance of GPR outperformed other methods in terms of all the measures.

Decomposition methods for WPP decompose the wind power time series into different components depending on different characteristics such as frequency, scale [64,65]. Next different prediction algorithms can be applied on these components for WPP. These decomposition methods can efficiently model the nonlinearities, however in many cases these components shows chaotic behaviour which degrades the prediction accuracy. To remove the uncertainty and low amplitude variations from these components and to improve the accuracy, singular spectrum analysis (SSA) [66,67] was found to be very useful. In [14], ensemble empirical mode decomposition (EEMD) is used to decomposed the time series data into different components. After determining the chaotic components, SSA is applied to remove the impact of the chaotic components on the accuracy. The proposed method has two stages—a decomposition stage and a prediction stage. The first stage consists of EEMD, chaotic TS analysis, and SSA and is referred to as multi-scale singular spectrum analysis (MSSSA). In the next stage, the authors used LSSVM-based framework as a prediction algorithm and developed an iterative multi-step short-term WPP method. Due to chaotic TS analysis and iterative multi-step algorithm, the accuracy of the prediction for both chaotic as well as non-chaotic components increases. The proposed method is evaluated on historical data from farms located in Spain and Canada.

A short-term wind power prediction method with high accuracy is presented in [68]. The hybrid prediction method combines empirical mode decomposition and kernel ridge regression (KRR). Mutual effects in different components of time series data are isolated using EMD. They further combined RVFL and ELM with EMD and comparative analysis of EMD-KRR, EMD-RVFL and EMD-ELM is also presented. To reduce computation complexity and improve the training time, an improved version of EMD-KRR is also presented in [68]. The proposed algorithm is evaluated on four different prediction horizons, i.e., 10 min, 30 min, 1 h and 3 h ahead and comparable improvement in accuracy and computation time is reported. In order to avoid limitations of EMD (mentioned in an earlier section), Ref. [48] combined VMD with multi-kernel ridge regression (MKRR) instead of EMD. Improvement in the performance is reported over its EMD counterpart.

Wavelet decomposition is widely used to decompose a signal into different frequency bands. Use of the wavelet kernels as an activation function of the CNN is recently trending in wind power prediction algorithms. In [69], wavelet kernel is used in LSTM and achieved 30% improvement in performance compared to existing wind power prediction methods. Gaussian, Morelet, Ricker and Shannon are four different wavelets that are used as activation functions. LSTM composed of four layers is trained using Rmsprop optimizer for wind power prediction. Data from seven farms in the European region are used for the evaluation of the work, results of four different wavenets (wavelet + LSTM network) on the data from seven wind farms are reported in terms of MSE, MAE, MAER, MAPE, and R2. The lowest prediction errors are observed in the case of all four networks. .

Depending on the weather conditions, the wind speed varies and hence wind power generated. Based on the wind speed there exists different wind grades such as breeze, cool breeze, strong wind etc. Fuzzy k-means clustering is applied in [70] to classify the historic time series data into these wind classes. Each class corresponds to different speed hence the wind power data corresponding to each class and amount of power generated by each class will be different. Therefore instead of learning single function that can fit all these classes, separate SVR is trained for each class. The optimization of various SVR models is performed using enhanced harmony search (EHS) algorithm. The authors also presented the uncertainty analysis in terms of confidence interval using EHS-based QR approach. The proposed multiple SVR-based method provides 3 h-ahead 15 min wind power forecasts.

Wind power series is characterised by long memory characteristics and strong unpredictability. The forecasting method should be able to capture both the characteristics. Hybrid approach in [71] combines autoregressive fractionally integrated moving average ARFIMA to capture long memory characteristic and LSSVR to capture nonlinearities in the data. Such integration of linear and nonlinear component for wind power forecast improves the performance. Experimental analysis and comparison with ARFIMA, LSSVM and hybrid ARFIMA-BP demonstrates the superiority of the ARFIMA-LSSVM model. Combination of time series and ML method are computationally efficient solution for short-tern wind power prediction.

Two optimization algorithms are employed in [72] to optimize ANFIS [73] for short-term wind power prediction. The initial parameters of ANFIS are randomly initialized and fuzzy c-means (FCM) clustering is used to generate fuzzy inference structure (FIS). Two optimizers GA and the PSO run simultaneously and independently and optimal model parameters are selected based on RMSE. The GA–PSO hybrid algorithm performs better than BPNN, GA-BPNN, and NF-based forecast models. Hybrid approach composed of DWT, seasonal autoregressive integrated moving average (SARIMA), and LSTM is proposed in [74]. First, the input data are cleaned using data pre-processing methods such as isolation forest, re-sampling, and interpolation. Next, the pair of DWT and IDWT is applied on the cleaned data to decompose data into different components and to remove noise. Next the approximation and detail components are analysed by SARIMA model. SARIMA being sensitive to seasonal components is more suitalbe than ARIMA for non-stationary datasets to improve the prediction accuracy. Finally each decomposed band is processed through LSTM for power prediction. Combined effect of DWT, SARIMA and LSTM has shown drastic improvement in the prediction accuracy.

In the case of ML and DL models, it has been observed that accuracy largely depends on hyperparameters and therefore optimization methods play an important role. In [75], training an LSTM novel optimization method is proposed. The hybrid approach in [75], ARIMA and LSTM are combined for short-term WPP. After data preprocessing and assessing stationarity, three different optimization approaches are applied for WPP. Grid search is applied to find optimum hyperparameters of ARIMA, LSTM. Along with grid search Optuna optimizer is proposed to accelerate the process of hyperparameter search. The integration of preprocessing, outlier removal, imputation, resampling and optimizer along with ARIMA and LSTM has resulted into significant improvements in results. Hybrid methods for short-term WPP are listed in Table 8.

Table 8. Hybrid methods for short-term WPP.

Sr. No.	Articles	Models	Evaluation Metric	Evaluation Value
1	Liang et al., 2016 [76]	SVM + ELM	NMAE, NRMSE	0.0417, 0.0621
2	Cornejo et al., 2017 [63]	Hybrid ML Methods	RMSE, MAE	5.3066, 3.9519
3	Huang et al., 2017 [70]	SVM + k-means clustering	RMSE, MRE	57.0628, 2.0112%
4	Yuan et al., 2017 [71]	Hybrid AR + LSSVM	RMSE, MAPE, MAE	114.80, 8.33%, 85.56
5	Safari et al., 2017 [14]	EEMD + SSA + LSSVM	NRMSE NMAE	5.9671 3.4262
6	Naik et al., 2018 [68]	EMD + KRR	RMSE, MAPE	1.0674, 7.68%
7	Zheng et al., 2018 [72]	GA + PSO+ ANFIS	MAE, NMAE, MAPE	45.73, 1.83, 6.64
8	Liu et al., 2019 [77]	LSTM + wavelet	MAE, MAPE, RMSE	10.12, 3.01%, 14.22
9	Son et al., 2019 [78]	LSTM + ANN	RMSE, MAPE	3.67, 5.04%
10	Mishra et al., 2019 [79]	ANN + RBF	MAPE, RMSE	8.1043, 0.8394
11	Zhang et al., 2019 [80]	CEEMD + LLE + ELM	NRMSE, NMAE, R	7.372, 6.124 91.958
12	Shahid et al., 2020 [69]	LSTM + Wavelet	MSE, MAE MAER, MAPE, R^2	0.0089, 0.0644 0.0416, 0.4512, 0.9221
13	Shahid et al., 2020 [69]	GP + ANNs + Axp-GPNN	RMSE, SDE, MAE	0.0845, 0.0580, 0.0841
14	Zhao et al., 2020 [81]	SSA+ Temporal CNN	RMSE, SMAPE, R^2	188.79, 23.07, 0.9804
15	Yan et al., 2020 [82]	EWT + KLEM + GRUNN	RMSE, MAE, GRA	33.75, 27.44 0.95
16	Liu et al., 2020 [83]	Wavelet + LSTM	MAE, MAPE, RMSE	49.896, 5.831, 63.991

Table 8. *Cont.*

Sr. No.	Articles	Models	Evaluation Metric	Evaluation Value
17	Han et al., 2020 [84]	wavelet WPD + VMD + SSA	NMAE, NRMSE, MAPE, R^2	0.66, 0.84, 2.3, 0.99
18	Duan et al., 2021 [85]	VMD + LSTM neural network	MAE,RMSE,TIC	41.10, 58.77, 0.0047
19	Shahid et al., 2021 [62]	LSTM + GA	MSE, MAE, RMSE, EV, R2Score	0.00924, 0.87413, 0.09615 0.07271, 0.87656
20	Huang et al., 2022 [86]	VMD + BiLSTM–CNN–WGAN GP	MAE, MAPE, RMSE	0.28, 1.26%, 0.33
21	Zhang et al., 2022 [74]	DWT + SARIMA + LSTM	Prediction Accuracy	0.99
22	Hanifi et al., 2022 [75]	ARIMA + LSTM	RMSE	484.3

5. Ultra Short-Term Wind Power Prediction

5.1. Machine Learning

A multi-linear regression algorithm is presented in [87] for ultra short-term WPP. Initially, the dimensionality of the data is reduced and only relevant parameters from the NWP data are selected. Next, phase space reconstruction is performed using a covariance matrix and eigen values. Further, state variables of the regressive model are extracted from the proposed phase space. Finally the multivariate regression model provides the predicted wind power. Performance of this approach is compared ARIMA, BPNN, LSSVE and single-variable phase space reconstruction and proposed model found more accurate and fast.

Ultra short-term (10 min) wind power prediction is presented in [88] using ELM wherein the weights are optimized using the Salp Swarm Algorithm. The input dataset consists of wind speed, wind direction, temperature and other climate factors. The ELM has single hidden layer, the weights and bias of this network are first optimized by SSA using historic wind data. SSA helps to avoid overfitting and improves generalization ability of ELM. This method is compared with other variants of ELM and found better in terms of accuracy. However, the performance of this method degrades in the presence of outliers.

An efficient yet low-complexity algorithm based on k-nearest neighbour classifier is proposed in [89] for very short-term wind power prediction. The proposed method utilizes the power of information that lies in different parameters of meteorological data. Instead of using highly complex ML method or an ensemble of them, in [89], a simple but efficient KNN classifier is trained using multidimensional data. The authors selected wind speed, wind power, wind direction, air temperature and barometric pressure time series as input data. The combined and individual influence of each of these parameters and different distance measures on the prediction accuracy is also analysed. Through this analysis, wind speed and barometric pressure are found to be most influential parameters for WPP whereas, wind direction and air pressure are decided as ineffective for WPP. Although this method is simple and effective its performance is not compared with existing methods.

5.2. Deep Learning

The prediction horizon of ultra short-term wind power prediction ranges from a few minutes toa few hours. Therefore, the prediction algorithms need to capture spatial as well as temporal variations in the data. Existing deep learning methods captures nonlinear relation between the input parameters and predicted power using spatial features. For accurate ultra short-term wind power prediction, Ref. [90] proposed combination of spatio-temporal correlation model (STCM) and LSTM. CNN is used for spatial feature extraction and LSTM extracts the temporal relation between input and output. Performance of the combined model is compared with individual CNN and LSTM and better results are reported in terms of MAE, MAPE, RMSE and NRMSE. Ref. [91] also explored the spatio-temporal relationship for ultra short-term WPP. Wherein, attention mechanism that automatically calculates the contribution of input in the output is used for feature selection. In general, the convolutional network carries only spatial information; in [91], a temporal

CNN is introduced for spatial as well as temporal feature extraction. The performance of TCN for ultra short-term WPP is improved by incorporating a self-attention mechanism in the TCN. The proposed SATCN extracts temporal features that improve the performance of the LSTM connected to it. Performance evaluation of the combined TCN-LSTM system is carried out using meteorological data and wind power data of full year. Combined feature extraction and prediction scheme shows better results than other methods. For ultra short-term WPP prediction—spatial as well as temporal—both the trends are important. Therefore, for this category of WPP, the combined effect of spatio-temporal features and a learning algorithm seems to be a promising future dimension.

5.3. Hybrid Methods

An improved EMD (IEMD) is proposed in [92] to overcome shortcomings of EMD through analysis and improvement of sifting process of EMD. IEMD decomposes non-stationary data into stationary components. Depending on the fluctuations in the data, a series of intrinsic mode functions (IMFs) is obtained from the IEMD. Large fluctuations degrade the prediction accuracy while a moderate one improves it [13]. That means depending on the available fluctuations in the data one can change the prediction model, i.e., for moderate variations linear prediction can provide required accuracy and for large fluctuations we need complex models. The authors used ANN for high frequency and separate SVM for mid frequency, low frequency and and trend item. Validation is performed using two different datasets and results are compared with only ANN and EMD.

A hybrid method combining k-means clustering and an adaptive neuro-fuzzy inference system (ANFIS) is proposed in [73] for ultra short-term wind power prediction. In this approach, phase space variables are first obtained from PSR; next, optimal input variables are selected using a feature selection method. Selected input variables are categorised into different subsets using k-means clustering and ANFIS is trained using these clusters. Parameters of the ANFIS are optimized using PSO.

In [93], a hybrid approach combining LSTM, wavelet transform and PCA was utilized to forecast ultra-short-term wind power. Initially, for signal decomposition and feature extraction, wavelet transform and PCA are applied on the time series. Further these features are applied as input to the LSTM network. Next the authors used normal condition distribution to find the prediction error of the wind power.

A deep learning based hybrid approach is proposed in [94] for ultra short-term (5 min) wind power prediction. Feature extraction is performed using CNN and the extracted features are used to train gated recurrent units. Long term trends in the data across the time steps are captured using GRU. Next, a fully connected NN is used to forecast wind power generation. Comparison with existing advanced prediction methods such as RNN, LSTM, Bi-LSTM, GRU, ARIMA and SVM is presented to show the effectiveness of the proposed hybrid scheme. The authors presented a fair comparison by separately tuning parameters of all the compared methods to their best setting. The results show that performance of proposed scheme is close to ARIMA and SVM in terms of MAE, RMSE and MAPE. Similar approach using GRU, CNN and LSTM is proposed in [95] for ultra short-term (5 min) wind power prediction. Authors used CNN for feature extraction, GRU to learn long-term variations and LSTM for prediction. However, parameter tuning has been carried out using Harris Hawks Optimization algorithm [96]. This combined approach outperformed all the compared method with large gap in terms of MAPE. Table 9 shows ultra short-term WPP methods reviewed in this article.

Table 9. Ultra short-term wind power prediction.

Sr. No.	Articles	Models	Evaluation Metric	Evaluation Value	Prediction Horizon
1	Yesilbudak et al., 2017 [89]	KNN classifier	e_{NRMSE}, e_{NMAE} e_{NMAPE}	2.475%, 15.839 1.158%	10 min
2	Liu et al., 2018 [87]	PSR and MLR	NMAE	5%	less than 4 h
			NRMSE	9.21%	
3	Yang et al., 2018 [92]	IEMD + ANN + SVM	DMAP	93.82%	4 h
			DMQP	91.91%	
4	Dong et al., 2018 [73]	mRMR + IVS	NRMSE, NMAE	6.67, 4.10	1 h to 4 h
			MAE	0.09	
5	Tan et al., 2020 [88]	SSA_ELM	MAPE	0.47	15 min to 4 h
			RMSE	0.16	
6	Sun et al., 2020 [93]	LSTM + WT	MAPE, RMSE	1.54%, 3.66	2 h
7	Hossain et al., 2021 [94]	LSTM + GRUNN	MAE, RMSE, MAPE	2.45, 3.85, 9.80%	5 min
8	Hossain et al., 2021 [95]	CNN + GRU + LSTM	MAPE	17.9	5 min and 10 min

6. Discussion

In this paper, we have presented a selective review of state-of-the-art wind power prediction methods. We do not aim to compare different methods and reported results rather, we highlight recent developments and benchmarks in this field. We presented three different classes of WPP based on prediction horizon and for each class, detail discussions on prediction methodology or algorithms are also presented.

From the presented review it has been observed that, in recent times, relatively few publications report on time series methods for WPP. Time series models are not competent enough to capture high degree of nonlinearity and stochastic behaviour of wind. Higher order polynomials can model nonlinear behaviour, but complexity increases with the degree of the polynomial and finding global minima not guaranteed.

Machine learning methods are suitable for all the prediction horizon. Variants of BPNN and ELM are proposed with different optimization methods. It is noted that, same network with different optimizers produce different results, since datasets are not same. Variants of Gaussian process regression and ELM with advanced optimizers shown improved results. Along with the accuracies, GP based approaches provide the confidence interval of the results. Being a nonparametric method, GP-based approaches do not require cross validation. Instead of using a single ML method, ensemble learning methods are widely used for WPP in different horizons. Through ensemble learning different base learner effectively models the non-stationary behaviour of wind. Ensemble of different learning algorithms have shown improvement in the accuracies.

With increased computing power and ability to model complex nonlinear functions deep learning models can provide accurate predictions than shallow machine learning methods. The deep learning models extract optimal features as well as learn a regression function. Each of these models perform well individually, combination of time series and ML and deep learning models substantially improves the results.

Rightly pointed out in [8], in recent years, a substantial increase in hybrid approaches has been observed. It is effectively applicable to all prediction horizons. Variants of EMD are combined with different ML, LSTM and deep learning models to improve the prediction accuracy. Decomposition power of wavelets are incorporated in deep learning by using wavelet as an activation function and substantial improvement is observed. Combination of LSTM, CNN and decomposition methods drastically improves accuracy of short-term and ultra short-term WPP. However, these methods become computationally very heavy. It is noted that appropriate feature selection, data cleaning and optimizers and network selection are key to improving the accuracy of WPP.

It has been observed that the selection of an optimizer is a very crucial step in the case of ML and DL models. Recently developed hybrid approaches make use of LSTM and CNN along with different decomposition methods; however, the combination of decomposition methods and deep learning models does not perform well unless hyperparameters are properly tuned. In our opinion, with sufficiently large development in the deep learning area, efficient optimization algorithms are needed, and researchers need to focus on this aspect as well.

Data preprocessing is an important step in WPP; it can be seen that, in [74,75], data preprocessing such as outlier removal, anomaly detection and removal, resampling and interpolation substantially improved the performance of the algorithm. Therefore, along with the algorithmic development data, preprocessing is also an important factor. The R package for data cleaning and preprocessing is presented in [97]. Researchers can used such tools for data preprocessing and also to analyze data at various scales and resolutions to find relevant features.

7. Conclusions

This paper presents a selective review of wind power forecasting methods. In this paper, WPP methods are classified based on the prediction horizon and for each category we investigated time series, machine learning, deep learning and hybrid approaches for WPP. Among these four categories, recent developments are skewed towards hybrid methods. This paper focuses on a comparison of existing state-of-the-art methods based on pre-processing, feature extraction, algorithm and performance. Compared to long term approaches, due to the high requirement for stable dispatching of the power grid, short-term forecasting methods are gaining more attention. A combination of feature extraction, time series decomposition and learning algorithms improves the forecasting accuracy. Investigations in this paper favour the hybrid methods, which show high performance for all three prediction horizons. It is noted that there is a large variation in databases, related NWP data and performance measures; therefore common datasets and parameters are needed for bench-marking. The discussions in this paper provide guidelines about current achievements and future requirements.

Author Contributions: Conceptualization, M.S. and R.P.; methodology, M.S., T.S., S.N. (Shreyas Nagle), S.C. and S.N. (Shivang Negi); software, M.S., T.S., S.N. (Shreyas Nagle), S.C. and S.N. (Shivang Negi); validation, M.S., R.P. and N.D.B.; formal analysis, M.S., T.S., S.N. (Shreyas Nagle), S.C. and S.N. (Shivang Negi); investigation, M.S., T.S., S.N. (Shreyas Nagle), S.C. and S.N. (Shivang Negi); resources, M.S., R.P. and N.D.B.; data curation, M.S., T.S., S.N. (Shreyas Nagle), S.C. and S.N. (Shivang Negi); writing—original draft preparation, M.S., R.P., T.S., S.N. (Shreyas Nagle), S.C., S.N. (Shivang Negi) and N.D.B.; writing—review and editing, M.S., R.P., T.S., S.N. (Shreyas Nagle), S.C., S.N. (Shivang Negi) and N.D.B.; visualization, M.S. and N.D.B.; supervision, M.S., R.P. and N.D.B.; project administration, M.S., R.P. and N.D.B.; funding acquisition, N.D.B. All authors have read and agreed to the published version of the manuscript.

Funding: This research received no external funding.

Data Availability Statement: Not applicable.

Conflicts of Interest: The authors declare no conflict of interest.

Abbreviations

The following abbreviations are used in this manuscript:

ANFIS	Adaptive Neuro-Fuzzy Inference System
ANN	Artificial Neural Network
AR	Autoregressive
ARX	Autoregressive with exogenous variable
ARIMA	Autoregressive Integrated Moving Average
ARMA	Autoregressive Moving Average
AWNN	Adaptive Wavelet Neural Network
BP	Back Propagation

BPNN	Back-Propagation Neural Network
CNN	Convolutional Neural Network
DBN	Deep Belief Network
ELM	Extreme Learning Machine
ENN	Elman Neural Network
EVS	Explained Variance Score
FFNN	Feed Forward Neural Network
GA	Genetic Algorithm
GFS	Global Forecasting System
GMM	Gaussian Mixture Model
GP	Gaussian Process
GPR	Gaussian Process Regression
IF	Isolation Forest
LSSVM	Least Square Support Vector Machine
LSTM	Long Short-Term Memory
MA	Moving Average
MAPE	Mean Absolute Percentage Error
MARE	Mean Absolute Relative Error
MDN	Mixture Density Neural Network
MLP	Multilayer Perceptron
MSE	Mean Square Error
MSLE	Mean Squared Logarithmic Error
NAAE	Normalised Absolute Average Error
NMAE	Normalised Mean Absolute Error
NMBE	Normalized Mean Bias Error
NMSE	Normalized Mean Square Error
NN	Neural Network
NRMSE	Normalised Root Mean Square Error
NWP	Numerical Weather Prediction
PSO	Particle Swarm Optimisation
R^2	R-Square
RBFNN	Radial Basis Function Neural Network
RF	Random Forest
RMSE	Root Mean Square Error
RVM	Relative Vector Machine
SDE	Standard Deviation Error
SNMAE	Square Normalized Mean Bias Error
SVM	Support Vector Machine
SVR	Support Vector Regression
WNN	Wavelet Neural Network

References

1. Zhang, Y.; Liu, K.; Qin, L.; An, X. Deterministic and probabilistic interval prediction for short-term wind power generation based on variational mode decomposition and machine learning methods. *Energy Convers. Manag.* **2016**, *112*, 208–219. [CrossRef]
2. Bokde, N.; Tranberg, B.; Andresen, G.B. A graphical approach to carbon-efficient spot market scheduling for Power-to-X applications. *Energy Convers. Manag.* **2020**, *224*, 113461. [CrossRef]
3. Tao, S.; Xu, Q.; Feijóo, A.; Kuenzel, S.; Bokde, N. Integrated wind farm power curve and power curve distribution function considering the wake effect and terrain gradient. *Energies* **2019**, *12*, 2482. [CrossRef]
4. Bokde, N.; Feijóo, A.; Villanueva, D. Wind turbine power curves based on the weibull cumulative distribution function. *Appl. Sci.* **2018**, *8*, 1757. [CrossRef]
5. Sawant, M.; Thakare, S.; Rao, A.P.; Feijóo-Lorenzo, A.E.; Bokde, N.D. A review on state-of-the-art reviews in wind-turbine-and wind-farm-related topics. *Energies* **2021**, *14*, 2041. [CrossRef]
6. Bokde, N.D.; Yaseen, Z.M.; Andersen, G.B. ForecastTB—An R package as a test-bench for time series forecasting—Application of wind speed and solar radiation modeling. *Energies* **2020**, *13*, 2578. [CrossRef]

7. Hanifi, S.; Liu, X.; Lin, Z.; Lotfian, S. A critical review of wind power forecasting methods—Past, present and future. *Energies* **2020**, *13*, 3764. [CrossRef]
8. Bokde, N.; Feijóo, A.; Villanueva, D.; Kulat, K. A review on hybrid empirical mode decomposition models for wind speed and wind power prediction. *Energies* **2019**, *12*, 254. [CrossRef]
9. Landberg, L. A mathematical look at a physical power prediction model. *Wind Energy Int. J. Prog. Appl. Wind Power Convers. Technol.* **1998**, *1*, 23–28. [CrossRef]
10. Catalao, J.; Pousinho, H.; Mendes, V. Hybrid wavelet-PSO-ANFIS approach for short-term wind power forecasting in Portugal. *IEEE Trans. Sustain. Energy* **2010**, *2*, 50–59. [CrossRef]
11. Bhaskar, K.; Singh, S.N. AWNN-assisted wind power forecasting using feed-forward neural network. *IEEE Trans. Sustain. Energy* **2012**, *3*, 306–315. [CrossRef]
12. Chen, N.; Qian, Z.; Nabney, I.T.; Meng, X. Wind power forecasts using Gaussian processes and numerical weather prediction. *IEEE Trans. Power Syst.* **2013**, *29*, 656–665. [CrossRef]
13. Sideratos, G.; Hatziargyriou, N.D. An advanced statistical method for wind power forecasting. *IEEE Trans. Power Syst.* **2007**, *22*, 258–265. [CrossRef]
14. Safari, N.; Chung, C.; Price, G. Novel multi-step short-term wind power prediction framework based on chaotic time series analysis and singular spectrum analysis. *IEEE Trans. Power Syst.* **2017**, *33*, 590–601. [CrossRef]
15. Tian, Z.; Li, S.; Wang, Y.; Wang, X. Wind power prediction method based on hybrid kernel function support vector machine. *Wind Eng.* **2018**, *42*, 252–264. [CrossRef]
16. Erdem, E.; Shi, J. ARMA based approaches for forecasting the tuple of wind speed and direction. *Appl. Energy* **2011**, *88*, 1405–1414. [CrossRef]
17. Liu, H.; Tian, H.Q.; Li, Y.F. Comparison of two new ARIMA-ANN and ARIMA-Kalman hybrid methods for wind speed prediction. *Appl. Energy* **2012**, *98*, 415–424. [CrossRef]
18. Suykens, J.A.; Vandewalle, J. Least squares support vector machine classifiers. *Neural Process. Lett.* **1999**, *9*, 293–300. [CrossRef]
19. Rasmussen, C.E. Gaussian processes in machine learning. In *Summer School on Machine Learning*; Springer: Berlin/Heidelberg, Germany, 2003; pp. 63–71.
20. Karakuş, O.; Kuruoğlu, E.E.; Altınkaya, M.A. One-day ahead wind speed/power prediction based on polynomial autoregressive model. *IET Renew. Power Gener.* **2017**, *11*, 1430–1439. [CrossRef]
21. Hong, T.; Pinson, P.; Fan, S. Global energy forecasting competition 2012. *Int. J. Forecast.* **2014**, *30*, 357–363. [CrossRef]
22. Ouyang, T.; Huang, H.; He, Y. Ramp events forecasting based on long-term wind power prediction and correction. *IET Renew. Power Gener.* **2019**, *13*, 2793–2801. [CrossRef]
23. Cui, M.; Ke, D.; Sun, Y.; Gan, D.; Zhang, J.; Hodge, B.M. Wind power ramp event forecasting using a stochastic scenario generation method. *IEEE Trans. Sustain. Energy* **2015**, *6*, 422–433. [CrossRef]
24. Lee, D.; Baldick, R. Short-term wind power ensemble prediction based on Gaussian processes and neural networks. *IEEE Trans. Smart Grid* **2013**, *5*, 501–510. [CrossRef]
25. Li, T.; Li, Y.; Liao, M.; Wang, W.; Zeng, C. A new wind power forecasting approach based on conjugated gradient neural network. *Math. Probl. Eng.* **2016**, *2016*, 8141790. [CrossRef]
26. Ouyang, T.; Zha, X.; Qin, L.; Kusiak, A. Optimisation of time window size for wind power ramps prediction. *IET Renew. Power Gener.* **2017**, *11*, 1270–1277. [CrossRef]
27. Jin, H.; Shi, L.; Chen, X.; Qian, B.; Yang, B.; Jin, H. Probabilistic wind power forecasting using selective ensemble of finite mixture Gaussian process regression models. *Renew. Energy* **2021**, *174*, 1–18. [CrossRef]
28. Fang, S.; Chiang, H.D. A high-accuracy wind power forecasting model. *IEEE Trans. Power Syst.* **2016**, *32*, 1589–1590. [CrossRef]
29. Bochenek, B.; Jurasz, J.; Jaczewski, A.; Stachura, G.; Sekuła, P.; Strzyżewski, T.; Wdowikowski, M.; Figurski, M. Day-ahead wind power forecasting in Poland based on numerical weather prediction. *Energies* **2021**, *14*, 2164. [CrossRef]
30. Tasnim, S.; Rahman, A.; Oo, A.M.T.; Haque, E. Autoencoder for wind power prediction. *Renew. Wind. Water Sol.* **2017**, *4*, 6. [CrossRef]
31. Abedinia, O.; Bagheri, M.; Naderi, M.S.; Ghadimi, N. A new combinatory approach for wind power forecasting. *IEEE Syst. J.* **2020**, *14*, 4614–4625. [CrossRef]
32. Zhang, H.; Yue, D.; Dou, C.; Li, K.; Hancke, G.P. Two-step wind power prediction approach with improved complementary ensemble empirical mode decomposition and reinforcement learning. *IEEE Syst. J.* **2021**. [CrossRef]
33. Dong, L.; Wang, L.; Khahro, S.F.; Gao, S.; Liao, X. Wind power day-ahead prediction with cluster analysis of NWP. *Renew. Sustain. Energy Rev.* **2016**, *60*, 1206–1212. [CrossRef]
34. Abedinia, O.; Lotfi, M.; Bagheri, M.; Sobhani, B.; Shafie-Khah, M.; Catalão, J.P. Improved EMD-based complex prediction model for wind power forecasting. *IEEE Trans. Sustain. Energy* **2020**, *11*, 2790–2802. [CrossRef]
35. Shi, X.; Lei, X.; Huang, Q.; Huang, S.; Ren, K.; Hu, Y. Hourly day-ahead wind power prediction using the hybrid model of variational model decomposition and long short-term memory. *Energies* **2018**, *11*, 3227. [CrossRef]
36. Heinermann, J.; Kramer, O. Machine learning ensembles for wind power prediction. *Renew. Energy* **2016**, *89*, 671–679. [CrossRef]
37. Lee, J.; Wang, W.; Harrou, F.; Sun, Y. Wind power prediction using ensemble learning-based models. *IEEE Access* **2020**, *8*, 61517–61527. [CrossRef]

38. Yan, J.; Li, K.; Bai, E.; Yang, Z.; Foley, A. Time series wind power forecasting based on variant Gaussian Process and TLBO. *Neurocomputing* **2016**, *189*, 135–144. [CrossRef]
39. Liu, T.; Wei, H.; Zhang, K. Wind power prediction with missing data using Gaussian process regression and multiple imputation. *Appl. Soft Comput.* **2018**, *71*, 905–916. [CrossRef]
40. Lahouar, A.; Slama, J.B.H. Hour-ahead wind power forecast based on random forests. *Renew. Energy* **2017**, *109*, 529–541. [CrossRef]
41. Wang, Q.; Lei, Y.; Cao, H. Wind power prediction based on nonlinear partial least square. *Math. Probl. Eng.* **2018**, *2018*, 6829274. [CrossRef]
42. Li, N.; He, F.; Ma, W.; Wang, R.; Zhang, X. Wind power prediction of kernel extreme learning machine based on differential evolution algorithm and cross validation algorithm. *IEEE Access* **2020**, *8*, 68874–68882. [CrossRef]
43. An, G.; Jiang, Z.; Cao, X.; Liang, Y.; Zhao, Y.; Li, Z.; Dong, W.; Sun, H. Short-term wind power prediction based on particle swarm optimization-extreme learning machine model combined with AdaBoost algorithm. *IEEE Access* **2021**, *9*, 94040–94052. [CrossRef]
44. An, J.; Yin, F.; Wu, M.; She, J.; Chen, X. Multisource wind speed fusion method for short-term wind power prediction. *IEEE Trans. Ind. Inf.* **2020**, *17*, 5927–5937. [CrossRef]
45. Yuan, X.; Chen, C.; Jiang, M.; Yuan, Y. Prediction interval of wind power using parameter optimized Beta distribution based LSTM model. *Appl. Soft Comput.* **2019**, *82*, 105550. [CrossRef]
46. Watts, D.J.; Strogatz, S.H. Collective dynamics of 'small-world'networks. *Nature* **1998**, *393*, 440–442. [CrossRef]
47. Zhang, Y.; Sun, H.; Guo, Y. Wind power prediction based on PSO-SVR and grey combination model. *IEEE Access* **2019**, *7*, 136254–136267. [CrossRef]
48. Xie, W.; Zhang, P.; Chen, R.; Zhou, Z. A nonparametric Bayesian framework for short-term wind power probabilistic forecast. *IEEE Trans. Power Syst.* **2018**, *34*, 371–379. [CrossRef]
49. Bokde, N.; Asencio-Cortés, G.; Martínez-Álvarez, F.; Kulat, K. PSF: Introduction to R Package for Pattern Sequence Based Forecasting Algorithm. *R J.* **2017**, *9*, 324–333. [CrossRef]
50. Shende, M.K.; Salih, S.Q.; Bokde, N.D.; Scholz, M.; Oudah, A.Y.; Yaseen, Z.M. Natural Time Series Parameters Forecasting: Validation of the Pattern-Sequence-Based Forecasting (PSF) Algorithm; A New Python Package. *Appl. Sci.* **2022**, *12*, 6194. [CrossRef]
51. Bokde, N.; Asencio-Cortes, G.; Martinez-Alvarez, F. PSF: Forecasting of Univariate Time Series Using the Pattern Sequence-Based Forecasting (PSF) Algorithm. R Package Version 0.4. 2017. Available online: https://cran.r-project.org/web/packages/PSF/PSF.pdf (accessed on 10 June 2022).
52. Bokde, N.; Troncoso, A.; Asencio-Cortés, G.; Kulat, K.; Martínez-Álvarez, F. Pattern sequence similarity based techniques for wind speed forecasting. In Proceedings of the International Work-Conference on Time Series, Granada, Spain, 18–20 September 2017; pp. 18–20.
53. Bokde, N.; Feijóo, A.; Villanueva, D.; Kulat, K. A novel and alternative approach for direct and indirect wind-power prediction methods. *Energies* **2018**, *11*, 2923. [CrossRef]
54. Ghoushchi, S.J.; Manjili, S.; Mardani, A.; Saraji, M.K. An extended new approach for forecasting short-term wind power using modified fuzzy wavelet neural network: A case study in wind power plant. *Energy* **2021**, *223*, 120052. [CrossRef]
55. Wang, S.X.; Li, M.; Zhao, L.; Jin, C. Short-term wind power prediction based on improved small-world neural network. *Neural Comput. Appl.* **2019**, *31*, 3173–3185. [CrossRef]
56. Qureshi, A.S.; Khan, A.; Zameer, A.; Usman, A. Wind power prediction using deep neural network based meta regression and transfer learning. *Appl. Soft Comput.* **2017**, *58*, 742–755. [CrossRef]
57. Wang, K.; Qi, X.; Liu, H.; Song, J. Deep belief network based k-means cluster approach for short-term wind power forecasting. *Energy* **2018**, *165*, 840–852. [CrossRef]
58. Zhang, H.; Liu, Y.; Yan, J.; Han, S.; Li, L.; Long, Q. Improved deep mixture density network for regional wind power probabilistic forecasting. *IEEE Trans. Power Syst.* **2020**, *35*, 2549–2560. [CrossRef]
59. Ding, M.; Zhou, H.; Xie, H.; Wu, M.; Nakanishi, Y.; Yokoyama, R. A gated recurrent unit neural networks based wind speed error correction model for short-term wind power forecasting. *Neurocomputing* **2019**, *365*, 54–61. [CrossRef]
60. Schmidhuber, J.; Hochreiter, S. Long short-term memory. *Neural Comput.* **1997**, *9*, 1735–1780.
61. Wang, S.; Li, B.; Li, G.; Yao, B.; Wu, J. Short-term wind power prediction based on multidimensional data cleaning and feature reconfiguration. *Appl. Energy* **2021**, *292*, 116851. [CrossRef]
62. Shahid, F.; Zameer, A.; Muneeb, M. A novel genetic LSTM model for wind power forecast. *Energy* **2021**, *223*, 120069. [CrossRef]
63. Cornejo-Bueno, L.; Cuadra, L.; Jiménez-Fernández, S.; Acevedo-Rodríguez, J.; Prieto, L.; Salcedo-Sanz, S. Wind power ramp events prediction with hybrid machine learning regression techniques and reanalysis data. *Energies* **2017**, *10*, 1784. [CrossRef]
64. Bokde, N.; Feijóo, A.; Kulat, K. Analysis of differencing and decomposition preprocessing methods for wind speed prediction. *Appl. Soft Comput.* **2018**, *71*, 926–938. [CrossRef]
65. Bokde, N.; Feijóo, A.; Al-Ansari, N.; Tao, S.; Yaseen, Z.M. The hybridization of ensemble empirical mode decomposition with forecasting models: Application of short-term wind speed and power modeling. *Energies* **2020**, *13*, 1666. [CrossRef]
66. Xie, H.B.; Guo, T.; Sivakumar, B.; Liew, A.W.C.; Dokos, S. Symplectic geometry spectrum analysis of nonlinear time series. *Proc. R. Soc. A Math. Phys. Eng. Sci.* **2014**, *470*, 20140409. [CrossRef]

67. Ghanbarzadeh, M.; Aminghafari, M. A new hybrid-multiscale SSA prediction of non-stationary time series. *Fluct. Noise Lett.* **2016**, *15*, 1650005. [CrossRef]
68. Naik, J.; Satapathy, P.; Dash, P. Short-term wind speed and wind power prediction using hybrid empirical mode decomposition and kernel ridge regression. *Appl. Soft Comput.* **2018**, *70*, 1167–1188. [CrossRef]
69. Shahid, F.; Zameer, A.; Mehmood, A.; Raja, M.A.Z. A novel wavenets long short term memory paradigm for wind power prediction. *Appl. Energy* **2020**, *269*, 115098. [CrossRef]
70. Huang, C.M.; Kuo, C.J.; Huang, Y.C. Short-term wind power forecasting and uncertainty analysis using a hybrid intelligent method. *IET Renew. Power Gener.* **2017**, *11*, 678–687. [CrossRef]
71. Yuan, X.; Tan, Q.; Lei, X.; Yuan, Y.; Wu, X. Wind power prediction using hybrid autoregressive fractionally integrated moving average and least square support vector machine. *Energy* **2017**, *129*, 122–137. [CrossRef]
72. Zheng, D.; Semero, Y.K.; Zhang, J.; Wei, D. Short-term wind power prediction in microgrids using a hybrid approach integrating genetic algorithm, particle swarm optimization, and adaptive neuro-fuzzy inference systems. *IEEJ Trans. Electr. Electron. Eng.* **2018**, *13*, 1561–1567. [CrossRef]
73. Dong, W.; Yang, Q.; Fang, X. Multi-step ahead wind power generation prediction based on hybrid machine learning techniques. *Energies* **2018**, *11*, 1975. [CrossRef]
74. Zhang, W.; Lin, Z.; Liu, X. Short-term offshore wind power forecasting-A hybrid model based on Discrete Wavelet Transform (DWT), Seasonal Autoregressive Integrated Moving Average (SARIMA), and deep-learning-based Long Short-Term Memory (LSTM). *Renew. Energy* **2022**, *185*, 611–628. [CrossRef]
75. Hanifi, S.; Lotfian, S.; Zare-Behtash, H.; Cammarano, A. Offshore Wind Power Forecasting—A New Hyperparameter Optimisation Algorithm for Deep Learning Models. *Energies* **2022**, *15*, 6919. [CrossRef]
76. Liang, Z.; Liang, J.; Wang, C.; Dong, X.; Miao, X. Short-term wind power combined forecasting based on error forecast correction. *Energy Convers. Manag.* **2016**, *119*, 215–226. [CrossRef]
77. Liu, Y.; Guan, L.; Hou, C.; Han, H.; Liu, Z.; Sun, Y.; Zheng, M. Wind power short-term prediction based on LSTM and discrete wavelet transform. *Appl. Sci.* **2019**, *9*, 1108. [CrossRef]
78. Son, N.; Yang, S.; Na, J. Hybrid forecasting model for short-term wind power prediction using modified long short-term memory. *Energies* **2019**, *12*, 3901. [CrossRef]
79. Mishra, S.; Dash, P. Short-term prediction of wind power using a hybrid pseudo-inverse Legendre neural network and adaptive firefly algorithm. *Neural Comput. Appl.* **2019**, *31*, 2243–2268. [CrossRef]
80. Zhang, C.; Ding, M.; Wang, W.; Bi, R.; Miao, L.; Yu, H.; Liu, L. An improved ELM model based on CEEMD-LZC and manifold learning for short-term wind power prediction. *IEEE Access* **2019**, *7*, 121472–121481. [CrossRef]
81. Zhao, Y.; Jia, L. A short-term hybrid wind power prediction model based on singular spectrum analysis and temporal convolutional networks. *J. Renew. Sustain. Energy* **2020**, *12*, 056101. [CrossRef]
82. Yan, H.; Wu, Z. A hybrid short-term wind power prediction model combining data processing, multiple parameters optimization and multi-intelligent models apportion strategy. *IEEE Access* **2020**, *8*, 227126–227140. [CrossRef]
83. Liu, B.; Zhao, S.; Yu, X.; Zhang, L.; Wang, Q. A novel deep learning approach for wind power forecasting based on WD-LSTM model. *Energies* **2020**, *13*, 4964. [CrossRef]
84. Han, Y.; Tong, X. Multi-step short-term wind power prediction based on three-level decomposition and improved grey wolf optimization. *IEEE Access* **2020**, *8*, 67124–67136. [CrossRef]
85. Duan, J.; Wang, P.; Ma, W.; Tian, X.; Fang, S.; Cheng, Y.; Chang, Y.; Liu, H. Short-term wind power forecasting using the hybrid model of improved variational mode decomposition and Correntropy Long Short-term memory neural network. *Energy* **2021**, *214*, 118980. [CrossRef]
86. Huang, L.; Li, L.; Wei, X.; Zhang, D. Short-term prediction of wind power based on BiLSTM–CNN–WGAN-GP. *Soft Comput.* **2022**, *26*, 10607–10621. [CrossRef]
87. Liu, R.; Peng, M.; Xiao, X. Ultra-short-term wind power prediction based on multivariate phase space reconstruction and multivariate linear regression. *Energies* **2018**, *11*, 2763. [CrossRef]
88. Tan, L.; Han, J.; Zhang, H. Ultra-short-term wind power prediction by salp swarm algorithm-based optimizing extreme learning machine. *IEEE Access* **2020**, *8*, 44470–44484. [CrossRef]
89. Yesilbudak, M.; Sagiroglu, S.; Colak, I. A novel implementation of kNN classifier based on multi-tupled meteorological input data for wind power prediction. *Energy Convers. Manag.* **2017**, *135*, 434–444. [CrossRef]
90. Wu, Q.; Guan, F.; Lv, C.; Huang, Y. Ultra-short-term multi-step wind power forecasting based on CNN-LSTM. *IET Renew. Power Gener.* **2021**, *15*, 1019–1029. [CrossRef]
91. Xiang, L.; Liu, J.; Yang, X.; Hu, A.; Su, H. Ultra-short term wind power prediction applying a novel model named SATCN-LSTM. *Energy Convers. Manag.* **2022**, *252*, 115036. [CrossRef]
92. Yang, M.; Chen, X.; Du, J.; Cui, Y. Ultra-short-term multistep wind power prediction based on improved EMD and reconstruction method using run-length analysis. *IEEE Access* **2018**, *6*, 31908–31917. [CrossRef]
93. Sun, Y.; Wang, P.; Zhai, S.; Hou, D.; Wang, S.; Zhou, Y. Ultra short-term probability prediction of wind power based on LSTM network and condition normal distribution. *Wind Energy* **2020**, *23*, 63–76. [CrossRef]
94. Hossain, M.A.; Chakrabortty, R.K.; Elsawah, S.; Ryan, M.J. Very short-term forecasting of wind power generation using hybrid deep learning model. *J. Clean. Prod.* **2021**, *296*, 126564. [CrossRef]

95. Hossain, M.A.; Chakrabortty, R.K.; Elsawah, S.; Gray, E.M.; Ryan, M.J. Predicting wind power generation using hybrid deep learning with optimization. *IEEE Trans. Appl. Supercond.* **2021**, *31*, 1–5. [CrossRef]
96. Heidari, A.A.; Mirjalili, S.; Faris, H.; Aljarah, I.; Mafarja, M.; Chen, H. Harris hawks optimization: Algorithm and applications. *Future Gener. Comput. Syst.* **2019**, *97*, 849–872. [CrossRef]
97. Shende, M.K.; Feijoo-Lorenzo, A.E.; Bokde, N.D. cleanTS: Automated (AutoML) Tool to Clean Univariate Time Series at Microscales. *Neurocomputing* **2022**, *500*, 155–176. [CrossRef]

energies

MDPI

Review

Evaluation Metrics for Wind Power Forecasts: A Comprehensive Review and Statistical Analysis of Errors

Paweł Piotrowski [1,*], Inajara Rutyna [1,2], Dariusz Baczyński [1] and Marcin Kopyt [1]

1 Electrical Power Engineering Institute, Warsaw University of Technology, Koszykowa 75 Street, 00-662 Warsaw, Poland
2 Ideas NCBR Sp. z o.o, Nowogrodzka Street 47A, 00-695 Warsaw, Poland
* Correspondence: pawel.piotrowski@pw.edu.pl; Tel.: +48-22-234-7255

Abstract: Power generation forecasts for wind farms, especially with a short-term horizon, have been extensively researched due to the growing share of wind farms in total power generation. Detailed forecasts are necessary for the optimization of power systems of various sizes. This review and analytical paper is largely focused on a statistical analysis of forecasting errors based on more than one hundred papers on wind generation forecasts. Factors affecting the magnitude of forecasting errors are presented and discussed. Normalized root mean squared error (nRMSE) and normalized mean absolute error (nMAE) have been selected as the main error metrics considered here. A new and unique error dispersion factor (EDF) is proposed, being the ratio of nRMSE to nMAE. The variability of EDF depending on selected factors (size of wind farm, forecasting horizons, and class of forecasting method) has been examined. This is unique and original research, a novelty in studies on errors of power generation forecasts in wind farms. In addition, extensive quantitative and qualitative analyses have been conducted to assess the magnitude of forecasting error depending on selected factors (such as forecasting horizon, wind farm size, and a class of the forecasting method). Based on these analyses and a review of more than one hundred papers, a unique set of recommendations on the preferred content of papers addressing wind farm generation forecasts has been developed. These recommendations would make it possible to conduct very precise benchmarking meta-analyses of forecasting studies described in research papers and to develop valuable general conclusions concerning the analyzed phenomena.

Keywords: forecasting error; evaluation criteria metrics; wind power forecasting; wind turbine; wind farm; statistical analysis of errors; hybrid methods; ensemble methods; machine learning; deep neural network

Citation: Piotrowski, P.; Rutyna, I.; Baczyński, D.; Kopyt, M. Evaluation Metrics for Wind Power Forecasts: A Comprehensive Review and Statistical Analysis of Errors. *Energies* **2023**, *15*, 9657. https://doi.org/10.3390/en15249657

Academic Editor: Frede Blaabjerg

Received: 16 November 2022
Accepted: 16 December 2022
Published: 19 December 2022

1. Introduction

The forecasting of power generation in wind farms has been an extensively explored research topic [1–8]. The growing significance of renewable energy sources (RES) and the remarkably dynamic growth of wind farms in most countries has highlighted the importance of accurate power generation forecasts due to, e.g., increased wind farm contribution to the overall power system. Cost-efficient and optimized management of a power system requires RES generation forecasts of the best possible accuracy. System operation optimization processes include scheduling the operation of fossil-based sources, scheduling maintenance works in the power grid, and preventive and remedial maintenance of RES themselves. In addition to obtaining forecasts with the best accuracy, the estimation of errors in these forecasts also proves to be important, as it translates into maintaining appropriate safety margins.

Forecasting purposes vary by time horizon [4,5,7]. Time horizon, also called planning horizon, is a fixed point in the future at which a certain process will be evaluated or assumed to have ended. In wind energy forecasting, time horizon affects the choice of forecasting techniques, as they are classified into four types: very short-term (seconds), short-term (minutes to hours), medium-term (months), and long-term forecasts (months/years) [9]. Based on this classification, each approach is defined in Table 1.

Table 1. Forecasting horizons.

Horizon	Applications	Duration
Very short-term	Turbine regulation, control strategies, electricity market billing, real-time grid operation, network stability, voltage control	10 s
Short-term	Economic load dispatch planning, load amount, reversibility of power management, electric market reliability in day-ahead, load decisions for increments	5 min, 10 min, 15 min, 1 h, 24 h
Medium-term	Unit commitment, reserved requirement, generation operator, operation scheduling	1 month
Long-term	Maintenance scheduling, wind farm design, electricity market restructuring, optimization of operating costs	1 month, 1 year

Each time horizon tends to use different types of data. Very short-term forecasts are most often based on time series data. Short-term forecasts very often use online measurement data from a meteorological station, NWP, or combination of both as input data, with the expectation that weather conditions will remain the same in a short time.

1.1. Major Factors Affecting Wind Power Forecasts

Magnitudes of forecasting errors for wind power generation can vary widely. The quality of power generation forecasts is affected by a large number of independent factors based on quantities or classes. The final forecasting error can be seen as the output of a non-linear function that uses all the factors presented in Table 2 as inputs. Obviously it is impossible to create a formal formula for such a function, although it is possible to verify how selected factors affect the magnitude of error. In some cases, it is possible to determine the functional relationship between the error and the factor, but, unfortunately, it requires a very large pool of research samples. In Table 2 modifiable and fixed factors are described. When the wind farm is already active and generating energy, the following factors are fixed: site, landscape, and size of the system.

Table 2. Description of major factors affecting wind power forecasts.

Factor	Description/Influence on Error Level
Time horizon	With increasing forecasting horizon, forecasting errors grow significantly, mainly due to the falling quality of NWP forecasts [5,6,8,10,11]
Forecasting method (complexity)	Complex (ensemble or hybrid) models have typically lower forecasting errors than single methods (more details in Section 2) [1–5,8]
Size of system	The inertia of power production usually grows with system size. This translates into more predictable production, especially in shorter horizons.
Site (onshore/offshore)	Forecasting errors for offshore wind farms should be less than for onshore wind farms due to distinctive characteristics of weather conditions (more stable and higher wind speeds).
Landscape	Forecasting errors for farms located in rough terrain can depend on the local landscape and meteorological features (e.g., a forest, hills, or a lake in a direct neighborhood). The best if the terrain has as little roughness as possible—this guarantees the optimal generation of wind energy [12]
Location of NWP forecasting points	NWP forecasts from points more distant from the wind farm can generate larger forecasting errors. The topic of optimum selection of NWP forecasting points and their location on the farm is subject to studies [12]

Table 2. *Cont.*

Factor	Description/Influence on Error Level
Types and quantities of input data	The more information related to power generation can be used in the model, the more accurate generation forecasts can be expected. In particular, using NWP as input data is especially important (if absent, generation forecast errors grow very sharply), and with horizons of more than several hours, NWPs are virtually indispensable [1,2,4,13]
NWP data sources	NWP data can vary by quality (forecasting accuracy)—the more accurate NWP forecasts, the lower power generation forecasting error [2]
Quantity of training sets	The model must use training data encompassing at least one year. With a growing number of years, the model uses more information and better represents the seasonality and daily variability of the process (a smaller forecasting error can be expected)
Data preprocessing	Properly conducted steps to clean up and process raw input data can reduce forecasting error [3]
Data postprocessing	Elimination of impossible situations, e.g., negative power forecasts or forecasts determined as unlikely for specific input data. Elimination of such cases reduces the error [13]
Measurement data availability lag	Using up-to-date, current data provides for identification and correction of errors "on the fly," e.g., using switchable forecasting models

More attention should be paid to forecast model inputs due to the fact that, whereas some factors, such as location, size of the system, or forecasting horizon, cannot be changed, the biggest reduction in error can be achieved by appropriate selection of input data or such selection that encompasses as much information as possible related to the forecast power generation time series.

Regarding the selection of input data itself, both statistical analysis and a semi-machine approach can be applied. Statistical analysis using various tests can help to draw conclusions on data interdependencies, however, the time required for this makes it impractical for big datasets. If larger quantities of data are available, expert selection of a pool of input data combinations and their subsequent review can prove to be a more practical approach. The choice of solution depends significantly on the tool—the time required for statistical analysis can bring more benefits for tools that usually require greater optimization, e.g., ANN models.

Forecasting models require input data to predict wind power generation. The data format used by forecasting models need to be relevant to the model itself, i.e., it must consider which external phenomena have a direct impact on wind generation. This data can be divided into NWP and time series [14].

NWP is a multivariate dataset based on a set of physical models used to simulate conditions in the atmosphere; these models are available both on local and global scales. NWP dataset contains information generated by power metering and prediction of several meteorological variables, such as wind speed, wind direction, temperature, humidity, air pressure, time of the day, day of the year, etc. [15,16]. As NWP is a general dataset, the main factor that affects wind-generated power is wind speed [17,18].

The time series is a univariate dataset of a wind speed or wind power that is measured at timestamps over a certain period. To obtain a wind speed time series, a mast is usually installed at the wind farm, with an anemometer mounted at the hub height.

Decomposition methods often used during the forecasting process are based on the premise that the wind power time series contain different frequency signals with different characteristics, and that modeling each of the decomposed series separately can lead to an overall improvement in forecast quality [8]. Popular techniques are discrete wavelet transform (DWT), empirical mode decomposition (EMD) [3,8], ensemble empirical mode decomposition (EEMD) [6], variational mode decomposition (VMD) [4,8], or wavelet packet transform (WPT).

Machine learning models can use NWP data as features, and/or time series as inputs, where NWPs are features with information related to the expected output wind speed. Input matrix X contains the following: historical wind inputs, weather information, and time data. Output vector y contains the time series and multiple values of predicted wind power with a changing prediction horizon [19].

When NWP data is not available to be used as input data, common practice is to use autoregression of the output variable. In some works, measured values of other variables from the recent past are also used—e.g., measured instantaneous rotor speed or measured weather parameters for a few hours in the past [20,21]. This allows us to take into account recent generation trends and a farm's short-term generation inertia. A practical drawback of this appeal would be that it cannot be used for forecasts with horizons longer than a few hours ahead. Other possible approaches would be using measures from one farm to forecast another one [22]. In this case, the final result would depend on the weather similarity between used and destination farms and tools used to translate one generation to another.

When only the wind speed time series is known, a technique called feature engineering can be used to fabricate new features. This technique's goal is to fabricate features by executing simple calculations based on the known feature, the wind speed time series. These calculations are described by standard deviation, average, minimum, or maximum wind speed for a period of time [20,23].

In practice, the choice of the source and types of NWP forecasts strongly depends on the *data cost-to-quality* ratio, nevertheless, it is worthwhile to maximize both the number of NWP models with various densities of forecasting points and the number of weather parameters derived from them [12]. In some cases, the application of various models, or even bundles of models, is recommended due to the diverse information content of different models. For instance, for long-term forecasts, NWP climatic models are different from each other, and identification of the best one can be not only difficult, but virtually impossible, and drawing any conclusion can require aggregation analysis of various scenarios. On the macro scale, equally important is to select a proper forecasting point, from which meteorological variables would be derived. Hence, the growing trend of extraction of spatial information using various tools, e.g., CNN, was applied by some research papers [24–26].

1.2. Objective and Contribution

The main objectives of this paper can be summarized as follows:

- classify wind power forecasting techniques;
- provide unique description of major factors affecting wind power forecasts;
- describe the performance of forecasting models;
- conduct comprehensive review (quantitative analysis) based on more than one hundred papers;
- conduct statistical analysis of errors (qualitative analysis).

Below are listed selected contributions of this paper:

- proposal for a novel, unique ratio, called EDF;
- analysis of variability of the new EDF ratio depending on selected characteristics (size of wind farm, forecasting horizon, and class of forecasting method) and the original, novel conclusions drawn from the analysis;
- development of a unique list of recommended content of papers addressing wind farm generation forecasts (the application of these recommendations would make it possible to conduct very accurate meta-analyses that would compare various forecasting studies).

The remainder of this paper is organized as follows: Section 2 presents the classification of wind power forecasting techniques. Section 3 describes the performance of the forecasting model. Section 4 is the main part of the paper and includes comprehensive statistical analysis (quantitative and qualitative analysis). Discussion is provided in Sections 5 and 6 draws the main conclusions. References are listed at the close of this paper.

2. Classification of Wind Power Forecasting Techniques

The following alternative methodologies are applied to wind power forecasting: Naive, Physical, Statistical, and AI/ML methods (see Table 3).

Table 3. Classification of forecasting techniques.

Technique	Description	Advantage	Disadvantage
Naive/Persistence Methods	Simple approach that evaluates the last period t, benchmark for more complex methods	Fast results, applicable only for short-term forecasts	Does not consider correlations between input data and the results are not reliable for the subsequent steps in time
Physical Methods	Physical mathematical model of the power turbine	No need for training or historical data, enables understanding of physical behavior	Computationally complex, it relies on fresh data from several features, extensive setup effort
Statistical/Multivariate Methods (time series)	Mapping of relations between features, generally through recursive techniques, no NWP data as an input	Models can be built easily and with small computational effort	Quick loss of accuracy with time
AI/ML methods (LR, GBT, RF, BPNN, KNNR, LSTM, CNN, DBN, GRU, EDCNN, RBF, SVR)	Techniques that enable the computer to mimic human behavior without a predefined mathematical model ("black box")	Highest accuracy and can learn more complex nonlinear relations	Slower convergence speed, risk of overfitting, computational complexity

Statistical models have high precision in very short-term prediction [27]. The most used statistical model for wind forecasting is the times series model, due to the fact that future levels of wind power depend on weather features, but they also can depend on the prior value of wind power generated. The amount of wind power produced in the current hour affects the amount of wind power generation in the next hour. These models can determine conditions in time based on relationships between parameters. However, they depend on pre-set coefficient values.

AI and ML models are suitable for systems that are more complex to model, as they attempt to discover underlying relationships, and are widely used to accurately predict wind. Without an a priori structural hypothesis that relates wind power to several historical meteorological variables, they have a strong generalization and fast speed [18,28].

Each approach mentioned above can have a high forecasting error due to inherent weaknesses, especially when wind speeds have significant non-linear characteristics, as volatility causes complex fluctuations. In particular, the conventional single ANN model has the drawback of falling into local minimum and overfitting, and its performance can be influenced by the initial parameters. These weaknesses cannot be easily remedied with a single method. To reduce forecasting error and obtain advanced models that can achieve higher accuracy, a combination of methods described in Table 4 is incorporated.

Ensemble forecasting methods are generated through the application of various machine learning techniques and then by merging the outputs, which reduces the risk of overestimation and is aimed at preserving the diversity of models. The ensemble technique is known to be applied in both cooperative and competitive styles.

In a cooperative ensemble, the dataset is divided into data subsets, each subset being forecast individually and then aggregated with other sub-forecasts [29]. This technique is computationally lightweight due to less need for parameter tuning and is in general used for very short-term or short-term forecasting.

Table 4. Classification of complex forecasting techniques.

Technique	Description	Advantage	Disadvantage
Ensemble methods	Aggregate a combination of results from different methods	Usually perform better than a single method	Larger computational cost of running individual methods; the combination of results can obscure problems inherent in the methods
Hybrid methods	Combination of different methods connected in series to create hybrid prediction structures.	Usually perform better than a single method	Larger computational cost of running individual methods; require large quantities of data

Competitive ensembles build individual forecasting models with different parameters and initial values, and the results are obtained by aggregation of forecasts by different techniques, such as the Bayesian model average. This technique, used by [30], can cover a larger dataset and is used to achieve early detection of a large wind ramp before the changes in the wind speed propagate to other locations. However, it is considered computationally expensive and is mostly used in medium-term and long-term forecasting.

To obtain an advanced model with higher accuracy, hybrid forecasting models combine the advantages of different methods with individual superior features [31]. Overall forecasting effectiveness of hybrid methods can be improved, since hybrid methods can overcome the limitations and take advantage of the merits of individual models by integrating two or more types of models [28].

A neural network can be used in different steps of the algorithm, for example, a CNN-based model using transfer learning is used to address the problem of some newly constructed farms not having sufficient historical wind speed data to train a well-performing model by producing synthetic data [32]. In [26], the CNN is trained in layers to extract local features and relationships between the nodes, and the output layer of CNN is set in multiple dimensions to directly forecast future wind speed.

The most common approach is to adopt the machine learning algorithm as the main forecasting tool and to perform data treatment using general techniques as shown by [33], which consist of variational mode decomposition (VMD) of raw wind power series into a certain number of sub-layers with different frequencies; the K-means as a data mining approach being executed for splitting the data into an ensemble of components with a similar fluctuant level of each sub-layer; and LSTM is adopted as the principal forecasting engine for capturing unsteady characteristics of each component.

Some authors also combine both hybrid and ensemble approaches into one [34], using a hybrid technique of intelligent and heuristic algorithms that include neural networks, wavelet transform, diverse heuristic algorithms, and fuzzy logic. The hybrid technique uses wavelet transform to filter distortions and noise in wind power signals, the radial neural networks (RBF) technique being used as a preliminary predictor to find local solutions. With the local solution, an ensemble combining three neural networks of MLP using various learning methods along with heuristic WIPSO is used for the final prediction and modeling of the non-linear behavior of the wind power curve.

3. Performance of Forecasting Model

3.1. RMSE, MAE, and MAPE as Frequently-Used Metrics

The root mean square error (RMSE), given by Formula 1, is a quadratic scoring rule that estimates the average magnitude of error. It is the most standard function used to calculate the difference between predicted and observed values, since it reflects the level of differences between the actual and forecast values, in other words, the absolute magnitude of prediction error [35]. However, RMSE is sensitive to outliers, so its outcome can be biased if the data is not clean [36].

$$\text{RMSE} = \sqrt{\frac{1}{N}\sum_{i=1}^{N}(y_i - \hat{y}_i)^2} \quad (1)$$

where $\hat{y}_i$ is predicted value, y_i is the actual value, and N is the number of prediction points or number of samples. A smaller RMSE means that the proposed model performs better.

The mean absolute error (MAE), Equation (2), corresponds to the estimated level of absolute error. This level indicates the average magnitude of the actual value and the predicted value [37].

$$\text{MAE} = \frac{1}{N}\sum_{i=1}^{N}|y_i - \hat{y}_i| \quad (2)$$

where $\hat{y}_i$ is predicted value, y_i is the actual value, and N is the number of prediction points or number of samples. MAE is not susceptible to outliers and can better reflect the actual status of predicted errors [38]. The model is deemed to be accurate when MAE is close to zero.

The mean absolute percentage error (MAPE), Equation (3), calculates the percentage error relative to the actual value, which is stated as the average ratio, and is also commonly used to compare different models [36,39].

$$\text{MAPE} = \frac{1}{N}\sum_{i=1}^{N}\left|\frac{y_i - \hat{y}_i}{y_i}\right| \cdot 100\% \quad (3)$$

where $\hat{y}_i$ is predicted value, y_i is the actual value, and N is the number of prediction points or number of samples.

Although RMSE is usually used to express the dispersion of the results, MAE and MAPE can indicate the deviation of the prediction [17]. The smaller the values of RMSE, MAE, and MAPE, the more accurate the forecasting model.

3.2. MSE, nMAE, nRMSE, and R2 As Occasionally Used Metrics

The mean squared error (MSE), Equation (4), simply averages the mean squared difference between the estimated and original parameters [40], which can avoid the problem that the errors cancel each other out, and accurately reflects the actual prediction error [35].

$$MSE = \frac{1}{N}\sum_{i=1}^{N}(y_i - \hat{y}_i)^2 \quad (4)$$

where $\hat{y}_i$ is predicted value, y_i is the actual value, and N is the number of prediction points or number of samples.

Sometimes authors need to normalize the MAE and RMSE to quantitatively examine the prediction performances of some models, their norms being given by the normalized mean absolute error (nMAE), Equation (5), and normalized root mean squared error (nRMSE), Equation (6).

$$nRMSE = \sqrt{\frac{1}{N}\sum_{i=1}^{N}\left(\frac{y_i - \hat{y}_i}{C_i}\right)^2} \quad (5)$$

$$nMAE = \frac{1}{N}\sum_{i=1}^{N}\left|\frac{y_i - \hat{y}_i}{C_i}\right|, \quad (6)$$

where C_i is the operating capacity of time point i, $\hat{y}_i$ is predicted value, y_i is the actual value, and N is the number of prediction points or number of samples. In general, smaller values of these metrics indicate that the corresponding solution offers less deviation of prediction performance [41].

The R-square or coefficient of determination (R2), Equation (7), is the proportion of the variance in the dependent variable that is predictable from independent variable(s) [42,43]. It indicates the level of correlation between predicted value and the actual value, and it helps to select the best model with highest forecasting accuracy [44]. It is mostly used in datasets of large amplitudes [17].

$$R2 = 1 - \frac{\sum_{i=1}^{N}(y_i - \hat{y}_i)^2}{\sum_{i=1}^{N}(y_i - \overline{y})^2},\ for\ \overline{y} = \frac{1}{N}\sum_{i=1}^{N} y_i \tag{7}$$

where $\overline{y}$ is the average of actual values, $\hat{y}_i$ is predicted value, y_i is the actual value, and N is the number of prediction points or number of samples. An R2 closer to one indicates more accurate forecasting. It can also be displayed as negative to denote an arbitrarily worse predicting model [20].

3.3. *R, PICP, PINAW, sMAPE, MRE, and TIC As Seldom Used Metrics*

The Pearson linear correlation coefficient (R or CC), Equation (8), is a metric that determines the relationship between inputs and outputs by determining the linear dependence between results and observations [9,37].

$$R = \frac{\sum_{i=1}^{N}(x_i - \hat{x}_i)(y_i - \hat{y}_i)}{\sqrt{\sum_{i=1}^{N}(x_i - \hat{x}_i)^2 \sum_{i=1}^{N}(y_i - \hat{y}_i)^2}} \tag{8}$$

where $\hat{y}_i$ and $\hat{x}_i$ are predicted values, y_i and x_i are the actual values, and N is the number of prediction points or number of samples. The possible R score range can vary between 1 and -1, with 1 representing the biggest correlation, and -1 the lowest one [45].

Prediction interval coverage probability (PICP), Formula (9), measures the ability of the constructed confidence interval to cover the target values for prediction intervals.

$$\text{PICP} = \frac{1}{N}\sum_{i=1}^{N} \rho_i,\ \text{for}\ \rho_i = \begin{cases} 1, & if\ y_i \in [L_i, U_i] \\ & else\ 0 \end{cases} \tag{9}$$

where L_i and U_i are the lower bound and the upper bound, respectively, of the prediction values, y_i is the actual value, and N is the number of prediction points or number of samples. The greater the PICP, the more reliable the prediction values [46,47].

Prediction interval normalized average width (PINAW), Equation (10), is used to measure the width of the PIs for a given length of the prediction interval.

$$\text{PINAW} = \frac{1}{N}\sum_{i=1}^{N} \frac{(U_i - L_i)}{t_{max} - t_{min}} \tag{10}$$

where t_{min} and t_{max} are the maximum and minimum values of the predicted values, and L and U_i are the lower bound and the upper bound, respectively, of the prediction values [48].

The Symmetric Mean Absolute Percentage Error (sMAPE) metric, Formula (11), a variation of MAPE, is used to describe the relative error of a set of forecasts and their labels as a percentage [36,37].

$$\text{sMAPE} = \frac{1}{N}\sum_{i=1}^{N}\left|\frac{y_i - \hat{y}_i}{y_i + \hat{y}_i}\right| \cdot 200\% \tag{11}$$

where $\overline{y}$ is the average of actual values, $\hat{y}_i$ is predicted value, y_i is the actual value, and N is the number of prediction points or number of samples.

Table 5. Summary of errors obtained for onshore wind power forecasting models.

No.	Title/Reference	Horizon	System Nominal Power ***	Method Details	Error Metric		Input Data Details
		-	-	-	nRMSE	nMAE	-
1.	Wind power forecasting based on daily wind speed data using machine learning algorithms [23] Proposed Method—	1 year	1 MW	RF (ensemble)	0.0302 *	0.0070 *	Daily wind speed, mean wind speed, standard deviation, total generated wind power values—Turkish State Meteorological Service
				XGBoost (ensemble)	0.0344 *	0.0065 *	
2.	Short-term wind power forecasting based on support vector machine with improved dragonfly algorithm [17] Proposed Method—IDA-SVM	48 h	2050 KW	IDA-SVM (hybrid)	0.0524	0.0404	Wind power, wind speed, wind direction, and temperature—La Haute Borne wind farm
				DA-SVM (hybrid)	0.0617	0.0515	
3.	A cascaded deep learning wind power prediction approach based on a two-layer of mode decomposition [24] Proposed Method—EMD-VMD-CNN-LSTM	1, 2, 3 steps	? MW	EMD-VMD-CNN-LSTM (Dataset #1, avg, hybrid)	0.0329 *	0.0227 *	Three sets of hourly averaged wind power, wind speed, and wind direction time series, Sotavento Galicia wind farm
				EMD-VMD-CNN (Dataset #1, avg, hybrid)	0.0396 *	0.0321 *	
4.	A Model Combining Stacked Auto Encoder and Back Propagation Algorithm for Short-Term Wind Power Forecasting [57] Proposed Method—SAE_BP	1, 2, . . . , 9 steps	1500 MW	SAE_BP (avg, hybrid/ensemble)	0.0941 *	0.0738 *	Wind generation from Ireland all island—EirGrid Group
				SVM (avg, single)	0.1396 *	0.1185 *	
5.	Deep belief network based k-means cluster approach for short-term wind power forecasting [16] Proposed Method—DBN	10 min	17.56 MW	DBN (single)	0.0322	0.0236	Wind speed, wind direction, temperature, humidity, pressure, history wind speed, history wind power—NWP meteogalicia and Sotavento wind farm
				BP (single)	0.0580	0.0446	
6.	Multi-distribution ensemble probabilistic wind power forecasting [29] Proposed Method—MDE	1, 6, 24 h	16 MW	Q-learning (6 h, single)	0.2355	0.1841	Meteorological information (many features), synthetic actual wind power, and wind power forecasts generated by the Weather Research and Forecasting (WRF) model—MDE probabilistic forecasting framework, from the Wind Integration National Dataset (WIND) Toolkit
				NWP (24 h, single)	0.1817	0.1337	
7.	Forecasting energy consumption and wind power generation using deep echo state network [36] Proposed Method—DeepESN—Stacked Hierarchy of Reservoirs	10 steps	6910 million KWh	DeepESN (hybrid/ensemble)	0.0326 *	0.0247 *	?—historical WPG data in Inner Mongolia
				BP (single)	0.0957 *	0.0764 *	
				NAÏVE (single)	0.2033 *	0.1593 *	
8.	Feature Extraction of NWP Data for Wind Power Forecasting Using 3D-Convolutional Neural Networks [25] Proposed Method—3D-CNN—Three Dimensional Convolutional Neural Network	30 min—72 h	Normalized data	3D-CNN (20 h, single)	0.210 **	0.150 **	Wind power from every 10 s, from April 2015 to July 2017 + NWP—wind farm in Tohoku region
				2D-CNN (20 h, single)	0.212 **	0.156 **	
				NAÏVE (20 h, single)	0.212 **	0.156 **	
9.	A gated recurrent unit neural networks based wind speed error correction model for short-term wind power forecasting [58] Proposed Method—Gated Recurrent Unit Neural Networks	24 h	?	Proposed (single)	0.1345	0.0687	Wind speed, wind power, and NWP wind speed data, sampled at a period of 15 min—wind farm in Sichuan Province
				ANN (single)	0.1350	0.0680	

Table 5. *Cont.*

No.	Title/Reference	Horizon	System Nominal Power ***	Method Details	Error Metric		Input Data Details
		-	-	-	nRMSE	nMAE	-
10.	Medium-term wind power forecasting based on multi-resolution multilearner ensemble and adaptive model selection [9] Proposed Method—MRMLE-AMS—Multilearner Ensemble and Adaptive Model Selection	6, 12, 18 h Step = 6 h	200 MW	MRMLE (dataset #4, step 3, hybrid)	0.1254 *	0.0943 *	Actual wind power series in Canada as experimental datasets
				Hao's model (dataset #4, step 3, hybrid/ensemble)	0.1554 *	0.1245 *	
				NAÏVE (dataset #4, step 3, single)	0.4218 *	0.3226 *	
11.	Short-term wind power prediction based on improved small-world neural network [59] Proposed Method—SWBP—Small-World BP neural network	15 min	Normalized data	SWBP (single)	0.0820	0.0601	Wind speed and power, air temperature, barometric pressure, relative humidity, wind velocity, and direction at the height of 10, 30, 50, and 70 m—Jiangsu wind farm
				BP (single)	0.1072	0.0841	
12.	Wind power forecasting based on singular spectrum analysis and a new hybrid Laguerre neural network [31] Proposed Method—OTSTA-SSA-HLNN—opposition transition state transition algorithm-SSA-HLNN	1, 3, 6, 12 steps	Approx. 1500 KW (graph)	OTSTA-SSA-HLNN (experiment1, hybrid)	0.0612	0.0923	?—Wind farm in Xinjiang
				STA-SSA-HLNN (experiment1, hybrid/ensemble)	0.0727	0.1101	
13.	Short-term wind power forecasting using the hybrid model of improved variational mode decomposition and Correntropy Long Short-term memory neural network [50] Proposed Method—IVMD-SE-MCC-LSTM—Improved VMD-SE-MCC-LSTM	?	1402.7 KW	IVMD-SE-MCC-LSTM (hybrid)	0.0347 *	0.0241 *	Wind power series from 1 to 4 parameters—Gansu and Dingbian wind farm
				EMD-SE-MCC-LSTM (hybrid)	0.0370 *	0.0270 *	
14.	Short term wind power forecasting using hybrid variational mode decomposition and multi-kernel regularized pseudo inverse neural network [44] Proposed Method—VAPWCA-VMD-RMKRPINN—VMD based kernel regularized pseudo inverse neural network	10, 30 min and 1, 3 h	Approx. 30 MW (graph)	VAPWCA-VMD-RMKRPINN (3h, 150 samples, hybrid)	0.0310 *	0.0261 *	Wind power—Wyoming wind farm
				VAWCA-EMD-RMKRPINN (3h, 150 samples, hybrid)	0.0337 *	0.0274 *	
15.	Short-term wind power forecasting using long-short term memory based recurrent neural network model and variable selection [27] Proposed Method—LSTM-RNN	1 to 24 h	17.56 MW	LSTM-RNN (24 h, single)	0.1043	0.0765	Surface pressure, temperature, wind speed at 10 M, 35 M, 100 M, 170 M, wind direction at 35 M, 170 M—Sotavento Galicia wind farm NWP from Weather & Energy PROGnoses (WEPROG)

Table 5. *Cont.*

No.	Title/Reference	Horizon	System Nominal Power ***	Method Details	Error Metric		Input Data Details
		-	-	-	nRMSE	nMAE	-
16.	A novel hybrid model for short-term wind power forecasting [35] Proposed Method—The Proposed Hybrid Model	1, 2, 3 steps	2.5 KW	The Proposed Hybrid Model (set 1, 3 steps, single)	0.0773 *	0.0570 *	Wind speed—Sotavento Galicia wind farm
				EMD-ENN (set 1, 3 steps, single)	0.0793 *	0.0625 *	
				NAÏVE (set 1, 3 steps, single)	0.1280 *	0.1020 *	
17.	A novel hybrid technique for prediction of electric power generation in wind farms based on WIPSO, neural network and wavelet transform [34] Proposed Method—RBF + HNN + WT + WIPSO	24, 72 h	254 MW	RBF + HNN + WT + WIPSO (72 h, winter, hybrid)	0.0201	0.0173	Wind power, temperature, wind direction, wind speed, humidity, and air pressure—wind farms in Alberta, Canada
				RBF + HNN + WT + MPSO (72 h, winter, hybrid)	0.0250	0.0200	
18.	Wind Speed Prediction of IPSO-BP Neural Network Based on Lorenz Disturbance [43] Proposed Method—LD-PCA-IPSO-BPNN	10 min ?	?	LD-PCA-IPSO-BPNN (hybrid)	0.0095 *	0.0082 *	Wind speed, wind direction, temperature, air pressure, specific volume, specific humidity, and surface roughness—Sotavento wind farm
				PCA-IPSO-BPNN (hybrid)	0.0269 *	0.0235 *	
19.	Multi-step Ahead Wind Power Forecasting Based on Recurrent Neural Networks [53] Proposed Method—LSTM/GRU with wind speed correction	1, 2, 3, 4 steps step: 10	3 MW	LSTM (wind speed correction step *, hybrid)	0.0181 *	0.0090 *	Wind turbine state, rotation rate, generating capacity, sine of wind direction, cosine of wind direction, pitch angle, temperature, and wind power—wind turbine in North China and NWP
				GRU (wind speed correction step *, hybrid)	0.0185 *	0.0096 *	
20.	A New Prediction Model Based on Cascade NN for Wind Power Prediction [60] Proposed Method—Proposed	10, 20, . . . , 50, 60 min 1, 2, . . . , 7 days	17.56 MW	Proposed (7 days, hybrid)	0.1942 *	0.2762 *	Hourly wind speed and power forecast—Sotavento wind farm
				MLP (7 days, single)	0.4704 *	0.3519 *	
				NAÏVE (7 days, single)	0.7204 *	0.5108 *	

Table 5. *Cont.*

No.	Title/Reference	Horizon	System Nominal Power ***	Method Details	Error Metric		Input Data Details
		-	-	-	nRMSE	nMAE	-
21.	A new short-term wind speed forecasting method based on fine-tuned LSTM neural network and optimal input sets [61] Proposed Method—Optimized WT-FS-LSTM by CSA and by PSO	1 step Step: approx. 240 (graph)	Normalized data	Optimized WT-FS-LSTM by CSA (hybrid)	0.1536	0.1217	Hourly wind speed—Sotavento wind farm
				Optimized WT-FS-LSTM by PSO (hybrid)	0.1621	0.1306	
				WT-FS-LSTM (hybrid)	0.1655	0.1311	
22.	Wind speed forecasting method based on deep learning strategy using empirical wavelet transform, long short term memory neural network and Elman neural network [54] Proposed Method—EWT-LSTM-Elman	1, 2, 3 steps	?	EWT-LSTM-Elman (3 steps, hybrid)	0.065 *	0.049 *	Wind speed series
				EWT-BP (3 steps, hybrid)	0.077 *	0.058 *	
23.	A Comparative Study of ARIMA and RNN for Short Term Wind Speed Forecasting [62] Proposed Method——	?	?	RNN (dataset 1, single)	0.0942 *	0.0750 *	Hourly mean wind speed—Coimbatore meteoblue weather website
				ARIMA (dataset 1, single)	0.1058 *	0.0917 *	
24.	Can we improve short-term wind power forecasts using turbine-level data? A case study in Ireland [63] Proposed Method—VMD-ELM with WT-data	1,2, . . . , 8 h	24,923 KW	VMD-ELM with WT-data	0.0773	0.0564	Wind power, WT-data, or WF-data—wind farm in West and Southwest of Ireland
				VMD-ELM with WF-data	0.0811	0.0605	
25.	Short term forecasting based on hourly wind speed data using deep learning algorithms [64] Proposed Method——	?	Normalized data	LSTM (October–December, single)	1.236	0.8742	Hourly wind speed, air temperature, pressure, dew point, and humidity—metrological wind dataset from the location
				ELM (October–December, single)	0.9215	0.3978	
26.	Hybrid Approach for Short Term Wind Power Forecasting [65] Proposed Method—Proposed Method	24 h	2100 KW	Proposed method (December, hybrid)	0.0162 *	0.0077 *	Wind speed—Jodhpur wind farm
27.	Very Short-Term Spatial and Temporal Wind Power Forecasting: A Deep Learning Approach [66] Proposed Method—CST-WPP—convolution-based spatial-temporal wind power predictor	5—30 min	Normalized data	CST-WPP (30 min)	0.0749	0.0494	5 min mean wind power data of 28 wind farms—Australian Energy Market Operator
				VAR (30 min, single)	0.0861	0.0651	
				NAÏVE (30 min, single)	0.1174	0.0974	

Table 5. *Cont.*

No.	Title/Reference	Horizon	System Nominal Power ***	Method Details	Error Metric		Input Data Details
		-	-	-	nRMSE	nMAE	-
28.	Short-Term Wind Speed Forecasting Using Ensemble Learning [67] Proposed Method—	48 h	?	Bagged trees (December, ensemble)	0.138 *	0.107 *	10 min wind speed data—modern-era retrospective analysis for research and applications, version 2 (MEERA-2)
				Boosted trees (December, ensemble)	0.147 *	0.113 *	
29.	Seasonal Self-evolving Neural Networks Based Short-term Wind Farm Generation Forecast [68] Proposed Method—SSEN—Seasonal Self-Evolving Neural Networks	10 min, 1 h	300.5 MW	SSEN (10 min, ensemble)	0.0302	0.0170	Power outputs, wind speed and direction—National Renewable Energy Laboratory (NREL) and Xcel Energy
				LSTM (10 min, single)	0.0307	0.0180	
30.	Markov Models based Short Term Forecasting of Wind Speed for Estimating Day-Ahead Wind Power [69] Proposed Method—	24 h	2100 KW	Markov chain model (Group 1, single)	0.0714 *	0.0620 *	Wind speed—Dark Sky API
31.	Day-Ahead Wind Power Forecasting Based on Wind Load Data Using Hybrid Optimization Algorithm [49] Proposed Method—VMD-mRMR-FA-LSTM	24 h	Approx. 85 MW (graph)	VMD-mRMR-FA-LSTM (hybrid)	0.0358 *	0.0348 *	Wind power, load, temperature, air pressure, humidity, wheel height, wind speed and wind direction at 10m, 30m, 50m and 70m—Beijing Lumingshan Wind Power Plant
				mRMR-FA-LSTM (hybrid)	0.0388 *	0.0379 *	
32.	Short-Term Wind Power Prediction Using GA-BP Neural Network Based on DBSCAN Algorithm Outlier Identification [70] Proposed Method—DBSCAN-GA-BP	10 min, 1 h	?	DBSCAN-GA-BP (single)	0.0646	0.0524	Wind speed, temperature, humidity, cosine value of wind direction, air pressure and power
				GA-BP (single)	0.0978	0.0798	
33.	Ultra-Short-Term Wind-Power Forecasting Based on the Weighted Random Forest Optimized by the Niche Immune Lion Algorithm [71] Proposed Method—WD-NILA-WRF	5 min	49.5 MW	WD-NILA-WRF (WF A, hybrid)	28.5797	24.2484	Wind power, wind speed, wind direction, temperature, humidity, air density, air pressure, ground roughness, and other factors—wind farm in Inner Mongolia A
				NILA-RF (WF A, hybrid)	70.9289	54.4785	
34.	Wind Power Short-Term Prediction Based on LSTM and Discrete Wavelet Transform [72] Proposed Method—DWT_LSTM	1, 2, 3, 4, 5 steps	Approx. 1500 MW (graph)	DWT_LSTM (WF 3—5 steps)	0.024 *	0.016 *	Wind power—wind farm in Yunnan, China
				DWT_BP (WF 3—5 steps)	0.027 *	0.018 *	
35.	Short-Term Wind Power Forecasting: A New Hybrid Model Combined Extreme-Point Symmetric Mode Decomposition, Extreme Learning Machine and Particle Swarm Optimization [73] Proposed Method—ESMD-PSO-ELM	15 min?	49.5 MW	ESMD-PSO-ELM (hybrid)	2.7	2.23	Wind power—wind farm in Yunnan, China
				EMD-PSO-ELM (hybrid)	3.25	2.88	

Table 5. *Cont.*

No.	Title/Reference	Horizon	System Nominal Power ***	Method Details	Error Metric		Input Data Details
		-	-	-	nRMSE	nMAE	-
36.	Hourly Day-Ahead Wind Power Prediction Using the Hybrid Model of Variational Model Decomposition and Long Short-Term Memory [74] Proposed Method—Direct-VMD-LSTM	24 h	Approx. 30 MW (graph)	D-VMD-LSTM (hybrid)	0.1287 *	0.1070 *	Hourly wind power—wind farm located in Henan
				VMD-BP (hybrid)	0.1487 *	0.1927 *	
37.	Wind Power Prediction Based on Extreme Learning Machine with Kernel Mean p-Power Error Loss [51] Proposed Method—ELM-KMPE	Short-term?	6 MW	ELM-KMPE (hybrid)	1.3643 *	0.6369 *	Wind speed, wind direction, temperature, atmospheric humidity, and atmospheric pressure + weather data—NWP and SCADA
				ELM (single)	1.3648 *	0.6703 *	
38.	TRSWA-BP Neural Network for Dynamic Wind Power Forecasting Based on Entropy Evaluation [75] Proposed Method—TRSWA-BP	1, 4, 6, 24 h	?	TRSWA-BP (24h, hybrid)	0.1703 *	0.1018 *	Wind power
				BP (24h, single)	0.1747 *	0.1132 *	
39.	Ultra Short-Term Wind Power Forecasting Based on Sparrow Search Algorithm Optimization Deep Extreme Learning Machine [76] Proposed Method—SSA-DELM	Ultra-short-term ?	3600 KW	SSA-DELM (hybrid)	0.0321 *	0.0194 *	Wind Power, wind Speed, wind Direction—SCADA + wind farm data
				WA-DELM (hybrid)	0.0323 *	0.0201 *	
40.	Numerical Weather Prediction and Artificial Neural Network Coupling for Wind Energy Forecast [77] Proposed Method—? Hybrid	> 6 h	?	Hybrid (hybrid)	0.163	0.105	?—NWP
				Baseline (single)	0.214	0.137	
41.	A Hybrid Framework for Short Term Multi-Step Wind Speed Forecasting Based on Variational Model Decomposition and Convolutional Neural Network [26] Proposed Method—VMD-CNN	16 h	?	VMD-CNN (hybrid)	0.0655 *	0.0515 *	Wind speed—Sotavento Galicia wind farm
				VMD-NN (hybrid)	0.0959 *	0.0764 *	
42.	Forecasting Models for Wind Power Using Extreme-Point Symmetric Mode Decomposition and Artificial Neural Networks [55] Proposed Method—Proposed model	24 h	49.5 MW	Proposed model (hybrid)	0.0251	0.0191	Wind power, humidity (30 m), pressure (50 m), temperature (100 m), wind direction (50 m), wind speed (50 m)—wind farm, China
				EEMD (ensemble)	0.0281	0.0218	
43.	Multi-Step Ahead Wind Power Generation Prediction Based on Hybrid Machine Learning Techniques [41] Proposed Method—mRMR-IVS	1, 2, 3, 4 h	94 MW	mRMR-IVS (Xuqiao, hybrid)	0.0971	0.0762	Power generation + wind speed, trigonometric wind direction, temperature, humidity, and atmospheric pressure—Anhui, Xuqiao wind farm + NWP
				PSR-IVS (Xuqiao, hybrid)	0.1041	0.087	
44.	An improved residual-based convolutional neural network for very short-term wind power forecasting [37] Proposed Method—Ours	1, 2, 3 h	2.5 MW	Ours (jan, single)	0.0326	0.0234	Wind power, wind speed, and wind direction—wind farm located in southeastern Turkey
				VGG-16 (jan, single)	0.0942	0.07	
45.	A novel hybrid approach based on variational heteroscedastic Gaussian process regression for multi-step ahead wind speed forecasting [45] Proposed Method—CEEMDAN-VHGPR (CVHGPR)	1, 2, 3 steps	?	CVHGPR (W1, 3-step)	0.0244 *	0.0189 *	Wind speed—Sotavento Galicia wind farm
				CGPR (W1, 3-step)	0.0250 *	0.0193 *	

Table 5. *Cont.*

No.	Title/Reference	Horizon	System Nominal Power ***	Method Details	Error Metric		Input Data Details
		-	-	-	nRMSE	nMAE	-
46.	A novel reinforced online model selection using Q-learning technique for wind speed prediction [78] Proposed Method—OMS-QL	1 h	?	OMS-QL (hybrid)	0.0635 *	0.0348 *	Wind speed—Colorado wind farm
				CLSTM (M9, hybrid)	0.0937 *	0.0670 *	
47.	Hybrid and Ensemble Methods of Two Days Ahead Forecasts of Electric Energy Production in a Small Wind Turbine [1] Proposed Method—The ensemble method based on hybrid methods	48 h	5 KW	The ensemble method based on hybrid methods (ensemble/hybrid)	0.0936 *	0.0563 *	Generated power, wind speed at different levels, blades pitch angle, nacelle orientation, etc.—SCADA and Levenmouth Demonstration Turbine (LDT)
				Physical model+LSTM (hybrid)	0.0940 *	0.0565 *	
				NAÏVE (single)	0.2005 *	0.1267 *	
48.	Advanced Ensemble Methods Using Machine Learning and Deep Learning for One-Day-Ahead Forecasts of Electric Energy Production in Wind Farms [2] Proposed Method—Ensemble Averaging Without Extremes (many single and ensemble methods)	24 h	50 MW	Ensemble Averaging Without Extremes (many single and ensemble methods) (ensemble/ hybrid)	0.1618	0.1130	Many input data, lagged values of forecasted time series, meteo forecast—wind speed, Indictor of variability of the daily electric energy production—NWP UM, south of Poland
				GBT (XGBOOST (ensemble)	0.1636	0.1185	
				NAÏVE (single)	0.3833	0.2877	
49.	A novel hybrid model based on nonlinear weighted combination for short-term wind power forecasting [79] Proposed Method—Proposed	Short-term ?	1112.40 KW	Proposed (hybrid)	0.0386 *	0.0318 *	Wind speed—Shaanxi Dingbian wind farm
				PSO-DBN (hybrid)	0.1565 *	0.1004 *	
50.	Ultra-short term wind power prediction applying a novel model named SATCN-LSTM [80] Proposed Method—SATCN-LSTM	1 step step: 16	Normalized data	SATCN-LSTM (Q1st, hybrid)	0.9070	0.4970	Wind speed, air density, wind direction, temperature, and surface pressure—wind farm California
				TCN-LSTM (Q1st, hybrid)	0.9160	0.5060	

* recalculated from RMSE or MAE and nominal system power, ** approximated values from graphs, *** nominal system power only applies to the error for a specific case rather than all cases.

Table 6. Summary of errors obtained for offshore wind power forecasting models.

No.	Title/Reference	Horizon	System Nominal Power ***	Method Details	Error Metric		Input Data Details
		-	-	-	nRMSE	nMAE	-
1.	Wind power forecasting of an offshore wind turbine based on high-frequency SCADA data and deep learning neural network [20] Proposed Method——	?	7 MW	Tensor Flow with feature engineering (single)	0.07789 *	0.0572 *	Active power, wind speeds, blade pitch angles, nacelle orientation, ambient temperature, and yaw error, from 1 July 2018 to 30 June 2019—SCADA database from Offshore Renewable Energy (ORE) Catapult
				Tensor Flow with ReLU activation (single)	0.07390 *	0.0534 *	
2.	A hybrid deep learning-based neural network for 24-h ahead wind power forecasting [42] Proposed Method—CNN-RBFNN-DGF	24 h	2,000,800 KW	CNN-RBFNN-DGF (hybrid)	0.0385 *	-	Wind power generation—Changgong
				CNN-RBFNN (hybrid)	0.0697 *	-	
3.	Exploiting Deep Learning for Wind Power Forecasting Based on Big Data Analytics [15] Proposed Method—EDCNN	24 h	900 MW	EDCNN (ensemble)	0.085	-	Wind speed, dew point temperature, dry bulb temperature, past lagged values of wind power, and wavelet packet decomposed wind power—NWP and 100 weather station in West Texas Mesonet (WTM)"
				SELU CNN (ensemble)	0.12	-	
4.	Wind power forecasting using attention-based gated recurrent unit network [19] Proposed Method—AGRU + Attention	5, 15, 13 min, 1h, 2 h	16 MW	AGRU + Attention (2 h, ensemble)	0.1007		Historical wind power sequences and NWP data—National Renewable Energy Laboratory (NREL)
				AGRU + RFE (2 h, ensemble)	0.1147		
5.	Offshore wind power forecasting-A new hyperparameter optimization algorithm for deep learning models [81] Proposed Method—LSTM optimized by Optuna	?	6.5 MW	LSTM optimised by Optuna (Data1, single)	0.0950 *	-	Generated power, wind speed at different levels, blades pitch angle, nacelle orientation, etc.—SCADA and wind farm, Levenmouth Demonstration Turbine (LDT)
				LSTM (Data1, single)	0.0963 *	-	
				NAÏVE (Data1, single)	0.0979 *	-	
6.	Short-term multi-hour ahead country-wide wind power prediction for Germany using gated recurrent unit deep learning [82] Proposed Method—RNN-GRU, RNN-LSTM	1, 3, 5, 12 h	33,626 MW	RNN-GRU (1 h)	0.0017 *	0.0009 *	Wind speeds, air density, air pressure, power price index, wind speed V1, V10, and V50 (2 m above displacement height), h2 (10 m above displacement height), and h3 (50 m above ground), surface roughness length, temperature 2 m—OPSD Time Series, 2019
				RNN-LSTM (1 h)	0.0026 *	0.0012	
				SVR-RBF (1 h)	0.2968 *	0.2333 *	

* recalculated from RMSE or MAE and nominal system power, *** nominal system power only applies to the error for a specific case rather than all cases.

In addition to the papers mentioned in Tables 5 and 6, in Section 4.1 (*Comprehensive Quantitative Review*) below we also use in certain analyses information from those papers which did not provide nRMSE and nMAE error values or these values could not be calculated due to absence of information on the rated power of the system. These papers are the following: [26,83–122].

4.1. Comprehensive Quantitative Review

Based on data from 116 papers, statistical analysis was conducted to determine, among other things: the frequency of use of various error metrics, classes of forecasting methods, distinct types of input variables for forecasting models, scopes of rated powers of the systems subject to forecasting, location of the systems subject to forecasting, and typical forecasting horizons. The analysis in this subparagraph excludes papers that provided less reliable or no data.

Figure 1 presents the outcome of statistical analysis of the number of forecasting studies concerning wind power generation in particular regions of the world based on the research papers analyzed here. What is remarkable is the very uneven distribution of studies across regions of the world. Special attention must be drawn to China—by far the largest number of papers addressing wind farm generation forecasting. The second best is the United States.

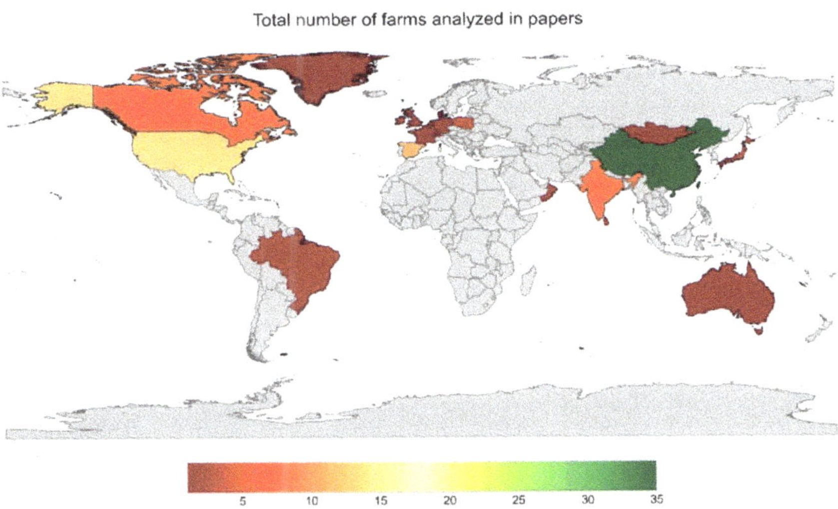

Figure 1. Total number of farms analyzed in 116 papers.

The performance of a wind power forecasting model is measured with different statistical metrics. These metrics quantify the prediction error of a model, providing the accuracy between the predicted values and the measured data [65]. It is difficult to make a comprehensive evaluation using the single error index, and, as Figure 2 shows, studies can consider up to six different metrics to evaluate, compare performance and quantify forecasting errors [20,40]. However, in general, only 2 or 3 statistical metrics are used in model validation. In some cases, authors also do not specify the metric used to evaluate the performance of the model. The combination of statistical metrics, presented by Figure 2, varies by study.

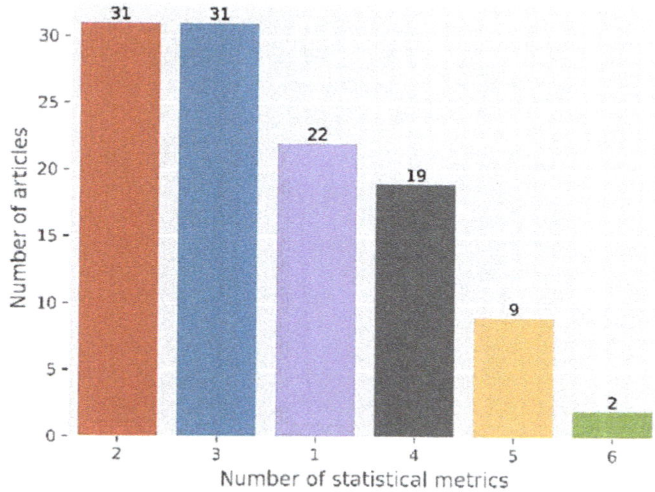

Figure 2. Number of statistical metrics used per article to evaluate the performance of forecasting models (evaluation of 114 articles).

It means that the metrics used to evaluate the performance of each model are different for each study. Figure 3 shows a summary of quantifiers used in the studies analyzed, and these metrics are split into four groups (see Table 7).

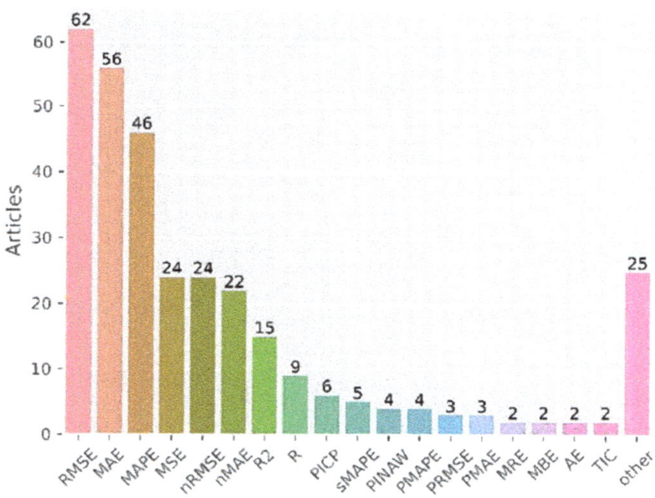

Figure 3. Common ways to measure and evaluate the error of models predicting quantitative data.

Table 7. Frequency of use of model performance evaluation metrics.

Use	Metric	Number of Times Used
Frequent	RMSE, MAE, and MAPE	46–62
Occasional	MSE, nRMSE, nMAE, and R2	15–24
Seldom	R, PICP, sMAPE, MBE, MRE, . . . , TIC	2–9
Rare	Other Metrics	1

RMSE, MAE, and MAPE are popular accuracy metrics due to their ease of interpretability by decision-makers and participants of energy markets. Those metrics, unlike mean bias errors, do not falsify the average quality of forecasts by compensating over-forecasting with under-forecasting. Moreover, they give a decent estimation of the average error one can expect from forecasts for each prediction step [123]. Because of squaring the error, RMSE is more sensitive to the detection of high values of errors in error time series, which make it a good metric for detecting extreme error values. In turn, MAE does not additionally magnify extreme values of errors and is the closest to the most naturally expected type of error—mean error. Unlike the two previous metrics, MAPE is not dependent on the scale of values in the data, which makes it useful for comparing data of different scales—e.g., errors for prosumer wind turbines and very big wind farms. It also allows us to find how an accurate model is in the scale of changing the momentary real value. This metric is, however, susceptible to zero/small values of generation appearing in its denominator [124]. The result can be either an indefinite expression or a substantial error at a given step, and as a consequence, its value is reflected in the final average value. Because of the aforementioned, we recommend not using MAPE. In Figure 3 one can see the rare use of metrics other than the three most frequent ones. Usually, they are root or derivative metrics of those three and are used to solve previous ones' drawbacks, e.g., nRMSE and nMAE add an aspect of comparability between objects of different scales, which cannot be easily conducted without normalization of time series, used in MAPE for example. A coefficient of determination is also relatively frequently used. It serves as a means of describing not how well a model predicts but how much of the modeled process is actually modeled.

Forecasting methods were classified into single methods, ensemble methods, and hybrid methods, and calculations were conducted on how frequently each of those methods was the best method (lowest nRMSE error) in each of the studies described in the papers reviewed here. Figure 4 presents the outcome of our analysis. A hybrid method was usually the best class of forecasting methods (almost 44%). Quite a substantial percentage (almost 25%) of studies in which a single method was the best is surprising. This can be explained by the fact that some papers proposed single methods only, without comparing them to other classes (ensemble, hybrid). Additionally, note that, in some cases, ensemble and hybrid methods have the characteristics of both classes. For instance, the general hybrid model also contains model(s) from the ensemble class. Comparison of forecast quality of the single, ensemble, and hybrid methods are presented in Section 4.2.2. Analysis of errors and EDF depending on the class of forecasting methods.

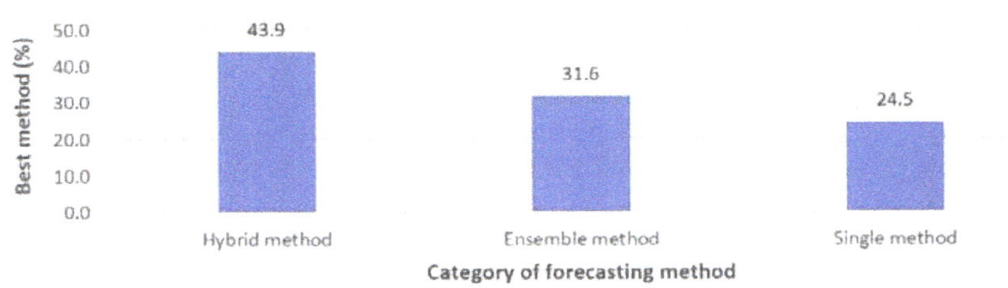

Figure 4. Frequency of using forecasting methods (quantity statistics).

Based on information from papers (for which rated powers were provided), the percentage of farms for which generation forecasts were conducted was calculated for specific ranges of rated powers. By far most frequently studied were systems sized from more than 10 MW to 100 MW, with the second largest group of systems being those sized up to 10 MW (Figure 5). Domination of the former range is probably due to the fact that it is the most frequent range of powers in wind farms, and, on the other hand, for very small (prosumer) systems, power generated from wind turbines is forecast much less frequently.

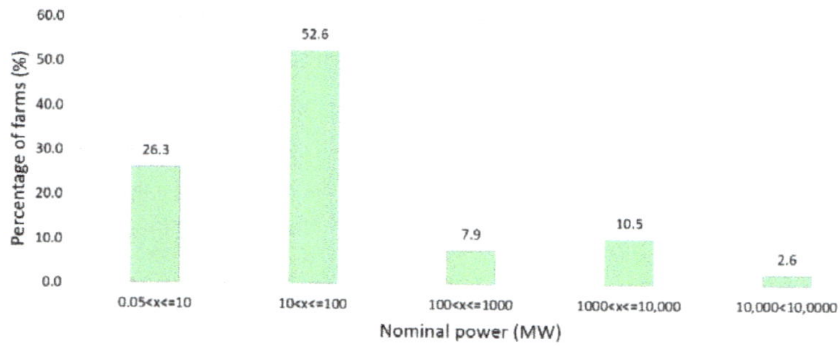

Figure 5. Ranges of rated powers of wind farms.

The percentage of farms for which generation forecasts have been conducted was calculated by forecasting horizon (Figure 6). By far the most frequent forecasting horizon is "24 h" and "few steps", with "one step" (5 min, 10, min, 15 min, or 1 h) being slightly less frequent. Forecasts with horizons of more than 24 h are clearly rare. On the one hand, the reason may be more difficult access to NWP with such horizons and the awareness of the loss of quality of such forecasts, especially as compared to horizons with few steps ahead.

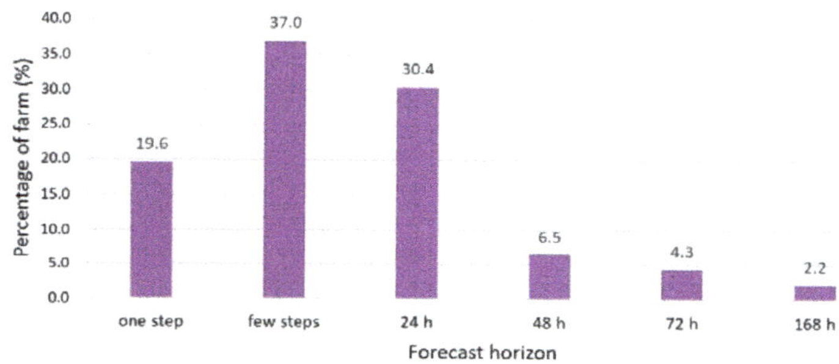

Figure 6. Frequency of forecasts from different forecasting horizons.

The frequency of use of various sets of input data in the forecasting models described in the papers was calculated (Figure 7). Lagged generation values of the forecast time series are used clearly most frequently. NWP and weather measurements are used only slightly less frequently. Other types of input data are used at least ten times less frequently than the three input data mentioned above (or incidentally). Such infrequent use of input data such as lagged NWP, time variables and generation stats (statistics on the forecast times series) in forecasting models is surprising.

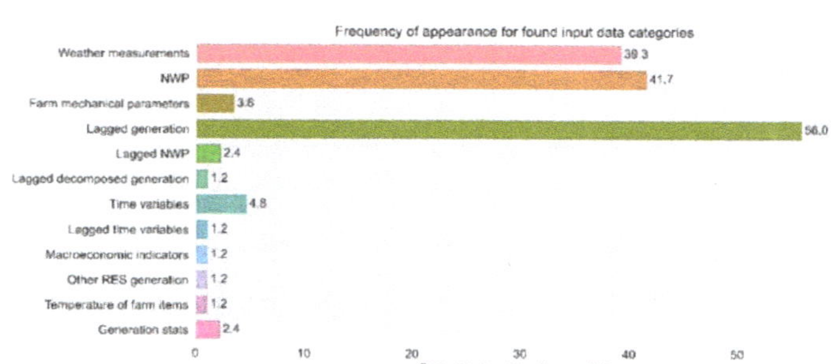

Figure 7. Input data categories by frequency of use.

4.2. Comprehensive Error Analysis

Out of 116 papers, error analysis was conducted on those which applied both nRMSE and nMAE errors and which could calculate these two error metrics based on the available rated power of the system and the values of RMSE and MAE errors (errors pre-normalization). In addition, quotients of nRMSE to nMAE errors were calculated. A new, unique EDF (error dispersion factor) metric has thus been introduced to analyses, described by Formula (20). Therefore, EDF is a combination of two frequently used error metrics. Statistical analyses in Section 4.2, Section 4.2.1 and Section 4.2.2 apply, among others, to the potential usefulness of EDF in analyses of wind power forecasts.

$$EDF = \frac{RMSE}{MAE} = \frac{nRMSE}{nMAE} = \frac{\sqrt{\frac{1}{N}\sum_{i=1}^{N}(y_i - \hat{y}_i)^2}}{\frac{1}{N}\sum_{i=1}^{N}|y_i - \hat{y}_i|} \quad (20)$$

The analysis in this subparagraph excludes papers that provided less reliable data (abnormal errors, abnormal error quotients)—abnormal phenomena are addressed in the Section 5. Table 8 presents basic statistics, and Figure 8 visualizes selected statistics.

Table 8. Descriptive statistics of errors and error quotients.

Descriptive Statistics	nRMSE	nMAE	EDF
Mean	0.0787	0.0566	1.4160
Standard deviation	0.0578	0.0407	0.2307
Minimum	0.0162	0.0077	1.0283
Maximum	0.2762	0.1942	2.1129
Quotient of maximum/minimum	17.05	25.22	1.75
The 10th percentile	0.0244	0.0174	1.1841
The 25th percentile (lower quartile)	0.0324	0.0236	1.2740
The 50th percentile (median)	0.0654	0.0504	1.3603
The 75th (upper quartile)	0.1007	0.0756	1.5019
The 90 percentile	0.1618	0.1070	1.7741
Variance	0.0033	0.0017	0.0562
Skewness	1.3448	1.2811	1.2939
Kurtosis	1.9858	1.8489	1.5423

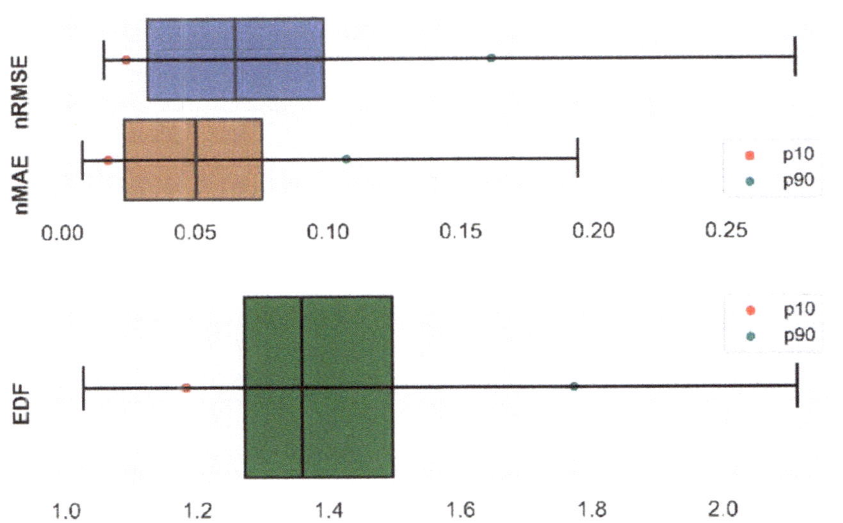

Figure 8. Selected statistics of errors and error quotients.

The averages in Table 8 are slightly larger than the medians both for nRMSE, nMAE, and EDF. The dispersion of errors is remarkably high—the maximum/minimum quotient for nRMSE error metric is more than 17, and for nMAE errors, the quotient is more than 25. Such large dispersion of values can be partly justified by different forecasting horizons (from 10 min to 72 h).

4.2.1. Analysis of Errors by Forecasting Horizon

Figures 9 and 10 present nRMSE and nMAE errors, respectively, in ascending order, based on the papers in which these error metrics were provided (also considering those which did not provide the rated power of the system). In addition, information on the forecasting horizon is provided. Forecasts with longer horizons display significantly much larger nRMSE errors, which is unsurprising (the accuracy of wind speed forecasts decreases with increasing forecasting horizon).

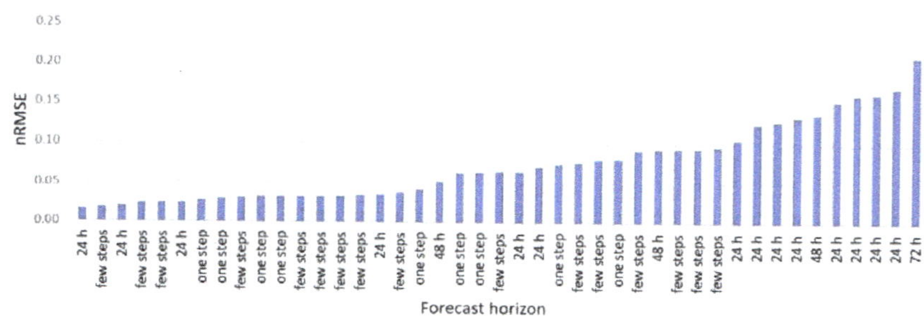

Figure 9. nRMSE errors with a note on forecasting horizon, in ascending order.

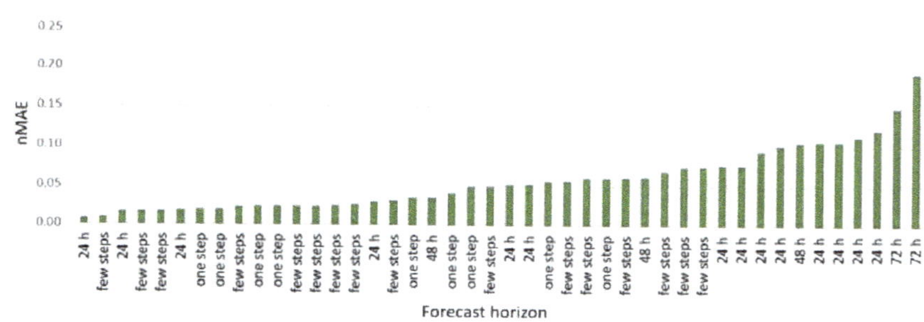

Figure 10. nMAE errors with a note on forecasting horizon, in ascending order.

Figure 11 presents how the amount of error depends on the forecasting horizon. This figure summarizes information from Figures 9 and 10—average values for both metrics were calculated for selected forecasting horizons. In general, average errors grow with increasing forecasting horizon, although, for the 24 h horizon, average errors are slightly more than for the 48 h horizon. This is probably due to the fact that there were significantly fewer papers describing forecasts with a 48 h horizon than with a 24 h horizon (random element of lower errors from a small number of samples). By far the largest were the average errors for the 72 h horizon—more than two-and-a-half larger than for 24 h and 48 h horizons. For the "one step" horizon, average errors are two times smaller than average errors for the 24 h horizon. This information has large practical significance—it shows what magnitude of normalized errors should be expected from the respective forecasting horizon. Please note that the averages calculated for the 48 h and 72 h horizons may not be fully representative due to a small number of samples.

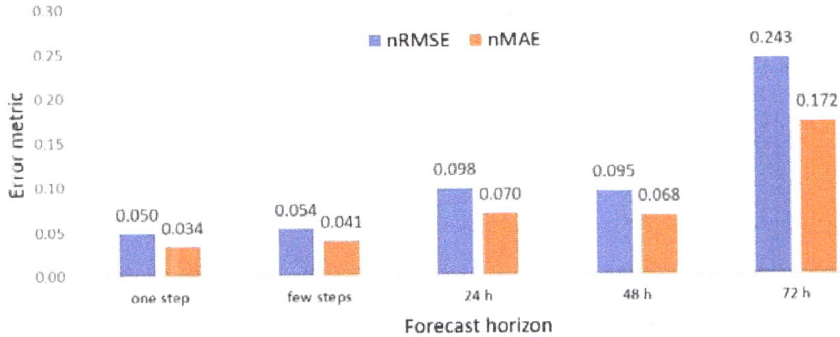

Figure 11. Magnitudes of error by forecasting horizon.

To determine precisely whether there is a statistically significant relationship between the forecasting horizon and error magnitudes, numerical forecasting horizons were selected (1/6 h, 1/4 h, 1/2 h, 1 h, 6 h, 12 h, 24 h, 48 h, and 72 h), which enabled us to calculate Pearson linear correlation. The statistical analysis concluded a statistically significant (5% level of significance) positive linear correlation between the forecasting horizon (multiples of 1) and the magnitude of nRMSE error (R = −0.347). nRMSE error grows with increasing forecasting horizon. Figure 12 presents how the magnitude of nRMSE error depends on the forecasting horizon.

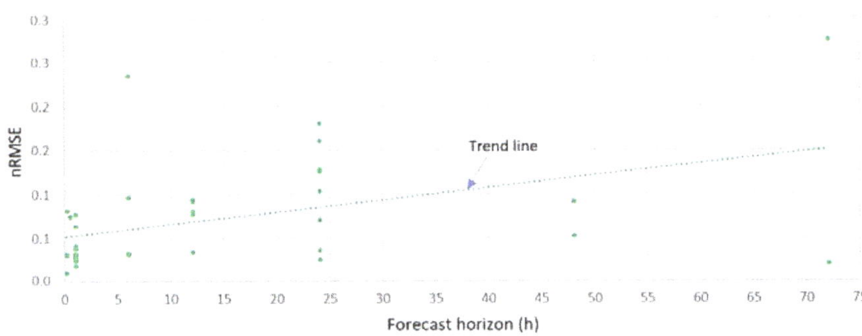

Figure 12. Dependence of nRMSE on forecasting horizon.

The statistical analysis concluded a statistically significant (5% level of significance) positive linear correlation between the forecasting horizon (multiple of 1) and the magnitude of nRMSE error (R = 0.410). nMAE error grows with increasing forecasting horizons. Figure 13 presents how the magnitude of nMAE error depends on the forecasting horizon. It is worthwhile to emphasize that the linear correlation between the forecasting horizon and the magnitude of error is slightly larger for the nRMSE error metric than for nMAE.

The statistical analysis concluded a statistically insignificant (5% level of significance) negative linear correlation between the forecasting horizon and EDF (R = −0.196). The EDF slightly decreases with increasing forecasting horizon. Figure 14 presents how EDF depends on the forecasting horizon.

For forecasts with very short horizons (from 10 min to 1 h), the average EDF is 1.422, and 1.3163 for 6 h, and it falls to 1.2724 for the 24 h horizon. For the 48 h and 72 h horizons, the samples are too few to calculate reliable averages.

In addition, statistical analysis omitting the 48 h and 72 h horizons concluded a negative correlation. Not a very large one, but statistically significant (15% level of significance), between the forecasting horizon and EDF (R = −0.283).

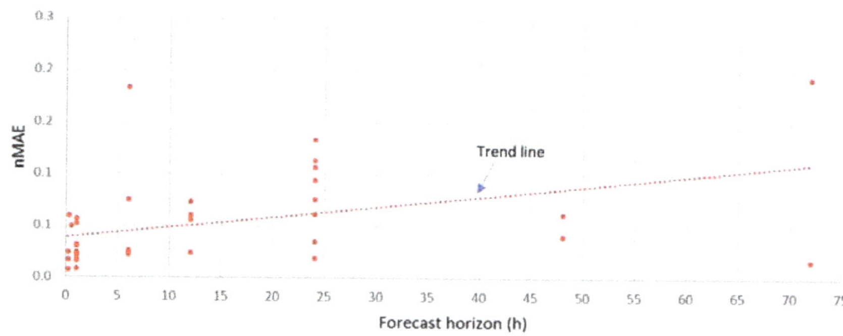

Figure 13. Dependence of nMAE errors on forecasting horizon.

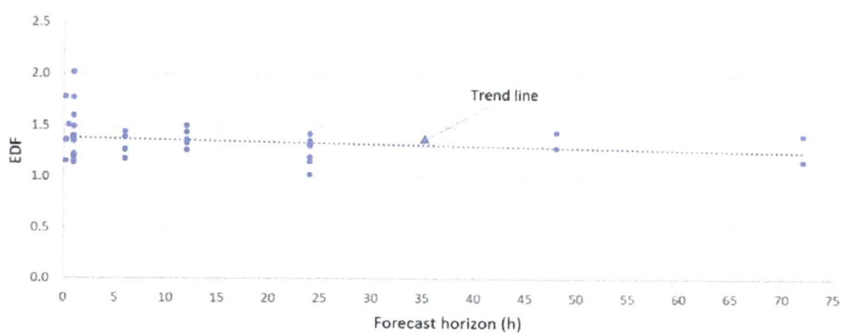

Figure 14. Dependence of EDF on forecasting horizon.

The EDF (Figure 14) and (Formula (20) show the average variability of the moduli of error regardless of the magnitude of the error. If absolute errors on all samples are the same, this ratio reaches its minimum value of 1. The larger the error deviation on particular samples from the average, the larger the ratio. This resembles the behavior of standard deviation determined for the moduli of error for samples, however, with the difference that standard deviation reaches a minimum value equal to zero, and the dynamics of that ratio is much larger—significantly dependent on particular samples. For the EDF, the dynamics of values are smaller, which better illustrates the variability of errors across the sample pool. It should also be mentioned that the EDF in fact shows the ratio of the second moment of error to the first moment of error.

The decreasing levels of EDF with a rising forecasting horizon means that the variability of error decreases with an increasing forecasting horizon. On the one hand, it is probably due to the growing error, and, on the other, the averaging nature of the forecasting models for longer horizons, which stabilizes errors around certain values.

It is worthwhile to note that statistical analysis of hourly values of wind speed presented in [1] concluded that the variance of wind speed forecasts for horizons ranging from 1 to 24 h was 3.121, and for 25- to 48-h horizons, it was 3.063, which is less.

4.2.2. Analysis of Errors and EDF Depending on the Class of Forecasting Methods

Some of the 116 papers analyzed here provide the forecasting error of a method from the "single method" class. The primary objective of the analysis was to investigate percentage error reduction achieved by the best (proposed) method from the ensemble or hybrid class relative to the single method with the largest forecasting error (excluding the outcome of the naive method). Figure 15 presents, in descending order, percentage reductions of nRMSE and percentage reductions of nMAE of the best methods relative to single methods. What is remarkable is a very wide dispersion of percentage reductions of error. For nRMSE, the largest percentage reduction of error is 80.02%, and the smallest is 2.76%. Similar observations apply to the dispersion of nMAE.

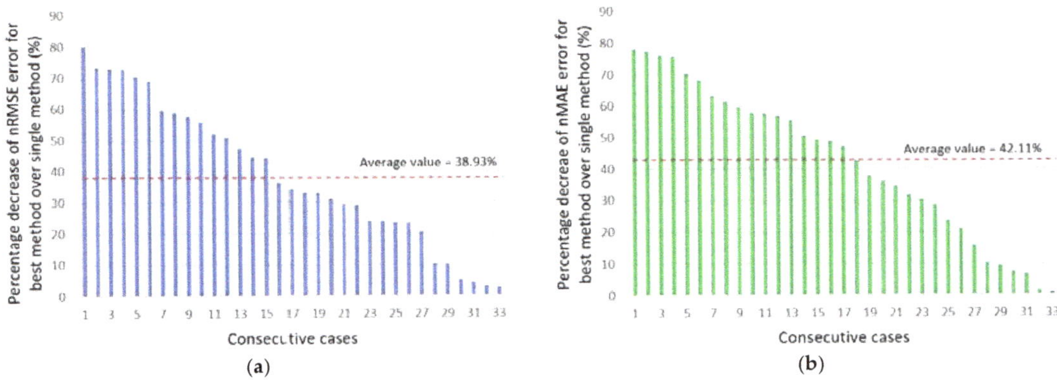

Figure 15. (**a**) Percentage reduction in nRMSE for the best method relative to the single method; (**b**) percentage reduction in nMAE for the best method relative to the single method.

Figure 16 presents the average percentage improvement of the hybrid methods and ensembles method relative to the single method for nRMSE and nMAE error metrics. The percentage improvement of error metrics is much bigger for hybrid methods in comparison to ensemble methods however the number of cases (19 for hybrid methods and 14 for ensemble methods) is too small to generalize this fact.

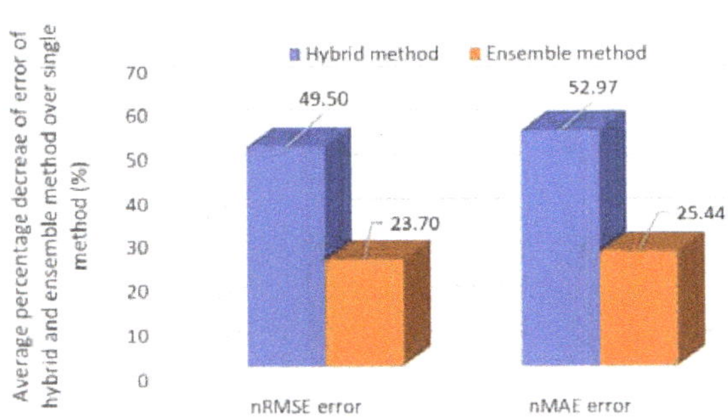

Figure 16. Average percentage improvement of the hybrids method and ensembles method relative to the single method for nRMSE and nMAE error metrics.

Unfortunately, a small proportion of the papers reviewed here provide forecasting error using a naive (persistence) method—such error would be the best benchmark for the level of improvement achieved by other proposed methods, including single methods. The forecasting methodology assumes that a forecasting method is valuable if its error is less than the error of the naive method. Six papers provide errors for the naive method. Figure 17 presents, in descending order, percentage reductions of nRMSE relative to the naive method for six cases (pairs of nRMSE and nMAE)

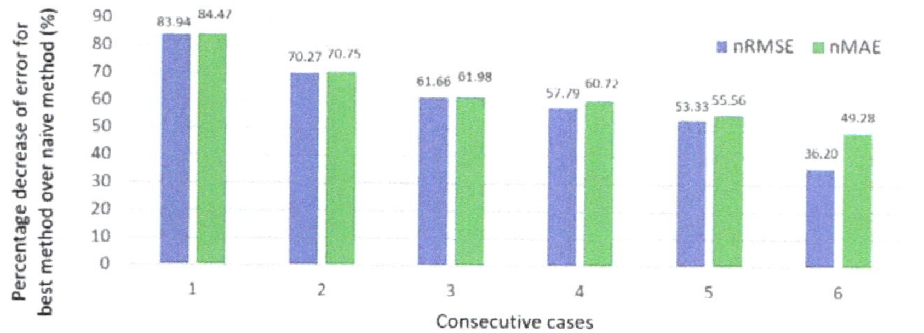

Figure 17. Percentage improvement of the best method relative to the naive method for nRMSE and nMAE error metrics.

The average percentage reduction calculated for six cases is 60.53% for nRMSE and 63.79% for nMAE. Therefore, both average percentages are much larger than similar values calculated for nRMSE and nMAE reductions when errors of best methods are compared to single methods. Nevertheless, in a small number of cases, percentage reductions of nRMSE and nMAE for the best method relative to single methods are large and similar to the best method compared to the naive method. It means that some single methods referred to in literature are only marginally better than naive methods.

The second objective of our analysis is to compare EDF for the best (proposed: ensemble or hybrid) method and a single method. Based on 33 cases (pairs of ratios), we have determined that in 77% of cases, the EDF for the best method is larger than the EDF for the single method used in the respective study—this is more frequently observed for larger values of those ratios. The Pearson coefficient of linear correlation (R) between the ratios for the best method and the ratios for the single method is 0.737.

Figure 18 presents pairs of EDF sorted in descending order by the level of ratios for the best method.

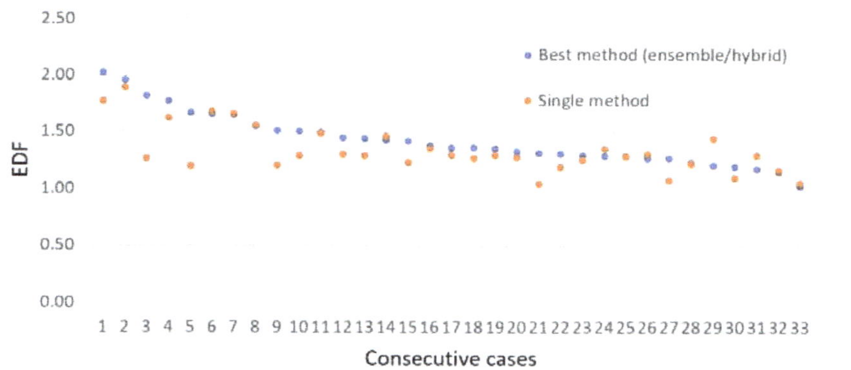

Figure 18. Pairs of EDF sorted in descending order by the level of quotients for the best method.

The average EDF for the best method is 1.432 and the median EDF is 1.364. The average EDF for the single method is 1.352, and the median EDF is 1.294. Therefore, both the average and median levels are clearly larger for the best method. Both series do not have normal distribution—Shapiro–Wilk test was conducted. Wilcoxon signed-rank test has been therefore applied to the analysis of pairs, which concluded that there are statistically significant differences between pairs in both series (they have different expected values). Therefore, differences between medians are statistically significant, and not without reason.

An interesting conclusion can be drawn based on our analysis—the variability of the moduli of errors in the best methods (smallest forecasting errors) is typically larger than for the "single method" class (much larger forecasting errors). The moduli of errors in the "single method" class are much larger and much closer to each other than in the best (hybrid or ensemble) method. In some studies, a single method could also use slightly less information (a different set of input data), which can also affect the characteristics of errors (magnitude and variability level).

4.2.3. Analysis of Errors Based on System Size

Our analysis covered the studies which provided nRMSE and nMAE and the size of the system. Statistical analysis did not reveal a statistically significant (5% level of significance) linear correlation between the size of the system (rated power) and nRMSE and nMAE errors (R = −0.110, R = −0.111, respectively). In theory, errors should grow with increasing size of the system due to much less uniform weather conditions (wind speed) in wind farms occupying extensive areas, the fact of using usually point-based meteorological forecasts, the wake effect, and other factors affecting the farm which are more difficult to represent if they overlap in the same space. The Pearson coefficient of linear correlation (R) between nRMSE and nMAE is 0.994 (5% level of significance). It means that these error metrics are very similar to each other. The details are presented in Figure 19. In addition, there is a large dispersion of the magnitudes of error for systems of similar sizes. This can be due to different sets of input data (different quality of information).

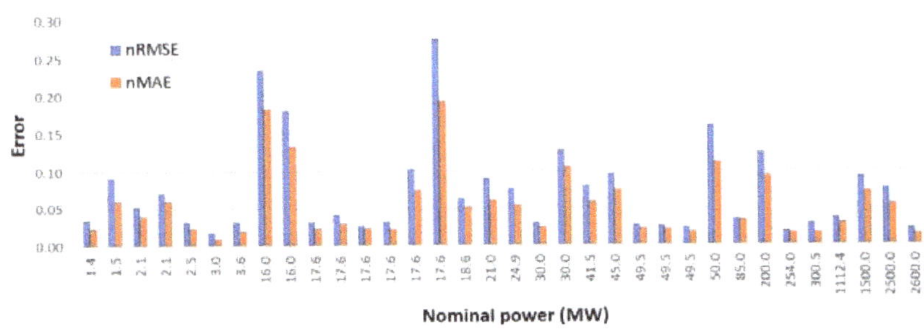

Figure 19. Magnitude of error depending on rated power of the system.

The statistical analysis concluded an insignificant (5% level of significance), marginally positive linear correlation between the size of the system (rated power) and EDF (R = 0.039). The EDF usually varies between 1 and 2. The average value of the EDF is 1.35. The details are presented in Figure 20.

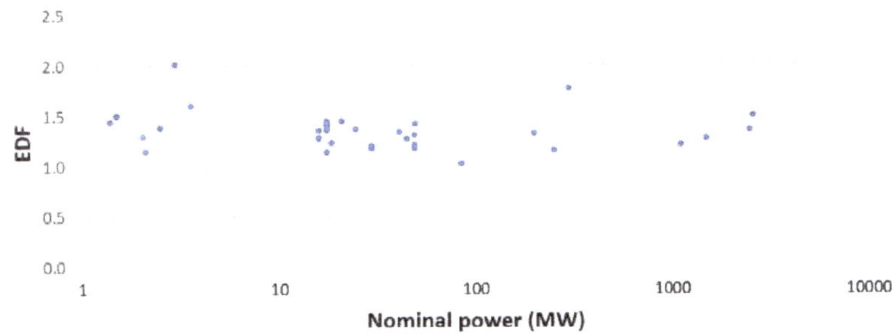

Figure 20. Magnitude of EDF depending on rated power of the system.

4.2.4. Analysis of Error Based on System Location (Onshore v. Offshore)

The number of papers addressing forecasting for offshore farms is small, as they constitute less than 6% of the 116 papers subject to this analysis. Only six papers (Table 6) provide nRMSE or nMAE, which is too little to conduct an accurate statistical analysis. Figure 21 compares nRMSEs for two forecasting horizons (average of the errors provided in the papers) for offshore and onshore farms.

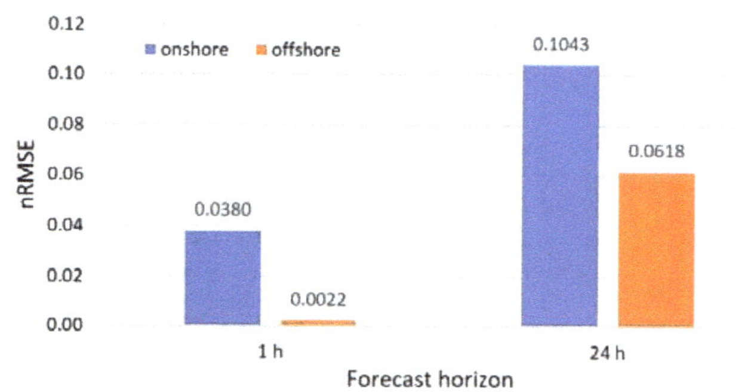

Figure 21. Magnitude of nRMSEs depending on farm location and forecasting horizon.

Offshore farms have smaller forecasting errors than onshore farms. This is expected, as it results from more stable and stronger winds at offshore farms. In addition, these are typically very large systems. For a 1 h horizon, such a sizable difference may result from the fact that the average for offshore farms was based on only two values of error and the fact that some onshore forecasts did not use meteorological forecasts of wind speed (larger forecasting errors occur in such cases). In real terms, at 24 h horizon, nRMSE at onshore farms can be about twice as large (assuming that meteorological forecasts are used in both locations).

5. Discussion

A comprehensive review and statistical analysis of errors based on an extensive selection of 116 papers allowed us to conclude, using actual figures, a correlation between the magnitude of error and selected factors. The quantitative analysis is provided for the aggregate assessment of how frequently various categories of (quite diverse) forecasting methods are applied, what typical input data are (meteorological forecasts are typically used for horizons above 6 h), how often various forecasting horizons have been used (typical horizons being in the range of 1 h to 24 h).

The analyses concluded that some papers used incomplete data that prevented them from being used in an aggregate meta-analysis of studies, which applies, in particular, to error metrics (nRMSE and nMAE).

In addition, several untypical (extreme) nRMSE and nMAE error levels have been identified, which, due to extreme dissimilarity to the remaining data of the same class (forecasting error being too large or too small) by expert judgment have been excluded from the analyses presented in Section 4.2. *Comprehensive Error analysis*. Figure 22 presents the variability of nRMSE error in the papers reviewed here.

A novel, unique ratio called EDF has been explored. The EDF shows the average variability of the moduli of error regardless of the magnitude of error. The analysis of variability of the new EDF ratio depending on selected characteristics (size of wind farm, forecasting horizon, and class of forecasting method) has been performed. There is a small negative correlation but statistically significant between the forecasting horizon and EDF. Additionally, the EDF for the best forecasting method is larger than the EDF for the single forecasting method. The analysis concluded an insignificant, marginally positive linear correlation between the size of the system (rated power) and EDF.

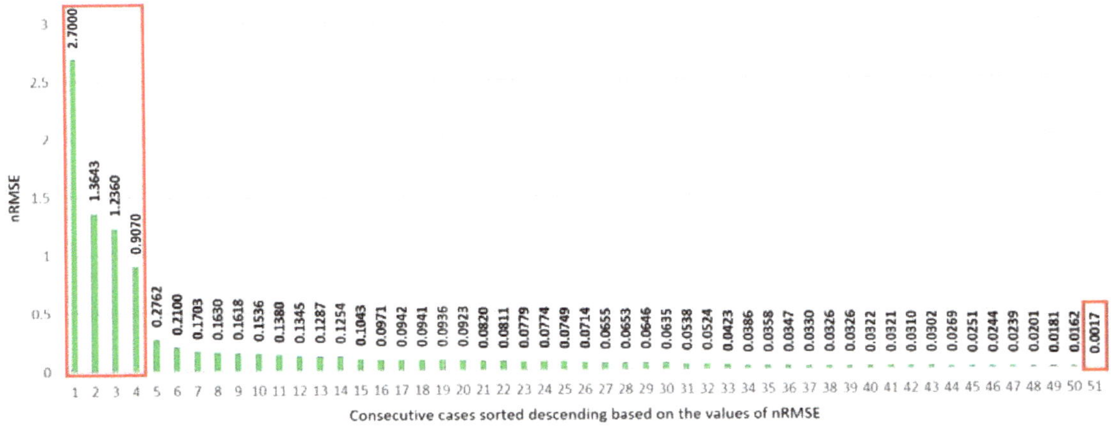

Figure 22. Variability of nRMSEs in the reviewed papers, with red-marked less reliable values.

Statistical analysis in one paper concluded, in addition, an untypical value of EDF. Statistical data from the reviewed papers for which EDF could be calculated shows that EDF levels range from 1.028 to 7.478, although a vast majority of EDF levels range between 1 and 2 (this range seems to be most credible—minimum value of the ratio is 1). The outcome of our analysis is presented in Figure 23.

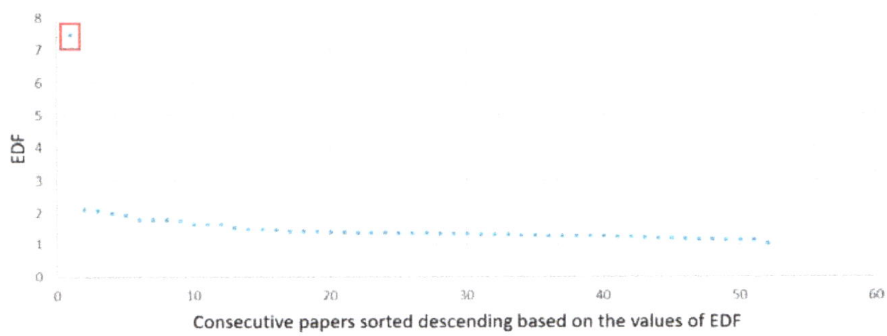

Figure 23. Variability of EDF in reviewed papers, with red-marked less credible value.

Based on our analysis of papers, in our subjective assessment, to maximize the quality of aggregate meta-analysis of studies addressing power generation forecasting in wind farms, a research paper should contain the following items (our recommendations):

- mandatory use of normalized error metrics for the assessment of forecasting quality: nRMSE Formula (5) and nMAE Formula (6), accompanied by a description of the normalization method (recommended normalization using the rated power of the system), which would enable comparative assessments of the quality of studies of systems of various sizes (or regardless of how big the wind farm is);
- we do not recommend the use of MAPE metric, which is susceptible to substantial error for small or zero generations;
- mandatory use of a forecast conducted by naive (persistence) method to enable the assessment of the quality of the best model proposed by the authors of the paper relative to the reference model; in such case calculation of the skill score metric by Formula (17), (18), or (19) is recommended in addition;

- provide strict, precise information on forecasting horizon(s) and information if the forecast is, e.g., "One Step," e.g., from 1 h to 24 h, or one step ahead with follow-up predictions (in which case forecasting errors are typically smaller);
- provide strict, precise information on the set(s) of input data for the proposed model(s);
- provide the source of meteo forecasts (GFS, ECMWF, other sources) if used in the forecasting model;
- explicit statement on whether the forecasting model uses meteo forecasts and/or on-site measurements of weather conditions at the wind farm;
- provide the range of training, validation, and testing data, and how the range of data is divided into the identified subsets;
- provide details of the location, unless confidential (location and size of the wind farm, and landscape features prevalent at the location).

6. Conclusions

This paper is the outcome of a comprehensive review and statistical analysis of errors using more than one hundred research papers. The quantitative analyses allowed us to assess the distribution of frequency of application of selected parameters in research studies (including the number and type of error metrics, forecasting horizon, rated power of the system, classes of forecasting methods, and location of the forecast systems).

Our qualitative analyses allowed us to provide an aggregate assessment of power generation forecasting in wind farms, including how selected factors affect the magnitude of forecasting errors. In addition, the rationale for using complex (ensemble, hybrid) forecasting methods instead of single methods was verified, by examining how this improves the quality of forecasts.

Notably, only 6 of 116 papers addressed power generation forecasts in offshore farms—it means that such research should intensify going forward, although it is in part due to a significantly smaller number of such systems than of onshore farms. The offshore location of a farm involves a number of distinct characteristics (such as a surface with exceptionally low roughness, significantly higher wind speeds, and more stable power generation). The magnitude of forecasting errors is significantly smaller. Due to a small number of offshore-related papers, our analysis was much more constrained.

In our view, research on topics related to aggregate statistical analyses (meta-analyses) should continue. We are planning to increase the number of reviewed papers at least two- or three-fold in the future. Such a number will enable us to conduct a more precise statistical assessment of a large number of factors affecting the magnitude of forecasting error, and expand the analyses related to the EDF factor proposed by us. In our view, it is crucial that published papers on generation forecasts in wind farms contain information from our recommended list, to enable conducting the necessary analyses.

Author Contributions: Conceptualization, P.P. and D.B.; methodology, P.P., D.B., I.R. and M.K.; investigation, P.P., D.B., I.R. and M.K.; supervision, P.P.; validation, P.P. and M.K.; writing, P.P., D.B., I.R. and M.K.; visualization P.P., I.R., M.K. and D.B.; project administration, P.P. All authors have read and agreed to the published version of the manuscript.

Funding: This research was funded by the 2021 edition of the competition for grant of the Scientific Council for the Discipline Automatic Control, Electronics and Electrical Engineering of the Warsaw University of Technology (to P.P., D.B., M.K.).

Institutional Review Board Statement: Not applicable.

Informed Consent Statement: Not applicable.

Data Availability Statement: Not applicable.

Conflicts of Interest: The authors declare no conflict of interest.

Abbreviations

The following abbreviations are used throughout this manuscript:

ANN	Artificial Neural Network
ARIMA	Autoregressive Integrated Moving Average
BP	Back Propagation
BPNN	BP Neural Network
CNN	Convolutional Neural Network
CSA	Crow Search Algorithm
DA	Dragonfly Algorithm
DBN	Deep Belief Network
DBSCAN	Density-Based Spatial Clustering of Applications with Noise
DGF	Double Gaussian Function
DWT	Discrete Wavelet Transform
EDCNN	Efficient Deep Convolution Neural Network
EDF	Error Dispersion Factor
EEMD	Ensemble Empirical Mode Decomposition
EMD	Empirical Mode Decomposition
EWT	Empirical Wavelet Transformation
FS	Feature Selection
GBT	Gradient Boosted Trees
GRU	Gated Recurrent Unit
HLNN	Laguerre Neural Network
HNN	Hybrid Neural Network
IPSO	Improved Particle Swarm Optimization
KNNR	K-Nearest Neighbors Regression
LD	Lorenz Disturbance System
LR	Linear Regression
LSTM	Long-Short-Term Memory
MAAPE	Mean Arctangent Absolute Percentage Error
MAE	Mean Absolute Error
MAPE	Mean Absolute Percentage Error
MBE	Mean Bias Error
MCC	Maximum Correntropy Criterion
MKRPINN	Multi-Kernel Regularized Pseudo Inverse Neural Network
MPSO	Modified Particle Swarm Optimization
MRE	Mean Relative Error
mRMR	Maximum Relevance and Minimum Redundancy Algorithm
MSE	Mean Squared Error
nMAE	Normalized Mean Absolute Error
NN	Neural Network
nRMSE	Normalized Root Mean Squared Error
NWP	Numerical Weather Prediction
OMS-QL	Online Model Selection using Q-learning
P	Promoting Percentages
PCA	Principal Component Analysis
PICP	Prediction Interval Coverage Probability
PINAW	Prediction Interval Normalized Average Width
PSO	Particle Swarm Optimization
R or CC	Pearson Linear Correlation Coefficient
R2	R-square or Coefficient of Determination
RBF	Radial Basis Function
RES	Renewable Energy Sources
RF	Random Forest
RMSE	Root Mean Square Error
SAE	Stacked Auto-Encoders
SE	Sample Entropy
sMAPE	Symmetric Mean Absolute Percentage Error
SS	Skill Score
SSA	Singular Spectrum Analysis
STA	State Transition Algorithm
SVM	Support Vector Machine

SVR	Support Vector Regression
TIC	Theil's Inequality Coefficient
VMD	Variational Mode Decomposition
WF	Wind Farm
WIPSO	Weight Improved Particle Swarm Optimization
WPT	Wavelet Packet Transform
WT	Wavelet Transform
xGBoost	eXtreme Gradient Boost

References

1. Piotrowski, P.; Kopyt, M.; Baczyński, D.; Robak, S.; Gulczynski, T. Hybrid and Ensemble Methods of Two Days Ahead Forecasts of Electric Energy Production in a Small Wind Turbine. *Energies* **2021**, *14*, 1225. [CrossRef]
2. Piotrowski, P.; Baczyński, D.; Kopyt, M.; Gulczyński, T. Advanced Ensemble Methods Using Machine Learning and Deep Learning for One-Day-Ahead Forecasts of Electric Energy Production in Wind Farms. *Energies* **2022**, *15*, 1252. [CrossRef]
3. Wang, Y.; Zou, R.; Liu, F.; Zhang, L.; Liu, Q. A review of wind speed and wind power forecasting with deep neural networks. *Appl. Energy* **2021**, *304*, 117766. [CrossRef]
4. Hanifi, S.; Liu, X.; Lin, Z.; Lotfian, S. A Critical Review of Wind Power Forecasting Methods-Past, Present and Future. *Energies* **2020**, *13*, 3764. [CrossRef]
5. Santhosh, M.; Venkaiah, C.; Vinod Kumar, D. Current advances and approaches in wind speed and wind power forecasting for improved renewable energy integration: A review. *Eng. Rep.* **2020**, *2*, e12178. [CrossRef]
6. González-Sopeña, J.; Pakrashi, V.; Ghosh, B. An overview of performance evaluation metrics for short-term statistical wind power forecasting. *Renew. Sustain. Energy Rev.* **2021**, *138*, 110515. [CrossRef]
7. Nazir, M.S.; Alturise, F.; Alshmrany, S.; Nazir, H.M.J.; Bilal, M.; Abdalla, A.N.; Sanjeevikumar, P.; Ali, Z.M. Wind Generation Forecasting Methods and Proliferation of Artificial Neural Network: A Review of Five Years Research Trend. *Sustainability* **2020**, *12*, 3778. [CrossRef]
8. Tawn, R.; Browell, J. A review of very short-term wind and solar power forecasting. *Renew. Sustain. Energy Rev.* **2022**, *153*, 111758. [CrossRef]
9. Chen, C.; Liu, H. Medium-term wind power forecasting based on multi-resolution multi-learner ensemble and adaptive model selection. *Energy Convers. Manag.* **2020**, *206*, 112492. [CrossRef]
10. Mishra, S.P.; Dash, P.K. Short-term prediction of wind power using a hybrid pseudo-inverse Legendre neural network and adaptive firefly algorithm. *Neural Comput. Appl.* **2019**, *31*, 2243–2268. [CrossRef]
11. Wang, H.; Liu, Y.; Zhou, B.; Li, C.; Cao, G.; Voropai, N.; Barakhtenko, E. Taxonomy research of artificial intelligence for deterministic solar power forecasting. *Energy Convers. Manag.* **2020**, *214*, 112909. [CrossRef]
12. Piotrowski, P.; Baczyński, D.; Kopyt, M.; Szafranek, K.; Helt, P.; Gulczyński, T. Analysis of forecasted meteorological data (NWP) for efficient spatial forecasting of wind power generation. *Electr. Power Syst. Res.* **2019**, *175*, 105891. [CrossRef]
13. Piotrowski, P.; Parol, M.; Kapler, P.; Fetliński, B. Advanced Forecasting Methods of 5-Minute Power Generation in a PV System for Microgrid Operation Control. *Energies* **2022**, *15*, 2645. [CrossRef]
14. Yousuf, M.U.; Al-Bahadly, I.; Avci, E. Current Perspective on the Accuracy of Deterministic Wind Speed and Power Forecasting. *IEEE Access* **2019**, *7*, 159547–159564. [CrossRef]
15. Mujeeb, S.; Alghamdi, T.A.; Ullah, S.; Fatima, A.; Javaid, N.; Saba, T. Exploiting Deep Learning for Wind Power Forecasting Based on Big Data Analytics. *Appl. Sci.* **2019**, *9*, 4417. [CrossRef]
16. Wang, K.; Qi, X.; Liu, H.; Song, J. Deep belief network based k-means cluster approach for short-term wind power forecasting. *Energy* **2018**, *165*, 840–852. [CrossRef]
17. Li, L.L.; Zhao, X.; Tseng, M.L.; Tan, R.R. Short-term wind power forecasting based on support vector machine with improved dragonfly algorithm. *J. Clean. Prod.* **2020**, *242*, 118447. [CrossRef]
18. Lopez, E.; Valle, C.; Allende, H.; Gil, E.; Madsen, H. Wind Power Forecasting Based on Echo State Networks and Long ShortTerm Memory. *Energies* **2018**, *11*, 526. [CrossRef]
19. Niu, Z.; Yu, Z.; Tang, W.; Wu, Q.; Reformat, M. Wind power forecasting using attention-based gated recurrent unit network. *Energy* **2020**, *196*, 117081. [CrossRef]
20. Lin, Z.; Liu, X. Wind power forecasting of an offshore wind turbine based on high-frequency SCADA data and deep learning neural network. *Energy* **2020**, *201*, 117693. [CrossRef]
21. Kisvari, A.; Lin, Z.; Liu, X. Wind power forecasting A data-driven method along with gated recurrent neural network. *Renew. Energy* **2021**, *163*, 1895–1909. [CrossRef]
22. Ju, Y.; Sun, G.; Chen, Q.; Zhang, M.; Zhu, H.; Rehman, M.U. A Model Combining Convolutional Neural Network and LightGBM Algorithm for Ultra-Short-Term Wind Power Forecasting. *IEEE Access* **2019**, *7*, 28309–28318. [CrossRef]
23. Demolli, H.; Dokuz, A.S.; Ecemis, A.; Gokcek, M. Wind power forecasting based on daily wind speed data using machine learning algorithms. *Energy Convers. Manag.* **2019**, *198*, 111823. [CrossRef]

24. Yin, H.; Ou, Z.; Huang, S.; Meng, A. A cascaded deep learning wind power prediction approach based on a two-layer of mode decomposition. *Energy* **2019**, *189*, 116316. [CrossRef]
25. Higashiyama, K.; Fujimoto, Y.; Hayashi, Y. Feature Extraction of NWP Data for Wind Power Forecasting Using 3D-Convolutional Neural Networks. *Energy Procedia* **2018**, *155*, 350–358. [CrossRef]
26. Pan, J.S.; Hu, P.; Chu, S.C. Novel Parallel Heterogeneous Meta-Heuristic and Its Communication Strategies for the Prediction of Wind Power. *Processes* **2019**, *7*, 845. [CrossRef]
27. Cali, U.; Sharma, V. Short-term wind power forecasting using long-short term memory based recurrent neural network model and variable selection. *Int. J. Smart Grid Clean Energy* **2019**, *8*, 103–110. [CrossRef]
28. Jiang, P.; Li, R.; Li, H. Multi-objective algorithm for the design of prediction intervals for wind power forecasting model. *Appl. Math. Model.* **2019**, *67*, 101–122. [CrossRef]
29. Sun, M.; Feng, C.; Zhang, J. Multi-distribution ensemble probabilistic wind power forecasting. *Renew. Energy* **2020**, *148*, 135–149. [CrossRef]
30. Du, P. Ensemble Machine Learning-Based Wind Forecasting to Combine NWP Output with Data from Weather Station. *IEEE Trans. Sustain. Energy* **2019**, *10*, 2133–2141. [CrossRef]
31. Wang, C.; Zhang, H.; Ma, P. Wind power forecasting based on singular spectrum analysis and a new hybrid Laguerre neural network. *Appl. Energy* **2020**, *259*, 114139. [CrossRef]
32. Wang, Z.; Zhang, J.; Zhang, Y.; Huang, C.; Wang, L. Short-Term Wind Speed Forecasting Based on Information of Neighboring Wind Farms. *IEEE Access* **2020**, *8*, 16760–16770. [CrossRef]
33. Sun, Z.; Zhao, S.; Zhang, J. Short-Term Wind Power Forecasting on Multiple Scales Using VMD Decomposition, K-Means Clustering and LSTM Principal Computing. *IEEE Access* **2019**, *7*, 166917–166929. [CrossRef]
34. Nourani Esfetang, N.; Kazemzadeh, R. A novel hybrid technique for prediction of electric power generation in wind farms based on WIPSO, neural network and wavelet transform. *Energy* **2018**, *149*, 662–674. [CrossRef]
35. Du, P.; Wang, J.; Yang, W.; Niu, T. A novel hybrid model for short-term wind power forecasting. *Appl. Soft Comput.* **2019**, *80*, 93–106. [CrossRef]
36. Hu, H.; Wang, L.; Lv, S.X. Forecasting energy consumption and wind power generation using deep echo state network. *Renew. Energy* **2020**, *154*, 598–613. [CrossRef]
37. Yildiz, C.; Acikgoz, H.; Korkmaz, D.; Budak, U. An improved residual-based convolutional neural network for very short-term wind power forecasting. *Energy Convers. Manag.* **2021**, *228*, 113731. [CrossRef]
38. Zheng, H.; Wu, Y. A XGBoost Model with Weather Similarity Analysis and Feature Engineering for Short-Term Wind Power Forecasting. *Appl. Sci.* **2019**, *9*, 3019. [CrossRef]
39. Bhatt, G.A.; Gandhi, P.R. Statistical and ANN based prediction of wind power with uncertainty. In Proceedings of the 3rd International Conference on Trends in Electronics and Informatics (ICOEI), Tirunelveli, India, 23–25 April 2019; pp. 622–627. [CrossRef]
40. Shahid, F.; Zameer, A.; Mehmood, A.; Raja, M.A.Z. A novel wavenets long short term memory paradigm for wind power prediction. *Appl. Energy* **2020**, *269*, 115098. [CrossRef]
41. Dong, W.; Yang, Q.; Fang, X. Multi-Step Ahead Wind Power Generation Prediction Based on Hybrid Machine Learning Techniques. *Energies* **2018**, *11*, 1975. [CrossRef]
42. Hong, Y.Y.; Rioflorido, C.L.P.P. A hybrid deep learning-based neural network for 24-h ahead wind power forecasting. *Appl. Energy* **2019**, *250*, 530–539. [CrossRef]
43. Zhang, Y.; Chen, B.; Zhao, Y.; Pan, G. Wind Speed Prediction of IPSO-BP Neural Network Based on Lorenz Disturbance. *IEEE Access* **2018**, *6*, 53168–53179. [CrossRef]
44. Naik, J.; Dash, S.; Dash, P.; Bisoi, R. Short term wind power forecasting using hybrid variational mode decomposition and multi-kernel regularized pseudo inverse neural network. *Renew. Energy* **2018**, *118*, 180–212. [CrossRef]
45. Zhang, C.; Peng, T.; Nazir, M.S. A novel hybrid approach based on variational heteroscedastic Gaussian process regression for multi-step ahead wind speed forecasting. *Int. J. Electr. Power Energy Syst.* **2022**, *136*, 107717. [CrossRef]
46. Shen, Y.; Wang, X.; Chen, J. Wind Power Forecasting Using Multi-Objective Evolutionary Algorithms for Wavelet Neural Network-Optimized Prediction Intervals. *Appl. Sci.* **2018**, *8*, 185. [CrossRef]
47. Yuan, X.; Chen, C.; Jiang, M.; Yuan, Y. Prediction interval of wind power using parameter optimized Beta distribution based LSTM model. *Appl. Soft Comput.* **2019**, *82*, 105550. [CrossRef]
48. Li, R.; Jin, Y. A wind speed interval prediction system based on multi-objective optimization for machine learning method. *Appl. Energy* **2018**, *228*, 2207–2220. [CrossRef]
49. Qin, G.; Yan, Q.; Zhu, J.; Xu, C.; Kammen, D.M. Day-Ahead Wind Power Forecasting Based on Wind Load Data Using Hybrid Optimization Algorithm. *Sustainability* **2021**, *13*, 1164. [CrossRef]
50. Duan, J.; Wang, P.; Ma, W.; Tian, X.; Fang, S.; Cheng, Y.; Chang, Y.; Liu, H. Short-term wind power forecasting using the hybrid model of improved variational mode decomposition and Correntropy Long Short-term memory neural network. *Energy* **2021**, *214*, 118980. [CrossRef]
51. Li, N.; He, F.; Ma, W. Wind Power Prediction Based on Extreme Learning Machine with Kernel Mean p-Power Error Loss. *Energies* **2019**, *12*, 673. [CrossRef]

52. Amila, T.; Peiris, J.J.; Rathnayake, U. Forecasting Wind Power Generation Using Artificial Neural Network: "Pawan Danawi" A Case Study from Sri Lanka. *J. Electr. Comput. Eng.* **2021**, *2021*, 5577547. [CrossRef]
53. Fu, Y.; Hu, W.; Tang, M.; Yu, R.; Liu, B. Multi-step Ahead Wind Power Forecasting Based on Recurrent Neural Networks. In Proceedings of the IEEE PES Asia-Pacific Power and Energy Engineering Conference (APPEEC), Kota Kinabalu, Malaysia, 7–8 October 2018; pp. 217–222. [CrossRef]
54. Liu, H.; Mi, X.W.; Li, Y.F. Wind speed forecasting method based on deep learning strategy using empirical wavelet transform, long short term memory neural network and Elman neural network. *Energy Convers. Manag.* **2018**, *156*, 498–514. [CrossRef]
55. Zhou, J.; Xu, X.; Huo, X.; Li, Y. Forecasting Models for Wind Power Using Extreme-Point Symmetric Mode Decomposition and Artificial Neural Networks. *Sustainability* **2019**, *11*, 650. [CrossRef]
56. Bazionis, I.K.; Georgilakis, P.S. Review of Deterministic and Probabilistic Wind Power Forecasting: Models, Methods, and Future Research. *Electricity* **2021**, *2*, 13–47. [CrossRef]
57. Jiao, R.; Huang, X.; Ma, X.; Han, L.; Tian, W. A Model Combining Stacked Auto Encoder and Back Propagation Algorithm for Short-Term Wind Power Forecasting. *IEEE Access* **2018**, *6*, 17851–17858. [CrossRef]
58. Ding, M.; Zhou, H.; Xie, H.; Wu, M.; Nakanishi, Y.; Yokoyama, R. A gated recurrent unit neural networks based wind speed error correction model for short-term wind power forecasting. *Neurocomputing* **2019**, *365*, 54–61. [CrossRef]
59. Wang, S.X.; Li, M.; Zhao, L.; Jin, C. Short-term wind power prediction based on improved small-world neural network. *Neural Comput. Appl.* **2019**, *31*, 3173–3185. [CrossRef]
60. Torabi, A.; Mousavy, S.A.K.; Dashti, V.; Saeedi, M.; Yousefi, N. A New Prediction Model Based on Cascade NN for Wind Power Prediction. *Comput. Econ.* **2019**, *53*, 1219–1243. [CrossRef]
61. Memarzadeh, G.; Keynia, F. A new short-term wind speed forecasting method based on fine-tuned LSTM neural network and optimal input sets. *Energy Convers. Manag.* **2020**, *213*, 112824. [CrossRef]
62. Shivani; Sandhu, K.; Ramachandran Nair, A. A Comparative Study of ARIMA and RNN for Short Term Wind Speed Forecasting. In Proceedings of the 10th International Conference on Computing, Communication and Networking Technologies (ICCCNT), Kanpur, India, 6–8 July 2019; pp. 1–7. [CrossRef]
63. Sopena, J.M.G.; Pakrashi, V.; Ghosh, B. Can we improve short-term wind power forecasts using turbine-level data? A case study in Ireland. In Proceedings of the IEEE Madrid PowerTech, Madrid, Spain, 28 June–2 July 2021; pp. 1–6. [CrossRef]
64. Saini, V.K.; Kumar, R.; Mathur, A.; Saxena, A. Short term forecasting based on hourly wind speed data using deep learning algorithms. In Proceedings of the 3rd International Conference on Emerging Technologies in Computer Engineering: Machine Learning and Internet of Things (ICETCE), Jaipur, India, 7–8 February 2020; pp. 1–6. [CrossRef]
65. Reddy, V.; Verma, S.M.; Verma, K.; Kumar, R. Hybrid Approach for Short Term Wind Power Forecasting. In Proceedings of the 9th International Conference on Computing, Communication and Networking Technologies (ICCCNT), Bengaluru, India, 10–12 July 2018; pp. 1–5. [CrossRef]
66. Hu, T.; Wu, W.; Guo, Q.; Sun, H.; Shi, L.; Shen, X. Very short-term spatial and temporal wind power forecasting A deep learning approach. *CSEE J. Power Energy Syst.* **2020**, *6*, 434–443. [CrossRef]
67. Karthikeyan, M.; Rengaraj, R. Short-Term Wind Speed Forecasting Using Ensemble Learning. In Proceedings of the 7th International Conference on Electrical Energy Systems (ICEES), Chennai, India, 11–13 February 2021; pp. 502–506. [CrossRef]
68. Liu, Y.; Ghasemkhani, A.; Yang, L.; Zhao, J.; Zhang, J.; Vittal, V. Seasonal Self-evolving Neural Networks Based Short-term Wind Farm Generation Forecast. In Proceedings of the IEEE International Conference on Communications, Control, and Computing Technologies for Smart Grids (SmartGridComm), Tempe, AZ, USA, 11–13 November 2020; pp. 1–6. [CrossRef]
69. Verma, S.M.; Reddy, V.; Verma, K.; Kumar, R. Markov Models Based Short Term Forecasting of Wind Speed for Estimating Day-Ahead Wind Power. In Proceedings of the International Conference on Power, Energy, Control and Transmission Systems (ICPECTS), Chennai, India, 22–23 February 2018; pp. 31–35. [CrossRef]
70. Zhang, P.; Wang, Y.; Liang, L.; Li, X.; Duan, Q. Short-Term Wind Power Prediction Using GA-BP Neural Network Based on DBSCAN Algorithm Outlier Identification. *Processes* **2020**, *8*, 157. [CrossRef]
71. Niu, D.; Pu, D.; Dai, S. Ultra-Short-Term Wind-Power Forecasting Based on the Weighted Random Forest Optimized by the Niche Immune Lion Algorithm. *Energies* **2018**, *11*, 1098. [CrossRef]
72. Liu, Y.; Guan, L.; Hou, C.; Han, H.; Liu, Z.; Sun, Y.; Zheng, M. Wind Power Short-Term Prediction Based on LSTM and Discrete Wavelet Transform. *Appl. Sci.* **2019**, *9*, 1108. [CrossRef]
73. Zhou, J.; Yu, X.; Jin, B. Short-Term Wind Power Forecasting: A New Hybrid Model Combined Extreme-Point Symmetric Mode Decomposition, Extreme Learning Machine and Particle Swarm Optimization. *Sustainability* **2018**, *10*, 3202. [CrossRef]
74. Shi, X.; Lei, X.; Huang, Q.; Huang, S.; Ren, K.; Hu, Y. Hourly Day-Ahead Wind Power Prediction Using the Hybrid Model of Variational Model Decomposition and Long Short-Term Memory. *Energies* **2018**, *11*, 3227. [CrossRef]
75. Wang, S.; Zhao, X.; Li, M.; Wang, H. TRSWA-BP Neural Network for Dynamic Wind Power Forecasting Based on Entropy Evaluation. *Entropy* **2018**, *20*, 284. [CrossRef]
76. An, G.; Jiang, Z.; Chen, L.; Cao, X.; Li, Z.; Zhao, Y.; Sun, H. Ultra Short-Term Wind Power Forecasting Based on Sparrow Search Algorithm Optimization Deep Extreme Learning Machine. *Sustainability* **2021**, *13*, 10453. [CrossRef]
77. Donadio, L.; Fang, J.; Porté-Agel, F. Numerical Weather Prediction and Artificial Neural Network Coupling for Wind Energy Forecast. *Energies* **2021**, *14*, 338. [CrossRef]

78. Kosana, V.; Teeparthi, K.; Madasthu, S.; Kumar, S. A novel reinforced online model selection using Q-learning technique for wind speed prediction. *Sustain Energy Technol. Assess.* **2022**, *49*, 101780. [CrossRef]
79. Duan, J.; Wang, P.; Ma, W.; Fang, S.; Hou, Z. A novel hybrid model based on nonlinear weighted combination for short-term wind power forecasting. *Int. J. Electr. Power Energy Syst.* **2022**, *134*, 107452. [CrossRef]
80. Xiang, L.; Liu, J.; Yang, X.; Hu, A.; Su, H. Ultra-short term wind power prediction applying a novel model named SATCN-LSTM. *Energy Convers. Manag.* **2022**, *252*, 115036. [CrossRef]
81. Hanifi, S.; Lotfian, S.; Seyedzadeh, S.; Zare-Behtash, H.; Cammarano, A. Offshore wind power forecasting—A new hyperparameter optimisation algorithm for deep learning models. *Energies* **2022**, *15*, 6919. [CrossRef]
82. Farah, S.; David, A.W.; Humaira, N.; Aneela, Z.; Steffen, E. Short-term multi-hour ahead country-wide wind power prediction for Germany using gated recurrent unit deep learning. *Renew. Sustain. Energy Rev.* **2022**, *167*, 112700. [CrossRef]
83. Kosovic, B.; Haupt, S.E.; Adriaansen, D.; Alessandrini, S.; Wiener, G.; Delle Monache, L.; Liu, Y.; Linden, S.; Jensen, T.; Cheng, W.; et al. A Comprehensive Wind Power Forecasting System Integrating Artificial Intelligence and Numerical Weather Prediction. *Energies* **2020**, *13*, 1372. [CrossRef]
84. Yu, R.; Gao, J.; Yu, M.; Lu, W.; Xu, T.; Zhao, M.; Zhang, J.; Zhang, R.; Zhang, Z. LSTM-EFG for wind power forecasting based on sequential correlation features. *Future Gener. Comput. Syst.* **2019**, *93*, 33–42. [CrossRef]
85. Yu, R.; Liu, Z.; Li, X.; Lu, W.; Ma, D.; Yu, M.; Wang, J.; Li, B. Scene learning: Deep convolutional networks for wind power prediction by embedding turbines into grid space. *Appl. Energy* **2019**, *238*, 249–257. [CrossRef]
86. Oh, E.; Wang, H. Reinforcement-Learning-Based Energy Storage System Operation Strategies to Manage Wind Power Forecast Uncertainty. *IEEE Access* **2020**, *8*, 20965–20976. [CrossRef]
87. Wang, X.; Li, P.; Yang, J. Short-term Wind Power Forecasting Based on Two-stage Attention Mechanism. *IET Renew. Power Gener.* **2019**, *14*, 297–304. [CrossRef]
88. Dorado-Moreno, M.; Navarin, N.; Gutierrez, P.; Prieto, L.; Sperduti, A.; Salcedo-Sanz, S.; Hervas-Martinez, C. Multi-task learning for the prediction of wind power ramp events with deep neural networks. *Neural Netw.* **2020**, *123*, 401–411. [CrossRef]
89. Ma, Y.J.; Zhai, M.Y. A Dual-Step Integrated Machine Learning Model for 24h-Ahead Wind Energy Generation Prediction Based on Actual Measurement Data and Environmental Factors. *Appl. Sci.* **2019**, *9*, 2125. [CrossRef]
90. Khosravi, A.; Machado, L.; Nunes, R. Time-series prediction of wind speed using machine learning algorithms: A case study Osorio wind farm, Brazil. *Appl. Energy* **2018**, *224*, 550–566. [CrossRef]
91. Wang, R.; Li, C.; Fu, W.; Tang, G. Deep Learning Method Based on Gated Recurrent Unit and Variational Mode Decomposition for Short-Term Wind Power Interval Prediction. *IEEE Trans. Neural Netw. Learn. Syst.* **2020**, *31*, 3814–3827. [CrossRef] [PubMed]
92. Han, L.; Jing, H.; Zhang, R.; Gao, Z. Wind power forecast based on improved Long Short Term Memory network. *Energy* **2019**, *189*, 116300. [CrossRef]
93. Sharifian, A.; Ghadi, M.J.; Ghavidel, S.; Li, L.; Zhang, J. A new method based on Type-2 fuzzy neural network for accurate wind power forecasting under uncertain data. *Renew. Energy* **2018**, *120*, 220–230. [CrossRef]
94. He, Y.; Li, H. Probability density forecasting of wind power using quantile regression neural network and kernel density estimation. *Energy Convers. Manag.* **2018**, *164*, 374–384. [CrossRef]
95. Shao, H.; Deng, X.; Jiang, Y. A novel deep learning approach for short-term wind power forecasting based on infinite feature selection and recurrent neural network. *J. Renew. Sustain. Energy* **2018**, *10*, 043303. [CrossRef]
96. Dong, D.; Sheng, Z.; Yang, T. Wind Power Prediction Based on Recurrent Neural Network with Long Short-Term Memory Units. In Proceedings of the 2018 International Conference on Renewable Energy and Power Engineering (REPE), Toronto, ON, Canada, 24–26 November 2018; pp. 34–38. [CrossRef]
97. Yang, X.; Zhang, Y.; Yang, Y.; Lv, W. Deterministic and Probabilistic Wind Power Forecasting Based on Bi-Level Convolutional Neural Network and Particle Swarm Optimization. *Appl. Sci.* **2019**, *9*, 1794. [CrossRef]
98. Ak, R.; Li, Y.F.; Vitelli, V.; Zio, E. Adequacy assessment of a wind-integrated system using neural network-based interval predictions of wind power generation and load. *Int. J. Electr. Power Energy Syst.* **2018**, *95*, 213–226. [CrossRef]
99. Sun, G.; Jiang, C.; Cheng, P.; Liu, Y.; Wang, X.; Fu, Y.; He, Y. Short-term wind power forecasts by a synthetical similar time series data mining method. *Renew. Energy* **2018**, *115*, 575–584. [CrossRef]
100. Sitharthan, R.; Parthasarathy, T.; Sheeba Rani, S.; Ramya, K.C. An improved radial basis function neural network control strategy-based maximum power point tracking controller for wind power generation system. *Trans. Inst. Meas. Control* **2019**, *41*, 3158–3170. [CrossRef]
101. Abedinia, O.; Lotfi, M.; Bagheri, M.; Sobhani, B.; Shafiekhah, M.; Catalao, J.P.S. Improved EMD-Based Complex Prediction Model for Wind Power Forecasting. *IEEE Trans. Sustain. Energy* **2020**, *11*, 2790–2802. [CrossRef]
102. Hammerschmitt, B.; Lucchese, F.C.; Silva, L.N.; Abaide, A.R.; Rohr, J.A.D.A. Short-Term Generation Forecasting Against the High Penetration of the Wind Energy. In Proceedings of the 8th International Conference on Modern Power Systems (MPS), Cluj-Napoca, Romania, 21–23 May 2019; pp. 1–6. [CrossRef]
103. Abesamis, K.; Ang, P.; Bisquera, F.I.; Catabay, G.; Tindogan, P.; Ostia, C.; Pacis, M. Short-Term Wind Power Forecasting Using Structured Neural Network. In Proceedings of the IEEE 11th International Conference on Humanoid, Nanotechnology, Information Technology, Communication and Control, Environment, and Management (HNICEM), Laoag, Philippines, 29 November–1 December 2019; pp. 1–4. [CrossRef]

104. Bhushan Sahay, K.; Srivastava, S. Short-Term Wind Speed Forecasting of Lelystad Wind Farm by Using ANN Algorithms. In Proceedings of the International Electrical Engineering Congress (iEECON), Krabi, Thailand, 7–9 March 2018; pp. 1–4. [CrossRef]
105. Vaitheeswaran, S.S.; Ventrapragada, V.R. Wind Power Pattern Prediction in time series measuremnt data for wind energy prediction modelling using LSTM-GA networks. In Proceedings of the 10th International Conference on Computing, Communication and Networking Technologies (ICCCNT), Kanpur, India, 6–8 July 2019; pp. 1–5. [CrossRef]
106. Wang, S.; Sun, Y.; Zhai, S.; Hou, D.; Wang, P.; Wu, X. Ultra-Short-Term Wind Power Forecasting Based on Deep Belief Network. In Proceedings of the Chinese Control Conference (CCC), Guangzhou, China, 27–30 July 2019; pp. 7479–7483. [CrossRef]
107. Medina, S.V.; Ajenjo, U.P. Performance Improvement of Artificial Neural Network Model in Short-term Forecasting of Wind Farm Power Output. *J. Mod. Power Syst. Clean Energy* **2020**, *8*, 484–490. [CrossRef]
108. Pang, D.; Li, G. Research on Wind Power Output Reconstruction Technology Based on Multiple Time Scales. In Proceedings of the 2nd IEEE Conference on Energy Internet and Energy System Integration (EI2), Beijing, China, 20–22 October 2018; pp. 1–5. [CrossRef]
109. Prabha, P.P.; Vanitha, V.; Resmi, R. Wind Speed Forecasting using Long Short Term Memory Networks. In Proceedings of the 2019 2nd International Conference on Intelligent Computing, Instrumentation and Control Technologies (ICICICT), Kannur, India, 5–6 July 2019; Volume 1, pp. 1310–1314. [CrossRef]
110. Musbah, H.; Aly, H.H.; Little, T.A. A Novel Approach for Seasonality and Trend Detection using Fast Fourier Transform in Box-Jenkins Algorithm. In Proceedings of the IEEE Canadian Conference on Electrical and Computer Engineering (CCECE), London, ON, Canada, 30 August–2 September 2020; pp. 1–5. [CrossRef]
111. Viet, D.T.; Phuong, V.V.; Duong, M.Q.; Tran, Q.T. Models for Short-Term Wind Power Forecasting Based on Improved Artificial Neural Network Using Particle Swarm Optimization and Genetic Algorithms. *Energies* **2020**, *13*, 2873. [CrossRef]
112. Liu, T.; Huang, Z.; Tian, L.; Zhu, Y.; Wang, H.; Feng, S. Enhancing Wind Turbine Power Forecast via Convolutional Neural Network. *Electronics* **2021**, *10*, 261. [CrossRef]
113. Tang, B.; Chen, Y.; Chen, Q.; Su, M. Research on Short-Term Wind Power Forecasting by Data Mining on Historical Wind Resource. *Appl. Sci.* **2020**, *10*, 1295. [CrossRef]
114. Liu, B.; Zhao, S.; Yu, X.; Zhang, L.; Wang, Q. A Novel Deep Learning Approach for Wind Power Forecasting Based on WD-LSTM Model. *Energies* **2020**, *13*, 4964. [CrossRef]
115. Bochenek, B.; Jurasz, J.; Jaczewski, A.; Stachura, G.; Sekuła, P.; Strzyżewski, T.; Wdowikowski, M.; Figurski, M. Day-Ahead Wind Power Forecasting in Poland Based on Numerical Weather Prediction. *Energies* **2021**, *14*, 2164. [CrossRef]
116. Li, Y.; Yang, F.; Zha, W.; Yan, L. Combined Optimization Prediction Model of Regional Wind Power Based on Convolution Neural Network and Similar Days. *Machines* **2020**, *8*, 80. [CrossRef]
117. Gouveia, H.T.V.; De Aquino, R.R.B.; Ferreira, A.A. Enhancing Short-Term Wind Power Forecasting through Multiresolution Analysis and Echo State Networks. *Energies* **2018**, *11*, 824. [CrossRef]
118. Ahmadi, A.; Nabipour, M.; Mohammadi-Ivatloo, B.; Amani, A.M.; Rho, S.; Piran, M.J. Long-Term Wind Power Forecasting Using Tree-Based Learning Algorithms. *IEEE Access* **2020**, *8*, 151511–151522. [CrossRef]
119. Liu, H.; Chen, C.; Lv, X.; Wu, X.; Liu, M. Deterministic wind energy forecasting: A review of intelligent predictors and auxiliary methods. *Energy Convers. Manag.* **2019**, *195*, 328–345. [CrossRef]
120. Al-Zadjali, S.; Al Maashri, A.; Al-Hinai, A.; Al Abri, R.; Gajare, S.; Al Yahyai, S.; Bakhtvar, M. A Fast and Accurate Wind Speed and Direction Nowcasting Model for Renewable Energy Management Systems. *Energies* **2021**, *14*, 7878. [CrossRef]
121. Ding, C.; Zhou, Y.; Ding, Q.; Li, K. Integrated Carbon-Capture-Based Low-Carbon Economic Dispatch of Power Systems Based on EEMD-LSTM-SVR Wind Power Forecasting. *Energies* **2022**, *15*, 1613. [CrossRef]
122. Yang, H.; Jiang, P.; Wang, Y.; Li, H. A fuzzy intelligent forecasting system based on combined fuzzification strategy and improved optimization algorithm for renewable energy power generation. *Appl. Energy* **2022**, *325*, 119849. [CrossRef]
123. Madsen, H.; Pinson, P.; Kariniotakis, G.; Nielsen, H.A.; Nielsen, T.S. A Protocol for Standardazing the Performance Evaluation of Short Term Wind Power Prediction Models. *Wind Eng.* **2005**, *29*, 475–489. [CrossRef]
124. Hyndman, R.J.; Koehler, A.B. Another Look at Measures of Forecast Accuracy. *Int. J. Forecast.* **2006**, *22*, 679–688. [CrossRef]

energies

Article

Analysis and Forecasting of Monthly Electricity Demand Time Series Using Pattern-Based Statistical Methods

Paweł Pełka

Electrical Engineering Faculty, Czestochowa University of Technology, 42-200 Czestochowa, Poland; pawel.pelka@pcz.pl

Abstract: This article provides a solution based on statistical methods (ARIMA, ETS, and Prophet) to predict monthly power demand, which approximates the relationship between historical and future demand patterns. The energy demand time series shows seasonal fluctuation cycles, long-term trends, instability, and random noise. In order to simplify the prediction issue, the monthly load time series is represented by an annual cycle pattern, which unifies the data and filters the trends. A simulation study performed on the monthly electricity load time series for 35 European countries confirmed the high accuracy of the proposed models.

Keywords: medium-term load forecasting; pattern-based forecasting; time-series preprocessing

1. Introduction

Forecasting loads on power systems is an integral activity embedded in the long-term system operation planning processes and in the processes of the ongoing control of its operation. The system cannot function without accurate forecasts. This is because electricity cannot be stored in large quantities. The demand must be covered on an ongoing basis with production, with the limitations resulting from the flexibility of the production units and the requirements of the reliability and safety of the system operation. The accuracy of forecasts translates into the costs of production, transmission, and the degree of reliability of electricity supply to consumers. Inflated forecasts lead to the maintenance of too many generating units in order to meet the safety requirements to ensure an adequate margin of reserve capacity. Underestimated forecasts have the opposite effect—too few generating units are planned, which are not able to cover the actual demand. In such a situation, additional units with quick start-up are intervened in the traffic, generating additional operating costs [1].

Medium-term forecasting most often concerns forecasting the monthly electricity load (MEL). MEL time series contain components of nonlinear trend, annual seasonality and random disturbances. They show a significant variation in the variance and shape of the annual cycle over time. MEL is highly dependent on economic and socioeconomic as well as climatic and weather variables. The factors disturbing the MEL include unpredictable economic events, extreme weather changes and political decisions [2]. The importance of MEL forecasting in the power sector and the complexity of the problem encourage the search for forecasting models that will meet the requirements of the specificity of the task and generate accurate forecasts.

Medium-term electricity demand forecasting is a very well-researched issue. There is a great deal of solutions used to settle this issue, including classical/statistical methods [3,4], neural networks (NNs) [5–7], deep learning [8–12] or similarity-based methods [13–18]. Traditional strategies were first launched for electricity load forecasting. Linear regression, ARIMA, and exponential smoothing (ETS) methods have been widely used [19]. The restricted versatile capacities of these strategies and their linear nature have brought about an expanded interest in artificial intelligence techniques [5]. Neural networks in [5] are utilized to predict the trend of the time series of MELs and the Fourier series are included to

Citation: Pełka, P. Analysis and Forecasting of Monthly Electricity Demand Time Series Using Pattern-Based Statistical Methods. *Energies* **2023**, *16*, 827. https://doi.org/10.3390/en16020827

Academic Editor: Surender Reddy Salkuti

Received: 13 December 2022
Revised: 4 January 2023
Accepted: 9 January 2023
Published: 11 January 2023

forecast the seasonal component. Then, at that point, the two gauges, trend and occasional changes, are totaled. In [20], a deep short-term memory network (LSTM) was carried out for the probabilistic estimating of client load profiles. The LSTM method is also used for predicting electricity prices in the article [21] with good accuracy.

The simplest classical models include naive models, which assume a selected historical load value as a forecast. In series that exhibit seasonality, such as monthly load series, this is a twelve-month shifted value. Random fluctuations and trend may negatively affect the forecast results.

Linear regression models allow for taking into account the trend (only linear), but the implementation of seasonal cycles in the model requires additional operations, e.g., decomposition of the series into individual months. An example of the application of a linear model for medium- and long-term forecasting of the electricity loads can be found in [22]. The model uses strong daily (24 h) and annual (52 weeks) correlations to forecast daily load profiles in the horizon from several weeks to several years. The forecast results are corrected for annual load increments. For forecasting with a one-year horizon, the MAPE errors were obtained with values not greater than 3.8%. In [3], the operation of the linear regression model and the ARIMA model in the task of forecasting monthly peak loads up to 12 months in advance was compared. The models are powered by the same set of inputs, including historical peak load data, weather data, and economic data. About twice the accuracy of the ARIMA model was demonstrated experimentally. For non-stationary time series with an irregular periodic trend, [23] proposed a linear regression model extended with periodic components implemented by the sine function of different frequencies.

The spatial autoregression model for medium-term load forecasting is described in [24]. The authors noted a strong correlation between the load on the system and GDP in the analyzed thirty Chinese provinces. To forecast the load in a given province, they used not only the local dependence of the load on GDP, but also the relationships identified for neighboring provinces. This allows to reduce forecast errors from 5.2–5.4% to 3.5–3.9%. Ref. [25] described ARIMA and ETS prognostic methods in combination with bootstrap aggregation. The time series were initially processed using the Box-Cox transform (a procedure often used to compensate for variances) and decomposed. The ARMA model was used for the generation of bootstrap tests, and the ARIMA and ETS models for their forecasting.

Classical methods also use Fourier series. In [5], they were used to model the seasonal component of the time series of monthly loads. Based on spectral analysis, six fundamental frequencies were identified, and a Fourier series was created for them. The forecast of the seasonal component calculated by the Fourier series was added to the trend forecast. Markov chains were used in [26] to analyze the input data and select the best forecasting model. This approach is especially useful when the upward trend of the time series is unstable. The classic models also include the model described in [27], which is based on a simple logistic function. The input variables are time, maximum atmospheric temperature and the social factor, taking into account religious holidays (the model was developed for Arab countries).

In this work, statistical methods for estimating month-to-month power demand are utilized. What differentiates these models from other traditional strategies is that they use pattern representation of seasonal cycles of the time series. The patterns allow us to unify the data and filter out the trend. The input and output variables in the pattern space are characterized by a less complex relationship contrasted with the original space. Thus, due to pattern representation, the classical forecasting method has an easier task to perform. Using time-series preprocessing by calculating yearly patterns constitutes the direct novelty of the proposed models.

Patterns represent fragments of time series and extract information about the shape of these fragments, filter out the trend and normalize the variance. By using a pattern-based representation, the relationship between the input and output variables is simplified, and hence the predictive model is simplified. Unlike parametric systems, which are most often "black box" models, the statistical methods' principle of operation is understandable, which is important in practical industrial applications and translates to a greater degree of confidence in the forecast.

2. MEL Time Series Analysis

A time series is a set of observations of a certain variable ordered in the time domain: $\{z_t, t = 1, 2, \ldots, N\}$. Formally, the observations z_t are implementations of a sequence of random variables $\{Z_t, t = 1, 2, \ldots, N\}$ having certain specified cumulative distribution. These definitions are applied to a discrete variable when the time variable t takes values at equal intervals (seconds, hours, days, months).

The purpose of the time series analysis is to detect and describe the regularities affecting the phenomenon expressed in the form of a time series. The following components of time series are distinguished, which are the effect of the influence of various factors on the studied phenomenon [28,29]:

- Development tendency (trend), which is revealed by one-way and systematic changes in the level of a given phenomenon that takes place over a long-term period;
- Periodic fluctuations, i.e., rhythmic fluctuations with a specific cycle;
- Business cycle fluctuations, i.e., systemic wave fluctuations observed in longer periods;
- Random fluctuations, i.e., irregular random disturbances.

Periodic fluctuations, also known as seasonal fluctuations [30], are characterized by a constant period, as opposed to a cyclical fluctuations in which there are no fixed period fluctuations.

A time series is called stationary if its statistical properties do not change over time. That is, a time series of $\{Z_t\}$ has the same properties as time series shifted in time $\{Z_{t+l}\}$, for each value shift l. In practice, weak stationarity is tested (so-called wide-sense stationarity), which requires immutability mean level, variance and time correlation. It can be written as follows [28]:

- $EZ_t = \text{const}$—expected value is constant (none trend);
- $\text{Var}(Z_t) = \text{const}$—the variance is homogeneous over time;
- $\text{Corr}(Z_r, Z_s) = \text{Corr}(Z_{r+t}, Z_{s+t}) = \rho(s - r) = \rho(l)$—the correlation between observations is solely dependent on time shift l.

The stationarity of the time series plays a key role in a forecasting model selection and the construction of forecasts. Establishing forecasts for the stationery time series, where the basic properties do not change over time, is much easier than for non-stationary series. Therefore, in practice, it is often sought to transform the time series into a stationary form before starting the forecasting process.

A series of MELs for Poland with a box diagram, showing the distribution of monthly values in the following years, is shown in Figure 1. The variability of the median and variance is clearly visible over time. The last year of observation is characterized by almost a three times smaller interquartile range (box height) than the first one. Figure 2 shows the MEL time series for 35 European countries. The time series vary in length, from 60 to 288 months, ending with a year 2014. As we can see, these time series are non-stationary, and they show variability of mean value and variance. They are characterized by strong annual seasonality and non-linear trends.

Other forms of graphical presentation of MEL time series are shown in Figures 3 and 4. Figure 3 presents the trends set in the subsequent months of the year. Strong upward trends are observed in the summer months, which is not the case for the winter months. Figure 4) allows us to observe distinctive features of annual cycles and assess the shape similarity of these cycles over the years. Despite the reduction in the amplitude of the yearly fluctuations in subsequent years, the shapes of the annual cycles show similarities. The demand in the winter months is significantly greater than the demand in the summer months. It is observed that there is understated demand in February with the following months (which cannot only be explained by the smaller number of days this month) and overstated demand in March and October. The shape similarities of the annuals cycles are particularly important in the nonparametric regression models [31]. The seasonal graph can be compiled in polar coordinates (Figure 4)).

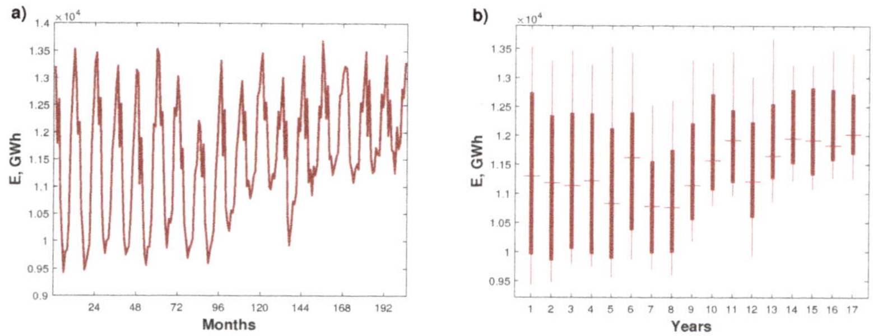

Figure 1. Graph of the monthly electricity demand time series for Poland (**a**) and the box diagram showing the medians and annual dispersions (**b**).

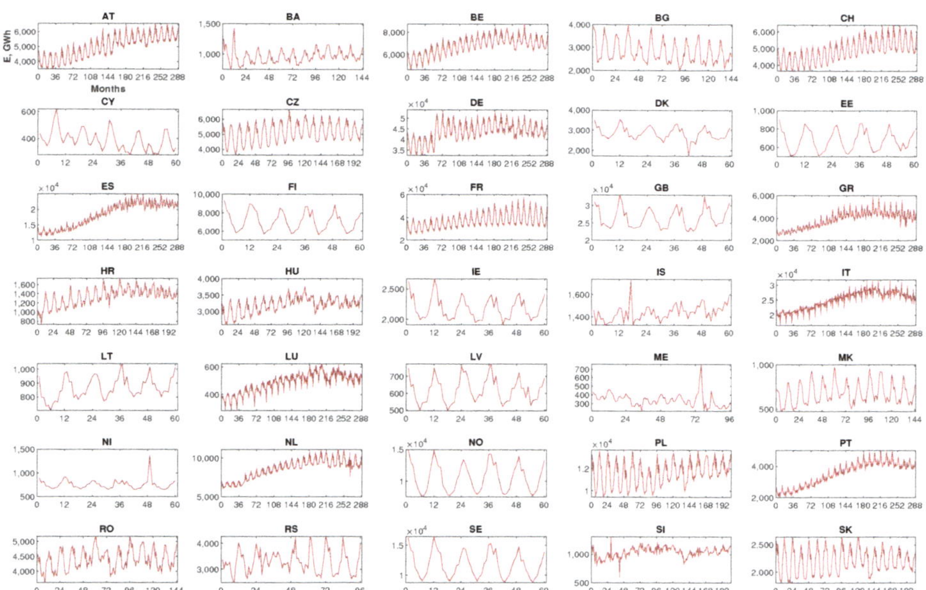

Figure 2. Monthly electricity demand time series for European countries.

Relationship between observations of the time series distant by l time units are assessed using the autocorrelation function (ACF) and the partial autocorrelation (PACF). The former has the form [28]:

$$ACF(l) = \frac{\widehat{\gamma}(l)}{\widehat{\gamma}(0)}, l = 0, 1, \ldots, N-1, \tag{1}$$

where $\widehat{\gamma}(h)$ is a sample autocovariance function:

$$\widehat{\gamma}(l) = \frac{1}{N}\sum_{j=1}^{N-l}\left(Z_{j+l} - \overline{Z}\right)\left(Z_j - \overline{Z}\right), \tag{2}$$

where $\overline{Z}$ is the mean value of the time series.

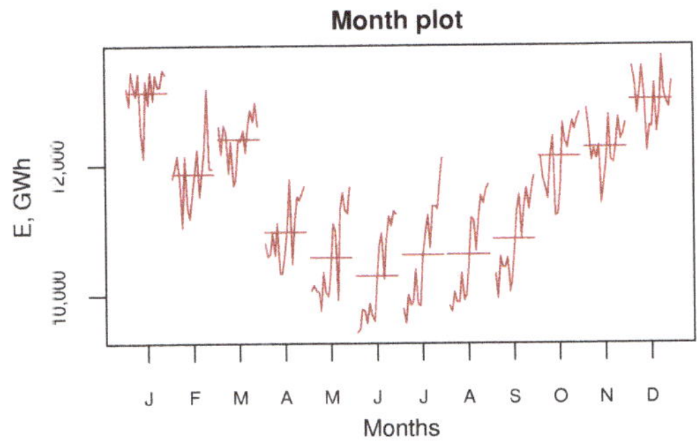

Figure 3. Month plot of energy load for Poland.

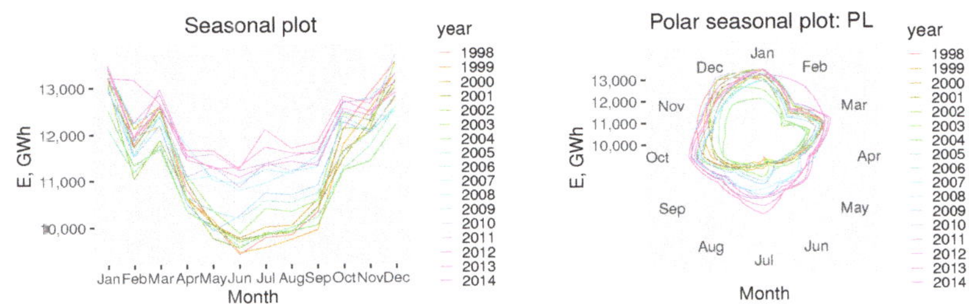

Figure 4. Seasonal MEL plots for Poland.

In the case of PACF, when determining the autocorrelation between Z_t and Z_{t+l}, the influence of intermediate observations on this is eliminated dependency, $Z_{t+1}, \ldots, Z_{t+l-1}$. Sample partial autocorrelation is an estimate of a theoretical PACF function of a given pattern:

$$\alpha(l) = \begin{cases} \mathrm{Corr}(Z_{t+1}, Z_t), l = 1 \\ \mathrm{Corr}(Z_{t+l} - P_{t,l}(Z_{t+l}), Z_t - P_{t,l}(Z_t)), l > 1' \end{cases} \tag{3}$$

where $P_{t,l}(Z_t)$ is the orthogonal projection operator on linear subspace spanning over variables $Z_{t+1}, \ldots, Z_{t+l-1}$.

Examples of ACF and PACF charts for the MEL time series are shown in Figure 5. The horizontal lines designate the confidence intervals, which allow a conclusion to be drawn about the statistical significance of the autocorrelation. The 95% confidence interval has the form $\left[-\frac{1.96}{\sqrt{N}}, \frac{1.96}{\sqrt{N}}\right]$. Autocorrelations outside this range are considered to be statistically significant. On the ACF chart, we can see very strong oscillations, proving a clear seasonality of the series. According to what is expected, the strongest autocorrelation occurs for lags which are multiples of twelve, which is confirmed by the annual cycle. The big PACF value for delay $l = 1$, close to 1, can signal the presence of an uptrend.

Seasonal fluctuations can also be identified using the harmonic (spectral, spectral) analysis [32]. It leads to creating a model consisting of the sum of the sine and cosine functions different frequency. A time series of length n is recorded using the Fourier series as follows [33,34]:

$$f(t) = a_0 + \sum_{i=1}^{N/2}\left[a_i sin\left(\frac{2\pi it}{N}\right) + b_i \cos\left(\frac{2\pi it}{N}\right)\right], \tag{4}$$

where a_0, a_i and b_i are the coefficients which are determined from patterns:

$$\begin{gathered} a_0 = \frac{1}{N}\sum_{t=1}^{N} Z_t \\ a_i = \frac{2}{N}\sum_{t=1}^{N} Z_t \sin\left(\frac{2\pi i t}{N}\right), i = 1,2,\ldots,\frac{N}{2}-1 \\ b_i = \frac{2}{N}\sum_{t=1}^{N} Z_t \cos\left(\frac{2\pi i t}{N}\right), i = 1,2,\ldots,\frac{N}{2}-1, \\ a_{N/2} = 0 \\ b_{N/2} = \frac{1}{N}\sum_{t=1}^{N} Z_t cos(\pi t) \end{gathered} \quad (5)$$

The magnitudes of the amplitudes for successive harmonics are as follows:

$$A_i = \sqrt{a_i^2 + b_i^2}. \quad (6)$$

The amplitude of i-th harmony, A_i, testifies to the participation of this harmony in explaining the variance of the considered variable. This share is expressed by the following formula [33]:

$$u_i = \frac{A_i^2}{2Var(Z_t)}, \quad (7)$$

The amplitudes of successive harmonics for the MEL time series for Poland are shown in a periodogram (Figure 6). As can be seen, the dominant period in this series is a year.

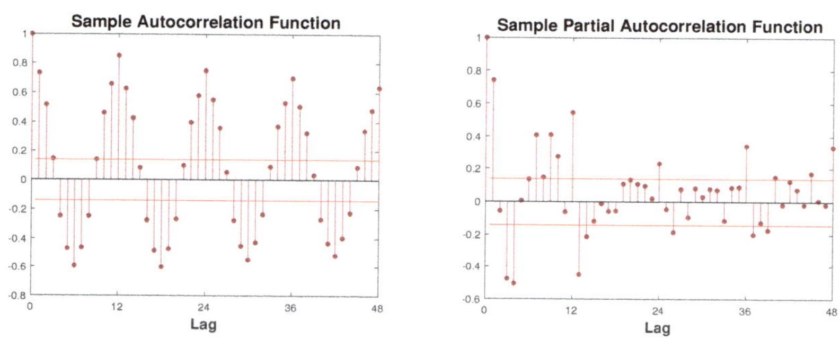

Figure 5. Graphs of autocorrelation and partial correlation of the MEL time series for Poland (*lag* means delay *l*).

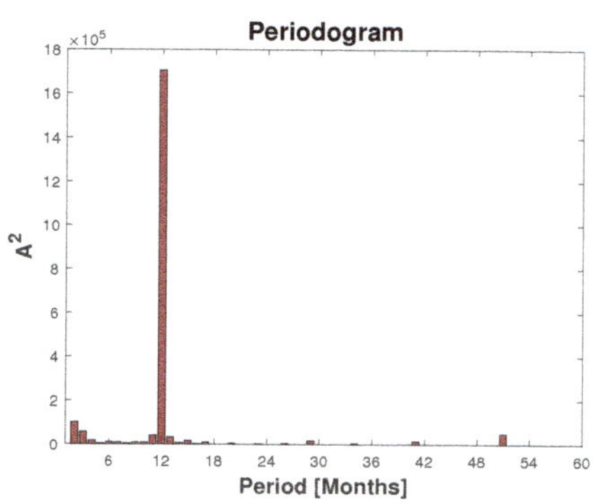

Figure 6. Periodogram of the MEL time series for Poland.

3. Forecasting Model

Let us consider the monthly electricity demand time series beginning from January and finishing off with December: $E = \{E_t : t = 1,2,\ldots,N\}$. We divide each time series into yearly fragments $E_i = \{E_t : t = 12(i-1)+1, 12(i-1)+2,\ldots,12(i-1)+12\}$, $i = 1,2,\ldots,N/12$. Every fragment can be described by a vector $\mathbf{E}_i = [E_{i,1} E_{i,2} \ldots E_{i,12}]^T$.

Let us formulate an x-pattern $\mathbf{x}_i = [x_{i,1}x_{i,2}\cdots x_{i,12}]^T$ as a vector, which represents a yearly fragment E_i. A function, which transforms time series points into patterns, is chosen, including the character of the time series, such as trend, variance, and seasonalities. We proposed a few definitions of this function in [35]. In this study, x-patterns are defined as [8]

$$x_{i,j} = \frac{E_{i,j} - \overline{E}_i}{\sigma_i}, \tag{8}$$

where $j = 1,2,\ldots,12$, $\overline{E}_i$ is a mean of sequence E_i, and $\sigma_i = \sqrt{\sum_{j=1}^{n}(E_{i,j} - \overline{E}_i)^2}$ is a measure of the sequence E_i dispersion.

The X-pattern is a normalized $\mathbf{E}_i$ vector. Note that yearly fragments expressed by $\mathbf{E}_i$ have a different mean and dispersion. After the normalization process, time-series fragments are also unified: all x-patterns have unity length, the same variance and also the mean value of these fragments equals zero. x-patterns carry information about the shapes of the yearly fragments. Then, the new time series composed of x-patterns representing successive yearly periods are created: $x = \{x_i : i = 1,2,\ldots,N/12\} = \{x_{1,1}, x_{1,2}, \ldots, x_{N/12,12}\}$. Note that it is distinguished by regular character and stationarity.

The forecasting procedure using the time series composed of x-patterns requires determining the demand forecast using the x-pattern forecast. After generating the x-pattern by the forecasting model, the MELs in the forecasted yearly period are computed from the forecasted x-pattern using transformed Equation (8) (this is called decoding):

$$\hat{E}_{i,j} = \hat{x}_{i,j} * \hat{\sigma}_i + \hat{\overline{E}}_i, j = 1,2,\ldots,12. \tag{9}$$

However, in this equation, the coding variables, $\overline{E}_i$ and σ_i, are not known because they are the mean and dispersion of the future fragment, which is forecasted. So, the coding variables must be forecasted based on their historical values. ARIMA and ETS models are used for this purpose in this work in every model and for preprocessed time-series forecasting, when we use these two models [8]. Figure 7 presents the idea of pattern-based forecasting using a block diagram of the proposed methodology presenting data structures and illustrating data flow.

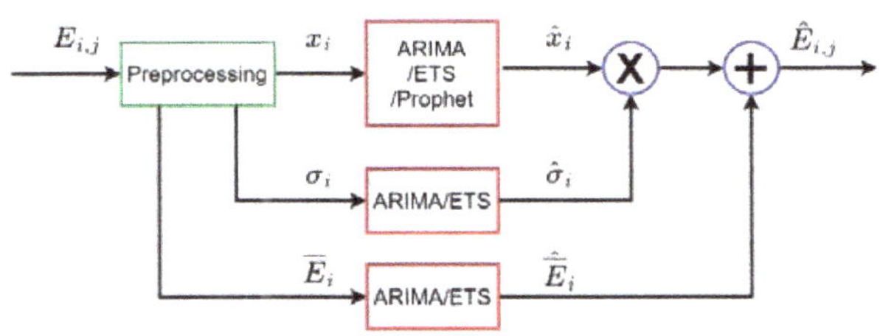

Figure 7. A block diagram of pattern-based MEL forecasting methodology.

3.1. *Autoregressive Integrated Moving Average ARIMA Model*

The autoregressive integrated moving average ARIMA model $ARIMA(p,d,q)(P,D,Q)_m$ [30,36] was used to model the MEL time series:

$$\Phi(B^m)\phi(B)(1-B)^D(1-B)^d z_t = c + \Theta(B^m)\theta(B)\xi_t, \tag{10}$$

where z_t are the terms of the time series (in the considered case of the series $\{E_t, t = 1,2,\ldots,N\}$), m is the length of the annual cycle (m = 12), B is the backward shift operator, D and d are the orders of seasonal differentiation and of the ordinary, respectively, $\phi(.)$, $\Phi(.)$, $\theta(.)$, and $\Theta(.)$ are polynomials of degree p, g, P and Q, respectively, c denotes a constant, and ξ_t is a white noise process with zero mean and the variance of σ^2.

The ARIMA model was also used to construct predictive models for the code variables $\hat{\overline{E}}_i$ and $\hat{\sigma}_i$ needed to decode the pattern x. In this case, the ARIMA model can be simplified to the non-seasonal ARIMA form (p, d, q):

$$\phi(B)(1 - B^d)z_t = c + \theta(B)\xi_t, \tag{11}$$

where z_t represents the terms of the time series $\{\overline{E}_1, \overline{E}_2, \ldots, \overline{E}_{N-1}\}$ or $\{\sigma_1, \sigma_2, \ldots, \sigma_{N-1}\}$.

3.2. Exponential Smoothing ETS

The seasonal exponential smoothing method (Holt–Winters method), known since the 1960s, is, next to the ARIMA method, one of the most frequently used in practice. The idea of ETS is to assign exponentially declining weights to observations from previous periods. Thus, observations from recent periods have a greater impact on the forecast. The Holt–Winters model is based on three smoothing equations that represent the level of the forecast variable, its increment and seasonality. There are two types of Holt–Winters methods, which depend on how seasonality is modeled. The additive version of the Holt–Winters method is used when seasonal fluctuations are independent of the trend. On the other hand, the multiplicative version is used when there is a proportional relationship between seasonal fluctuations and the trend [28].

The equations for the additive version of the Holt–Winters method are [36]

$$\begin{array}{c} Level:\ l = \alpha(y - s_{-m}) + (1-\alpha)(l_{-1} - b_{-1}) \\ Increment:\ b = \beta(l - l_{-1}) + (1-\beta)b_{-1} \\ Seasonality:\ s = \gamma(y - l_{-1} - b_{-1}) + (1-\gamma)s_{-m}{}' \\ \text{Forecast}:\ \hat{y}_{t+h|t} = l + bh + s_{-m+h_m^+} \end{array} \tag{12}$$

where m is the period of seasonal fluctuations ($m = 12$ for the MEL series), h is the forecast horizon, $h_m^+ = [(h-1)\ mod\ m] + 1$, and α, β, γ are the smoothing coefficients from the range $(0, 1)$.

Before using the model, the initial values of the states l_0, b_0, $s_{1-m}, \ldots, s_0$ and the smoothing parameters α, β and γ should be given. All these values are then estimated from the observed data.

In [36], ETS models were defined by state equations and classified into 30 types. These types differ in the ways in which the model includes the trend components (the sum of the level and the increment), seasonality and error. Components can be expressed as additive or multiplicative, and the trend can be further suppressed.

The ETS model was also used to build forecasting models for the code variables $\hat{\overline{E}}_i$ and $\hat{\sigma}_i$.

3.3. Prophet

Prophet is a time series forecasting method designed by Facebook for direct use in business applications [37]. It is distinguished by a completely automatic forecasting procedure with an intuitive selection of parameter values that can be adjusted without knowing the details of the base model. Prophet is immune to data deficiencies and trend changes. It usually copes well with outliers. The model implementation is publicly available in the Python and R environments.

Prophet is an additive model with three components—non-linear trend, seasonality, and a component that represents holidays:

$$y(t) = g(t) + s(t) + h(t) + \epsilon_t, \tag{13}$$

where $g(t)$ is a trend function modeling non-periodic changes in the value of the time series, $s(t)$ represents seasonal changes (e.g., weekly and annual seasonality), $h(t)$ represents holiday effects, and ϵ_t is a residual component with a normal distribution.

The model specification is similar to the generalized additive model (GAM), with nonlinear smoothing and time t as the sole regressor. The trend is modeled in two ways: using a piecewise-linear model or a limited-growth model. In the latter case, the logistic curve is used:

$$g(t) = \frac{C(t)}{1 + exp(-k(t)(t - q(t)))}, \quad (14)$$

where $C(t)$ is the carrying capacity, $k(t)$ is the increment function, and $q(t)$ is the offset function.

In (14) both the carrying capacity and the increment and offset are functions of time. They vary depending on the characteristics of the time series. This makes it possible to flexibly shape the trend function.

The seasonal component is modeled using the Fourier series:

$$s(t) = \sum_{i=1}^{N} \left(a_i cos\left(\frac{2\pi it}{m}\right) + b_i \sin\left(\frac{2\pi it}{m}\right) \right), \quad (15)$$

where m is the length of the seasonal cycle, and a_i and b_i are the coefficients.

The component representing the effects of public holidays, $h(t)$, is not applicable in forecasting MEL time series.

4. Simulation Study

In this section, we analyze the monthly electricity demand time series for 35 European countries and verify our proposed forecasting model on these time series. The data were downloaded from the ENTSO-E repository, (www.entsoe.eu, accessed on 12 April 2016).

4.1. MEL Time Series Analysis Results

Table 1 shows the statistics and parameters describing the analyzed time series:

- Median—median as a measure of the average level of the series,
- IQR—average of the annual interquartile ranges as a measure of the annual dispersion of the series,
- $\text{iqr}_{\%}$—mean relative annual dispersion as mean ratio of annual interquartile ranges to annual medians:

$$\text{iqr}_{\%} = \frac{100}{M} \sum_{i=1}^{M} \frac{\text{IQR}_i}{\text{Median}_i}, \quad (16)$$

where M is the length of the series in years, IQR_i and Median_i is the interquartile range and the median of the year i,

- $ACF(12)$—value of the autocorrelation function for the delay $l = 12$,
- u^{12}—annual-period harmonics share in the series variance (7).

The countries classified as the largest, most developed and with the largest number of inhabitants in Europe have the greatest demand for electricity, i.e., Germany, France, Italy and Great Britain. The greatest relative annual volatility $\text{iqr}_{\%}$ is characterized in the following order: Norway, Sweden, Montenegro, Bulgaria, Estonia and France. In these cases, the interval of the median is over 25% of the median. The least annual volatility is observed for Italy and Iceland (less than 7%).

The strongest autocorrelation for the annual delay $l = 12$ ($ACF(12) \geq 0.9$), signaling the clearest annual cycles, occurs for the time series of the MEL of Switzerland, Spain, Portugal, France and Italian. The MEL time series of Montenegro, Northern Ireland and Iceland show the lowest values $ACF(12)$.

The annual-period harmonics share is the highest for Norway, Sweden, Finland and Estonia. There is a high correlation u^{12} with mean relative annual dispersion $\text{iqr}_{\%}$. The harmonic analysis in some cases gives very low values u^{12} despite the high values of the annual autocorrelation, e.g., for Italy, Greece and Spain. Periodograms for these countries show a high bar for the year period but many lower bars for higher frequencies. This means that there are disturbances in the series of those countries with the lower value of u^{12}, masking the picture of the annual cycle.

Table 1. Statistics and describing parameters analyzed MEL time series.

Symbol	Median	IQR	iqr%
AT	4978.00 ± 641.00	720.00 ± 164.00	15.00 ± 4.36
BA	955.00 ± 49.00	141.00 ± 58.00	14.91 ± 5.66
BE	6981.00 ± 609.00	848.00 ± 96.00	12.66 ± 1.93
BG	2654.00 ± 152.00	740.00 ± 200.00	27.69 ± 6.64
CH	4885.00 ± 425.00	862.00 ± 120.00	18.15 ± 2.87
CY	375.00 ± 35.00	87.00 ± 12.00	23.33 ± 3.41
CZ	5134.00 ± 243.00	1105.00 ± 203.00	21.98 ± 4.82
DE	44,140.00 ± 2883.00	5921.00 ± 1411.00	13.79 ± 3.81
DK	2824.00 ± 86.00	431.00 ± 86.00	15.46 ± 2.95
EE	672.00 ± 16.00	181.00 ± 42.00	27.37 ± 6.66
ES	19,294.00 ± 3778.00	1466.00 ± 437.00	8.11 ± 1.49
FI	7052.00 ± 89.00	1636.00 ± 204.00	23.33 ± 2.84
FR	36,860.00 ± 2948.00	9116.00 ± 2822.00	25.01 ± 6.69
GB	25,751.00 ± 869.00	4559.00 ± 750.00	17.66 ± 2.80
GR	4094.00 ± 623.00	467.00 ± 161.00	12.03 ± 3.05
HR	1346.00 ± 158.00	196.00 ± 40.00	15.24 ± 3.86
HU	3224.00 ± 180.00	303.00 ± 92.00	9.70 ± 3.51
IE	2157.00 ± 12.00	290.00 ± 42.00	13.48 ± 1.91
IS	1413.00 ± 41.00	92.00 ± 9.00	6.48 ± 0.71
IT	25,913.00 ± 2767.00	1290.00 ± 246.00	5.15 ± 0.86
LT	864.00 ± 38.00	102.00 ± 9.00	11.99 ± 0.91
LU	509.00 ± 64.00	46.00 ± 11.00	9.53 ± 2.44
LV	622.00 ± 18.00	105.00 ± 19.00	17.01 ± 3.23
ME	345.00 ± 42.00	77.00 ± 46.00	22.25 ± 11.75
MK	641.00 ± 41.00	190.00 ± 38.00	29.60 ± 6.47
NI	742.00 ± 8.00	109.00 ± 14.00	14.79 ± 1.90
NL	8880.00 ± 1099.00	837.00 ± 243.00	9.78 ± 2.00
NO	10,339.00 ± 397.00	3827.00 ± 460.00	37.03 ± 4.90
PL	11,299.00 ± 429.00	1774.00 ± 517.00	15.65 ± 4.86
PT	3764.00 ± 716.00	292.00 ± 70.00	8.37 ± 1.20
RO	4357.00 ± 142.00	543.00 ± 104.00	12.53 ± 2.64
RS	3201.00 ± 104.00	632.00 ± 289.00	19.82 ± 9.55
SE	11,600.00 ± 299.00	3511.00 ± 462.00	30.55 ± 3.64
SI	1046.00 ± 64.00	86.00 ± 28.00	8.43 ± 3.00
SK	2165.00 ± 59.00	383.00 ± 117.00	17.70 ± 5.66

Figure 8 shows pie charts presenting the shares of each components of decomposition, i.e., trend, periodic fluctuations and random fluctuations, in the total variance of the MEL time series. The results were used as seasonal–trend decomposition using LOESS (STL). The shares were calculated from the formulas

$$u_T = \frac{\mathrm{Var}(T_t)}{\mathrm{Var}(E_t)}, u_S = \frac{\mathrm{Var}(S_t)}{\mathrm{Var}(E_t)}, u_R = \frac{\mathrm{Var}(R_t)}{\mathrm{Var}(E_t)}, \tag{17}$$

where T_t, S_t, R_t is the component of the trend, respectively, seasonal and random fluctuations, and E_t is the MEL time series.

By analyzing pie charts, one can divide the MEL time series from due to the dominant component. Countries whose time series have the highest content of the trend component, above 80%, are Spain, Portugal, the Netherlands and Italy. The highest share of the seasonal component, over 90%, distinguishes Norway, Finland, Estonia, Sweden and Ireland. Countries whose MEL time series show the dominance of the seasonal component comprise the most numerous group. It should be noted that the results of these analyzes are dependent on the length of the time series. The trend shines through the time series longer ones, while in short ones, the seasonal component dominates. Montenegro is the only country whose series includes the disturbance component as dominant (53%). Other countries with a high proportion of random fluctuations, over 30% are in the following order: Northern Ireland, Serbia, Iceland and Slovenia [31].

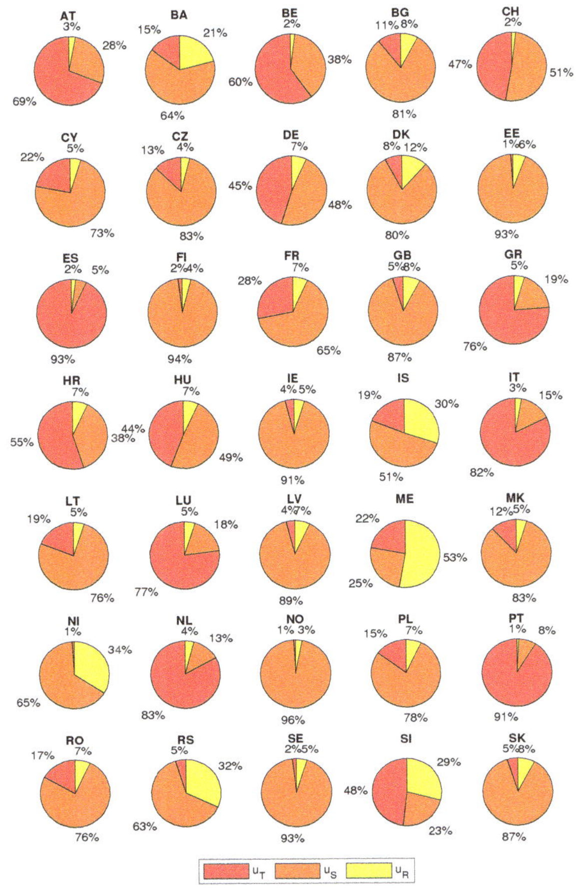

Figure 8. Shares of individual components decomposition in the total variance of the MEL time series.

4.2. MEL Time Series Forecasting Results

It follows that the time series show extensive contrasts and permit us to dependably assess predictive models. The forecasting issue is to create the multi-step-ahead forecasts for the each month of 2014 (last year of data) utilizing the information from the past period for training. For hyperparameter selection, the models were trained and validated on data up to 2013.

In this work, classical statistical models in connection with preprocessing are proposed:

- ARIMA — ARIMA$(p, d, q)(P,D,Q)_{12}$ model used in function `auto.arima` in R environment (package `forecast`). This function uses automatic ARIMA modeling. It combines unit root tests, maximum likelihood estimation and minimization of the Akaike information criterion (AICc) to obtain the optimal ARIMA model [38].
- ETS — exponential smoothing state space model [4] used in function `ets` (R package `forecast`). This implementation uses many types of ETS models depends on the consideration of the trend, seasonal and error components. It can be expressed multiplicatively or additively, and the trend could be damped or not. Similar to the

case of auto.arima, ETS returns the optimal model estimated by the model parameters using AICc [38].
- Prophet — modular additive regression model with nonlinear trend and seasonal components [37] implemented in function `Prophet` in R environment (package `prophet`).

The proposed models were here compared with other computational intelligence models as well as classical statistical models. They include the following:

- *k*-NN—*k* nearest neighbor is an ML-type model designed for the medium-term load forecasting (MTLF) [13]. It learns from patterns defined by (8).
- N-WE—Nadaraya–Watson estimator [39] for MTLF using pattern representation (8).
- LSTM—Long short-term memory (LSTM) network developed for MEL forecasting in [8]. It works on patterns (8).
- MLP—Multilayer perceptron proposed in [6]. This model is developed for MEL forecasting. It learns from patterns defined by (8).
- ANFIS—adaptive neuro-fuzzy inference system proposed for MTLF in [40]. It works on patterns (8).
- SVM—Support vector machine proposed for MTLF in [41]. It works on patterns (8).
- ES-RNN—A hybrid method of exponential smoothing and recurrent neural networks [42–44].

The length of the x-patterns is the one of the main hyperparameters for k-NN, N-WE, LSTM, MLP, ANFIS, and SVM models. Despite the natural choice for this hyperparameter, which is equal to the seasonal cycle length, i.e., 12 for the monthly type of time series, the optimal value of n, in a range from 3 to 24, for each model and each time series was selected in leave-one-out procedure using historical data.

The parameters of the ARIMA, ETS, and Prophet models were selected in the optimization procedures implemented in the `auto.arima`, `ets`, and `prophet` functions, respectively. These functions ensure a fully automatic selection of model structure and parameters for each time series individually. The MLP model learned from x patterns. A separate MLP network was trained for each time series and each month of the forecast period (2014). A single-hidden layer network with sigmoid activation functions was used. The networks were trained using the Levenberg–Marquardt method with Bayesian regularization, which helped prevent overfitting. The number of hidden nodes was selected from 1 to 10, individually for each time series and each forecasted month. The adaptive-network-based fuzzy inference system, ANFIS, like MLP, learned from x patterns. A separate ANFIS model was trained for each time series and month of the forecast period. The initial parameters of the Gaussian membership functions in the premise parts of the rules were selected using fuzzy c-means clustering. ANFIS was trained with a hybrid method that uses a combination of the least squares method to estimate the consequent parameters and the backpropagation gradient descent method to select the premise parameters. The number of fuzzy rules M was selected from 2 to 13. SVM, like MLP and ANFIS, is learned from x patterns. For each time series and each month of the forecast period, a separate SVM model was trained with kernels in the form of a dot product $K(x_i, x_j) = x_i^T x_j$. The length of the input pattern was selected for each series. The remaining hyperparameters (BoxConstraint, KernelScale, and Epsilon) were selected in the automatic optimization procedure implemented in the `fitrsvm` function from the Statistics and Machine Learning Toolbox in the Matlab environment. LSTM networks in all variants were trained using the Adam optimization algorithm (adaptive moment estimation). The number of hidden units was selected for each time series individually from the set $\{1, 2, \ldots, 10, 15, \ldots, 50, 60, \ldots, 200\}$. The remaining hyperparameters assumed default values: number of epochs, 250; initial value of the learning rate, 0.005; and a threshold value of the gradient (to prevent gradient explosion), 1. In the middle of the learning process, the learning rate was reduced to 0.001. The ETS+RD-LSTM model learned on all-time series simultaneously (cross-learning). The optimal values of the hyperparameters of this model for the set of 35-time series were as follows: number of epochs, 16; learning rate, 0.001; length of the cell and hidden state, 40; pinball loss, 0.4; regularization parameter, 50; and ensembling parameters, $L = 5$, $K = 3$, and $R = 3$.

In the case of the k-NN, MLP, ANFIS, SVM, and LSTM models, the optimal values of their hyperparameters were selected individually for each of the 35 time series. The selection of parameters was carried out separately for each of the variants of the models, V1, or (V2, and V3). The selection of hyperparameters was performed on the training set in the grid search procedure using cross-validation—the leave-one-out cross-validation method or by treating the last year of training data (2013) as validation data (SVM and LSTM models). The training set contained all terms of a given series from the historical period, up to and including 2013. The model with the optimal set of hyperparameter values was used to forecast the time series in the test period, which covered 12 months in 2014. Optimization procedures for ARIMA, ETS, and Prophet models are fully automatic, built into the functions that implement these models (`auto.arima, ets, prophet`). The ETS+RD-LSTM model was optimized on training data, treating the last year of training data (2013) as validation data.

The *k*-NN, N-WE, ARIMA and ETS are deterministic models, and they return the same results for the same data. NN-based models, for example, MLP, ANFIS, LSTM, and SVM, return various outcomes for the same data because of the stochastic type of the learning system. In this study, these models were run 100 times, and the final errors were averaged from 100 independent trials.

Taking into account the *x*-pattern encoding variants described in Section 3, three variants of each models (which also work with pattern representation usage) are considered:

- V1. The basic variant, where the coding variables for *x*-patterns are the mean and dispersion of previous sequence X_{i-1} for *k*-NN, N-WE, LSTM, MLP, ANFIS, and SVM. No patterns and coding variables are used for ES-RNN, ARIMA, ETS, LSTM, and Prophet in this case. We can use this variant to forecast the MEL from (8) without additional forecasting for coding variables.
- V2. The variant, for which the mean and dispersion of sequence X_i serve as the coding variables. Using ARIMA model, they are both independently forecasted for the query pattern based on their previous values. This variation broadens the denotations by "+AR", e.g., "k-NN + AR", and "ANFIS + AR".
- V3. The mean and dispersion of sequence X_i serve as the coding variables, as in variant V2. However, in this instance, they are predicted using ETS for the query pattern. "+ETS" is used to extend the denotations in this variant, such as "k-NN + ETS", and "ANFIS + ETS".

The following measures are used to assess the quality of forecasts and forecasting models:

- Percentage error (PE):

$$PE = \frac{E - \widehat{E}}{E} \cdot 100, \tag{18}$$

where E is the actual value and $\widehat{E}$ is the forecasted value.

- Mean percentage error (MPE):

$$MPE = \frac{1}{N} \sum_{i=1}^{N} PE_i. \tag{19}$$

- Absolute percentage error (APE) :

$$APE = |PE| \, . \tag{20}$$

- Mean absolute percentage error (MAPE):

$$MAPE = \frac{100}{N} \sum_{i=1}^{N} \left| \frac{E_i - \widehat{E}_i}{E_i} \right|. \tag{21}$$

- Interquartile range of absolute percentage error (IQR):

$$IQR(APE) = Q3(APE) - Q1(APE), \tag{22}$$

where $Q1(APE)$, $Q3(APE)$ are the lower and upper quartiles, respectively.
The quarter range allows you to assess the variability of the APE error. It includes 50% of all observations located centrally in the distribution.

- Root mean squared error (RMSE):

$$RMSE = \sqrt{\frac{1}{N}\sum_{i=1}^{N}(E_i - \widehat{E}_i)^2}. \tag{23}$$

- Standard deviation of percentage errors ($stdPE$):

$$Std(PE) = \sqrt{\frac{1}{N}\sum_{i=1}^{N}(PE_i - MPE)^2}. \tag{24}$$

- Coefficient of asymmetry (skewness) of the distribution of percentage errors (skewPE):

$$Skew(PE) = \frac{\frac{1}{N}\sum_{i=1}^{N}(PE_i - MPE)^3}{Std(PE)^3}, \tag{25}$$

The coefficient of asymmetry is zero for symmetric distributions, negative values for left asymmetric distributions (most of the population is below average) and positive for right asymmetric distributions (most of the population is above average).

- Kurtosis of percentage error distribution (kuPE):

$$ku(PE) = \frac{\frac{1}{N}\sum_{i=1}^{N}(PE_i - MPE)^4}{Std(PE)^4} - 3, \tag{26}$$

kuPE is a measure of the clustering of PE errors around the mean value of MPE. The higher the kurtosis value, the more slender the distribution of errors and the greater the concentration of their values around the mean.

The median absolute percentage error (MdAPE), mean absolute percentage error (MAPE), interquartile range of absolute percentage error (APE) as a measure of the forecast dispersion, and root mean square error (RMSE) are all shown in Table 2. The RMSE error measure, as can be seen from this table, indicates that Prophet + ETS is the most accurate model when compared to its rivals. In cases of MdAPE and MAPE metrics, Prophet + ETS results are very close to the state-of-the-art model ETS-RNN + ETS, but the predictions are fully interpretable, and the model has a small number of parameters to estimate.

Table 2. Results comparison among proposed and comparative models.

Model	MdAPE	MAPE	IQR	RMSE
k-NN	3.11	5.19	4.17	385.68
k-NN + AR	2.88	4.71	4.25	352.42
k-NN + ETS	2.72	4.58	3.58	333.27
N-WE	2.84	5.00	3.97	352.01
N-WE + AR	2.85	4.59	3.95	340.26
N-WE + ETS	2.68	4.37	3.36	320.51
LSTM	3.73	6.11	4.50	431.83
LSTM + AR	3.43	5.28	4.79	392.47
LSTM + ETS	3.08	5.19	4.54	366.45
MLP	2.97	5.27	3.84	378.81
MLP + AR	3.12	4.83	4.26	362.03
MLP + ETS	3.11	4.80	4.12	358.07

Table 2. *Cont.*

Model	MdAPE	MAPE	IQR	RMSE
ANFIS	3.56	6.18	4.87	488.75
ANFIS + AR	3.66	6.05	5.07	473.80
ANFIS + ETS	3.54	6.32	4.26	464.29
SVM	2.80	5.41	3.97	382.60
SVM + AR	3.14	4.91	4.09	348.52
SVM + ETS	2.85	4.74	3.60	330.94
ETS-RNN	2.74	4.48	3.55	347.24
ETS-RNN + AR	2.58	4.23	3.47	332.74
ETS-RNN + ETS	2.64	4.09	3.13	314.01
ARIMA	3.32	5.65	5.24	463.07
ARIMA + AR	2.99	4.64	3.95	357.84
ARIMA + ETS	2.85	4.52	3.61	339.49
ETS	3.50	5.05	4.80	374.52
ETS + AR	2.94	4.50	3.68	345.37
ETS + ETS	2.76	4.30	3.19	326.94
Prophet	3.08	4.72	4.37	349.01
Prophet + AR	2.99	4.39	3.65	334.46
Prophet + ETS	2.68	4.15	3.44	311.56

Figure 9 provides more specific results, such as MAPE for each nation. It is important to note that Prophet + ETS is frequently one of the most precise models. The model rankings, based on MAPE and RMSE, are shown in Figure 10. They display the models' average positions in the country-by-country rankings. Take note of Prophet + ETS's top spot in both rankings.

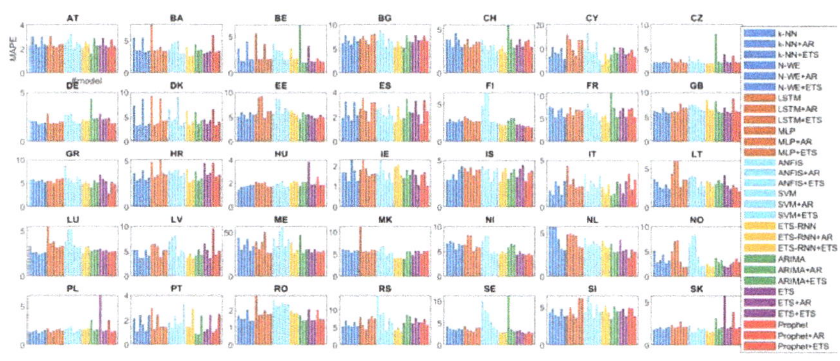

Figure 9. MAPE for each country.

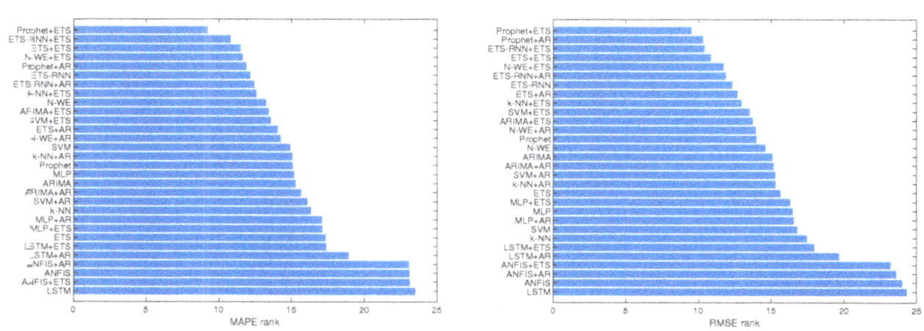

Figure 10. Rankings of the models.

In Table 3, the forecast errors and the descriptive statistics for the percentage errors are shown. The bias of the forecasts can be estimated from the average value of the PE error. For all models, the mPE is negative, which means that the forecast is overstated. The least loaded forecasts are made by the Prophet model (mPE = −0.70), and the most by the LSTM model (mPE = −3.12). The distribution of PE errors is described by the statistics: medPE, stdPE), skewPE, and kurtPE. The distribution of Prophet + ETS errors (medPE = 0.00) is the closest to zero, and the most distant LSTM (medPE = −1.81). The PE errors for the SVM model show the greatest dispersion around the mean (stdPE = 16.76), and the Prophet + AR errors the smallest (stdPE = 7.02). The most slender and centered around the mean value are the error distributions MLP + AR, MLP + ETS, ANFIS + AR, ARIMA + AR, ETS, ETS + AR, Prophet, Prophet + AR (kurtPE > 15), and the most flattened is error distribution MLP + ETS (kurtPE = 11.83). The negative skewness values, skewPE, which characterize the PE distributions of all models, indicate the left-hand skewness (the greater part of the population has values below the average). The most flattened distribution shows the greatest symmetry, obtained for ANFIS + ETS (skewPE = 0.96), and the smallest, the most slender distribution, obtained for SVM (skew (PE) < −13.40).

Table 3. Descriptive statistics of percentage errors among proposed and comparative models.

Model	mPE	medPE	stdPE	skewPE	kuPE
k-NN	−1.96	−1.27	10.83	−4.88	49.39
k-NN + AR	−1.76	−0.75	8.10	−2.66	20.96
k-NN + ETS	−1.26	−0.20	9.11	−4.47	38.22
N-WE	−1.91	−1.18	10.82	−5.41	48.94
N-WE + AR	−1.75	−0.85	7.82	−2.68	21.38
N-WE + ETS	−1.26	−0.17	8.68	−4.63	40.75
LSTM	−3.12	−1.81	9.49	−2.86	22.21
LSTM + AR	−1.86	−0.78	8.66	−2.75	21.20
LSTM + ETS	−1.41	−0.55	10.15	−5.35	50.04
MLP	−1.37	−0.68	11.88	−7.52	109.64
MLP + AR	−1.64	−0.92	7.45	−1.64	12.16
MLP + ETS	−1.71	−1.03	7.32	−1.55	11.83
ANFIS	−2.51	−1.43	11.37	−4.35	34.93
ANFIS + AR	−1.94	−0.65	9.63	−1.67	13.29
ANFIS + ETS	−1.30	−0.40	12.65	−0.96	39.37
SVM	−2.22	−0.91	16.76	−13.40	229.68
SVM + AR	−1.66	−0.61	8.58	−3.44	28.95
SVM + ETS	−1.28	−0.09	11.06	−7.84	95.91
ETS-RNN	−1.11	−0.27	10.07	−6.37	63.61
ETS-RNN + AR	−0.86	−0.20	7.30	−2.75	24.75
ETS-RNN + ETS	−0.32	0.47	8.48	−5.36	52.51
ARIMA	−2.35	−1.03	13.62	−9.01	119.20
ARIMA + AR	−1.69	−0.77	7.43	−1.65	12.67
ARIMA + ETS	−1.15	−0.26	8.40	−3.76	30.82
ETS	−1.04	−0.31	7.97	−1.89	13.52
ETS + AR	−1.71	−0.90	7.20	−1.73	13.02
ETS + ETS	−1.17	−0.28	8.20	−4.06	33.64
Prophet	−0.70	−0.18	7.47	−1.37	12.27
Prophet + AR	−1.30	−0.55	7.02	−1.66	12.84
Prophet + ETS	−0.76	0.00	7.86	−3.79	30.82

Figure 11 shows the MAPE errors broken down into individual months of the forecast period (2014). Attention should be paid to errors that are lower for months 8–10 and higher for months 1–4 and 12 [18]. The lowest errors were most often achieved by models with the forecast of code variables using ETS. Examples of forecasts generated by the models with the + ETS representation for several European countries are shown in Figure 12. The forecast errors of the PL series, except for the ETS model, do not exceed 2%, which should be considered a very good result. DE, ES, and IT ranks are forecast with a slightly greater error. In the case of DE, there are greater deviations in the forecasts set by ANFIS + ETS for September. The GB series is projected well below the actual mileage. This is due to the unexpected increase in demand in the UK system in 2014, despite the downward trend observed in 2010–2013 (see Figure 5). The opposite situation for FR, ES, and IT resulted in a slight overestimation of forecasts.

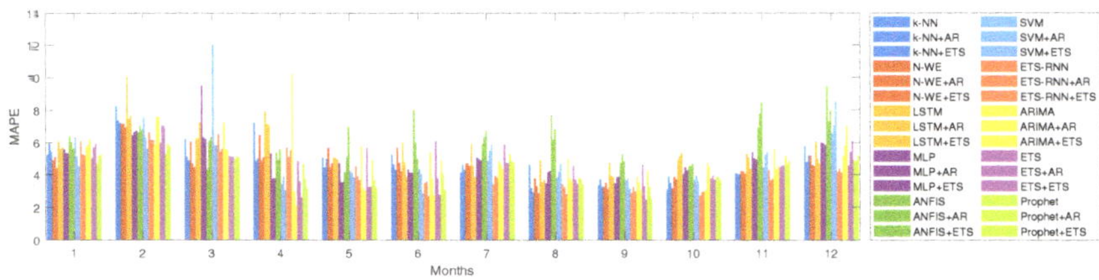

Figure 11. MAPE for each month.

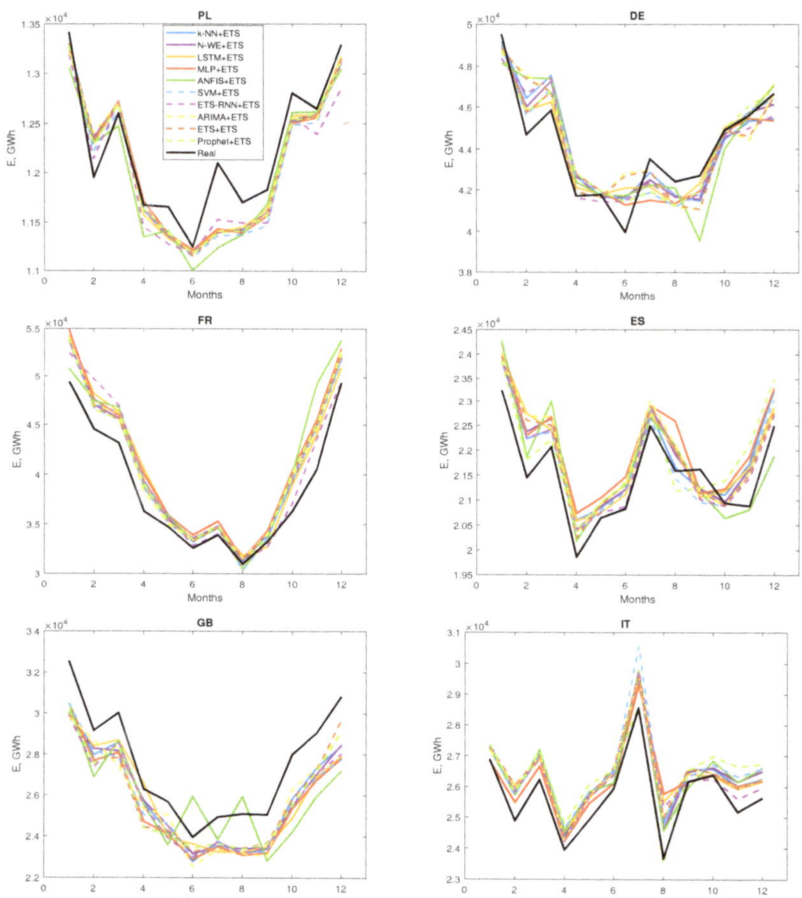

Figure 12. Examples of forecasts of MOD.

MAPE and RMSE results are presented in Figure 13. The time series of code variables and their forecasts for six sample countries are shown in Figure 14). Figure 15 shows Wilcoxon tests and the distribution of APE errors. The models were ranked from the lowest to the highest median (APE) value. The first two positions are occupied by the ETS-RNN models using the forecast of code variables. It should be noted that the proposed statistical models achieve higher positions in the case of using ETS to forecast code variables. ANFIS models take final positions to confirm the statistical significance of the differences in APE errors. Wilcoxon tests were performed for each pair of models. A white diagram element means that the models intersecting this element do not differ statistically in terms of APE. A yellow element means that the model pointed to by this element on the OY axis has reached a smaller error than the model indicated on the OX axis. A red element means that the model pointed to by this element on the OY axis has reached a greater error than the model pointed to on the OX axis. Figure 15 shows that models without prediction of code variables in many cases bring greater errors than hybrid models that contain predictions of these variables. The results of the Wilcoxon test for the ETS-RNN + ETS model, which is characterized by the lowest MAPE error (see Table 3), is similar to the results of the Prophet + ETS model. Both of these models show advantages over at least twenty-three other models (statistically significant difference in errors) and are as accurate as the other models. The ETS-RNN + ETS model shows the greatest advantage: in terms of error, the APE is more accurate than twenty-five other models.

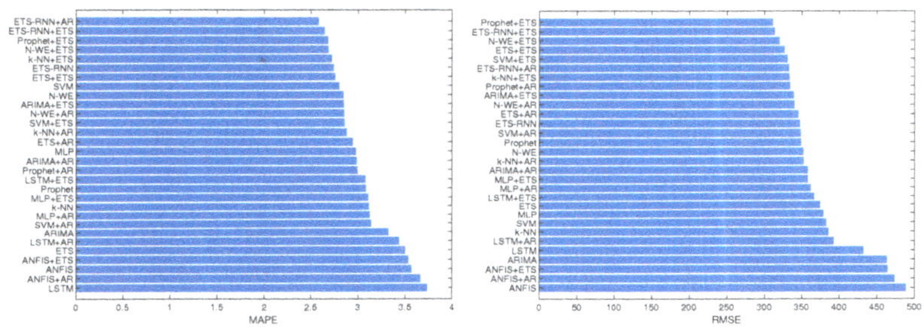

Figure 13. Errors of the models.

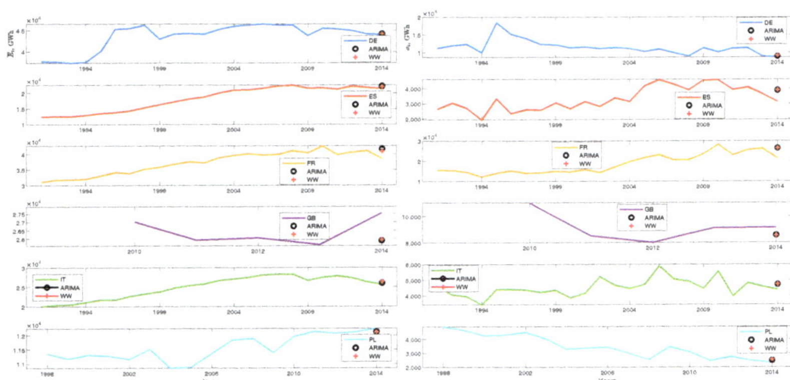

Figure 14. Forecasts of coding variables.

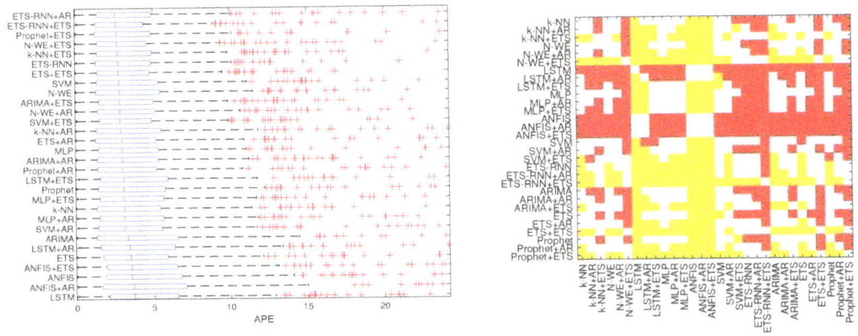

Figure 15. APE and Willcoxon test results for each model.

Summarizing the preliminary results, it should be noted that the performance of the prediction model is completely dependent on proper TS preprocessing. The introduction of initial normalization and predictive coding variables through ETS in the classic models significantly improves performance.

5. Discussion

In the ARIMA and ETS models, the optimization is global, i.e., their parameters are adjusted to ensure the lowest error for all terms of the time series (however, the length of the series can be restricted to the last terms so that the model takes into account only the specificity of the last historical period). The memory of the ARIMA model is limited to the last values of the time series (the size of the memory is determined by the rows of the AR and MA processes). Thus, the predicted value is "constructed" from the last terms. The ETS model is based on all values, but their influence on the predicted value decreases exponentially with time, i.e., more distant points have lower weights. Unlike these models, for example, k-NN constructs a local model individually for each query pattern. For the construction of the forecast, it uses all terms of the time series, not limited to the last period, as in ARIMA, and without introducing weights depending on the time distance, as in ETS. To construct a local regression function, k-NN considers values that may be distant in time from the query pattern if such distant cases are similar in shape to the query pattern. It is worth noting, however, that the time distance information can be easily introduced into, for example, N-WE by additional weights (decreasing in time) assigned to the x output patterns in the model. We can also enter weights based on seasonality. Outliers in the time series interfere with the selection of ARIMA and ETS parameters, leading to suboptimal models. In, for example, k-NN, as previously stated, outliers have a reduced impact in part. An additional distinction between statistical models and, for example, N-WE, is that the former generates forecasts one step ahead. Forecasts with longer horizons are achieved recursively, taking the the prediction for the preceding time step as an input for the prediction of the following time step. In contrast, for example, k-NN predicts the x-pattern, representing the entire predicted sequence, in one step.

6. Conclusions

In this paper, statistical methods for mid-term load forecasting were proposed. The input data for the models represent the normalized annual seasonal cycle of a load time series with filtered trend and unified variance. They express shapes of the yearly cycles. The proposed approach uses statistical methods for forecasting annual patterns and also for forecasting coding variables for decoding these patterns. Due to the pattern representation of the time series, forecasting models do not need to grasp the essence of the complex time series, which simplifies the predictive problem.

In the experimental part of the work, classical models with pattern representation were tested on MEL predicting issues for 35 European countries. The outcomes showed the advantages of the proposed approach over an alternative approach without forecasting coding variables. For statistical models and also for many comparative models, the proposed approach improved the accuracy. For example, in the cases of Prophet + ETS, ETS + ETS, and ARIMA + ETS usage, it leads to outperform predecessors Prophet, ETS, and ARIMA about 13.7%, 17.4%, and 25% in case of MAPE error. The proposed models can be further implemented for short-term load forecasting.

Funding: This research received no external funding.

Institutional Review Board Statement: Not applicable.

Informed Consent Statement: Not applicable.

Data Availability Statement: We use real-world data collected from www.entsoe.eu (accessed on 12 April 2016).

Conflicts of Interest: The author declares no conflict of interest.

Abbreviations

The following abbreviations are used in this manuscript:

ML	Machine Learning	ES	Spain
MEL	Monthly Energy Load	FI	Finland
MTLF	Mid-term Load Forecasting	FR	France
ARIMA	Autoregressive Integrated Moving Average	GB	Great Britain
ETS	Exponential Smoothing	GR	Greece
NN	Neural Network	HR	Croatia
ACF	Autocorrelation Function	HU	Hungary
PACF	Partial Autocorrelation Function	IE	Ireland
k-NN	k Nearest Neighbor	IS	Iceland
N-WE	Nadaraya–Watson estimator	IT	Italy
LSTM	Long Short-Term Memory	LT	Lithuania
MLP	Multilayer Perceptron	LU	Luxembourg
ANFIS	Adaptive Neuro-Fuzzy Inference System	LV	Latvia
SVM	Support Vector Machine	ME	Montenegro
ETS-RNN	Exponential Smoothing and Recurrent Neural Networks	MK	Macedonia
AT	Austria	NI	Northern Ireland
BA	Bosnia and Herzegovina	NL	Netherlands
BE	Belgium	NO	Norway
BG	Bulgaria	PL	Poland
CH	Switzerland	PT	Portugal
CY	Cyprus	RO	Romania
CZ	Czech Republic	RS	Serbia
DE	Germany	SE	Sweden
DK	Denmark	SI	Slovenia
EE	Estonia	SK	Slovakia

References

1. Pełka, P. Przegląd Metod Wykorzystywanych Do Średnioterminowego Prognozowania Obciążeń Elektroenergetycznych. *Przegląd Elektrotechniczny* **2019**, *95*, 155–159. [CrossRef]
2. Dogan, E. Are Shocks to Electricity Consumption Transitory or Permanent? Sub-National Evidence from Turkey. *Utilities Policy* **2016**, *41*, 77–84. [CrossRef]
3. Bunnoon, P.; Chalermyanont, K.; Limsakul, C. Mid Term Load Forecasting of the Country Using Statistical Methodology: Case Study in Thailand. In Proceedings of the 2009 International Conference on Signal Processing Systems, Singapore, 15–17 May 2009; pp. 924–928.
4. Hyndman, R.; Koehler, A.B.; Ord, J.K.; Snyder, R.D. *Forecasting with Exponential Smoothing: The State Space Approach*; Springer Science & Business Media: Berlin/Heidelberg, Germany, 2008.

5. González-Romera, E.; Jaramillo-Morán, M.A.; Carmona-Fernández, D. Monthly Electric Energy Demand Forecasting with Neural Networks and Fourier Series. *Energy Convers. Manag.* **2008**, *49*, 3135–3142. [CrossRef]
6. Pełka, P.; Dudek, G. Pattern-Based Forecasting Monthly Electricity Demand Using Multilayer Perceptron. In *Artificial Intelligence and Soft Computing*; Springer: Cham, Switzerland, 2019; pp. 663–672.
7. Gao, T.; Niu, D.; Ji, Z.; Sun, L. Mid-Term Electricity Demand Forecasting Using Improved Variational Mode Decomposition and Extreme Learning Machine Optimized by Sparrow Search Algorithm. *Energy* **2022**, *E261*, 125328. [CrossRef]
8. Pełka, P.; Dudek, G. Pattern-Based Long Short-Term Memory for Mid-Term Electrical Load Forecasting. In Proceedings of the 2020 International Joint Conference on Neural Networks (IJCNN), Glasgow, UK, 19–24 July 2020; pp. 1–8.
9. Peng, L.; Wang, L.; Xia, D.; Gao, Q. Effective Energy Consumption Forecasting Using Empirical Wavelet Transform and Long Short-Term Memory. *Energy* **2022**, *238*, 121756. [CrossRef]
10. Sharma, K.; Dwivedi, Y.K.; Metri, B. Incorporating Causality in Energy Consumption Forecasting Using Deep Neural Networks. *Ann. Oper.* **2022**, 1–36. [CrossRef]
11. Li, J.; Wei, S.; Dai, W. Combination of Manifold Learning and Deep Learning Algorithms for Mid-Term Electrical Load Forecasting. *IEEE Trans. Neural Netw. Learn. Syst.* **2021**. [CrossRef]
12. Xu, A.; Tian, M.-W.; Firouzi, B.; Alattas, K.A.; Mohammadzadeh, A.; Ghaderpour, E. A New Deep Learning Restricted Boltzmann Machine for Energy Consumption Forecasting. *Sustainability* **2022**, *14*, 10081. [CrossRef]
13. Dudek, G.; Pełka, P. Prognozowanie Miesięcznego Zapotrzebowania Na Energię Elektryczną Metodą k Najbliższych Sąsiadów. *Przeglad Elektrotechniczny* **2017**, *1*, 64–67. [CrossRef]
14. Pełka, P.; Dudek, G. Prediction of Monthly Electric Energy Consumption Using Pattern-Based Fuzzy Nearest Neighbour Regression. *ITM Web Conf.* **2017**, *15*, 02005. [CrossRef]
15. Pełka, P.; Dudek, G. Medium-Term Electric Energy Demand Forecasting Using Generalized Regression Neural Network. In *International Conference on Information Systems Architecture and Technology*; Springer: Cham, Switzerland, 2018; pp. 218–227.
16. Pełka, P.; Dudek, G. Ensemble Forecasting of Monthly Electricity Demand Using Pattern Similarity-Based Methods. In *International Conference on Artificial Intelligence and Soft Computing*; Springer: Cham, Switzerland, 2020; pp. 712–723.
17. Castán-Lascorz, M.A.; Jiménez-Herrera, P.; Troncoso, A.; Asencio-Cortés, G. A New Hybrid Method for Predicting Univariate and Multivariate Time Series Based on Pattern Forecasting. *Inf. Sci.* **2022**, *586*, 611–627. [CrossRef]
18. Dudek, G.; Pełka, P. Pattern Similarity-Based Machine Learning Methods for Mid-Term Load Forecasting: A Comparative Study. *Appl. Soft Comput.* **2021**, *104*, 107223. [CrossRef]
19. Papalexopoulos, A.D., Hesterberg, T.C.: A regression-based approach to short-term system load forecasting. *IEEE Trans. Power Syst.* **1990**, *5*, 1535–1547. [CrossRef]
20. Wang, Y.; Gan, D.; Sun, M.; Zhang, N.; Lu, Z.; Kang, C. Probabilistic Individual Load Forecasting Using Pinball Loss Guided LSTM. *Appl. Energy* **2019**, *235*, 10–20. [CrossRef]
21. Miseta, T.; Fodor, A.; Vathy-Fogarassy, Á. Energy Trading Strategy for Storage-Based Renewable Power Plants. *Energy* **2022**, *250*, 123788. [CrossRef]
22. Al-Hamadi, H.M.; Soliman, S.A. Long-Term/Mid-Term Electric Load Forecasting Based on Short-Term Correlation and Annual Growth. *Electr. Power Syst. Res.* **2005**, *74*, 353–361. [CrossRef]
23. Barakat, E.H. Modeling of Nonstationary Time-Series Data. Part II. Dynamic Periodic Trends. *Int. J. Electr. Power Energy Syst.* **2001**, *23*, 63–68. [CrossRef]
24. Cai, G.; Yang, D.; Jiao, Y.; Pan, C. The Characteristic Analysis and Forecasting of Mid-Long Term Load Based on Spatial Autoregressive Model. In Proceedings of the 2009 International Conference on Sustainable Power Generation and Supply, Nanjing, China, 6–7 April 2009; pp. 1–5.
25. de Oliveira, E.M.; Oliveira, F.L.C. Forecasting Mid-Long Term Electric Energy Consumption through Bagging ARIMA and Exponential Smoothing Methods. *Energy* **2018**, *144*, 776–788. [CrossRef]
26. Zhang, D.-L.; Yan, J.; Wang, W.-H.; Yang, X.-L. Mid-Long Term Load Forecasting of the Unstable Growth Sequence Based on Markov Chains Screening Combination Forecasting Models. In Proceedings of the 2016 China International Conference on Electricity Distribution (CICED), Xi'an, China, 10–13 August 2016; pp. 1–5.
27. Barakat, E.E.; Al-Rashed, S.A. Long Range Peak Demand Forecasting under Conditions of High Growth. *IEEE Trans. Power Syst.* **1992**, *7*, 1483–1486. [CrossRef]
28. Zagdański, A.; Suchwałko, A. *Analiza i Prognozowanie Szeregów Czasowych: Praktyczne Wprowadzenie Na Podstawie Środowiska R*; Wydawnictwo Naukowe PWN: Warsaw, Poland, 2016.
29. Jóźwiak, J.; Podgórski, J. *Statystyka Od Podstaw*; Wyd. PWE: Warszawa, Poland, 2001.
30. Hyndman, R.J.; Athanasopoulos, G. *Forecasting: Principles and Practice*; OTexts: Melbourne, Australia, 2013; ISBN 978-0-9875071-0-5.
31. Pełka, P. Nonparametric Machine Learning Methods for Mid-Term Forecasting of Power System Loads. Ph.D. Thesis, Czestochowa University of Technology, Częstochowa, Poland, 2021.
32. Chatfield, C. *The Analysis of Time Series: An Introduction*, 6th ed.; Chapman and Hall/CRC: Boca Raton, FL, USA, 2003; ISBN 978-1-58488-317-3.
33. Dittmann, P. *Prognozowanie w Przedsiebiorstwie*; Oficyna Ekonomiczna: Kraków, Poland, 2004.
34. Dudek, G. Wybrane Metody Analizy Szeregów Czasowych Obciążeń Elektroenergetycznych. In Proceedings of the Materiały VII Konferencji Naukowej Prognozowanie w elektroenergetyce, Częstochowa, Poland, 26–27 September 2004; pp. 116–125.

35. Dudek, G. Pattern Similarity-Based Methods for Short-Term Load Forecasting–Part 1: Principles. *Appl. Soft Comput.* **2015**, *37*, 277–287. [CrossRef]
36. Hyndman, R.J.; Khandakar, Y. Automatic Time Series Forecasting: The Forecast Package for R. *J. Stat. Softw.* **2008**, *27*, 1–22. [CrossRef]
37. Taylor, S.J.; Letham, B. Forecasting at Scale. *Am. Stat.* **2018**, *72*, 37–45. [CrossRef]
38. Hyndman, R.J.; Athanasopoulos, G.; Bergmeir, C.; Caceres, G.; Chhay, L.; O'Hara-Wild, M.; Petropoulos, F.; Razbash, S.; Wang, E.; Yasmeen, F. *Forecast: Forecasting Functions for Time Series and Linear Models*; R Package: Melbourne, Australia, 2018.
39. Dudek, G.; Pełka, P. Medium-Term Electric Energy Demand Forecasting Using Nadaraya-Watson Estimator. In Proceedings of the 2017 18th International Scientific Conference on Electric Power Engineering (EPE), Kouty nad Desnou, Czech Republic, 17–19 May 2017; pp. 1–6.
40. Pełka, P.; Dudek, G. Neuro-Fuzzy System for Medium-Term Electric Energy Demand Forecasting. In *International Conference on Information Systems Architecture and Technology*; Springer: Cham, Switzerland, 2017; pp. 38–47.
41. Pełka, P. Pattern-Based Forecasting of Monthly Electricity Demand Using Support Vector Machine. In Proceedings of the 2021 International Joint Conference on Neural Networks (IJCNN), Shenzhen, China, 18–22 July 2021.
42. Smyl, S. A Hybrid Method of Exponential Smoothing and Recurrent Neural Networks for Time Series Forecasting. *Int. J. Forecast.* **2020**, *36*, 75–85. [CrossRef]
43. Dudek, G., Pełka, P., Smyl, S.: A Hybrid Residual Dilated LSTM and Exponential Smoothing Model for Midterm Electric Load Forecasting. *IEEE Trans. Neural Netw. Learn. Syst.* **2022**, *33*, 2879–2891. [CrossRef] [PubMed]
44. Dudek, G.; Pełka, P.; Smyl, S. 3ETS+ RD-LSTM: A New Hybrid Model for Electrical Energy Consumption Forecasting. In Proceedings of the International Conference on Neural Information Processing, Bangkok, Thailand, 23–27 November 2020; Springer: Berlin/Heidelberg, Germany, 2020; pp. 519–531.

energies

Article

Six Days Ahead Forecasting of Energy Production of Small Behind-the-Meter Solar Sites

Hugo Bezerra Menezes Leite and Hamidreza Zareipour *

Department of Electrical and Software Engineering, University of Calgary, Calgary, AB T2N 1N4, Canada
* Correspondence: hzareipo@ucalgary.ca

Abstract: Due to the growing penetration of behind-the-meter (BTM) photovoltaic (PV) installations, accurate solar energy forecasts are required for a reliable economic energy system operation. A new hybrid methodology is proposed in this paper with a sequence of one-step ahead models to accumulate 144 h for a small-scale BTM PV site. Three groups of models with different inputs are developed to cover 6 days of forecasting horizon, with each group trained for each hour of the above zero irradiance. In addition, a novel dataset preselection is proposed, and neighboring solar farms' power predictions are used as a feature to boost the accuracy of the model. Two techniques are selected: XGBoost and CatBoost. An extensive assessment for 1 year is conducted to evaluate the proposed method. Numerical results highlight that training the models with the previous, current, and 1 month ahead from the previous year referenced by the target month can improve the model's accuracy. Finally, when solar energy predictions from neighboring solar farms are incorporated, this further increases the overall forecast accuracy. The proposed method is compared with the complete-history persistence ensemble (CH-PeEn) model as a benchmark.

Keywords: photovoltaic (PV); forecast; behind-the-meter (BTM); spatio-temporal; strategic training

Citation: Bezerra Menezes Leite, H.; Zareipour, H. Six Days Ahead Forecasting of Energy Production of Small Behind-the-Meter Solar Sites. *Energies* **2023**, *16*, 1533. https://doi.org/10.3390/en16031533

Academic Editor: Jayanta Deb Mondol

Received: 19 December 2022
Revised: 26 January 2023
Accepted: 26 January 2023
Published: 3 February 2023

1. Introduction

A deployment of 138 GW of rooftop photovoltaic (PV) systems has been identified between 2020 and 2021 [1]. PV systems deployment did not slow down even during the COVID-19 pandemic and all of the related health and logistic limitations. The growth of behind-the-meter (BTM) solar sites makes net demand forecasting challenging as it introduces additional uncertainty in net demand patterns [2]. Net demand is the critical input in both short-term and long-term planning of power systems [3]. To carry this out, net demand must be forecasted, accounting for a modified shape pattern between the morning hours of the day and the end of the afternoon [4]. Net demand can be predicted directly or indirectly by subtracting the BTM PV power forecast from the demand. Therefore, forecasting models with enhanced accuracy for small BTM PV sites are important to support net demand forecasting in power systems.

One-step ahead forecasters were the most common between 2010 and 2019, while more recently, multi-step forecast methods are gaining momentum [5]. Even though many hybrid approaches have been proposed in the literature, they are limited mainly by intra-day horizons or, in some cases, limited to 3 days ahead [5]. For example, in [6], a hybrid method called physical hybrid artificial neural network (PHANN) is proposed to predict up to 72 h ahead. Finally, in [7], a hybrid method with artificial neural network (ANN) and an analog ensemble (AnEn) is proposed to generate 72 h forecasts of power.

In the solar energy forecasting domain, there are two primary training practices. The first approach, Generalization, is when research works use a significant amount of data as possible to train a unique model to forecast any hour of the day, month, or season. The second approach, Classification, is when different models are built based on categories. For example, the first two full years of data were used for training in [8], and in [9], the authors used the total accumulated historical data to train a model. On the other hand, an optimization using a selection training dataset with a standard setting consisting of days with different cloud conditions from half a year of data was applied in [10]. Whereas, in [11], three models were trained with a dataset divided into three categories: Sunny, cloudy, and overcast days, according to the mean irradiance.

The spatio-temporal correlation is attracting attention in solar energy forecasts. Some works have applied historical data from neighboring solar sites and weather stations to forecast solar energy. For example, five nearby solar irradiance stations, with distances from 0 to 200 km to predict from 5 min to 24 h ahead; first, solar irradiance, converted to solar power, was accounted in [12]. More recently, in [13], the authors considered the collaborative data from 44 rooftop-scale solar units located in a Portuguese city in their model to produce 6 h ahead of solar power forecasts.

In distinction to the common practices, we propose a new hybrid methodology with three groups of models and different inputs to cover a forecasting horizon of 6 days. Each group is trained for each hour of above zero irradiance. In addition, the method includes a monthly pattern preselection approach where the most recent and the most likely future weather patterns are present. We select similar months based on the target month we want to predict, reducing the available dataset to only the months with very similar characteristics to the target month. Namely, the previous month, the current month, and the next month ahead from the previous year are selected and referenced by the forecasting origin. This strategy balances enough generalization and similar days classification in a reduced dataset which is more correlated with the target forecast. Moreover, we propose the reinforcement of the methodology by benefiting from spatio-temporal correlation using publicly available regional aggregated solar power predictions (RASPP) as a feature. This specific feature helps the proposed method to take advantage of other forecasts for solar energy generation within the same neighboring region.

In summary, the main contributions of this work are as follows: First, we propose a horizontally cascaded set of models to extend the forecasting horizon of short-term solar energy forecasting to 6 days. The forecasting horizon of 6 days is divided into three groups of models with different inputs, and, within each group, a separate model for each hour of the day is proposed. Moreover, a novel classification and training strategy is proposed for the enhancement of forecast accuracy. Second, we propose the use of publicly available regional aggregated solar power predictions (RASPP) as an input to the model to benefit from potential spatio-temporal correlations between the power production at the target site and the general solar energy production patterns in the same geographical region.

The remainder of this paper is organized according to the following sections: Section 2 presents a literature review of solar power forecasting. Section 3 describes the proposed solar power forecasting methodology. Section 4 presents the numerical results and discussions, followed by Section 5, which summarizes this work, and suggests directions for potential future work.

2. Literature Review

The solar energy forecasting literature can be classified into subdomains regarding the spatial horizon [14], time horizon [15], methods [5], techniques [14], inputs [14], benchmarks [16], and level of uncertainty [17,18]. Regarding methods, some strategies consider numerical and probabilistic methods, physical models, and artificial intelligence (AI) techniques, including machine learning (ML), deep learning (DL), and hybrid methods [5]. In terms of time horizon [14], these works can be categorized into four subdomains: Intra-hour or nowcasting, intra-day, i.e., 6 h to day-ahead, or multi-days ahead or more prolonged (2 days and longer). Regarding modelling inputs [14], forecasting strategies consider endogenous inputs [8] and [19], exogenous, or both [20], including numerical weather predictions (NWP), sky cameras [17], satellite imagery [21], neighboring PV plants [22], adjacent weather stations [12], and other predictions.

A review of the literature showed that some works used a significant amount of data as possible to train a model. They based their models on Generalization. For example, in [8], where the first two full years of data were used for training; or in [9], the authors used the total accumulated historical data to train a model with no specific selection of days; and in [23], the authors used one whole year of data to train six models. However, in terms of Classification, different models are built based on categories. Some works targeted the pattern of the preselected data to train a model in order to focus on the similar pattern to be predicted, as seen in [10], in which an optimization using a selection training dataset with a standard setting consisting of days with different cloud conditions from half a year of data was applied; or in [11], the authors trained three models with the dataset divided into three categories: Sunny, cloudy, and overcast days, according to the mean irradiance; or in [24], the researchers developed a weather scenario based on generation, in which a copula was adopted to model the correlation among weather variables, including the data from local weather stations and historical NWP, through a high-dimensional joint distribution; or in [25], three selection methods were presented for training purposes: First, the previous 30 days, or second, the 30 days according to the absolute difference between the clearness index of the day to be predicted and each day included in the database, or finally the third strategy, which considered the 30 days according to the similarity between the empirical distribution function of the irradiance forecast for the day to be predicted and for each day included in the database. The first group of works considered that a generalization strategy would assist in increasing the quality of their models. On the other hand, the second group argued that a classification strategy would perform better. It is very unlikely that a model trained with data from Winter would be helpful to predict something during Summer. Therefore, we understand that a generalization strategy is ineffective in this case. Meanwhile, the second group relied on the forecasted weather in order to select the respective model to predict accurately. However, if the weather forecast is wrong, it is very likely that the prediction, based on the classified day would not perform well. For example, an overcast day would be predicted, but no clouds are above the solar panels and then, an underestimated production would be expected. In this case, a methodology that considers enough generalization and classification is needed. Therefore, a reduced dataset also divided by each hour of the day is a potential solution. In conclusion, to the best of our knowledge, the gap related to the month pattern preselection exists. As a result, it merits investigation in this paper.

Some works have explored the application of using historical data from neighboring solar sites or weather stations in the context of solar energy forecasting. For example, the authors of [26] accounted for 80 distributed rooftop PV plants in the Arizona region as a network of irradiance sensors to predict cloud speed and solar power; or the authors of [27] used historical information from five neighboring rooftop PV plants in the Netherlands and one meteorological station to predict solar power of a 500 W PV system; or the authors of [12] accounted for five nearby solar irradiance stations, with distances from 0 to 200 km for prediction, first, of solar irradiance, and then conversion to solar power; or the authors of [22] developed an individual model for each hour for each of three utility-scale solar farms (Solar Farms A, B, and C) to predict with an hourly resolution, accounting for the inclusion of independent variables from the adjacent solar farms (for example, for Solar Farm A, the neighboring solar farms included in the model were B and C, etc.). More recently, the authors of [13] considered the collaborative data from 44 rooftop-scale solar units located in a Portuguese city in their model to produce solar power forecasts. However, none of the existing works had explored the possibility of using publicly available regional aggregated solar power predictions (RASPP) for improving the forecasts of small BTM solar facilities. In addition, only one publication [27] focused on intra-hour, intra-day, day-ahead, and longer horizons as 7 days ahead, 15 days ahead, 20 days ahead, and 1 month ahead. The remaining works concentrated their findings on the intra-hour forecast, in [26], which focused on predictions from 15 to 90 min ahead; on the intra-day forecast, in [13], which focused on 6 h ahead; or on intra-day to day-ahead forecast, in [12], which focused on predictions from 5 min to 24 h ahead, and finally in [22], which focused on 24 h ahead. Therefore, the presented works used historical data from neighboring solar sites or weather stations to forecast solar irradiance or solar power and none of them used regional aggregated solar power predictions from adjacent solar farms. Moreover, most of them limited their works to forecast up to 1 day ahead horizons. As a result, a hybrid methodology to predict solar power forecast accounting for publicly available regional aggregated solar power predictions and covering from intra-day to 6 days ahead will be explored in this paper.

3. Proposed Solar Power Forecasting Methodology

The proposed method comprises relevant inputs, data preprocessing steps, and training of three groups of separate one-step ahead model for each hour of the day, including two regression models per hour to produce a set of deterministic forecasts. The final step considers data postprocessing. These components together form the proposed solar power forecasting framework, presented in Figure 1, which will be described in the following sections.

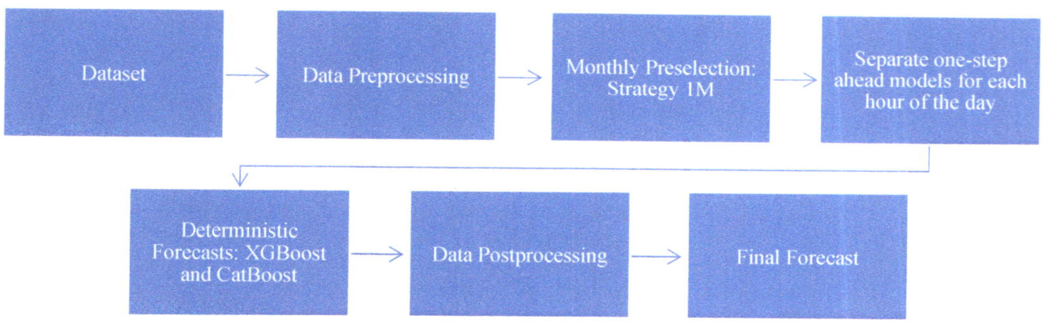

Figure 1. The proposed solar power forecasting methodology.

3.1. Dataset

A dataset with an hourly resolution of PV power output, including lagged power production from the previous 15, 30, 45, 60, and 75 min, and lagged power production from the previous 1, 24, 48, 72, 96, 120, and 144 h from 11 March 2019 to 31 July 2022 are used in this work. Moreover, the following is aggregated in this dataset: Global horizontal irradiance (GHI), zenith, and azimuth. GHI is the total power of solar radiation per unit area, measured in W/m^2 at a horizontal surface to the ground during the absence of visible clouds across the sky. It provides the maximum irradiance under clear sky conditions. In addition, numerical weather predictions (NWP) are included in the dataset, such as ambient temperature, solar irradiance, wind speed, wind direction, relative humidity, cloud cover, dew point, gust speed, and pressure, in which forecasts are performed by an external source. The solar irradiance forecast considers the likelihood of clouds and their effects on the availability of solar irradiance. In a nearby geographical location of the target PV system, in the City of Medicine Hat, Alberta, Canada, there are more than 20 utility-scale solar farms. Therefore, the historical regional aggregated solar power output (RASPO), and the publicly available regional aggregated solar power predictions (RASPP), which are provided by the independent system operator (ISO) are also added to the dataset. Lagged features or past power production improve the forecast quality since time dependency is considered in the forecasting problem [28]. Data from NWP in solar power forecasting are most applicable for day-ahead forecasting [29]. NWP uses mathematical models of the atmosphere and oceans based on measured conditions to predict with excellent forecast skill up to 6 days ahead and a relatively accurate forecast up to 14 days ahead.

3.2. Data Preprocessing

The standard scaler normalizes all features to remove the mean and scale to unit variance. A correlation analysis is a statistical summary that assists in identifying the strength and direction of the relationship between two variables. Spearman's correlation is applicable for nonlinear relationships and non-Gaussian distribution, which assumes that the relationship between variables is monotonic and tends to move in the same relative direction but not at a constant rate. An autocorrelation analysis was performed to explore the relevance of lagged features. In addition, a correlation among other variables was carried out. Therefore, only the most relevant and correlated features are used in this study.

3.3 Monthly Preselection

We propose a new training strategy for this work. The strategy is based on the similarity of seasonal weather and the general solar power production for each month in hourly resolution. For example, the objective is to avoid training with a database from Winter months, while the target is to predict Summer. Two main benefits can be observed: First, the correlation among features is increased, which directly impacts the accuracy of the predictions. Second, the training process speed is increased, which reduces the computational cost.

Herein, primary and secondary strategies are proposed in a simplified way to train a model weekly. Strategy 1M targets the dataset selection of the previous month, the current month, 1 month ahead from the previous year, and the previous month and the last weeks of the current month of the current year, always referenced by the forecasting origin. On the other hand, strategy 3M targets the dataset selection of the previous 3 months of the previous and current year, and the last weeks of the current month of the current year. For example, Figure 2 presents a dataset containing data from 1 January 2020 to 31 December 2021, and the forecast origin is 1 August 2021. For strategy 1M, the preselected months to train the model are from July to September 2020 as well as July 2021. For strategy 3M and the same forecast origin, the preselected months to train the model are from May to August 2020, as well as May to July 2021. Namely, the proposed training strategy 1M considers the typical previous 30 days in the current and previous years, as well as the following 60 days in the previous year. The numerical results in Section 4 highlight the fact that the proposed method can improve the model's accuracy.

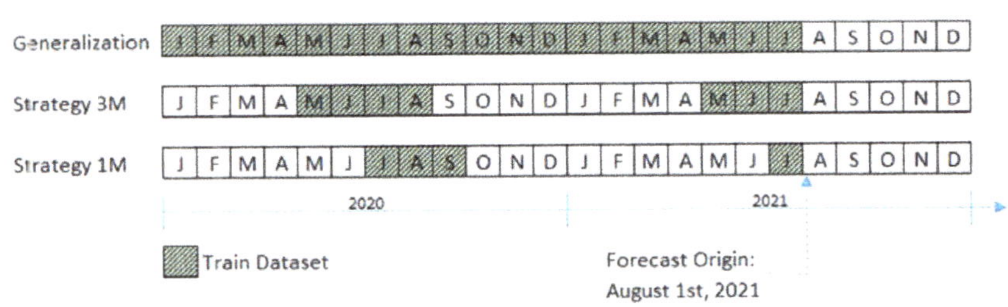

Figure 2. The proposed training strategy.

3.4. Separate One-Step Ahead Models for Each Hour of the Day

In this paper, we propose the development of three groups of separate one-step ahead deterministic models, with each group trained for each hour of the day and receiving different inputs. During the Summer, from 4:00 a.m. to 11:00 p.m., there are a total of 20 h of above zero solar irradiance. The strategy does not consider the remaining 4 h of the day since they have zero solar irradiance. Therefore, the proposed methodology trains 20 separate models. In addition, two techniques are selected to fit the models: XGBoost and CatBoost. As a result, the number of separate models per hour multiplied by the number of methods accumulates 40 models per group, and finally, multiplied by three groups results in a total of 120 models. Different combinations of inputs and time horizons are proposed in this work to extract the best outcomes from each available feature and reflect the accuracy gained for each step ahead in the forecasting horizon.

3.4.1. Group A: One-Step Ahead for the 1st Hour Ahead Framework

The most impactful prediction horizon is 1 h ahead since it is critical for monitoring and dispatching purposes. For example, in [30], the authors identified that lagged observations are more important for shorter forecasting horizons than weather forecasts. Group A is a set of one-step models for the 1st hour ahead, with separate models for each hour of the day. It leverages the most recent observations of the target PV system's power (p) from the previous 15, 30, 45, 60, and 75 min to increase the accuracy of the next hour ahead prediction. In addition, exogenous inputs from NWP, such as GHI, ambient temperature (T), and solar irradiance (SI) are considered, according to Equation (1). For a set of models, M^A with h hours of the day is limited to 4:00 a.m. $\leq$ h $\leq$ 11:00 p.m. and two techniques (XGBoost, CatBoost), herein Equation (2) represents Group A, as follows:

$$(p_{t+1}) = f(p_{t-15}, p_{t-30}, p_{t-45}, p_{t-60}, p_{t-75}, GHI_{t+1}, T_{t+1}, SI_{t+1}) \tag{1}$$

$$M^A = \left\{ \left(XGB_4^A, CTB_4^A\right), \ldots \left(XGB_h^A, CTB_h^A\right) \right\}_{h=4AM, \ldots 11PM} \tag{2}$$

3.4.2. Group B: One-Step Ahead for 2nd to 56th Hour Ahead Framework

Intra-day and day-ahead forecasts are relevant for scheduling the spinning reserve capacity. Group B is a set of one-step ahead models for recursive predictions from the 2nd to the 56th hour ahead, with separate models for each hour of the day. Group B leverages the information from NWP and the solar farms' power predictions to increase the accuracy of the target PV power output model. Since these forecasts are available with a forecasting horizon limited to the 56th hour ahead, Group B is also determined by the same horizon. The set of inputs considered in this group of models are past power output (p), GHI, ambient temperature (T), solar irradiance (SI), wind speed (WS), relative humidity (RH), and the publicly available regional aggregated solar power predictions (RASPP), according to Equation (3) with k = (2, ... , 56). The forecasting engine uses a recursive forecasting strategy, i.e., to keep the properties of the time series, the outputs from Group A are used as inputs to Group B and then from Group B to Group C [31]. For a set of models, M^B with h hours of the day is limited to 4:00 a.m. $\leq$ h $\leq$ 11:00 p.m. and two techniques (XGBoost, CatBoost), herein Equation (4) represents Group B, as follows:

$$\begin{aligned}(p_{t+2h}, \ldots, p_{t+k}) \\ = f(p_{t-1h}, p_{t-24h}, p_{t-48h}, p_{t-72h}, p_{t-96h}, p_{t-120h}, p_{t-144h}, GHI_{t+k}, T_{t+k}, SI_{t+k}, WS_{t+k}, RH_{t+k}, RASPP_{t+k})\end{aligned} \tag{3}$$

$$M^B = \left\{ \left(XGB_4^B, CTB_4^B\right), \ldots \left(XGB_h^B, CTB_h^B\right) \right\}_{h=4AM, \ldots 11PM} \tag{4}$$

3.4.3. Group C: One-Step Ahead for 57th to 144th Hour Ahead Framework

The following days ahead forecasting is essential for managing the grid operations. Group C is a set of one-step ahead models for recursive predictions from 57th to 144th hour ahead, with separate models for each hour of the day. Group C models rely on the NWP and the most recent forecasts from Group B, following the same recursive forecasting strategy. The input variables are past power output (p), GHI, ambient temperature (T), solar irradiance (SI), wind speed (WS), and relative humidity (RH), according to Equation (5) with k = (57, ... , 144). For a set of models, M^C with h hours of the day limited to 4:00 a.m. $\leq$ h $\leq$ 11:00 p.m. and two techniques (XGBoost, CatBoost), herein Equation (6) represents Group C, as follows:

$$(p_{t+57h}, \ldots, p_{t+k}) = f(p_{t-1h}, p_{t-24h}, p_{t-48h}, p_{t-72h}, p_{t-96h}, p_{t-120h}, p_{t-144h}, GHI_{t+k}, T_{t+k}, SI_{t+k}, WS_{t+k}, RH_{t+k}) \tag{5}$$

$$M^C = \left\{ \left(XGB_4^C, CTB_4^C\right), \ldots \left(XGB_h^C, CTB_h^C\right) \right\}_{h=4AM, \ldots 11PM} \tag{6}$$

Figure 3 shows an example of the forecasting strategy and how Groups A, B, and C accumulate outputs for the forecasting horizon of 144 h ahead. The instance considers the issuing time of 1 August 2021 at 4:20 a.m.. Therefore, the forecasting origin is 1 August 2021 at 5:00 a.m., which is one step ahead of forecasting. From a set of 20 pairs of models, Group A will select the appropriate model specific for 5:00 a.m. to forecast the first step. It will provide predictions for XGBoost and CatBoost. Next, Group B will predict the second step by selecting the 6:00 a.m. models for the specific hour. Then, Group B sets the next pair of models recursively for 7:00 a.m. until it reaches 56 h ahead of forecast. Similarly, Group C will provide the following one-step forecasting from 57 to 144 h ahead, when the last prediction of the forecasting horizon is reached. In summary, a serial sequence of one-step ahead or 1 h ahead models will be selected from 1 to 144 h ahead.

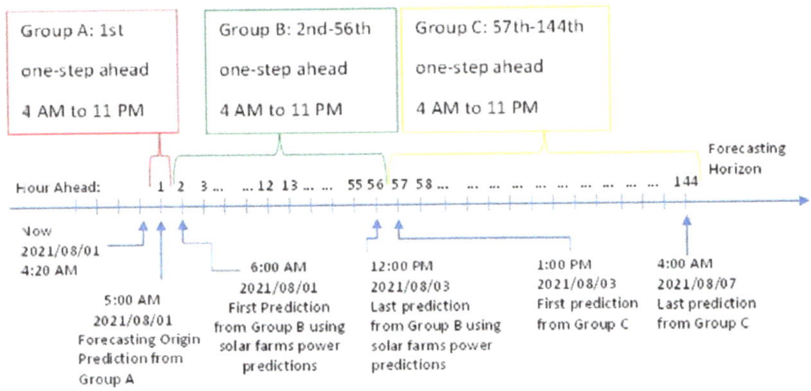

Figure 3. Group A, B, and C models to predict 1-144 h ahead.

2.5. Deterministic Forecast

Point forecasts, deterministic forecasts, or single-value forecasts are all synonyms. They can be used to define that the predictions or forecasts made by this class of models can output only one value for each instance or each time stamp. Two deterministic models will be presented and then, individual performances will be evaluated. Producing probabilistic forecasts are left for future works.

- XGBoost (XGB) [32], or the eXtreme Gradient Boosting, is an evolution implementation of the gradient tree boosting (GB), which is a technique first introduced in 2000 by the authors of [33]. XGBoost gained recognition in several data mining challenges and machine learning competitions. For example, in 2017, one of the five best teams in The Global Energy Forecasting Competition 2017 (GEFCom2017) used XGBoost to solve a hierarchical probabilistic load forecasting problem. The technique is a gradient boosted tree algorithm, a supervised learning method capable of fitting generic non-parametric predictive models. For XGBoost, a search for the hyperparameters with RandomizedSearchCV class and GridSearchCV class from Scikitlearn is performed.
- CatBoost (CTB), or categorical boosting [34], is an open-source machine learning tool developed in Germany in 2017. The authors claim that this updated method outperforms the existing state-of-the-art implementations of gradient-boosted decision trees XGBoost. CatBoost proposes ordered boosting, a modification of the standard gradient boosting algorithm that avoids both a target leakage and prediction shift, with a new algorithm for processing categorical features. It presents three main advantages: First, it can integrate data types, such as numerical, images, audio, and text features. Second, it can simplify the feature engineering process since it requires minimal categorical feature transformation. Finally, it has a built-in hyperparameter optimization, which simplifies the learning process while increasing the overall speed of the model.

4. Numerical Results

4.1. Evaluation Criteria

The most common and accepted deterministic forecasting accuracy measures are the root mean squared error (RMSE) and its respective RMSE skill score [35,36]. The RMSE is a common error metric used in point forecasting due to its squared error, which is more sensitive to outliers [37]. The RMSE is calculated according to Equation (7). It is measured with the same unit as the target forecasting, in kilowatts (kW). For n = 144, t ∈ (1, 2, . . . 144), $\hat{y}_t$ is the forecast at time t, and y_t is the observed PV power at time t. The best way to measure the accuracy gain of a proposed forecasting method is to calculate the forecast skill score using the RMSE as the base metric and compare the results from the proposed method versus a benchmark method [35,38]. The RMSE skill score is calculated using Equation (8) and is measured in percent (%), as follows:

$$RMSE = \sqrt{\frac{1}{n}\sum_{t=1}^{n}(\hat{y}_t - y_t)^2} \tag{7}$$

$$Skill\ Score_{RMSE} = 1 - \frac{RMSE_{Proposed\ Method}}{RMSE_{Benchmark}} \tag{8}$$

4.2. Benchmark

Solar energy forecasts, including irradiance and power, strongly depend on data, location, resolution, and horizon. Therefore, according to the author in [35], without a universal benchmark, it is sometimes impossible to interpret the quality of a solar energy forecast model. The complete-history persistence ensemble (CH-PeEn) was proposed in 2019 to be a universal benchmarking method for probabilistic solar forecasting [16]. The historical PV power output with hourly resolution used to calculate this model is from 11 March 2019 to 31 July 2022. This paper will use the percentile 50% or the mean as a benchmark. Figure 4 shows the target PV system's mean and probability distribution of power.

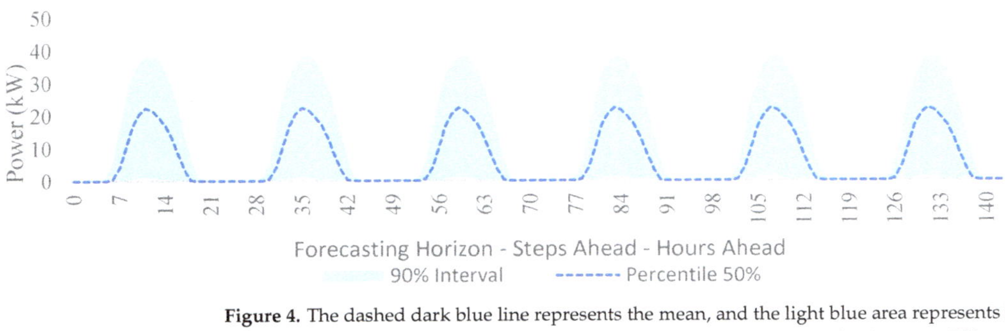

Figure 4. The dashed dark blue line represents the mean, and the light blue area represents the 90% interval probability distribution of power for the CH-PeEn benchmark for the target PV system for six consecutive days.

4.3. Test Design

In this paper, the following six scenarios will be performed to demonstrate the quality of each strategy: Models S, 3M, 1M, N, S3M, and S1M, covering from 1 h to 6 days ahead. First, Model S will verify whether the hybrid method with publicly available regional aggregated solar power predictions can be effectively applied to increase the accuracy of the small-scale BTM PV model. Second, Model 3M will verify whether solely selecting the current and the previous 3 months related to a targeted pattern of historical data can increase the forecasting model accuracy. Third, Model 1M will examine the accuracy improvement from the solely 1M strategy. Fourth, Model N will verify whether a model considering none of the proposed strategies is better and can increase the forecasting model accuracy. Fifth, Model S3M will explore whether a model assuming a combination of Models S and 3M can improve the model's accuracy. Sixth, Model S1M, the proposed model of this paper, will consider including Models S and 1M. A CH-PeEn model benchmark is also performed to compare the forecasting skill of each model. The training data available are from 11 March 2019 to 31 July 2021. Each model is retrained every week, adding one previous week. The solar power forecasting methodology is deployed to predict a total of 1 year of testing from 1 August 2021 to 31 July 2022.

The skill score RMSE of each model per month and season is presented in Tables 1 and 2, respectively, for models XGBoost and CatBoost. Table 1 shows that from October to February, the skill score ranges from 73% to 96% compared with the CH-PeEn benchmark, in which the skill score is more significant than in the other months. This result aligns with the baseline in Section 4.2 since it has only one shape for all days of the year, represented by the mean curve in Figure 4. For example, in Table 1, the results for S1M-CatBoost show that the highest skill score occurred in December 2021 and the lowest in May and July 2022. This indicates that compared with the benchmark in Figure 4, the general shape and magnitude of the solar power predictions are more similar to months, such as May and July than December. According to the authors of [35], due to an effective model considering a station in the upwind direction to the target point, higher forecast skill scores can be found between 50% and 70% during specific periods and not year round.

Table 1. Average skill score RMSE (%) per month per model.

INDEX	202108	202109	202110	202111	202112	202201	202202	202203	202204	202205	202206	202207
S-XGBoost	58	69	80	85	91	88	78	67	51	39	52	42
S-CatBoost	59	69	81	86	91	88	79	68	52	41	52	44
3M-XGBoost	52	60	75	82	90	84	77	57	42	47	45	34
3M-CatBoost	53	63	75	84	92	85	78	59	45	46	47	35
1M-XGBoost	72	72	84	91	95	94	89	74	70	55	60	54
1M-CatBoost	72	74	85	91	96	94	90	77	70	57	61	57
N-XGBoost	53	53	73	84	92	85	81	61	48	45	48	35
N-CatBoost	55	56	75	84	92	86	81	62	49	46	51	38
S3M-XGBoost	58	66	77	84	92	84	80	63	48	44	49	43
S3M-CatBoost	58	68	79	85	93	85	80	65	49	45	50	47
S1M-XGBoost	74	74	85	91	95	94	89	76	72	56	61	57
S1M-CatBoost	73	75	85	91	95	94	90	77	70	58	62	58

Table 2. Average skill score RMSE (%) per season per model.

INDEX	SPRING	SUMMER	FALL	WINTER	AVG Year	Ranking
S-XGBoost	52	51	78	86	66.7	7
S-CatBoost	54	52	79	86	67.5	5
3M-XGBoost	48	44	72	84	62.2	12
3M-CatBoost	50	45	74	85	63.6	10
1M-XGBoost	67	62	82	93	76.0	4
1M-CatBoost	68	63	83	93	76.8	3
N-XGBoost	52	45	70	86	63.2	11
N-CatBoost	52	48	72	86	64.7	9
S3M-XGBoost	52	50	76	85	65.6	8
S3M-CatBoost	53	52	77	86	67.0	6
S1M-XGBoost	68	64	83	93	76.9	2
S1M-CatBoost	68	64	84	93	77.3	1

In Table 2, it can be observed that strategies S, 3M, N, and S3M presented skill score RMSE results below 67.5%. Moreover, this table shows how Models S1M and 1M are more relevant than the other suggested models. When Model N is compared with Model S1M, it can be observed that Model S1M could leverage the combination of both strategies: S and 1M, indicating that the solar farms' power forecasts and the reduced dataset assisted in increasing the forecasting skill. The best individual technique was S1M-CatBoost, with an average per year of 77.3%, followed by S1M-XGBoost with 76.9%.

In Table 2, the best deterministic model is CatBoost, since it outperformed XGBoost on all scenarios. Therefore, an analysis of the average RMSE and the skill score RMSE of hours ahead predictions of the six solar scenarios will be presented only for the CatBoost scenarios. Figure 5 shows that Model 3M presents the highest RMSE, while Model 1M is the second best, which indicates that a reduced dataset using strategy 1M leveraged the highest correlation to improve the model's accuracy. Next, Model N was easily outperformed by Models S3M and S when the publicly available regional aggregated solar power predictions were added as a feature. Finally, the proposed Model S1M consistently outperformed from 2 to 144 h ahead. For Model S1M, the error increases more noticeably from 1 to 2 h ahead, and the average error of the following steps increases very slightly. Moreover, an improvement is observed between steps 91 and 97. It could be related to the recursive forecasting strategy and the lagged power production as inputs for the model, highlighting a higher correlation.

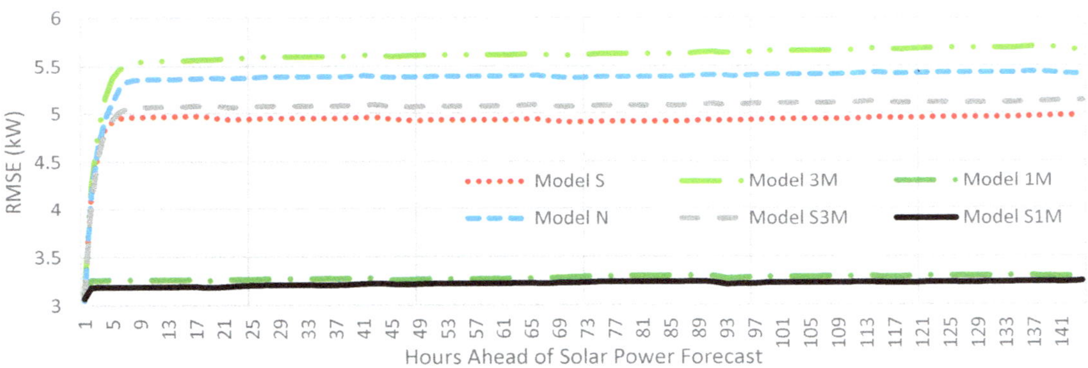

Figure 5. RMSE of hours ahead predictions of the six CatBoost scenarios.

Figure 6 shows that all CatBoost scenarios outperformed the mean of the benchmark CH-PeEn with skill score RMSE ranging from 79% to 62%. Although Models S1M, 1M, S, and S3M were outperformed by Model N via a small margin in the 1st hour ahead, all models outperformed Model N consistently for all of the remaining 143 steps ahead. Models 1M and 3M outperformed the benchmark, but the latter did not outperform Model N any step forward. Finally, Models 1M and 3M were improved when the solar farms' power predictions were added as a feature, later identified as Models S1M and S3M, proving the relevance of this feature.

Since the best deterministic model is the S1M-CatBoost, an analysis of the average RMSE and skill score RMSE for each hour of the day will be presented. The average RMSE and the respective skill score RMSE for each hour of the day in the four seasons are shown as two main characteristics in Figures 7 and 8. First, each season has a particular magnitude, and second, each has a specific duration, from sunrise to sunset. For example, during Summer, the highest magnitude of RMSE is around 3.8 kW at 4:00 p.m., and the daylight hours are from 6:00 a.m. to 9:00 p.m., with an average skill score of 65%. However, during Winter, the lowest magnitude of RMSE is found since it is directly proportional to the lower availability of solar irradiance, around 1.6 kW at 9:00 a.m., with daylight hours from 8:00 a.m. to 5:00 p.m., and an average skill score of 93%.

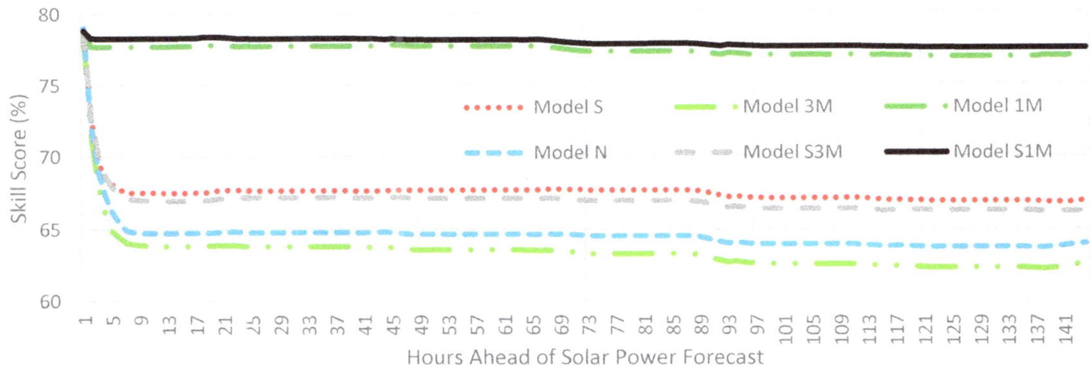

Figure 6. Skill score RMSE of hours ahead predictions of the six CatBoost scenarios.

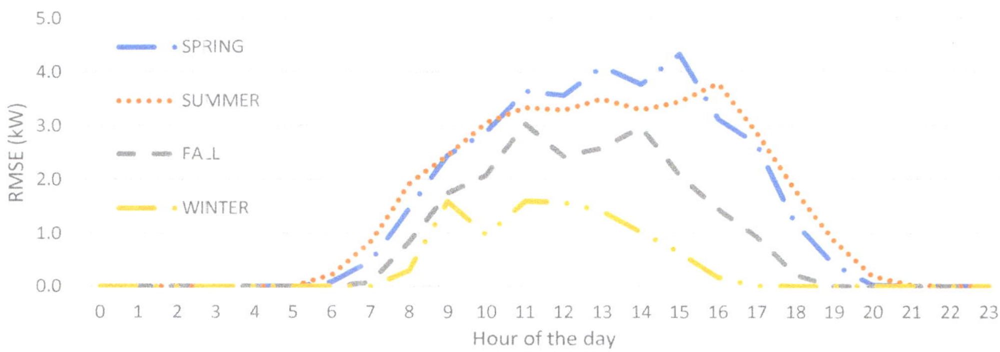

Figure 7. Average RMSE for each hour of the day for the proposed Model S1M.

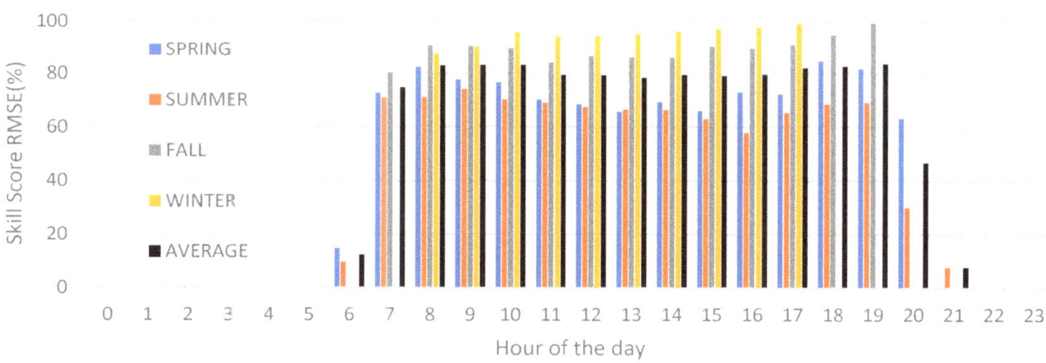

Figure 8. Average skill score RMSE for each hour of the day for the proposed Model S1M.

In Figure 8, the average skill score identified from 7:00 a.m. to 7:00 p.m. for all seasons are similar, but two exceptions can be found. During Spring and Summer, at 6:00 a.m. and 8:00 p.m., and during Summer, at 9:00 p.m., the magnitude of the skill score is lower than the season average. This situation occurs due to the higher errors found, especially during sunrise and sunset, which are harder to predict even though the magnitude of the PV output is significantly lower than the daylight peak hours.

5. Conclusions

The contributions of this paper are three-fold: First, a new hybrid methodology was proposed with a sequence of one-step models to forecast 6 days ahead for a small-scale BTM PV site with three groups of models with different inputs—each group was trained for each hour of above zero irradiance. The best technique identified was CatBoost and the proposed method was S1M. Second, a novel dataset preselection was presented, named 1M, and individual results proved the method's efficiency against a benchmark and other scenarios. Third, applying neighboring solar farm predictions as a feature boosted the model's accuracy. Therefore, to the author's knowledge, no other research work has developed a simplified targeted training strategy, such as the one presented or used publicly available regional aggregated solar power predictions from neighboring utility-scale solar farms to improve the quality of the small-scale BTM PV system forecasts.

Future Work

Probabilistic forecasts describe the embedded variability and assist in the decision-making process [18]. Therefore, a bootstrap strategy can be implemented by creating N different scenarios with an increased bandwidth of predictions to output a probabilistic forecast. For example, the authors of [24] used 20,000 weather scenarios and averaged the estimates with quantile regression averaging (QRA) to produce probabilistic forecasts.

Author Contributions: Conceptualization, H.Z.; methodology, H.B.M.L.; software, H.B.M.L.; validation, H.B.M.L.; formal analysis, H.B.M.L.; investigation, H.B.M.L.; resources, H.B.M.L.; data curation, H.B.M.L.; writing—original draft, H.B.M.L.; writing—review and editing, H.Z.; visualization, H.B.M.L.; supervision, H.Z.; funding acquisition, H.Z. All authors have read and agreed to the published version of the manuscript.

Funding: This research was supported by funding from Canada NSERC Discovery Grants.

Data Availability Statement: Not applicable.

Acknowledgments: The authors would like to thank NRGStream for providing complimentary access to their data warehouse.

Conflicts of Interest: The authors declare no conflict of interest.

References

1. Masson, G. International Energy Agency Snapshot. 2022. Available online: https://iea-pvps.org/snapshot-reports/snapshot-2022/ (accessed on 11 August 2022).
2. Haupt, S.E.; Dettling, S.; Williams, J.K.; Pearson, J.; Jensen, T.; Brummet, T.; Kosovic, B.; Wiener, G.; McCandless, T.; Burghardt, C. Blending Distributed Photovoltaic and Demand Load Forecasts. *Sol. Energy* **2017**, *157*, 542–551. [CrossRef]
3. Chu, Y.; Pedro, H.T.C.; Kaur, A.; Kleissl, J.; Coimbra, C.F.M. Net Load Forecasts for Solar-Integrated Operational Grid Feeders. *Sol. Energy* **2017**, *158*, 236–246. [CrossRef]
4. Kaur, A.; Nonnenmacher, L.; Pedro, H.T.C.; Coimbra, C.F.M. Benefits of Solar Forecasting for Energy Imbalance Markets. *Renew. Energy* **2016**, *86*, 819–830. [CrossRef]
5. Mellit, A.; Pavan, A.M.; Ogliari, E.; Leva, S.; Lughi, V. Advanced Methods for Photovoltaic Output Power Forecasting: A Review. *Appl. Sci.* **2020**, *10*, 487. [CrossRef]
6. Dolara, A.; Grimaccia, F.; Leva, S.; Mussetta, M.; Ogliari, E. A Physical Hybrid Artificial Neural Network for Short Term Forecasting of PV Plant Power Output. *Energies* **2015**, *8*, 1138–1153. [CrossRef]
7. Cervone, G.; Clemente-Harding, L.; Alessandrini, S.; Delle Monache, L. Short-Term Photovoltaic Power Forecasting Using Artificial Neural Networks and an Analog Ensemble. *Renew. Energy* **2017**, *108*, 274–286. [CrossRef]
8. Yagli, G.M.; Yang, D.; Srinivasan, D. Automatic Hourly Solar Forecasting Using Machine Learning Models. *Renew. Sustain. Energy Rev.* **2019**, *105*, 487–498. [CrossRef]
9. Zhang, X.; Li, Y.; Lu, S.; Hamann, H.F.; Hodge, B.M.; Lehman, B. A Solar Time Based Analog Ensemble Method for Regional Solar Power Forecasting. *IEEE Trans. Sustain. Energy* **2019**, *10*, 268–279. [CrossRef]
10. Wolff, B.; Kühnert, J.; Lorenz, E.; Kramer, O.; Heinemann, D. Comparing Support Vector Regression for PV Power Forecasting to a Physical Modeling Approach Using Measurement, Numerical Weather Prediction, and Cloud Motion Data. *Sol. Energy* **2016**, *135*, 197–208. [CrossRef]

11. Mellit, A.; Massi Pavan, A.; Lughi, V. Short-Term Forecasting of Power Production in a Large-Scale Photovoltaic Plant. *Sol. Energy* **2014**, *105*, 401–413. [CrossRef]
12. Yang, C.; Xie, L. A Novel ARX-Based Multi-Scale Spatio-Temporal Solar Power Forecast Model. In Proceedings of the 2012 North American Power Symposium, NAPS 2012, Champaign, IL, USA, 9–11 September 2012. [CrossRef]
13. Goncalves, C.; Bessa, R.J.; Pinson, P. Privacy-Preserving Distributed Learning for Renewable Energy Forecasting. *IEEE Trans. Sustain. Energy* **2021**, *12*, 1777–1787. [CrossRef]
14. Antonanzas, J.; Osorio, N.; Escobar, R.; Urraca, R.; Martinez-de-Pison, F.J.; Antonanzas-Torres, F. Review of Photovoltaic Power Forecasting. *Sol. Energy* **2016**, *136*, 78–111. [CrossRef]
15. Yang, D. Standard of Reference in Operational Day-Ahead Deterministic Solar Forecasting. *J. Renew. Sustain. Energy* **2019**, *11*, 053702. [CrossRef]
16. Yang, D. A Universal Benchmarking Method for Probabilistic Solar Irradiance Forecasting. *Sol. Energy* **2019**, *184*, 410–416. [CrossRef]
17. Pedro, H.T.C.; Coimbra, C.F.M.; David, M.; Lauret, P. Assessment of Machine Learning Techniques for Deterministic and Probabilistic Intra-Hour Solar Forecasts. *Renew. Energy* **2018**, *123*, 191–203. [CrossRef]
18. van der Meer, D. Comment on "Verification of Deterministic Solar Forecasts": Verification of Probabilistic Solar Forecasts. *Sol. Energy* **2020**, *210*, 41–43. [CrossRef]
19. Pedro, H.T.C.; Coimbra, C.F.M. Assessment of Forecasting Techniques for Solar Power Production with No Exogenous Inputs. *Sol. Energy* **2012**, *86*, 2017–2028. [CrossRef]
20. Panamtash, H.; Zhou, Q.; Hong, T.; Qu, Z.; Davis, K.O. A Copula-Based Bayesian Method for Probabilistic Solar Power Forecasting. *Sol. Energy* **2020**, *196*, 336–345. [CrossRef]
21. Marquez, R.; Pedro, H.T.C.; Coimbra, C.F.M. Hybrid Solar Forecasting Method Uses Satellite Imaging and Ground Telemetry as Inputs to ANNs. *Sol. Energy* **2013**, *92*, 176–188. [CrossRef]
22. Huang, J.; Perry, M. A Semi-Empirical Approach Using Gradient Boosting and k-Nearest Neighbors Regression for GEFCom2014 Probabilistic Solar Power Forecasting. *Int. J. Forecast.* **2016**, *32*, 1081–1086. [CrossRef]
23. Pierro, M.; Gentili, D.; Liolli, F.R.; Cornaro, C.; Moser, D.; Betti, A.; Moschella, M.; Collino, E.; Ronzio, D.; van der Meer, D. Progress in Regional PV Power Forecasting: A Sensitivity Analysis on the Italian Case Study. *Renew. Energy* **2022**, *189*, 983–996. [CrossRef]
24. Sun, M.; Feng, C.; Zhang, J. Probabilistic Solar Power Forecasting Based on Weather Scenario Generation. *Appl. Energy* **2020**, *266*. [CrossRef]
25. Pinho, M.; Muñoz, M.; De, I.; Perpiñán, O. Comparative Study of PV Power Forecast Using Parametric and Nonparametric PV Models. *Sol. Energy* **2017**, *155*, 854–866. [CrossRef]
26. Lonij, V.P.A.; Brooks, A.E.; Cronin, A.D.; Leuthold, M.; Koch, K. Intra-Hour Forecasts of Solar Power Production Using Measurements from a Network of Irradiance Sensors. *Sol. Energy* **2013**, *97*, 58–66. [CrossRef]
27. Vaz, A.G.R.; Elsinga, B.; van Sark, W.G.J.H.M.; Brito, M.C. An Artificial Neural Network to Assess the Impact of Neighbouring Photovoltaic Systems in Power Forecasting in Utrecht, the Netherlands. *Renew. Energy* **2016**, *85*, 631–641. [CrossRef]
28. Henze, J.; Schreiber, J.; Sick, B. *Representation Learning in Power Time Series Forecasting*; Springer International Publishing: Kassel, Germany, 2020; ISBN 9783030317607.
29. Zhang, G.; Yang, D.; Galanis, G.; Androulakis, E. Solar Forecasting with Hourly Updated Numerical Weather Prediction. *Renew. Sustain. Energy Rev.* **2022**, *154*, 111768. [CrossRef]
30. Persson, C.; Bacher, P.; Shiga, T.; Madsen, H. Multi-Site Solar Power Forecasting Using Gradient Boosted Regression Trees. *Sol. Energy* **2017**, *150*, 423–436. [CrossRef]
31. ben Taieb, S.; Hyndman, R.J. Recursive and Direct Multi-Step Forecasting: The Best of Both Worlds. *Int. J. Forecast.* Available online: https://www.monash.edu/business/ebs/research/publications/ebs/wp19-12.pdf (accessed on 11 August 2022).
32. Chen, T.; Guestrin, C. XGBoost: A Scalable Tree Boosting System. In Proceedings of the 22nd ACM SIGKDD International Conference on Knowledge Discovery and Data Mining, San Francisco, CA, USA, 13–17 August 2016; ACM: New York, NY, USA, 2016; Volume 42, pp. 785–794.
33. Friedman, J.; Tibshirani, R.; Hastie, T. Additive Logistic Regression: A Statistical View of Boosting (With Discussion and a Rejoinder by the Authors). *Ann. Stat.* **2000**, *28*, 337–407. [CrossRef]
34. Prokhorenkova, L.; Gusev, G.; Vorobev, A.; Dorogush, A.V.; Gulin, A. Catboost: Unbiased Boosting with Categorical Features. *Adv. Neural Inf. Process. Syst.* **2018**, 6638–6648.
35. Yang, D. A Guideline to Solar Forecasting Research Practice: Reproducible, Operational, Probabilistic or Physically-Based, Ensemble, and Skill (ROPES). *J. Renew. Sustain. Energy* **2019**, *11*. [CrossRef]
36. Pedro, H.T.C.; Larson, D.P.; Coimbra, C.F.M. A Comprehensive Dataset for the Accelerated Development and Benchmarking of Solar Forecasting Methods. *J. Renew. Sustain. Energy* **2019**, *11*. [CrossRef]

37. van der Meer, D.W.; Widén, J.; Munkhammar, J. Review on Probabilistic Forecasting of Photovoltaic Power Production and Electricity Consumption. *Renew. Sustain. Energy Rev.* **2018**, *81*, 1484–1512. [CrossRef]
38. Nonnenmacher, L.; Kaur, A.; Coimbra, C.F.M. Day-Ahead Resource Forecasting for Concentrated Solar Power Integration. *Renew Energy* **2016**, *86*, 866–876. [CrossRef]

energies

MDPI

Article

Short-Term Occupancy Forecasting for a Smart Home Using Optimized Weight Updates Based on GA and PSO Algorithms for an LSTM Network

Sameh Mahjoub, Sami Labdai, Larbi Chrifi-Alaoui *, Bruno Marhic and Laurent Delahoche

Laboratory of Innovative Technology (LTI, UR-UPJV 3899), University of Picardie Jules Verne, 80000 Amiens, France
* Correspondence: larbi.alaoui@u-picardie.fr

Abstract: In this work, we provide a smart home occupancy prediction technique based on environmental variables such as CO_2, noise, and relative temperature via our machine learning method and forecasting strategy. The proposed algorithms enhance the energy management system through the optimal use of the electric heating system. The Long Short-Term Memory (LSTM) neural network is a special deep learning strategy for processing time series prediction that has shown promising prediction results in recent years. To improve the performance of the LSTM algorithm, particularly for autocorrelation prediction, we will focus on optimizing weight updates using various approaches such as Genetic Algorithm (GA) and Particle Swarm Optimization (PSO). The performances of the proposed methods are evaluated using real available datasets. Test results reveal that the GA and the PSO can forecast the parameters with higher prediction fidelity compared to the LSTM networks. Indeed, all experimental predictions reached a range in their correlation coefficients between 99.16% and 99.97%, which proves the efficiency of the proposed approaches.

Keywords: deep neural networks; LSTM; time series prediction; optimisation; GA; PSO

Citation: Mahjoub, S.; Labdai, S.; Chrifi-Alaoui, L.; Marhic, B.; Delahoche, L. Short-Term Occupancy Forecasting for a Smart Home Using Optimized Weight Updates Based on GA and PSO Algorithms for an LSTM Network. *Energies* **2023**, *16*, 1641. https://doi.org/10.3390/en16041641

Academic Editors: Paweł Piotrowski, Grzegorz Dudek and Dariusz Baczyński

Received: 24 December 2022
Revised: 21 January 2023
Accepted: 24 January 2023
Published: 7 February 2023

1. Introduction

One of the most efficient systems to save energy is to reduce a building's heating and cooling load, which is mostly caused by heat transfer over its envelope. Smart buildings are required to provide permanent, healthy and comfortable indoor environments, independent of exterior weather conditions [1,2]. Indeed, the major part of energy in such buildings is used by Heating, Ventilation, and Air Conditioning (HVAC) systems, which have a significant influence on both home comfort and the environment. Therefore, managing these systems in residential structures should be tackled in order to increase energy efficiency through improved energy planning [3]. One of the most essential features of smart buildings is their ability to self-control the systems used to maintain the comfort of the inside atmosphere while also minimizing energy use. Because HVAC systems are the primary source of energy consumption in buildings, intelligent HVAC system control is a current trend in research studies that necessitates the insertion of occupancy information into the control process [4]. Moreover, the rise of smart buildings, as well as the pressing need to reduce energy use, has rekindled interest in building energy demand prediction. Intelligent controls are a solution for optimizing power consumption in buildings without reducing interior comfort [5]. For example, in [6], a Model Predictive Control (MPC) is developed to obtain a hybrid HVAC control with energy savings while maintaining of thermal comfort. Building energy consumption prediction strives to achieve various goals such as evaluating the impact of energy-saving interventions and assume energy demands based on regular requirements. It can anticipate the fluctuations in power consumption of certain events at specfic times that may modify the systems' customary energy usage [7]. Furthermore, based on detailed and extensive studies, it was concluded that occupant behavior is one of the most significant elements affecting residential structure energy use. Occupancy behavior includes activities such as turning on and off lights, switching on and off heating and cooling systems, and regulating the temperature.

Previous research has shown that various occupant demands and behaviors necessitate specific technological solutions, which may induce or change behavior patterns, and that occupant behavior affects the flexibility and deployment of technologies. However, the lack of comprehensive knowledge of occupant behaviors in residential building leads to misunderstanding and inaccurate decisions in both technical design and policy making [8]. The context of our research is energy efficiency. In recent years, energy efficiency has been realized by improving the thermal performances of the building envelope's insulation layer. The research strategies aim to permanently adjust the comfort conditions to the living situation, as well as to ensure greater energy supervision and management within the smart buildings. To achieve this, it is important to automatically characterize the activities of the building's residents. The significant challenge in today's new technical design for smart buildings is understanding customer behaviors [9]. In the future, our occupancy prediction approach will guarantee energy savings in a smart building environment. Ambient intelligence is an important prerequisite for improving human quality of life.

The rest of this work is structured as follows: Section 2 explains the technique employed in this project. Firstly, it offers the overall framework of the LSTM forecasting model. Next, it presents, step-by-step, the implementation procedure of the suggested technique; it includes descriptions of database processing, the parameters, and the assessment indicators. Section 3 features experimental details, as well as an analysis of the results. Finally, Section 4 provides some conclusions and future works.

2. Related Works

Building energy consumption is influenced by the thermal insulation, heating, ventilation, air conditioning, lighting, and occupants' behaviors [10]. Characterising human activity has become an increasingly prominent application of machine learning in a many disciplinary fields. Indeed, for the past two decades, researchers from several application fields have investigated activity recognition by developing a variety of methodologies and techniques for each of these key tasks. The prediction of human behaviour represents a key challenge, and many approaches have already been proposed in the industrial, medical, home care, and energy efficiency domains, and many others [11]. For example, in [12], an end-to-end technique for forecasting multi-zone interior temperatures using LSTM-based sequence to sequence has been introduced. The goal of this prediction is to improve the building's energy efficiency while maintaining occupant comfort. Authors in [13] also proposed implementing simple XGBoost machine learning methods to predict the interior room temperature, relative humidity, and CO_2 concentration in a commercial structure. The proposed technique presents a practical option because it does not require a large data set for training. Additionally, these models eliminate the necessity for multiple sensors, which create sophisticated and expensive networks. In [14], a short-term load consumption forecasting approach for nonresidential buildings using artificial occupancy attributes and based on Support Vector Machines (SVM) has been developed. However, the determination of human behaviour in this work is imprecise. The authors in [15], present a load forecasting model for office buildings based on artificial intelligence and regression analysis to effectively extract the cooling and heating load characteristics. However, the model assumes that the building's internal disturbing influences are steady. In [16], an optimal deep learning LSTM model for forecasting electricity consumption utilizing feature selection and a Genetic Algorithm (GA) is implemented. The goal of this suggested technique is to determine the optimal time delays and number of layers for LSTM architecture's predictive performance optimisation, as well as to minimize overfitting, resulting in more accurate and consistent forecasting. Furthermore, recently, machine learning approaches based on Artificial Neural Networks (ANNs) have been widely used to forecast the thermal behavior of modern buildings for modeling HVAC systems. As an example, in [17], four comparative models have been developed and refined to forecast the inside temperature of a public building. These proposed techniques can be adapted to various scenarios. However, we must keep in mind that the adoption of an online technique such as OMLP (Online MultiLayer Perceptron) might be influenced by outliers. The authors also in [18] tackle a non-linear autoregressive neural network methodology for forecasting interior air temperature in the short and medium terms. Realistic artificial temperature data are used to train the proposed model. The goal of this strategy is to make up for the lack of real-world data collected by sensors in energy experiments. Thus, an improved technique integrating real-time information and addressing possible noise or missing data is necessary to prove the reliability of the proposed strategy in real scenarios. Differently from previous research solutions, which typically rely on a basic and simple LSTM model, we designed an optimised architecture exploiting GA and PSO algorithms to update weights and select the optimal values that give the best prediction precision and reduce model overfitting. As a matter of fact, these two methods (PSO and GA) were chosen due to their good reputation in the literature, and they add a stochastic approach to the neural network that resulted in better performance. We compared our results with the LSTM method, which is considered the best neural approach in time series forecasting, as proven in previously conducted works based on LSTM. As an example, Ref. [19] introduces comprehensive comparative studies that include several deep learning methods used in forecasting extra-short-term Plug-in Electric Vehicle (PEV) charging loads such as ANN, RNN, LSTM, gated recurrent units (GRU), and bi-directional long short-term memory (Bi-LSTM). Among these approaches, the LSTM model outperforms the others, and it is competent in giving satisfactory results.

3. Materials and Methodologies

3.1. Data Description

A year of data were collected from a smart home between 1 January 2018 and 31 December 2018 with a resolution of 10 min. Each room of the house was equipped with several sensors, including set points of the room temperature, CO_2 concentration, pressure, noise, lighting, and occupancy:

- CO_2 values of a floor of house;
- Outdoor temperature;
- Noise values.
- Pressure values.

The concentration of these factors varied depending on the room; for example, the concentration of CO_2 in the living room differed from that in the office or the kitchen. Moreover, the CO_2 variable does not have a direct relationship with the interior temperature. However, because CO_2 is a strong predictor of room occupancy, it may have a direct impact on the indoor temperature during the cold season. The variation in the CO_2, the noise, and the temperature are given by Figures 1–3, respectively.

3.2. Data Pre-Processing

The prediction of building energy use based on an occupant behavior assessment is a multivariate time series issue in which sensors create data that may contain uncertainty, redundancy, missing values, non-unified time intervals, noise, and so on. Traditional machine learning techniques struggle to reliably anticipate power usage due to unpredictable trend components and seasonal trends. The collection of suitable data contributes to efficiently addressing prediction challenges. As a result, several considerations should be made [20]. So, numerous techniques have been proposed to obtain meaningful inferences and insights; nevertheless, these solutions are still in the early phases of development. Therefore, current research is focusing on improving the procedures for processing and cleaning the collected data in order to produce accurate prediction [21].

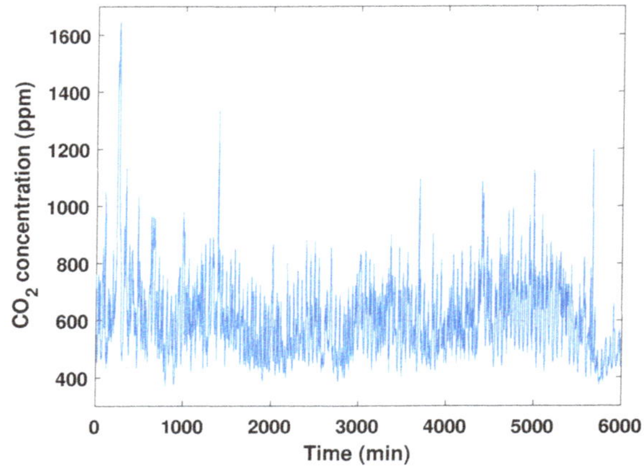

Figure 1. Overview of the CO_2 set points.

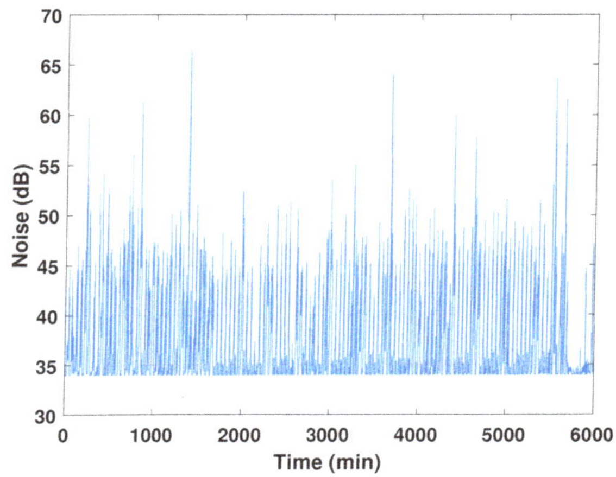

Figure 2. Overview of the room noise set points.

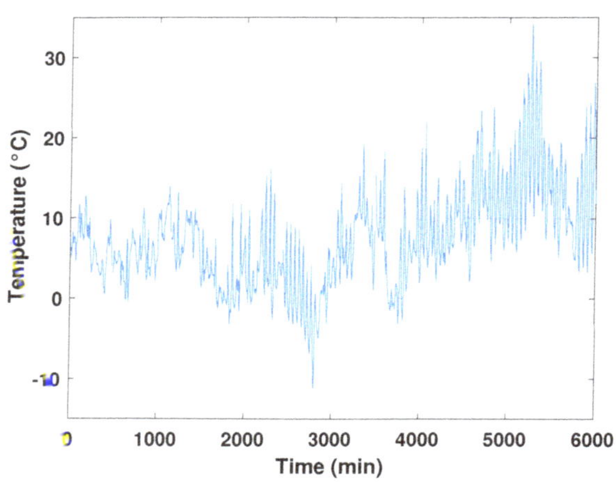

Figure 3. Overview of the room temperature set points.

3.2.1. Missing Values

Many real-world datasets may include missing values for various reasons. So, training a model using a dataset that has a large number of missing values can have a considerable influence on the machine learning model's quality. To prevent information leakage, missing data were interpolated using Exponential Moving Average (EMA). This method is described in [22].

3.2.2. Normalisation

The data for a sequence prediction problem probably need to be normalised to the range of [−1, +1] when training a neural network such as a long short-term memory recurrent neural network. When a network is fit on unscaled data, it is possible for large inputs to slow down the learning and convergence of that network and, in some cases, prevent the network from effectively learning the problem. The Z-score is used for the normalization, and the formula is given as [23]:

$$Z_{Score} = \frac{x - x_{mean}}{x_{\sigma}} \tag{1}$$

where:

$$x_{\sigma} = \sqrt{\frac{1}{n-1}\sum_{i=1}^{n}(x_i - x_{mean})^2} \tag{2}$$

$$x_{mean} = \frac{1}{n}\sum_{i=1}^{n} x_i \tag{3}$$

and n is the number of time periods.

4. Modeling Approaches

The main aim of this research is to investigate the performance of various occupancy forecasting strategies to identify the most accurate ones. In fact, we choose three distinct methods, based on a deep learning method: GA-LSTM and PSO-LSTM as optimiser based-models and LSTM as a simple deep learning technique.

4.1. LSTM Architecture

Recurrent Neural Networks (RNNs) struggle with learning long-term dependencies. LSTM-based models are an extension of RNNs that can solve the vanishing gradient problem and exploding gradient problem of RNNs and which perform more favorably than RNN on longer sequences. LSTM models basically expand the memory of RNNs to allow them to maintain and learn long-term input dependencies properly. This memory expansion can recall data for a longer amount of time, allowing them to read, write, and delete information from their memories. The LSTM memory is referred to as a "gate" structure because it has the power to decide whether to keep or discard memory information [24,25]. A gate is a way of transferring information selectively that includes a sigmoid neural network layer and a bitwise multiplication operations. The LSTM process and mathematical representation consists mostly of the four phases listed below [26]:

1. Deciding to remove useless information:

$$f_t = \sigma(w_f[h_{t-1}, X_t] + b_f) \tag{4}$$

where f_t represents the forget gate and σ is the sigmoid activation function and it can be defined as:

$$\sigma(x) = (1 + e^{-x})^{-1} \tag{5}$$

This function is utilized for this gate to decide what information should be removed from the LSTM's memory. This decision is mainly dependent on the values of the previous hidden layer output h_{t-1} and the input x_t. The output f_t takes a value between 0 and 1, where 0 means fully discard the learned value and 1 means preserve the entire value. w_f is the recurrent weight matrix, while b_f is the bias term.

2. Updating information:

$$i_t = \sigma(w_i[h_{t-1}, X_t] + b_i) \tag{6}$$

$$\tilde{c_t} = tanh(w_c[h_{t-1}, X_t] + b_c) \tag{7}$$

in which i_t is the input gate and denotes if the value needs to be updated or not and $\tilde{c_t}$ designates a vector of new candidate values that will be added into the LSTM memory. Indeed, the sigmoid layer determines which values require updating, and the tanh layer generates a vector of new candidate values.

3. Updating the cell status:

$$c_t = f_t * c_{t-1} + i_t * \tilde{c_t} \tag{8}$$

where c_t and c_{t-1} represent the current and previous memory states, respectively. This phase is carried out by updating the previous cell's state, multiplying the old value by f_t, deleting the information to be forgotten, and adding $i_t * \tilde{c_t}$ to generate a new candidate value.

4. Outputting information:

$$o_t = \sigma(w_o[h_{t-1}, X_t] + b_o) \tag{9}$$

$$h_t = o_t * tanh(c_t) \tag{10}$$

where o_t is the output gate and h_t is the current hidden layer outputs whose representations are a value between −1 and 1. This step defines the ultimate result. To begin, a sigmoid layer, represented by o_t, selects which part of the cell state will be output. The cell state is then processed by the tanh activation function and multiplied by the sigmoid layer output to create the output.

A typical LSTM network is seen in Figure 4. LSTM layers are composed of memory blocks rather than neurons. These memory blocks are interconnected across the layers, and each block may contain one or more recurrently connected memory elements or cells. As indicated in this figure (yellow shaded area), the flow of information is managed by three types of gates: the forget gate (f_t), the input gate (I_t), and the output gate (O_t).

4.2. LSTM Model Settings and Optimisation

Optimizing an LSTM model entails establishing a set of model parameters that yields the best model performance. The number of units and hidden layers and the optimiser, activation function, batch size, and learning rate are typical examples of such elements. So, the choice of a suitable algorithm is critical to success in addressing any type of optimisation issue. Wolpert and Macready demonstrated this in their "no free lunch" theorem, which states that no method is perfect for solving every type of optimisation issue. As a result, the basic idea is to select an effective optimisation approach to solve a given hand-in optimisation problem with less computational effort and a greater rate of convergence [27].

4.2.1. Genetic Algorithm (GA)

Genetic algorithms (GAs) have been around for over four decades. GAs are heuristic search algorithms that provide answers to optimisation and search problems. The name "GA" is derived from the biological terminology of natural selection, crossing, and mutation. In reality, GAs simulate natural evolutionary processes [28]. Thus, a literature review provides many instances of using GA in the analysis and optimisation of various elements from many sectors, such as energy systems. Moreover, GA can be used for the optimisation of ANN predictions or for the optimisation of ANN architecture [29]. GAs provide a general and global optimisation process. Since the GA is a global search technique, it will be less vulnerable to local search flaws such as back-propagation. The GA may be used to design the network's architecture as well as its weight. There have been various attempts to utilise GAs to determine the architecture of a neural network and the link weights for a fixed architecture network. Many attempts have been made to use a GA to determine the architecture as well as the link weights.

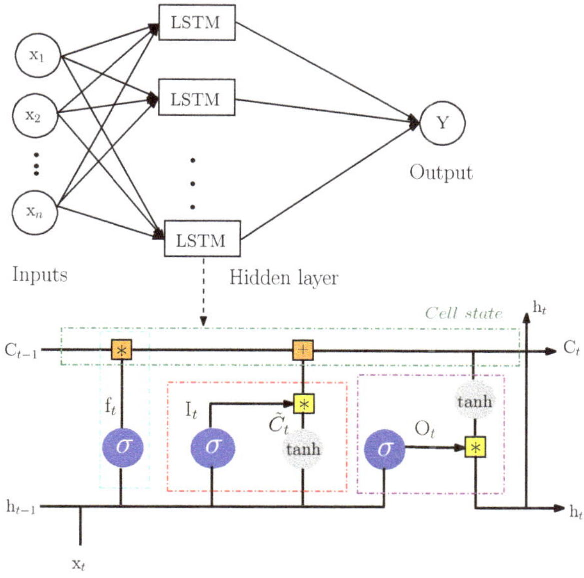

Figure 4. A typical Long Short-Term Memory (LSTM) network topology.

4.2.2. Particle Swarm Optimization (PSO)

The particle swarm optimisation (PSO) method is a swarm-based stochastic optimisation approach introduced by Eberhart and Kennedy (1995). This technique replicates the social behavior of birds inside a flock to reach the food objective. A swarm of birds approaches their food goal using a combination of personal and communal experience. They constantly update their position based on their best position as well as the best position of the entire swarm, and reunite themselves to form an ideal configuration [30]. This nature-inspired method is becoming increasingly popular due to its reliability and easy implementation. In addition, classical neural networks do not operate well when forecasting parameters within short intervals. Moreover, because of their dependability, hybrid ANNs based on particle swarm optimisation have been frequently advocated in literature reviews. The PSO method, like the GA, is used as an optimisation technique within neural networks to optimise ANN forecasts or ANN architecture (the number of layers, neurons, etc.) [31]. Thus, we use this algorithm to optimise the weights.

4.3. LSTM Network Parameters

The network's trainable parameters, known as the trainable weights, influence the network's complexity. They are represented in LSTMs via connections between the input, hidden, and output layers, as well as internal connections. The following formula is used to calculate the Number of Trainable Weights (NTW) of a neural network with x inputs, y outputs, and z LSTM cells in the hidden layer:

$$NTW = 4xz + 4zz + 4z + yz + y \tag{11}$$

where:

- $4xz$: the connection weights between the input layer and the hidden layer;
- $4zz$: the hidden layer's recursive weights;
- $4z$: the hidden layer's bias;
- yz: the connection weights between the hidden layer and the output layer;
- y: the output layer's bias.

Choosing ideal neural network settings can frequently imply the difference between mediocre and peak performance. However, there is limited information in the literature on the selection of different neural network parameters x, y, and z; it requires the expertise of professionals.

4.4. Train–Validation–Test dataset

The one-year target variables were divided into three datasets: the first served as the training set, the second served as the test set, and depending on the length of the output sequence, random samples drawn from the last part served as the validation set. So, for the validation, we use cross-validation, which is a popular data resampling approach for estimating the true forecasting prediction error of models and tuning model parameters. This technique evaluates the generalization capabilities of prediction models and prevents over-fitting. It is the process of generating numerous train–test splits from the training data, which are then applied to adjust the model [32] . k-fold cross-validation is identical to repeated random sub-sampling, but the sampling is performed in such a manner that no two test sets overlap. The available learning set is divided into k disjoint subsets of about equivalent size. Indeed, each time, one of the k subsets is utilised as the validation/test batch, while the remaining $(k-1)$ subsets are combined to form the training set. The total efficacy of the model is calculated by averaging the error estimation over all k trials. Each sample is placed in a validation/test set precisely once and in the training set $(k-1)$ times [33]. Figure 5 illustrates this process as a popular evaluation mechanism in machine learning.

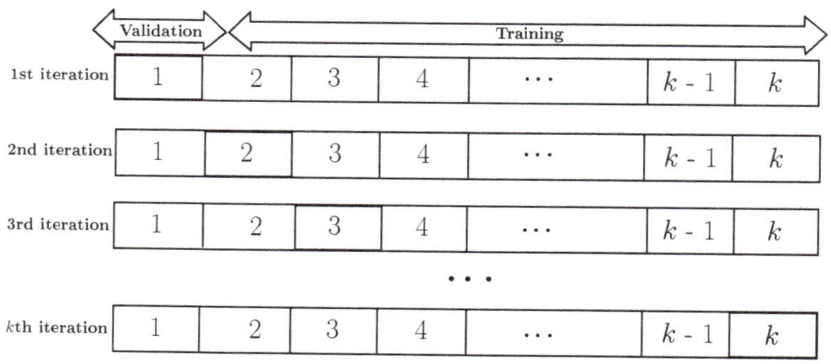

Figure 5. k-fold cross-validation.

We train the LSTM with various architectures for 12-h forecasting of thermal parameters such as CO_2, noise, and temperature. As a result, the window size of the input and output parameters is determined by the time scale of the chosen parameter prediction. We apply the ADAM optimiser, which is one of the optimisation methods employed in deep learning. The learning rate is fixed to 0.01 and gradually drops after every 50 epochs. We train the LSTM with 60, 60, and 100 hidden units for the forecasting of the CO_2, the noise, and the temperature, respectively. The window size of the input and output parameters depends on the time scale of the load prediction. The validation and training results of each parameter are illustrated in Figures 6–8.

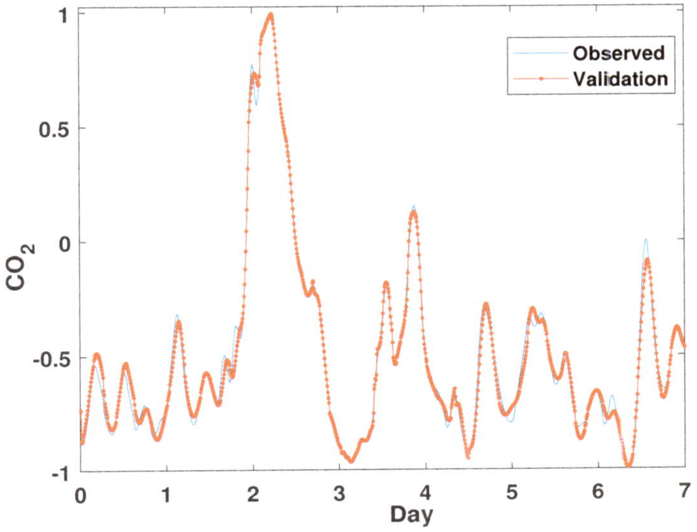

Figure 6. Training and validation of the CO_2 data.

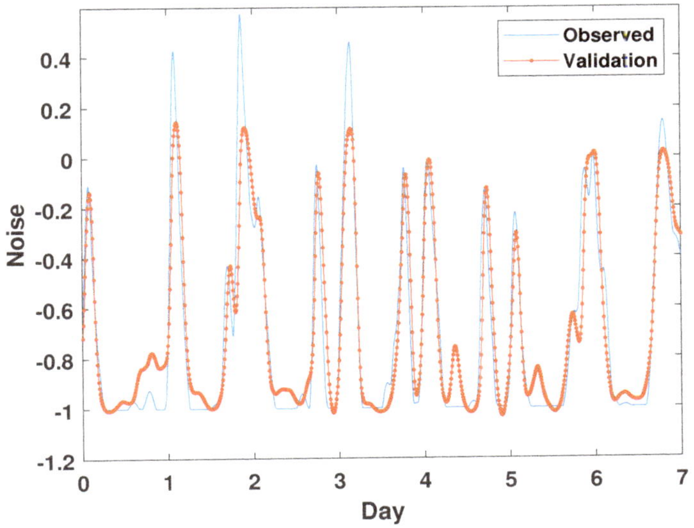

Figure 7. Training and validation of the noise data.

4.5. Evaluation Metrics

This study uses the Root Mean Square Error ($RMSE$) as the loss function and the Mean Absolute Error (MAE) and the Correlation Coefficient (CC) to evaluate the various performance measures. These indicators are measurements of the anticipated value's departure from the actual data, and they indicate the prediction's overall inaccuracy. The corresponding definition of each indicator is given by the following as [34]:

$$RMSE = \sqrt{\frac{1}{N}\sum_{i=1}^{N}(y_i - \tilde{y}_i)^2} \tag{12}$$

$$MAE = \frac{\sum_{i=1}^{N} |y_i - \tilde{y}_i|}{N} \quad (13)$$

$$CC = \frac{\sum_{i=1}^{N} (\bar{y}_i - \bar{p})(y_i - \tilde{y}_i)}{\sqrt{\sum_{i=1}^{N} (\bar{y}_i - \bar{p})^2 \sum_{i=1}^{N} (y_i - \tilde{y}_i)^2}} \quad (14)$$

where y_i and $\tilde{y}_i$ represent the real value and the forecasted value at the time t, N denotes the total time step, and $\bar{y}_i$ and $\bar{p}$ are the average of the real value and the forecasted value, respectively. The smaller the values of $RMSE$ and MAE, the smaller the deviation of the projected outcomes from the actual values. A value of CC closer to 1 indicates lower errors and a more accurate prediction.

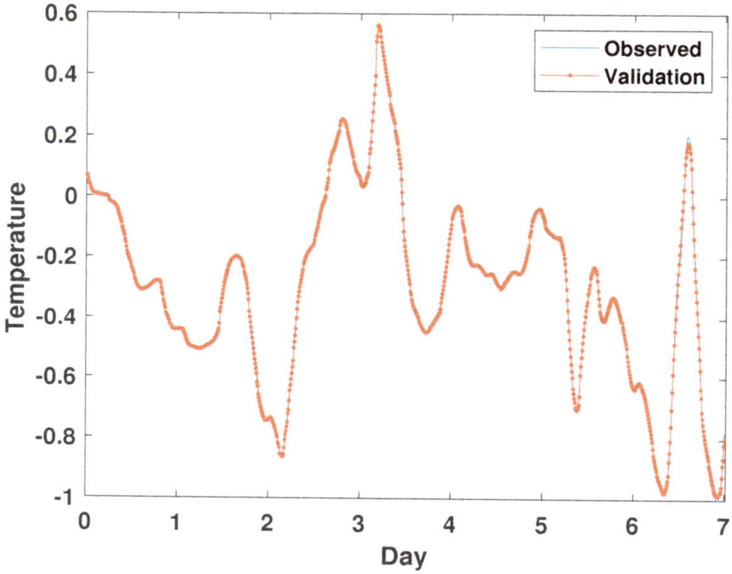

Figure 8. Training and validation of the temperature data.

5. Experimental Results

5.1. Parameters Forecasting

We show in this research a forecast of the thermal characteristics of a smart house outfitted with various types of sensors. The fundamental architecture of LSTM networks is predetermined and immutable; each LSTM unit has a vector input of n values, including the current value of the specified parameters (CO_2, noise, and temperature) at time $t = 0$ as well as the past values. We create three neural networks with various designs, each one adapted to the predicting parameter. After 10 min, these neural networks can forecast. We can anticipate the full period of the required horizon by repeating the process and selecting the appropriate parameters for these models.

5.2. CO_2 Forecasting

In the first experiment, we give the CO_2 prediction of a house for 12 h. Figures 9–11 show the predicted results obtained by the LSTM, the GA-LSTM, and the PSO-LSTM algorithms, respectively. As shown, the predicted results are closer to the real data values and the $RMSE$ of each technique is quite low, which proves the forecasting performance of the suggested strategies.

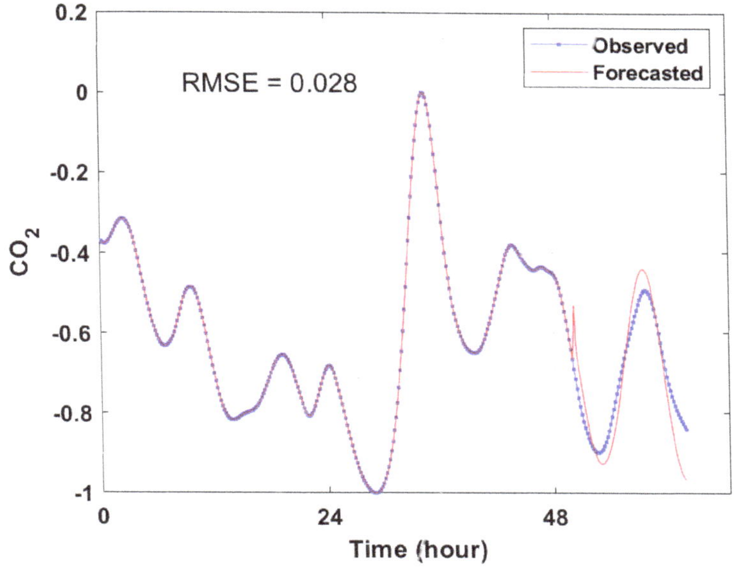

Figure 9. CO_2 forecasting by LSTM.

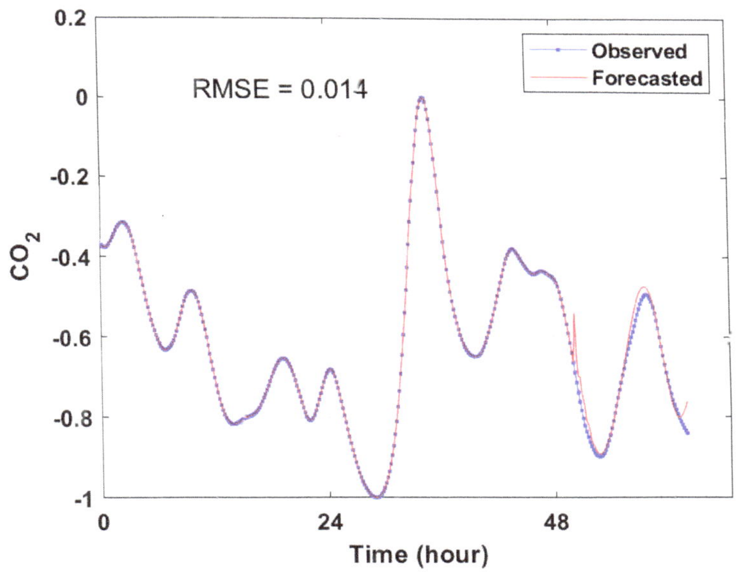

Figure 10. CO_2 forecasting by GA-LSTM.

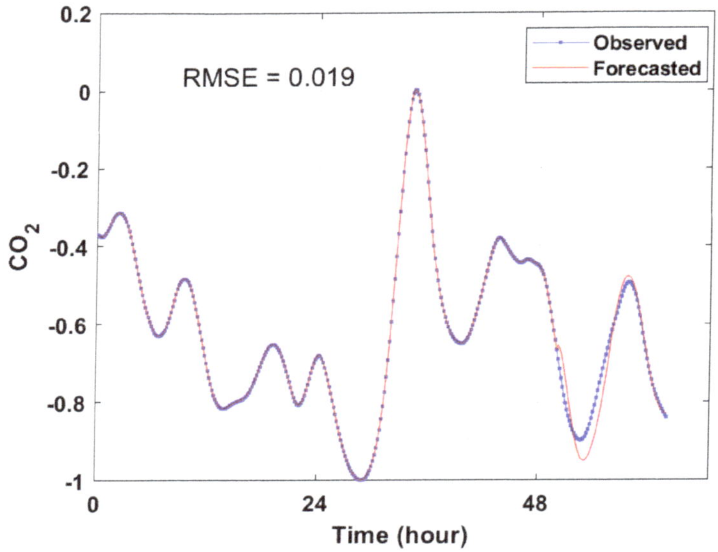

Figure 11. CO_2 forecasting by PSO-LSTM.

5.3. Noise Forecasting

The second experiment also illustrates the noise prediction results for 12 h. Figures 12–14 show the findings with the error rate of the LSTM, the GA-LSTM, and the PSO-LSTM models. It appears that each model's curve prediction retains the shape of the real data curve.

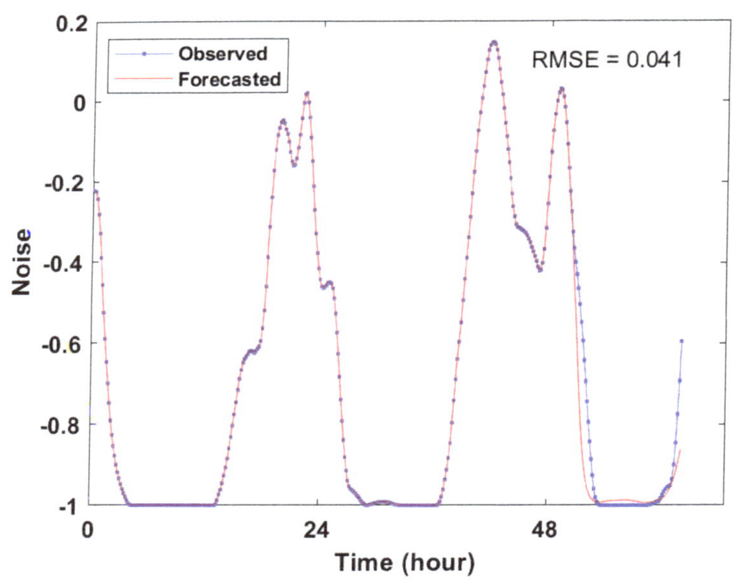

Figure 12. Noise forecasting by LSTM.

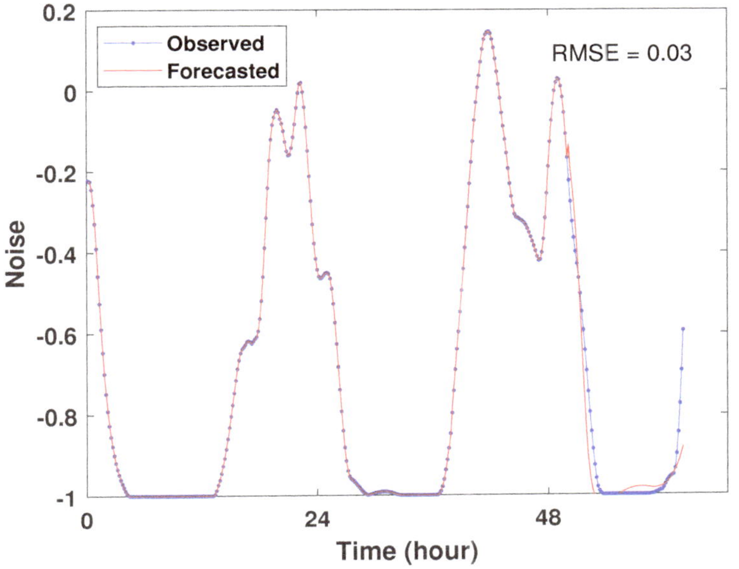

Figure 13. Noise forecasting by GA-LSTM.

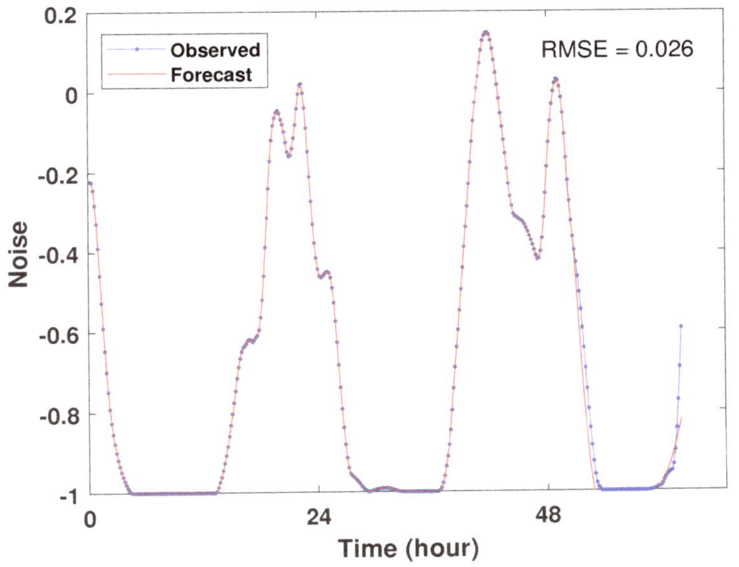

Figure 14. Noise forecasting by PSO-LSTM.

5.4. Temperature Forecasting

The third experiment shows the temperature forecasted results for 12 h. Figures 15–17 depict the results with the *RMSE* value of the LSTM, the GA-LSTM, and the PSO-LSTM approaches. Likewise, each model's curve prediction looks to keep the form of the real data curve.

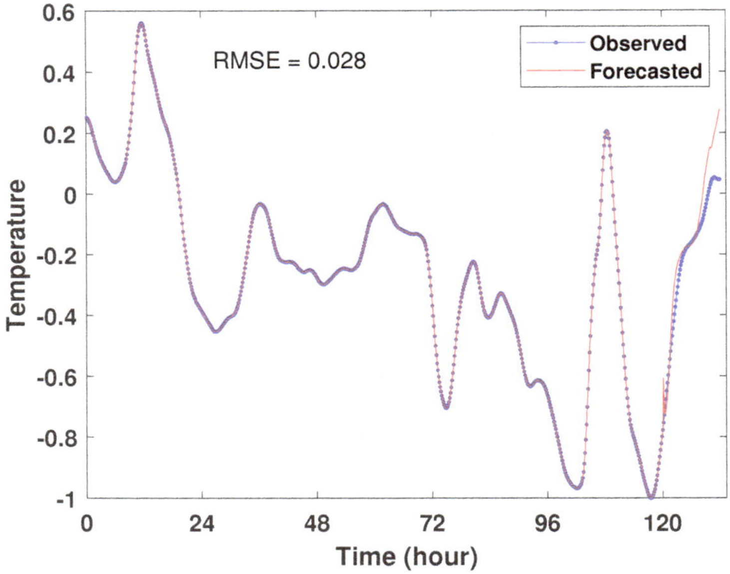

Figure 15. Temperature forecasting by LSTM.

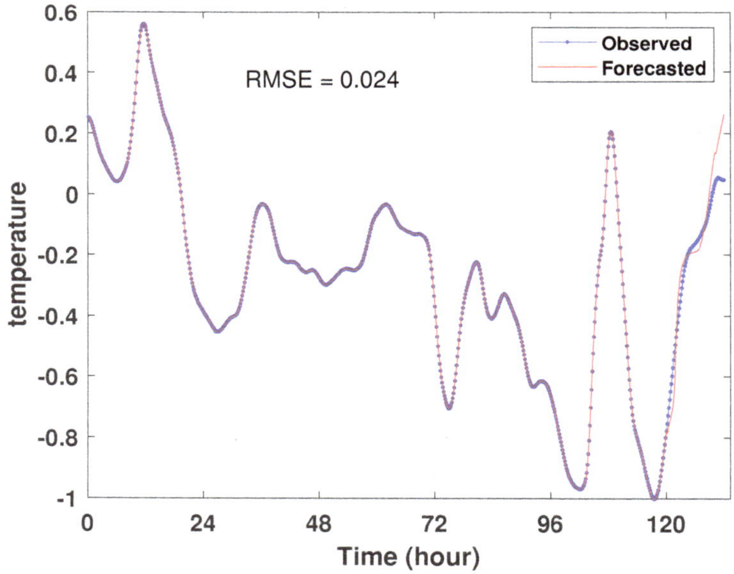

Figure 16. Temperature forecasting by GA-LSTM.

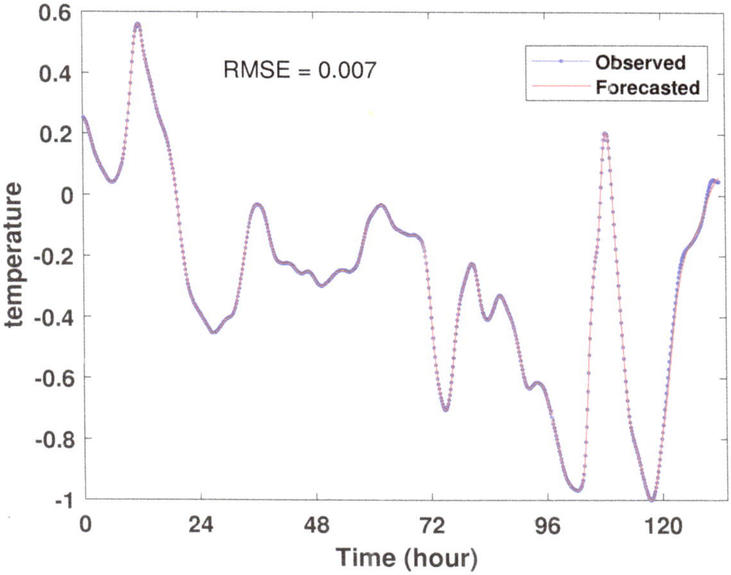

Figure 17. Temperature forecasting by PSO-LSTM.

5.5. Analysis of Results

This work basically assesses the performance of the suggested model from two angles: precision and running time. Tables 1–3 provide the various performance measures for testing predictions on the studied building.

We can see that the implemented approaches produce quite excellent results, and the predicted findings are precise and dependable.

Tables 1–3 reveal that the two performance metrics, $RMSE$ and MAE, have small values. These predictions are fairly close and representative to the real data. The correlation coefficient (CC) is also very close to 1, which proves the high precision of the forecasting strategies. As indicated in the tables and figures of forecasting results, the simple LSTM model without optimisation gives the worst results compared with the GA-LSTM and the PSO-LSTM techniques. We emphasize that the experimental results of the CO_2 prediction show that the GA-LSTM outperforms the PSo-LSTM and the LSTM models with $RMSE$s of 0.0135, 0.0185, and 0.0281 and CCs of 99.80%, 99.62%, and 99.16% for GA-LSTM, PSO-LSTM, and LSTM, respectively. For noise and temperature prediction, the performance of the PSO-LSTM outperforms the GA-LSTM in terms of $RMSE$ and CC. Overall, we have successfully shown that the proposed optimisation techniques (GA-LSTM and PSO-LSTM networks) may successfully extract relevant information from noisy human behavior data.

The statistical analysis of the obtained results shows that the proposed model tuned by the two evolutionary metaheuristic search algorithms (GA and PSO) provides more precise results than the benchmark LSTM model, whose parameters were established through limited experience and a discounted number of experiments.

Table 1. Performance criteria of the CO_2 prediction.

Algorithms	LSTM	GA-LSTM	PSO-LSTM
$RMSE$	0.0281	0.0135	0.0185
MAE	0.0102	0.0039	0.0061
CC	0.9916	0.9980	0.9962

Table 2. Performance criteria of the noise prediction.

Algorithms	LSTM	GA-LSTM	PSO-LSTM
RMSE	0.0405	0.0290	0.0256
MAE	0.0097	0.0070	0.0080
CC	0.9942	0.9970	0.9978

Table 3. Performance criteria of the temperature prediction.

Algorithms	LSTM	GA-LSTM	PSO-LSTM
RMSE	0.0275	0.0243	0.0070
MAE	0.0063	0.0075	0.0017
CC	0.9968	0.9974	0.9997

6. Conclusions

In this work, we have proposed two optimised metaheuristic algorithms based on the LSTM architecture for dealing with occupancy forecasting in the context of smart buildings. The GA-LSTM and PSO-LSTM models give very satisfactory prediction results with a high level of precision and reliability compared with the LSTM forecasting results. The choice of these two methods (PSO and GA) is based on their reputation in literature. A comparison shows that the implementation of the two metaheuristic algorithms (GA and PSO) for the optimal configuration of occupancy forecasting derived an optimal LSTM model that performs significantly better than the benchmark models, including other machine learning approaches such as the basic LSTM model. The predicted values have been used to check the presence of residents and then control real electrical consumption. This was carried out to prove that the optimised LSTM can decrease power consumption, improve security, and maintain comfort for the occupants. A potential field for future research would be to perform thermal parameters forecasting, using recurrent neural networks, for various construction such as hospitals, hotels, and public establishments. It would be worthwhile to investigate whether a recurrent neural network can maintain such a high accuracy to forecast thermal features and room occupancy rates in a smart building. Thus, future studies will also focus on the deployment and integration of various hybrid optimisation algorithms in recurrent neural networks such as the LSTM model in order to select the best architecture, weights, and learning rate in order to achieve greater energy savings in the building energy management system. As a result, our findings provide a solid foundation for future research aimed at providing a more accurate assessment of building occupancy. Nonetheless, the current findings will provide a basis for occupancy prediction, which might be used to enhance our context-driven approaches for managing active building systems such as the HVAC, lighting, and shading systems. Again, a forecasting model for thermal characteristics and room occupancy rates with a low estimation error would help energy producers in making operational, tactical, and strategic decisions. Finally, better building load forecasting allows the implementation of the real-time management of smart buildings.

Author Contributions: Conceptualization, L.C.-A. and B.M.; methodology, S.M., B.M. and L.D.; software, S.M. and S.L.; validation, S.M. and L.C.-A.; formal analysis, L.D. and L.C.-A.; investigation, S.M. and S.L.; resources, B.M. and L.D.; data curation, B.M. and L.D.; writing—original draft preparation, S.M.; writing—review and editing, L.C.-A. and S.L.; visualization, S.M.; supervision, L.C.-A. and L.D.; project administration, B.M., L.D. and L.C.-A.; funding acquisition, B.M. All authors have read and agreed to the published version of the manuscript.

Funding: This research received no external funding.

Institutional Review Board Statement: Not applicable.

Informed Consent Statement: Not applicable.

Conflicts of Interest: The authors declare no conflict of interest.

References

1. Stazi, F. *Thermal Inertia in Energy Efficient Building Envelopes*; Butterworth-Heinemann: Oxford, UK, 2017.
2. Lotfabadi, P.; Hancer, P. A comparative study of traditional and contemporary building envelope construction techniques in terms of thermal comfort and energy efficiency in hot and humid climates. *Sustainability* **2019**, *11*, 3582. [CrossRef]
3. Alawadi, S.; Mera, D.; Fernández-Delgado, M.; Alkhabbas, F.; Olsson, C.M.; Davidsson, P. A comparison of machine learning algorithms for forecasting indoor temperature in smart buildings. *Energy Syst.* **2020**, *13*, 1–17. [CrossRef]
4. Dai, X.; Liu, J.; Zhang, X. A review of studies applying machine learning models to predict occupancy and window-opening behaviours in smart buildings. *Energy Build.* **2020**, *223*, 2020. [CrossRef]
5. Boodi, A.; Beddiar, K.; Amirat, Y.; Benbouzid, M. Building Thermal-Network Models: A Comparative Analysis, Recommendations, and Perspectives. *Energies* **2022**, *15*, 1328. [CrossRef]
6. Turley, C.; Jacoby, M.; Pavlak, G.; Henze, G. Development and evaluation of occupancy-aware HVAC control for residential building energy efficiency and occupant comfort. *Energies* **2020**, *13*, 5396. [CrossRef]
7. Fotopoulou, E.; Zafeiropoulos, A.; Terroso-Saenz, F.; Simsek, U.; Gonzalez-Vidal, A.; Tsiolis, G.; Gouvas, P.; Liapis, P.; Fensel, A.; Skarmeta, A. Providing Personalized Energy Management and Awareness Services for Energy Efficiency in Smart Buildings. *Sensors* **2017**, *17*, 2054. [CrossRef]
8. Hu, S.; Yan, D.; Guo, S.; Cui, Y.; Dong, B. A survey on energy consumption and energy usage behavior of households and residential building in urban China. *Energy Build.* **2017**, *148*, 366–378. [CrossRef]
9. Cao, T.D.; Delahoche, L.; Marhic, B.; Masson, J.B. Occupancy Forecasting using two ARIMA Strategies. In Proceedings of the ITISE 2019, Granada, Spain, 25–27 September 2019; Volume 2.
10. Mariano-Hernández, D.; Hernández-Callejo, L.; García, F.S.; Duque-Perez, O.; Zorita-Lamadrid, A.L. A review of energy consumption forecasting in smart buildings: Methods, input variables, forecasting horizon and metrics. *Appl. Sci.* **2020**, *10*, 8323. [CrossRef]
11. Scheurer, S.; Tedesco, S.; Brown, K.N.; O'Flynn, B. Using domain knowledge for interpretable and competitive multi-class human activity recognition. *Sensors* **2020**, *20*, 1208. [CrossRef]
12. Fang, Z.; Crimier, N.; Scanu, L.; Midelet, A.; Alyafi, A.; Delinchant, B. Multi-zone indoor temperature prediction with LSTM-based sequence to sequence model. *Energy Build.* **2021**, *245*, 111053. [CrossRef]
13. Kaligambe, A.; Fujita, G.; Keisuke, T. Estimation of Unmeasured Room Temperature, Relative Humidity, and CO_2 Concentrations for a Smart Building Using Machine Learning and Exploratory Data Analysis. *Energies* **2022**, *15*, 4213. [CrossRef]
14. Massana, J.; Pous, C.; Burgas, L.; Melendez, J.; Colomer, J. Short-term load forecasting for non-residential buildings contrasting artificial occupancy attributes. *Energy Build.* **2016**, *130*, 519–531. [CrossRef]
15. Zhao, J.; Liu, X. A hybrid method of dynamic cooling and heating load forecasting for office buildings based on artificial intelligence and regression analysis. *Energy Build.* **2018**, *174*, 293–308. [CrossRef]
16. Bouktif, S.; Fiaz, A.; Ouni, A.; Serhani, M.A. Optimal deep learning lstm model for electric load forecasting using feature selection and genetic algorithm: Comparison with machine learning approaches. *Energies* **2018**, *11*, 1636. [CrossRef]
17. Alawadi, S.; Mera, D.; Fernandez-Delgado, M.; Taboada, J.A. Comparative study of artificial neural network models for forecasting the indoor temperature in smart buildings. In *International Conference on Smart Cities*; Springer: Cham, Switzerland, 2017; pp. 29–38.
18. Aliberti, A.; Bottaccioli, L.; Macii, E.; Di Cataldo, S.; Acquaviva, A.; Patti, E. A non-linear autoregressive model for indoor air-temperature predictions in smart buildings. *Electronics* **2019**, *8*, 979. [CrossRef]
19. Zhu, J.; Yang, Z.; Mourshed, M.; Guo, Y.; Zhou, Y.; Chang, Y.; Wei, Y.; Feng, S. Electric Vehicle Charging Load Forecasting: A Comparative Study of Deep Learning Approaches. *Energies* **2019**, *12*, 2692. [CrossRef]
20. Liu, Z.; Wu, D.; Liu, Y.; Han, Z.; Lun, L.; Gao, J.; Jin, G.; Cao, G. Accuracy analyses and model comparison of machine learning adopted in building energy consumption prediction. *Energy Explor. Exploit.* **2019**, *37*, 1426–1451. [CrossRef]
21. Saleem, T.J.; Chishti, M.A. Data analytics in the Internet of Things: A survey. *Scalable Comput. Pract. Exp.* **2019**, *20*, 607–630. [CrossRef]
22. Mahjoub, S.; Chrifi-Alaoui, L.; Marhic, B.; Delahoche, L. Predicting Energy Consumption Using LSTM, Multi-Layer GRU and Drop-GRU Neural Networks. *Sensors* **2022**, *22*, 4062. [CrossRef]
23. Mahjoub, S.; Chrifi-Alaoui, L.; Marhic, B.; Delahoche, L.; Masson, J.B.; Derbel, N. Prediction of energy consumption based on LSTM Artificial Neural Network. In Proceedings of the 2022 19th International Multi-Conference on Systems, Signals Devices (SSD), Setif, Algeria, 6–10 May 2022; pp. 521–526.
24. Siami-Namini, S.; Tavakoli, N.; Namin, A.S. The performance of LSTM and BiLSTM in forecasting time series. In Proceedings of the 2019 IEEE International Conference on Big Data (Big Data), Los Angeles, CA, USA, 9–12 December 2019; pp. 3285–3292.
25. Apaydin, H.; Feizi, H.; Sattari, M.T.; Colak, M.S.; Shamshirband, S.; Chau, K.W. Comparative analysis of recurrent neural network architectures for reservoir inflow forecasting. *Water* **2020**, *12*, 1500. [CrossRef]
26. Zhang, X.; Zhao, M.; Dong, R. Time-series prediction of environmental noise for urban IoT based on long short-term memory recurrent neural network. *Appl. Sci.* **2020**, *10*, 1144. [CrossRef]
27. Jain, M.; Saihjpal, V.; Singh, N.; Singh, S.B. An Overview of Variants and Advancements of PSO Algorithm. *Appl. Sci.* **2022**, *12*, 8392. [CrossRef]
28. Hoschel, K.; Lakshminarayanan, V. Genetic algorithms for lens design: A review. *J. Opt.* **2019**, *48*, 134–144. [CrossRef]

29. Lorencin, I.; Andelic, N.; Mrzljak, V.; Car, Z. Genetic algorithm approach to design of multi-layer perceptron for combined cycle power plant electrical power output estimation. *Energies* **2019**, *12*, 4352. [CrossRef]
30. Wang, D.; Tan, D.; Liu, L. Particle swarm optimization algorithm: An overview. *Soft Comput.* **2018**, *22*, 387–408. [CrossRef]
31. Aljanad, A.; Tan, N.M.; Agelidis, V.G.; Shareef, H. Neural network approach for global solar irradiance prediction at extremely short-time-intervals using particle swarm optimization algorithm. *Energies* **2021**, *14*, 1213. [CrossRef]
32. Berrar, D. Cross-validation. In *Encyclopedia of Bioinformatics and Computational Biology*; Elsevier: Amsterdam, The Netherlands, 2019; Volume 1, pp. 542–545. [CrossRef]
33. Xiong, Z.; Cui, Y.; Liu, Z.; Zhao, Y.; Hu, M.; Hu, J. Evaluating explorative prediction power of machine learning algorithms for materials discovery using k-fold forward cross-validation. *Comput. Mater. Sci.* **2020**, *171*, 109203. [CrossRef]
34. Xiao, Y.; Yin, H.; Zhang, Y.; Qi, H.; Zhang, Y.; Liu, Z. A dual-stage attention-based Conv-LSTM network for spatio-temporal correlation and multivariate time series prediction. *Int. J. Intell. Syst.* **2021**, *36*, 2036–2057. [CrossRef]

MDPI
St. Alban-Anlage 66
4052 Basel
Switzerland
www.mdpi.com

Energies Editorial Office
E-mail: energies@mdpi.com
www.mdpi.com/journal/energies

www.ingramcontent.com/pod-product-compliance
Lightning Source LLC
Chambersburg PA
CBHW041015270926
51850CB00024B/196
9783036590806